INTERGOVERNMENTAL OCEANOGRAPHIC COMMISSION

1984

ARCTIC OCEAN   OCÉAN ARCTIQUE

5·00

ENCYCLOPEDIA
OF
OCEAN SCIENCES

ENCYCLOPEDIA
OF
OCEAN SCIENCES

Editor-in-Chief

JOHN H. STEELE

Editors

STEVE A. THORPE

KARL K. TUREKIAN

ACADEMIC PRESS

A Harcourt Science and Technology Company

San Diego   San Francisco   New York   Boston
London   Sydney   Tokyo

Academic Press
A Harcourt Science and Technology Company
Harcourt Place, 32 Jamestown Road, London NW1 7BY, UK
http://www.academicpress.com

Academic Press
A Harcourt Science and Technology Company
525 B Street, Suite 1900, San Diego, California 92101-4495, USA
http://www.academicpress.com

ISBN 0-12-227430-X

Library of Congress Catalog Number: 2001092473

A catalogue record for this book is available from the British Library

Access for a limited period to an on-line version of the Encyclopedia of Ocean Sciences is
included in the purchase price of the print edition.

This on-line version has been uniquely and persistently identified by the Digital Object Identifier
(DOI)

10.1006/rwos.2001

By following the link

http://dx.doi.org/10.1006/rwos.2001

from any Web Browser, buyers of the Encyclopedia of Ocean Sciences
will find instructions on how to register for access.

If you have any problems with accessing the on-line version, e-mail:
idealreferenceworks@harcourt.com

Typeset by Macmillan India Limited, Bangalore, India
Printed and bound in Spain by Grafos SA Arte Sobre Papel, Barcelona
01 02 03 04 05 06 GF 9 8 7 6 5 4 3 2 1

Editors

EDITOR-IN-CHIEF

John H. Steele
Woods Hole Oceanographic Institution
Mail Stop 41
Woods Hole
MA 02543, USA

EDITORS

Steve A. Thorpe
Southampton Oceanography Centre
University of Southampton
Waterfront Campus
European Way
Southampton
SO14 3ZH, UK

Karl K. Turekian
Yale University
Department of Geology and Geophysics
New Haven
CT 06520-8109, USA

Editorial Advisory Board

Preface

In 1942, a monumental volume was published on *The Oceans* by H. U. Sverdrup, M. W. Johnson, and R. H. Fleming. It was comprehensive and covered the knowledge at that time of the scientific study of the oceans. This seminal book helped to initiate the tremendous burgeoning of marine research that occurred during the following decades. The *Encyclopedia of Ocean Sciences* aims to embody the great growth of knowledge in a major new reference work.

There have been remarkable new approaches to the study of the oceans that blur the distinctions between the physical, chemical, biological, and geological disciplines. New theories and technologies have expanded our knowledge of ocean processes. For example, plate tectonics has revolutionized our view not only of the geology and geophysics of the seafloor but also of ocean chemistry and biology. Satellite remote sensing provides a global vision as well as detailed understanding of the close coupling of ocean physics and biology at local and regional scales.

Exploration, fishing, warfare, and the impact of storms have driven the past study of the seas, but we now have a great public awareness of and concern with broader social and economic issues affecting the oceans. For this reason, we have invited articles explicitly on marine policy and environmental topics, as well as encouraging authors to address these aspects of their particular subjects. We believe the Encyclopedia should be of use to those involved with policy and management as well as to students and researchers.

Over 400 scientists have contributed to this description of what we now know about the oceans. They are distinguished researchers who have generously shared their knowledge of this ever-growing body of science. We are extremely grateful to all these authors, whose ability to write concisely on complex subjects has generated a perspective on our science that we, as editors, believe will enhance the appreciation of the oceans, their uses, and the research ahead.

It has been a major challenge for the members of the Editorial Advisory Board to cover such a heterogeneous subject. Their knowledge of the diverse areas of research has guaranteed comprehensive coverage of the ocean sciences. The Board contributed significantly by suggesting topics, persuading authors to contribute, and reviewing drafts. Many of them wrote Overviews that give broad descriptions of major parts of the ocean sciences. Clearly it was the dedicated involvement of the Editorial Advisory Board that made this venture successful.

Such a massive enterprise as a multivolume encyclopedia would not be possible without the long-term commitment of the staff of the Major Reference Works team at Academic Press. In particular, we are very grateful for the consistent support of our Senior Developmental Editor, Colin McNeil, who has worked so well with us throughout the whole process. Also, we are very pleased that new technology permits enhanced search and retrieval through the Internet. We believe this will make the Encyclopedia much more accessible to individual researchers and students.

In Memoriam

Tragically, one member of the Editorial Board, Professor Michael Mullin of Scripps Institute of Oceanography, died before the Encyclopedia was completed. His efforts contributed significantly to the articles on biological oceanography. We all mourn his loss.

John H. Steele, Steve A. Thorpe, Karl K. Turekian
Editors

Guide to Use of the Encyclopedia

Introductory Points

In devising the vision and structure for the Encyclopedia, the Editors have striven to unite and interrelate all current knowledge that can be designated "Ocean Sciences". To aid users of the Encyclopedia, this new reference work offers intuitive searching and extensive cross-linking of content. These features are explained in more detail below.

Structure of the Encyclopedia

The material in the Encyclopedia is arranged as a series of entries in alphabetical order. Some entries comprise a single article, whilst entries on more diverse subjects consist of several articles that deal with various aspects of the topic. In the latter case the articles are arranged in a logical sequence within an entry.

To help you realize the full potential of the material in the Encyclopedia we have provided three features to help you find the topic of your choice.

1. Contents Lists

Your first point of reference will probably be the contents list. The complete contents list appearing in each volume will provide you with both the volume number and the page number of the entry. On the opening page of an entry a contents list is provided so that the full details of the articles within the entry are immediately available.

Alternatively you may choose to browse through a volume using the alphabetical order of the entries as your guide. To assist you in identifying your location within the Encyclopedia a running headline indicates the current entry and the current article within that entry.

You will find 'dummy entries' where obvious synonyms exist for entries or where we have grouped together related topics. Dummy entries appear in both the contents list and the body of the text. For example, a dummy entry appears for Kelvin Waves which directs you to Coastal Trapped Waves, where the material is located.

Example

If you were attempting to locate material on Ecosystems via the contents list.

Ecosystems   *See* Fiordic Ecosystems; Large Marine Ecosystems; Ocean Gyre Ecosystems: Polar Ecosystems; Upwelling Ecosystems

At the appropriate location in the contents list, the page numbers for these articles are given.

If you were trying to locate the material by browsing through the text and you looked up Ecosystems then the following information would be provided.

ECOSYSTEMS

See **FIORDIC ECOSYSTEMS; LARGE MARINE ECOSYSTEMS; OCEAN GYRE ECOSYSTEMS; POLAR ECOSYSTEMS; UPWELLING ECOSYSTEMS**

2. Cross References

All of the articles in the encyclopedia have been extensively cross referenced. The cross references, which appear at the end of each article, have been provided at three levels:

i. To indicate if a topic is discussed in greater detail elsewhere.

> **ACOUSTICS, ARCTIC**
> *See also:* Acoustics in Marine Sediments. Acoustic Noise. Acoustics, Shallow Water. Arctic Basin Circulation. Autonomous Underwater Vehicles (AUVs). Bioacoustics. Ice–Ocean Interaction. Nepheloid Layers. North Atlantic Oscillation (NAO). Satellite Passive Microwave Measurements of Sea Ice. Sea Ice: Overview; Variations in Extent and Thickness. Seals. Seismic Structure. Thermohaline Circulation. Tomography. Under-ice Boundary Layer. Water Types and Water Masses.

ii. To draw the reader's attention to parallel discussions in other articles.

> **ACOUSTICS, ARCTIC**
> *See also:* Acoustics in Marine Sediments. Acoustic Noise. Acoustics, Shallow Water. Arctic Basin Circulation. Autonomous Underwater Vehicles (AUVs). Bioacoustics. Ice–Ocean Interaction. Nepheloid Layers. North Atlantic Oscillation (NAO). Satellite Passive Microwave Measurements of Sea Ice. Sea Ice: Overview; Variations in Extent and Thickness. Seals. Seismic Structure. Thermohaline Circulation. Tomography. Under-ice Boundary Layer. Water Types and Water Masses.

iii. To indicate material that broadens the discussion.

> **ACOUSTICS, ARCTIC**
> *See also:* Acoustics in Marine Sediments. Acoustic Noise. Acoustics, Shallow Water. Arctic Basin Circulation. Autonomous Underwater Vehicles (AUVs). Bioacoustics. Ice–Ocean Interaction. Nepheloid Layers. North Atlantic Oscillation (NAO). Satellite Passive Microwave Measurements of Sea Ice. Sea Ice: Overview; Variations in Extent and Thickness. Seals. Seismic Structure. Thermohaline Circulation. Tomography. Under-ice Boundary Layer. Water Types and Water Masses.

3. Index

The index will provide you with the page number where the material is to be located, and the index entries differentiate between material that is a whole article, is part of an article or is data presented in a table. The page numbers contained within each volume of the Encyclopedia are listed in a footer on all pages of the index. On the opening page of the index detailed notes are provided.

4. Appendices

In addition to the articles that form the main body of the encyclopedia, there are a number of appendices which provide bathymetric charts and lists of data used throughout the encyclopedia.

The appendices are located in volume 6, before the index.

5. Contributors

A full list of contributors appears at the beginning of each volume.

Contributors

Adams, E
Massachusetts Institute of Technology
Department of Civil and Environmental Engineering
77 Massachusetts Avenue
Cambridge, MA 02139, USA

Aiken, J
Plymouth Marine Laboratory
Prospect Place
West Hoe
Plymouth, PL1 3DH, UK

Ainley, D G
HT Harvey Associates
3150 Almaden Expressway Suite 145
San Jose, CA 95118, USA

Akal, T
NATO SACLANT Undersea Research Centre
Viale San Bartolomeo 400
La Spezia, Italy

Alldredge, A
University of California, Santa Barbara
Department of Ecology, Evolution and Marine Biology
Santa Barbara, CA 93106, USA

Alpers, W
University of Hamburg
Institute of Oceanography
Troplowitzstr 7
D-22529 Hamburg, Germany

Anderson, D M
Woods Hole Oceanographic Institution
Biology Department
Woods Hole, MA 02543, USA

Anderson, O R
Lamont-Doherty Earth Observatory
of Columbia University
PO Box 1000
61 Route 9W
Palisades, NY 10964, USA

Andrews, J T
University of Colorado
Institute of Arctic and Alpine Research
and Department of Geological Sciences
Box 450
Boulder, CO 80309, USA

Apel, J R
Global Ocean Associates
PO Box 12131
Silver Spring, MD 20908, USA

Arnold, G P
Centre for Environment, Fisheries and
Aquaculture Science
Pakefield Road
Lowestoft, Suffolk, NR33 0HT, UK

Arp, A J
Romberg Tiburon Center for Environment Studies
3152 Paradise Drive
Tiburon, CA 94920, USA

Askew, T
Harbor Branch Oceanographic Institution
5600 US 1 North
Fort Pierce, FL 34946, USA

Baggeroer, A B
Massachusetts Institute of Technology
Department of Ocean Engineering
77 Massachusetts Avenue
Cambridge, MA 02139, USA

Bailey, K M
Alaska Fisheries Science Center
7600 Sand Point Way NE
Seattle, WA 98115, USA

Baines, P G
CSIRO Division of Atmospheric Research
Aspendale, VIC 3195, Australia

Baker, J M
Clock Cottage
Ruyton-XI-Towns
Shrewsbury, SY4 1LA, UK

Balance, L T
National Marine Fisheries Service (NOAA)
Southwest Fisheries Science Center
PO Box 271
La Jolla, CA 92037, USA

Baldauf, J G
Texas A&M University
Department of Oceanography
College Station, TX 77845, USA

Ballard, R D
Institute for Exploration
55 Coogan Boulevard
Mystic, CT 06355, USA

Bannister, J L
The Western Australian Museum
Francis Street
Perth, Western Australia, Australia

Barber, R T
Duke University Marine Laboratory
135 Duke Marine Laboratory Road
Beaufort, NC 28516, USA

Barnabé, G
Université de Montpellier II
Laboratoire d'Ecologie Marine
1 Quai de la Daurade
34200 Sete, France

Barnes, R S K
University of Cambridge
Department of Zoology
Downing Street
Cambridge, CB2 3EJ, UK

Barton, E D
University of Wales, Bangor
School of Ocean Sciences
Menai Bridge
Anglesey, LL59 5RH, UK

Bartram, J
World Health Organization
Avenue Appia 20
CH-1211 Geneva 27, Switzerland

Bates, N R
Bermuda Biological Station for Research
Ferry Reach, St George's GE01, Bermuda

Batiza, R
National Science Foundation
Marine Geology and Geophysics
Division of Ocean Sciences
4201 Wilson Boulevard
Arlington, VA 22230, USA

Beckmann, A
Alfred-Wegener-Institut für Polar-
   und Meeresforschung
Postfach 12 01 61
27515 Bremerhaven, Germany

Bellingham, J
Massachusetts Institute of Technology
292 Main Street
Cambrige, MA 02139, USA

Benfield, M C
Louisiana State University
Department of Oceanography and Coastal Sciences
218 Wetland Resources
Baton Rouge, LA 70803, USA

Berger, W H
Scripps Institution of Oceanography
University of California, San Diego
9500 Gilman Drive
La Jolla, CA 92093, USA

Bergstad, O A
Institute of Marine Research
Flødevigen Research Station
N-4817 His, Norway

Bewers, J M
Bedford Institute of Oceanography
PO Box 1006
Dartmouth
Nova Scotia, B2Y 4A2, Canada

Birnbaum, G
Alfred-Wegener-Institut für Polar-
   und Meeresforschung
Postfach 12 01 61
27515 Bremerhaven, Germany

Black, K D
Scottish Association for Marine Science
Dunstaffnage Marine Laboratory
Oban, Argyll, PA34 4AD, UK

Blanton, B O
The University of North Carolina at Chapel Hill
Department of Marine Sciences
12-7 Venable Hall
CB 3300
Chapel Hill, NC 27599, USA

Blaxter, J H S
Scottish Association for Marine Science
Dunstaffnage Marine Laboratory
Oban, Argyll, PA34 4AD, UK

Blough, N V
University of Maryland
Department of Chemistry and Biochemistry
College Park, MD 20742, USA

Bodkin, J L
US Geological Survey
Alaska Biological Science Center
1011 E. Tudor Road
Anchorage, AK 99503, USA

Bogden, P S
Gulf of Maine Ocean Observing System (GoMOOS)
PO Box 4919
Portland, ME 04112, USA

Boyd, I L
British Antarctic Survey
High Cross, Madingley Road
Cambridge, CB3 0ET, UK

Boyd, T J
Oregon State University
College of Oceanic and Atmospheric Sciences
Corvallis, OR 97331, USA

Boyle, E A
Massachusetts Institute of Technology
Department of Earth, Atmospheric
  and Planetary Physics
77 Massachusetts Avenue
Cambridge, MA 02139, USA

Boyle, P
University of Aberdeen
Department of Zoology
Tillydrone Avenue
Aberdeen, AB24 2TZ, UK

Branch, G M
University of Cape Town
Marine Biology Research Institute
Zoology Department
Rondebosch 7701
Cape Town, Republic of South Africa

Brander, K
International Council for the Exploration of the Sea (ICES)
Palægade 2-4
DK 1261 Copenhagen, Denmark

Brown, A C
University of Cape Town
Department of Zoology
Rondebosch 7701
Cape Town, Republic of South Africa

Burger, J
Rutgers University
Institute of Marine and Coastal Sciences
604 Allison Road
Piscataway, NJ 08854, USA

Burkill, P H
Plymouth Marine Laboratory
Prospect Place
West Hoe
Plymouth, PL1 3DH, UK

Bush, D M
State University of West Georgia
Department of Geosciences
Carrollton, GA 30118, USA

Caldeira, K
Lawrence Livermore National Laboratory
Climate and Carbon Cycle Modeling Group
Livermore, CA 94550, USA

Camphuysen, C J
Netherlands Institute for Sea Research
PO Box 59, 1790 AB Den Burg
Texel, The Netherlands

Caputi, N
Fisheries WA Research Division
PO Box 20
North Beach, WA 6020, Australia

Carbotte, S M
Lamont-Doherty Earth Observatory
  of Columbia University
PO Box 1000
61 Route 9W
Palisades, NY 10964, USA

Carlotti, F
CNRS/Université Bordeaux 1
Laboratoire Oceanographie Biologique, UMR 5805
2, rue du Professeur Jolyet
33120 Arachon, France

Carlson, C A
University of California, Santa Barbara
Department of Ecology, Evolution and Marine Biology
Santa Barbara, CA 93106, USA

Casciotti, K L
Princeton University
Department of Geosciences
Princeton, NJ 08544, USA

Chamley, H
Université de Lille 1
Sedimentologie et Geodynamique
59655 Villeneuve d'Ascq, France

Chandler, G T
University of South Carolina
School of Public Health
Columbia, SC 29208, USA

Cheney, R E
NOAA Laboratory for Satellite Altimetry
1315 East-West Highway
Silver Spring, MD 20910, USA

Chereskin, T K
Scripps Institution of Oceanography
University of California, San Diego
9500 Gilman Drive
La Jolla, CA 92093, USA

Chester, R
Liverpool University
Oceanography Laboratories
Department of Earth Sciences
Liverpool, L69 3BX, UK

Christensen, V
University of British Columbia
Fisheries Centre
2204 Main Mall
Vancouver, BC V6T 1Z4, Canada

Church, J A
Antarctic CRC and CSIRO Marine Research
GPO Box 1538, Hobart
Tasmania 7001, Australia

Clarke, A
British Antarctic Survey
High Cross, Madingley Road
Cambridge, CB3 0ET, UK

Coale, K H
Moss Landing Marine Laboratories
PO Box 450
Moss Landing, CA 95039, USA

Cochran, J K
State University of New York, Stony Brook
Marines Sciences Research Center
Stony Brook, NY 11794, USA

Coffin, M F
University of Texas at Austin
Institute for Geophysics
4412 Spicewood Springs Road
Building 600
Austin, TX 78759, USA

Collar, P
Southampton Oceanography Centre
University of Southampton
Waterfront Campus
European Way
Southampton, SO14 3ZH, UK

Collie, J S
Danish Institute for Fisheries Research
Charlottenlund Castle
DK-2920 Charlottenlund, Denmark

Corkeron, P J
James Cook University
School of Tropical Environment Studies
  and Geography
Townsville, QLD 4811, Australia

Coull, B C
University of South Carolina
School of the Environment
Columbia, SC 29208, USA

Cowen, R
Rosenstiel School of Marine and Atmospheric Science
University of Miami
4600 Rickenbacker Causeway
Miami, FL 33149, USA

Cresswell, G
CSIRO Marine Research
GPO Box 1538
Hobart, Tasmania 7001, Australia

Cronan, D S
Imperial College of Science, Technology and Medicine
TH Huxley School of Environment,
  Earth Sciences and Engineering
Prince Consort Road
London, SW7 2BP, UK

Cronin, M F
NOAA Pacific Marine Environmental Laboratory
7600 Sand Point Way NE
Seattle, WA 98115, USA

Cullen, J J
Dalhousie University
Department of Oceanography
Halifax
Nova Scotia, B3H 4J1, Canada

Cushing, D H
The Centre for Environmental Fisheries
  and Aquaculture Science
Pakefield Road
Lowestoft, Suffolk, NR33 0HT, UK

Cuthbert, R J
University of Otago
Department of Zoology
Box 56
Dunedin, New Zealand

Cutter, G A
Old Dominion University
Ocean Earth and Atmospheric Sciences
Norfolk, VA 23529, USA

Dacey, J W
Woods Hole Oceanographic Institution
Biology Department
Mail Stop 38
Woods Hole, MA 02543, USA

Damm, K L V
University of New Hampshire
Institute for the Study of Earth, Oceans and Space
39 College Road
Durham, NH 03824, USA

Davenport, J
University College Cork
Department of Zoology and Animal Ecology
Lee Maltings
Prospect Row, Cork, Ireland

David, A R J
3 Maynard Park
Bere Alston
Devon PL20 7AR, UK

Davis, L S
University of Otago
Department of Zoology
Box 56
Dunedin, New Zealand

de Angelis, M A
Humboldt State University
Department of Oceanography
Arcata, CA 95521, USA

de Jonge, V N
Groningen University
Department of Marine Biology
PO Box 14
9750 AE Haren, The Netherlands

DeMaster, D J
North Carolina State University
Department of Marine, Earth
and Atmospheric Sciences
Raleigh, NC 27695, USA

Denman, K L
University of Victoria
Canadian Centre for Climate Modelling
and Analysis
Victoria, BC, Canada

Deuser, W
Woods Hole Oceanographic Institution
Mail Stop 25
Woods Hole, MA 02543, USA

Dickey, T D
University of California, Santa Barbara
Ocean Physics Laboratory
6487 Calle Real Suite A
Santa Barbara, CA 93117, USA

Diemand, D
Coriolis, PO Box 284
Shoreham, VT 05770, USA

Dinsmore, R P
Woods Hole Oceanographic Institution
Mail Stop 37
Woods Hole, MA 02543, USA

Divoky, G J
University of Alaska, Fairbanks
Institute of Arctic Biology
Fairbanks, AK 99775, USA

Doake, C S M
British Antarctic Survey
High Cross, Madingley Road
Cambridge, CB3 0ET, UK

Donat, J
Old Dominion University
Department of Chemistry and Biochemistry
Norfolk, VA 23529, USA

Doney, S C
Climate and Global Dynamics
National Center for Atmospheric Research
Boulder, CO 80027, USA

Donlon, C J
Joint Research Centre of the European Commission
Space Applications Institute
Marine Environment Unit
Ispra, Italy

Dorman, L M
Scripps Institution of Oceanography
University of California, San Diego
9500 Gilman Drive
La Jolla, CA 92093, USA

Douglas, R H
City University
Department of Optometry and Visual Science
Northampton Square
London, EC1V 0HB, UK

Doumenge, F
Museé Océanographique de Monaco
Avenue Saint-Martin
MC 98000, Monaco

Dover, C L Van
College of William and Mary
Biology Department
Williamsburg, VA 23187, USA

Dower, J F
University of British Columbia
Department of Earth and Ocean Sciences
Vancouver, BC, Canada

Draxler, S
Karl-Franzens-Universität Graz
Institut für Experimentalphysik
Universitatsplatz 5
A-8010 Graz, Austria

Dryden, C
Old Dominion University
Department of Chemistry and Biochemistry
Norfolk, VA 23529, USA

Duce, R A
Texas A&M University
Department of Oceanography
and Department of Atmospheric Sciences
College Station, TX 77843, USA

Ducklow, H W
The Virginia Institute of Marine Science
College of William and Mary
PO Box 1346
Gloucester Point, VA 23062, USA

Duffy-Anderson, J T
Alaska Fisheries Science Center
7600 Sand Point Way NE
Seattle, WA 98115, USA

Dufour, A
United States Environmental Protection Agency
26 West Martin Luther King Drive
Cincinnati, OH 45268, USA

Dyer, I
9 Cliff Street
Marblehead, MA 01945, USA

Dyer, K
University of Plymouth
Institute of Marine Studies
Drake Circus
Plymouth, PL4 8AA, UK

Dyrssen, D W
Gothenburg University
Department of Analytical and Marine Chemistry
Gothenburg, SE-412 96, Sweden

Edmonds, H N
University of Texas at Austin
Marine Science Institute
750 Channel View Drive
Port Aransas, TX 78373, USA

Edson, J B
Woods Hole Oceanographic Institution
Mail Stop 12
Woods Hole, MA 02543, USA

Edwards, C A
University of Connecticut
Department of Marine Science
1084 Shennecossett Road
Groton, CT 06340, USA

Edwards, M
SAHFOS
1 Walker Terrace
The Hoe
Plymouth, PL1 3BN, UK

Eglinton, T I
Woods Hole Oceanographic Institution
Mail Stop 4
Woods Hole, MA 02543, USA

Eldholm, O
University of Oslo
Department of Geology
PO Box 1047
Blindern, N-0316 Oslo, Norway

Elliott, M
University of Hull
Department of Biological Sciences
Institute of Estuarine and Coastal Studies
Hull, HU6 7RX, UK

Ellis, A E
Marine Laboratory
Victoria Road
Aberdeen, AB11 9DB, UK

Emery, W J
University of Colorado
Colorado Center for Astrodynamics Research (CCAR)
Boulder, CO 80309, USA

England, M H
University of New South Wales
Centre for Environmental Modelling
    and Prediction (CEMAP)
School of Mathematics
Sydney, NSW 2052, Australia

Ettwein, V
University College London
Department of Geography
Environmental Change Research Centre
26 Bedford Way
London, WC1H 0AP, UK

Evans, S M
Newcastle University
Department of Marine Sciences
    and Coastal Management
Dove Marine Laboratory
Cullercoats
Tyne and Wear, NE30 4PZ, UK

Everson, I
British Antarctic Survey
High Cross, Madingley Road
Cambridge, CB3 0ET, UK

Fahrbach, E
Alfred-Wegener-Institut für Polar-
    und Meeresforschung
Postfach 12 01 61
27515 Bremerhaven, Germany

Farmer, D M
Institute of Ocean Sciences
9860 West Saanich Road, Sidney
British Columbia, V8L 4B2, Canada

Farrington, J W
Woods Hole Oceanographic Institution
Mail Stop 31
Woods Hole, MA 02543, USA

Farrow, S
Carnegie Mellon University
Center for the Study and Improvement of Regulation
Pittsburgh, PA 15213, USA

Fieux, M
Université Pierre et Marie Curie
LODYC, Tour 14-2ê4 Place de Jussieu
75252 Paris, France

Fine, R A
Rosenstiel School of Marine and Atmospheric Science
University of Miami
4600 Rickenbacker Causeway
Miami, FL 33149, USA

Flather, R A
Proudman Oceanographic Laboratory
Bidston Observatory
Birkenhead
Wirral, CH43 7RA, UK

Fogarty, M J
National Marine Fisheries Service (NOAA)
166 Water Street
Woods Hole, MA 02543, USA

Fonteyne, R
Agricultural Research Centre
Ghent Sea Fisheries Department
Ankerstraat 1, B-8400 Oostende, Belgium

Foote, K G
Woods Hole Oceanographic Institution
Department of Applied Ocean Physics
   and Engineering
Mail Stop 12
Woods Hole, MA 02543, USA

Fornari, D J
Woods Hole Oceanographic Institution
Department of Geology and Geophysics
Mail Stop 22
Woods Hole, MA 02543, USA

Forteath, N
Seahorse Australia
Inspection Head Wharf
Beauty Point
Tasmania 7270, Australia

Føyn, L
Institute of Marine Research
PO Box 1870
Nordnes, N-5024 Bergen, Norway

François, L
University of Liege
Laboratory for Planetary
   and Atmospheric Physics
Liege, Belgium

Friedrichs, M A M
Old Dominion University
Center for Coastal Physical Oceanography
Norfolk, VA 23529, USA

Fuhrman, J
University of Southern California
Department of Biological Sciences
Los Angeles, CA 90089, USA

Gage, J D
Scottish Association for Marine Science
Dunstaffnage Marine Laboratory
Oban, Argyll, PA34 4AD, UK

Gallienne, C P
Plymouth Marine Laboratory
Prospect Place
West Hoe
Plymouth, PL1 3DH, UK

Ganachaud, A
IFREMER UM/LPO
BP 70, 29280 Plouzané, France

Garcia, S M
Food and Agriculture Organization of
   the United Nations
Fishery Resources Division
Viale delle Terme di Caracalla
00100 Rome, Italy

Garrett, C
University of Victoria
Department of Physics
PO Box 3055
Victoria, British Columbia V8W 3P6, Canada

Gaston, T
National Wildlife Research Centre
Canadian Wildlife Service
100 Gamelin Boulevard Hull
Quebec, K1A 0H3, Canada

Gibson, C H
Scripps Institution of Oceanography
University of California, San Diego
9500 Gilman Drive
La Jolla, CA 92093, USA

Gibson, R N
Scottish Association for Marine Science
Dunstaffnage Marine Laboratory
Oban, Argyll, PA34 4AD, UK

Giese, G S
Woods Hole Oceanographic Institution
Mail Stop 22
Woods Hole, MA 02543, USA

Glover, D M
Woods Hole Oceanographic Institution
Marine Chemistry and Geochemistry Department
Woods Hole, MA 02543, USA

Gochfeld, M
Robert Wood Johnson Medical School
Environmental and Community Medicine
UMDNJ
Piscataway, NJ 08854, USA

Goddéris, Y
University of Liege
Laboratory for Planetary
  and Atmospheric Physics
Liege, Belgium

Godfrey, S
CSIRO Division of Marine Research
GPO Box 1538
Hobart 7001, Tasmania, Australia

Gooday, A J
Southampton Oceanography Centre
University of Southampton
Waterfront Campus
European Way
Southampton, SO14 3ZH, UK

Goodbred Jr, S L
State University of New York, Stony Brook
Marine Sciences Research Center
Stony Brook, NY 11794, USA

Gordon, A L
Lamont-Doherty Earth Observatory
  of Columbia University
PO Box 1000
61 Route 9W
Palisades, NY 10964, USA

Gordon, J D M
Scottish Association for Marine Science
Dunstaffnage Marine Laboratory
Oban, Argyll, PA34 4AD, UK

Gorlov, A M
Northeastern University
Boston, MA 02115, USA

Grassle, J F
Rutgers University
Institute of Marine and Coastal Sciences
71 Dudley Road
New Brunswick, NJ 83901, USA

Gray, J S
University of Oslo
Biological Institute
Pb 1064
Blindern, 0316 Oslo, Norway

Gregory, J M
The Meteorological Office
Hadley Centre
London Road
Bracknell
Berkshire, RG12 2SY, UK

Griffiths, G
Southampton Oceanography Centre
University of Southampton
Waterfront Campus
European Way
Southampton, SO14 3ZH, UK

Grottoli, A G
University of Pennsylvania
Department of Earth
  and Environmental Science
240 South 33rd Street
162 Hayden Hall
Philadelphia, PA 19104, USA

Hall, S J
Flinders University
School of Biological Sciences
Adelaide, SA 5001, Australia

Hamer, K C
University of Durham
Department of Biological Sciences
South Road
Durham, DH1 3LE, UK

Hammond, D
University of Southern California
Department of Earth Sciences
Los Angeles, CA 90089, USA

Hansell, D A
Rosenstiel School of Marine and Atmospheric Science
University of Miami
4600 Rickenbacker Causeway
Miami, FL 33149, USA

Haq, B U
10 Vendome Court
Bethesda, MD 20817, USA

Harbison, G R
Woods Hole Oceanographic Institution
Mail Stop 38
Woods Hole, MA 02543, USA

Harding, A
Scripps Institution of Oceanography
University of California, San Diego
9500 Gilman Drive
La Jolla, CA 92093, USA

Harding Jr, L W
University of Maryland Center for
  Environmental Science
Horn Point Laboratory
  and Sea Grant College
0112 Skinner Hall
College Park, MD 20742, USA

Hart, P J B
University of Leicester
Zoology Department
University Road
Leicester, LE1 7RH, UK

Harris, R
Plymouth Marine Laboratory
Prospect Place
West Hoe
Plymouth, PL1 3DH, UK

Hay, W W
GEOMAR, Christian-Albrechts University
Wischhofstrasse 1-3
D-24148 Kiel, Germany

Haymon, R M
University of California, Santa Barbara
Department of Geological Sciences
Santa Barbara, CA 93106, USA

Heath, J W
Louisiana State University
Coastal Fisheries Institute
CCEER
Baton Rouge, LA 70803, USA

Hebert, D L
University of Rhode Island
Graduate School of Oceanography
Narragansett Bay Campus, RI 02882, USA

Hedgecock, D
University of California
Bodega Marine Laboratory
Bodega Bay, CA 94923, USA

Helfrich, K R
Woods Hole Oceanographic Institution
Mail Stop 21
Woods Hole, MA 02543, USA

Helmond, I
CSIRO Division of Marine Research
GPO Box 1538, Hobart
Tasmania, 7000, Australia

Hemleben, C
Tübingen University
Institute and Museum of Geology
   and Paleontology
Sigwartstrasse 10
72076 Tübingen, Germany

Herbert, T D
Brown University
Geological Sciences
Box 1846
Providence, RI 02912, USA

Herring, P J
Southampton Oceanography Centre
University of Southampton
Waterfront Campus
European Way
Southampton, SO14 3ZH, UK

Herzen, R P V
Woods Hole Oceanographic Institution
Department of Geology and Geophysics
Woods Hole, MA 02543, USA

Herzog, H
Massachusetts Institute of Technology
Energy Laboratory
77 Massachusetts Avenue
Cambridge, MA 02139, USA

Hey, R
University of Hawaii at Manoa
Hawaii Institute of Geophysics
   and Planetology Scientific
2525 Correa Road
Honolulu, HI 96822, USA

Heyning, J E
The Natural History Museum of
   Los Angeles County
900 Exposition Boulevard
Los Angeles, CA 90007, USA

Hickey, B M
University of Washington
School of Oceanography
Box 357940
Seattle, WA 98195, USA

Higgs, D M
University of Maryland
Department of Biology
College Park, MD 20742, USA

Hixon, M A
Oregon State University
Department of Zoology
Corvallis, OR 97331, USA

Hoagland, P
Woods Hole Oceanographic Institution
Mail Stop 41
Woods Hole, MA 02543, USA

Hofmann, E E
Old Dominion University
Center for Coastal Physical Oceanography
Norfolk, VA 23529, USA

Holman, R A
Oregon State University
College of Ocean and Atmospheric Science
Corvallis, OR 97331, USA

Holt, J T
Proudman Oceanographic Laboratory
Bidston Observatory
Birkenhead
Wirral, CH43 7RA, UK

Hong, S-Y
Ministry of Maritime Affairs
  and Fisheries (MOMAF)
Vice Minister
139 Chungjong-No 3, Seodaemun-Gu
Seoul, 120-715, South Korea

Honjo, S
Woods Hole Oceanographic Institution
Department of Geology and Geophysics
Woods Hole, MA 02543, USA

Hood, M
Intergovernmental Oceanographic Commission
UNESCO, 1 rue Miollis
75732 Paris Cedex 15, France

Hooker, S K
British Antarctic Survey
High Cross, Madingley Road
Cambridge, CB3 0ET, UK

Hotta, H
Japan Marine Science Technology Center (JAMSTEC)
2–15 Natsushimacho Yokosuka
237-0061, Japan

Houde, E D
University of Maryland
Center for Environmental Science
Chesapeake Biological Laboratory
PO Box 38
Solomons, MD 20688, USA

Howarth, M J
Proudman Oceanographic Laboratory
Bidston Observatory
Birkenhead
Wirral, CH43 7RA, UK

Hunt Jr, G L
University of California, Irvine
Department of Ecology and Evolutionary Biology
Irvine, CA 92697, USA

Hurrell, J W
National Center for Atmospheric Research
PO Box 3000
Boulder, CO 80307, USA

Hutchinson, P
North Atlantic Salmon Conservation Organization
11 Rutland Square
Edinburgh, EH1 2AS, UK

Huthnance, J M
Proudman Oceanographic Laboratory
Bidston Observatory
Birkenhead
Wirral, CH43 7RA, UK

Ierley, G R
Scripps Institution of Oceanography
University of California, San Diego
9500 Gilman Drive
La Jolla, CA 92093, USA

Jacoby, J
Woods Hole Oceanographic Institution
Marine Policy Center
Mail Stop 41
Woods Hole, MA 02543, USA

Jähne, B
University of Heidelberg
Research Group Image Processing
Im Neuenheimer Feld 368
69120 Heidelberg, Heidelberg, Germany

Jahnke, R A
Skidaway Institut of Oceanography
10 Ocean Science Circle
Savannah, GA 31411, USA

Jenkins, W J
Southampton Oceanography Centre
University of Southampton
Waterfront Campus
European Way
Southampton, SO14 3ZH, UK

Jensen, F B
North Atlantic Treaty Organisation
SACLANT Undersea Research Centre
Viale San Bartolomeo 400
19138 La Spezia, Italy

John, A
SAHFOS
The Laboratory
Citadel Hill
Plymouth
Devon, PL1 2PB, UK

Joseph, J
2790 Palomino Circle
La Jolla, CA 92037, USA

Karl, D M
University of Hawaii at Manoa
Department of Oceanography
1000 Pope Road
Honolulu, HI 96822, USA

Katsaros, K B
Atlantic Oceanographic and
  Meteorological Laboratory (NOAA)
4301 Rickenbacker Causeway
Miami, FL 33149, USA

Kemp, A E S
Southampton Oceanography Centre
University of Southampton
Waterfront Campus
European Way
Southampton, SO14 3ZH, UK

Kennedy, V S
University of Maryland
Horn Point Laboratory
Center for Environmental Science
Box 775
Cambridge, MD 21613, USA

Key, R M
Princeton University
Atmospheric and Oceanic Sciences Program
Department of Geoscience
Princeton, NJ 08540, USA

Killworth, P D
Southampton Oceanography Centre
University of Southampton
Waterfront Campus
European Way
Southampton, SO14 3ZH, UK

Kingston, P F
Heriot-Watt University
Department of Biological Sciences
Riccarton
Edinburgh, EH14 4AS, UK

Kite-Powell, H L
Woods Hole Oceanographic Institution
Marine Policy Center
Mail Stop 41
Woods Hole, MA 02543, USA

Klinck, J
Old Dominion University
Center for Coastal Physical Oceanography
Crittenton Hall
Norfolk, VA 23529, USA

Klinger, B
Center for Ocean-Land-Atmosphere
  Studies (COLA)
4041 Powder Mill Road
Suite 302
Calverton, MD 20705, USA

Kominz, M A
Western Michigan University
Department of Geology
Kalamazoo, MI 49008, USA

Kooyman, G L
Scripps Institution of Oceanography
University of California, San Diego
9500 Gilman Drive
La Jolla, CA 92093, USA

Kope, R
Northwest Fisheries Science Center
2725 Montlake Boulevard East
Seattle, WA 98112, USA

Kraemer, D R B
The Johns Hopkins University
Civil Engineering Department
108 Latrobe Hall
3400 N. Charles Street
Baltimore, MD 21218, USA

Krauss, W
Universität Kiel
Institut für Meereskunde, Kiel, Germany

Krijgsman, W
University of Utrecht
Faculty of Earth Sciences
Budapestlaan 17
3584 HD Utrecht, The Netherlands

Krishnaswami, S
Physical Research Laboratory
Navrangpura
Ahmedabad, 380 009, India

Kristoffersen, J B
University of Bergen
Department of Fisheries and Marine Biology
Bergen High Technology Centre
PO Box 7800
N-5020 Bergen, Norway

Kunze, E L
Applied Physics Laboratory
University of Washington, 1013 NE 40th
Seattle, WA 98105, USA

Kuperman, W A
Scripps Institution of Oceanography
University of California, San Diego
9500 Gilman Drive
La Jolla, CA 92093, USA

Lagerloef, G
Earth and Space Research
1910 Fairview Avenue E, Suite 102
Seattle, WA 98102, USA

Laird, L M
Aberdeen University
Department of Zoology
Aberdeen, AB24 2TZ, UK

Lal, D
Scripps Institution of Oceanography
University of California, San Diego
9500 Gilman Drive
La Jolla, CA 92093, USA

Lambeck, K
Australian National University
Research School of Earth Sciences
Canberra, ACT, Australia

Lampitt, R S
Southampton Oceanography Centre
University of Southampton
Waterfront Campus
European Way
Southampton, SO14 3ZH, UK

Landry, M
University of Hawaii at Manoa
Department of Oceanography
1000 Pope Road
Honolulu, HI 96822, USA

Lane-Serff, G F
UMIST
Department of Civil and Structural Engineering
PO Box 88
Manchester, M60 1QD, UK

Langereis, C G
University of Utrecht
Faculty of Earth Sciences
Budapestlaan 17
3584 HD Utrecht, The Netherlands

Langford, T E L
University of Southampton
Centre for Environmental Sciences
Building 46
Highfield
Southampton, SO41 0RJ, UK

Larson, N G
Sea-Bird Electronics Inc
1808-136th Place NE
Bellevue, WA 98005, USA

Lascaratos, A
University of Athens
Department of Applied Physics
Oceanography Group
University Campus, Building PHYS-V
Athens, 15784, Greece

Law, C S
Plymouth Marine Laboratory
Prospect Place
West Hoe
Plymouth, PL1 3DH, UK

Lawson, K
Sea-Bird Electronics Inc
1808-136th Place NE
Bellevue, WA 98005, USA

Lazier, J R N
Bedford Institute of Oceanography
PO Box 1006
Dartmouth
Nova Scotia, B2Y 4A2, Canada

Ledwell, J R
Woods Hole Oceanographic Institution
Mail Stop 12
Woods Hole, MA 02543, USA

Lee, C-W
Kyungnam University
Department of Civil and Environmental Engineering
449 Wolyoung-dong
Masan, South Korea

Leibovich, S
Cornell University
Department of Mechanical
   and Aerospace Engineering
105 Upson Hall
Ithaca, NY 14853, USA

Lermusiaux, P F J
Harvard University
Division of Engineering and Applied Sciences
Department of Earth and Planetary Sciences
20 Oxford Street
Cambridge, MA 02138, USA

Leslie, W G
Harvard University
Division of Engineering and Applied Sciences
Department of Earth and Planetary Sciences
20 Oxford Street
Cambridge, MA 02138, USA

Lindstrom, E J
NASA, Code YS
Washington, DC 20546, USA

Lippitsch, M E
Karl-Franzens-Universität Graz
Institut für Experimentalphysik
Universitatsplatz 5
A-8010 Graz, Austria

Liu, A K
NASA Goddard Space Flight Center
Oceans and Ice Branch
Greenbelt, MD 20771, USA

Longhurst, A
Place de l'Eglise
46160 Cajarc, France

Lueck, R
University of Victoria
School of Earth and Ocean Sciences
PO Box 1700, Victoria
British Columbia, V8W 2YX, Canada

Lukas, R
University of Hawaii at Manoa
Department of Oceanography
1000 Pope Road
Honolulu, HI 96822, USA

Lupton, J E
NOAA Pacific Marine Environmental Laboratory
Hatfield Marine Science Center
Newport, OR 97365, USA

Lutjeharms, J R E
University of Cape Town
Department of Oceanography
Rondebosch 7700, Republic of South Africa

Lutz, R A
Rutgers University
Institute of Marine and Coastal Sciences
71 Dudley Road
New Brunswick, NJ 08901, USA

Macdonald, A M
Woods Hole Oceanographic Institution
Mail Stop 21
Woods Hole, MA 02543, USA

Macdonald, K C
University of California
Department of Geological Sciences
Santa Barbara
CA 93106, USA

Mackenzie, F T
University of Hawaii at Manoa
Department of Oceanography
1000 Pope Road
Honolulu, HI 96822, USA

Madin, L P
Woods Hole Oceanographic Institution
Mail Stop 38
Woods Hole, MA 02543, USA

Malanotte-Rizzoli, P
Massachusetts Institute of Technology
Department of Earth, Atmospheric
   and Planetary Sciences
77 Massachusetts Avenue
Cambridge, MA 02139, USA

Martin, S
University of Washington
School of Oceanography
Box 357940
Seattle, WA 98195, USA

Maslin, M
University College London
Department of Geography
Environmental Change Research Centre
26 Bedford Way
London, WC1H 0AP, UK

Masutani, S M
University of Hawaii at Manoa
Hawaii Natural Energy Institute
School of Ocean and Earth Science
   and Technology
2540 Dole Street, Holmes Hall 246
Honolulu, HI 96822, USA

Matano, R P
Oregon State University
College of Oceanic Atmospheric Sciences
Corvallis, OR 97331, USA

Matsumoto, R
University of Tokyo
Department of Earth and Planetary Science
Graduate School of Science
Hongo, Tokyo 113, Japan

Maul, G A
Florida Institute of Technology
Department of Marine
   and Environmental Systems
150 West University Boulevard
Melbourne, FL 32901, USA

McCave, I N
University of Cambridge
Department of Earth Sciences
Cambridge, CB2 3EQ, UK

McCay, B J
Rutgers University
Department of Human Ecology at Cook College
55 Dudley Road
New Brunswick, NJ 08901, USA

McCleave, J D
University of Maine
School of Marine Sciences
5741 Libby Hall
Orono, ME 04469, USA

McClain, C
NASA Goddard Space Flight Center
Earth Sciences Directorate
Greenbelt, MD 20771, USA

McCormick, M E
The Johns Hopkins University
Civil Engineering Department
108 Latrobe Hall
3400 N. Charles Street
Baltimore, MD 21218, USA

McGillicuddy Jr, D J
Woods Hole Oceanographic Institution
Department of Applied Ocean Physics
   and Engineering
Woods Hole, MA 02543, USA

McIntyre, A D
University of Aberdeen
Aberdeen, AB15 5BW, UK

McManus, J W
University of Miami
National Center for Atlantic and
   Caribbean Reef Research
Rosenstiel School of Marine
   and Atmospheric Science
4600 Rickenbacker Causeway
Miami, FL 33149, USA

McMurtry, G M
University of Hawaii at Manoa
Department of Oceanography
1000 Pope Road
Honolulu, HI 96822, USA

McNeil, C
Academic Press
32 Jamestown Road
London, NW1 7BY, UK

McNutt, M
MBARI
7700 Sandholdt Road
PO Box 628
Moss Landing, CA 95039, USA

McPhee, M
McPhee Research Company
Naches, WA 98937, USA

Melville, W K
Scripps Institution of Oceanography
University of California, San Diego
9500 Gilman Drive
La Jolla, CA 92093, USA

Melville-Smith, R
Fisheries WA Research Division
PO Box 20
North Beach, WA 6020, Australia

Merrin, C L
University of British Columbia
Department of Earth and Ocean Sciences
6270 University Boulevard
Vancouver
British Columbia V6T 1Z4, Canada

Michaels, A F
University of Southern California
Wrigley Institute for Environmental Sciences
Los Angeles, CA 90089, USA

Middleton, J H
The University of New South Wales
Department of Aviation
Sydney 2052, New South Wales, Australia

Mikhalevsky, P N
Science Applications International Corporation
Ocean Sciences Division
1700 Goodridge Drive
McLean, VA 22102, USA

Miller, W D
University of Maryland
Center for Environmental Science
Horn Point Laboratory
0112 Skinner Hall
College Park, MD 20742, USA

Milliman, J D
The Virginia Institute of Marine Science
College of William and Mary
PO Box 1346
Gloucester Point, VA 23062, USA

Mills, D
Atlantic Salmon Trust
Moulin, Pitlochry
Perthshire, PH16 5JQ, UK

Millward, G E
University of Plymouth
Department of Environmental Sciences
Plymouth, Devon PL4 8AA, UK

Minchin, D
Marine Organism Investigations
3 Marina Village, Ballina
Killaloe, Co. Clare, Ireland

Minnett, P J
Rosenstiel School of Marine and
   Atmospheric Science
University of Miami
4600 Rickenbacker Causeway
Miami, FL 33149, USA

Mobley, C D
Sequoia Scientific Inc
Westpark Technical Center
15317 NE 90th Street
Redmond, WA 98052, USA

Momma, H
Japan Marine Science Technology Center (JAMSTEC)
2–15 Natsushimacho Yokosuka
237-0061, Japan

Monahan, E C
University of Connecticut
Department of Marine Sciences
1084 Shennecossett Road
Groton, CT 06340, USA

Montevecchi, W A
Memorial University of Newfoundland
Biopsychology Programme
Departments of Psychology and Biology
St John's
Newfoundland, A1B 3X9, Canada

Morel, A
Université Pierre et Marie Curie, Paris 6
Laboratoire Physique et Chimie Marine
Quai de la Darse BP8, F-06238
Villefranche-sur-mer, France

Moore, C
WET Labs, Inc
PO Box 518
620 Applegate Street
Philomath, OR 97370, USA

Moore, J C
University of California at Santa Cruz
Earth Sciences Department
Santa Cruz, CA 95064, USA

Moran, K
Joint Oceanographic Institutions
1755 Massachusetts Avenue NW
Suite 800
Washington, DC 20036, USA

Moreno, I D L
Food and Agriculture Organization of
   the United Nations
Fishery Resources Division
Viale delle Terme di Caracalla
00100 Rome, Italy

Morison, J H
University of Washington
Applied Physics Laboratory, Polar Science Center
1013 NE 40th Street
Seattle, WA 98105, USA

Morreale, S J
Cornell University
Department of Natural Resources
Ithaca, NY 14853, USA

Moum, J N
Oregon State University
College of Oceanic and Atmospheric Sciences
Corvallis, OR 97331, USA

Mullin, M M[†]
Scripps Institution of Oceanography
University of California, San Diego
9500 Gilman Drive
La Jolla, CA 92093, USA

Munk, W
University of California San Diego
Institute of Geophysics and Planetary Physics
La Jolla, CA 92093, USA

Murphy, E J
British Antarctic Survey
High Cross, Madingley Road
Cambridge, CB3 0ET, UK

Murray, L A
Fisheries Laboratory
Remembrance Avenue
Burnham-on-Crouch
Essex, CM0 8HA, UK

Naidu, P D
National Institute of Oceanography
Dona Paula 403 004, Goa, India

Narayanaswamy, R
UMIST
Department of Instrumentation
   and Analytical Science
PO Box 88
Manchester, M60 1QD, UK

Neal, W J
Grand Valley State University
Department of Geology
Allendale, MI 49401, USA

Neilson, J D
Department of Fisheries and Oceans
Marine Fish Division, Biological Station
St Andrews, New Brunswick E0G 2X0, Canada

Nicholls, K W
British Antarctic Survey
High Cross, Madingley Road
Cambridge, CB3 0ET, UK

[†]deceased

Niitsuma, N
Shizuoka University
836 Ohya, Shizuoka
422-8529, Japan

Nowlin Jr, W D
Texas A&M University
Department of Oceanography
305 Arguello Drive
College Station, TX 77843, USA

Nozaki, Y
University of Tokyo
The Ocean Research Institute
1-15-1 Minamidai
Nakano-Ku, Tokyo 164, Japan

Oakey, N S
Bedford Institute of Oceanography
PO Box 1006
Dartmouth
Nova Scotia B2Y 4A2, Canada

O'Shea, T J
United States Geological Survey
Southern Rocky Mountain Ecosystems Section
4512 McMurry Avenue
Fort Collins, CO 80525, USA

Olson, D B
Rosenstiel School of Marine and Atmospheric Science
University of Miami
4600 Rickenbacker Causeway
Miami, FL 33149, USA

O'Neil Baringer, M O
NOAA-AOML/PHOD
4301 Rickenbacker Causeway
Key Biscayne, FL 33149, USA

Orians, K J
University of British Columbia
Department of Earth and Ocean Sciences
6270 University Boulevard
Vancouver, British Columbia V6T 1Z4, Canada

Osterkamp, T E
University of Alaska, Fairbanks
Geophysical Institute
Fairbanks, AK 99775, USA

Paladino, F V
Indiana-Purdue University at Fort Wayne
Department of Biology
Fort Wayne, IN 46805, USA

Parijs, S M V
Norwegian Polar Institute
Tromsø 9296, Norway

Paris, C
Rosenstiel School of Marine and Atmospheric Science
University of Miami
4600 Rickenbacker Causeway
Miami, FL 33149, USA

Parkinson, C L
NASA Goddard Space Flight Center
Oceans and Ice Branch
Code 971
Greenbelt, MD 20771, USA

Paulson, C A
Oregon State University
College of Oceanic and Atmospheric Sciences
104 Ocean Admin Bldg
Corvallis, OR 97331, USA

Pauly, D
Fisheries Centre, University of British Columbia
2204 Main Hall
Vancouver, British Columbia V6T 1Z4, Canada

Pearson, A
Woods Hole Oceanographic Institution
Mail Stop 4
Woods Hole, MA 02543, USA

Pegau, W S
Oregon State University
College of Oceanic and Atmospheric Sciences
104 Ocean Admin Bldg
Corvallis, OR 97331, USA

Penn, J W
Fisheries WA Research Division
PO Box 20
North Beach, WA 6020, Australia

Perfit, M R
University of Florida
Department of Geological Sciences
PO Box 112120
Gainesville, FL 32611, USA

Perry, R I
Department of Fisheries and Oceans
Pacific Biological Station
Nanaimo, British Colombia
V9R 5K6, Canada

Peterson, L C
Rosenstiel School of Marine and Atmospheric Science
University of Miami
4600 Rickenbacker Causeway
Miami, FL 33149, USA

Philander, S G
Princeton University
Department of Geosciences
Princeton, NJ 08540, USA

Pilcher, N J
Universiti Malaysia Sarawak
Institute of Biodiversity and Environmental Conservation
94300 Kota Samarahan
Sarawak, Malaysia

Pilkey, O H
Duke University
Division of Earth and Ocean Sciences
Nicholas School of the Environment
Durham, NC 27708, USA

Piola, A R
Servicio de Hidrografa Naval
Departamento Oceanografa
Avenida Montes de Oca 2124
1271 Buenos Aires, Argentina

Pitcher, T J
University of British Columbia
Fisheries Centre
2204 Main Mall
Vancouver V6T 1Z4, Canada

Plant, W J
University of Washington
Applied Physics Laboratory
Seattle, WA 98105, USA

Platt, T
Dalhousie University
Oceanography Department
Halifax
Nova Scotia, B3H 4J1, Canada

Plueddemann, A J
Woods Hole Oceanographic Institution
Mail Stop 29
Woods Hole, MA 02543, USA

Polovina, J J
National Marine Fisheries Service (NOAA)
Honolulu Laboratory
2570 Dole Street
Honolulu, HI 96822, USA

Popper, A N
University of Maryland
Department of Biology
College Park, MD 20742, USA

Powell, J A
Florida Fish and Wildlife Conservation Commission
Florida Marine Research Institute
100 Eighth Avenue SE
St Petersburg, FL 33701, USA

Price, J F
Woods Hole Oceanographic Institution
Mail Stop 29
Woods Hole, MA 02543, USA

Prien, R D
Southampton Oceanography Centre
University of Southampton
Waterfront Campus
European Way
Southampton, SO14 3ZH, UK

Proctor, R
Proudman Oceanographic Laboratory
Bidston Observatory
Birkenhead
Wirral, CH43 7RA, UK

Pugh, D T
Southampton Oceanography Centre
University of Southampton
Waterfront Campus
European Way
Southampton, SO14 3ZH, UK

Qiu, B
University of Hawaii at Manoa
Department of Oceanography
1000 Pope Road
Honolulu, HI 96822, USA

Quadfasel, D
Niels Bohr Institute
Geophysics Department
Juliane Maries Vej 30
DK-2100 Copenhagen, Denmark

Rahmstorf, S
Potsdam Institute for Climate Impact Research
PO Box 601203
D-14412 Potsdam, Germany

Raven, J A
University of Dundee
Biological Sciences
Dundee, DD1 4HN, UK

Ravizza, G E
Woods Hole Oceanographic Institution
Mail Stop 22
Woods Hole, MA 02543, USA

Ray, R D
NASA Goddard Space Flight Center
Laboratory for Terrestrial Physics
Code 926
Greenbelt, MD 20771, USA

Reeve, M R
National Science Foundation
4201 Wilson Boulevard
Room 752N
Arlington, VA 22230, USA

Reeves, R R
Okapi Wildlife Associates
27 Chandler Lane
Hudson
Quebec, JOP 1HO, Canada

Reid, P C
SAHFOS
1 Walker Terrace
West Hoe, Plymouth
Devon, PL1 3BN, UK

Reverdin, G
LEGOS
14 Av. E. Belin
31401, Toulouse Cedex, France

Reysenbach, A-L
Portland State University
Department of Biology
PO Box 751
Portland, OR 97207, USA

Rhines, P B
University of Washington
School of Oceanography
Box 357940
Seattle, WA 98195, USA

Richardson, P L
Woods Hole Oceanographic Institution
Mail Stop 29
Woods Hole, MA 02543, USA

Robinson, A R
Harvard University
Division of Engineering and Applied Sciences
Department of Earth and Planetary Sciences
20 Oxford Street
Cambridge, MA 02138, USA

Roman, C T
University of Rhode Island
United States Geological Survey
Patuxent Wildlife Research Center
Narragansett, RI 02882, USA

Rossby, H T
University of Rhode Island
Graduate School of Oceanography
Kingston, RI 02881, USA

Royer, T C
Old Dominion University
Center for Coastal Physical Oceanography
Department of Ocean, Earth and
    Atmospheric Sciences
Norfolk, VA 23508, USA

Rozwadowski, H M
Georgia Institute of Technology
School of History, Technology, and Society
Atlanta, GA 30332, USA

Rubega, M
University of Connecticut
Department of Ecology
    and Evolutionary Biology
75 North Eagleville Rd. U-3043
Storrs, CT 06269, USA

Rudels, B
Finnish Institute of Marine Research
PO Box 33
Helsinki, FIN-00931, Finland

Rudnick, D L
Scripps Institution of Oceanography
University of California, San Diego
9500 Gilman Drive
La Jolla, CA 92093, USA

Ruttenberg, K C
Woods Hole Oceanographic Institution
Mail Stop 8
Woods Hole, MA 02543, USA

Salas, H
CEPIS/HEP/Pan American Health Organization
Lima, Peru

Salvanes, A G V
University of Bergen
Department of Fisheries and Marine Biology
Bergen High Technology Centre
PO Box 7800
N-5020 Bergen, Norway

Sathyendranath, S
Dalhousie University
Oceanography Department
Halifax
Nova Scotia, B3H 4J1, Canada

Sayer, M D J
Scottish Association for Marine Sciences
Dunstaffnage Marine Laboratory
Post Box 3
Oban, Argyll, PA34 4AD, UK

Schiebel, R
Tübingen University
Institute and Museum of Geology and Paleontology
Sigwartstrasse 10
72076 Tübingen, Germany

Schmitt, R W
Woods Hole Oceanographic Institution
Mail Stop 21
Woods Hole, MA 02543, USA

Schumacher, M E
Woods Hole Oceanographic Institution
Marine Policy Center
Mail Stop 41
Woods Hole, MA 02543, USA

Scott, J
DERA Winfrith
A22 Building West Tech Centre, Winfrith
Newburgh, Dorchester
Dorset, DT2 8XJ, UK

Scranton, M I
State University of New York, Stony Brook
Marine Sciences Research Center
Stony Brook, NY 11794, USA

Seki, M P
National Marine Fisheries Service (NOAA)
Honolulu Laboratory
2570 Dole Street
Honolulu, HI 96822, USA

Shannon, L V
University of Cape Town
Department of Oceanography
Rondebosch 7701, Cape Town, Republic of South Africa

Sharples, J
Southampton Oceanography Centre
University of Southampton
Waterfront Campus
European Way
Southampton, SO14 3ZH, UK

Shay, L K
Rosenstiel School of Marine and Atmospheric Science
University of Miami
Division of Meteorology and Physical Oceanography
4600 Rickenbacker Causeway
Miami, FL 33149, USA

Shepherd, K
Canadian Scientific Submersible Facility
c/o Institute of Ocean Sciences
9680 West Saanich Road
PO Box 6000, Sidney
British Columbia, V8L 4B2, Canada

Sherman, K
National Marine Fisheries Service (NOAA)
Narragansett Laboratory
28 Tarzwell Drive
Narragansett, RI 02882, USA

Short, A D
University of Sydney
Coastal Studies Unit
School of Geosciences F09
Sydney, NSW 2006, Australia

Siegel-Causey, D
Harvard University
Museum of Comparative Zoology
26 Oxford Street
Cambridge, MA 02138, USA

Sigman, D M
Princeton University
Department of Geosciences
Princeton, NJ 08544, USA

Simpson, J H
University of Wales, Bangor
School of Ocean Sciences
Menai Bridge
Anglesey, LL59 5RH, UK

Sissenwine, M P
National Marine Fisheries Service (NOAA)
Northeast Fisheries Science Center
166 Water Street
Woods Hole, MA 02543, USA

Smedbol, R K
Dalhousie University
Department of Oceanography
Halifax
Nova Scotia, B3H 4J1, Canada

Smith, T P
National Marine Fisheries Service (NOAA)
Northeast Fisheries Science Center
166 Water Street
Woods Hole, MA 02543, USA

Smyth, W D
Oregon State University
College of Oceanic Atmospheric Sciences
Corvallis, OR 97331, USA

Snelgrove, P V R
Memorial University of Newfoundland
Fisheries and Marine Institute
St John's
Newfoundland, A1C 5R3, Canada

Soloviev, A
Nova Southeastern University
Oceanographic Center
8000 North Ocean Drive
Dania Beach, FL 33004, USA

Solow, A R
Woods Hole Oceanographic Institution
Mail Stop 41
Woods Hole, MA 02543, USA

Spalding, M D
UNEP World Conservation Monitoring Centre
219 Huntingdon Road
Cambridge, CB3 0DL, UK

Spall, M A
Woods Hole Oceanographic Institution
Mail Stop 21
Woods Hole, MA 02543, USA

Spear, L B
HT Harvey & Associates
3150 Almaden Expressway, Suite 145
San Jose, CA 95118, USA

Sprintall, J
Scripps Institution of Oceanography
University of California, San Diego
9500 Gilman Drive
La Jolla, CA 92093, USA

Steele, J H
Woods Hole Oceanographic Institution
Mail Stop 41
Woods Hole, MA 02543, USA

Stein, C A
University of Illinois at Chicago
Department of Earth and
   Environmental Sciences
845 W, Taylor Street
Chicago, IL 60607, USA

Steinberg, D K
College of William and Mary
Virginia Institute of Marine Science
PO Box 1346
Gloucester Point, VA 23062, USA

Stephenson, R L
St Andrews Biological Station
531 Brandy Cove Road, St Andrews
New Brunswick, E5B 2L9, Canada

Stickley, C
University College London
Department of Geography
Environmental Change Research Centre
26 Bedford Way
London, WC1H 0AP, UK

Stigebrandt, A
Gothenburg University
Department of Oceanography
Box 460
S-40530 Gothenburg, Sweden

Stow, D A V
Southampton Oceanography Centre
University of Southampton
Waterfront Campus
European Way
Southampton SO14 3ZH, UK

Stramma, L
University of Kiel
Institut fur Meereskunde
Dusternbrooker Weg 20
D-24105 Kiel, Germany

Sumaila, U R
University of British Columbia
Fisheries Center
2204 Main Hall
Vancouver, British Columbia V6T 1Z4, Canada

Swift, R N
NASA Goddard Space Flight Center
Wallops Flight Facility
Wallops Island, VA 23337, USA

Takagawa, S
Japan Marine Science Technology Center (JAMSTEC)
2–15 Natsushimacho Yokosuka 237-0061, Japan

Takahashi, P K
University of Hawaii at Manoa
Department of Oceanography
1000 Pope Road
Honolulu, HI 96822, USA

Takahashi, T
Lamont-Doherty Earth Observatory
   of Columbia University
PO Box 1000
61 Route 9W
Palisades, NY 10964, USA

Talley, L D
Scripps Institution of Oceanography
University of California, San Diego
9500 Gilman Drive
La Jolla, CA 92093, USA

Tande, K S
University of Tromsø
Norwegian College of Fishery Science
Department of Marine and Freshwater Biology
9037 Tromsø, Norway

Taylor, P K
Southampton Oceanography Centre
University of Southampton
Waterfront Campus
European Way
Southampton SO14 3ZH, UK

Teal, J M
Woods Hole Oceanographic Institution
Woods Hole, MA 02543, USA

Theocharis, A
National Centre for Marine Research (NCMR)
Aghios Kosmas
Hellinikon 16604
Athens, Greece

Thomas, E
Yale University
Department of Geology and Geophysics
Box 208109
New Haven, CT 06520, USA

Thorpe, S A
Southampton Oceanography Centre
University of Southampton
Waterfront Campus
European Way
Southampton SO14 3ZH, UK

Ticco, P C
Massachusetts Maritime Academy
101 Academy Drive
Buzzards Bay, MA 02532, USA

Toggweiler, J R
Geophysical Fluid Dynamics Laboratory
NOAA
PO Box 308
Princeton, NJ 08542, USA

Tomczak, M
Flinders University of South Australia
School of Chemistry, Physics and Earth Sciences
GPO Box 2100
Adelaide, SA 5001, Australia

Trask, R P
Woods Hole Oceanographic Institution
Mail Stop 29
Woods Hole, MA 02543, USA

Trenberth, K E
National Center for Atmospheric Research
PO Box 3000
Boulder, CO 80307, USA

Trites, A W
University of British Columbia
Marine Mammal Research Unit
Fisheries Centre
Vancouver
British Columbia, V6T 1Z4, Canada

Turekian, K K
Yale University
Department of Geology and Geophysics
PO Box 208109
New Haven, CT 06520, USA

Turner, A
University of Plymouth
Department of Environmental Sciences
Plymouth, Devon PL4 8AA, UK

Tyack, P L
Woods Hole Oceanographic Institution
Biology Department
Mail Stop 34
Woods Hole, MA 02543, USA

Tyrrell, T
Southampton Oceanography Centre
University of Southampton
Waterfront Campus
European Way
Southampton, SO14 3ZH, UK

Underwood, G J C
University of Essex
Department of Biological Sciences
Wivenhoe Park, Colchester, CO4 3SQ, UK

van der Loeff, M M R
Alfred-Wegener-Institut für Polar-
   und Meeresforschung
Postfach 120161
27515 Bremerhaven, Germany

Ver, L M
University of Hawaii at Manoa
Department of Oceanography
1000 Pope Road
Honolulu, HI 96822, USA

Videler, J J
Rijksuniversiteit Groningen
Department of Marine Biology, Kerklaan 30
9750 AA Haren, The Netherlands

Vine, F J
University of East Anglia
School of Environmental Sciences
Norwich, NR4 7TJ, UK

von Blanckenburg, F
Universität Bern
Mineralogisch-Petroghraphisches Institut
Abteilung Isotopengeochemie
Erlachstrasse 9a
3012 Bern, Switzerland

Vivian, C M G
The Centre for Environment, Fisheries
    and Aquaculture Science
Burnham Laboratory
Essex, CM0 8HA, UK

Wadhams, P
University of Cambridge
Scott Polar Research Institute
Lensfield Road
Cambridge, CB2 1ER, UK

Walsh, J J
University of South Florida
Department of Marine Science
140 7th Avenue South
St Petersburg, FL 33701, USA

Warwick, R M
Plymouth Marine Laboratory
Prospect Place
West Hoe
Plymouth, PL1 3DH, UK

Watson, A J
University of East Anglia
School of Environmental Sciences
Norwich, NR4 7TJ, UK

Weeks, W F
University of Alaska Fairbanks
Department of Geology and Geophysics
Fairbanks, AK 99775, USA

Wefer, G
Universität Bremen
Fachbereich Geowissenschaften
28334 Bremen, Germany

Weller, R A
Woods Hole Oceanographic Institution
Mail Stop 29
Woods Hole, MA 02543, USA

Wells, N C
Southampton Oceanography Centre
University of Southampton
Waterfront Campus
European Way
Southampton, SO14 3ZH, UK

Werner, F E
The University of North Carolina at Chapel Hill
Department of Marine Sciences
12-7 Venable Hall, CB 3300
Chapel Hill, NC 27599, USA

Whitehead, J A
Woods Hole Oceanographic Institution
Mail Stop 21
Woods Hole, MA 02543, USA

Widder, E A
Harbor Branch Oceanographic Institution
5600 US 1N
Fort Pierce, FL 34946, USA

Wiebe, P H
Woods Hole Oceanographic Institution
Mail Stop 33
Woods Hole, MA 02543, USA

Wijesekera, H
Oregon State University
College of Oceanic and Atmospheric Sciences
Corvallis, OR 97331, USA

Wijffels, S
CSIRO Marine Research
GPO Box 1538
Hobart, Tasmania 7001, Australia

Wildish, D J
Fisheries and Oceans Canada
Biological Station
531 Brandy Cove Road
St Andrews, New Brunswick E5B 2L9, Canada

Wilkinson, M
Heriot-Watt University
Department of Biological Science
Riccarton, Edinburgh, EH14 4AS, UK

Williams, R G
University of Liverpool
Oceanography Laboratories
Liverpool, L69 7ZL, UK

Wilson III, C A W
Louisiana State University
Department of Oceanography and Coastal Sciences
Baton Rouge, LA 70803, USA

Wilson, D C
Danish Institute for Fisheries Research
North Sea Centres
PO Box 101
9850 Hirtshals, Denmark

Wilson, S
NOAA HCHB 5224
14th Const Ave NW
Washington, DC 20230, USA

Wiltshire, J C
University of Hawaii at Manoa
Department of Oceanography
1000 Pope Road
Honolulu, HI 96822, USA

Windsor, M
North Atlantic Salmon Conservation Organization
11 Rutland Square
Edinburgh, EH1 2AS, UK

Woodroffe, C
University of Wollongong
Northfields Avenue
Wollongong, NSW 2522, Australia

Woolf, D K
Southampton Oceanography Centre
University of Southampton
Waterfront Campus
European Way
Southampton, SO14 3ZH, UK

Worcester, P F
Scripps Institution of Oceanography
University of California, San Diego
9500 Gilman Drive
La Jolla, CA 92093, USA

Wright, C W
NASA Goddard Space Flight Center
Wallops Flight Facility
Wallops Island, VA 23337, USA

Wright, J D
Rutgers University
Department of Geology
610 Taylor Road
Piscataway, NJ 08854, USA

Wu, S Y
NASA Goddard Space Flight Center
Oceans and Ice Branch
Greenbelt, MD 20771, USA

Wunsch, C
Massachusetts Institute of Technology
Department of Earth, Atmospheric and
    Planetary Sciences
Cambridge, MA 02139, USA

Zenk, W
Universität Kiel
Institut fur Meereskundean
Dusternbrooker Weg 20
24105 Kiel, Germany

Zemmelink, H J
Rijbsuniversiteit Groningen
Department of Marine Biology
9751 NN Haren, The Netherlands

Contents

Volume 1

A

B

C

Volume 2

D

E

Volume 3

I

K

L

M

Volume 4

N

Volume 5

S

Volume 6

T

SALINITY

See **SATELLITE MEASUREMENTS OF SALINITY. WATER TYPES AND WATER MASSES.**

SALMON FISHERIES

Atlantic

P. Hutchinson and M. Windsor,
North Atlantic Salmon Conservation Organization,
Edinburgh, UK

Copyright © 2001 Academic Press

doi:10.1006/rwos.2001.0471

Introduction

Migrating animals, concentrated in space and time, represent readily harvestable resources that have a long history of exploitation by humans. The anadromous Atlantic salmon (*Salmo salar*) is no exception. Cave paintings and stone carvings dating back 25 000 years from the Dordogne region of France confirm its long association with, and importance to, humans. Throughout its range in the North Atlantic, the Atlantic salmon has been and continues to, be exploited by a variety of gear in rivers, lakes, estuaries, and the sea, providing employment and recreation, and generating considerable economic benefits, often in remote rural areas. The Atlantic salmon also has cultural, ceremonial, and symbolic significance, but it is difficult to ascribe a value to these important facets of the resource. Throughout the history of exploitation of Atlantic salmon by humans, there have been many changes in the nature and scale of the fisheries.

Description of the Salmon Fisheries

Although it is a matter of conjecture, the most ancient method of harvesting Atlantic salmon was probably by hand in rivers where adults returning to spawn may well have been an important component of the diet prior to the development of agriculture and techniques for animal husbandry. Apart from the use of clubs or stones, the spear or harpoon was probably the first fishing gear used for salmon. The snare, hook, and dip net probably followed. The earliest method of harvesting salmon in quantity was probably the fishing weir.

Spears, hooks, nets, and weirs thought to have been used for catching salmon at least 8000 years ago have been discovered in Sweden. In eastern Canada, the first harvesting of Atlantic salmon is thought to have started about 8800 BC when Amerindians arrived in the area. The spear was the preferred implement.

Documentary evidence of the use of salmon weirs is available from the eleventh century. The Battle of Clontarf in Ireland in 1014 was known as the Battle of the Salmon Weir. Use of weirs and nets (probably hand nets, seine, and gill nets) by North American Indians was documented in the sixteenth century. The seine net is known to have been used for catching salmon in Scotland and Ireland in the seventeenth century and probably much earlier than that. Amerindians practiced a primitive form of angling. Angling for salmon as a hobby is known from at least the seventeenth century in some countries, although it was introduced to Norway only in the nineteenth century.

While a considerable variety of types of salmon fishing gear has been developed, on comparison they appear to be based on a few basic methods of capture, which have been categorized under four general headings: fixed gears or traps (e.g., bag nets, stake nets, set gill nets); floating gears (e.g., long-lines and drift nets); seine or draft nets; and rod and line (using a variety of artificial flies, baits, and lures).

There have been many improvements to these fishing methods over the period of their deployment. One of the most significant has been the development of synthetic twines that made the gear (including recreational fishing gear) easier to handle and less visible.

Salmon fisheries are often categorized as 'recreational' and 'commercial' to distinguish between sport fishing with rod and line and fishing with other gears with the intention of selling the harvest. However, the distinction is sometimes blurred. For example: 'recreational' licenses may be issued to fish for salmon with gill nets for local consumption purposes in Greenland; in some countries the sale of rod-caught salmon is permitted; and rod-and-line fisheries may be let or sold for considerable sums of money. For the purposes of this article, recreational fisheries are considered to be sport fisheries using rod and line and a variety of artificial flies, baits, and lures; commercial fisheries are those fisheries conducted with a variety of other gears where the intention is to sell the harvest. A third category, 'subsistence fisheries', is conducted with the intention of using the harvest of salmon for consumption by the local community; for example, the fisheries by native people in Canada, Finland and Greenland. Salmon fishing may also be conducted for research purposes, in some cases using methods that would ordinarily be prohibited.

Some countries have only recreational fisheries. For example, all netting of salmon was prohibited in Spain and the salmon fishery was dedicated entirely to recreational fishing in 1942 following the Civil War, during which the salmon populations had been heavily exploited for food. Similarly, in the United States all commercial exploitation of Atlantic salmon ceased in 1948. Other countries have a mixture of commercial and recreational fisheries (e.g., Norway, United Kingdom, Ireland, France, Iceland, and Russia). In Iceland there is no coastal netting and commercial fisheries are conducted in only two rivers in the south of the island (A. Isaksson, personal communication). Canada had a major commercial fishery, but management measures introduced since the mid-1960s have progressively reduced the fishery and in 2000 no commercial licenses were issued, with the effect that the Canadian salmon fishery is now recreational and subsistence in nature. In Russia, the fisheries were mainly commercial and angling for salmon was prohibited

in all but three rivers, where it was strictly controlled by restrictions on the number of licenses issued. However, since the mid-1980s, recreational fisheries have developed in the rivers of the Kola peninsula and are popular with foreign anglers. Greenland and the Faroe Islands have only one and five salmon rivers, respectively, so the opportunities for recreational fishing are limited and fishing has mainly been either commercial or subsistence in Greenland, and commercial or research in the Faroe Islands.

Salmon fisheries have been described as single or mixed stock on the basis of whether they exploit a significant number of salmon from one or from more than one river stock, respectively. Some mixed stock fisheries may exploit salmon originating in different countries. Mixed stock fisheries have also been referred to as interception fisheries and the term is often applied specifically to the Greenland and Faroes fisheries. However, prior to the closure of the commercial fisheries in Newfoundland and Labrador in Canada, there was concern about the harvest of US fish by this fishery. Similarly, in the North-East Atlantic area there are harvests in the fisheries of one country of salmon originating in the rivers of another country (**Table 1**). Thus, many salmon fisheries are interceptory in nature, but these interceptions have declined in recent years as a result of international agreements in the North Atlantic Salmon Conservation Organization (NASCO), national or regional regulations, economic factors and other reasons.

The terms 'home water' and 'distant water' are also used in relation to salmon fisheries. Since 1983,

Table 1 Origin of salmon caught in home water fisheries in the North-east Atlantic in 1992

Origin of stock	Catch by country									
	Russia	Finland	Norway	Sweden	England and Wales	Scotland	Northern Ireland	Ireland	France	Iceland
Wild										
Russia	100%	—	+	—	—	—	—	—	—	—
Finland	—	99%	+	—	—	—	—	—	—	—
Norway	—	+	75%	6%	—	—	—	+	—	—
Sweden	—	—	1%	46%	—	—	—	—	—	—
England and Wales	—	—	—	—	62%	+	+	10%	—	—
Scotland	—	—	—	—	38%	95%	3%	5%	—	—
Northern Ireland	—	—	—	—	+	+	92%	5%	—	—
Ireland	—	—	—	—	+	+	+	80%	—	—
France	—	—	—	—	+	+	+	+	100%	—
Iceland	—	—	—	—	—	—	—	—	—	28%
Reared										
Escapees	—	<1%	23%	2%	—	5%	1%	—	—	—
Ranched	—	—	1%	46%[a]	—	—	3%	<1%	—	72%

Source: Report of the ICES Advisory Committee on Fishery Management 1994. NASCO Council document CNL(94)13.
[a]Fish released for mitigation purposes and not expected to contribute to spawning.
+, Catches thought to occur but contribution not estimated.
−, Catches occur rarely or not at all.

with the implementation of the Convention for the Conservation of Salmon in the North Atlantic Ocean, which prohibits fishing beyond areas of fisheries' jurisdiction, all salmon fisheries are in effect home water fisheries. The term home water fishery is therefore more correctly used to indicate fisheries within the jurisdiction of the state of origin of the salmon (i.e., in the country in whose rivers the salmon originated), as opposed to distant water fisheries, which harvest salmon outside the jurisdiction of the state of origin.

The Resource

Limits on production during the freshwater phase of the life cycle constrain the abundance of Atlantic salmon and result in catches that are low compared to those of Pacific salmon and pelagic marine fish species such as herring and mackerel.

A wide range of factors has already affected this freshwater production capacity, including urbanization, land drainage, overgrazing, forestry practices, infrastructure developments, water abstraction, sewage and industrial effluents, hydroelectricity generation, and the introduction of nonindigenous species. Many salmon rivers were damaged as a result of the Industrial Revolution. For example, in Canada, there has been a net loss of productive capacity of salmon of 16% since 1870, and in the state of Maine, USA, about two-thirds of the historic salmon habitat had been lost by the mid-1980s. Early attempts at enhancement through stocking programs date to the middle of the nineteenth century. These stocking programs continue and in 1999 more than 30 million Atlantic salmon eggs and juveniles were stocked in rivers around the North Atlantic. With the decline of many heavy industries there have been improvements in salmon habitat and in England and Wales, for example, there are now more salmon-producing rivers than there were 150 years ago. Much progress has also been made in recent years in improving fish passage facilities at dams. The effects of the Industrial Revolution are, however, still being felt today, through the continuing problem of acidification of rivers and lakes, for example. As the human population continues to increase, pressures on salmon habitat from domestic, industrial, and agricultural demands will increase.

Catch statistics compiled for the North Atlantic region by the International Council for the Exploration of the Sea (ICES) are available for the period from 1960, during which the total reported catch has ranged from approximately 2200 tonnes in 1999 to approximately 12 500 tonnes in 1973 (**Figure 1**).

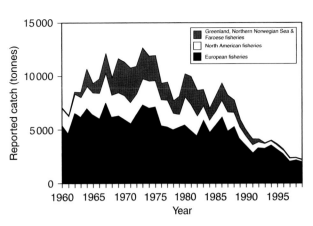

Figure 1 Reported catch of Atlantic salmon (tonnes) from the North Atlantic area, 1960–1999.

The mean reported catch in tonnes by country for each of the four decades 1960–69 to 1990–99 is shown in **Table 2**. There has been a steady decline in the total reported North Atlantic catch of salmon since the early 1970s. **Figure 2** shows for four major states of origin that there is some degree of synchronicity in the trend in catches over the 40-year period from 1960 when expressed as the percentage difference from the long-term mean reported catch. While catches in all four countries were above or close to the 40-year mean in the period from the 1960s to the late 1980s, the last decade of the twentieth century was characterized by below-average catches. Although some of the reduction in catches was the result of the introduction of management measures, which have reduced fishing effort, the abundance of both European and North American Atlantic salmon stocks has declined since the 1970s, particularly the multi-sea-winter components. This decline in abundance appears to be related to reduced survival at sea.

In addition to the reported catches illustrated in **Figure 1**, catches may go unreported for a variety of reasons. These include the absence of a requirement for statistics to be collected; suppression of information thought to be unfavorable; and illegal fishing. Estimates of unreported catch for the North Atlantic region for the period from 1987 have ranged between approximately 800 and 3200 tonnes, or 29–51% of the reported catch. Illegal fishing appears to be a particular problem in some countries. Associated with all forms of fishing gear is mortality generated directly or indirectly by the gear but which is not included in reported catches. This mortality may be associated with predation, discards, and escape from the gear. For salmon fishing gear the contribution of most sources of this

Table 2 Mean reported catch (in tonnes) by country during the four decades 1960–1969 to 1990–1999

Country	1960–1969	1970–1979	1980–1989	1990–1999
Canada	2053	2142	1638	395
Denmark	138	491	152	1
England and Wales	325	384	370	224
Faroe Islands	64	152	606	47
Finland	—	42	54	57
France	—	14	23	13
Germany	2	3	—	—
Greenland	773	1300	816	119
Iceland	131	197	176	138
Ireland	1329	1676	1263	616
Northern Ireland	291	174	114	82
Norway	1822	1745	1453	840
Russia	690	559	520	158
Scotland	1684	1437	1058	468
Spain	36	27	21	8
St. Pierre and Miquelon	—	—	3	2
Sweden	50	39	35	34
USA	1	2	3	1

Notes: (1) The catch for Iceland excludes returns to commercial ranching stations. (2) The catches for Norway, Sweden, and Faroe Islands include harvests at West Greenland and in the Northern Norwegian Sea fishery. (3) The catch for Finland includes harvests in the Northern Norwegian Sea fishery. (4) The catch for Denmark includes catches in the Faroese zone, in the Northern Norwegian Sea fishery, and at West Greenland. (5) The catch for Germany is from the Northern Norwegian Sea fishery.

mortality is estimated to be low (0–10%) but highly variable.

By-catch of nontarget species in salmon fishing gear is thought to be generally low. Drift nets may have a by-catch associated with their use, but this has not been fully quantified. However, as this gear is often tended by the fishermen, there may be an opportunity to release sea birds and marine mammals from the nets. 'Ghost fishing' by lost or abandoned nets is not thought to be a problem associated with salmon fishing gear. By-catch of salmon in gear set for species such as bass, lumpsucker, mackerel, herring, and cod is known to occur but it is not generally a problem. In some countries regulations have been introduced to protect salmon from capture in coastal fisheries for other species. There is, however, concern about the possible by-catch of salmon post-smolts in pelagic fisheries for herring and mackerel in the Norwegian Sea, which overlap spatially and temporally with European-origin post-smolt migration routes.

In addition to exploitation of Atlantic salmon in the North Atlantic region, there are fisheries in the Baltic Sea. Catches since 1972 have ranged from approximately 2000 to 5600 tonnes. These fisheries, which are based to a large extent on hatchery smolts released to compensate for loss of habitat following hydroelectric development, are described in detail by Christensen *et al.* (see Further Reading).

Economic Value

A wide variety of techniques have been used to assess the economic value of Atlantic salmon and, in the absence of a standardized approach, assessment of the economic value of salmon fisheries on a North Atlantic basis is not possible. Many assessments concern the expenditure associated with salmon fishing. Economic value, however, reflects willingness to pay for use of the resource, and as willingness to pay must at least be equal to actual expenditure, many assessments underestimate the full economic value. However, it is clear that throughout its range the Atlantic salmon generates considerable economic benefits that may have impacts on a regional basis or, where visiting anglers from other countries are involved or where the harvest is exported, impacts on national economies. The following examples serve to highlight the considerable economic value of salmon fisheries.

The total net economic value of salmon fisheries, both recreational and commercial, in Great Britain was estimated in 1988 to be £340 million, of which

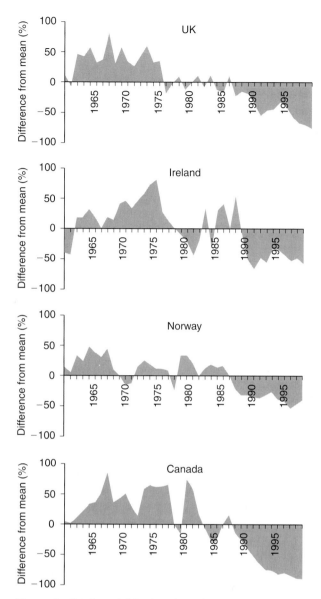

Figure 2 Catches of Atlantic salmon (tonnes), expressed as the percentage difference from the 40-year mean, for four major states of origin.

the recreational fisheries accounted for approximately £327 million.

In Canada, recreational anglers spent Can$39 million on salmon fishing in 1985 with a further Can$45 million invested on major durables and property.

In Greenland, the salmon fishery in 1980 was a substantial source of income (30–35% of total annual income to the fishermen), and 50% of fishermen could not have met their current vocational and domestic expenses at that time without the salmon fishery. Many people other than the fishermen depended on the salmon fishery for gear and equipment sales and repair and shore processing.

The expenditure by recreational salmon fishermen visiting one major Scottish salmon river, the Tweed, was estimated to be £9 million in 1996, with a total economic impact of more than £12 million. Approximately 500 full-time job equivalents depended on this activity. This is for one river and there are more than 2000 salmon rivers in the North Atlantic region, with fisheries that bring economic benefits, often to remote areas where job creation is otherwise very difficult.

In addition to the economic value associated with the fisheries, individuals are willing to contribute to salmon conservation even though they have no interest in fishing. Sixty percent of the New England population was found to 'care' about the Atlantic salmon restoration program and in 1987 their willingness to pay was estimated to exceed the cost of the restoration program. Economic assessments that fail to take these non-user aspects into account will considerably underestimate the economic value of the resource. The salmon has a special place in human perception and there are many nongovernment organizations dedicated to its conservation.

Management of the Fisheries

Legislation regulating the operation of salmon fisheries is known to have been introduced in Europe as early as the twelfth century. In Scotland, for example, legislation was introduced to establish a weekly close time and to prevent total obstruction of rivers by fishing weirs. Similarly, in the middle of the thirteenth century, legislation establishing close seasons was introduced in Spain. Since these early conservation measures were enacted, a wide variety of laws and regulations concerning the salmon fisheries have been developed by each North Atlantic country. These laws and regulations include those that permit or prohibit certain methods of fishing; specify permitted times and places of fishing; restrict catch by quota; prohibit the taking of young salmon and kelts; restrict or place conditions on the trade in salmon; and ensure the free passage of salmon.

The last quarter of the twentieth century witnessed dramatic changes in the exploitation of Atlantic salmon. Commercial fisheries have been greatly reduced, partly as a result of management measures taken in response to concern about abundance and partly as a result of the growth of salmon farming. Production of farmed Atlantic salmon has increased from less than 5000 tonnes in 1980 to more than 620 000 tonnes in 1999 (**Figure 3**). This rapidly growing industry has had a marked impact on the profitability of commercial fisheries for salmon. While it has been argued that the growth of

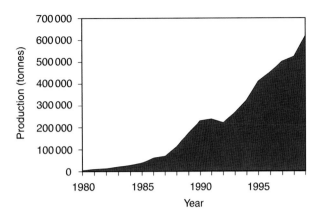

Figure 3 Production of farmed Atlantic salmon (tonnes) in the North Atlantic area.

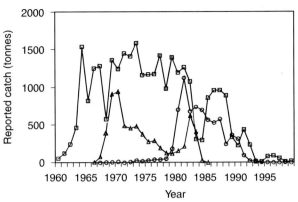

Figure 4 Reported catches (tonnes) of Atlantic salmon in the fisheries at West Greenland (□), in the Northern Norwegian Sea (△) and in the Faroese zone (○).

salmon farming, which in 1999 produced about 300 times the harvest of the fisheries, has reduced exploitation pressure on the wild stocks, there are concerns about the genetic, disease, parasite, and other impacts the industry may be having on the wild Atlantic salmon. In some countries, escaped farm salmon frequently occur in fisheries for wild salmon and in spawning stocks.

Distant Water Fisheries

Prior to the 1960s, management of salmon fisheries in the North Atlantic region was at a local, regional, or national level. During the 1960s and early 1970s, however, distant water fisheries developed at West Greenland (harvesting both European and North American origin salmon) and in the Northern Norwegian Sea and, later, in the Faroese zone (harvesting predominantly European-origin salmon). The rational management of these fisheries required international cooperation, the forum for which was created in 1984 with the establishment of the intergovernment North Atlantic Salmon Conservation Organization (NASCO). The development and subsequent regulation of these fisheries in terms of reported catch are illustrated in **Figure 4**. The Newfoundland and Labrador commercial fishery in Canada, which before its closure harvested US-origin salmon in addition to salmon returning to Canadian rivers, was also subject to a regulatory measure agreed in NASCO.

West Greenland Salmon Fishery The presence of salmon off West Greenland was first reported in the late eighteenth century and a fishery for local consumption purposes has probably been conducted since the beginning of the twentieth century. From 1960 to 1964 the landings by Greenlandic vessels using fixed gill nets increased from 60 tonnes to

more than 1500 tonnes and increased further from 1965 when vessels from Denmark, Sweden, the Faroe Islands, and Norway joined the fishery and monofilament gill nets were introduced. From 1975 the fishery was restricted to Greenlandic vessels. The salmon harvested at West Greenland are almost exclusively one-sea-winter salmon that would return to rivers in North America (principally Canada, but harvests of US salmon were significant in comparison to the number of fish returning to spawn) and Europe (particularly the United Kingdom and Ireland) as multi-sea-winter salmon. During the 1990s, the proportion of salmon of North American origin in the catch has increased, comprising 90% of samples in 1999.

International agreement on regulation of the harvests at West Greenland first occurred in 1972 when the International Commission for the Northwest Atlantic Fisheries (ICNAF) endorsed a US–Danish bilateral agreement to limit the catch to 1100 tonnes (adjusted to 1191 tonnes in 1974). This quota, with small adjustments to take account of delays in the start of the seasons in 1981 and 1982, applied until 1984, since when regulatory measures have been developed within NASCO. Details of these measures are given in **Table 3**.

Northern Norwegian Sea Fishery Seven years after the start of the West Greenland fishery, a salmon fishery involving, at different times, vessels from Denmark, Norway, Sweden, Finland, Germany, and the Faroe Islands commenced in the Northern Norwegian Sea. Initially drift nets were used, but the vessels soon changed to longlines. Prior to 1975, the fishery was conducted over a large geographical area between 68°N and 75°N and between the Greenwich meridian and 20°E. However, following

Table 3 Regulatory measures agreed by NASCO for the West Greenland salmon fishery

Year	Allowable catch (tonnes)	Comments/other measures
1984	870	
1985	–	Greenlandic authorities unilaterally established quota of 852 t.
1986	850	Catch limit adjusted for season commencing after 1 August.
1987	850	Catch limit adjusted for season commencing after 1 August.
1988–1990	2520	Annual catch in any year not to exceed annual average (840 t) by more than 10%. Catch limit adjusted for season commencing after 1 August.
1991	–	Greenlandic authorities unilaterally established quota of 840 t.
1992	–	No TAC imposed by Greenlandic authorities but if the catch in first 14 days of the season had been higher compared to the previous year, a TAC would have been imposed.
1993	213	
1994	159	
1995	77	
1996	–	Greenlandic authorities unilaterally established a quota of 174 t.
1997	57	
1998	Internal consumption fishery only	Amount for internal consumption in Greenland has been estimated to be 20 t.
1999	Internal consumption fishery only	Amount for internal consumption in Greenland has been estimated to be 20 t.
2000	Internal consumption fishery only	Amount for internal consumption in Greenland has been estimated to be 20 t.

TAC, total allowable catch.

the extension of fishery limits to 200 nautical miles, the fishery shifted westward to the area between the Norwegian fishery limit and Jan Mayen Island. The catch peaked at almost 950 tonnes in 1970. In response to the rapid escalation of this fishery, the North-East Atlantic Fisheries Commission (NEAFC) adopted a variety of measures intended to stabilize harvests, although a proposal to prohibit high-seas salmon fishing failed to obtain unanimous approval. However, this fishery ceased to exist in 1984 as a result of the prohibition on fishing for salmon beyond areas of fisheries jurisdiction in the Convention for the Conservation of Salmon in the North Atlantic Ocean (the NASCO Convention).

In the period 1989–94 vessels were identified fishing for salmon in international waters. These vessels were based mainly in Denmark and Poland and some had re-registered to Panama in order to avoid the provisions of the NASCO Convention. On the basis of information on the number of vessels, the number of trips per year, and known catches, ICES has provided estimates of the harvest (tonnes) as follows:

1990	1991	1992	1993	1994
180–350	25–100	25–100	25–100	25–100

Following diplomatic initiatives by NASCO and its Contracting Parties there have been no sightings of vessels fishing for salmon in international waters in the North-East Atlantic since 1994. NASCO is cooperating with coastguard authorities in order to coordinate and improve surveillance activities.

Faroes Salmon Fishery During the period 1967–78 exploratory fishing for salmon was conducted off the Faroe Islands using floating longlines. During this period no more than nine Faroese vessels were involved in the fishery and the catches, which were mainly of one-sea-winter salmon, did not exceed 40 tonnes. During the period 1978–85 Danish vessels also participated in the fishery and in 1980 and 1981 there was a marked increase in fishing effort and catches. As the fishery developed, it moved farther north and catches were dominated by two-sea-winter salmon. The salmon caught in the fishery are mainly of Norwegian and Russian origin. Initially negotiations on regulatory measures for the Faroese fishery were conducted on a bilateral basis between the Faroese authorities and the European Commission. Since 1984, the fishery has been regulated through NASCO. Details of these measures are given in **Table 4.**

Compensation Arrangements In the period 1991–98 the North Atlantic Salmon Fund (NASF) entered into compensation arrangements with the Faroese salmon fishermen. Similar arrangements were in place at West Greenland in 1993 and 1994. Under these arrangements the fishermen in these

Table 4 Regulatory measures agreed by NASCO for the Faroese salmon fishery

Year	Allowable catch (tonnes)	Comments/other measures
1984/85	625	
1986	–	
1987–1989	1790	Catch in any year not to exceed annual average (597 t) by more than 5%.
1990–1991	1100	Catch in any year not to exceed annual average (550 t) by more than 15%.
1992	550	
1993	550	
1994	550	
1995	550	
1996	470	No more than 390 t of the quota to be allocated if fishing licenses issued.
1997	425	No more than 360 t of the quota to be allocated if fishing licenses issued.
1998	380	No more than 330 t of the quota to be allocated if fishing licenses issued.
1999	330	No more than 290 t of the quota to be allocated if fishing licenses issued.
2000	300	No more than 260 t of the quota to be allocated if fishing licenses issued.

Note: The quotas for the Faroe Islands detailed above were agreed as part of effort limitation programs (limiting the number of licenses, season length, and maximum number of boat fishing days) together with measures to minimize the capture of fish less than 60 cm in length. The measure for 1984/85 did not set limits on the number of licenses or the number of boat fishing days.

countries were paid not to fish the quotas agreed within NASCO. As a result of the permanent closure of the Northern Norwegian sea fishery, regulatory measures agreed by NASCO, and compensation arrangements, the proportion of the total North Atlantic catch taken in the distant water fisheries declined from an average of 21% in the 1970s to an average of only 4% in the 1990s.

Home Water Fisheries

Management measures introduced in home water fisheries partly for domestic reasons and partly under the process of 'putting your own house in order before expecting others to make or continue to make sacrifices' have also resulted in major changes in the level and pattern of exploitation of Atlantic salmon.

In Canada, approximately Can$80 million was invested in the period 1972–99 to reduce the number of commercial salmon fishing licenses. No commercial salmon fishing licenses were issued in the year 2000. Drift netting for salmon was prohibited in Scotland in 1962 and in Norway in 1989. Between 1970 and 1999, the number of fixed commercial gears in Norway has been reduced by 68%. Similarly, in the United Kingdom and Ireland there have been reductions in netting effort. In Scotland the reduction in effort between 1970 and 1999 was 83%. In England and Wales there has been a 53%

reduction in the number of salmon netting licenses issued over the last 25 years. It is the UK government's policy to phase out fisheries in coastal waters that exploit stocks from more than one river. In Ireland, there has been a reduction in netting effort of at least 20% since 1997.

Recreational fisheries have also been subject to restrictive management measures. In the United States, the recreational fishery was restricted to catch-and-release fishing and in the year 2000 closed completely with the exception of a fishery in the Merrimack River based on releases of surplus hatchery broodstock. In Canada, daily and seasonal catch limits have been reduced, mandatory catch-and-release has been introduced, and where conditions require, individual rivers have been closed to fishing. In England and Wales, measures were introduced in 1999 to protect early running salmon by delaying the start of the netting season to 1 June and by requiring anglers to return salmon to the water before 16 June. Catch-and-release fishing is becoming increasingly commonplace. In 1999, 100%, 77%, 49%, 44% and 29% of the total rod catch in the United States, Russia, Canada, England and Wales, and Scotland respectively, was released.

While these examples highlight the severe nature of the restrictions on fisheries, all countries around the North Atlantic have introduced measures

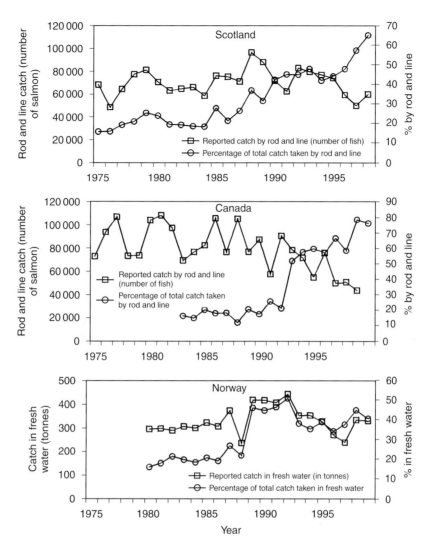

Figure 5 Reported catches (tonnes) of Atlantic salmon by rod and line (Scotland and Canada) or in fresh water (Norway) expressed as number or weight of fish and as a percentage of the total catch. Source of data: Scottish Rural Affairs Department, Edinburgh; Canadian Department of Fisheries and Oceans, Ottawa; Norwegian Directorate for Nature Management, Trondheim.

designed to conserve the resource. One result of these measures has been to change the pattern of exploitation, with rod fisheries taking an increasing proportion of the total catch. This trend is illustrated for Canada, Norway and Scotland in **Figure 5**. Exploitation is, therefore, increasingly occurring in fresh water rather than at sea, and is focused more on individual river stocks.

Management of Salmon Fisheries Under a Precautionary Approach

Concern about the status of salmon stocks in the North Atlantic has given rise to the adoption of a precautionary approach to salmon management by NASCO and its Contracting Parties. This ap-

proach, which will guide management of North Atlantic salmon fisheries in the twenty-first century, means that there is a need for caution when information is uncertain, unreliable, or inadequate and that the absence of adequate scientific information should not be used as a reason for postponing or failing to take conservation and management measures. The precautionary approach requires, *inter alia*, consideration of the needs of future generations and avoidance of changes that are not potentially reversible; prior identification of undesirable outcomes and of measures that will avoid them or correct them; and that priority be given to conserving the productive capacity of the resource where the likely impact of resource use is uncertain. A decision structure for the management of North

1. **Is the stock threatened by external factors (e.g., acidification, disease)?**
   If yes, take special management action as appropriate (e.g., establish gene bank).
   If no, go to (2).

2. **Assess status of the stock (abundance and diversity)**
   (a) Have age-specific conservation limits been set?
      (i) If yes, is the conservation limit being exceeded according to agreed compliance criteria (e.g., 3 out of 4 years)?
      (ii) If no, assess other measures of abundance.
   (b) Is the stock meeting other diversity criteria?

3. **If either abundance or diversity are unsatisfactory, then seek to identify the reasons**
   (a) Immediately implement pre-agreed procedures to introduce appropriate measures to address reasons for failure (including stock rebuilding programs).
   (b) Monitor the effect of the measures and take the results into account in future management and assessment; include identification of information gaps, process, and timeframe for resolution.

4. **If both abundance and diversity are satisfactory**
   (a) Implement pre-agreed management actions to permit harvest of the surplus taking into account uncertainty (where appropriate use management targets to establish the exploitable surplus).
   (b) Monitor the effect of the measures and take the results into account in future management and assessment; include identification of information gaps, process and timeframe for resolution.

Figure 6 Decision structure for implementing the precautionary approach to management of single stock salmon fisheries.

1. **Identify river stocks that are available to the fishery**

2. **Identify stock components that are exploited by the fishery**

3. **Assess abundance and diversity of individual stocks contributing to the fishery**

4. **Are abundance and diversity satisfactory (consider the percentage of stocks that are unsatisfactory and the extent of failure for each stock)?**
   (a) If yes, go to (5).
   (b) If no, consider closing the fishery (taking into account socioeconomic factors). If the decision is made not to close the fishery, then continue to (5).

5. **Are the combined conservation limit(s) for all stocks subject to the fishery being exceeded?**
   (a) If yes, implement pre-agreed procedures for the management of the fishery based on effort or quota control:
      ● *Quota control*
         – define management target based on an assessment of risk of failing conservation limits
         – predict prefishery abundance
         – determine exploitable surplus
         – apply pre-agreed rules on setting quotas
      ● *Effort control* (and quota control in the absence of management targets and/or prediction of pre-fishery abundance)
         – evaluate effectiveness of previous effort control measures and apply appropriate changes.
   (b) If no, consider closing the fishery, taking into account socioeconomic factors. If the decision is made not to close the fishery, apply pre-agreed reserve measures to minimize exploitation.

6. **Monitor the effect of the measures and take the results into account in future management and assessment; include identification of information gaps, process, and timeframe for resolution**

Figure 7 Decision structure for implementing the precautionary approach to mixed stock salmon fisheries.

Atlantic salmon fisheries has been adopted by NASCO on a preliminary basis. This decision structure is shown in **Figures 6** and **7** for single stock (i.e., exploiting salmon from one river) and mixed stock (i.e., exploiting salmon from more than one river) fisheries, respectively.

In short, salmon fisheries changed greatly in the last four decades of the twentieth century and the development of salmon farming had a marked effect on these fisheries. There is great concern about the future of the wild stocks and the fisheries continue to undergo critical re-examination.

See also

Fishery Management. Fishing Methods and Fishing Fleets. Salmonid Farming. Salmonids.

Further Reading

Anon (1991) *Salmon Net Fisheries: Report of a Review of Salmon Net Fishing in the Areas of the Yorkshire and Northumbria Regions of the National Rivers Authority and the Salmon Fishery Districts from the River Tweed to the River Ugie*. London: HMSO.

Ayton W (1998) *Salmon Fisheries in England and Wales*. Moulin, Pitlochry: Atlantic Salmon Trust.

Barbour A (1992) *Atlantic Salmon, An Illustrated History*. Edinburgh: Canongate Press.

Baum ET (1997) *Maine Atlantic Salmon. A National Treasure*. Herman, ME: Atlantic Salmon Unlimited.

Dunfield RW (1985) *The Atlantic Salmon in the History of North America*. Canadian Special Publication of Fisheries and Aquatic Sciences 80. Ottawa: Department of Fisheries and Oceans.

Hansen LP and Bielby GH (1988) *Salmon in Norway*. Moulin, Pitlochry: Atlantic Salmon Trust.

Jakupsstovu SHi (1988) Exploitation and migration of salmon in Faroese waters. In: Mills D and Piggins D (eds) *Atlantic Salmon: Planning for the Future*, pp. 458–482. Beckenham: Croom Helm.

Mills DH (1983) *Problems and Solutions in the Management of Open Seas Fisheries for Atlantic Salmon*. Moulin, Pitlochry: Atlantic Salmon Trust.

Mills DH (1989) *Ecology and Management of Atlantic Salmon*. London: Chapman and Hall.

Møller Jensen J (1988) Exploitation and migration of salmon on the high seas in relation to Greenland. In: Mills D and Piggins D (eds) *Atlantic Salmon: Planning for the Future*, pp. 438–457. Beckenham: Croom Helm.

NASCO (1991) *Economic Value of Atlantic Salmon*. Council document CNL(91)29. Edinburgh: North Atlantic Salmon Conservation Organization.

Taylor VR (1985) *The Early Atlantic Salmon Fishery in Newfoundland and Labrador*. Canadian Special Publication of Fisheries and Aquatic Sciences 76.

van Brandt A (1964) *Fish Catching Methods of the World*. London: Fishing News (Books) Ltd.

Vickers K (1988) *A Review of Irish Salmon and Salmon Fisheries*. Moulin, Pitlochry: Atlantic Salmon Trust.

Went AEJ (1955) *Irish Salmon and Salmon Fisheries*. The Buckland Lectures. London: Edward Arnold.

Williamson R (1991) *Salmon Fisheries in Scotland*. Moulin, Pitlochry: Atlantic Salmon Trust.

Pacific

R. G. Kope, Northwest Fisheries Science Center, Seattle, WA, USA

doi:10.1006/rwos.2001.0452

Introduction

Pacific salmon comprise six species of anadromous salmonids that spawn in fresh water from central California in North America across the North Pacific Ocean to Korea in Asia: chinook salmon (*Oncorhynchus tshawytscha*), coho salmon (*O. kisutch*), sockeye salmon (*O. nerka*), chum salmon (*O. keta*), pink salmon (*O. gorbuscha*), and masu or cherry salmon (*O. masou*). Pacific salmon spawn in rivers, streams, and lakes where they die soon after spawning. Most juveniles migrate to the ocean as smolts, where they spend a significant portion of their life cycle. The length of freshwater and marine residence varies by species and the life span ranges from 2 years for pink salmon to as much as 7 or 8 years for some chinook salmon populations. Spawning runs of adult salmon have contributed an important source of protein for human cultures as well as a large influx of marine nutrients into terrestrial ecosystems. Large runs of mature fish returning from the sea every year have been highly visible to people living near rivers and salmon have historically assumed a role in the lives of people that extends beyond subsistence and commerce. Salmon became part of the social fabric of the cultures with which they interacted, and this significance continues today.

History

Salmon played an important role in the lives of people long before the arrival of Europeans on the

Pacific rim. The predictable appearance of large runs of fish in the rivers emptying into the North Pacific Ocean provided a readily available source of high quality protein that could be harvested in large quantities and preserved for consumption in the winter when other sources of food were scarce. In the coastal areas of Washington, British Columbia, and south-east Alaska, Pacific salmon were a staple in the diets of tribal people and supported a level of human population density, commerce, and art unrivaled elsewhere on the continent. In Asia, salmon were harvested for subsistence by native people in Siberia, and the Japanese have harvested and dry-salted salmon in Kamchatka and Sakhalin at least since the seventeenth century.

When Europeans arrived on the Pacific rim, they were quick to take advantage of the abundant salmon. In the late eighteenth century Russian fur traders in Alaska caught and preserved salmon to provision native trappers. Distant markets developed for fresh, dried, salted, and pickled salmon, but the industry was hampered because the methods of preserving fish were inadequate, and spoilage was a recurrent problem. Commercial fisheries for Pacific salmon did not expand to an industrial scale in North America until the introduction of canning in the 1860s.

Pacific salmon were first canned on the Sacramento River in California in 1864. The canning industry then rapidly spread to the Columbia River, Puget Sound, British Columbia, and Alaska. Within 20 years canneries were established along the entire west coast from Monterey in California to Bristol Bay in Alaska, and commercial fisheries for Pacific salmon were conducted on an industrial scale. By the beginning of the twentieth century a few canneries had also been established in Asia, but the principle product produced in Asian salmon fisheries remained dry-salted salmon preferred by Japanese consumers.

With the advent of powered fishing boats in the early twentieth century, new fishing gears became prevalent and a trend developed to intercept salmon further and further from their natal streams. Two factors contributing to this trend were the possibility of harvesting salmon at times when local stocks were unavailable because of run timing or depletion, and the harvesting of fish before they became available to other fisheries. Off the coast of the United States troll fisheries targeting chinook and coho salmon developed (largely to avoid the closures imposed on river fisheries to protect stocks) and purse seine gear came into use in pink, chum, and sockeye fisheries. The Japanese began using drift gill nets which enabled the development of a high-seas

mothership fleet off the coast of Kamchatka, and shore-based fisheries in the Kuril Islands.

Increasing catch of salmon stocks offshore led to a period, in the latter half of the twentieth century, of increasing international collaboration on research and management of Pacific salmon fisheries. One of the underlying principles of international management in the twentieth century has been that salmon belong to the countries in which they originate. Because of this principle, much of the research has focused on the migration and distribution of different stocks, and on methods to determine the origin of salmon encountered on the high seas. This principle of ownership has also encouraged the escalation of hatchery production.

Gear Types

Aboriginal fisheries traditionally used a variety of gear to harvest salmon. Weirs and traps were employed throughout North America and Asia and were probably the most common and efficient means of harvest. Spears and dipnets were also widely used. North American tribal fishermen also used hook and line, bow and arrow, gill nets, seines, and an elaborate gear called a reef net. The reef net was employed in northern Puget Sound to target primarily Fraser River sockeye salmon. It consisted of a rectangular net, suspended between two canoes in shallow water in the path of migrating salmon and was fished on a flood tide. Leads helped to direct fish into the net, which was raised when fish were seen swimming over it.

Commercial fisheries in Asia have primarily utilized traps in coastal waters, and seines and weirs in the rivers. In the early twentieth century the Japanese began using drift gill nets largely to avoid the uncertainty of retaining access to trap sites in Kamchatka. This allowed the Japanese to harvest salmon stocks originating from Kamchatka and the Sea of Okhotsk outside of Russian territorial waters and ultimately led to the development of the Japanese high-seas mothership fishery. Today traps remain the most prevalent and efficient gear used in Asian coastal fisheries.

In North America, early commercial fisheries primarily used haul seines and gill nets. Traps were employed very effectively in the Columbia River, Puget Sound, and Alaska, and fishwheels were also used on the Columbia River and in Alaska. Traps are no longer permitted in North America, and fishwheels are only used in Alaskan subsistence fisheries and as experimental gear in tribal fisheries in British Columbia. As internal combustion engines came into use by the fishing fleet, troll and purse seine gears were developed to target salmon in coastal waters.

Catch

Since 1950, the catch of Pacific salmon has been dominated by pink and chum salmon, followed by sockeye, coho, and chinook salmon (**Figure 1**). Masu salmon are relatively uncommon, and spawn only in Japan and rivers entering the Seas of Japan and Okhotsk. They account for 1000–3000 tonnes per year, nearly all of that in Japan. Total harvest of Pacific salmon was relatively stable at around 400 000 tonnes during the 1950s, 1960s, and 1970s, but has increased dramatically since the 1970s, reaching a peak of more than 1 million tonnes in 1995. This has been due to increases in harvest of chum salmon in Japan, and in pink and sockeye salmon, primarily in Alaska (**Figure 2**). It is noteworthy that these three species have fundamentally different marine distributions to chinook and coho salmon. Chinook and coho salmon tend to have more coastal distributions while sockeye, pink, and chum undergo extensive marine migrations and have more offshore distributions. These differences

in marine distributions have apparently contributed to differences in abundance trends in response to environmental changes like El Niño-Southern Oscillation (ENSO) events and regime shifts occurring on decadal scales.

Chinook Salmon

Harvest of chinook salmon in the North Pacific has been variable but has been relatively stable since 1950, at least until the 1990s (**Figure 3**). Until the last couple of years, total production has varied between 15 000 and 30 000 tonnes with the bulk of production coming from North America. Average annual total harvest from the North Pacific for the 5 years period from 1994 through 1998 has been 18 000 tonnes including production from aquaculture, with approximately 10 000 tonnes coming from capture fisheries.

Chinook salmon originate primarily from large river systems, with the Columbia, Fraser, Sacramento, and Yukon Rivers being the largest producers. Historically the majority of the chinook harvest came from the USA, but recent environmental conditions have impacted southern stocks and the

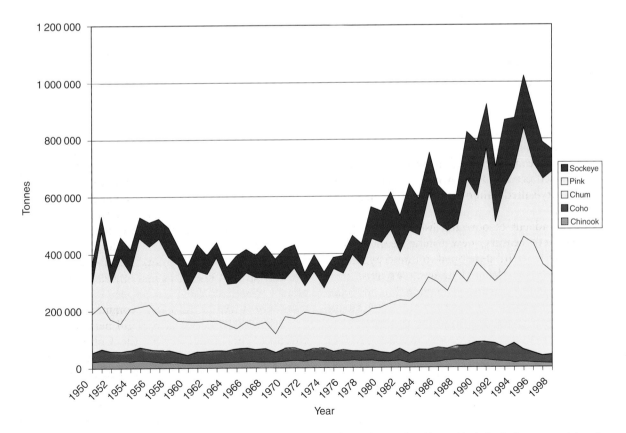

Figure 1 Commercial harvest of Pacific salmon in the North Pacific Ocean by species. Harvest includes both commercial capture fisheries and aquaculture. (Data from FAO Fishstat database.)

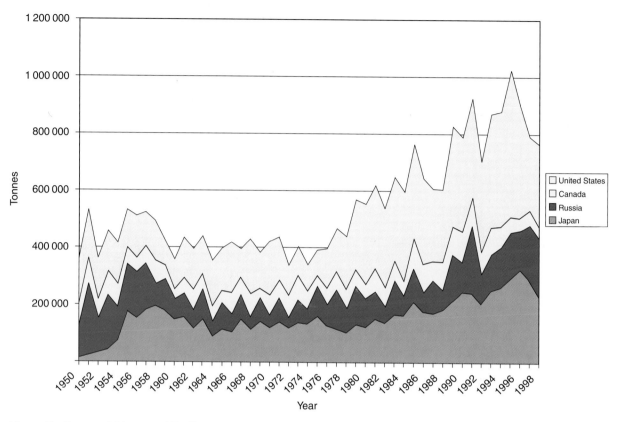

Figure 2 Commercial harvest of Pacific salmon by major fishing nations. (Data from FAO Fishstat database.)

harvest in Canada has surpassed that of the USA. Two pronounced dips in production, in early 1960s and 1980s, immediately followed strong ENSO events in 1959 and in the winter of 1982–83. The decline in production in the 1990s has also been associated with a series of ENSO events.

A trend which is not apparent in **Figure 3** is the shift from natural to hatchery production and aquaculture. During the latter half of the twentieth century there was a transition from predominantly naturally produced chinook salmon to production that is dominated by hatchery fish. Annual releases of hatchery chinook salmon have increased from approximately 50 million juveniles in 1950 to an average of 317 million from 1993 to 1995. The increase in Canadian landings in the late 1980s is coincident with the rapid expansion of a Canadian hatchery program throughout the 1980s. In recent years, poor marine survival, and conservation concerns for natural stocks have led to reductions in harvest in both the USA and Canada.

In the last decade, there has been a rapid increase in production of chinook salmon through aquaculture in net pens (**Figure 4**). While Canada is a major producer of pen-reared chinook salmon, the bulk

of this production now comes from the Southern Hemisphere, primarily from Chile and New Zealand.

Coho Salmon

Many aspects of the history of coho salmon production are very similar to that of chinook salmon production. Like chinook salmon, harvest of coho salmon in the North Pacific has been variable, but relatively stable until the last few years, with most of the production coming from North America (**Figure 5**). Production has varied from 30 000 to 60 000 tonnes, dipping below 30 000 tonnes following the strong ENSO events in 1959 and 1982. North American coho stocks have also suffered from poor marine survival in recent years, and conservation concerns have prompted reductions in harvest. Average annual production from the North Pacific from 1994 through 1998 have been 38 000 tonnes, with 24 000 tonnes coming from capture fisheries.

Like chinook there has also been a shift from natural production to hatchery production and aquaculture. Hatchery releases have increased from approximately 20 million juveniles in 1950 to an

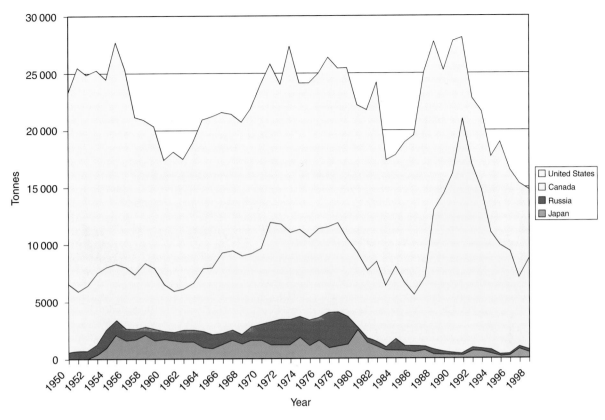

Figure 3 Commercial harvest of chinook salmon by major fishing nations in the North Pacific Ocean. Harvest includes both capture fisheries and aquaculture. (Data from FAO Fishstat database.)

average of 111 million juveniles annually from 1993 to 1995. Much of the increase in production by Japan in the 1980s was from aquaculture. The shift from capture fisheries to aquaculture production has been even more dramatic for coho salmon than for chinook (**Figure 6**). While Japan was the largest producer of pen-reared coho in the late 1980s and early 1990s, Chilean production has rapidly expanded in the 1990s and now exceeds all other production combined.

Sockeye Salmon

Sockeye harvest was relatively stable in the 1950s, varying between 50 000 and 100 000 tonnes, but reached a low point in the 1970s of < 40 000 tonnes (**Figure 7**). In the 1980s and 1990s, a combination of favorable marine conditions and harvest management that allowed sufficient spawning escapement led to dramatic increases in sockeye production in Alaska and the Fraser River in British Columbia. This has led to all-time record harvests in Alaska and the highest harvests in British Columbia since a blockage at Hell's Gate on the Fraser River devastated sockeye salmon runs in 1913 and 1914.

The peak harvest exceeded 230 000 tonnes in 1993, and the average harvest from 1994 to 1998 was 148 000 tonnes annually.

Sockeye salmon have the most complex life history of any Pacific salmon. Most sockeye salmon rear in lakes for 1–3 years. They then migrate to the ocean where they spend from 1 to 4 years. Despite this variability, most spawning runs are dominated by fish of a single total age, usually either 4 or 5 years. The extended freshwater rearing period allows for interactions between successive year-classes within populations which contribute to cycles with periods of 4–5 years. While there is some evidence of cyclic dominance in the catch record, it is largely masked by aggregation of stocks in the fisheries. Differences in spawner abundance between peak and off-peak years in individual populations can be greater than an order of magnitude, and cycles can persist for decades.

Unlike chinook and coho salmon, most sockeye production is the result of natural spawning. While artificial production from 1993 to 1995 has averaged 326 million juveniles annually, 75% of this artificial production has been from natural spawning in artificial spawning channels.

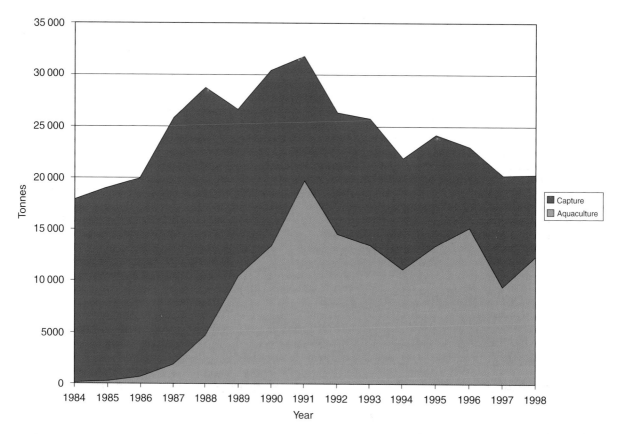

Figure 4 Total commercial production of chinook salmon in the Pacific basin. Most aquaculture production is in the Southern Hemisphere. (Data from FAO Fishstat database.)

Pink Salmon

Pink salmon are the most abundant species of Pacific salmon, but they also have the smallest average body size. Unlike other species of salmon, pink salmon all have a fixed 2 year life span. Because of this, the even-year and odd-year brood lines are genetically and demographically isolated, and tend to fluctuate independently with either the even-year or odd-year being dominant for long periods of time. In some streams only one of the brood lines is present. This is readily apparent in their harvest history (**Figure 8**), especially in Asian stocks where the odd-year brood line has been dominant.

The historical pattern of pink salmon harvest is similar to that of sockeye; landings were relatively stable in the 1950s and 1960s but declined slightly to reach a low in the early 1970s (**Figure 8**). Since the mid-1970s production has increased to recent record levels, reaching a peak harvest of > 430 000 tonnes in 1991. This increase has occurred in Russian and Alaskan harvested while Canadian harvest has been stable and Japanese harvest has declined. The average annual harvest of pink salmon from the

North Pacific from 1994 through 1998 was 327 000 tonnes.

Russia and Alaska are also where most pink salmon originate. While natural production is the major source of pink salmon, Alaska, Russia, and Japan have significant hatchery programs. Japan has had a long-standing hatchery program that has increased gradually, while Alaska did not begin hatchery production of pink salmon until the 1970s, but has expanded rapidly to surpass all others combined.

Chum Salmon

Like pink and sockeye salmon, the harvest of chum salmon declined in the 1950s and 1960s, and has increased to record levels in the 1990s (**Figure 9**). However, there is a fundamental difference between the production history of chum salmon and those of pink and sockeye salmon. North American production has been relatively stable and natural production in Asia has declined while there has been a large increase in Japanese hatchery production. The decline in Asian natural production is greater than is apparent from the catch history of Russian fisheries because the Japanese catch in

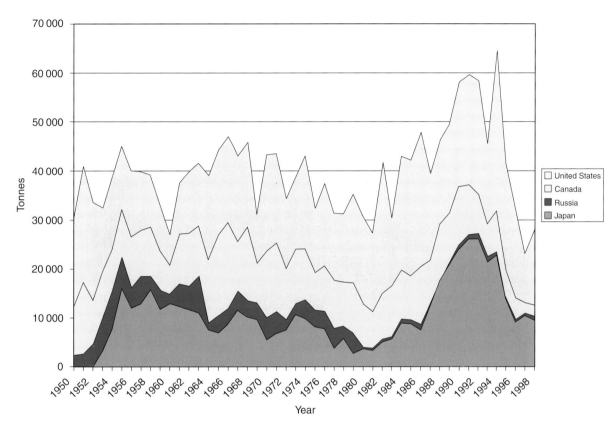

Figure 5 Commercial harvest of coho salmon by major fishing nations in the North Pacific Ocean. Harvest includes both capture fisheries and aquaculture. (Data from FAO Fishstat database.)

the 1950s and 1960s was primarily from high-seas fisheries, while the recent Japanese catch has been predominantly from coastal fisheries targeting returning hatchery fish from Hokkaido and northern Honshu.

Issues

Hatcheries

Pacific salmon fisheries are unique among marine commercial and recreational fisheries in the scale of their dependence on artificial propagation. Because freshwater habitat is often the factor most limiting to salmon production, massive artificial propagation programs have been implemented for all species of Pacific salmon to enhance fisheries, and as mitigation for freshwater habitat losses.

While hatcheries have augmented the abundance of salmon, there is increasing concern over potentially deleterious effects of hatcheries on natural salmon populations. Mixed stock fisheries targeting abundant hatchery stocks can overharvest less productive natural stocks that are intermingled with the hatchery fish. Hatchery stocks often differ genetically from the local natural stocks because the

original broodstock for the hatchery may have been obtained from a nonlocal population or hatchery, hatchery breeding practices which unintentionally exert selection pressure for particular traits, and because of domestication through adaptation to hatchery rearing conditions. Stray spawners from the hatcheries can make natural stocks appear more productive than they really are, and interbreeding between natural and hatchery stocks can further reduce the productivity of natural stocks through outbreeding depression.

While artificial propagation is widespread, the focus has been on different species in different areas.

In the contiguous United States, the focus of artificial propagation has been on chinook and coho salmon. Hatchery programs exist in most rivers that support chinook and coho populations, and hatchery releases from 1993 through 1995 have averaged approximately 71 million coho and 252 million chinook annually. Alaska and Canada also have hatchery programs, but their combined releases in the same period have averaged about 38 million coho salmon and 64 million chinook salmon annually. Hatchery fish account for the majority of the harvest of these two species in the contiguous United States.

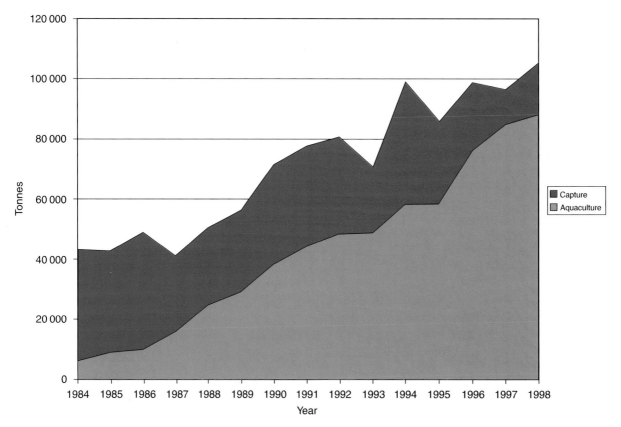

Figure 6 Total commercial production of coho salmon in the Pacific basin. Most aquaculture production is in the Southern Hemisphere. (Data from FAO Fishstat database.)

Canada has the largest program for artificial production of sockeye salmon, but the majority of this is in the form of artificial spawning channels in the Fraser River system. In these artificial channels adults spawn naturally and juveniles emigrate to lakes for rearing without being artificially fed. This program has developed rapidly since 1960 and the 1993–95 average annual production was approximately 244 million fry. Alaska has the largest hatchery program for sockeye smolts with annual production that has averaged approximately 69 million.

Alaska also has the largest hatchery program for pink salmon. This program has expanded rapidly since the 1970s. Annual releases of hatchery fry averaged 843 million from 1993 to 1995, and returns from this program have contributed substantially to the increases in landings of pink salmon in recent years. Over the same period, Japanese hatcheries have released an annual average of 132 million, and Russian hatcheries, 264 million pink salmon fry.

The largest Pacific salmon hatchery program is operated by the Japanese for chum salmon. Japan's chum salmon hatchery program dates back to 1888 when the national salmon hatchery was built in

Chitose on Hokkaido. This program released 200 000–500 000 juveniles annually in the 1950s, but expanded rapidly in the 1970s. From 1993 to 1995 Japanese hatcheries released an average of more than 2 billion chum salmon fry annually. During the same period, Canada and the Russian Federation each released an average of 220 million hatchery chum salmon annually, and annual releases from US hatcheries have averaged more than 500 million hatchery chum salmon fry. Combined, these programs have been releasing nearly 3 billion chum salmon juveniles per year, nearly 60% of the hatchery production of all Pacific salmon combined.

The recent and rapid increase in hatchery production of salmon, coupled with the increase in total landings and a concurrent decline in the average size of individuals of all salmon species has raised concerns, and prompted research into the possibility that ocean carrying capacity is currently limiting production of salmon in the North Pacific Ocean.

International Management

One of the challenges in management of Pacific salmon fisheries is the direct result of their

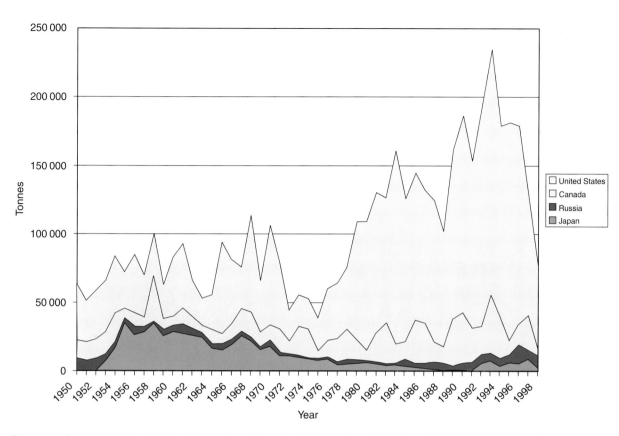

Figure 7   Commercial harvest of sockeye salmon by major fishing nations. (Data from FAO Fishstat database.)

anadromous life history. Although stocks are genetically distinct, and segregate at spawning, they undergo extensive marine migrations and co-mingle in the ocean. This characteristic makes stocks vulnerable to interception in foreign fisheries. Countries harvesting Pacific salmon have long recognized the need for international coordination of harvest management, but cooperation has been an ongoing challenge complicated by international relationships that have been less than cordial at times.

Japanese have fished for salmon along the coasts of Sakhalin, Kamchatka, and the Kuril Islands at least since the early seventeenth century. In the early nineteenth century, Russia levied a tariff on salmon exported to Japan, and in the 1890s Russia began to restrict Japanese access to trap sites. Through the early part of the twentieth century relations between Japan and Russia became increasingly strained and the Japanese fishing industry generally had increasing difficulty accessing traditional shore-based trap sites in Russia. This encouraged the development and expansion of Japanese high-seas drift gill net fisheries in the western Bering Sea and southern Sea of Okhotsk and of shore-based trap and gill net fisheries in the Kuril Islands targeting salmon stocks originating primarily from the Sea of Okhotsk. Dur-

ing World War II Japan lost all access to trap sites in the Soviet Union, and the Japanese high-seas fishery ceased. At the end of the war, the Soviet Union took possession of the Kuril Islands and Japanese fisheries based in the Kuril Islands ceased as well.

In 1952, the governments of Canada, Japan, and the USA signed the International Convention for the High-Seas Fisheries of the North Pacific Ocean which established the International North Pacific Commission (INPFC). Under the terms of this Convention, Japan resumed and rapidly expanded their high-seas salmon fishery in the northern Pacific Ocean, Bering Sea, and the southern portion of the Sea of Okhotsk. The primary management action of the INPFC was to set an eastern limit on the extent of high-seas fishing, but it also embarked on a large-scale research program directed at addressing issues relevant to the Commission with much of the focus on stock identification, distribution, and migration patterns. Because the Japanese high-seas fishery targeted Asian stocks, mostly from the Soviet Union, the Soviet Union responded by excluding the Japanese high-seas fishery from the Sea of Okhotsk and a large portion of the western Bering Sea. This action led to the negotiation of a Convention between Japan and the USSR in 1957 which

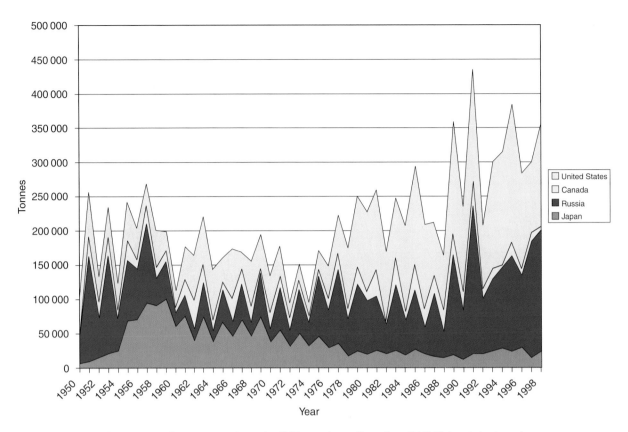

Figure 8   Commercial harvest of pink salmon by major fishing nations. (Data from FAO Fishstat database.)

established the Soviet–Japan Fisheries Commission to regulate the harvest of Asian salmon stocks on the high-seas.

In the late 1970s, leading up to adoption of the United Nations Convention of Law of the Sea in 1982, the USA, Canada, and the Soviet Union established 200 mile fishery zones off their coasts. This further reduced the areas accessible to high-seas fisheries. In 1993 the North Pacific Anadromous Fisheries Commission (NPAFC) replaced the INPFC with Japan, the Russian Federation, Canada, and the USA as members. Between the NPAFC Convention, which prohibits high-seas salmon fishing and trafficking in illegally caught salmon, and the United Nations General Assembly Resolution 46/215, prohibiting large-scale pelagic drift gill netting, high-seas fisheries for Pacific salmon have been eliminated except for illegal fishing.

In North America, much of the focus of international management has been on the Fraser River run of sockeye salmon. This was the largest run of sockeye in North America, and one of the most profitable for the canning industry in the late nineteenth century and early twentieth century. Returning adult Fraser River sockeye have two migratory approaches to the Fraser River: the southern

approach through the Strait of Juna de Fuca of the south of Vancouver Island where they are vulnerable to American fisheries, and the northern approach through the Johnstone and Georgia Straits which lies entirely within Canadian waters. In most years the majority of the run has taken the southern approach where it was intercepted by US fisheries, causing persistent friction between the two countries.

The need for cooperative management of the fisheries in these southern boundary waters was recognized by both countries. Beginning in the 1890s, a series of commissions were formed to study the problem and make recommendations. In 1913 the Fraser River was obstructed by debris from railroad construction at Hell's Gate in the Fraser River Canyon. This passage problem was exacerbated by a rock slide caused by the construction in 1914 which effectively destroyed the largest, most productive sockeye run in North America. This had devastating impacts on inland native tribes who depended on salmon as a dietary staple, and had lasting impacts on the commercial fisheries as well. The desire to restore Fraser River sockeye production increased the incentive for cooperation, but it has been a long and difficult process.

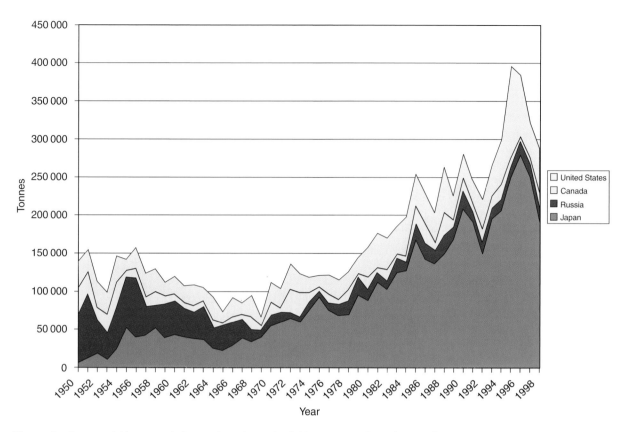

Figure 9 Commercial harvest of chum salmon by major fishing nations. (Data from FAO Fishstat database.)

Bilateral treaties were signed in 1908, 1919, and 1929, but enabling legislation was never passed by the US Congress. The 1929 treaty was amended and again signed in 1930 but was not ratified until 1937, 45 years after negotiations had started. The treaty established the International Pacific Salmon Fisheries Commission (IPSFC) to study and restore Fraser River sockeye salmon, and manage sockeye fisheries in Convention waters of the southern approach to the Fraser River, but postponed any regulatory authority for another 8 years. The IPSFC constructed many fish passage structures in the Fraser River basin, and its authority was extended to include pink salmon with the addition of a pink salmon protocol in 1957.

In 1985 a new treaty was signed and ratified that replaced the IPSFC with the Pacific Salmon Commission (PSC). The PSC has expanded authority, which includes all five species of Pacific salmon harvested in North America, and encompasses marine boundary waters of the northern and southern Canada–US borders, and transboundary rivers passing through Canada and south-east Alaska. However, the effectiveness of the PSC has been hampered by the inability to come to unanimous agreement on allocation issues in some years.

Endangered Species

Because of their dependence on freshwater habitat, anadromous fish are vulnerable to habitat degradation resulting from human activities in inland areas. The requirements of Pacific salmon for spawning gravel free of fine sediments and cold oxygenated water for spawning and juvenile rearing make them particularly vulnerable to habitat loss and degradation from nearly all human activities where salmon and people co-occur. Construction of impassable dams for flood control, hydropower, and irrigation has eliminated access to much historic habitat. Stream channelization and levee construction for flood control has reduced habitat availability and complexity. Logging has resulted in scoured stream channels, increased runoff and sediment loads, and has diminished the availability of shade and large woody debris which provides rearing habitat. Mining has directly destroyed stream channels and choked streams with sediment, as well as contaminating the water with heavy metals. Water diversions for agricultural, industrial, and domestic use have reduced the water available and directly removed juvenile salmon. Grazing has increased sedimentation and destroyed streamside vegetation.

Home construction and urbanization have contributed to sedimentation and chemical pollution, and have increased the amount of impervious surface, increasing the variability in stream flow.

Collectively these impacts have eliminated many populations of Pacific salmon and have compromised the productivity of many remaining ones. Reduced productivity of natural stocks has increased their susceptibility to overharvest, and the construction of hatcheries to mitigate the impacts of water development projects and enhance fisheries has often exacerbated the problem by increasing competition between natural and hatchery fish and increasing the harvest pressure on all fish in the attempt to harvest hatchery fish. As a result, many natural populations are at critically low abundance where they are at higher risk of extirpation from random environmental and demographic variability or from catastrophic events. As a result, a number of distinct population segments of Pacific salmon and steelhead trout in the contiguous United States have been listed as threatened or endangered under the US Endangered Species Act. At the time of writing (2000) the listings include 17 distinct population segments of chinook, coho, chum, and sockeye salmon, and another 10 distinct population segments of steelhead trout. The additional regulatory complexities of dealing with listed species has greatly complicated the management of Pacific salmon fisheries in the USA.

In response to these listings, and critically low abundance of some Canadian stocks, harvest impacts on depressed stocks have been substantially reduced in the contiguous United States and British Columbia. Efforts are being made to reduce the combined negative impacts of habitat loss and degradation, overharvest, and the negative impacts of hatchery production. Fishery scientists and managers are exploring changes in harvest practices to allow more selective harvest of hatchery stocks and healthy natural stocks while reducing impacts on listed stocks.

See also

El Niño Southern Oscillation (ENSO). Fishing Methods and Fishing Fleets. International Organizations. Law of the Sea. Ocean Ranching. Salmonid Farming. Salmonids.

Further Reading

Cobb JN (1917) *Pacific Salmon Fisheries*. Bureau of Fisheries Document No. 839. Washington, DC: Government Printing Office.

Groot C and Margolis L (eds) (1991) *Pacific Salmon Life Histories*. Vancouver, BC: UBC Press.

McNeil WJ (ed.) (1988) *Salmon Production, Management and Allocation*. Corvallis, OR: Oregon State University Press.

NMFS (1999) *Our Living Oceans. Report on the Status of US Living Marine Resources, 1999*. US Department of Commerce, NOAA Technical Memo. NMFS-F/SPO-41.

Shepard MP, Shepard CD and Argue AW (1985) *Historic Statistics of Salmon Production Around the Pacific Rim*. Canadian Manuscript Report of Fisheries and Aquatic Sciences No. 1819, 297 pages.

SALMONIDS

D. Mills, Atlantic Salmon Trust, UK

doi:10.1006/rwos.2001.0018

Introduction

The Atlantic and Pacific salmon and related members of the Salmonidae are anadromous fish, breeding in fresh water and migrating to sea as juveniles at various ages where they feed voraciously and grow fast. Survival at sea is dependent on exploitation, sea surface temperature, ocean climate and predation. Their return migration to breed reveals a remarkable homing instinct based on various guidance mechanisms. Some members of the family, however, are either not anadromous or have both anadromous and nonanadromous forms.

Taxonomy

The Atlantic salmon (*Salmo salar*) and the seven species of Pacific salmon (*Oncorhynchus*) are members of one of the most primitive superorders of the teleosts, namely the Protacanthopterygii. The family Salmonidae includes the Atlantic and Pacific salmon, the trout (*Salmo* spp.), the charr (*Salvelinus* spp.) and huchen (*Hucho* spp.). The anatomical features that separate the genera *Salmo* and *Oncorhynchus* from the genus *Salvelinus* are the positioning of the teeth. In the former two genera the teeth form a double or zigzag series over the whole of the vomer bone, which is flat and not boat-shaped, whereas in the latter the teeth are restricted to the front of a boat-shaped vomer. In the genus *Salmo* there is only a small gap between the vomerine and palatine teeth but this gap is wide in adult *Oncorhynchus* and not in *Salmo*. A specialization occurring in *Oncorhynchus* and not in *Salmo* is the simultaneous ripening of all the germ cells so that these fish can only spawn once (semelparity). There are a number of anatomical features that help in the identification of the various species of *Salmo* and *Oncorhynchus*; these include scale and fin ray counts, the number and shape of the gill rakers on the first arch and the length of the maxilla in relation to the eye.

Origin

There has been much debate as to whether the Salmonidae had their origins in the sea or fresh water. Some scientists considered that the Salmonidae had a marine origin with an ancestor similar to the Argentinidae (argentines) which are entirely marine and, like the salmonids and smelts (Osmeridae), bear an adipose fin. Other scientists considered the salmonids to have had a freshwater origin, supporting their case by suggesting that since the group has both freshwater-resident and migratory forms within certain species there has been recent divergence. Furthermore, there are no entirely marine forms among modern salmonids so they can not have had a marine origin. The Salmonidae have been revised as relatively primitive teleosts of probable marine pelagic origin whose specializations are associated with reproduction and early development in fresh water. The hypothesis of the evolution of salmonid life histories through penetration of fresh water by a pelagic marine fish, and progressive restrictions of life history to the freshwater habitat, involves adaptations permitting survival, growth and reproduction there. The salmonid genera show several ranges of evolutionary progression in this direction, with generally greater flexibility among *Salmo* and *Salvelinus* than among *Oncorhynchus* species (**Table 1**). Evidence for this evolutionary progression is perhaps

Table 1 Examples of flexibility of life history patterns in salmonid genera: anadromy implies emigration from fresh water to the marine environment as juveniles and return to fresh water as adults; nonanadromy implies a completion of the life cycle without leaving fresh water, although this may involve migration between a river and a lake habitat

Genus	Species		
	Anadromous form only	Both anadromous and nonanadromous forms	Nonanadromous form only
Oncorhynchus	gorbuscha (pink salmon) keta (chum salmon) tschawytscha (chinook salmon)	nerka (sockeye salmon) kisutch (coho salmon) masou (cherry salmon) rhodurus (amago salmon) mykiss (steelhead/rainbow trout) clarki (cut throat trout)	aguabonito (golden trout)
Salmo	none	salar (Atlantic salmon) trutta (sea/brown trout)	
Salvelinus	none	alpinus (arctic charr) fontinalis (brook trout) malma (Dolly Varden) leucomanis	namaycush (lake trout)
Hucho	none	perryi	hucho (Danube salmon)

Adapted from Thorpe (1988).

even greater if one starts with the Argentinidae and Osmeridae which are basically marine coastal fishes. Some enter the rivers to breed, some live in fresh water permanently, and others such as the capelin (*Mallotus villosus*) spawn in the gravel of the seashore.

Life Histories

Members of the Salmonidae have a similar life history pattern but with varying degrees of complexity. A typical life history involves the female excavating a hollow in the river gravel into which the large yolky eggs are deposited and fertilized by the male. Because of egg size and the protection afforded them in the gravel the fecundity of the Salmonidae is low when compared with species such as the herring and the cod which are very fecund but whose eggs have no protection. On hatching the salmonid young (alevins) live on their yolk sac within the gravel for some weeks depending on water temperature. On emerging the fry may remain in the freshwater environment for a varying length of time (**Table 2**) changing as they grow into the later stages of parr and then smolt, at which stage they go to sea (**Figure 1**). Not all the Salmonidae have a prolonged freshwater life before entering the sea, and the juveniles of some species such as the pink salmon (*Oncorhynchus gorbuscha*) and chum salmon (*O. keta*) migrate to sea on emerging from the gravel. Others such as the sockeye salmon (*O. nerka*) have specialized freshwater requirements, namely the need for a lacustrine environment to which the young migrate on emergence (**Table 2**).

Distribution

Atlantic Salmon

The Atlantic salmon occurs throughout the northern Atlantic Ocean and is found in most countries whose rivers discharge into the North Atlantic Ocean and Baltic Sea from rivers as far south as Spain and Portugal to northern Norway and Russia and one river in Greenland. It has been introduced to some countries in the Southern Hemisphere, including New Zealand where they only survive as a land-locked form.

Sea Trout

The marine distribution of the anadromous form of *S. trutta* is confined to coastal and near-offshore waters and is not found in the open ocean. It has a more limited distribution than the salmon being confined mainly to the eastern seaboard of the North Atlantic, although it has been introduced includes Iceland, and the Faroe Islands, Scandinavia, the Cheshkaya Gulf in the north, throughout the Baltic and down the coast of Europe to northern Portugal. It occurs as a subspecies in the Black Sea (*S. trutta labrax*) and Caspian Sea (*S. trutta caspius*). It has been introduced to countries in the Southern Hemisphere including Chile and the Falkland Islands.

Sockeye Salmon

The natural range of the sockeye, as other species of Pacific salmon, is the temperate and subarctic waters of the North Pacific Ocean and the northern

Table 2 Life histories

Species	Length of freshwater life (years)	Particular features
Oncorhynchus nerka	1–3	Lake environment required for juveniles
O. gorbuscha	Migrate to estuarine waters on emergence	Tend to spawn closer to sea than other oncorhynchids, and may frequent smaller river systems
O. keta	Migrate to estuarine waters on emergence	Spawning takes place in lower reaches of rivers
O. tschawytscha	Some migrate to estuary on emergence, others remain in fresh water for one or more years	Two races: *anadromous*, long freshwater residence *semelparous*, short freshwater residence
O. kisutch	1–2	Tend to utilize small coastal streams
O. masou	1–2	Large parr become smolts and go to sea Small to medium-sized parr remain in fresh water
Salmo salar	1–7	A small percentage spawn more than once. 'Land-locked' forms live in lakes and spawn in afferent or efferent rivers
S. trutta	1–4 (anadromous form)	May spawn frequently, the anadromous form after repeat spawning migrations
S. alpinus	2–6	Anadromous form only occurs in rivers lying north of 60°N

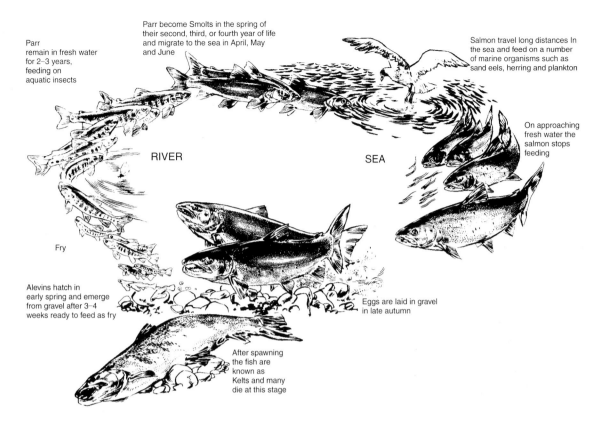

Parr become Smolts in the spring of their second, third, or fourth year of life and migrate to the sea in April, May and June

Parr remain in fresh water for 2–3 years, feeding on aquatic insects

Salmon travel long distances In the sea and feed on a number of marine organisms such as sand eels, herring and plankton

RIVER

SEA

On approaching fresh water the salmon stops feeding

Fry

Alevins hatch in early spring and emerge from gravel after 3–4 weeks ready to feed as fry

Eggs are laid in gravel in late autumn

After spawning the fish are known as Kelts and many die at this stage

Figure 1 Life cycle of the Atlantic salmon. (Reproduced from Mills, 1989.)

adjoining Bering Sea and Sea of Okhotsk. However, because sockeye usually spawn in areas associated with lakes, where their juveniles spend their freshwater existence before going to sea, their spawning distribution is related to north temperate rivers with lakes in their systems. The Bristol Bay watershed in southwestern Alaska and the Fraser River system are therefore the major spawning areas for North American sockeye.

Pink Salmon

The natural freshwater range of pink salmon embraces the Pacific coast of Asia and North America north of 40°N and during the ocean feeding and maturation phase they are found throughout the North Pacific north of 40°N. The pink is the most abundant of the Pacific salmon species followed by the chum and sockeye in that order. Pink salmon have been transplanted outside their natural range to the State of Maine, Newfoundland, Hudson Bay, the North Kola Peninsula and southern Chile.

Chum Salmon

The chum occurs throughout the North Pacific Ocean in both Asian and North American waters north of 40°N to the Arctic Ocean and along

the western and eastern arctic seaboard to the Lena River in Russia and the Mackenzie River in Canada.

Chinook Salmon

The chinook has a more southerly distribution than the other Pacific salmon species, extending as far south as the Sacramento–San Joaquin River system in California as well as into the northerly waters of the Arctic Ocean and Beaufort Sea. It has been transplanted to the east coast of North America, the Great Lakes, south Chile and New Zealand.

Coho Salmon

The coho is the least abundant of the Pacific salmon species but has a similar distribution as the other species. It has been introduced to the Great Lakes, some eastern states of North America and Korea and Chile.

Masu Salmon

This species only occurs in Far East Asia, in the Sea of Japan and Sea of Okhotsk. Some have been transported to Chile. The closely related amago salmon mainly remain in fresh water but some do migrate to coastal waters but few to the open sea.

Arctic Charr

The anadromous form of this species has a wide distribution throughout the subarctic and arctic regions of the North Atlantic north of 40°N. It enters rivers in the late summer and may remain in fresh water for some months.

Migrations

Once Atlantic salmon smolts enter the sea and become postsmolts they move relatively quickly into the ocean close to the water surface. Patterns of movement are strongly influenced by surface water currents, wind direction and tidal cycle. In some years postsmolts have been caught in the near-shore zone of the northern Gulf of St Lawrence throughout their first summer at sea, and in Iceland some postsmolts, mainly maturing males, forage along the shore following release from salmon-ranching stations. These results indicate that the migratory behavior of postsmolts can vary among populations. Pelagic trawl surveys conducted in the Iceland/ Scotland/Shetland area during May and June and in the Norwegian Sea from 62°N to 73°N in July and August, have shown that postsmolts are widely distributed throughout the sampled area, although they do not reach the Norwegian Sea until July and August. Over much of the study area, catches of postsmolts are closely linked to the main surface currents, although north of about 64°N, where the current systems are less pronounced, postsmolts appear to be more diffusely distributed.

Young salmon from British Columbia rivers descend in discontinuous waves, and it has been suggested that this temporal pattern has evolved in response to short-term fluctuations in the availability of zooplankton prey. Although zooplankton production along the British Columbia coast is adequate to meet this seasonal demand, there is debate about the adequacy of the Alaska coastal current to support the vast populations of growing salmon in the summer. Density-dependent growth has been shown for the sockeye salmon populations at this time, suggesting that food can be limiting.

The movements of salmon in offshore waters are complex and affected by physical factors such as season, temperature and salinity and biological factors such as maturity, age, size and food availability and distribution of food organisms and stock-of-origin (i.e. genetic disposition to specific migratory patterns). Through sampling of stocks of the various Pacific salmon at various times of the year over many years scientists have been able to construct oceanic migration patterns of some of the major stocks of North American sockeye, chum and pink salmon (**Figure 2**) as well as for stocks of chinook and masu salmon. Similarly, as a result of ocean surveys and tagging experiments it has been possible to determine the migration routes of North American Atlantic salmon very fully (**Figures 3 and 4**). A picture of the approximate migration routes of Atlantic salmon in the North Atlantic area as a whole has been achieved from tag recaptures (**Figure 5**).

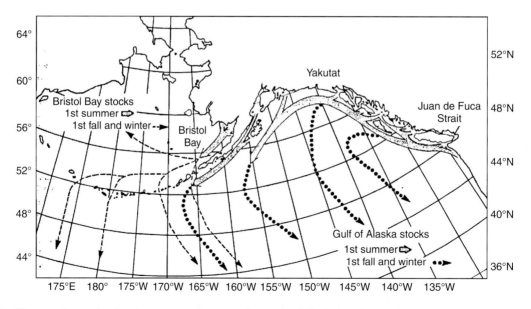

Figure 2 Diagram of oceanic migration patterns of some major stocks of North American sockeye, chum and pink salmon during their first summer at sea, plus probable migrations during their first fall and winter. (Reproduced from Burgner, 1991.)

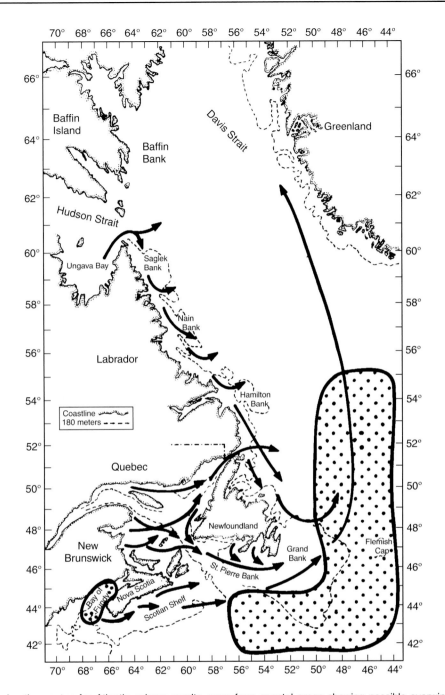

Figure 3 The migration routes for Atlantic salmon smolts away from coastal areas showing possible overwintering areas and movement of multi-sea winter salmon into the West Greenland area. Arrows indicate the path of movement of the salmon, and dotted area indicates the overwintering area. (Reproduced from Reddin, 1988.)

Movements

Vertical and horizontal movements of Atlantic salmon have been investigated using depth-sensitive tags and data storage tags. Depth records show that salmon migrate mostly in the uppermost few meters and often show a diel rhythm in vertical movements. The salmon are closest to the surface at mid-day and go deeper at night. They have been recorded diving to a depth of 110 m.

Available information from research vessels and operation of commercial Pacific salmon fisheries suggest that Pacific salmon generally occur in near-surface waters.

Details of rates of travel are given in **Table 3**.

Food

The marine diet of both Pacific and Atlantic salmon comprises fish and zooplankton. The proportion of fish and zooplankton in their diet varies with season, availability and area. Among the fish species sockeye eat capelin (*Mallotus villosus*), sand eels (*Ammodytes hexapterus*), herring (*Clupea harengus pallasi*) and pollock (*Theragra chalcogramma*).

Zooplankton organisms include euphausiids, squid, copepods and pteropods.

The diet of pink salmon includes fish eggs and larvae, squid, amphipods, euphausiids and copepods, whereas chum salmon were found to take pteropods, salps, euphausiids and amphipods. Chinook salmon eat herring, sand eels, pilchards and anchovies and in some areas zooplankton never exceeds 6% of the diet. However, the diet varies

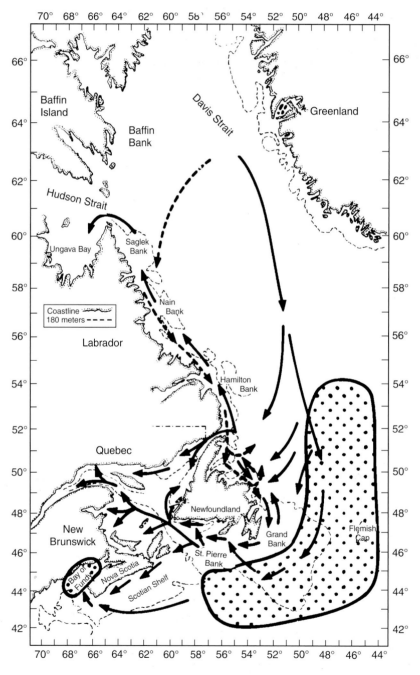

Figure 4 The migration routes of salmon from West Greenland and overwintering areas on return routes to rivers in North America. Solid arrows indicate migration in mid-summer and earlier; broken arrows indicate movement in late summer and fall; dotted areas are the wintering areas. (Reproduced from Reddin, 1988.)

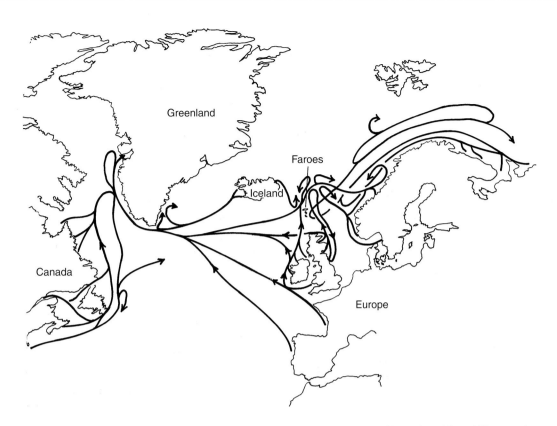

Figure 5 Approximate migration routes of Atlantic salmon in the North Atlantic area. (Reproduced from Mills, 1989.)

Table 3 Rates of travel in the sea of Atlantic and Pacific salmon

Species	Rate of travel (km day^{-1})	Conditions
Atlantic salmon	19.5–24	Icelandic postsmolts from ranching stations
	15.2–20.7	Postsmolts based on smolt tag recaptures
	22–52	Grilse and large salmon
	32	Average for maturing salmon of all sea ages
	26	For previous spawners migrating to Newfoundland and Greenland
	28.8–43.2	Icelandic coastal waters
	10–50	Baltic Sea
Sockeye salmon	46–56	During their final 30–60 days at sea
Pink salmon	17.2–19.8	Juveniles during first 2–3 months at sea
	45–54.3	Fish recovered at sea and in coastal waters
	43.3–60.2	For eastern Kamchatka stocks tagged in Aleutian Island passes or the Bering Sea
Coho salmon	30	
	55	Could be maintained over long distances

considerably from area to area and up to 21 different taxonomic groups have been recorded in this species' diet. Similarly, the diet of coho salmon is a varied one, with capelin, sardines, lantern fish (myctophids), other coho salmon, being eaten along with euphausiids, squid, goose barnacles and jellyfish.

Masu salmon eat mainly small-sized fish and squid and large zooplankton such as amphipods and euphausiids. Fish species taken include capelin, herring, Dolly Varden charr, Japanese pearlside (*Maurolicus japonicus*), saury (*Cololabis saira*), sand eels, anchovies, greenlings (*Hexagrammos otakii*) and sculpins (*Hemilepidotus* spp.).

There is no evidence of selective feeding among sockeye, pink and chum salmon.

The food of Atlantic salmon postsmolts is mainly invertebrate consisting of chironomids and gammarids in the early summer in inshore waters and in the late summer and autumn the diet changes to one of small fish such as sand eels and herring larvae. A major dietary study of 4000 maturing Atlantic salmon was undertaken off the Faroes and it confirmed the view that salmon forage opportunistically, but that they demonstrate a preference for fish rather than crustaceans when both are available. They are also selective when feeding on crustaceans, preferring hyperiid amphipods to euphausiids. Feeding intensity and feeding rate of Atlantic salmon north of the Faroes have been shown to be lower in the autumn than in the spring, which might suggest that limited food is available at this time of year. Similar results have been found for Atlantic salmon in the Labrador Sea and in the Baltic. Salmon in the north Atlantic also rely on amphipods in the diet in the autumn, whereas at other times fish are the major item. Fish taken include capelin, herring, sprat (*Clupea sprattus*), lantern fishes, barracudinas and pearlside (*Maurolicus muelleri*).

Predation

Marine mammals recorded predating on Pacific salmon include harbour seal (*Phoca vitulina*), fur seal (*Callorhinus uresinus*), Californian sea lion (*Zalophus grypus*), humpbacked whale (*Megaptera novaeangliae*) and Pacific white-sided dolphin (*Lagenorhynchus obliquideus*). Pinniped scar wounds on sockeye salmon, caused by the Californian sea lion and harbor seal, increased from 2.8% in 1991 to 25.9% in 1996 and on spring-run chinook salmon they increased from 10.5% in 1991 to 31.8% in 1994.

Predators of Atlantic salmon postsmolts include gadoids, bass (*Dicentrarchus labrax*), gannets (*Sula bassana*), cormorants (*Phalacrocorax carbo*) and Caspian terns (*Hydroprogne tschegrava*). Adult fish are taken by a number of predators including the grey seal (*Halichoerus grypus*), the common seal (*Phoca vitulina*), the bottle-nosed dolphin (*Tursiops truncatus*), porbeagle shark (*Lamna cornubica*), Greenland shark (*Somniosus microcephalus*) and ling (*Molva molva*).

Environmental Factors

Surface Salinity

Salinity may have an effect on fish stocks and it has been shown that the great salinity anomaly of the 1970s in the North Atlantic adversely affected the spawning success of eleven of fifteen stocks of fish whose breeding grounds were traversed by the anomaly. In the North Pacific there was found to be little relation between the high seas distribution of sockeye salmon and surface salinity. Sockeye are distributed across a wide variety of salinities, with low salinities characterizing 'salmon waters' of the Subarctic Pacific Region.

Sea Surface Temperature

Sockeye salmon are found over a wide variety of conditions. Sea surface temperature (SST) has not been found to be a strong and consistent determinant of sockeye distribution, but it definitely influences distribution and timing of migrations. Sockeye tend to prefer cooler water than the other Pacific salmon species.

Temperature ranges of waters yielding catches of various salmon species in the northwest Pacific in winter are: sockeye, 1.5–6.0°C; pink, 3.5–8.5°C; chum, 1.5–10.0°C; coho, 5.5–9.0°C.

A significant relationship was found for SST and Atlantic salmon catch rates in the Labrador Sea, Irminger sea and Grand Banks (**Figure 6**), with the greatest abundance of salmon being found in SSTs between 4 and 10°C. This significant relationship suggests that salmon may modify their movements at sea depending on SST. It has also been suggested that the number of returning salmon is linked to environmental change and that the abundance of salmon off Newfoundland and Labrador was linked to the amount of water of < 0°C on the shelf in summer. In years when the amount of cold water was large and the marine climate tended to be cold, there were fewer salmon returning to coastal waters. A statistically significant relationship has been found between the area of ice off Labrador and northern Newfoundland and the number of returning salmon

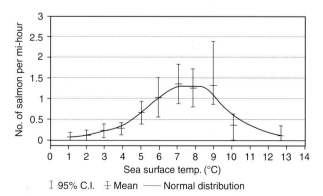

Figure 6 The relationship between sea surface temperatures and salmon catch rates from Labrador Sea, Irminger Sea and Grand Banks, 1965–91. (Reproduced from Reddin and Friedland, 1993.)

in one of the major rivers on the Atlantic coast of Nova Scotia. The timing and geographical distribution of Atlantic salmon along the Newfoundland and Labrador coasts have been shown to be dependent on the arrival of the 4°C water, salmon arriving earlier during warmer years.

In two Atlantic salmon stocks that inhabit rivers confluent with the North Sea a positive correlation was found between the area of 8–10°C water in May and the survival of salmon. An analysis of SST distribution for periods of good versus poor salmon survival showed that when cool surface waters dominated the Norwegian coast and the North Sea during May salmon survival has been poor. Conversely, when the 8°C isotherm has extended northward along the Norwegian coast during May, survival has been good.

Temperature may also be linked to sea age at maturity. An increase in temperature in the northeast Atlantic subarctic was found to be associated with large numbers of older (multi-sea winter) salmon and fewer grilse (one-sea winter salmon) returning to the Aberdeenshire Dee in Scotland.

Ocean Climate

Ocean climate appears to have a major influence on mortality and maturation mechanisms in salmon. Maturation, as evidenced by returns and survival of salmon of varying age, has been correlated with a number of environmental factors. The climate over the North Pacific is dominated by the Aleutian Low Pressure System. The long-term pattern of the Aleutian Low Pressure System corresponds with trends in salmon catch, with copepod production and with other climatic indices, indicating that climate and the marine environment may play an important role in salmon production. Survival of Pacific salmon species varies with fluctuations in large-scale circulation patterns such as El Niño Southern Oscillation (ENSO) events and more localized upwelling circulation that would be expected to affect local productivity and juvenile salmon growth. Runs of Pacific salmon in rivers along the western margin of North America stretching from Alaska to California vary on a decadal scale. When catches of a species are high in one region (e.g. Oregon) they may be low in another (e.g. Alaska). These changes are in part caused by marked interdecadal changes in the size and distribution of salmon stocks in the northeast Pacific, which are in turn associated with important ecosystem shifts forced by hydroclimatic changes linked to El Niño and possibly climate change.

Similarly, in the North Atlantic there have been annual and decadal changes in the North Atlantic Oscillation (NAO) index. Associated with these changes is a wide range of physical and biological responses, including effects on wind speed, ocean circulation, sea surface temperature, prevalence and intensity of Atlantic storms and changes in zooplankton production. For example, during years of positive NAO index, the eastern and western North Atlantic display increases in temperature while temperatures in central North Atlantic and in the Labradon Basin decline. The extremely low temperatures in the Labrador Sea during the 1980s coincided with a decline in salmon abundance. Similarly, in Europe, years of low NAO index were associated with high catches, whereas stocks have declined dramatically during high index years.

In the northeast Atlantic significant correlations were obtained between the variations in climate and hydrography with declines in primary production, standing crop of zooplankton and with reduced abundance and altered distribution of pelagic forage fishes and salmon catches.

By analogy with the North Pacific it is likely that changes in salmon abundance in the northeast Atlantic are linked to alterations in plankton productivity and/or structural changes in trophic transfer, each forced by hydrometeorological variability and possibly climate change as a result of the North Atlantic Oscillation and the Gulf Stream indexes.

Homing

It is suggested that homeward movement involves directed navigation. Fish can obtain directional information from the sun, polarized light, the earth's magnetic field and olfactory clues. Chinook and sockeye salmon have been shown capable of detecting changes in magnetic field. It has been suggested that migration from the feeding areas at sea to the natal stream must be accomplished without strictly retracing the outward migration using a variety of geopositioning mechanisms including magnetic and celestial navigation coordinated by an endogenous clock. Olfactory and salinity clues take over from geopositioning (bi-coordinate navigation) once in proximity of the natal stream because even small changes in declination during the time at sea correspond to a large search area for the home stream.

See also

El Niño Southern Oscillation (ENSO). Fish Feeding and Foraging. Fish Larvae. Fish Migration, Horizontal. Geophysical Heat Flow. North Atlantic Oscillation (NAO). Plankton.

Further Reading

Burgner RL (1991) Life history of sockeye salmon (*Oncorhynchus nerka*). In: Groot C and Margolis L (eds) *Pacific Salmon Life Histories*, pp. 2–117. Vancouver: UBC Press.

Foerster RE (1968) The sockeye salmon, *Oncorhynchus nerka*. *Bulletin of the Fisheries Research Board of Canada* 162: 422.

Groot C and Margolis L (eds) (1991) *Pacific Salmon Life Histories*. Vancouver: UBC Press.

Healey MC (1991) Life history of chinook salmon (*Oncorhynchus tschawytscha*). In: Groot C and Margolis L (eds) *Pacific Salmon Life Histories*, pp. 312–393. Vancouver: UBC Press.

Heard WR (1991) Life history of pink salmon (*Oncorhynchus gorbuscha*). In: Groot C and Margolis L (eds) *Pacific Salmon Life Histories*, pp. 120–230. Vancouver: UBC Press.

Kals F (1991) Life histories of masu and amago salmon (*Oncorhynchus masou and Oncorhynchus rhodurus*). In: Groot C and Margolis L (eds) *Pacific Salmon Life Histories*, pp. 448–520. Vancouver: UBC Press.

McDowell RM (1988) *Diadromy in Fishes*. London: Croom Helm.

Mills DH (1989) *Ecology and Management of Atlantic Salmon*. London: Chapman and Hall.

Mills DH (ed.) (1993) *Salmon in the Sea and New Enhancement Strategies*. Oxford: Fishing News Books.

Mills DH (ed.) (1999) *The Ocean Life of Atlantic Salmon*. Oxford: Fishing News Books.

Reddin D (1988) Ocean Life of Atlantic Salmon (*Salmo salar* L.) in the northwest Atlantic. In: Mills D and Piggins D (eds) *Atlantic Salmon: Planning for the Future*, pp. 483–511. London and Sydney: Croom Helm.

Reddin D and Friedland K (1993) Marine environmental factors influencing the movement and survival of Atlantic salmon. In: Mills D (ed.) *Salmon in the Sea and New Enhancement Strategies*, pp. 79–103. Oxford: Fishing News Books.

Salo EO (1991) Life history of chum salmon (*Oncorhynchus keta*). In: Groot C and Margolis L (eds) *Pacific Salmon Life Histories*, pp. 232–309. Vancouver: UBC Press.

Sandererock FK (1991) Life history of coho salmon (*Oncorhynchus kisutch*). In: Groot C and Margolis L (eds) *Pacific Salmon Life Histories*, pp. 396–445. Vancouver: UBC Press.

Thorpe JE (1988) Salmon migration. *Science Progress (Oxford)* 72: 345–370.

SALMONID FARMING

L. M. Laird, Aberdeen University, Aberdeen, UK

doi:10.1006/rwos.2001.0475

Introduction

All salmonids spawn in fresh water. Some of them complete their lives in streams, rivers, or lakes but the majority of species are anadromous, migrating to sea as juveniles and returning to spawn as large adults after one or more years feeding. The farmed process follows the life cycle of the wild fish; juveniles are produced in freshwater hatcheries and smolt units and transferred to sea for ongrowing in floating sea cages. An alternative form of salmonid mariculture, ocean ranching, takes advantage of their accuracy of homing. Juveniles are released into rivers or estuaries, complete their growth in sea water and return to the release point where they are harvested.

The salmonids cultured in seawater cages belong to the genera *Salmo*, *Oncorhynchus*, and *Salvelinus*. The last of these, the charrs are currently farmed on a very small scale in Scandinavia; this article concentrates on the former two genera. The Atlantic salmon, *Salmo salar* is the subject of almost all production of fish of the genus *Salmo* (1997 worldwide production 640 000 tonnes) although a small but increasing quantity of sea trout (*Salmo trutta*) is produced (1997 production 7000 tonnes). Three species of *Oncorhynchus*, the Pacific salmon are farmed in significant quantities in cages, the chinook salmon (also known as the king, spring or quinnat salmon), *O. tshawytscha* (1997, 10 000 tonnes), the coho (silver) salmon, *O. kisutch* (1997, 90 000 tonnes) and the rainbow trout, *O. mykiss*. The rainbow trout (steelhead) was formerly given the scientific name *Salmo gairdneri* but following studies on its genetics and native distribution was reclassified as a Pacific salmon species. Much of the world rainbow trout production (1997, 430 000 tonnes) takes place entirely in fresh water although in some countries such as Chile part-grown fish are transferred to sea water in the same way as the salmon species.

Here, the history of salmonid culture leading to the commercial mariculture operations of today is reviewed. This is followed by an overview of the requirements for successful operation of marine salmon farms, constraints limiting developments and prospects for the future.

History

Salmonids were first spawned under captive conditions as long ago as the fourteenth century when Dom Pinchon, a French monk from the Abbey of Reome stripped ova from females, fertilized them with milt from males, and placed the fertilized eggs in wooden boxes buried in gravel in a stream. At that time, all other forms of fish culture were based on the fattening of juveniles captured from the wild. However, the large (4–7 mm diameter) salmonid eggs were much easier to handle than the tiny, fragile eggs of most freshwater or marine fish. By the nineteenth century the captive breeding of salmonids was well established; the main aim was to provide fish to enhance river stocks or to transport around the world to provide sport in countries where there were no native salmonids. In this way, brown trout populations have become established in every continent except Antarctica, sustaining game fishing in places such as New Zealand and Patagonia.

A logical development of the production of eggs and juveniles for release was to retain the young fish in captivity, growing them until a suitable size for harvest. The large eggs hatch to produce large juveniles that readily accept appropriate food offered by the farmer. In the early days, the fish were first fed on finely chopped liver and progressed to a diet based on marine fish waste. The freshwater rainbow trout farming industry flourished in Denmark at the start of the twentieth century only to be curtailed by the onset of World War I when German markets disappeared. The success of the Danish trout industry encouraged a similar venture in Norway. However, when winter temperatures in fresh water proved too low, fish were transferred to pens in coastal sea water. Although these pens broke up in bad weather, the practice of seawater salmonid culture had been successfully demonstrated. The next major steps in salmonid mariculture came in the 1950s and 1960s when the Norwegians developed the commercial rearing of rainbow trout and then Atlantic salmon in seawater enclosures and cages. Together with the development of dry, manufactured fishmeal-based diets this led to the industry in its present form.

Salmonid Culture Worldwide

Fish reared in seawater pens are subject to natural conditions of water quality and temperature. Optimum water temperatures for growth of most salmonid species are in the range 8–16°C. Such temperatures, together with unpolluted waters are found not only around North Atlantic and North Pacific coasts within their native range but also in the southern hemisphere along the coastlines of Chile, Tasmania, and New Zealand. Salmon and trout are thus farmed in seawater cages where conditions are suitable both within and outwith their native ranges.

Seawater cages are used for almost the entire sea water production of salmonids. A very small number of farms rear fish in large shore-based silo-type structures into which sea water is pumped. Such structures have the advantage of better protection against storms and predators and the possibility of control of environmental conditions and parasites such as sea lice. However, the high costs of pumping outweigh these advantages and such systems are generally now used only for broodfish that are high in value and benefit from controlled conditions.

The production figures for the four species of salmonid reared in seawater cages are shown in Table 1.

Norway, the pioneering country of seawater salmonid mariculture, remains the biggest producer of Atlantic salmon. The output figures for 1997 show the major producing countries to be Norway (331 367 t), Scotland, UK (99 422 t), Chile (96 675 t), Canada (51 103 t), USA (18 005 t). Almost the entire farmed production of chinook salmon comes from Canada and New Zealand with Chile producing over 70 000 tonnes of coho salmon (1997 figures). Most of the Atlantic salmon produced in Europe is sold domestically or exported to other European countries such as France and Spain. Production in Chile is exported to North America and to Japan.

Seawater Salmonid Rearing

Smolts

Anadromous salmonids undergo physiological, anatomical and behavioral changes that preadapt them for the transition from fresh water to sea water. At this stage one of the most visible changes in the young fish is a change in appearance from mottled brownish to silver and herring-like. The culture of farmed salmonids in sea water was made possible by the availability of healthy smolts, produced as

Table 1 Production of four species of salmonids reared in seawater cages

	1988 production (t)	1997 production (t)
Atlantic salmon	112 377	638 951
Rainbow trout[a]	248 010	428 963
Coho salmon	25 780	88 431
Chinook salmon	4 698	9 774

[a]Includes freshwater production

a result of the technological progress in freshwater units. Hatcheries and tank farms were originally operated to produce juveniles for release into the wild for enhancement of wild stocks, often where there had been losses of spawning grounds or blockage of migration routes by the construction of dams and reservoirs. One of the most significant aspects of the development of freshwater salmon rearing was the progress in the understanding of dietary requirements and the production of manufactured pelleted feed. The replacement of a diet based on wet trash fish with a dry diet also benefitted the freshwater rainbow trout farming industry, by improving growth and survival and reducing disease and the pollution of the watercourses receiving the outflow water from earth ponds.

It was found possible to transfer smolts directly to cages moored in full strength sea water. If the smolts are healthy and the transfer stress-free, survival after transfer is high and feeding begins within 1–3 days (Atlantic salmon).

The smolting process in salmonids is controlled by day length; natural seasonal changes regulate the physiological processes, resulting in the completion of smolting and seaward migration of wild fish in spring. For the first two decades of seawater Atlantic salmon farming, producers were constrained by the annual seasonal availability of smolts. These fish are referred to as 'S1s', being approximately one year post-hatching. This had consequences for the use of equipment (nonoptimal use of cages) and for the timing of harvest. Most salmon reached their optimum harvest size or began to show signs of sexual maturation at the same time of year; thus large quantities of fish arrived on the market together for biological rather than economic reasons. These fish competed with wild salmonids and missed optimum market periods such as Christmas and Easter.

Research on conditions controlling the smolting process enabled smolt producers to alter the timing of smolting by manipulating photoperiod. Compressing the natural year by shortening day length and giving the parr an early 'winter' results in S1/2s or S3/4s, smolting as early as six months after hatch. Similarly, by delaying winter, smolting can be postponed. Thus it is now possible to have Atlantic salmon smolts ready for transfer to sea water throughout the year. Although this benefits marketing it makes site fallowing (see below) more difficult than when smolt input is annual.

The choice of smolts for seawater rearing is becoming increasingly important with the establishment of controlled breeding programs. Few species of fish can be said to be truly domesticated. The only examples approaching domestication are

carp species and, to a lesser degree, rainbow trout. Other salmon species have been captive bred for no more (and usually far less than) ten generations; the time between successive generations of Atlantic salmon is usually a minimum of three years which prevents rapid progress in selection for preferred characters although this is countered by the fact that many thousand eggs are produced by each female. Trials carried out mainly in Norway have demonstrated that several commercially important traits can be improved by selective breeding. These include growth rate, age at sexual maturity, food conversion efficiency, fecundity, egg size, disease resistance and survival, adaptation to conditions in captivity and harvest quality, including texture, fat, and color. All of these factors can also be strongly influenced by environmental factors and husbandry.

Sexual maturation before salmonids have reached the desired size for harvest has been a problem for salmonid farmers. Pacific salmon species (except rainbow trout) die after spawning; Atlantic salmon and rainbow trout show increased susceptibility to disease, reduced growth rate, deterioration in flesh quality and changes in appearance including coloration. Male salmonids generally mature at a smaller size and younger age than the females. One solution to this problem, routinely used in rainbow trout culture, is to rear all-female stocks, produced as a result of treating eggs and fry of potential broodstock with methyl testosterone to give functional males which are in fact genetically female. When crossed with normal females, all-female offspring are produced as the Y, male, sex chromosome has been eliminated. Sexual maturation can be eliminated totally by subjecting all-female eggs to pressure or heat shock to produce triploid fish. This is common practice for rainbow trout but used little for other salmonids, partly because improvements in stock selection and husbandry are overcoming the problem but also because of adverse press comment on supposedly genetically modified fish. This same reaction has limited the commercial exploitation of fast-growing genetically modified salmon, produced by the incorporation into eggs of a gene from ocean pout.

Site Selection

The criteria for the ideal site for salmonid cage mariculture have changed with the development of stronger cage systems, use of automatic feeders with a few days storage capacity and generally bigger and stronger boats, cranes and other equipment on the farm. The small, wooden-framed cages (typically 6 m × 6 m frame, 300 m³ capacity) with polystyrene flotation required sheltered sites with protection from wind and waves greater than 1–2 m high.

Recommended water depth was around three times the depth of the cage to ensure dispersal of wastes. This led to the siting of cages in inshore sites such as inner sea lochs and fiords. These sheltered sites had several disadvantages, notably variable water quality caused by runoff of fresh water, silt, and wastes from the land and susceptibility to the accumulation of feces and waste feed on the seabed because of poor water exchange. In addition, cage groups were often sited near public roads in places valued for their scenic beauty, attracting adverse public reaction to salmon farming.

Cages in use today are far larger (several thousand m^3 volume) and stronger. Frames are made from either galvanized steel or flexible plastic or rubber and can be designed to withstand waves of 5 m or more. Flotation collars are stronger and mooring systems designed to match cages to sites. Such sites are likely to provide more constant water quality than inshore sites; an ideal salmonid rearing site has temperatures of 6–16°C and salinities of 32–35‰ (parts per thousand). Rearing is thus moving into deeper water away from sheltered lochs and bays. However, there are still advantages to proximity to the coast; these include ease of access from shore bases, proximity to staff accommodation, reduction in costs of transport of feed and stock and ability to keep sites under regular surveillance. Other factors to be taken into account in siting cage groups are the avoidance of navigation routes and the presence of other fish or shellfish farms. Maintaining a minimum separation distance from other fish farms is preferred to minimize the risk of disease transfer; if this is not possible, farms should enter into agreements to manage stock in the same way to reduce risk. Models have been developed in Norway and Scotland to determine the carrying capacity of cage farm sites.

Current speed is an important factor in site selection. Water exchange through the cage net ensures the supply of oxygen to the stock and removal of dissolved wastes such as ammonia as well as feces and waste feed. Salmon have been shown to grow and feed most efficiently in currents with speeds equivalent to 1–2 body lengths per second. In an ideal site this current regime should be maintained for as much of the tidal cycle as possible. At faster current speeds the salmon will use more energy in swimming and cage nets will tend to twist, sometimes forming pockets and trapping fish, causing scale removal.

Some ideal sites may be situated near offshore islands; access from the mainland may require crossing open water with strong tides and currents making access difficult on stormy days. However,

modern workboats and feeding barges with the capacity to store several days supply of feed make the operation of such sites possible.

The presence of predators in the vicinity is often taken as a criterion for site selection. Unprotected salmon cage farms are likely to be subject to predation from seals or, in Chile, sea lions, and birds, such as herons and cormorants. Such predators not only remove fish but also damage others, tear holes in nets leading to escapes and stress stock making it more susceptible to disease. Protection systems to guard against predators include large mesh nets surrounding cages or cage groups, overhead nets and acoustic scaring devices. When used correctly these can all be effective in preventing attacks. Attacks from predators are frequently reported to involve nonlocal animals, attracted to a food source. Because of this and the possibility of excluding and deterring predators, it seems that proximity to colonies is not necessarily one of the most important factors in determining site selection.

A further factor, which must be taken into account in the siting of cage salmonid farms, is the occurrence of phytoplankton blooms. Phytoplankton can enter surface-moored cages and can physically damage gills, cause oxygen depletion or produce lethal toxins that kill fish. Historic records may indicate prevalence of such blooms and therefore sites to be avoided although some cages are now designed to be lowered beneath the surface and operated as semisubmersibles, keeping the fish below the level of the bloom until it passes.

Farm Operation

Operation of the marine salmon farm begins with transfer of stock from freshwater farms. Where possible, transfer in disinfected bins suspended under helicopters is the method of choice as it is quick and relatively stress-free. For longer journeys, tanks on lorries or wellboats are used. The latter require particular vigilance as they may visit more than one farm and have the potential to transfer disease. Conditions in tanks and wellboats should be closely monitored to ensure that the supply of oxygen is adequate (minimum $6 \, mg \, l^{-1}$).

The numbers of smolts stocked into each cage is a matter for the farmer; some will introduce a relatively small number, allowing for growth to achieve a final stocking density of $10–15 \, kg^{-3}$ whereas others stock a greater number and split populations between cages during growth. This latter method makes better use of cage space but increases handling and therefore stress. Differential growth may make grading into two or three size groups necessary.

Stocking density is the subject of debate. It is essential that oxygen concentrations are maintained and that all fish have access to feed when it is being distributed. Fish may not distribute themselves evenly within the water column; because of crowding together the effective stocking density may therefore be a great deal higher than the theoretical one.

As with all farmed animals the importance of vigilance of behavior and health and the maintenance of accurate, useful records cannot be overemphasized. When most salmon farms were small, producing one or two hundred tonnes of salmon a year rather than thousands, hand feeding was normal; observation of stock during feeding provided a good indication of health. Today, fish are often fed automatically using blowers attached to feed storage systems. The best of these systems incorporate detectors to monitor consumption of feed and underwater cameras to observe the stock.

All of the nutrients ingested by cage-reared salmonids are supplied in the feed distributed. Typically, manufactured diets for salmonids will contain 40% protein (mainly obtained from fishmeal) and up to 30% oil, providing the source of energy, sparing protein for growth. Although very poorly digested by salmonids, carbohydrate is necessary to bind other components of the diet. Vitamins and minerals are also added, as are carotenoid pigments such as astaxanthin, necessary to produce the characteristic pink coloration of the flesh of anadromous salmonids. The feed used on marine salmon farms is nowadays almost exclusively a pelleted or extruded fishmeal-based diet manufactured by specialist companies. Feed costs make up the biggest component of farm operating costs, sometimes reaching 50%. It is therefore important to make optimum use of this valuable input by minimizing wastes. This is accomplished by ensuring that feed is delivered to the farm in good condition and handled with care to prevent dust formation, increasing the size of pellets as the fish grow and distributing feed to satisfy the appetites of the fish. Improvements in feed manufacture and in feeding practices have reduced feed conversion efficiency (feed input : increase in weight of fish) from 2 : 1 to close to 1 : 1. Such figures may seem improbable but it must be remembered that they represent the conversion of a nearly dry feed to wet fish flesh and other tissues.

The importance of maintaining a flow of water through the net mesh of the cages has been emphasized. Mesh size is generally selected to be the maximum capable of retaining all fish and preventing escapes. Any structure immersed in the upper few meters of coastal or marine waters will quickly be subjected to colonization by fouling organisms including bacteria, seaweeds, mollusks and sea squirts. Left unchecked, such fouling occludes the mesh, reducing water exchange and may place a burden on the cage reducing its resistance to storm damage. One of the most effective methods of preventing fouling of nets and moorings is to treat them with antifouling paints and chemicals prior to installation. However, one particularly effective treatment used in the early 1980s, tributyl tin, has been shown to have harmful effects on marine invertebrates and to accumulate in the flesh of the farmed fish; its use in aquaculture is now banned. Other antifoulants are copper or oil based; alternative, preferred methods of removing fouling organisms include lifting up sections of netting to dry in air on a regular basis or washing with high pressure hoses or suction devices to remove light fouling.

The aim of the salmonid farmer is to produce maximum output of salable product for minimum financial input. To do this, fish must grow efficiently and a high survival rate from smolt input to harvest must be achieved. Minimizing stress to the fish by reducing handling, maintaining stable environmental conditions and optimizing feeding practices will reduce mortalities. Causes of mortality in salmonid and other farms are reviewed elsewhere (*see* **Mariculture Diseases and Health**). It is vital to keep accurate records of mortalities; any increase may indicate the onset of an outbreak of disease. It is also important that dead fish are removed; collection devices installed in the base of cages are often used to facilitate this. Treatment of diseases or parasitic infestations such as sea lice (*Lepeophtheirus salmonis, Caligus elongatus*) is difficult in fish reared in sea cages because of their large volumes and the high numbers of fish involved. Some treatments for sea lice involve reducing the cage volume and surrounding with a tarpaulin so that the fish can be bathed in chemical. After the specified time the tarpaulin is removed and the chemical disperses into the water surrounding the cage. Newer treatments incorporate the chemicals in feed and are therefore simpler to apply. In the future, vaccines are increasingly likely to replace chemicals.

The health of cage-reared salmonids can be maintained by a site management system incorporating a period of fallowing when groups of cages are left empty for a period of at least three months and preferably longer. This breaks the life cycle of parasites such as sea lice and allows the seabed to recover from the nutrient load falling from the cages. Ideally a farmer will have access to at least three sites; at any given time one will be empty and the other two will contain different year classes, separated to prevent cross-infection.

Harvesting

Most of the farmed salmonids reared in sea water reach the preferred harvest size (3–5 kg) 10 months or more after transfer to sea water. Poor harvesting and handling methods can have a devastating effect on flesh quality, causing gaping in muscle blocks and blood spotting. After a period of starvation to ensure that guts are emptied of feed residues the fish are generally killed by one of two methods. One of these involves immersion in a tank of sea water saturated with carbon dioxide, the other an accurate sharp blow to the cranium. Both methods are followed by excision of the gill arches; the loss of blood is thought to improve flesh quality. It is important that water contaminated with blood is treated to kill any pathogens which might infect live fish.

Ocean Ranching

The anadromous behavior of salmonids and their ability to home to the point of release has been exploited in ocean ranching programs which have been operated successfully with Pacific salmon. Some of these programs are aimed at enhancing wild stocks and others are operated commercially. The low cost of rearing Pacific salmon juveniles, which are released into estuaries within weeks of hatching, makes possible the release of large numbers. In Japan over two billion juveniles are released annually; overall return rates have increased to 2%, 90% of which are chum (*Oncorhynchus keta*) and 8% pink (*Oncorhynchus gorbuscha*) salmon. The success of the operation depends on cooperation between those operating and financing the hatcheries and those harvesting the adult fish. The relatively high cost of producing Atlantic salmon smolts and the lack of control over harvest has restricted ranching operations.

See also

Mariculture Diseases and Health. Ocean Ranching. Open Ocean Convection. Salmon Fisheries: Atlantic; Pacific.

Further Reading

Anon (1999) *Aquaculture Production Statistics 1988–1997*. Rome: Food and Agriculture Organization.

Black KD and Pickering AD (eds) (1998) *Biology of Farmed Fish*. Sheffield Academic Press.

Heen K, Monahan RL and Utter F (eds) (1993) *Salmon Aquaculture*. Oxford: Fishing News Books.

Pennell W and Barton BA (eds) (1996) *Principles of Salmonid Culture*. Amsterdam: Elsevier.

Stead S and Laird LM (In press) *Handbook of Salmon Farming Praxis*. Chichester: Springer-Praxis.

Willoughby S (1999) *Manual of Salmonid Farming*. Oxford: Blackwell Science.

SALT MARSH VEGETATION

C. T. Roman, University of Rhode Island, Narragansett, RI, USA

doi:10.1006/rwos.2001.0088

Introduction

Coastal salt marshes are intertidal features that occur as narrow fringes bordering the upland or as extensive meadows, often several kilometers wide. They occur throughout the world's middle and high latitudes, and in tropical/subtropical areas they are mostly, but not entirely, replaced by mangrove ecosystems. Salt marshes develop along the shallow, protected shores of estuaries, lagoons, and behind barrier spits. Here, low energy intertidal mud and sand flats are colonized by halophytes, plants that are tolerant of saline conditions. The initial colonizers serve to enhance sediment accumulation and over time the marsh expands vertically and spreads horizontally, encroaching the upland or growing seaward. As salt marshes mature they become geomorphically and floristically more complex with establishment of creeks, pools, and distinct patterns or zones of vegetation.

Several interacting factors influence salt marsh vegetation patterns, including frequency and duration of tidal flooding, salinity, substrate, surface elevation, oxygen and nutrient availability, disturbance by wrack deposition, and competition among plant species. Moreover, the ability of individual flowering plant species to adapt to an environment with saline and waterlogged soils plays an important role in defining salt marsh vegetation patterns. Morphological and physiological adaptations that halophytes may possess to manage salt stress include a succulent growth form, salt-excreting glands, mechanisms to reduce water loss, such as few stomates and low surface area, and a C4 photosynthetic pathway to promote high water use efficiency. To deal with anaerobic soil conditions, many salt marsh plants have well-developed aerenchymal tissue that delivers oxygen to below-ground roots.

Salt Marsh Vegetation Patterns

Vegetation patterns often reflect the stage of maturation of a salt marsh. Early in the development, halophytes, such as *Spartina alterniflora* along the east coast of the United States, colonize intertidal flats. These initial colonizers are tolerant of frequent flooding. Once established, the plants spread vegetatively by rhizome growth, the plants trap sediments, and the marsh begins to grow vertically. As noted from A. C. Redfield's classic study of salt marsh development along a coast with a rising sea level, salt marshes often extend seaward over tidal flats, while also accreting vertically and encroaching the upland or freshwater tidal wetlands. With this vertical growth and maturity of the marsh, discrete patterns of vegetation develop: frequently flooded low marsh vegetation borders the seaward portion of the marsh and along creekbanks, while high marsh areas support less flood-tolerant species, such as *Spartina patens*. There is some concern that vertical growth or accretion of salt marshes may not be able to keep pace with accelerated rates of sea level rise, resulting in submergence or drowning of marshes. This has been observed in some areas of the world.

Plant species and patterns of vegetation that dominate the salt marsh vary from region to region of the world and it is beyond the scope of this article to detail this variability; however, the general pattern of low marsh and high marsh remains throughout. There is often variation in vegetation patterns from marshes within a region and even between marshes within a single estuary, but in general, the low marsh is dominated by a limited number of species, often just one. On the Atlantic coast of North America, *Spartina alterniflora* is the early colonizer and almost exclusively dominates the low marsh. In European marshes, *Puccinellia maritima* often dominates the intertidal low marsh. *Spartina anglica*, a hybrid of *Spartina alterniflora* and *Spartina maritima*, has invaded many of the muddier low marsh sites of European marshes over the past century. With increasing elevation of the high marsh, species richness tends to increase. *Spartina patens*, *Distichlis spicata*, *Juncus gerardi*, *Juncus roemerianus*, and short-form *Spartina alterniflora* occupy the US east coast high marsh, with each species dominating in patches or zones to form a mosaic vegetation pattern. In European marshes, the low marsh may give way to a diverse high marsh of *Halimione portulacoides*, *Limonium* sp., *Suaeda maritima*, and *Festuca* sp., among others.

Factors Controlling Vegetation Patterns

The frequency and duration of tidal flooding are mostly responsible for the low and high marsh delineation, but many other factors contribute to the wide variation found in salt marsh vegetation patterns (**Figure 1**). Soil salinity is relatively constant within the low marsh because of frequent tidal flooding, but extremes in soil salinity can occur on the less-flooded high marsh contributing to the vegetation mosaic. Concentrations in excess of 100 parts per thousand can occur, resulting in hypersaline pannes that remain unvegetated or are colonized by only the most salt-tolerant halophytes (e.g., *Salicornia*). Salt marshes of southern temperate and subtropical/tropical latitudes tend to have higher soil salinity because of more intense solar radiation and higher evaporation rates. At the other extreme, soil salinity of the high marsh can be dramatically depressed by rainfall or by discharge of groundwater near the marsh upland border. On salt marshes of the New England coast (USA), *Juncus gerardi* and the shrub, *Iva fructescens*, are less tolerant of salt, and thus, grow at higher elevations where tidal flooding is only occasional or near upland freshwater sources.

For successful growth in environments of high soil salinity, halophytes must be able to maintain a flow of water from the soil into the roots. Osmotic pressure in the roots must be higher than the surrounding soil for water uptake. To maintain this osmotic difference, halophytes have high concentrations of solutes in their tissues (i.e., high osmotic pressure), concentrations greater than the surrounding environment. This osmotic adjustment can be achieved by an accumulation of sodium and chloride ions or organic solutes. To effectively tolerate high internal salt concentrations, many halophytes have a succulent growth form (e.g., *Salicornia*, *Suaeda*). This high tissue water content serves to dilute potentially toxic salt concentrations. Other halophytes, such as *Spartina alterniflora*, have salt glands to actively excrete salt from leaves.

Waterlogged or anaerobic soil conditions also strongly influence the pattern of salt marsh vegetation. Plants growing in waterlogged soils must deal with a lack of oxygen at the rhizosphere and the accumulation of toxins resulting from biogeochemical soil processes (i.e., sulfate reduction). High concentrations of hydrogen sulfide are toxic to root metabolism, inhibiting nitrogen uptake and resulting in decreased plant growth. Many salt marsh plants are able to survive these soil conditions because they have well-developed aerenchyma tissue, or a

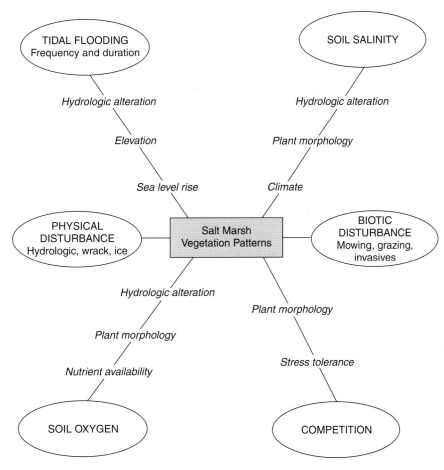

Figure 1   Conceptual model of factors controlling vegetation patterns in salt marshes. Ovals denote major factors and the key interacting parameters are shown in italics. For example, soil salinity responds to hydrologic alterations, climate (e.g., temperature/evaporation and precipitation) influences soil salinity, and a species response to soil salinity is dependent on morphological considerations (e.g., succulence, salt glands, etc.).

network of intercellular spaces that serve to deliver oxygen from above-ground plant parts to below-ground roots. Extensive research has been conducted on the relationships between sediment oxygen levels and growth of *Spartina alterniflora*. Aerenchyma transports oxygen to roots to support aerobic respiration, facilitating nitrogen uptake, and the aerated rhizosphere serves to oxidize reduced soil compounds, such as sulfide, thereby promoting growth. Under severely reducing soil conditions, the aerenchyma may not supply sufficient oxygen for aerobic respiration. *Spartina alterniflora* then has the ability to respire anaerobically, but growth is reduced. It is clear that the degree of soil aeration can dramatically influence salt marsh vegetation patterns. Plants that have the ability to form 1aerenchyma (e.g., *Spartina alterniflora*, *Juncus roemerianus*, *Puccinellia maritima*) or develop anaerobic respiration will be able to tolerate moderately reduced or waterlogged soils, whereas other species will be limited to better-drained areas of the marsh.

Coupled with the physical factors of salinity stress and waterlogging, competition is another process that controls salt marsh vegetation patterns. It has been suggested that competition is an important factor governing the patterning of vegetation near the upland boundaries of the high marsh. Plants that dominate the low marsh, like *Spartina alterniflora*, or forbs in salt pannes (e.g., *Salicornia*), can tolerate harsh environmental conditions and exist with few competitors. However, plant assemblages of the high marsh, and especially near the upland, may occur there because of their exceptional competitive abilities. For example, the high marsh plant, *Spartina patens*, has a dense growth form and is able to out-compete species such as *Distichlis spicata* and *Spartina alterniflora*, each with a diffuse or clumped morphology.

Natural and human-induced disturbances also influence salt marsh vegetation patterns. Vegetation can be killed and bare spots created following deposition of wrack on the marsh surface. The resulting bare areas may then become vegetated by early colonizers (e.g., *Salicornia*). In northern regions, ice scouring can dramatically alter the creek or bayfront of marshes. Also, large blocks of ice, laden with sediment, are often deposited on the high marsh creating a new microrelief habitat, or these blocks may transport plant rhizomes to mud or sand flats and initiate the process of salt marsh development.

Regarding human-induced disturbances, hydrologic alterations have dramatic effects on salt marsh vegetation patterns throughout the world. Some salt marshes have been diked and drained for agriculture. In others, extensive ditching has drained salt marshes for mosquito-control purposes. In yet another hydrologic alteration, water has been retained or impounded within salt marshes to alter wildlife habitat functions or to control mosquitoes. Impoundments generally restrict tidal inflow and retain fresh water, resulting in conversion from salt-tolerant vegetation to brackish or freshwater vegetation. Practices that drain the marsh, such as ditching, tend to lower the water-table level and aerate the soil, and the resulting vegetation may shift toward that typical of a high marsh. Another type of hydrologic alteration is the restriction of tidal flow by bridges, culverts, roads, and causeways. This is particularly common along urbanized shorelines, where soil salinity can be reduced and water-table levels altered resulting in vegetation changes, such as the conversion from *Spartina*-dominated salt marsh to *Phragmites australis*, as is most evident throughout the north-eastern US. The role of hydrologic alterations in controlling vegetation patterns is clearly identified in **Figure 1** as a key physical disturbance, and also as a variable that influences tidal flooding, soil salinity, and soil oxygen levels.

Grazing by domestic animals, mostly sheep and cattle, and mowing for hay are two practices that have been ongoing for many centuries on salt marshes worldwide, although they seem to be declining in some regions. Studies in Europe have demonstrated that certain plant species are favored by intensive grazing (e.g., *Puccinellia*, *Festuca rubra*, *Agrostis stolonifera*), whereas others become dominant on ungrazed salt marshes (e.g., *Halimione portulacoides*, *Limonium*, *Suaeda*).

Conclusions

Vegetation zones of salt marshes have been described as belts of plant communities from creekbank or bayfront margins to the upland border, or most appropriately, as a mosaic of communities along this elevation gradient. All salt marsh sites display some zonation but because of the complex of interacting factors that influence marsh vegetation patterns, there is extraordinary variability in zonation among individual marshes. Moreover, salt marsh vegetation patterns are constantly changing on seasonal to decadal timescales. Experimental research and long-term monitoring efforts are needed to further evaluate vegetation pattern responses to the myriad of interacting environmental factors that influence salt marsh vegetation. An ultimate goal is to model and predict the response of marsh vegetation as a result of natural or human-induced disturbance, accelerated rates of sea level rise, or marsh-restoration strategies.

See also

Coastal Circulation Models. Intertidal Fishes. Mangroves. Salt Marshes and Mud Flats. Sea Level Change.

Further Reading

Adam P (1990) *Saltmarsh Ecology*. Cambridge: Cambridge University Press.

Bertness MD (1999) *The Ecology of Atlantic Shorelines*. Sunderland, MA: Sinauer.

Chapman VJ (1960) *Salt Marshes and Salt Deserts of the World*. New York: Interscience.

Daiber FC (1986) *Conservation of Tidal Marshes*. New York: Van Nostrand Reinhold.

Flowers TJ, Hajibagheri MA and Clipson NJW (1986) Halophytes. *Quarterly Review of Biology* 61: 313–337.

Niering WA and Warren RS (1980) Vegetation patterns and processes in New England salt marshes. *BioScience* 30: 301–307.

Redfield AC (1972) Development of a New England salt marsh. *Ecological Monographs* 42: 201–237.

Reimold RJ and Queen WH (eds) (1974) *Ecology of Halophytes*. New York: Academic Press.

SALT MARSHES AND MUD FLATS

J. M. Teal, Woods Hole Oceanographic Institution, Rochester, MA, USA

Copyright © 2001 Academic Press

doi:10.1006/rwos.2001.0087

Structure

Salt marshes are vegetated mud flats. They are above mean sea level in the intertidal area where higher plants (angiosperms) grow. Sea grasses are an exception to the generalization about higher plants because they live below low tide levels. Mud flats are vegetated by algae.

Geomorphology

Salt marshes and mud flats are made of soft sediments deposited along the coast in areas protected from ocean surf or strong currents. These are long-term depositional areas intermittently subject to erosion and export of particles. Salt marsh sediments are held in place by plant roots and rhizomes (underground stems). Consequently, marshes are resistant to erosion by all but the strongest storms. Algal mats and animal burrows bind mud flat sediments, although, even when protected along tidal creeks within a salt marsh, mud flats are more easily eroded than the adjacent salt marsh plain.

Salinity in a marsh or mud flat, reported in parts per thousand (ppt), can range from about 40 ppt down to 5 ppt. The interaction of the tides and weather, the salinity of the coastal ocean, and the elevation of the marsh plain control salinity on a marsh or mud flat. Parts of the marsh with strong, regular tides (1 m or more) are flooded twice a day, and salinity is close to that of the coastal ocean. Heavy rain at low tide can temporarily make the surface of the sediment almost fresh. Salinity may vary seasonally if a marsh is located in an estuary where the river volume changes over the year. Salinity varies within a marsh with subtle changes in surface elevation. Higher marshes at sites with regular tides have variation between spring and neap tides that result in some areas being flooded every day while other, higher, areas are flooded less frequently. At higher elevations flooding may occur on only a few days each spring tide, while at the highest elevations flooding may occur only a few times a year.

Some marshes, on coasts with little elevation change, have their highest parts flooded only sea-sonally by the equinoctial tides. Other marshes occur in areas with small lunar tides where flooding is predominantly wind-driven, such as the marshes in the lagoons along the Texas coast of the United States. They are flooded irregularly and, between flooding, the salinity is greatly raised by evaporation in the hot, dry climate. The salinity in some of the higher areas becomes so high that no rooted plants survive. These are salt flats, high enough in the tidal regime for higher plants to grow, but so salty that only salt-resistant algae can grow there. The weather further affects salinity within marshes and mud flats. Weather that changes the temperature of coastal waters or varying atmospheric pressure can change sea level by 10 cm over periods of weeks to months, and therefore affect the areas of the marsh that are subjected to tidal inundation.

Sea level changes gradually. It has been rising since the retreat of the continental glaciers. The rate of rise may be increasing with global warming. For the last 10 000 years or so, marshes have been able to keep up with sea level rise by accumulating sediment, both through deposition of mud and sand and through accumulation of peat. The peat comes from the underground parts of marsh plants that decay slowly in the anoxic marsh sediments. The result of these processes is illustrated in **Figure 1**, in which the basement sediment is overlain by the accumulated marsh sediment. Keeping up with sea level rise creates a marsh plain that is relatively flat; the elevation determined by water level rather than by the geological processes that determined the original, basement sediment surface on which the marsh developed. Tidal creeks, which carry the tidal waters on and off the marsh, dissect the flat marsh plain.

Organisms

The duration of flooding and the salinities of the sediments and tidal waters control the mix of higher

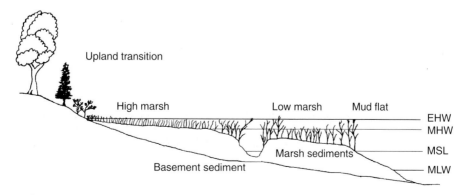

Figure 1 Cartoon of a typical salt marsh of eastern North America. The plants shown are mostly grasses and may differ in other parts of the world. MLW, mean low water; MSL, mean sea level; MHW, mean high water; EHW, extreme high water. The mud flat is shown as a part of the marsh but mud flats also exist independently of marshes.

vegetation. Competitive interactions between plants and interactions between plants and animals further determine plant distributions. Duration of flooding duration controls how saturated the sediments will be, which in turn controls how oxygenated or reduced the sediments are. The roots of higher plants must have oxygen to survive, although many can survive short periods of anoxia. Air penetrates into the creekbank sediments as they drain at low tide. Evapotranspiration from plants at low tide also removes water from the sediments and facilitates entry of air. Most salt marsh higher plants have aerenchyma (internal air passages) through which oxygen reaches the roots and rhizomes by diffusion or active transport from the above-ground parts. However, they also benefit from availability of oxygen outside the roots.

The species of higher plants that dominate salt marshes vary with latitude, salinity, region of the world, and tidal amplitude. They are composed of relatively few species of plants that have invested in the ability to supply oxygen to roots and rhizomes in reduced sediments and to deal with various levels of salt. Grasses are important, with *Spartina alterniflora* the dominant species from mid-tide to high-tide levels in temperate Eastern North America. *Puccinellia* is a dominant grass in boreal and arctic marshes. The less regularly flooded marshes of East Anglia (UK) support a more diverse vegetation community in which grasses are not dominant. The salty marshes of the Texas coast are covered by salt-tolerant *Salicornia* species. Adjacent to the upper, landward edge of the marsh lie areas flooded only at times when storms drive ocean waters to unusual heights. Some land plants can survive occasional salt baths, but most cannot. An extreme high-water even usually results in the death of plants at the marsh border.

Algae on the marsh and mud flat are less specialized. Depending upon the turbulence of the tidal water, macroalgae (seaweeds) may be present, but a diverse microalgal community is common. Algae live on or near the surface of the sediments and obtain oxygen directly from the air or water and from the oxygen produced by photosynthesis. Their presence on surface sediment is controlled by light. In highly turbid waters they are almost entirely limited to the intertidal flats. In clearer waters, they can grow below low-tide levels. Algae growing on the vegetated marsh plain and on the stems of marsh plants get less light as the plants mature. Production of these algae is greatest in early spring, before the developing vegetation intercepts the light.

Photosynthetic bacteria also contribute to marsh and mud flat production. Blue-green bacteria can be abundant enough to forms mats. Photosynthetic sulfur bacteria occupy a thin stratum in the sediment where they get light from above and sulfide from deeper reduced levels for their hydrogen source but are below the level of oxygen penetration that would kill them. These strange organisms are relicts from the primitive earth before the atmosphere contained oxygen.

Salt marsh animals are from terrestrial and marine sources; mud flat inhabitants are limited to marine sources. Insects, spiders, and mites live in marsh sediments and on marsh plants. Crabs, amphipods, isopods and shrimps, polychaete and oligochaete annelids, snails, and bivalves live in and on the sediments. Most of these marine animals have planktonic larval stages that facilitate movement between marshes and mud flats. Although burrowing animals, such as crabs that live at the water edge of the marsh, may be fairly large (2–15 cm), in general burrowers in marshes are smaller than those in mud flats, presumably because the root mats of the higher plants interfere with burrowing.

Fish are important faunal elements in regularly flooded salt marshes and mud flats. They can be characterized as permanent marsh residents; seasonal residents (species that come into the marsh at the beginning of summer as new post-larvae and live in the marsh until cold weather sets in); species that are primarily residents of coastal waters but enter the marshes at high tide; and predatory fish that come into marshes on the ebb tide to feed on the smaller fishes forced off the marsh plain and out of the smaller creeks by falling water levels.

A few mammals live in the marshes, including those that flee only the highest tides by retreating to land, such as voles, or those that make temporary refuges in tall marsh plants, such as raccoons. The North American muskrat builds permanent houses on the marsh from the marsh plants, although muskrats are typically found only in the less-saline marshes. Grazing mammals feed on marsh plants at low tide. In Brittany, lambs raised on salt marshes are specially valued for the flavor of their meat.

Many species of birds use salt marshes and mud flats. Shore bird species live in the marshes and/or use associated mud flats for feeding during migration. Northern harriers nest on higher portions of salt marshes and feed on their resident voles. Several species of rails dwell in marshes as do bitterns, ducks, and some wrens and sparrows. The nesting species must keep their eggs and young from drowning, which they achieve by building their nests in high vegetation, by building floating nests, and by nesting and raising their young between periods of highest tides.

Functions

Marshes, and to some extent mud flats, produce animals and plants, provide nursery areas for marine fishes, modify nutrient cycles, degrade organic chemicals, immobilize elements within their sediments, and modify wave action on adjacent uplands.

Production and Nursery

Plant production from salt marshes is as high as or higher than that of most other systems because of the ability of muddy sediments to serve as nutrient reservoirs, because of their exposure to full sun, and because of nutrients supplied by sea water. Although the plant production is food for insects, mites and voles, large mammalian herbivores that venture onto the marsh, a few crustacea, and other marine animals, most of the higher plant production is not eaten directly but enters the food web as detritus. As the plants die, they are attacked by fungi and bacteria that reduce them to small particles on the surface of the marsh. Since the labile organic matter in the plants is quickly used as food by the bacteria and fungi, most of the nutrient value of the detritus reaches the next link in the food chain through these microorganisms. These are digested from the plant particles by detritivores, but the cellulose and lignin from the original plants passes through them and is deposited as feces that are recolonized by bacteria and fungi. Besides serving as a food source in the marsh itself, a portion of the detritus–algae mixture is exported by tides to serve as a food source in the marsh creeks and associated estuary.

The primary plant production supports production of animals. Fish production in marshes is high. Resident fishes such as North American *Fundulus heteroclitus* live on the marsh plain during their first summer, survive low tide in tiny pools or in wet mud, feed on the tiny animals living on the detritus–algae–microorganism mix, and grow to migrate into small marsh creeks. At high tide they continue to feed on the marsh plain, where they are joined by the young-of-the-year of those species that use the marsh principally as a nursery area. The warm, shallow waters promote rapid growth and are refuge areas where they are protected from predatory fishes, but not from fish-eating herons and egrets.

The fishes are the most valuable export from the marshes to estuaries and coastal oceans. Some of the fishes are exported in the bodies of predatory fishes that enter the marsh on the ebb tide to feed. Many young fishes, raised in marshes, migrate offshore in the autumn after having spent the summer growing in the marsh.

Nutrient and Element Cycling

Nitrogen is the critical nutrient controlling plant productivity in marshes. Phosphorus is readily available in muddy salt marsh sediments and potassium is sufficiently abundant in sea water. Micro-nutrients, such as silica or iron, that may be limiting for primary production in deeper waters are abundant in marsh sediments. Thus nitrogen is the nutrient of interest for marsh production and nutrient cycling.

In marshes where nitrogen is in short supply, blue-green bacteria serve as nitrogen fixers, building nitrogen gas from the air into their organic matter. Nitrogen-fixing bacteria associated with the roots of higher plants serve the same function. Nitrogen fixation is an energy-demanding process that is absent where the supply is sufficient to support plant growth.

Two other stages of the nitrogen cycle occur in marshes. Organic nitrogen released by decomposition is in the form of ammonium ion. This can be oxidized to nitrate by certain bacteria that derive energy from the process if oxygen is present. Both nitrate and ammonium can satisfy the nitrogen needs of plants, but nitrate can also serve in place of oxygen for the respiration of another group of bacteria that release nitrogen gas as a by-product in the process called denitrification. Denitrification in salt marshes and mud flats is significant in reducing eutrophication in estuarine and coastal waters.

Phosphorus, present as phosphate, is the other plant nutrient that can be limiting in marshes, especially in regions where nitrogen is in abundant supply. It can also contribute to eutrophication of coastal waters, but phosphate is readily bound to sediments and so tends to be retained in marshes and mud flats rather than released to the estuary.

Sulfur cycling in salt marshes, while of minor importance as a nutrient, contributes to completing the production cycle. Sulfate is the second most abundant anion in sea water. In anoxic sediments, a specialized group of decomposing organisms living on the dead, underground portions of marsh plants can use sulfate as an electron acceptor – an oxygen substitute – in their respiration. The by-product is hydrogen sulfide rather than water. Sulfate reduction yields much less energy than respiration with oxygen or nitrate reduction, so these latter processes occur within the sediment surface, leaving sulfate reduction as the remaining process in deeper parts of the sediments. The sulfide carries much of the free energy not captured by the bacteria in sulfate reduction. As it diffuses to surface layers, most of the sulfide is oxidized by bacteria that grow using it as an energy source. A small amount is used by the photosynthetic sulfur bacteria mentioned above.

Pollution

Marshes, like the estuaries with which they tend to be associated, are depositional areas. They tend to accumulate whatever pollutants are dumped into coastal waters, especially those bound to particles. Much of the pollution load enters the coast transported by rivers and may originate far from the affected marshes. For example, much of the nitrogen and pesticide loading of marshes and coastal waters of the Mississippi Delta region of the United States comes from farming regions hundreds or thousands of kilometers upstream.

Many pollutants, both organic and inorganic, bind to sediments and are retained by salt marshes and mud flats. Organic compounds are often degraded in these biologically active systems, especially since many of them are only metabolized when the microorganisms responsible are actively growing on other, more easily degraded compounds. There are, unfortunately, some organics, the structures of which are protected by constituents such as chlorine, that are highly resistant to microbial attack. Some polychlorinated biphenyls (PCBs) have such structures, with the result that a PCB mixture will gradually lose the degradable compounds while the resistant components will become relatively more concentrated.

Metals are also bound to sediments and so may be removed and retained by marshes and mud flats. Mercury is sequestered in the sediment, while cadmium forms soluble complexes with chloride in sea water and is, at most, temporarily retained.

Since marshes and mud flats tend, in the long term, to be depositional systems, they remove pollutants and bury them as long as the sediments are not remobilized by erosion. Since mud flats are more easily eroded than marsh sediments held in place by plant roots and rhizomes, they are less secure long-term storage sites.

Storm Damage Prevention

While marshes and mud flats exist only in relatively protected situations, they are still subject to storm damage as are the uplands behind them. During storms, the shallow waters and the vegetation on the marshes offer resistance to water flow, making them places where wave forces are dissipated, reducing the water and wave damage to the adjoining upland.

Human Modifications

Direct Effects

Many marshes and mud flats in urban areas have been highly altered or destroyed by filling or by dredging for harbor, channel or marina development. Less intrusive actions can have large impacts. Since salt marshes and mud flats typically lie in indentations along the coast, the openings where tides enter and leave them are often sites of human modification for roads and railroads. Both culverts and bridges restrict flow if they are not large enough. Flow is especially restricted at high water unless the bridge spans the entire marsh opening, a rare situation because it is expensive. The result of restriction is a reduction in the amount of water that floods the marsh. The plants are submerged for a shorter period and to a lesser depth, and the floodwaters do not extend as far onto the marsh surface. The ebb flow is also restricted and the marsh may not drain as efficiently as in the unimpeded case. Poor drainage could freshen the marsh after a heavy rain and runoff. Less commonly, it could increase salinity after an exceptionally prolonged storm-driven high tide.

The result of the disturbance will be a change in the oxygen and salinity relations between roots and sediments. Plants may become oxygen-stressed and drown. Tidal restrictions in moist temperate regions usually result in a freshening of the sediment salinity. This favors species that have not invested in salt control mechanisms of the typical salt marsh plants. A widespread result in North America has been the spread of the common reed, *Phragmites australis*, a brackish-water and freshwater species. Common reed is a tall (3 m) and vigorous plant that can spread horizontally by rhizomes at 10 meters per year. Its robust stem decomposes more slowly than that of the salt marsh cord grass, *Spartina alterniflora* and as a result, it takes over a marsh freshened by tidal restrictions. Since its stems accumulate above ground and rhizomes below ground, it tends to raise the marsh level, fill in the small drainage channels, and reduce the value of the marsh for fish and wildlife. Although *Phragmites australis* is a valuable plant for many purposes (it is the preferred plant for thatching roofs in Europe), its takeover of salt marshes is considered undesirable.

The ultimate modification of tidal flow is restriction by diking. Some temperate marshes have been diked to allow the harvest of salt hay, valued as mulch because it lacks weed seeds. Since some diked marshes are periodically flooded in an attempt to maintain the desired vegetation, they are not completely changed and can be restored. Other marshes, such as those in Holland, have been diked and removed from tidal flow so that the land may be used for upland agriculture. Many marshes and mud flats have been modified to create salt pans for production of sea salt and for aquaculture. The latter is a greater problem in the tropics, where the impact is on mangroves rather than on the salt marshes of more temperate regions.

Indirect Effects

Upland diking The upper borders of coastal marshes were often diked to prevent upland flooding. People built close to the marshes to take advantage of the view. With experience they found that storms could raise the sea level enough to flood upland. The natural response was to construct a barrier to prevent flooding. Roads and railroads along the landward edges of marshes are also barriers that restrict upland flooding. They are built high enough to protect the roadbed from most flooding and usually have only enough drainage to allow rain runoff to pass to the sea. In both cases, the result is a barrier to landward migration of the marsh. As the relative sea level rises, sandy barriers that protect coastal marshes are flooded and, during storms, the sand is washed onto the marsh. As long as the marsh can also move back by occupying the adjacent upland it may be able to persist without loss in area, but if a barrier prevents landward transgression the marsh will be squeezed between the barrier and the rising sea and will eventually disappear. During this process, the drainage structures under the barrier gradually become flow restrictors. The sea will flood the land behind the barrier through the culverts, but these are inevitably too small to permit unrestricted marsh development. When flooding begins, the culverts are typically fitted with tide gates to prevent whatever flooding and marsh development could be accommodated by the capacity of the culverts.

Changes in sediment loading Increases in sediment supplies can allow the marshes to spread as the shallow waters bordering them are filled in. The plant stems further impede water movement and enhance spread of the marsh. This assumes that storms do not carry the additional sediment onto the marsh plain and raise it above normal tidal level, which would damage or destroy rather than extend the marsh.

Reduced sediment supply can destroy a marsh. In a river delta where sediments gradually de-water and consolidate, sinking continually, a continuous supply of new sediment combined with vegetation remains, accumulating as peat, and maintains the marsh level. When sediment supplies are cut off, the peat accumulation may be insufficient to maintain the marsh at sea level. Dams, such as the Aswan Dam on the Nile, can trap sediments. Sediments can be channeled by levées so that they flow into deep water at the mouth of the river rather than spreading over the delta marshes, as is happening in the Mississippi River delta. In the latter case, the coastline of Louisiana is retreating by kilometers a year as a result of the loss of delta marshlands.

Introduction of foreign species Dramatic changes in the marshes and flats of England and Europe occurred after *Spartina alterniflora* was introduced from the east coast of North America and probably hybridized with the native *S. maritima* to produce *S. anglica*. The new species was more tolerant of submergence than the native forms and turned many mud flats into salt marshes. This change reduced populations of mud flat animals, many with commercial value, and reduced the foraging area for shore birds that feed on mud flats. A similar situation has developed in the last decades on the north-west coast of the United States, where introduced *Spartina alterniflora* is invading mud flats and reducing the available area for shellfish.

Marsh restoration Salt marshes and mud flats may be the most readily restored of all wetlands. The source and level of water is known. The vascular plants that will thrive are known and can be planted if a local seed source is not available. Many of the marsh animals have planktonic larvae that can invade the restored marsh on their own. Although many of the properties of a mature salt marsh take time to develop, such as the nutrient-retaining capacity of the sediments, these will develop if the marsh is allowed to survive.

See also

Fish Feeding and Foraging. Fish Larvae. Mangroves. Nitrogen Cycle. Primary Production Processes. Salt Marsh Vegetation. Tides.

Further Reading

Adam P (1990) *Salt Marsh Ecology*. Cambridge: Cambridge University Press.

Mitsch WJ and Gosselink JG (1993) *Wetlands*. New York: Wiley.

Peterson CH and Peterson NM (1972) *The Ecology of Intertidal Flats of North Carolina: A Community Profile*. Washington, DC: US Fish and Wildlife Service, Office of Biological Services, FWS/OBS-79/39.

Streever W (1999) *An International Perspective on Wetland Rehabilitation*. Dordrecht: Kluwer.

Teal JM and Teal M (1969) *Life and Death of the Salt Marsh*. New York: Ballentine.

Weinstein MP and Kreeger DA (2001) *Concepts and Controversies in Tidal Marsh Ecology*. Dordrecht: Kluwer.

Whitlatch RB (1982) *The Ecology of New England Tidal Flats: A Community Profile*. Washington, DC: US Fish and Wildlife Service, Biological Services Program, FES/OBS-81/01.

SANDY BEACHES, BIOLOGY OF

A. C. Brown, University of Cape Town, Cape Town, Republic of South Africa

doi:10.1006/rwos.2001.0085

Introduction

Some 75% of the world's ice-free shores consists of sand. Nevertheless, biological studies on these beaches lagged behind those of other coastal marine habitats for many decades. This was probably because there are very few organisms to be seen on a beach at low tide during the day, when biological activity is at a minimum. A very different impression may be gained at night, but the beach fauna is then much more difficult to study. Even at night few species may be observed but these can be present in vast numbers. Most sandy-beach animals are not seen at all, as they are very tiny, virtually invisible to the naked eye, and they live between the sand grains (i.e., they are interstitial). A large number of these species, comprising the meiofauna, may be present. A further striking characteristic of ocean beaches is the total absence of attached plants intertidally and in the shallow subtidal. The animals are thus deprived of a resident primary food source, such as exists in most other habitats. We may, therefore, suspect that ecosystem functioning differs considerably from that found elsewhere and research has shown that this is, indeed, the case.

The Dominating Factor

In harsh, unpredictably varying environments, physical factors are far more important in shaping the ecosystem than are biological interactions; ocean beaches are no exception. Here the 'superparameter' controlling community structure, species diversity, and the modes of life of the organisms is water movement – waves, tides, and currents. Water movements determine the type of shore present and then interact with the sand particles so that a highly dynamic, unstable environment results. Instability reaches a peak during storms, when many tons of sand may be washed out to sea and physical conditions on the beach become chaotic. To exist in the face of such instability, beach animals must be very mobile and all must be able to burrow. Indeed the ability to burrow before being swept away by the next swash is critical for those animals that habitually emerge from the sand onto the beach face. The more exposed the beach to wave action, the more rapid does burrowing have to be. Furthermore, on most ocean beaches the sand is too unstable not only to support plant growth but also to allow permanent burrows to exist intertidally. Consequently animals such as the bloodworm, *Arenicola*, and the burrowing prawn, *Callianassa*, are found only on sheltered beaches. On exposed beaches, there is virtually nothing the aquatic fauna can do to modify their environment; they are entirely at the mercy of whatever the physical regime may dictate. During storms, behavior patterns must often change if the animals are to survive. Some animals simply dig themselves deeper into the sand, others may move offshore, and some semiterrestrial species (e.g., sand hoppers) move up into the dunes until the storm has passed. Notwithstanding these behavior patterns, mortality due to physical stresses often outweighs mortality due to predation.

Tidal Migrations and Zonation

Despite this extremely harsh environment, the essential mobility of the beach fauna permits exploitation of tidal rise and fall to an extent not available to more-sedentary animals. The meiofauna migrate up and down through the sand column in time with the tides, achieving the most optimal conditions available, and maximizing food resources and, in some cases, promoting photosynthesis. Many of the larger species (i.e., macrofauna) emerge from the sand and follow the tide up and down the slope of the beach face. The aquatic species tend to keep within the swash zone as it rises and falls; this is the area in which their food is likely to be most plentiful. As a bonus, this behavior also reduces predation, as the swash is too shallow for predatory fish to get to them, and most shore birds do not enter the water. Many semiterrestrial crustaceans, above the waterline, follow the falling tide down the slope, feed intertidally, and then migrate back to their positions above high water-mark as the tide rises. They do so only at night, remaining buried during the day, so that predation by birds is at a minimum.

If the macrofauna is sampled at low tide, a zonation of species is often apparent (**Figure 1**) and attempts have been made to relate these zones to those found on rocky shores or with changing physical conditions up the beach. However, the number of zones differs on different beaches and the zones tend to be blurred, with considerable overlap.

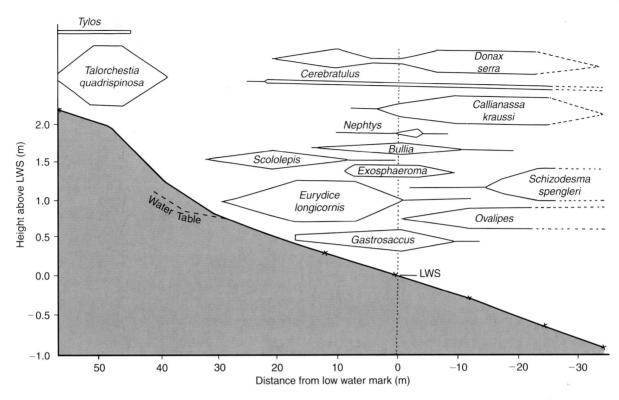

Figure 1 Low-tide distribution of the macrofauna up the beach slope at Muizenberg, near Cape Town. Some zonation is apparent. *Nephtys* and *Scololepis* are annelid worms; *Schizodesma* is a bivalve mollusk, larger than *Donax*; *Cerebratulus* is a nemertean worm; *Eurydice* and *Exosphaeroma* are isopods (pill bugs), *Ovalipes* is a swimming crab. The remaining genera are mentioned in the text. LWS, the level of Low Water of Spring tide.

Moreover, due to tidal migrations, the zones change as the tide rises. It has, therefore, been suggested that only two zones are apparent on all beaches at all states of the tide – a zone of aquatic animals and, higher up the shore, a zone of semiterrestrial air breathers; these are known as Brown's Zones.

Longshore Distribution

Not only is the distribution of the macrofauna up the beach slope not uniform but their longshore distribution is also typically discontinuous, or patchy. There is no one reason for this patchy distribution in all species. Variations in penetrability of the sand account for discontinuity in some species, breeding behavior or food maximization in others. In the semiterrestrial pill bug, *Tylos*, aggregations are apparently an incidental effect of their tendency to use existing burrows.

Diversity and Abundance: Meiofauna

The meiofauna consists of interstitial animals that will pass through a 1.0 mm sieve. They tend to be slender and worm-shaped, a necessary adaptation for gliding or wriggling between the grains. Nematoda (round worms) and harpacticoid copepods (small Crustacea, **Figure 2**) are usually dominant. However, most animal phyla are represented, including Platyhelminthes (flat worms), Nemertea (ribbon worms), Rotifera (wheel animalcules), Gastrotricha (**Figure 2**), Kinorhyncha, Annelida (segmented worms) and various Arthropoda such as mites and Collembola (spring tails). Rarer meiofaunal animals include an occasional cnidarian (hydroid), a few species of nudibranchs (sea slugs), tardigrades, a bryozoan, and even a primitive chordate. One whole phylum of animals, the Loricifera, is found only in beach meiofauna, as is the crustacean subclass Mystacocarida (**Figure 2**).

Both the diversity and abundance of the meiofauna are commonly highest on beaches intermediate between those with very gentle slopes (i.e., dissipative) and those with abrupt slopes (reflective). This is largely because these intermediate beaches also present an intermediate particle size range; the occurrence of most meiofaunal species is governed not by particle size itself but by the sizes of the pores between the grains. Pore size (porosity) also governs the depth to which adequate oxygen ten-

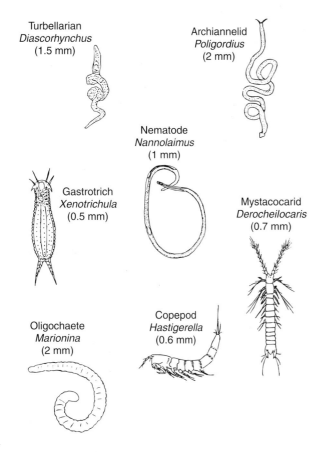

Figure 2 Some characteristic meiofaunal genera found in sandy beaches. (Reproduced from Brown and McLachlan, 1990 with permission from Elsevier Science.)

sions penetrate, this depth being least in very fine sand. This limits the vertical distribution of aerobic members of the meiofauna. Groups such as the Gastrotricha are completely absent from fine sands and harpacticoid copepods are relatively uncommon. As with the macrofauna, a zonation of the meiofauna is often apparent; maximum diversity and abundance are usually attained in the sand of the mid to upper part of the intertidal beach. Meiofaunal numbers typically average $10^6 \, \text{m}^{-2}$ but they may be as low as $0.05 \times 10^6 \, \text{m}^{-2}$ or as high as $3 \times 10^6 \, \text{m}^{-2}$. One of the few beaches whose meiofauna has been studied intensively is a beach on the island of Sylt, in the North Sea. No less than 652 meiofaunal species have been identified here.

Diversity and Abundance: Macrofauna

As with the meiofauna, the diversity and abundance of the macrofauna vary greatly with beach morphodynamics. Maximum numbers, species, and biomass are found on gently sloping beaches with wide surf zones (i.e., dissipative beaches). Important

criteria here, apart from slope, are the relatively fine sand, the extensive beach width and the long, essentially laminar flow of the swash, which results in minimum disturbance of sand, facilitates the transport of some species (see below) and renders burrowing easier. At the other extreme are abruptly sloping, reflective beaches, with waves breaking on the beach face itself, leading to turbulent swash, considerable sand movement and large particle size. Such beaches are inhospitable to virtually all species of aquatic macrofauna. There is, in fact, some correlation between diversity/abundance/biomass and beach slope, breaker height or the particle size of the sand, as these are all related. The mean individual size of the macrofaunal animals correlates best with particle size, whereas biomass correlates best with breaker height. Diversity and abundance are usually best correlated with Dean's Parameter (Ω) (**Figure 3**), which gives a good indication of the overall morphodynamic state of the beach:

$$\Omega = \frac{Hb}{Ws \, T}$$

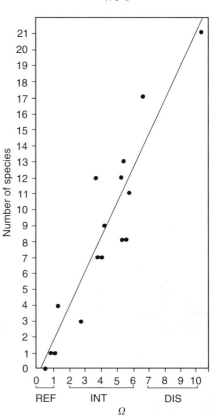

Figure 3 Relationship between macrofaunal diversity and beach morphodynamic state (Dean's parameter, Ω) for 23 beaches in Australia, South Africa, and the USA. REF, reflective; INT, intermediate; DIS, dissipative. (Reproduced from Brown and McLachlan, 1990 with permission from Elsevier Science.)

where Hb is average breaker height, Ws the mean fall velocity of the sand and T the wave period.

Beaches with extensive surf zones and gentle slopes may harbour over 20 intertidal species of aquatic macrofauna but this number decreases with increasing slope until fully reflective beaches may have no aquatic macrofauna at all.

Biomass varies greatly but on dissipative beaches averages about 7000 g dry mass per meter stretch of beach. One beach in Peru was found to have a biomass of no less than $25\,700\,\mathrm{g\,m^{-1}}$. Although diversity and abundance generally decrease with increasing wave exposure, some very exposed beaches display a surprisingly high biomass due to the larger individual sizes of the macrofauna. Species diversity and biomass tend to decrease from low to high tide marks but often increase again immediately above the high water level, especially if algal debris (e.g., kelp or wrack) is present.

On oceanic beaches, there is a gap in size between the meiofaunal animals and the macrofauna. This is probably because, although the meiofauna move between the grains, macrofaunal animals burrow by displacing the sand. They therefore have to be relatively robust and much larger than the sand grains. Common aquatic macrofauna include filter-feeding bivalve mollusks (clams) of several genera, of which the most wide-spread is *Donax*. Scavenging gastropod mollusks (whelks) may also occur; most is known about the plough-shell, *Bullia* (**Figure 4**). On relatively sheltered shores, a variety of annelid worms occurs, but these decrease in numbers and diversity with increasing wave action, small crustaceans becoming more dominant. On tropical and subtropical beaches, the mole-crabs *Hippa* and *Emerita* (**Figure 5**) may achieve dense populations, whereas on temperate beaches mysid shrimps (**Figure 6**) and aquatic isopods (pill bugs) may be much in evidence.

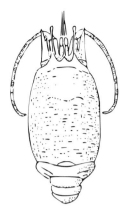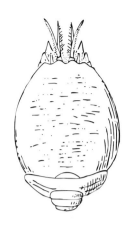

Figure 5 The mole-crabs *Emerita* and *Hippa* in dorsal view. (Reproduced from Brown and McLachlan, 1990 with permission from Elsevier Science.)

Above the water-line, semiterrestrial isopods, such as *Tylos* (**Figure 7**), are often abundant, and sand hoppers (or 'beach fleas') (*Talitrus*, **Figure 8**, *Orchestia* and *Talorchestia*) may occur in their millions on temperate beaches. On tropical and subtropical beaches, semiterrestrial crabs, such as the ghost crab, *Ocypode* (**Figure 9**), are important. In addition, insects may be present in some numbers; these include kelp flies and their relatives, staphylinids (rove beetles) and other beetles, and mole crickets.

Although the above forms are those most commonly encountered, some beaches have a quite different fauna. One beach has been found to be dominated by small Tanaidacea (related to pill bugs), another by picno-mole crickets (tiny insects). Seasonal visitors may also make considerable impact; people who know the beaches of the Atlantic seaboard of the United States or the beaches of Japan, Korea, or Malaysia will be familiar with the horseshoe crab, *Limulus* (actually not a crab at all), which may come ashore in vast numbers, and on some tropical beaches female turtles invade the beach to breed, their eggs and hatchlings providing a valuable food supply for a variety of marauding animals.

Figure 4 The 'plough shell' *Bullia*, about to surf.

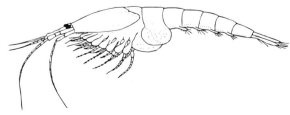

Figure 6 The mysid shrimp, *Gastrosaccus*.

Figure 7 The pill bug, *Tylos granulatus*, showing postcopulatory grooming. The male is mounted back-to-front on the female and is stroking her with his antennae. (Photo: Claudio Velasquez.) (Reproduced with permission from Brown and Odendaal, 1994.)

Food Relationships

As the instability of the substratum precludes the growth of attached plants in the intertidal area, the sandy-beach community is almost entirely dependent on imported food. A few diatoms may be present in the surface layers of sheltered beaches, but they are in general too sparse to be of much nutritional significance. Some potential food may be blown in from the land (plant debris, insects, etc.) and birds may defecate on the beach or die there; however, food washed up by the sea is of overwhelming importance. Three types of economy based on water-transported food may be distinguished, although they are not mutually exclusive.

One of these is based on detached algae (such as kelp), washed up onto the beach. In most cases, the rising tide pushes this material towards the top of the intertidal slope, where the semiterrestrial fauna make short work of it. However, not only are these animals messy feeders but their assimilation of the ingested material is poor, so that their feces are

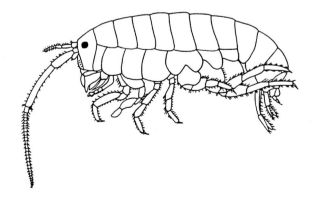

Figure 8 The semiterrestrial sand hopper, *Talitrus saltator*.

Figure 9 The ghost crab, *Ocypode*. (Reproduced from Brown and McLachlan, 1990 with permission from Elsevier Science.)

nutrient-rich. Algal fragments and feces find their way into the sand, where they form a food supply for bacteria and the meiofauna. From this community, mineralized nutrients leach back to sea, where they may support further algal growth – and so the cycle is completed. Some nutrients also pass to the land, because shore birds, and even birds such as passerines (e.g., swallows), feed on the semiterrestrial crustaceans and insects, as do some mammals and reptiles. On reflective beaches with a negligible aquatic macrofauna, this kelp-based economy may be the only nutrient cycle of significance.

A second type of food web depends on carrion, such as stranded jellyfish and animals detached from nearby rocky shores. Even dead seals, penguins, sea snakes, and fish may add to the food supply. Macrofaunal scavengers (swimming crabs, ghost crabs, and whelks such as *Bullia*) come into their own where carrion is plentiful. They will eat any animal matter and their assimilation is good, so they pass very little on to the bacteria and meiofauna. The acquatic scavengers are preyed on by the fishes and crabs of the surf zone, as the tide rises, and ghost crabs and other semiterrestrial crabs are taken by birds during the day and more particularly by small mammals and snakes which invade the beach at night.

The third type of economy is found on beaches with extensive surf zones displaying circulating cells of water. A number of phytoplankton (diatom) species are adapted to live in such cells (**Figure 10**), where they may achieve vast numbers, giving primary production rates of up to $10 \, \mathrm{g \, C \, m^{-3} \, h^{-1}}$. Some of this production is inevitably exported to sea but the remaining dissolved and particulate organic material supports three types of community (**Figure 11**). Firstly, it drives a 'microbial loop' in the surf, consisting of bacteria, which are eaten by flagellated protozoans, which in turn are consumed by microzooplankton of various kinds. Secondly,

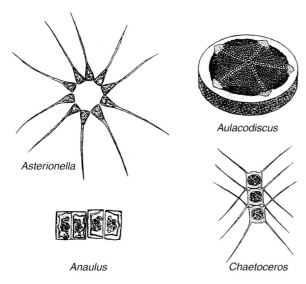

Figure 10 Common surf-zone diatoms (Reproduced from Brown and McLachlan, 1990 with permission from Elsevier Science.)

the production supports the interstitial meiofauna of both the surf zone and the intertidal beach, again largely through bacteria. Thirdly, the surf phytoplankton is eaten by a number of species of zooplankton (notably swimming prawns), which in turn are eaten by fish; the phytoplankton also supports the filter-feeders (*Donax*, *Emerita*, etc.) of the beach and these too are consumed by fish as the tide rises. The surf zone of such a beach is thus highly productive, displays considerable diversity and biomass, and forms an important nursery area for fish.

Adaptations of the Macrofauna

Locomotion

It has already been stressed that beach animals must be highly mobile and able to burrow into the sand. Two distinct types of burrowing are in evidence. The Crustacea, having hard exoskeletons, typically use their jointed walking legs as spades to dig themselves in, often with amazing rapidity; *Emerita* takes less than 1.5 s to completely bury itself. Soft-bodied invertebrates (annelid worms and mollusks) have to employ totally different methods (**Figure 12**). The sand surface is probed repeatedly, to liquefy it and so increase its permeability, and the head (of a worm) or the foot (of a clam or whelk) is inserted deep enough for it to swell to form a terminal anchor. With such an anchor in place, the rest of the body can be drawn down towards it. The swelling of the terminal anchor now subsides, while a new

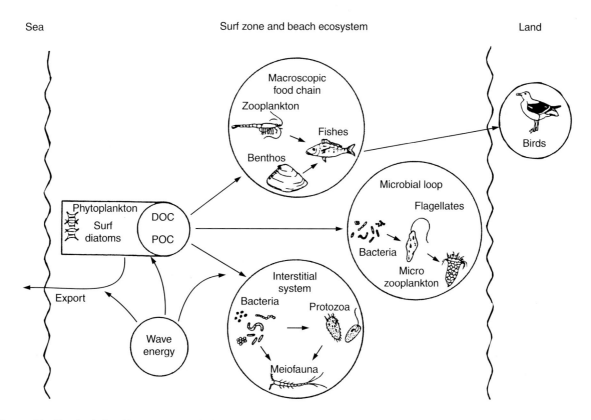

Figure 11 Food relationships on a typical dissipative beach with a rich surf-zone diatom flora, based on carbon flux (greatly simplified). DOC, dissolved organic carbon; POC, particulate organic carbon, including the diatoms themselves. (Reproduced from Brown and McLachlan, 1990 with permission from Elsevier Science.)

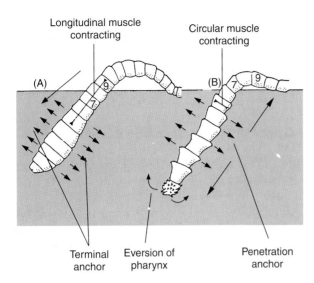

Figure 12 Principal stages of burrowing in the annelid worm *Arenicola*. (A) Anterior segments dilated to form a terminal anchor. (B) Penetration anchor formed by flanging. (Based on Trueman (1996) The mechanism of burrowing in the polychaete worm *Arenicola marina* (L.). *Biological Bulletin* 131: 369–377.

anchor appears, by swelling or flanging, closer to the surface, forming a penetration anchor. This allows the front part of the worm or the foot of the mollusk to penetrate deeper into the substratum; the cycle is repeated and so burrowing continues until the animal is completely buried.

Burrowing, crawling and swimming are common forms of locomotion, and a few semiterrestrial crustaceans (e.g., ghost crabs) run. However, some members of the macrofauna exploit not only the tides but also the waves in their quest for food. These are the surfers or swash-riders. Mollusks such as *Donax* and *Bullia* surf by maximally extending their feet and allowing the swash to transport them, whereas some crustaceans, such as juvenile *Tylos*, roll up into a ball when caught by the surf and get carried up the beach. Surfing has been studied in detail in *Bullia digitalis*. The whelk emerges from the sand in response to olfactory stimulation from stranded carrion or in response to increased water currents, spreads its foot to form a turgid underwater sail and then tacks at an angle to the direction of flow of the swash up the slope. As the whelk detects a decrease in olfactory stimulus (volatile amines), it flips over onto its other side and tacks in the opposite direction. It does this repeatedly and so reaches the carrion quickly and efficiently.

Nutrition and Energy Conservation

Many of the special adaptations of beach animals are concerned with making the best use of a highly erratic food supply (**Figure 13**). Thus many

members of the macrofauna have more than one method of feeding. The mysid shrimp *Gastrosaccus* filter-feeds while swimming, scoops up detritus deposited on the sand, and will also tear off pieces of carrion. The ghost crab, *Ocypode*, is both a scavenger and a voracious predator and, when all else fails, it sucks sand for any organic matter it may contain. *Bullia* will also turn predator in the absence of carrion; in addition, it grows an algal garden on its shell and uses its long proboscis to harvest the algae in a manner similar to mowing a lawn.

Not only must the fauna be opportunistic feeders and able to maximize the resource, they must also be able to conserve energy when no food at all is available. Some species remain inactive for long periods and allow their metabolic rates to decline to very low levels. *Tylos*, in its burrows during the day, has very modest rates of oxygen consumption, which increase dramatically during its short period of activity (**Figure 14**). *Bullia digitalis* also allows its metabolic rate to decline to low levels when inactive; in addition, its oxygen consumption is independent of temperature at all levels of activity, thus saving energy as the temperature rises.

Sensory Adaptations

The mobility of the macrofauna and the tidal migrations of many species up and down the shore, present considerable potential danger to them. If the animal moves too far down the slope it may be washed out to sea, and too far up the slope it will die of exposure (heat and desiccation). Sensory adaptations to ensure maintenance of position in the optimal area are thus essential. Some of these adaptations are relatively simple. For example, *Gastrosaccus* always swims against the current, facing the sea on an incoming swash and turning to face the land as the swash retreats; and the faster the current, the faster it swims. *Donax* and *Bullia* have

Figure 13 The swimming crab, *Ovalipes*, breaking into the shell of a living *Bullia*. (Photo: George Branch.)

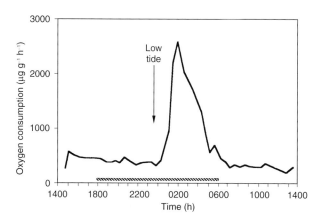

Figure 14 Respiratory rhythm of an individual of *Tylos granulatus*, showing a very low rate of oxygen consumption during the day but with a marked peak immediately after the time of low water at night. The horizontal bar denotes hours of darkness.

sensory cells in the foot, which detect the degree of saturation of the sand; as the animal migrates up the beach, it eventually finds itself in unsaturated sand and migrates no further upshore but allows itself to be carried down the slope.

In contrast, the sensory adaptations of semiterrestrial crustaceans are extremely complex and sophisticated. Most is known about the sand hopper, *Talitrus*. This little amphipod, with a 'brain' that can only be seen under the microscope, nevertheless has a whole suite of responses in its repertoire. It can sense where the shoreline of its particular beach lies relative to the sun and the moon, an orientation learnt while it was carried in the maternal brood pouch. Moreover, it has an internal clock which constantly adjusts this orientation as the sun or moon moves across the sky. It also senses beach slope and will move up the slope if it finds itself on wet sand (i.e., near the sea), or down the slope if the sand is very dry. It will react to broken horizons (e.g., dunes), which reinforces its sense of direction. Some have a magnetic compass and the animal may also orientate to polarized light patterns if the sun or moon is obscured. It also has a second internal clock geared to tidal rise and fall, which tells it when to emerge from the sand (on a falling tide) and when to retreat up the slope (on a rising tide). Some hoppers can detect an approaching storm (how is not known) and then reverse their normal response to slope, migrating as far away from the intertidal zone as possible. It is not surprising that these animals are so successful.

Plasticity of Behavior

In relatively constant environments, animals can often survive with a set of routine behavioral responses but in harsh, unpredictably changing habitats a much greater degree of flexibility (or plasticity) of behavior is called for. The sandy-beach macrofauna provides an outstanding opportunity to study this plasticity among invertebrates. Variations in behavior may in part be determined by genetic selection but the ability to learn plays a most important role in most cases. Both genetically determined plasticity and learning ability are necessary for populations of a species inhabiting different beaches, as no two beaches present exactly the same environment.

For example, *Tylos* typically moves down the intertidal slope to feed but on Mediterranean beaches it commonly moves up the slope after emergence, as food is more plentiful there. In the Eastern Cape Province of South Africa, some populations of *Tylos* have moved permanently into the dunes and have abandoned their tidal rhythm of emergence and reburial; this is also food-related. Another example, *Bullia digitalis* is typically an intertidal, surfing whelk, but in some localities where the beach is inhospitable to it, it occurs offshore, although it may return to the beach and surf if conditions become more favorable.

Perhaps the most remarkable example of the ability to adapt to circumstances concerns observations of *Tylos* on a beach in Japan. Detecting an approaching storm, the animals moved *en masse* towards the sea (i.e., towards the coming danger) and onto an artificial jetty, to which they clung until the storm had passed. This phenomenon has also been observed on a Mediterranean beach. Such behavioral flexibility plays a major role in the survival of the macrofauna and must have been rigorously selected for during the course of evolution.

Conclusion

All of the above phenomena can be linked directly or indirectly to the movements of the water. Waves, tides, and currents determine the type of shore, patterns of erosion, and deposition, the particle size distribution of the substratum, and the slope of the beach. The interaction of waves and sand results in an instability, which precludes attached plants, leading to a unique series of food webs and a series of faunal adaptations to deal with an erratic and unpredictable supply of imported food. Tides and waves are exploited to take maximum advantage of this food. Instability requires that the fauna be mobile and able to burrow, but the three-dimensional nature of the substratum also allows the development of a diverse meiofauna living between the grains. Being of necessity extremely mobile, the macrofauna require sophisticated sensory responses to maintain an optimum position on the beach and the ability to react appropriately to the conditions and circumstances they encounter on their particular beach. Finally, energy conservation is essential, so that metabolic, biochemical adaptations are ultimately dictated by an erratic food supply, again dependent on water movements.

See also

Beaches, Physical Processes Affecting. Diversity of Marine Species. Large Marine Ecosystems. Microbial Loops. Network Analysis of Food Webs. Ocean Gyre Ecosystems. Polar Ecosystems. Storm Surges. Tides. Upwelling Ecosystems.

Further Reading

Brown AC (1996) Behavioural plasticity as a key factor in the survival and evolution of the macrofauna on exposed sandy beaches. *Revista Chilena de Historia Natural* 69: 469–474.

Brown AC and McLachlan A (1990) *Ecology of Sandy Shores*. Amsterdam: Elsevier.

Brown AC and Odendaal FJ (1994) The biology of oniscid Isopoda of the genus *Tylos. Advances in Marine Biology* 30: 89–153.

Brown AC, Stenton-Dozey JME and Trueman ER (1989) Sandy-beach bivalves and gastropods: a comparison between *Donax serra* and *Bullia digitalis. Advances in Marine Biology* 25: 179–247.

Campbell EE (1996) The global distribution of surf diatom accumulations. *Revista Chilena de Historia Natural* 69: 495–501.

Little C (2000) *The Biology of Soft Shores and Estuaries*. Oxford: Oxford University Press.

McLachlan A and Erasmus T (eds) (1983) *Sandy Beaches as Ecosystems*. The Hague: W. Junk.

McLachlan A, De Ruyck A and Hacking N (1996) Community structure on sandy beaches: patterns of richness and zonation in relation to tide range and latitude. *Revista Chilena de Historia Natural* 69: 451–467.

SATELLITE ALTIMETRY

R. E. Cheney, Laboratory for Satellite Altimetry, Silver Spring, Maryland, USA

Copyright © 2001 Academic Press

doi:10.1006/rwos.2001.0340

Introduction

Students of oceanography are usually surprised to learn that sea level is not very level at all and that the dominant force affecting ocean surface topography is not currents, wind, or tides; rather it is regional variations in the Earth's gravity. Beginning in the 1970s with the advent of satellite radar altimeters, the large-scale shape of the global ocean surface could be observed directly for the first time. What the data revealed came as a shock to most of the oceanographic community, which was more accustomed to observing the sea from ships. Profiles telemetered back from NASA's pioneering altimeter, Geos-3, showed that on horizontal scales of hundreds to thousands of kilometers, the sea surface is extremely complex and bumpy, full of undulating hills and valleys with vertical amplitudes of tens to hundreds of meters. None of this came as a surprise to geodesists and geophysicists who knew that the oceans must conform to these shapes owing to spatial variations in marine gravity. But for the oceanographic community, the concept of sea level was forever changed. During the following two decades, satellite altimetry would provide exciting and revolutionary new insights into a wide range of earth science topics including marine gravity, bathymetry, ocean tides, eddies, and El Niño, not to mention the marine wind and wave fields which can also be derived from the altimeter echo. This chapter briefly addresses the technique of satellite altimetry and provides examples of applications.

Measurement Method

In concept, radar altimetry is among the simplest of remote sensing techniques. Two basic geometric measurements are involved. In the first, the distance between the satellite and the sea surface is determined from the round-trip travel time of microwave pulses emitted downward by the satellite's radar and reflected back from the ocean. For the second measurement, independent tracking systems are used to compute the satellite's three-dimensional position relative to a fixed Earth coordinate system. Combining these two measurements yields profiles of sea surface topography, or sea level, with respect to the reference ellipsoid (a smooth geometric surface which approximates the shape of the Earth).

In practice, the various measurement systems are highly sophisticated and require expertise at the cutting edge of instrument and modeling capabilities. This is because accuracies of a few centimeters

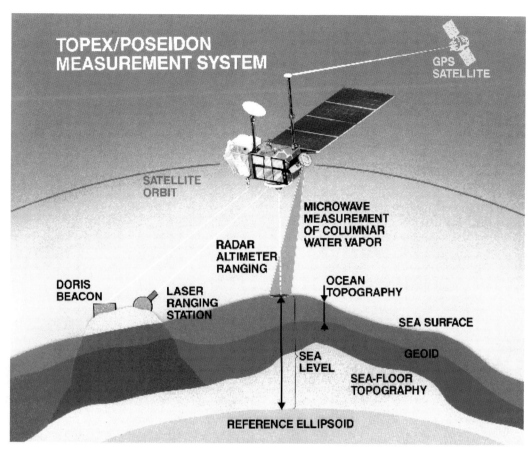

Figure 1 Schematic diagram of satellite radar altimeter system. Range to the sea surface together with independent determination of the satellite orbit yields profiles of sea surface topography (sea level) relative to the Earth's reference ellipsoid. Departures of sea level from the geoid drive surface geostrophic currents.

must be achieved to properly observe and describe the various oceanographic and geophysical phenomena of interest. **Figure 1** shows a schematic of the Topex/Poseidon (T/P) satellite altimeter system. Launched in 1992 as a joint mission of the American and French Space agencies (and still operating as of 2001), T/P is the most accurate altimeter flown to date. Its microwave radars measure the distance to the sea surface with a precision of 2 cm. Two different frequencies are used to solve for the path delay due to the ionosphere, and a downward-looking microwave radiometer provides measurements of the integrated water vapor content which must also be known. Meteorological models must be used to estimate the attenuation of the radar pulse by the atmosphere, and other models correct for biases created by ocean waves. Three different tracking systems (a laser reflector, a Global Positioning System receiver, and a 'DORIS' Doppler receiver) determine the satellite orbit to within 2 cm in the radial direction. The result of all these measurements is a set of global sea level observations with an absolute accuracy of 3–4 cm at intervals of 1 s, or about 6 km, along the satellite track. The altimeter footprint is exceedingly small – only 2–3 km – so regional maps or 'images' can only be derived by averaging data collected over a week or more.

Gravitational Sea Surface Topography

Sea surface topography associated with spatial variations in the Earth's gravity field has vertical amplitudes 100 times larger than sea level changes generated by tides and ocean currents. To first order, therefore, satellite altimeter data reveal information about marine gravity. Within 1–2% the ocean topography follows a surface of constant gravitational potential energy known as the geoid or the equipotential surface, shown schematically in **Figure 1**. Gravity can be considered to be constant in time for most purposes, even though slight changes do occur as the result of crustal motions, redistribution of terrestrial ice and water, and other slowly varying phenomena. An illustration of the

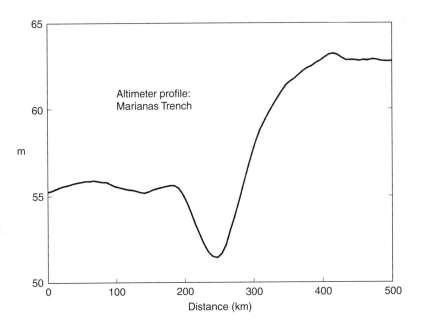

Figure 2 Sea surface topography across the Marianas Trench in the western Pacific measured by Topex/Poseidon. Heights are relative to the Earth's reference ellipsoid.

gravitational component of sea surface topography is provided in **Figure 2**, which shows a T/P altimeter profile collected in December 1999 across the Marianas Trench in the western Pacific. The trench represents a deficit of mass, and therefore a negative gravity anomaly, so that the water is pulled away from the trench axis by positive anomalies on either side. Similarly, seamounts represent positive gravity anomalies and appear at the ocean surface as mounds of water. The sea level signal created by ocean bottom topography ranges from ~ 1 m for seamounts to ~ 10 m for pronounced features like the Marianas Trench, and the peak-to-peak amplitude for the large-scale gravity field is nearly 200 m.

Using altimeter data collected by several different satellites over a period of years, it is possible to create global maps of sea surface topography with extraordinary accuracy and resolution. When these maps are combined with surface gravity measurements, models of the Earth's crust, and bathymetric data collected by ships, it is possible to construct three-dimensional images of the ocean floor – as if all the water were drained away (**Figure 3**). For many oceanic regions, especially in the Southern Hemisphere, these data have provided the first reliable maps of bottom topography. This new data set has many scientific and commercial applications, from numerical ocean modeling, which requires realistic bottom topography, to fisheries, which have been able to take advantage of new fishing grounds over previously uncharted seamounts.

Dynamic Sea Surface Topography

Because of variations in the density of sea water over the globe, the geoid and the mean sea surface are not exactly coincident. Departures of the sea surface with respect to the geoid have amplitudes of about 1 m and constitute what is known as 'dynamic topography'. These sea surface slopes drive the geostrophic circulation: a balance between the surface slope (or surface pressure gradient) and the Coriolis force (created by the Earth's rotation). The illustration in **Figure 4** shows an estimate of the global geostrophic circulation derived by combining a mean altimeter-derived topography with a geoid computed from independent gravity measurements. Variations are between − 110 cm (deep blue) and 110 cm (white). The surface flow is along lines of equal dynamic topography (red arrows). In the Northern Hemisphere, the flow is clockwise around the topography highs, while in the Southern Hemisphere, the flow is counter-clockwise around the highs. The map shows all the features of the general circulation such as the ocean gyres and associated western boundary currents (e.g. Gulf Stream, Kuroshio, Brazil/Malvinas Confluence) and the Antarctic Circumpolar Current.

At the time of writing, global geoid models are not sufficiently accurate to reveal significant new information about the surface circulation of the ocean. However, extraordinary gravity fields will soon be available from dedicated satellite missions

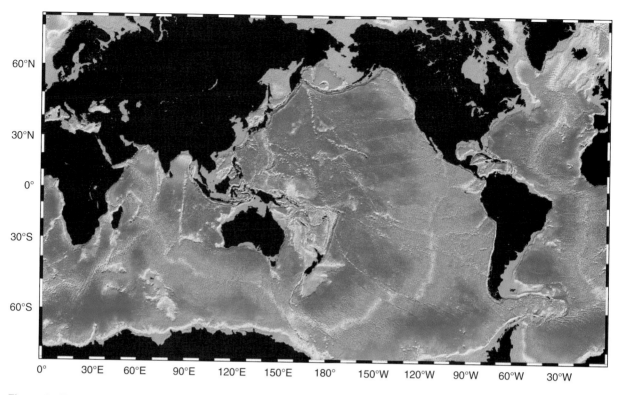

Figure 3 Topography of the ocean bottom determined from a combination of satellite altimetry, gravity anomalies, and bathymetric data collected by ships. (Courtesy of Walter H. F. Smith, NOAA, Silver Spring, MD, USA.)

such as the Challenging Minisatellite Payload (CHAMP: 2000 launch), the Gravity Recovery and Climate Experiment (GRACE: 2002 launch), and the Gravity Field and Steady-state Ocean Circulation Explorer (GOCE: 2005 launch). These satellite missions, sponsored by various agencies in the USA

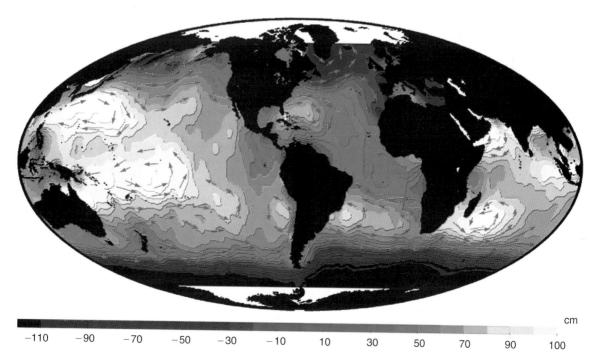

Figure 4 Surface geostrophic circulation determined from a combination of satellite altimetry and a model of the marine gravity field. (Courtesy of Space Oceanography Division, CLS, Toulouse, France.).

and Europe, will employ accelerometers, gravity gradiometers, and the Global Positioning System to virtually eliminate error in marine geoid models at spatial scales larger than 300 km and will thereby have a dramatic impact on physical oceanography. Not only will it be possible to accurately compute global maps of dynamic topography and geostrophic surface circulation, but the new gravity models will also allow recomputation of orbits for past altimetric satellites back to 1978, permitting studies of long, global sea level time-series. Furthermore, measurement of the change in gravity as a function of time will provide new information about the global hydrologic cycle and perhaps shed light on the factors contributing to global sea level rise. For example, how much of the rise is due simply to heating and how much to melting of glaciers? Together with complementary geophysical data, satellite gravity data represent a new frontier in studies of the Earth and its fluid envelope.

Sea Level Variability

At any given location in the ocean, sea level rises and falls over time owing to tides, variable geostrophic flow, wind stress, and changes in temperature and salinity. Of these, the tides have the largest signal amplitude, on the order of 1 m in mid-ocean. Satellite altimetry has enabled global

tide models to be dramatically improved such that mid-ocean tides can now be predicted with an accuracy of a few cm (see **Tides**). In studying ocean dynamics, the contribution of the tides is usually removed using these models so that other dynamic ocean phenomena can be isolated.

The map in **Figure 5** shows the variability of global sea level for the period 1992–98. It is derived from three satellite altimeter data sets: ERS-1, T/P, and ERS-2 (ERS is the European Space Agency Remote Sensing Satellite), from which the tidal signal has been removed. The map is dominated by mesoscale (100–300 km) variability associated with the western boundary currents, where the rms variability can be as high as 30 cm. This is due to a combination of current meandering, eddies, and seasonal heating and cooling. Other bands of relative maxima (10–15 cm) can be seen in the tropics where interannual signals such as El Niño are the dominant contributor. The smallest variability is found in the eastern portions of the major ocean basins where values are < 5 cm rms.

To examine a sample of the sea level signal more closely, **Figure 6** shows the record from the region of the Galapagos Islands in the eastern equatorial Pacific. The plot includes two time-series: one from the T/P altimeter and the other from an island tide gauge, both averaged over monthly time periods. These independent records agree at the level of

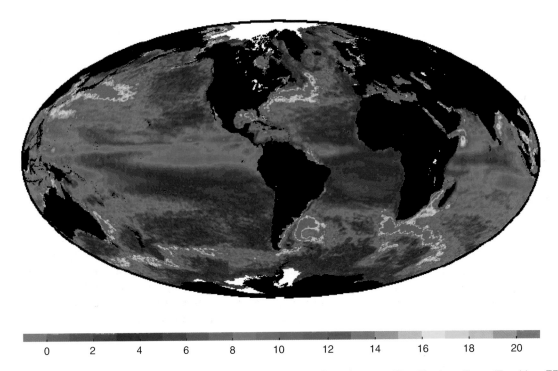

Figure 5 Variability of sea surface topography over the period 1992–98 from three satellite altimeters: Topex/Poseidon, ERS-1, and ERS-2. Highest values (cm) correspond to western boundary currents which meander and generate eddies. (Courtesy of Space Oceanography Division, CLS, Toulouse, France.).

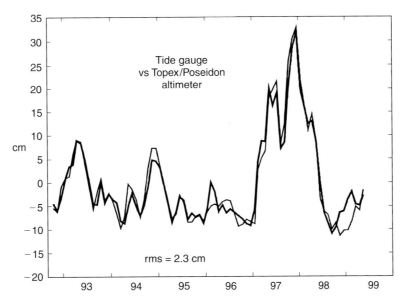

Figure 6 Monthly mean sea level deviation near the Galapagos Islands derived from tide gauge data and altimeter data. The ~2 cm agreement demonstrates the accuracy of altimetry for observing sea level variability. The effect of the 1997–98 El Niño is apparent.

2 cm, an indication of the remarkable reliability of satellite altimetry. The plot also illustrates changes associated with the El Niño event which took place during 1997–98. During El Niño, relaxation of the Pacific trade winds cause a dramatic redistribution of heat in the tropical oceans. In the eastern Pacific, sea level during this event rose to 30 cm above normal by December 1977 and fell by a corresponding amount in the far western Pacific. The global picture of sea level deviations observed by the T/P altimeter at this time is shown in **Figure 7**. Because sea level changes can be interpreted as changes in heat (and to a lesser extent, salinity) in the upper layers, altimetry provides important information for operational ocean models which are used for long-range El Niño forecasts (*see* **El Niño Southern Oscillation (ENSO)**).

Global Sea Level Rise

Tide gauge data collected over the last century indicate that global sea level is rising at about 1.8 mm y^{-1}. Unfortunately, because these data are relatively sparse and contain large interdecadal fluctuations, the observations must be averaged over 50–75 years in order to obtain a stable mean value. It is therefore not possible to say whether sea level rise is accelerating in response to recent global warming. Satellite altimeter data have the advantage of dense, global coverage and may offer new insights on the problem in a relatively short period of time. Based on T/P data collected since 1992, it

is thought that 15 years of continuous altimeter measurements may be sufficient to obtain a reliable estimate of the current rate of sea level rise. This will require careful calibration of the end-to-end altimetric system, not to mention cross-calibration of a series of two or three missions (which typically last only 5 years). Furthermore, in order to interpret and fully understand the sea level observations, the various components of the global hydrologic system must be taken into account, for example, polar and glacial ice, ground water, fresh water stored in man-made reservoirs, and the total atmospheric water content. It is a complicated issue, but one which may yield to the increasingly sophisticated observational systems that are being brought to bear on the problem. For additional information, *see* **Sea Level Change**.

Wave Height and Wind Speed

In addition to sea surface topography, altimetry provides indirect measurements of ocean wave height and wind speed (but not wind direction). This is made possible by analysis of the shape and intensity of the reflected radar signal: a calm sea sends the signal back almost perfectly, like a flat mirror, whereas a rough sea scatters and deforms it. Wave height measurements are accurate to about 0.5 m or 10% of the significant wave height, whichever is larger. Wind speed can be measured with an accuracy of about 2 m s^{-1}. For additional information,

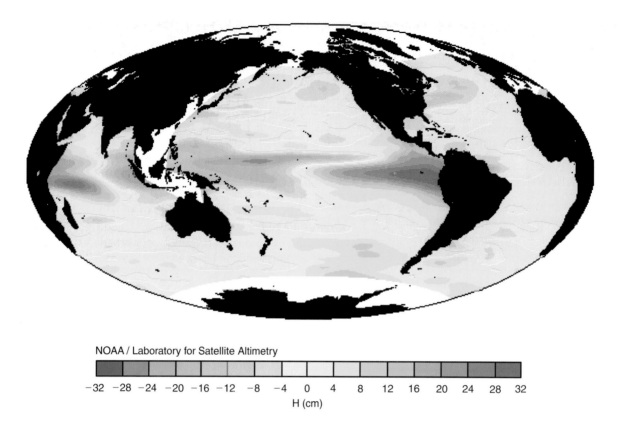

NOAA / Laboratory for Satellite Altimetry

-32 -28 -24 -20 -16 -12 -8 -4 0 4 8 12 16 20 24 28 32

H (cm)

Figure 7 Global sea level anomaly observed by the Topex/Poseidon altimeter at the height of the 1997–98 El Niño. High (low) sea level corresponds to areas of positive (negative) heat anomaly in the ocean's upper layers.

see Wave Generation by Wind and Surface, Gravity and Capillary Waves.

Conclusions

Satellite altimetry is somewhat unique among ocean remote sensing techniques because it provides much more than surface observations. By measuring sea surface topography and its change in time, altimeters provide information on the Earth's gravity field, the shape and structure of the ocean bottom, the integrated heat and salt content of the ocean, and geostrophic ocean currents. Much progress has been made in the development of operational ocean applications, and altimeter data are now routinely assimilated in near-real-time to help forecast El Niño, monitor coastal circulation, and predict hurricane intensity. Although past missions have been flown largely for research purposes, altimetry is rapidly moving into the operational domain and will become a routine component of international satellite systems during the twenty-first century.

See also

Elemental Distribution: Overview. El Niño Southern Oscillation (ENSO). El Niño Southern Oscillation (ENSO) Models. Heat Transport and Climate. Satellite Oceanography, History and Introductory Concepts. Satellite Remote Sensing Microwave Scatterometers. Satellite Remote Sensing SAR. Upper Ocean Time and Space Variability. Wind Driven Circulation.

Further Reading

Cheney RE (ed) (1995) TOPEX/POSEIDON: Scientific Results. *Journal of Geophysical Research* 100: 24 893–25 382.

Douglas BC, Kearney MS and Leatherman SP (eds) (2001) *Sea Level Rise: History and Consequences*. London: Academic Press.

Fu LL and Cheney RE (1995) Application of satellite altimetry to ocean circulation studies, 1987–1994. *Reviews of Geophysics* Suppl: 213–223.

Fu LL and Cazenave A (eds) (2001) *Satellite Altimetry and Earth Sciences*. London: Academic Press.

SATELLITE MEASUREMENTS OF SALINITY

G. Lagerloef, Earth and Space Research, Seattle, WA, USA

Copyright © 2001 Academic Press

doi:10.1006/rwos.2001.0345

Introduction

Surface salinity is an ocean state variable which controls, along with temperature, the density of sea water and influences surface circulation and formation of dense surface waters in the higher latitudes which sink into the deep ocean and drive the thermohaline convection. Although no satellite measurements are made at present, emerging new technology and a growing scientific need for global measurements are stimulating efforts to launch salinity observing satellite sensors within the present decade. Salinity remote sensing is possible because the dielectric properties of sea water which depend on salinity also affect the surface emission at certain microwave frequencies. Experimental heritage extends over the past 30 years, including laboratory studies, airborne sensors, and one instrument flown briefly in space on Skylab. Requirements for very low noise radiometers and large antenna structures have limited the advance of satellite systems, and are now being addressed. Science needs, primarily for climate studies, dictate a resolution requirement of 100 km spatial grid, observed monthly, with a 0.1‰, error (or 1 part in 10 000), which demand very precise radiometers and that several ancillary errors be accurately corrected. Measurements will be made in the 1.413 GHz astronomical hydrogen absorption band to avoid radio interference. The first satellite could be launched as early as 2005.

Definition and Theory

How Salinity is Defined and Measured

Salinity is the concentration of dissolved inorganic salts in sea water (grams of salt per kilogram sea water, or parts per thousand, and given by the symbol ‰). Oceanographers have developed methods based on the electrical conductivity of sea water which permit accurate measurement by use of automated electronic *in situ* sensors. Salinity is derived from conductivity, temperature, and pressure with a set of empirical equations known as the practical salinity scale. Accordingly, the literature sometimes quotes salinity measurements in practical salinity units (PSU), which is equivalent to ‰ or grams per kilogram salt. Salinity ranges from near zero adjacent to the mouths of major rivers to > 40‰ in the Red Sea. Aside from such extremes, open-ocean surface values away from coastlines generally fall between 32‰ and 37‰ (**Figure 1**).

This global mean surface salinity field has been compiled from all available oceanographic observations. A significant fraction of the 1° latitude longitude cells have no observations, requiring such maps to be interpolated and smoothed over several hundred kilometer scales. Seasonal to interannual salinity variations can only be resolved in very limited geographical regions where the sampling density is suitable. Data are most sparse over large regions of the Southern Hemisphere. Remote sensing from satellites will be able to fill this void and monitor multi-year variations globally.

Remote sensing theory Salinity remote sensing with microwave radiometry is likewise possible through the electrically conductive properties of sea water. A radiometric measurement of an emitting surface is given in terms of a brightness temperature (T_B), measured in degrees Kelvin (K). T_B is related to the true absolute surface temperature (T) through the emissivity coefficient (e):

$$T_B = eT$$

For sea water, e depends on the complex dielectric constant (ε), the viewing angle (Fresnel laws), and surface roughness (due to wind waves). The complex dielectric constant is governed by the Debye equation:

$$\varepsilon = \varepsilon_\infty + \frac{\varepsilon_S(S,T) - \varepsilon_\infty}{1 + i2\pi f\tau(S,T)} - \frac{iC(S,T)}{2\pi f\varepsilon_0}$$

and includes electrical conductivity (C), the static dielectric constant (ε_S), and the relaxation time (τ) which are all sensitive to salinity and temperature (S, T). The equation also includes radio frequency (f) and terms for permittivity at infinite frequency (ε_∞) which may vary weakly with T, and permittivity of free space (ε_0) which is a constant. The relation of electrical conductivity to salinity and temperature is determined through laboratory measurements or derived from the practical salinity scale. The static dielectric and time constants have been modeled by making laboratory measurements of ε at various

Global mean surface salinity

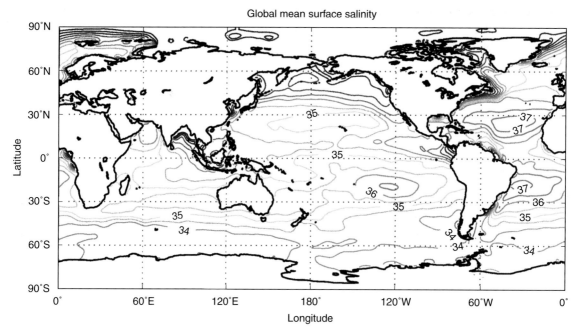

Figure 1   Contour map of the mean global surface salinity field based on the *World Ocean Atlas, 1998* (WOA98). Generated from data obtained from the US Department of Commerce, NOAA, National Oceanographic Data Center (www.nodc.noaa.gov).

frequencies, temperatures and salinities, and fitting ε_S and τ to polynomial expressions of (S, T) to match the ε data. Different models in the literature show similar variations with respect to (f, S, T).

Emissivity for the horizontal (H) and vertical (V) polarization state is related to ε by Fresnel reflection:

$$e_H = 1 - \left[\frac{\cos\theta - (\varepsilon - \sin^2\theta)^{1/2}}{\cos\theta + (\varepsilon - \sin^2\theta)^{1/2}}\right]^2$$

and

$$e_V = 1 - \left[\frac{\varepsilon\cos\theta - (\varepsilon - \sin^2\theta)^{1/2}}{\varepsilon\cos\theta + (\varepsilon - \sin^2\theta)^{1/2}}\right]^2$$

where θ is the vertical incidence angle from which the radiometer views the surface, and $e_H = e_V$ when $\theta = 0$. The above set of equations provides a physically based model function relating T_B to surface S, T, θ, and H or V polarization state for smooth water (no wind roughness). This can be inverted to retrieve salinity from radiometric T_B measurements provided the remaining parameters are known. The microwave optical depth is such that the measured emission originates in the top 1 cm of the ocean, approximately.

The rate at which T_B varies with salinity is sensitive to microwave frequency, achieving levels practical for salinity remote sensing at frequencies below about 3 GHz. Considerations for selecting

a measurement radio frequency include salinity sensitivity, requisite antenna size (see below), and radio interference from other (mostly man-made) sources. A compromise of these factors, dominated by the interference issue, dictates a choice of about a 20 MHz wide frequency band centered at 1.413 GHz, which is the hydrogen absorption band protected by international treaty for radio astronomy research. This falls within a frequency range known as L-band. Atmospheric clouds have a negligible effect, allowing observations to be made in all weather except possibly heavy rain. Accompanying illustrations are based on applying $f = 1.413$ GHz in the Debye equation and using a model that included laboratory dielectric constant measurements at the nearby frequency 1.43 GHz. Features of this model function and their influence on measurement accuracy are discussed in the section on resolution and error sources (below).

Antennas   Unusually large radiometer antennas will be required to be deployed on satellites to measure salinity. Radiometer antenna beam width varies inversely with both antenna aperture and radio frequency; 1.413 GHz is a significantly lower frequency than found on conventional satellite microwave radiometers, and large antenna structures are necessary to avoid excessive beam width and accordingly large footprint size. For example, a 50 km footprint requires about a 6 m aperture antenna, whereas conventional radiometer antennas are around 1–2 m. To decrease the footprint by a factor of two requires doubling the antenna size. Various filled and thinned array technologies for large antennas have now reached a development stage where application to salinity remote sensing is feasible.

History of Salinity Remote Sensing

The only experiment to date to measure surface salinity from space took place on the NASA Skylab mission during the fall and winter 1973–74, when a 1.413 GHz microwave radiometer with a 1 m antenna collected intermittent data. A weak correlation was found between the sensor data and surface salinity, after correcting for other influences. There was no 'ground truth' other than standard surface charts, and many of the ambient corrections were not as well modeled then as they could be today. Research leading up to the Skylab experiment began with several efforts during the late 1940s and early 1950s to measure the complex dielectric constant of saline solutions for various salinities, temperatures, and microwave frequencies. These relationships provide the physical basis for microwave remote sensing of the ocean as described above.

The first airborne salinity measurements were demonstrated in the Mississippi River outflow and published in 1970. This led to renewed efforts during the 1970s to refine the dielectric constants and governing equations. Meanwhile, a series of airborne experiments in the 1970s mapped coastal salinity patterns in the Chesapeake and Savannah river plumes and freshwater sources along the Puerto Rico shoreline. In the early 1980s, a satellite concept was suggested that might achieve an ideal precision of about 0.25‰ and spatial resolution of about 100 km. At that time, space agencies were establishing the oceanic processes remote sensing program around missions and sensors for measuring surface dynamic topography, wind stress, ocean color, surface temperature, and sea ice. For various reasons, salinity remote sensing from satellites was then considered only marginally feasible and lacked a strongly defined scientific need.

Interest in salinity remote sensing revived in the late 1980s with the development of a 1.4 GHz airborne Electrically Scanning Thinned Array Radiometer (ESTAR) designed primarily for soil moisture measurements. ESTAR imaging is done electronically with no moving antenna parts, thus making large antenna structures more feasible. The airborne version was developed as an engineering prototype and to provide the proof-of-concept that aperture synthesis can be extrapolated to a satellite design. The initial experiment to collect ocean data with this sensor consisted of a flight across the Gulf Stream in 1991 near Cape Hatteras. The change from 36‰ in the offshore waters to < 32‰ near shore was measured, along with several frontal features visible in the satellite surface temperature image from the same day. This Gulf Stream transect demonstrated that small salinity variations typical of the open ocean can be detected as well as the strong salinity gradients in the coastal and estuary settings demonstrated previously.

By the mid-1990s, a new airborne salinity mapper Scanning Low Frequency Microwave Radiometer (SLFMR) was developed for light aircraft and has been used extensively by NOAA and the US Navy to survey coastal and estuary waters on the US east coast and Florida. A version of this sensor is now being used in Australia, and a second generation model is presently being built for the US Navy.

In 1999 a satellite project was approved by the European Space Agency for the measurement of Soil Moisture and Ocean Salinity (SMOS) with projected launch in 2005. The SMOS mission design emphasizes soil moisture measurement requirements, which is done at the same microwave frequency for many of the same reasons as salinity. The T_B dynamic range is about 70–80 K for varying soil moisture conditions and the precision requirement is therefore much less rigid than for salinity. The greater requirement is for spatial resolution on the ground where SMOS will employ a large two-dimensional phased array system that will produce a minimum 35 km resolution around the center swath. The imaging system will yield measurements at various incidence angles and H and V polarization. Retrieval algorithms for salinity and the eventual measurement accuracy will be developed and evaluated prior to launch. Other satellite concepts with primary emphasis on ocean salinity rather than soil moisture are being investigated and may be proposed in 2001.

The focus will be on optimizing salinity accuracy and addressing the error sources presented in the following section.

Requirements for Observing Salinity from Satellites

Scientific Issues

Three broad scientific themes have been identified for a satellite salinity remote sensing program. These themes relate directly to the international climate research and global environmental observing program goals.

Improving seasonal to interannual climate predictions This focuses primarily on El Niño forecasting and involves the effective use of surface salinity data (1) to initialize and improve the coupled climate forecast models, and (2) to study and model the role of freshwater flux in the formation and maintenance of barrier layers and mixed-layer heat budgets in the tropics. Climate prediction models in which satellite altimeter sea level data are assimilated must be adjusted for steric height (sea level change due to ocean density) caused by the variations in upper layer salinity. If not, the adjustment for model heat content is incorrect and the prediction skill is degraded. Barrier layer formation occurs when excessive rainfall creates a shallow, freshwater stratified, surface layer which effectively isolates the deeper thermocline from exchanging heat with the atmosphere with consequences on the air–sea coupling processes that govern El Niño dynamics.

Improving ocean rainfall estimates and global hydrologic budgets Precipitation over the ocean is still poorly known and relates to both the hydrologic budget and to latent heating of the overlying atmosphere. Using the ocean as a rain gauge is feasible with precise surface salinity observations coupled with ocean surface current velocity data and mixed-layer modeling. Such calculations will reduce uncertainties in the surface freshwater flux on climate timescales and will complement satellite precipitation and evaporation observations to improve estimates of the global water and energy cycles.

Monitoring large-scale salinity events and thermohaline convection Studying interannual surface salinity variations in the subpolar regions, particularly the North Atlantic and Southern Oceans, is essential to long timescale climate prediction and modeling. These variations influence the rate of oceanic convection and poleward heat transport (thermohaline circulation) which are known to have been coupled to extreme global climate changes in the geologic record. Outside of the polar regions, salinity signals are stronger in the coastal ocean and marginal seas than in the open ocean in general, but large footprint size will limit near-shore applications of the data. Many of the larger marginal seas which have strong salinity signals might be adequately resolved nonetheless, such as the East China Sea, Bay of Bengal, Gulf of Mexico, Coral Sea/Gulf of Papua, and Mediterranean.

Science requirements The science themes lead to four specific study topics to benefit from satellite observations. Each topic has accuracy and spatial and temporal resolution requirements as follows. (1) Barrier layer effects on tropical Pacific heat flux: 0.2‰, 100 km, and 30 days. (2) Steric adjustment of heat storage from sea level: 0.2‰, 200 km, and 7 days. (3) North Atlantic thermohaline circulation: 0.1‰, 100 km, and 30 days. (4) Surface freshwater flux balance: 0.1‰, 300 km, and 30 days. Thermohaline circulation and convection in the North Atlantic and other subpolar seas have the most demanding requirements, and are the most technically challenging because of the reduced T_B/salinity ratio at low seawater temperatures (see below). This can serve as a prime satellite mission requirement, allowing for the others to be met by reduced mission requirements as appropriate. The suite of science requirements can be addressed with a spatial grid resolution of 100 km, observed monthly with a 0.1‰ error (or 1 part in 10 000).

Resolution and Error Sources

Model function Figure 2 shows that the dynamic range of T_B is about 4 K over the range of typical open-ocean surface salinity and temperature conditions. T_B gradients are greater with respect to salinity than to temperature. At a given temperature, T_B decreases as salinity increases, whereas the tendency with respect to temperature changes sign. The differential of T_B with respect to salinity ranges from -0.2 to -0.7 K per ‰. Corrected T_B will need to be measured to 0.02–0.07 K precision to achieve 0.1‰ resolution. The sensitivity is strongly affected by temperature, being largest at the highest temperatures and yielding better measurement precision in warm versus cold ocean conditions. Random error can be reduced by temporal and spatial averaging. The degraded measurement precision in higher latitudes will be somewhat compensated by the greater sampling frequency from a polar orbiting satellite.

The T_B variation with respect to temperature falls generally between ± 0.15 K $^\circ$C^{-1} and near zero over a broad S and T range. Knowledge of the surface temperature to within a few tenths of a degree Celsius will be adequate to correct T_B for temperature effects and can be obtained using data from other satellite systems. T_B values for the H and V polarizations have large variations with incidence angle and spacecraft attitude will need to be monitored very precisely.

Other errors Several other error sources will bias T_B measurements and must be either corrected or avoided. These include ionosphere and atmosphere effects, cosmic and galactic background radiations,

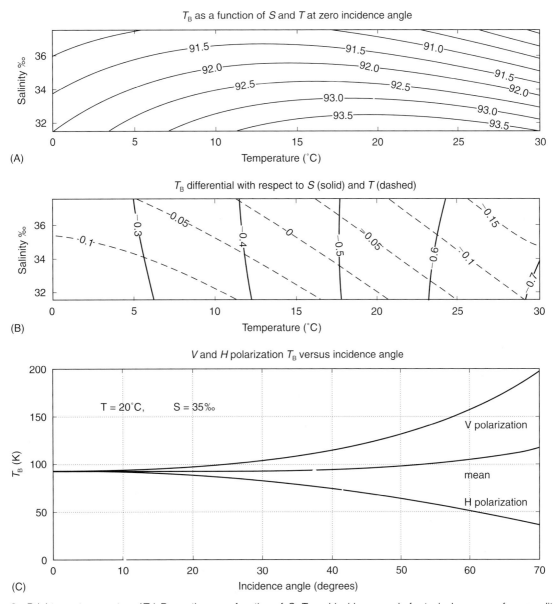

Figure 2 Brightness temperature (T_B) Properties as a function of S, T, and incidence angle for typical ocean surface conditions. (A) T_B contours. (B) T_B derivatives relative to S (solid curves) and T (dashed curves). (C) T_B variation versus incidence angle for H and V polarization. (Calculations based on formulas in Klein and Swift, 1977.)

surface roughness from winds, sun glint, and rain effects. Cosmic background and lower atmospheric adsorption are nearly constant biases easily corrected. An additional correction will be needed when the reflected radiation from the galactic core is in the field of view. The ionosphere and surface winds (roughness) have wide spatial and temporal variations and require ancillary data and careful treatment to avoid T_B errors of several K.

The ionosphere affects the measurement through attenuation and through Faraday rotation of the H and V polarized signal. There is no Faraday effect when viewing at nadir ($\theta = 0$) because H and

V emissivities are identical, whereas off-nadir corrections will be needed to preserve the polarization signal. Correction data can be obtained from ionosphere models and analyses but may be limited by unpredictable short-term ionosphere variations. Onboard correction techniques have been developed that require fully polarimetric radiometer measurements from which the Faraday rotation may be derived. Sun-synchronous orbits can be selected that minimize the daytime peak in ionosphere activity as well as sun glint off the surface.

The magnitude of the wind roughness correction varies with incidence angle and polarization, and ranges between 0.1 and $0.4\,K/(ms^{-1})$. Sea state conditions can change significantly within the few hours that may elapse until ancillary measurements are obtained from another satellite. Simultaneous wind roughness measurement can be made with an onboard radar backscatter sensor. A more accurate correction can be applied using a direct relationship between the radar backscatter and the T_B response rather than relying on wind or sea state information from other sensors.

Microwave attenuation by rain depends on rain rate and the thickness of the rain layer in the atmosphere. The effect is small at the intended microwave frequency, but for the required accuracy the effect must be either modeled and corrected with ancillary data, or the contaminated data discarded. For the accumulation of all the errors described here, it is anticipated that the root sum square will be reduced to 0.1‰ with adequate radiometer engineering, correction models, onboard measurements, ancillary data, and spatio-temporal filtering with methods now in development.

See also

Abrupt Climate Change. Aircraft Remote Sensing. Data Assimilation in Models. El Niño Southern Oscillation (ENSO). El Niño Southern Oscillation (ENSO) Models. Freshwater Transport and Climate. Ocean Circulation. Open Ocean Convection. Primary Production Distribution. Satellite Oceanography, History and Introductory Concepts. Satellite Altimetry. Satellite Passive Microwave Measurements of Sea Ice. Satellite Remote Sensing Microwave Scatterometers. Satellite Remote Sensing of Sea Surface Temperatures. Thermohaline Circulation. Upper Ocean Heat and Freshwater Budgets. Upper Ocean Mixing Processes. Upper Ocean Time and Space Variability. Water Types and Water Masses.

Further Reading

Blume H-JC, Kendall BM and Fedors JC (1978) Measurement of ocean temperature and salinity via microwave radiometry. *Boundary-Layer Meteorology* 13: 295–380.

Blume H-JC, Kendall BM and Fedors JC (1981) Multifrequency radiometer detection of submarine freshwater sources along the Puerto Rican coastline. *Journal of Geophysical Research* 86: 5283–5291.

Blume HC and Kendall BM (1982) Passive microwave measurements of temperature and salinity in coastal zones. *IEEE Transactions Geoscience and Remote Sensing* GE: 394–404.

Broecker WS (1991) The great ocean conveyer. *Oceanography* 4: 79–89.

Delcroix T and Henin C (1991) Seasonal and interannual variations of sea surface salinity in the tropical Pacific Ocean. *Journal of Geophysical Research* 96: 22135–22150.

Delworth T, Manabe S and Stouffer RJ (1993) Interdecadal variations of the thermohaline circulation in a coupled ocean–atmosphere model. *Journal of Climate* 6: 1993–2011.

Dickson RR, Meincke R, Malmberg S-A and Lee JJ (1988) The 'Great Salinity Anomaly' in the Northern North Atlantic, 1968–1982. *Progress in Oceanography* 20: 103–151.

Droppelman JD, Mennella RA and Evans DE (1970) An airborne measurement of the salinity variations of the Mississippi River outflow. *Journal of Geophysical Research* 75: 5909–5913.

Kendall BM and Blanton JO (1981) Microwave radiometer measurement of tidally induced salinity changes off the Georgia coast. *Journal of Geophysical Research* 86: 6435–6441.

Klein LA and Swift CT (1977) An improved model for the dielectric constant of sea water at microwave frequencies. *IEEE Transactions Antennas and Propagation* AP-25(#1): 104–111.

Lagerloef G, Swift C and LeVine D (1995) Sea surface salinity: the next remote sensing challenge. *Oceanography* 8: 44–50.

Lagerloef GSE (2000) Recent progress toward satellite measurements of the global sea surface salinity field. In: Halpern D (ed.) *Satellites, Oceanography and Society*, pp. 309–319. Elsevier Science.

Lerner RM and Hollinger JP (1977) Analysis of 1.4 GHz radiometric measurements from Skylab. *Remote Sensing Environment* 6: 251–269.

Le Vine DM, Kao M, Tanner AB, Swift CT and Griffis A (1990) Initial results in the development of a synthetic aperture radiometer. *IEEE Transactions Geoscience and Remote Sensing* 28: 614–619.

Miller J, Goodberlet M and Zaitzeff J (1998) Airborne salinity mapper makes debut in coastal zone. *EOS Transactions American Geophysical Union* 79: 173, 176–177.

Reynolds R, Ji M and Leetmaa A (1998) Use of salinity to improve ocean modeling. *Physics and Chemistry of the Earth* 23: 545–555.

Swift CT and McIntosh RE (1983) Considerations for microwave remote sensing of ocean-surface salinity. *IEEE Transactions Geoscience Remote Sensing* GE-21: 480–491.

Webster P (1994) The role of hydrological processes in ocean–atmosphere interactions. *Reviews of Geophysics* 32: 427–476.

SATELLITE OCEANOGRAPHY, HISTORY AND INTRODUCTORY CONCEPTS

S. Wilson, NOAA, HCHB 5224, Washington, DC, USA
J. R. Apel, Global Ocean Associates, Silver Spring, MD, USA
E. J. Lindstrom, NASA Code YS, Washington, DC, USA

doi:10.1006/rwos.2001.0335

Oceanography from a satellite – the words themselves sound incongruous and, to a generation of scientists accustomed to Nansen bottles and reversing thermometers, the idea may seem absurd.

Gifford C. Ewing, 1965

A Story of Two Communities

The history of oceanography from space is a story of the coming together of two communities – satellite remote sensing and traditional oceanography.

For over a century oceanographers have gone to sea in ships, learning how to sample beneath the surface, making detailed observations of the vertical distribution of properties. Giff Ewing noted that oceanographers had been forced to consider 'the class of problems that derive from the vertical distribution of properties at stations widely separated in space and time.'

With the introduction of satellite remote sensing in the 1970s, traditional oceanographers were provided with a new tool to collect synoptic observations of conditions at or near the surface of the global ocean. Since that time, there has been dramatic progress; satellites are revolutionizing oceanography. (The Appendix to this article provides a brief overview of the principles of satellite remote sensing.)

Yet much remains to be done. Traditional subsurface observations and satellite-derived observations of the sea surface – collected as an integrated set of observations and combined with state-of-the-art models – have the potential to yield estimates of the three-dimensional, time-varying distribution of properties for the global ocean. Neither a satellite nor an *in situ* observing system can do this on its own. Furthermore, if such observations can be collected over the long term, they can provide oceanographers with an observational capability conceptually similar to that which meteorologists use on a daily basis to forecast atmospheric weather.

Our ability to understand and forecast oceanic variability, how the oceans and atmosphere interact, critically depends on an ability to observe the three-dimensional global oceans on a long-term basis. Indeed, the increasing recognition of the role of the ocean in weather and climate variability compels us to implement an integrated, operational satellite and in situ observing system for the ocean now – so that it may complement the system which already exists for the atmosphere.

The Early Era

The origins of satellite oceanography can be traced back to World War II – radar, photogrammetry, and the V-2 rocket. By the early 1960s a few scientists had recognized the possibility of deriving useful oceanic information from the existing aerial sensors. These included (1) the polar-orbiting meteorological satellites, especially in the 10–12-μm thermal infrared band, and (2) color photography taken by astronauts in the Mercury, Gemini, and Apollo manned spaceflight programs. Examples of the kinds of data obtained from NASA flights collected in the 1960s are shown in **Figures 1** and **2**.

Such early imagery held the promise of deriving interesting and useful oceanic information from space, and led to three important conferences on space oceanography during the same time period. In 1964, NASA sponsored a conference at the Woods Hole Oceanographic Institution (WHOI) to examine the possibilities of conducting scientific research from space. The report from the conference, entitled *Oceanography from Space* (Ewing, 1965), summarized findings to that time; it clearly helped to stimulate a number of NASA projects in ocean observations and in sensor development. Moreover, with the exception of the synthetic aperture radar, all instruments flown through the 1980s used techniques described in this report. Dr Ewing has since become justifiably regarded as the father of oceanography from space.

A second important step occurred in 1969 when the 'Williamstown Conference' was held at Williams College in Massachusetts. The ensuing report (Kaula, 1969) set forth the possibilities for a space-based geodesy mission to determine the equipotential figure of the Earth using a combination of (a) accurate tracking of satellites and (b) the precision measurement of satellite elevation above the sea surface using radar altimeters. Dr William Von Arx of WHOI realized the possibilities for determining

Figure 1 Thermal infrared image of the US south-east coast showing warmer waters of the Gulf Stream and cooler slope waters closer to shore taken in the early 1960s. While the resolution and accuracy of the TV on Tiros were not ideal, they were sufficient to convince oceanographers of the potential usefulness of infrared imagery. The AVHRR scanner (see text) has improved images considerably. (Figure courtesy of NASA.)

large-scale oceanic currents with precision altimeters in space. The requirements for measurement precision of 10 cm height error in the elevation of the sea surface with respect to the geoid was articulated. NASA scientists and engineers felt that such accuracy could be achieved in the long run, and the agency initiated the 'Earth and Ocean Physics Applications Program,' the first formal oceans-oriented program to be established within the organization. The required accuracy was not to be realized until 1992 with TOPEX/Poseidon, which was reached only over a 25-year period of incremental progress that saw the flights of five US altimetric satellites of steadily increasing capabilities: Skylab, Geos-3, Seasat, Geosat, and TOPEX/Poseidon. (See **Figure 3** for representative satellites.)

A third conference, focused on sea surface topography from space, was convened by the National Oceanic and Atmospheric Administration (NOAA), NASA, and the US Navy in Miami in 1972, with 'sea surface topography' being defined as undulations of the ocean surface with scales ranging from approximately 5000 km down to 1 cm. The conference identified several data requirements in oceanography that could be addressed with space-based radar and radiometers. These included determination of surface currents, Earth and ocean tides,

the shape of the marine geoid, wind velocity, wave refraction patterns and spectra, and wave height. The conference established a broad scientific justification for space-based radar and microwave radiometers, and it helped to shape subsequent national programs in space oceanography.

The First Generation

Two first-generation ocean-viewing satellites, Skylab in 1973 and Geos-3 in 1975, had partially responded to concepts resulting from the first two of these conferences. Skylab carried not only several astronauts, but a series of sensors that included the S-193, a radar-altimeter/wind-scatterometer, a long-wavelength microwave radiometer, a visible/infrared scanner, and cameras. S-193, the so-called Rad/Scatt, was advanced by Drs Richard Moore and Willard Pierson. These scientists held that the scatterometer could return wind velocity measurements whose accuracy, density, and frequency would revolutionize marine meteorology. Later aircraft data gathered by NASA showed that there was merit to their assertions. Skylab's scatterometer was damaged during the opening of the solar cell panels and as a consequence, returned indeterminate results (except for passage over a hurricane), but the altimeter made observations of the geoid anomaly due to the Puerto Rico Trench.

Geos-3 was a small satellite carrying a dual-pulse radar altimeter whose mission was to improve the knowledge of the Earth's marine geoid, and coincidentally to determine the height of ocean waves via the broadening of the short transmitted radar pulse upon reflection from the rough sea surface. Before the end of its 4 year lifetime, Geos-3 was returning routine wave height measurements to the National Weather Service for inclusion in its Marine Waves Forecast. Altimetry from space had become a clear possibility, with practical uses of the sensor immediately forthcoming. The successes of Skylab and Geos-3 reinforced the case for a second generation of radar-bearing satellites to follow.

The meteorological satellite program also provided measurements of sea surface temperature using far-infrared sensors, such as the Visible and Infrared Scanning Radiometer (VISR), which operated at wavelengths near 10 μm, the portion of the terrestrial spectrum wherein thermal radiation at terrestrial temperatures is at its peak, and where coincidentally the atmosphere has a broad passband. The coarse, 5 km resolution of the VISR gave blurred temperature images of the sea, but the promise was clearly there. **Figure 1** is an early 1960s TV image of the south-eastern USA taken by the

Figure 2 Color photograph of the North Carolina barrier islands taken during the Apollo-Soyuz Mission (AS9-20-3128). Capes Hatteras and Lookout, shoals, sediment- and chlorophyll-bearing flows emanating from the coastal inlets are visible, and to the right, the blue waters of the Gulf Stream. Cloud streets developing offshore of the warm current suggest that a recent passage of a cold polar front has occurred, with elevated air–sea evaporative fluxes. Later instruments, such as the Coastal Zone Color Scanner (CZCS) on Nimbus-7 and the SeaWiFS imager have advanced the state of the art considerably. (Figure courtesy of NASA.)

NASA TIROS program, showing the Gulf Stream as a dark signal. While doubts were initially held by some oceanographers as to whether such data actually represented the Gulf Stream, nevertheless the repeatability of the phenomenon, the verisimilitude of the positions and temperatures with respect to conventional wisdom, and their own objective judgment finally convinced most workers of the validity of the data. Today, higher resolution, temperature-calibrated infrared imagery constitutes a valuable data source used frequently by ocean scientists around the world.

During the same period, spacecraft and aircraft programs taking ocean color imagery were delineating the possibilities and difficulties of determining sediment and chlorophyll concentrations remotely. **Figure 2** is a color photograph of the North Carolina barrier islands taken with a hand-held camera, with Cape Hatteras in the center. Shoals, sediment- and chlorophyll-bearing flows emanating from the coastal inlets are visible, and to the right, the blue waters of the Gulf Stream. Cloud streets

developing offshore of the warm Stream suggest a recent passage of a cold polar front and attendant increases in air–sea evaporative fluxes.

The Second Generation

The combination of the early data and advances in scientific understanding that permitted the exploitation of those data resulted in spacecraft sensors explicitly designed to look at the sea surface. Information returned from altimeters and microwave radiometers gave credence and impetus to dedicated microwave spacecraft. Color measurements of the sea made from aircraft had indicated the efficacy of optical sensors for measurement of near-surface chlorophyll concentrations. Infrared radiometers returned useful sea surface temperature measurements. These diverse capabilities came together when, during a 4 month interval in 1978, the USA launched a triad of spacecraft that would profoundly change the way ocean scientists would observe the sea in the future. On June 26, the first dedicated

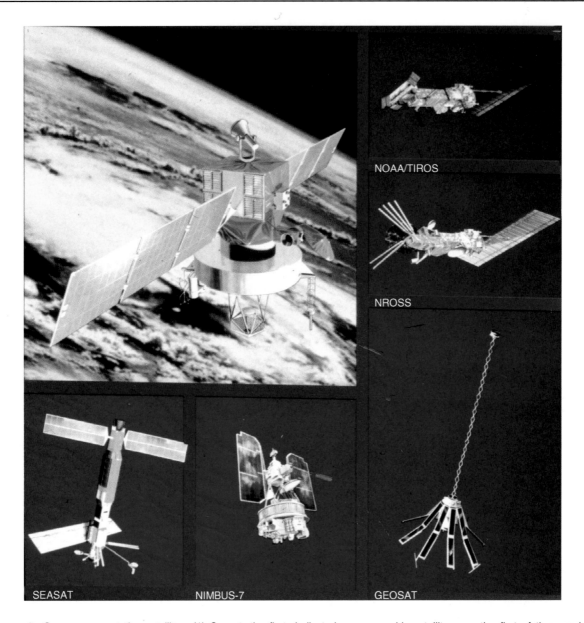

Figure 3 Some representative satellites: (1) Seasat, the first dedicated oceanographic satellite, was the first of three major launches in 1978; (2) the Tiros series of operational meteorological satellites carried the AVHRR surface temperature sensor; Tiros-N, the first of this series, was the second major launch in 1978; (3) Nimbus-7, carrying the CZCS color scanner, was the third major launch in 1978; (4) NROSS, an oceanographic satellite approved as an operational demonstration in 1985, was later cancelled; (5) Geosat, an operational altimetric satellite, was launched in 1985; and (6) this early version of TOPEX was reconfigured to include the French Poseidon; the joint mission, TOPEX/Poseidon, was launched in 1992. (Figure courtesy of NASA.)

oceanographic satellite, Seasat, was launched; on October 13, Tiros-N was launched immediately after the catastrophic failure of Seasat on October 10; and on October 24, Nimbus-7 was lofted. Collectively they carried sensor suites whose capabilities covered virtually all known ways of observing the oceans remotely from space.

This second generation of satellites would prove to be extraordinarily successful. They returned data that vindicated their proponents' positions on the measurement capabilities and utility, and they set the direction for almost all subsequent efforts in satellite oceanography.

In spite of its very short life of 99 days, Seasat demonstrated the great utility of altimetry by measuring the marine geoid to within a very few meters, by inferring the variability of large-scale ocean surface currents, and by determining wave heights. The wind scatterometer could yield oceanic surface wind velocities equivalent to 20 000 ship observations per day. The scanning multifrequency radiometer also provided wind speed and atmospheric water content data; and the synthetic aperture radar penetrated clouds to show features on the surface of the sea, including surface and internal waves, current boundaries, upwellings, and rainfall patterns. All of these measurements could be extended to basin-wide scales, allowing oceanographers a view of the sea never dreamed of before. Seasat stimulated several subsequent generations of ocean-viewing satellites, which outline the chronologies and heritage for the world's ocean-viewing spacecraft. Similarly, the early temperature and color observations have led to successor programs that provide large quantities of quantitative data to oceanographers around the world.

The Third Generation

The second generation of spacecraft would demonstrate that variables of importance to oceanography could be observed from space with scientifically useful accuracy. As such, they would be characterized as successful concept demonstrations. And while both first and second generation spacecraft had been exclusively US, international participation in demonstrating the utility of their data would lead to the entry of Canada, the European Space Agency (ESA), France, and Japan into the satellite program during this period. This, paper, however, will focus on the US effort.[1]

Partnership with Oceanography

Up to 1978, the remote sensing community had been the prime driver of oceanography from space and there were overly optimistic expectations. Indeed, the case had not yet been made that these observational techniques were ready to be exploited for ocean science. Consequently, in early 1979 the central task was establishing a partnership with the traditional oceanographic community. This meant involving them in the process of evaluating the performance of Seasat and Nimbus-7, as well as building an ocean science program at NASA

---

[1] Additional background on US third generation missions covering the period from 1980 through 1987 can be found in the series of Annual Reports for the Oceans Program: NASA Technical Memoranda 80233, 84467, 85632, 86248, 87565, 88987, and 4025. For information on missions in other countries, see Further Reading: Kawamura (2000) for Japan; Minster and Lefebvre (1997) for France; Guymer et al. (2001) for the UK; and Victorov (1996) and Cherny and Raizer (1998) for Russia, Ukraine, and the former Soviet Union.

Headquarters to complement the on-going remote sensing effort.

National Oceanographic Satellite System

This partnership with the oceanographic community was lacking in a notable and early false start on the part of NASA, Navy, and NOAA – the National Oceanographic Satellite System (NOSS). This was to be an operational system, with a primary and a back-up satellite, along with a fully redundant ground data system. NOSS was proposed shortly after the failure of Seasat, with a first launch expected in 1986. NASA formed a Science Working Group in 1980 under Francis Bretherton to define the potential that NOSS offered the oceanographic community, as well as to recommend sensors to constitute the 25% of its payload allocated for research. However, with oceanographers essentially brought in as junior partners, the job of securing a new start for NOSS fell to the operational community – which it proved unable to do. NOSS was canceled in early 1981. The prevailing and realistic view was that the greater community was not ready to implement such an operational system.

Science Working Groups

During this period, Science Working Groups (SWGs) were formed to look at each promising satellite sensing technique, assess its potential contribution to oceanographic research, and define the requirements for its future flight. The notable early groups were the TOPEX SWG formed in 1980 under Carl Wunsch for altimetry, Satellite Surface Stress SWG in 1981 under James O'Brien for scatterometry, and Satellite Ocean Color SWG in 1981 under John Walsh for color scanners. These SWGs were true partnerships between the remote sensing and oceanographic communities, developing consensus for what would become the third generation of satellites.

Partnership with Field Centers

Up to this time, NASA's Oceans Program had been a collection of relatively autonomous, in-house activities run by NASA Field Centers. In 1981 an overrun in the Space Shuttle program forced a significant budget cut at NASA Headquarters, including the Oceans Program. This in turn forced a re-prioritization and refocusing of NASA programs. This was a blessing in disguise, as it provided an opportunity to initiate a comprehensive, centrally led program – which would ultimately result in significant funding for the oceanographic, as well as remote sensing communities. Outstanding relationships with individuals like Mous Chahine in senior management at the Jet Propulsion Laboratory (JPL) enabled the partnership between NASA Headquarters and the two prime ocean-related Field Centers (JPL and the Goddard Space Flight Center) to flourish.

Partnerships in Implementation

A milestone policy-level meeting occurred on July 13, 1982 when James Beggs, then Administrator of NASA, hosted a meeting of the Ocean Principals Group – an informal group of leaders of the ocean-related agencies. A NASA presentation on opportunities and prospects for oceanography from space was received with much enthusiasm. However, when asked how NASA intended to proceed, Beggs told the group that – while NASA was the sole funding agency for space science and its missions – numerous agencies were involved in and support oceanography. Beggs said that NASA was willing to work with other agencies to implement an ocean satellite program, but that it would not do so on its own. Beggs' statement defined the approach to be pursued in implementing oceanography from space, namely, a joint approach based on partnerships.

Research Strategy for the Decade

As a further step in strengthening its partnership with the oceanographic community, NASA collaborated with the Joint Oceanographic Institutions Incorporated (JOI), a consortium of the oceanographic institutions with a deep-sea-going capability. At the time, JOI was the only organization in a position to represent and speak for the major academic oceanographic institutions. A JOI Satellite Planning Committee (1984) under Jim Baker examined SWG reports, as well as the potential synergy between the variety of oceanic variables which could be measured from space; this led to the idea of understanding the ocean as a system. (From this, it was a small leap to understanding the Earth as a system, the goal of NASA's Earth Observing System.)

The report of this Committee, *Oceanography from Space: A Research Strategy for the Decade, 1985–1995*, linked altimetry, scatterometry, and ocean color with the major global ocean research programs being planned at that time – the World Ocean Circulation Experiment (WOCE), Tropical Ocean Global Atmosphere program (TOGA), and Joint Global Ocean Flux Study (JGOFS). This strategy, still being followed today, served as a catalyst to engage the greater community, to identify the most important missions, and to develop an approach for their prioritization. Altimetry,

scatterometry, and ocean color emerged from this process as national priorities.

Promotion and Advocacy

The *Research Strategy* also provided a basis for promoting and building an advocacy for the NASA program. If requisite funding was to be secured to pay for proposed missions, it was critical that government policy makers, the Congress, the greater oceanographic community, and the public had a good understanding of oceanography from space and its potential benefits. In response to this need, a set of posters, brochures, folders, and slide sets was designed by Payson Stevens of Internetwork Incorporated and distributed to a mailing list which grew to exceed 3000. These award-winning materials – sharing a common recognizable identity – were both scientifically accurate and esthetically pleasing.

At the same time, dedicated issues of magazines and journals were prepared by the community of involved researchers. The first example was the issue of *Oceanus* (Wilson, 1981) which presented results from the second generation missions and represented a first step toward educating the greater oceanographic community in a scientifically useful and balanced way about realistic prospects for satellite oceanography.

Implementation Studies

Given the SWG reports taken in the context of the *Research Strategy*, the NASA effort focused on the following sensor systems. Listed with each are the various flight opportunities which were studied.

- Altimetry – the flight of a dedicated altimeter mission, first TOPEX as a NASA mission, and then TOPEX/Poseidon jointly with the French Centre Nationale d'Etudes Spatiales (CNES).
- Scatterometry – the flight of a NASA scatterometer (NSCAT), first on NOSS, then on the Navy Remote Ocean Observing Satellite (NROSS), and finally on the Advanced Earth Observing Satellite (ADEOS) of the Japanese National Space Development Agency (NASDA).
- Visible radiometry – the flight of a NASA color scanner on a succession of missions (NOSS, NOAA-H/-I, SPOT-3 (Systeme Pour l'Observation de la Terre), and Landsat-6) and finally the purchase of ocean color data from the SeaWiFS sensor to be flown by the Orbital Sciences Corporation.
- Microwave radiometry – a system to utilize data from the series of SSMI microwave radiometers

to fly on the Defense Meteorological Satellite Program satellites.

- Synthetic aperture radar (SAR) – a NASA ground station, the Alaska SAR Facility, to enable direct reception of SAR data from the ERS-1/-2, JERS-1, and Radarsat satellites of the European Space Agency, NASDA, and the Canadian Space Agency, respectively.

New Starts

Using the results of the studies listed above, the Oceans Program entered the new start process at NASA Headquarters numerous times attempting to secure funds for implementation of elements of the third generation. TOPEX was first proposed as a NASA mission in 1980. However, considering limited prospects for success, partnerships were sought and the most promising was with the French. CNES initially proposed a mission using a SPOT bus with a US launch. However, NASA rejected this because SPOT, constrained to be sun synchronous, would alias solar tidal components. NASA proposed instead a mission using a US bus capable of flying in a non-sun-synchronous orbit with CNES providing an Ariane launch. The NASA proposal was accepted for study in Fiscal Year (FY) 1983, and a new start was finally secured for the combined TOPEX/Poseidon in FY 1987.

In 1982 when the Navy first proposed NROSS, NASA offered to be a partner and provide a scatterometer. The Navy and NASA obtained new starts for both NROSS and NSCAT in FY 1985. However, NROSS suffered from a lack of strong support within the Navy, experienced a number of delays, and was finally terminated in 1987. Even with this termination, NASA was able to keep NSCAT alive until establishing the partnership with NASDA for its flight on their ADEOS mission.

Securing a means to obtain ocean color observations as a follow-on to the Coastal Zone Color Scanner (CZCS) was a long and arduous process, finally coming to fruition in 1991 when a contract was signed with the Orbital Sciences Corporation (OSC) to purchase data from the flight of their SeaWiFS sensor. By that time, a new start had already been secured for NASA's Earth Observing System (EOS), and ample funds were available in that program for the SeaWiFS data purchase.

Finally, securing support for the Alaska SAR Facility was straightforward; being small in comparison with the cost of flying space hardware, its funding had simply been included in the new start that NSCAT obtained in FY 1985. Also funding for utilization of SSMI data was small enough to be covered by the Oceans Program itself.

Implementing the Third Generation

With the exception of the Navy's Geosat, these third generation missions would take a very long time to come into being. As seen in **Figure 5**, TOPEX/Poseidon was launched in 1992 – 14 years after Seasat; NSCAT was launched on ADEOS in 1996 – 18 years after Seasat; and SeaWiFS was launched in 1997 – 19 years after Nimbus-7. In fact, these missions came so late that they had limited overlap with the field phases of the major ocean research programs (WOCE, TOGA, and JGOFS) they were to complement. Why did it take so long?

Understanding and Consensus

First, it took time to develop a physically unambiguous understanding of how well the satellite sensors actually performed, and this involved learning to cope with the data – satellite data rates being orders of magnitude larger than those encountered in traditional oceanography. For example, it was not until 3 years after the launch of Nimbus-7 that CZCS data could be processed as fast as collected by the satellite. And even with only a 3 month data set from Seasat, it took 4 years to produce the first global maps of variables such as those shown in **Figure 4**.

In evaluating the performance of both Seasat and Nimbus-7, it was necessary to have access to the data. Seasat had a free and open data policy; and after a very slow start, the experiment team concept (where team members had a lengthy period of exclusive access to the data) for the Nimbus-7 CZCS was replaced with that same policy. Given access to the data, delays were due to a combination of sorting out the algorithms for converting the satellite observations into variables of interest, as well as being constrained by having limited access to raw computing power.

In addition, the rationale for the third-generation missions represented a major paradigm shift. While earlier missions had been justified largely as demonstrations of remote sensing concepts, the third-generation missions would be justified on the basis of their potential contribution to oceanography. Hence, the long time it took to understand sensor performance translated into a delay in being able to convince traditional oceanographers that satellites were an important observational tool ready to be exploited for ocean science. As this case was made, it was possible to build consensus across the remote sensing and oceanographic communities.

Space Policy

Having such consensus reflected at the highest levels of government was another matter. The White

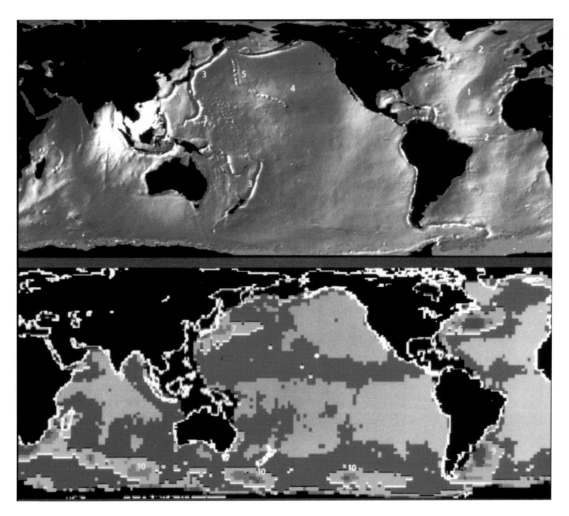

Figure 4 Global sea surface topography c. 1983. This figure shows results computed from the 70 days of Seasat altimeter data in 1978. Clearly visible in the mean sea surface topography, the marine geoid (upper panel), are the Mid-Atlantic Ridge (1) and associated fracture zones (2), trenches in the western Pacific (3), the Hawaiian Island chain (4), and the Emperor seamount chain (5). Superimposed on the mean surface is the time-varying sea surface topography, the mesoscale variability (lower panel), associated with the variability of the ocean currents. The largest deviations (10–25 cm), yellow and orange, are associated with the western boundary currents: Gulf Stream (6), Kuroshio (7), Agulhas (8), and Brazil/Falkland Confluence (9); large variations also occur in the West Wind Drift (10). (Figure courtesy of NASA.)

House Fact Sheet on US Civilian Space Policy of October 11, 1978 states, ' ... emphasizing space applications ... will bring important benefits to our understanding of earth resources, climate, weather, pollution ... and provide for the private sector to take an increasing responsibility in remote sensing and other applications.' Landsat was commercialized in 1979 as part of this space policy. As Robert Stewart explains, 'Clearly the mood at the presidential level was that earth remote sensing, including the oceans, was a practical space application more at home outside the scientific community. It took almost a decade to get an understanding at the policy level that scientific needs were also important, and that we did not have the scientific understanding necessary to launch an operational system

for climate.' The failures of NOSS, and later NROSS, were examples of an effort to link remote sensing directly with operational applications without the scientific underpinning.

The view in Europe was not dissimilar; governments felt that cost recovery was a viable financial scheme for ocean satellite missions, i.e. that the data have commercial value and the user would be willing to pay to help defray the cost of the missions.

Joint Satellite Missions

It is relatively straightforward to plan and implement missions within a single agency, as with NASA's space science program. However, implementing a satellite mission across different organizations, countries, and cultures is both challenging and time-consuming. An enormous amount of time and energy was invested in studies of various flight options, many of which fizzled out, but some were implemented. With the exception of the former Soviet Union, NASA's third-generation missions would be joint with each nation having a space program at that time, as well as with a private company.

The Geosat Exception

Geosat was the notable US exception, having been implemented so quickly after the second generation. It was approved in 1981 and launched in 1985 in order to address priority operational needs on the part of the US Navy. During the second half of its mission, data would become available within 1–2 days. As will be discussed below, Geosat shared a number of attributes with the meteorological satellites: it had a specific focus; it met priority operational needs for its user; experience was available for understanding and using the observations; and its implementation was done in the context of a single organization.

The Next Generation

In contrast to previous long delays, one only has to look at **Figure 5** to see that within the past few years ocean-related satellites are becoming more numerous, and the distinction between generations is getting blurred. In addition to TOPEX/Poseidon, there are altimeters on ERS-2, ENVISAT, and Jason-1; and CHAMP and GRACE are complementary gravity missions. In addition to Quikscat and ERS-2, there are scatterometers on ADEOS-2 and METOP-1; and in addition to SeaWiFS, there are color scanners on Terra, Aqua, ENVISAT, and ADEOS-2. With these observations, satellites will continue to revolutionize oceanography – not only further advancing our understanding of the ocean and how it interacts with the atmosphere, but also laying the basis for a long-term, routine ocean observing system.

The maturing science of oceanography sees the development of a suite of global oceanographic services being carried out in a manner similar to the development of weather services. The delivery of these services and their associated informational products will emerge as the result of the successes in ocean science (research push), as well as an increasing demand for ocean analyses and forecasts from a variety of sectors (user pull).

Integrated and Operational Observing Systems

The next generation of ocean remote sensing systems faces another major paradigm shift. From the research perspective, it is necessary to transition successfully demonstrated – experimental – observing techniques of the third generation into regular, long-term, systematic – operational – observing systems to meet a broad range of user requirements – while maintaining the capability to collect long-term, research-quality observations. From the operational perspective, it is necessary to implement proven, cost-effective observing systems capable of meeting specific societal needs, such as those associated with economic benefits or the protection of life and property. Meeting these sometimes competing, but quite complementary demands will be the challenge and legacy of the next generation of ocean remote sensing satellites.

An essential element of meeting the demands of both the research and a broader user community is stepping back from oceanography from space as a separate endeavor, and moving toward integrated observing systems. Such systems involve combinations of satellites and *in situ* instruments feeding observations into data processing systems capable of delivering a comprehensive view of one or more geophysical variables (sea level, surface temperature, winds, etc.).

Three examples help to illustrate the nature of integrated observing systems. First, the combination of the Jason-1 altimeter, its precision orbit determination system, and the suite of precision tide gauges around the globe is an integrated system which allows scientists to make an estimate of annual global sea level changes. Such information is critical for developing plans for our coastal zones. Second, global estimates of vector winds at the sea surface are produced from the Seawinds scatterometer on Quikscat, a global array of *in situ* surface buoys, and the Seawinds data processing system. Delivery of this product in real-time has significant potential to improve marine weather prediction. The third example concerns the Jason-1 altimeter together with the Argo global profiling float array. When combined in a sophisticated data assimilation system – using a state-of-the-art ocean model – these data enable the estimation of the physical state of the ocean as it changes through time. This information (the rudimentary weather map depicting the circulation of the oceans) is a critical component of climate models and provides the fundamental context for addressing a broad range of issues in chemical and biological oceanography.

Integrated observing systems serve the dual purpose of collecting the data needed as the foundation for the next generation of research in oceanography, while at the same time providing the product and customer focus needed for successfully establishing

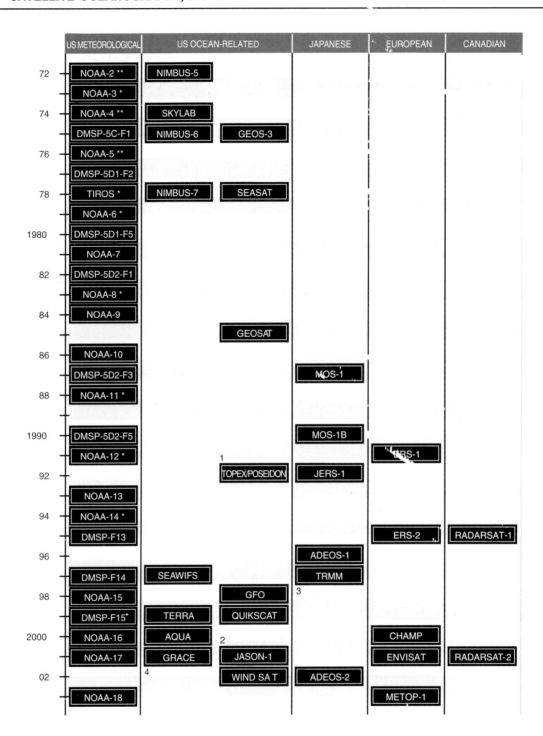

Figure 5 Approved US meteorological and international ocean-related satellite missions, arranged by the year of launch for the period 1972–2003. Column headings denote national sponsorship, except for the following joint missions: (1) US/France TOPEX/Poseidon, (2) France/US Jason-1, (3) Japan/US TRMM, and (4) US/German Grace. In addition, the Japanese ADEOS-1 included the US NSCAT in its sensor complement, and the US Aqua included the Japanese AMSR; the USA provided a launch for the Canadian Radarsat-1. The first column shows 38 polar-orbiting operational meteorological satellites launched from 1972 through 2000; each asterisk (*) in this column denotes an additional DMSP satellite launched during the year in question. For a chronology with mission lifetimes – including recent satellites of Brazil, China, India, Republic of Korea, and Taiwan – see Patzert and Van Woert (2000). Additionally, Masson (1991) presents a summary of missions, their payloads, how to access their data, and references.

operational oceanography. Thus, the next generation might most appropriately be characterized as oceanography from integrated observing systems.

Placing the next generation of ocean remote sensing satellites within the context of integrated observing systems necessitates new demands on space systems – long-term continuity of research-quality observations. These are needed to serve both operational oceanography – where the uninterrupted supply of real-time data is critical; and the research community – where long-term observations of subtle and slowly varying ocean phenomena are highly valued. This is a big challenge to be met by the space systems because it demands higher reliability and redundancy, while calling for stringent calibration and accuracy requirements. For example, the next generation of ocean altimetric satellites must incorporate the observations of ocean tides which vary on the order of a meter per day, along with accurate estimates of global sea level which varies on the order of a millimeter per year. The integrated system for winds requires the resolution of light, variable winds in climatically important regions like the western tropical Pacific, as well as high winds in hurricanes. Meeting these demands

and delivering the required products requires close cooperation between the research and operational, observational, and modeling communities.

The Meteorological Experience

Meteorologists have had a dramatically different experience than oceanographers with satellites, and it is useful to look at that history when considering ocean observing systems. With the launch in 1960 of the world's first meteorological satellite, the polar-orbiting Tiros-1 carrying two TV cameras, the value of the resulting imagery to the operational weather services was recognized immediately. The very next year a National Operational Meteorological System was implemented, with NASA to build and launch the satellites and the Weather Bureau to be the operator. The feasibility of using satellite imagery to locate and track tropical storms was soon demonstrated, and by 1969 this capability had become a regular part of operational weather forecasting. In 1985 Richard Hallgren, former Director of the US National Weather Service, stated that 'the use of satellite information simply permeates every aspect of the [forecast and warning] process and all this in a mere 25 years.'

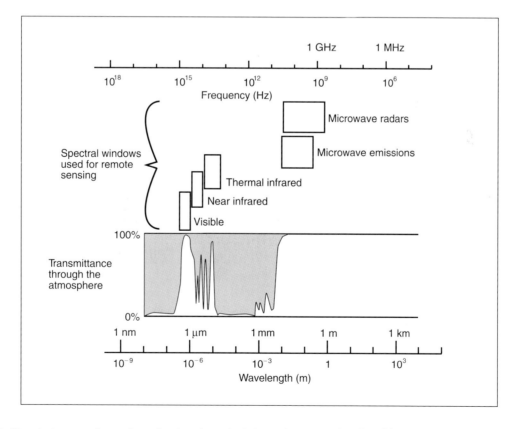

Figure 6 The electromagnetic spectrum showing atmospheric transmitance as a function of frequency and wavelength, along with the spectral windows used for remote sensing. Microwave bands are typically defined by frequency and the visible/infrared by wavelength. (After Robinson and Guymer, 1996.)

Since 1960, there has been a continuing series of 50 US operational, polar-orbiting satellites – 35 civilian and 15 military. If **Figure 5** were to show these satellites, it would have to begin in 1960 and would show slightly more than one satellite every year! Contrast that with the 16 US ocean-related satellites which have flown within the past three decades.

Why the dramatically different experience?

The first meteorological satellites had a specific focus – synoptic meteorology and weather forecasting. Initial image interpretation was straightforward (i.e. physically unambiguous), and there was a demonstrated value of observations to meet a societal need. Indeed, since 1960 satellites have ensured that no hurricane has gone undetected. In addition, the coupling between meteorology and remote sensing started very early. An institutional mechanism for transition from research to operations was established almost immediately. Finally, recognition of this endeavor extended to the highest levels of government, resulting in the financial commitment needed to enable success.

The challenge for oceanography Similar attributes are needed with regard to the ocean. What is the specific focus of the proposed long-term observing system? Koblinsky and Smith (2001) outline a growing international consensus for one such focus and the associated observational requirements. What is the demonstrated value of the resulting observations in terms of meeting a specific societal need? Addressing this question will help ensure an equivalent user pull to complement the research push. And unlike meteorology where there is a National Weather Service in each country to provide an institutional focus, ocean-observing systems have multiple user institutions whose interests must be reconciled. In the US, the dozen agencies with ocean-related responsibilities are using the National Oceanographic Partnership Program to provide a focus for reconciling such interests.

In 1994 – 34 years after the launch of TIROS-1 – a decision was made by President Clinton to merge separate civilian and military systems into the National Polar-orbiting Operational Environmental Satellite System (NPOESS) with its first satellite to

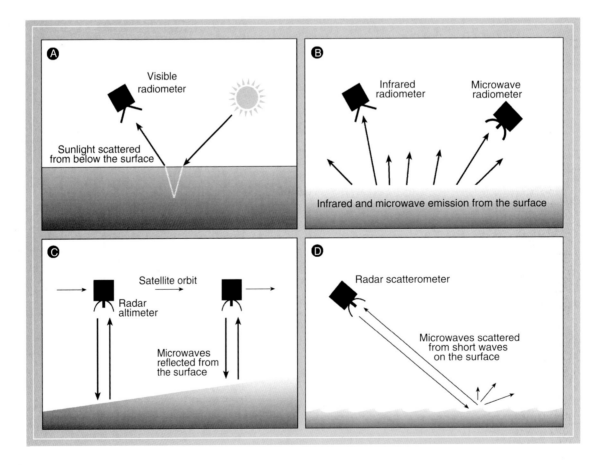

Figure 7 Four techniques for making oceanic observations from satellites: (A) visible radiometry, (B) infrared and microwave radiometry, (C) altimetry, and (D) scatterometry. (After Robinson and Guymer, 1996.)

be launched in ~ 2008. NPOESS will be an operational and an environmental satellite system, offering significant potential not just for the atmosphere, but also for the ocean and land, and not just for weather, but also for climate. As such, NPOESS is a target for transitioning a broad range of remote sensing capabilities to an operational footing. At the time of writing, the extent to which NPOESS will incorporate specific observational capabilities (including ocean surface topography, surface vector winds, and ocean color) is yet to be determined. Factors to be considered in addressing this issue of its mission configuration include: the demands individual sensors place on the satellite platform, the cost of those sensors, and the demonstrated value of resulting observations to meet specific societal needs.

In a major policy speech delivered to the American Geophysical Union on December 6, 1998, NASA Administrator Dan Goldin said NASA's

'role is to push the leading edge of remote sensing science and technology. We have an important but limited role in getting the benefits of new Earth science understanding into the hands of those who can make practical use of it … The next link in the chain is the operational satellite systems, those that can be counted on over the long term … it has become clear that the nation and the world needs an operational ocean observing system to pair with the atmospheric one now extant … NASA has proven the value and achievability of ocean topography, ocean color, ocean surface wind…measurements. The nation must have a plan to supply these and the corresponding in situ measurements on an operational basis.'

The paper by the Ocean Theme Team (2001) prepared under the auspices of the Integrated Global Observing Strategy Partnership www.igospartners.org represents how the space-faring nations are proceeding in this direction. IGOS partners include the major global research program sponsors, global observing systems, space agencies, and international organizations. Ongoing discussions in this forum are allowing for improved strategic planning and optimal use of resources in building a Global Ocean Observing System that will truly integrate space capabilities, *in situ* systems, and deliver the needed products to the greater user community.

Appendix: A Brief Overview of Satellite Remote Sensing

Unlike the severe attenuation in the sea, the atmosphere has 'windows' in which certain electro-magnetic (EM) signals are able to propagate. These windows, depicted in **Figure 6**, are defined in terms of atmospheric transmittance – the percentage of an EM signal which is able to propagate through the atmosphere – expressed as a function of wavelength or frequency.

Given a sensor onboard a satellite observing the ocean, it is necessary to understand and remove the effects of the atmosphere (such as scattering and attenuation) as the EM signal propagates through it. For passive sensors (**Figure 7(A)** and **(B)**), it is then possible to relate the EM signals collected by the

	PASSIVE SENSORS (RADIOMETERS)			ACTIVE SENSORS (MICROWAVE RADARS)		
SENSOR TYPE	VISIBLE	INFRARED	MICROWAVE	ALTIMETRY	SCATTEROMETRY	SYNTHETIC APERTURE RADAR
MEASURED PHYSICAL VARIABLE	Solar radiation backscattered from beneath the sea surface	Infrared emission from the sea surface	Microwave emission from the sea surface	Travel time, shape, and strength of reflected pulse	Strength of return pulse when illuminated from different directions	Strength and phase of return pulse
APPLICATIONS	Ocean color; chlorophyll; primary production; water clarity; shallow-water bathymetry	Surface temperature; ice cover	Ice cover, age and motion; sea surface temperature; wind speed	Surface topography for geostrophic currents and tides; bathymetry; oceanic geoid; wind and wave conditions	Surface vector winds; ice cover	Surface roughness at fine spatial scales; surface and internal wave patterns; bathymetric patterns; ice cover and motion

Figure 8 Measured physical variables and applications for both passive and active sensors, expressed as a function of sensor type.

sensor to the associated signals at the bottom of the atmosphere, i.e. the natural radiation emitted or reflected from the sea surface. Note that passive sensors in the visible band are dependent on the sun for natural illumination.

Active sensors, microwave radar (**Figure 7(C)** and **(D)**), provide their own source of illumination and have the capability to penetrate clouds, and to a certain extent, rain. Atmospheric correction must be done to remove effects for a round trip from the satellite to the sea surface.

With atmospheric corrections made, measurements of physical variables are available: emitted radiation for passive sensors, and the strength, phase, and/or travel time for active sensors. **Figure 8** shows typical measured physical variables for both types of sensors in their respective spectral bands, as well as applications or derived variables of interest – ocean color, surface temperature, ice cover, sea level, and surface winds. The companion articles on this topic address various aspects of **Figure 8** in more detail, so only this general overview is given here. (See also Further Reading: Robinson and Guymer (1996), Fu *et al.* (1990). Committee on Earth Sciences (1995) provides an overview of ocean-related satellites in the context of the Earth sciences.)

Acknowledgements

We would like to acknowledge contributions to this article from: Mary Cleave, Murel Cole, William Emery, Michael Freilich, Lee Fu, Rich Gasparovic, Trevor Guymer, Tony Hollingsworth, Hiroshi Kawamura, Michele Lefebvre, Jean-François Minster, Richard Moore, William Patzert, Willard Pierson, Jim Purdom, Keith Raney, Payson Stevens, Robert Stewart, Ted Strub, Tasuku Tanaka, William Townsend, Mike Van Woert, and Frank Wentz.

See also

Aircraft Remote Sensing. History of Ocean Sciences. IR Radiometers. Ocean Color from Satellites. Satellite Altimetry. Satellite Measurements of Salinity. Satellite Passive Microwave Measurements of Sea Ice. Satellite Remote Sensing Microwave Scatterometers. Satellite Remote Sensing SAR. Satellite Remote Sensing of Sea Surface Temperatures. Upper Ocean Time and Space Variability.

Further Reading

Apel JR (ed.) (1972) *Sea Surface Topography from Space*, vols 1 and 2. NOAA Technical Reports: ERL No. 228, AOML No. 7, Boulder: NOAA.

Cherny IV and Raizer VY (1998) *Passive Microwave Remote Sensing of Oceans*. Chichester: Praxis.

Committee on Earth Sciences (1995) *Earth Observations from Space: History, Promise, and Reality*. Washington: Space Studies Board, National Research Council.

Ewing GC (1965) *Oceanography from Space*. Proceedings of a Conference held in Woods Hole, August 24–28 1964, Woods Hole Oceanographic Institution Ref. No. 65–10.

Fu L-L, Liu WT and Abbott MR (1990) Satellite remote sensing of the ocean. In: *The Sea: Ocean Engineering Science*, vol. 9, pp. 1193–1236. New York: John Wiley & Sons.

Guymer TH, Challenor PG and Srokosz MA (2001) Oceanography from space: past success, future challenge. In: Deacon M *et al.* (eds) *Understanding the Oceans. A Century of Ocean Exploration*. London: UCL Press.

JOI Satellite Planning Committee (1984) *Oceanography from Space: A Research Strategy for the Decade, 1985–1995*, parts 1 and 2. Washington: Joint Oceanographic Institutions Incorporated.

Kaula WM (ed.) (1969) *The Terrestrial Environment: Solid-Earth and Ocean Physics*. Proceedings of a Conference held at William College, August 11–21, 1969. NASA CR-1579.

Kawamura H (2000) Era of ocean observations using satellites. *Sokko-Jiho 67*: S1–S9 (in Japanese).

Koblinsky CJ and Smith NR (eds) (2001) *Observing the Oceans in the 21st Century*. Melbourne: Global Ocean Data Assimilation Experiment and the Bureau of Meteorology. (In press).

Masson RA (1991) *Satellite Remote Sensing of Polar Regions*. London: Belhaven Press.

Minster JF and Lefebvre M (1997) TOPEX/Poseidon: satellite altimetry and the circulation of the oceans. In: Minster JF (ed.) *La Machine Océan*, pp. 111–135. Paris: Flammarion (in French).

Ocean Theme Team (2001) *An Ocean Theme for the IGOS Partnership*. Washington, DC: NASA.

Patzert WC and Van Woert ML (2000) *Ocean and Land Space Missions during the 1990s and Beyond* (http://airsea-www.jpl.nasa.gov/missions.html).

Purdom JFW and Menzel WP (1996) Evolution of satellite observations in the United States and their use in meteorology. In: Fleming JR (ed.) *Historical Essays on Meteorology: 1919–1995*, pp. 99–156. Boston: American Meteorological Society.

Robinson IS and Guymer T (1996) Observing oceans from space. In: Summerhayes CP and Thorpe SA (eds) *Oceanography: An Illustrated Guide*, pp. 69–87. Chichester: John Wiley & Sons.

Victorov SV (1996) *Regional Satellite Oceanography*. London: Taylor & Francis.

Wilson WS (ed.) (1981) Oceanography from space. *Oceanus 24*(3): 1–76.

SATELLITE PASSIVE MICROWAVE MEASUREMENTS OF SEA ICE

C. L. Parkinson, NASA Goddard Space Flight Center, Greenbelt, MD, USA

doi:10.1006/rwos.2001.0336

Introduction

Satellite passive-microwave measurements of sea ice have provided global or near-global sea ice data for most of the period since the launch of the Nimbus 5 satellite in December 1972, and have done so with horizontal resolutions on the order of 25–50 km and a frequency of every few days. These data have been used to calculate sea ice concentrations (percent areal coverages), sea ice extents, the length of the sea ice season, sea ice temperatures, and sea ice velocities, and to determine the timing of the seasonal onset of melt as well as aspects of the ice-type composition of the sea ice cover. In each case, the calculations are based on the microwave emission characteristics of sea ice and the important contrasts between the microwave emissions of sea ice and those of the surrounding liquid-water medium.

Background on Satellite Passive-Microwave Sensing of Sea Ice

Rationale

Sea ice is a vital component of the climate of the polar regions, insulating the oceans from the atmosphere, reflecting most of the solar radiation incident on it, transporting cold, relatively fresh water equatorward, and at times precipitating overturning in the ocean and even bottom water formation through its rejection of salt to the underlying water. Furthermore, sea ice spreads over vast distances, globally covering an area approximately the size of North America at any given time, and it is highly dynamic, experiencing a prominent annual cycle in both polar regions and many short-term fluctuations as it is moved by winds and waves, melted by solar radiation, and augmented by additional freezing. It is a major player in and indicator of the polar climate state, and consequently it is highly desirable to monitor the sea ice cover on a routine basis. In view of the vast areal coverage of the ice and the harsh polar conditions, the only feasible means of obtaining routine monitoring is through satellite observations. Visible, infrared, active-microwave, and passive-microwave satellite instruments are all proving useful for examining the sea ice cover, with the passive-microwave instruments providing the longest record of near complete sea ice monitoring on a daily or near daily basis.

Theory

The tremendous value of satellite passive-microwave measurements for sea ice studies results from the combination of the following factors:

1. Microwave emissions of sea ice differ noticeably from those of sea water, making sea ice generally readily distinguishable from liquid water on the basis of the amount of microwave radiation received by the satellite instrument. For example, **Figure 1** presents color-coded images of the data from one channel on a satellite passive-microwave instrument, presented in units (termed 'brightness temperatures') indicative of the intensity of emitted microwave radiation at that channel's frequency, 19.4 GHz. The ice edge, highlighted by the dashed white curve, is clearly identifiable from the brightness temperatures, with open-ocean values of 172–198 K outside the ice edge and sea ice values considerably higher, predominantly greater than 230 K, within the ice edge.

2. The microwave radiation received by earth-orbiting satellites derives almost exclusively from the earth system. Hence, microwave sensing does not require sunlight, and the data can be collected irrespective of the level of daylight or darkness. This is a major advantage in the high polar latitudes, where darkness lasts for months at a time, centered on the winter solstice.

3. Many of the microwave data are largely unaffected by atmospheric conditions, including the presence or absence of clouds. Storm systems can produce atmospheric interference, but the microwave signal from the ice–ocean surface can pass through most nonprecipitating clouds essentially unhindered. Hence, microwave sensing of the surface does not require cloud-free conditions.

4. Satellite passive-microwave instruments can obtain a global picture of the sea ice cover at least every few days with a resolution of 50 km or better, providing reasonable spatial resolution

and extremely good temporal resolution for most large-scale or climate-related studies.

Satellite Passive-Microwave Instruments

The first major satellite passive-microwave imager was the Electrically Scanning Microwave Radiometer (ESMR) launched on the Nimbus 5 satellite of the US National Aeronautics and Space Administration (NASA) in December 1972, preceded by a nonscanning passive-microwave radiometer launched on the Russian Cosmos satellite in September 1968. The ESMR was a single-channel instrument recording radiation at a wavelength of 1.55 cm and corresponding frequency of 19.35 GHz. It collected good-quality data for much of the 4-year period from January 1973 through December 1976, although with some major data gaps, including one that lasted for 3 months, from June through August 1975. Being a single-channel instrument, it did not allow some of the more advanced studies that have been done with subsequent instruments, but its flight was a highly successful proof-of-concept mission, establishing the value of satellite passive-microwave technology for observing the global sea ice cover. The ESMR data were used extensively in the determination and analysis of sea ice conditions in both the Arctic and the Antarctic over the four years 1973–1976. Emphasis centered on the determination of ice concentrations (percent areal coverages of ice) and, based on the ice concentration results, the calculation of ice extents (integrated areas of all grid elements with ice concentration ≥ 15%).

The Nimbus 5 ESMR was followed by a less successful ESMR on the Nimbus 6 satellite and then by the more advanced 10-channel Scanning Multichannel Microwave Radiometer (SMMR) on board NASA's Nimbus 7 satellite and a sequence of 7-channel Special Sensor Microwave Imagers (SSMIs) on board satellites of the United States Defense Meteorological Satellite Program (DMSP). The Nimbus 7 was launched in late October 1978, and the SMMR on board it was operational through mid-August 1987. The first of the DMSP SSMIs was operational as of early July 1987, providing welcome data overlap with the Nimbus 7 SMMR and thereby allowing intercalibration of the SMMR and SSMI data sets. SSMIs continue to operate into the twenty-first century. There was also a SMMR on board the short-lived Seasat satellite in 1978; and there was a 2-channel Microwave Scanning Radiometer (MSR) on board the Japanese Marine

Observation Satellites starting in February 1987. Each of these successor satellite passive-microwave instruments, after the ESMR, has been multi-channel, allowing both an improved accuracy in the ice concentration derivations and the calculation of additional sea ice variables, including ice temperature and the concentrations of separate ice types.

The Japanese have developed an Advanced Microwave Scanning Radiometer (AMSR) for the Earth Observing System's Aqua satellite (formerly named the PM-1 satellite), scheduled for launch by NASA in December 2001, and for the Advanced Earth Observing Satellite II, scheduled for launch by the Japanese National Space Development Agency in February 2002. The AMSR will allow sea ice measurements at a higher spatial resolution (12–25 km) than is currently available with the SSMI instruments (25–50 km resolution for the major derived sea ice products).

Sea Ice Determinations from Satellite Passive-Microwave Data

Sea Ice Concentrations

Ice concentration is among the most fundamental and important parameters for describing the sea ice cover. Defined as the percent areal coverage of ice, it is directly critical to how much impact the ice cover has on restricting exchanges between the ocean and atmosphere and on reflecting incoming solar radiation. Ice concentration is calculated at each ocean grid element, for whichever grid is being used to map or otherwise display the derived satellite products. A map of ice concentrations presents the areal distribution of the ice cover, to the resolution of the grid. In cases where ice thickness data are also available, the combination of ice concentration and ice thickness allows the calculation of ice volume.

With a single channel of microwave data, taken at a radiative frequency and polarization combination that provides a clear distinction between ice and water, approximate sea ice concentrations can be calculated by assuming a uniform radiative brightness temperature TB_w for water and a uniform radiative brightness temperature TB_I for ice, with both brightness temperatures being appropriate for the values received at the altitude of the satellite, i.e., incorporating an atmospheric contribution. Assuming no other surface types within the field of view, the observed brightness temperature TB is given by eqn [1].

$$TB = C_w TB_w + C_I TB_I \qquad [1]$$

C_w is the percent areal coverage of water and C_I is the ice concentration. With only the two surface types, $C_w + C_I = 1$, and eqn [1] can be expressed as eqn [2].

$$TB = (1 - C_I)TB_w + C_I TB_I \qquad [2]$$

This is readily solved for the ice concentration (eqn [3]).

$$C_I = \frac{TB - TB_w}{TB_I - TB_w} \qquad [3]$$

Equation [3] is the standard equation used for the calculation of ice concentrations from a single channel of microwave data, such as the data from the ESMR instrument. A major limitation of the formulation is that the polar ice cover is not uniform in its microwave emission, so that the assumption of a uniform TB_I for all sea ice is only a rough approximation, far less justified than the assumption of a uniform TB_w for sea water, although that also is an approximation.

Multichannel instruments allow more sophisticated, and generally more accurate, calculation of the ice concentrations. They additionally allow many options as to how these calculations can be done. To illustrate the options, two algorithms will be described, both of which assume two distinct ice types, thereby advancing over the assumption of a single ice type made in eqns [1]–[3], especially for the many instances in which two ice types dominate the sea ice cover. Assume then that the field of view contains only water and two ice types, type 1 ice and type 2 ice (see later for more on ice types), and that the three surface types have identifiable brightness temperatures, TB_w, TB_{I1}, and TB_{I2}, respectively. Labeling the concentrations of the two ice types as C_{I1} and C_{I2}, respectively, the percent coverage of water is $1 - C_{I1} - C_{I2}$, and the integrated observed brightness temperature is given by eqn [4].

$$TB = (1 - C_{I1} - C_{I2})TB_w + C_{I1} TB_{I1} + C_{I2} TB_{I2} \qquad [4]$$

With two channels of information, as long as appropriate values for TB_w, TB_{I1}, and TB_{I2} are known for each of the two channels, eqn [4] can be used individually for each channel, yielding two linear equations in the two unknowns C_{I1} and C_{I2}. These equations are immediately solvable for C_{I1} and C_{I2}, and the total ice concentration C_I is then given by eqn [5].

$$C_I = C_{I1} + C_{I2} \qquad [5]$$

Although the scheme described in the preceding paragraph is a marked advance over the use of a single-channel calculation (eqn [3]), most algorithms for sea ice concentrations from multichannel data make use of additional channels and concepts to improve further the ice concentration accuracies. A widely used algorithm (termed the NASA Team algorithm) for the SMMR data employs three of the ten SMMR channels, those recording horizontally polarized radiation at a frequency of 18 GHz and vertically polarized radiation at frequencies of 18 GHz and 37 GHz. The algorithm is based on both the polarization ratio (PR) between the 18 GHz vertically polarized data (abbreviated 18 V) and the 18 GHz horizontally polarized data (18 H) and the spectral gradient ratio (GR) between the 37 GHz vertically polarized data (37 V) and the 18 V data. PR and GR are defined in eqns [6] and [7].

$$PR = \frac{TB(18V) - TB(18H)}{TB(18V) + TB(18H)} \qquad [6]$$

$$GR = \frac{TB(37V) - TB(18V)}{TB(37V) + TB(18V)} \qquad [7]$$

Substituting into eqns [6] and [7] expanded forms of $TB(18\,V)$, $TB(18\,H)$, and $TB(37\,V)$ obtained from eqn [4], the result yields equations for PR and GR in the two unknowns C_{I1} and C_{I2}. Solving for C_{I1} and C_{I2} yields two algebraically messy but computationally straightforward equations for C_{I1} and C_{I2} based on PR, GR, and numerical coefficients determined exclusively from the brightness temperature values assigned to water, type 1 ice, and type 2 ice for each of the three channels (these assigned values are termed 'tie points' and are determined empirically). These are the equations that are then used for the calculation of the concentrations of type 1 and type 2 ice once the observations are made and are used to calculate PR and GR from eqns [6] and [7]. The total ice concentration C_I is then obtained from eqn [5]. The use of PR and GR in this formulation reduces the impact of ice temperature variations on the ice concentration calculations. This algorithm is complemented by a weather filter that sets to 0 all ice concentrations at any time and location with a GR value exceeding 0.07. The weather filter eliminates many of the erroneous calculations of sea ice presence arising from the influence of storm systems on the microwave data.

For the SSMI data, the same basic NASA Team algorithm is used, although 18 V and 18 H in

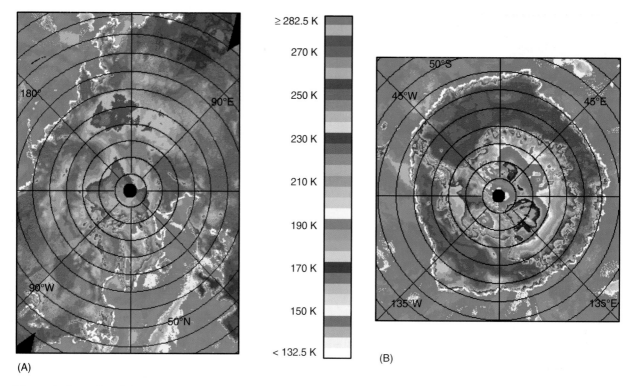

(A)

(B)

Figure 1 Late winter brightness temperature images of 19.4 GHz vertically polarized (19.4 V) data from the Defense Meteorological Satellite Program's Special Sensor Microwave Imager (SSMI) for (A) the north polar region on 15 March 1998, and (B) the south polar region on 15 September 1998, showing near-maximum sea ice coverage in each hemisphere. The dashed white curve has been added to indicate the location of the sea ice edge. Black signifies areas of no data; the absence of data poleward of 87.6° latitude results from the satellite's near-polar orbit and is consistent throughout the data set.

eqns [6] and [7] are replaced by 19.4 V and 19.4 H, reflecting the placement on the SSMI of channels at a frequency of 19.4 GHz rather than the 18 GHz channels on the SMMR. Also, because the data from the 19.4 GHz channels tend to be more contaminated by water vapor absorption/emission and other weather effects than the 18 GHz data, the weather filter for the SSMI calculations incorporates a threshold level for the gradient ratio calculated from the 22.2 GHz vertically polarized data and 19.4 V data as well as a threshold for the gradient ratio calculated from the 37 V and 19.4 V data. To illustrate the results of this ice concentration algorithm, **Figure 2** presents the derived sea ice concentrations for March 15, 1998 in the Northern Hemisphere and for September 15, 1998 in the Southern Hemisphere, the same dates as used in **Figure 1**.

As mentioned, there are several alternative ice concentration algorithms in use. Contrasts from the NASA Team algorithm just described include: use of different microwave channels; use of regional tie points rather than hemispherically applicable tie points; use of cluster analysis on brightness temperature data, without PR and GR formulations; use of

iterative techniques whereby an initial ice concentration calculation leads to refined atmospheric temperatures and opacities, which in turn lead to refined ice concentrations; use of iterative techniques involving surface temperature, atmospheric water vapor, cloud liquid water content, and wind speed; and use of a Kalman filtering technique in conjunction with an ice model. The various techniques tend to obtain very close overall distributions of where the sea ice is located, although sometimes with noticeable differences (up to 20%, and on occasion even higher) in the individual, local ice concentrations. The differences can often be markedly reduced by adjustment of tunable parameters, such as the algorithm tie points, in one or both of the algorithms being compared. However, the lack of adequate ground data often makes it difficult to know which tuning is most appropriate or which algorithm is yielding the better results.

Sea Ice Extents

Sea ice extent is defined as the area of coverage of sea ice of at least some set percentage ice concentration, with the set percentage most frequently being

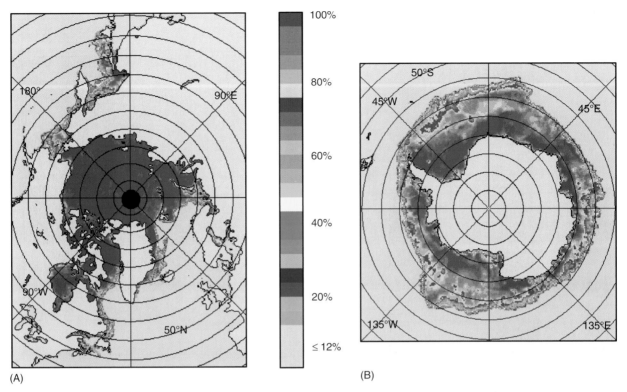

Figure 2 North and south polar sea ice concentration images for (A) 15 March 1998 and (B) 15 September 1998, respectively. The ice concentrations are derived from the data of the DMSP SSMI, including the 19.4 V data depicted in **Figure 1**.

15%. Sea ice extents are readily calculated from sea ice concentration maps by adding the areas of all grid elements in the region of interest having a calculated ice concentration of at least the predetermined cutoff (generally 15%). Ice extents are now regularly calculated from satellite passive-microwave data for the north polar region as a whole, for the south polar region as a whole, and for each of many subregions within the two polar domains.

A major early result in the use of satellite passive-microwave data was the detailed determination of the seasonal cycle of ice extents in each hemisphere. The Southern Ocean ice extents vary from about $2-4 \times 10^6 \, \text{km}^2$ in February to about $17-20 \times 10^6 \, \text{km}^2$ in September; the north polar ice extents vary from about $6-8 \times 10^6 \, \text{km}^2$ in September to about $14-16 \times 10^6 \, \text{km}^2$ in March. The ranges reflect the level of interannual variability observed from the satellite record over the course of the 1970s through the 1990s.

As the data sets lengthened, a major goal became the determination of trends in the ice extents and placement of these trends in the context of other climate variables and climate change. Because of the lack of a period of data overlap between the ESMR and the SMMR, matching of the ice extents derived from the ESMR data to those derived from the SMMR and SSMI data has been difficult and uncertain. Consequently, published results regarding trends found from the latter two data sets generally do not include the ESMR data. However, trends have been calculated for the combined SMMR/SSMI data set, and these calculations have been done for the data set as a whole, for individual seasons and months, and for individual regions. For the data set as a whole, the SMMR/SSMI record from late 1978 until the late 1990s indicates an overall decrease in Arctic ice extents of about 2–3% per decade and an overall increase in Antarctic ice extents of about 1–2% per decade. In both hemispheres, the trends are nonuniform with time and individual regions within the hemisphere display percent per decade trends of higher magnitude, some with positive signs and some with negative signs. These concepts are illustrated in **Figure 3A** with 20-year March time series for the Northern Hemisphere and two regions within the Northern Hemisphere and in **Figure 3B** with 20-year September time series for the Southern Hemisphere and two seas within the Southern Hemisphere. March and September are generally the months with the most sea ice coverage in the respective hemispheres.

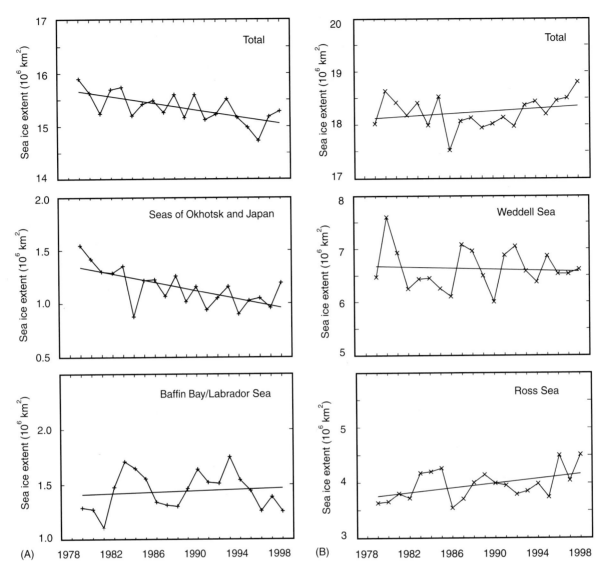

Figure 3 (A) Time series of monthly average 1979–1998 March sea ice extents for the Northern Hemisphere and two regions within the Northern Hemisphere. (B) Time series of monthly average 1979–1998 September sea ice extents for the Southern Hemisphere and two regions within the Southern Hemisphere. All ice extents are derived from the Nimbus 7 SMMR and DMSP SSMI data. The trend lines have slopes of $-31700\,km^2\,y^{-1}$ (-2.0% per decade) for the Northern Hemisphere total; $-20\,000\,km^2\,y^{-1}$ (-14.9% per decade) for the Seas of Okhotsk and Japan; $3100\,km^2\,y^{-1}$ (2.2% per decade) for Baffin Bay/Labrador Sea; $11\,900\,km^2\,y^{-1}$ (0.7% per decade) for the Southern Hemisphere total; $-5300\,km^2\,y^{-1}$ (-0.8% per decade) for the Weddell Sea; and $21\,300\,km^2\,y^{-1}$ (5.7% per decade) for the Ross Sea.

Sea Ice Types

The sea ice cover in both polar regions is a mixture of various types of ice, ranging from individual ice crystals to coherent ice floes several kilometers across. Common ice types include frazil ice (fine spicules of ice suspended in water); grease ice (a soupy layer of ice that looks like grease from a distance); slush ice (a viscous mass formed from a mixture of water and snow); nilas (a thin elastic sheet of ice 0.01–0.1 m thick); pancake ice (small, roughly circular pieces of ice, 0.3–3 m across and up to 0.1 m thick); first-year ice (ice at least 0.3 m thick that has not yet undergone a summer melt period); and multiyear ice (ice that has survived a summer melt).

Because different ice types have different microwave emission characteristics, once these differences are understood, appropriate satellite passive-microwave data can be used to distinguish ice types. The ice types most frequently distinguished with such data are first-year ice and multiyear ice in the

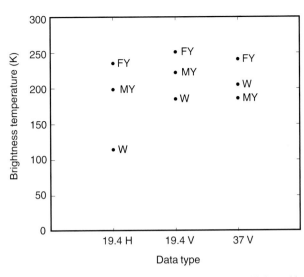

Figure 4 Brightness temperatures of first-year ice (FY), multi-year ice (MY), and liquid water (W) at three channels of SSMI data from the DMSP F13 satellite. These are the values used as tie points for the Arctic calculations in the NASA Team algorithm described in the text. (Data from Cavalieri *et al.* (1999).)

Other Sea Ice Variables: Season Length, Temperature, Melt, Velocity

Although sea ice concentrations, extents, and, to a lesser degree, types have been the sea ice variables most widely calculated and used from satellite passive-microwave data, several additional variables have also been obtained from these data, including the length of the sea ice season, sea ice temperature, sea ice melt, and sea ice velocity. The length of the sea ice season for any particular year is calculated directly from that year's daily maps of sea ice concentrations, by counting, at each grid element, the number of days with ice coverage of at least some predetermined (generally 15% or 30%) ice concentration. Trends in the length of the sea ice season from the late 1970s to the 1990s show coherent spatial patterns in both hemispheres, with a predominance of negative values (shortening of the sea ice season) in the Northern Hemisphere and a predominance of positive values in the Southern Hemisphere, consistent with the respective hemispheric trends in sea ice extents.

The passive-microwave-based ice temperature calculations generally depend on the calculated sea ice concentrations, empirically determined ice emissivities, a weighting of the water and ice temperatures within the field of view, and varying levels of sophistication in incorporating effects of the polar atmosphere and the presence of multiple ice types. The derived temperature is not the surface temperature but the temperature of the radiating portion of the ice, for whichever radiative frequency is being used. The ice temperature fields derived from passive-microwave data complement those derived from satellite infrared data, which have the advantages of having finer spatial resolution and more nearly approaching surface temperatures but the disadvantage of greater contamination by clouds. The passive-microwave and infrared data are occasionally used together, iteratively, for an enhanced ice temperature product.

The seasonal onset of melt on the sea ice and its overlying snow cover generally produces marked changes in microwave emissions, first increasing the emissions as liquid water emerges in the snow, then decreasing the emissions once the snow has melted and meltwater ponds cover the ice. Because of the emission changes, these events on the ice surface frequently become detectable through time series of the satellite passive-microwave data. The onset of melt in particular can often be identified, and hence yearly maps can be created of the dates of melt onset. Melt ponds, however, present greater difficulties, as they can have similar microwave emissions

Arctic Ocean. In fact, the NASA Team algorithm described earlier was initially developed specifically for first-year and multiyear ice, with the resulting calculations yielding the concentrations, C_{I1} and C_{I2}, of those two ice types. First-year and multiyear ice are distinguishable in their microwave signals because the summer melt process drains down through the ice some of its salt content, reducing the salinity of the upper layers of the ice and thereby changing the microwave emissions; these changes are dependent on the frequency and polarization of the radiation. To illustrate the differences, **Figure 4** presents a plot of the tie points used by the NASA Team algorithm in the Arctic for the three SSMI channels used to calculate ice concentrations prior to the application of the weather filter. The plot shows that while the transition from first-year to multiyear ice lowers the brightness temperatures for each of the three channels, the reduction is greatest for the 37 V data and least for the 19.4 V data. Examination of **Figure 4** further reveals that the polarization PR (eqn [6], revised for 19.4 GHz rather than 18 GHz data) increases from first-year to multiyear ice, though it remains well below the polarization of water, and that the gradient ratio GR (eqn [7], revised for 19.4 GHz data) is positive for water, slightly negative for first-year ice, and considerably more negative for multiyear ice. The differences allow the sorting out, either through the calculation of C_{I1} and C_{I2} as described or through alternative algorithms, of the first-year ice and multiyear ice percentages in the satellite field of view.

to those of the water in open spaces between ice floes, so that a field of heavily melt-ponded ice can easily be confused in the microwave data with a field of low-concentration ice. The ambiguities can be reduced through analysis of the passive-microwave time series and comparisons with active-microwave, visible, and infrared data.

The calculation of sea ice velocities from satellite data has in general relied upon data with fine enough resolution to distinguish individual ice floes, such as visible data and active-microwave data rather than the much coarser resolution passive-microwave data. However, in the 1990s several groups devised methods of determining ice velocity fields from passive-microwave data, some using techniques based on cross-correlation of brightness temperature fields and others using wavelet analysis. These techniques have yielded ice velocity maps on individual dates for the entire north and south polar sea ice covers. Comparisons with buoy and other data have been quite encouraging regarding the potential of using the satellite data for long-term records and monitoring of ice motions.

Looking Toward the Future

Monitoring of the polar sea ice covers through satellite passive-microwave technology is ongoing with the operational United States SSMI instruments and should be further enhanced with the Japanese AMSR instruments scheduled for launch in 2001 and 2002. The resulting lengthening sea ice records provide an improved basis with which scientists can examine trends in the sea ice cover and interactions between the sea ice and other elements of the climate system. For instance, the lengthened records will be essential to answering many of the questions raised concerning whether the negative overall trends found in Arctic sea ice extents for the first two decades of the SMMR/SSMI record will continue and how these trends relate to temperature trends, in particular to possible climate warming, and to oscillations within the climate system, in particular the North Atlantic Oscillation, the Arctic Oscillation, and the Southern Oscillation. In addition to covering a longer period, other expected improvements in the satellite passive-microwave record of sea ice include further algorithm developments, following additional analyses on the microwave properties of sea ice and liquid water. Such analyses are likely to lead both to improved algorithms for the variables already examined and to the development of techniques for calculating additional sea ice variables (such as sea ice thickness) from the satellite data.

Glossary

Brightness temperature Unit used to express the intensity of emitted microwave radiation received by the satellite, presented in temperature units (K) following the Rayleigh–Jeans approximation to Planck's law, whereby the radiation emitted from a perfect emitter at microwave wavelengths is proportional to the emitter's physical temperature.
Sea ice concentration Percent areal coverage of sea ice.
Sea ice extent Integrated area of all grid elements with sea ice concentration $\geq 15\%$.

See also

Acoustics, Arctic. Antarctic Circumpolar Current. Arctic Basin Circulation. Bottom Water Formation. Ice–Ocean Interaction. Current Systems in the Southern Ocean. Marine Mammal Migrations and Movement Patterns. Millenial Scale Climate Variability. North Atlantic Oscillation (NAO). Okhotsk Sea Circulation. Polar Ecosystems. Polynyas. Satellite Oceanography, History and Introductory Concepts. Satellite Remote Sensing Microwave Scatterometers. Satellite Remote Sensing SAR. Seabird Migration. Sea Ice: Overview. Sea Ice: Variations in Extent and Thickness. Seals. Weddell Sea Circulation.

Further Reading

Barry RG, Maslanik J, Steffen K et al. (1993) Advances in sea-ice research based on remotely sensed passive microwave data. *Oceanography* 6(1): 4–12.

Carsey FD (ed.) (1992) *Microwave Remote Sensing of Sea Ice*. Washington, DC: American Geophysical Union.

Cavalieri DJ, Parkinson CL, Gloersen P, Comiso JC and Zwally HJ (1999) Deriving long-term time series of sea ice cover from satellite passive-microwave multisensor data sets. *Journal of Geophysical Research* 104(C7): 15803–15814.

Gloersen P, Campbell WJ, Cavalieri DJ et al. (1992) *Arctic and Antarctic Sea Ice, 1978–1987: Satellite Passive-Microwave Observations and Analysis*. Washington, DC: National Aeronautics and Space Administration.

Gurney RJ, Foster JL and Parkinson CL (eds) (1993) *Atlas of Satellite Observations Related to Global Change*. Cambridge: Cambridge University Press.

Jeffries MO (ed.) (1998) *Antarctic Sea Ice: Physical Processes, Interactions and Variability*. Washington, DC: American Geophysical Union.

Massom R (1991) *Satellite Remote Sensing of Polar Regions: Applications, Limitations and Data Availability*. London: Belhaven Press.

Parkinson CL (1997) *Earth from Above: Using Color-Coded Satellite Images to Examine the Global Environment*. Sausalito, CA: University Science Books.

Parkinson CL (2000) Variability of Arctic sea ice: the view from space, an 18-year record. *Arctic* 53(4): 341–358.

Smith WO Jr and Grebmeier JM (eds) (1995) *Arctic Oceanography: Marginal Ice Zones and Continental Shelves*. Washington, DC: American Geophysical Union.

Ulaby FT, Moore RK and Fung AK (1986) Monitoring sea ice. In: *Microwave Remote Sensing: Active and Passive, Vol. III: From Theory to Applications*, pp. 1478–1521. Dedham, MA: Artech House.

SATELLITE REMOTE SENSING MICROWAVE SCATTEROMETERS

W. J. Plant, Applied Physics Laboratory, University of Washington, Seattle, WA, USA

doi:10.1006/rwos.2001.0337

Introduction

Microwave scatterometers are instruments that transmit low-power pulses of radiation toward the Earth's surface at intermediate incidence angles and measure the intensity of the signals scattered back at the same angles from surface areas a few kilometers on a side. Satellite scatterometers operate continuously and therefore scatter from land and ice as well as the ocean. Useful information is available in the signals from land and ice, but will not be discussed here. This article will concentrate on the primary goal of satellite scatterometers: the measurement of near-surface wind speed and direction over the ocean.

Scatterometers achieve this goal by measuring the intensity, or cross-section, of microwave signals backscattered from the ocean surface. Common frequencies of the transmitted signals for satellite scatterometers are near 5.3 GHz (C-band) on European instruments and 14 GHz (Ku-band) on US ones. At these frequencies, microwaves penetrate only a few millimeters into sea water, so all backscatter originates at the surface and is caused by the roughness of the surface; a perfectly calm sea surface produces no detectable scattering in the direction of the incident radiation. Changes in the average roughness of the ocean surface over scales of several kilometers are caused primarily, but not exclusively, by changes in the wind speed or direction at the ocean surface. Standard assumptions of scatterometry are that the backscatter cross-section, usually called σ_o, over such scales depends only on parameters of the scatterometer and on the mean wind, increases with wind speed, is a maximum when the antenna looks upwind, and is a minimum when the antenna looks nearly perpendicular to the wind, or crosswind. These assumptions allow the wind speed and direction to be determined from cross-sections measured for the same patch of the ocean, but with the antenna directed at several different azimuth angles. For satellite scatterometry a given patch of ocean can be viewed from several different directions only by allowing the scatterometer to sweep its antenna beams across the patch, a process that requires as much as 4 min. Thus an additional assumption of scatterometry is that average winds over kilometer-scale patches of ocean surface are stationary for several minutes.

With these assumptions and an adequate definition of the wind being measured (discussed below), satellite scatterometers have proven to be able to measure winds over the ocean with accuracies as good as or better than *in situ* measurement techniques. Because oceans cover most of the Earth, this means that microwave scatterometers carried on satellites can monitor the wind field over most of the globe every few days. **Table 1** gives typical specifications that a satellite scatterometer can be expected to meet.

The spatial coverage offered by satellite scatterometry is far better than can be achieved by *in situ* measurements. It allows scatterometers to provide data to study global weather patterns,

Table 1 Expected specifications of the NSCAT satellite scatterometer

Parameter	Value	Accuracy/comment
Wind speed	3–30 m s^{-1}	2 m s^{-1} or 10%
Wind direction	3–30 m s^{-1}	20°
Spatial resolution	50 km	Wind cells
Location accuracy	25/10 km	Absolute/relative
Coverage	90% of ice-free ocean	Every 2 d
Mission duration	3 y	Includes check out

(Data from Naderi *et al.* 1991.)

Table 2 Scatterometers in space to date and planned for the future

Satellite	Country/agency	Scatterometer	Launch date	Status
Seasat	USA/NASA	SASS	June 1978	Failed, October 1978
ERS-1	Europe/ESA	CSCAT(AMI)	July 1991	Standby, June 1996
ERS-2	Europe/ESA	CSCAT(AMI)	April 1995	Operational
ADEOS-I	USA/NASA Japan/NASDA	NSCAT	August 1996	ADEOS failed June 1997
QuikSCAT	USA/NASA	SeaWinds-1	June 1999	Operational
ADEOS-II	USA/NASA Japan/NASDA	SeaWinds-2	November 2001	Approved
ASCAT	Europe/ESA	Adv.CSCAT	2003	Proposed

(Adapted from Patzert and Van Woert, private communication.)

monitor storm intensities, improve meteorological forecasts, impact global ocean circulation models, facilitate climate prediction, and much more. In addition to introducing the basics of scatterometry, this article will provide examples of these benefits of satellite scatterometry and indicate how continued improvements in the technique may be expected to provide even better results in the future.

Satellite Scatterometers

Other instruments such as radar altimeters that look straight down and real and synthetic aperture radars (RARs and SARs) that image surface scenes at resolutions of meters to kilometers have been operated in space and are capable of measuring wind speed or direction, but not both simultaneously and routinely. Microwave radiometers, passive instruments that measure the naturally occurring radiation from the ocean surface, are presently being developed as spaceborne anemometers capable of measuring wind speed and direction simultaneously. However, only microwave scatterometers, active instruments that both transmit and receive radiation, have a history of wind vector measurement from space. Because they are microwave, scatterometers can make their measurements both day and night and in most kinds of weather. Only very heavy rainfall, as discussed below, can hinder a scatterometer's view of the surface.

For these reasons, microwave scatterometers have been the instruments of choice for measuring near-surface winds from space. **Table 2** lists the scatterometers that have been in space to date and those that are planned for the future. As the table shows, the first scatterometer in space specifically designed to measure winds was the one on Seasat in 1978. An earlier microwave radar on Skylab viewed the ocean surface at intermediate incidence angles from space in 1973, but did not produce multiple looks at a single ocean patch from which wind vector information could be obtained. With the launch of the C-band scatterometer on the first European Remote Sensing Satellite (ERS-1) in 1991, a continuous series of global wind vector measurements was begun. As **Table 2** shows, this series has continued to the present through the launch of ERS-2 and the subsequent decommissioning of ERS-1. If present plans are carried out, at least two microwave scatterometers (and one or more microwave radiometers) will continue to produce global wind vector information into the foreseeable future.

While most scatterometers in space have consisted of multiple, fixed waveguide (stick) antennas, more recent US scatterometers have used rotating parabolic antennas. All satellite scatterometers to date have been in orbits with inclinations near 98.5°, except for Seasat which was 108°. Spatial resolutions have usually been 50×50 km, although 25×25 km is becoming more common. **Table 3**

Table 3 Principal characteristics of satellite scatterometers

Satellite	Type	Frequency (GHz)	Polarization	Incidence angle (°)	Swath (km)	Altitude (km)
Seasat	4 stick	14.6	4 VV, 4 HH	25–55 0–4	475, 475 140	800
ERS-1	3 stick	5.3	3 VV	18–57	500	785
ERS-2	3 stick	5.3	3 VV	18–57	500	785
ADEOS-I	6 stick	14.0	6 VV, 2 HH	15–63	600, 600	797
QuikSCAT	1 rotating	13.4	1 VV, 1 HH	46(H), 54(V)	1800	803
ADEOS-II	1 rotating	13.4	1 VV, 1 HH	46(H), 54(V)	1800	803
METOP	3 stick	–	–	–	–	835

addresses the primary characteristics of the scatterometers listed in **Table 2**. As **Table 3** shows, ERS scatterometers have had about half the spatial coverage of their NASA counterparts because their antennas view only one side of the subsatellite path. As discussed below, however, uniformly reliable wind vectors are not obtained over the whole swath of the SeaWinds instruments. Also a gap is present between the two swaths on either side of the subsatellite path

of the NASA stick-type scatterometers because the response of the cross-section to the wind vector is weak at low incidence angles. The Seasat scatterometer did have a mode that observed the surface at very low incidence angles, but could not get wind direction in this swath without extrapolation from the wider swaths. **Figure 1** shows the fixed-stick ERS-1 scatterometer (**Figure 1A**) and the rotating-antenna QuickSCAT scatterometer (**Figure 1B**).

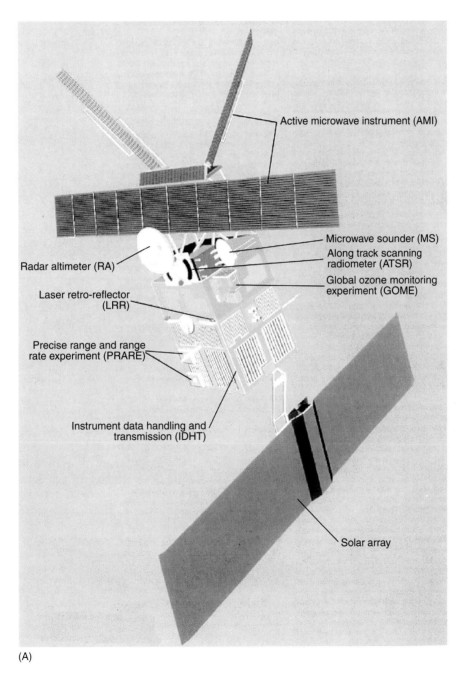

(A)

Figure 1 (A) ERS-1 satellite with two scatterometer waveguide antennas directed upwards. (B) QuikSCAT satellite with parabolic antenna with dual beams pointed downwards.

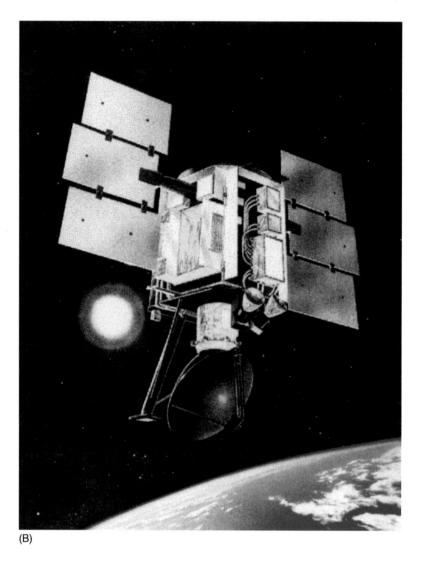

(B)

Figure 1 *Continued*

The Normalized Radar Cross-section of the Sea

The basis of a scatterometer's ability to measure the near-surface wind vector is the dependence of the normalized radar cross-section of the sea, σ_o, on this vector. This cross-section is defined through the radar equation as follows:

$$P_r = \frac{P_t G^2 \lambda^2 A \sigma_o}{(4\pi)^3 R^4} + P_n \tag{1}$$

where P_r is the received power, P_t is the transmitted power, G is antenna gain, λ is microwave length, A is the total illuminated area on the sea surface, and R is the range to the surface. P_n is the noise signal due to thermal noise in the components of the system and natural radiation of the earth to the receiving antenna. In a real scatterometer, the received power is reduced below the above level by system losses that must be taken into account. Also, for greater accuracy, the variation of G and R over the scatterometer footprint are taken into account by integration over the footprint. By relating σ_o rather than P_r to the wind vector, the importance of system-specific parameters, P_t, G, R, and λ, are greatly reduced. In fact, σ_o is independent of the first three of these parameters and depends on λ only because the mechanism of backscatter from the sea surface is weakly dependent on λ. Two other system parameters upon which σ_o depends due to the surface scattering mechanism are the incidence angle and the polarization. The strength of backscatter from the ocean surface increases with decreasing incidence angle and depends on the direction of the electric field of the incident radiation, its polarization.

The radar equation must be solved in order to obtain σ_o. This requires a knowledge of P_n, which is usually obtained by sampling only noise signals on a small fraction of the transmitted pulses, typically about 15%. The noise level obtained in this manner is subtracted from P_r and the difference is multiplied or divided by the other quantities which are known. This yields σ_o values that are accurate down to very low signal-to-noise ratios, but that can become negative due to sampling variability. This is an important consideration in the measurement of very low wind speeds.

Figure 2 shows the measured dependance of σ_o on polarization and incidence angle as well as on the wind vector. Two polarizations are indicated in the figure: the electric field vertical on both transmission and reception (VV) and the electric field horizontal on both (HH). These are the two polarizations of importance in scatterometry. The characteristics of σ_o that make it useful for wind vector measurement are obvious from this figure: for any given incidence angle and polarization, it depends on both wind speed and direction. In general the slope of σ_o versus wind speed is smaller at lower incidence angles. The dependence of σ_o on azimuth angle, χ, defined as the angle between the horizontal direction the antenna is pointing and the direction from which the wind comes, has generally been found to fit a three-term Fourier cosine series in χ very well. The relationship between σ_o and incidence angle, θ_i, polarization, p, wind speed, U, and wind direction, χ, is called the geophysical model function. Its general form is taken

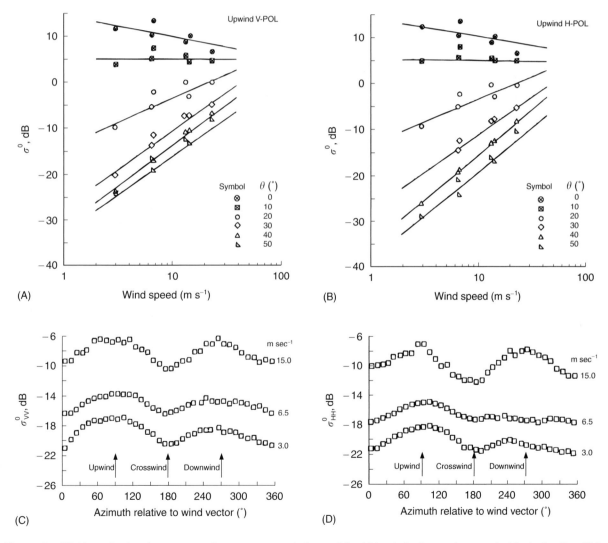

Figure 2 (A) Normalized radar cross section, σ_0 versus wind speed for VV polarization and an upwind look direction (θ is incidence angle); (B) same as (A) for HH polarization; (C) azimuthal dependence of σ_0 for VV polarization and a 30° incidence angle; (D) same as (C) for HH polarization. (Reproduced with permission from Jones WL *et al.* Aircraft measurements of the microwave scattering signature of the ocean. *IEEE Journal of Oceanic Engineering* © 1977 IEEE.)

to be the following:

$$\sigma_o = A_o(U, \theta_i, p) + A_1(U, \theta_i, p)\cos\chi$$
$$+ A_2(U, \theta_i, p)\cos2\chi \qquad [2]$$

Usually the A coefficients are specified in tabular form.

Some characteristics of the dependence of σ_o on χ can be easily discerned in **Figure 2**. Cross-sections measured with the antenna looking nearly perpendicular to the wind direction ($\chi = 90°$ or $270°$) are always lower than those measured with the antenna directed upwind ($\chi = 0°$) or downwind ($\chi = 180°$). Furthermore, σ_o is larger when the antenna looks upwind than when it looks downwind, except perhaps for VV polarization at small incidence angles.

Spaceborne Scatterometer Wind Measurements

Given the behavior of σ_o shown in **Figure 2**, a scatterometer fixed on Earth can easily yield the wind

vector. If the antenna is rotated and the direction of maximum σ_o is determined, then the level of σ_o in this direction yields the wind speed. In most cases, the direction of maximum σ_o would be the wind direction. The exception to this might be at VV polarization for some incidence angles where the downwind look direction could yield the maximum σ_o. Thus a $180°$ ambiguity might exist in the wind direction.

Unfortunately this simple technique will not work from satellites because of their high speeds. By the time the antenna rotates one revolution, the satellite has travelled several kilometers, so the footprint on the surface samples many different, widely separated areas. The solution to this problem is to use σ_o values measured in different directions at different times so that nearly the same surface area is illuminated in each direction. **Figure 3** indicates how this might be accomplished in the case of three, fixed stick antennas (**Figure 3A**) and in the case of a rotating antenna with two beams (**Figure 3B**). The first case corresponds to the ERS scatterometers

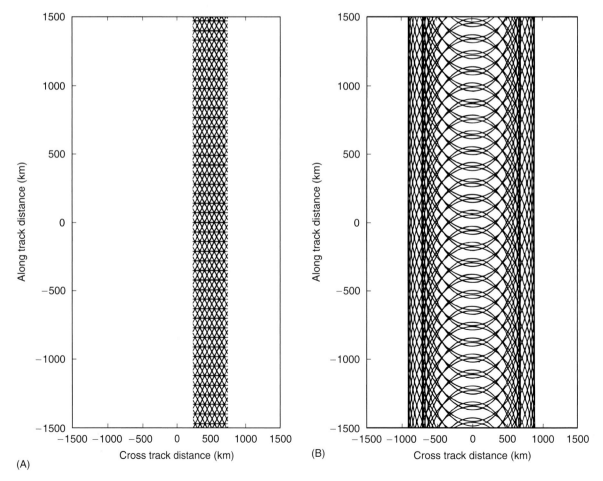

(A)   (B)

Figure 3   Lines on which scatterometer surface footprints lie showing crossing where multiple look directions occur. Along track distance expanded by a factor of 5. (A) ERS-1/2, (B) SeaWinds.

while the latter corresponds to SeaWinds whose single rotating antenna transmits beams at two different incidence angles. **Figure 3** shows the lines on which the surface resolution cells are located. For clarity, the vertical distance traveled by the satellite has been expanded by a factor of 5 in **Figure 3**. Thus crossings of the lines, where cross-sections can be measured from different directions, are more frequent than indicated in the figure. The two parts of **Figure 3** have the same horizontal scale to emphasize the much wider swath of the SeaWinds scatterometer compared with ERS. For regions of the swath near the center, however, the azimuth angles at which the SeaWinds beam views the surface are nearly opposite or parallel. Furthermore, near the edges, only two nearly parallel looks at a given surface area can be obtained. Estimation of wind speed and direction is consequently more difficult near the center and edge of the swath due to the ambiguous nature of the model function.

The ambiguities inherent in the form of the model function are illustrated in **Figure 4**. The figure shows the angular dependence of σ_o for winds coming from $80°$, $180°$, $260°$, and $345°$ with respect to north. Each curve maximizes when the antenna is directed into the wind direction. The circles in the figure represent measurements of the cross-section made by an ERS-type scatterometer where the three beams are $45°$ apart. As the figure shows, winds from $180°$ and $345°$ fit the measurements equally well. If the scatterometer had only two beams that were $90°$ apart, like the Seasat scatterometer, then the other two wind directions, $80°$ and $260°$ would also fit the data, provided that the wind speeds were slightly higher for the $260°$ direction and slightly lower for the $80°$ direction. Ambiguous wind vectors always exist in the output of scatterometers. Attempts to resolve these ambiguities have ranged from using human analysts to insure consistency of the wind fields to using median filters in which a given wind vector is forced to be in the direction

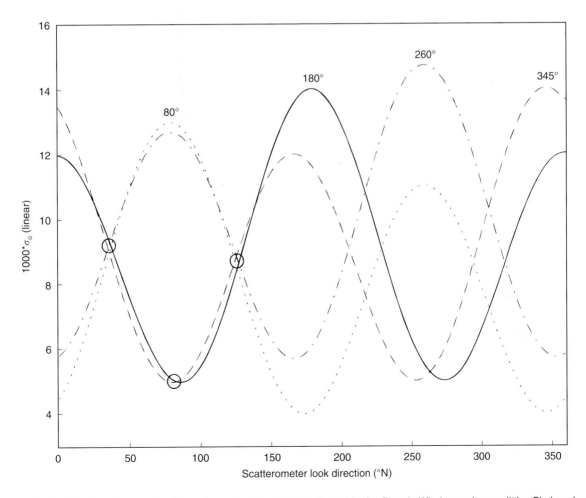

Figure 4 Angular dependence of σ_0 for various wind directions (indicated in the figure). Wind speeds vary little. Circles show possible data points; curves crossing at data points produce ambiguities in wind direction.

of the median of its surrounding vectors. While automatic techniques such as median filtering are now producing skills upwards of 90% in selecting the correct ambiguity, incorrect assignment of vector directions in some locations is still a problem for scatterometry.

Retrieval of the ambiguous wind vectors is accomplished in practice by noting that any single measurement of σ_o is an average of measurements from several scatterometer pulses and therefore has a probability distribution near Gaussian. The variance, δ^2, of this distribution may be determined from the characteristics of the scatterometer, especially its noise level, along with σ_o. Thus the probability of measuring a given σ_o value, call it $\hat{\sigma}_o$, is given by

$$P(\hat{\sigma}_o) = \frac{1}{\sqrt{2\pi\delta^2}} e^{(\hat{\sigma}_o - \sigma_o)^2/2\delta^2} \qquad [3]$$

where σ_o is the true normalized cross-section. Since this is unknown in practice, the value of σ_o given by the geophysical model function is used instead.

If all the measured values of $\hat{\sigma}_o$ from every view of the surface that is within the specified resolution of the system (for the NASA scatterometer (NSCAT) with a 50×50 km resolution this could be up to 24) then the joint probability of measuring these values is the product of many Gaussian distributions such as eqn [3]. The wind speed and direction, and therefore σ_o, are then varied to maximize this joint probability. This is the same as minimizing the following objective function:

$$J(U, \chi) = \sum_{i=1}^{N} \left[\ln \delta_i^2 + \left(\frac{\hat{\sigma}_{oi} - \sigma_{oi}(U, \chi)}{\delta_i} \right)^2 \right] \qquad [4]$$

Good knowledge of both the geophysical model function and the variances associated with the different looks is essential for a good retrieval.

Calibration

In principle, σ_o values obtained by scatterometers need not be calibrated in an absolute sense; they could simply be related to the surface wind that produced them. In order to cross-check the operation of different scatterometers and to further the development of rough surface scattering theories, however, all scatterometers flown in space have produced normalized radar cross-sections that have been calibrated against a standard target. Usually signals from a geophysical target that has been calibrated by airborne measurements and found

to be isotropic are used for calibration of σ_o from spaceborne scatterometers. The Amazon Rain Forest is the most common geophysical target chosen, although areas of Antarctica have also been used.

Calibration and verification of scatterometer winds require an exact definition of the wind that a scatterometer measures. Since a scatterometer really responds to surface roughness, the suggestion has often been made that a scatterometer measures the stress of the wind on the ocean surface rather than the wind vector itself. In fact, a geophysical model function has been developed to relate σ_o directly to the friction velocity, the square root of the stress divided by the density of air. However, to date this procedure has not been adopted by any agency operating a scatterometer. Rather, official winds produced by scatterometers are the winds measured at 10 m above the ocean surface that would yield the same surface stress under neutrally buoyant atmospheric stratification. The only exception was the Seasat scatterometer, whose output winds were specified at a height of 19.5 m above the surface; these are approximately 6% higher than winds at 10 m. Thus wind fields produced by scatterometers are not necessarily the wind fields that a fixed array of *in situ* anemometers would measure, even if they were all at a height of 10 m. Not only are scatterometer wind fields the neutrally buoyant equivalent winds, but evidence is growing that these wind fields are the ones that would be measured under neutral conditions by an anemometer drifting along with the ambient ocean current.

Although physical models of backscatter from the wind-roughened ocean have been developed, they have not proven to be sufficiently accurate to be used as geophysical model functions. Therefore wind retrieval from scatterometers flown to date has been achieved using empirical model functions developed by various means. Experiments have been mounted to relate cross-sections measured by airborne scatterometers to *in situ* measurements of winds converted to neutral conditions. The Europeans noted that if σ_o depends on no other environmental variables than U and χ, then σ_o values measured by three antennas looking in three different directions must fall on a well-defined surface when σ_o values from the three separate antennas are plotted on three orthogonal axes. This observation has allowed them to determine the properties of the geophysical model function to within a constant calibration factor, which was obtained from comparisons with buoy measurements. US geophysical model functions have also been developed by comparison of cross-sections with buoy measurements and with winds measured by other remote sensing instruments, such as microwave radiometers. Most recently, however, US geophysical model functions have been developed by binning satellite scatterometer σ_o values according to U and χ values produced by numerical weather prediction models. The idea is that errors in the numerical model will cancel out in the mean so that the correct model function can be obtained even in the presence of these errors.

After a model function has been developed by some means, a period of validation of the scatterometer wind fields always follows. Generally scatterometer wind fields are compared with those measured by buoys moored in fixed locations whose anemometer readings have been corrected to 10 m and neutral conditions. As an indication of the accuracy that can be obtained by scatterometers, **Figure 5** shows a comparison of wind speeds and directions produced by NSCAT with winds measured by anemometers on buoys operated by the US National Data Buoy Center (NDBC). To produce the wind speed comparison in **Figure 5A**, buoy wind speeds have been binned into 0.5 m s^{-1} bins and the corresponding NSCAT winds measured near the buoy have been averaged for each bin. The nonzero values of NSCAT wind speed at zero buoy wind speed are known to be a result of comparing magnitudes of vectors whose components are Gaussian distributed so the figure shows good agreement. **Figure 5B** shows the distribution of differences between NSCAT wind directions and buoy wind directions for several different wind speeds and again agreement is good.

Applications of Scatterometry

Further indications of the accuracy and usefulness of data on winds from satellite scatterometers come from studies that compare these winds with, or use them in, atmospheric, oceanic, or climate models to assess the improvements possible through the use of scatterometer winds. Comparisons between wind fields from satellite scatterometers and numerical models have shown that significant differences in these fields often exist. Cyclones predicted by numerical models have been found to disagree in location and intensity with those observed in satellite scatterometry, sometimes by 300 km and 10 mbar or more. The location of the Intertropical Convergence Zone (ITCZ) has been shown to be 1°–2° farther south in NSCAT wind fields than in wind fields predicted by the numerical model of the European Centre for Medium-Range Weather Fore-

casts (ECMWF). Furthermore, the ITCZ observed by NSCAT was stronger and narrower than that predicted by ECMWF. An example of the differences that can occur between scatterometer wind vectors and those of numerical models is given in **Figure 6A**. Here NSCAT measurements in the South China Sea are compared with predictions of the numerical model of the National Centers for Environmental Prediction (NCEP). Obviously the fields can be quite different.

In the South China Sea, the Princeton Ocean Model (POM) was run using wind fields from NSCAT and from NCEP, some of which are shown in **Figure 6A**. The results indicate significant differences in the output of the POM depending on the wind field used, although these differences were generally smaller than those of the wind fields themselves. This is illustrated in **Figure 6B**, which shows surface currents predicted by POM for the two different wind fields and their differences. While this study did not assess which POM prediction was the most accurate, other studies have indicated that the predictions of ocean models and coupled atmosphere–ocean models using scatterometer wind fields agreed with observations better than those using other wind fields. Comparisons of sea levels predicted by the modular ocean model (MOM) with those of the TOPEX/Poseidon radar altimeter have indicated that more accurate predictions are achieved using ERS-1 wind fields than using NCEP fields. Similarly, prediction of the 1997–98 El Niño by the Lamont-Doherty Earth Observatory model has been shown to be improved by using NSCAT wind fields rather than the surface observations collected by Florida State University (FSU). This is illustrated in **Figure 7**. Here sea surface temperature anomalies (with respect to FSU climatology) in the eastern tropical Pacific predicted by the model with FSU and NSCAT wind fields are compared with those actually observed in 1997 and 1998. NSCAT wind fields have improved the prediction.

Limitations and Improvements

The examples above indicate the usefulness of satellite scatterometry in oceanic and atmospheric modeling. However, limitations still exist in the satellite wind fields which make them less useful than they could be. Ambiguities have been mentioned earlier. The temporal and spatial sampling patterns inherent in the data collection of individual satellites can obscure geophysical effects that occur on spatial scales less than about 200 km and temporal scales less than a day or two. This effectively limits the

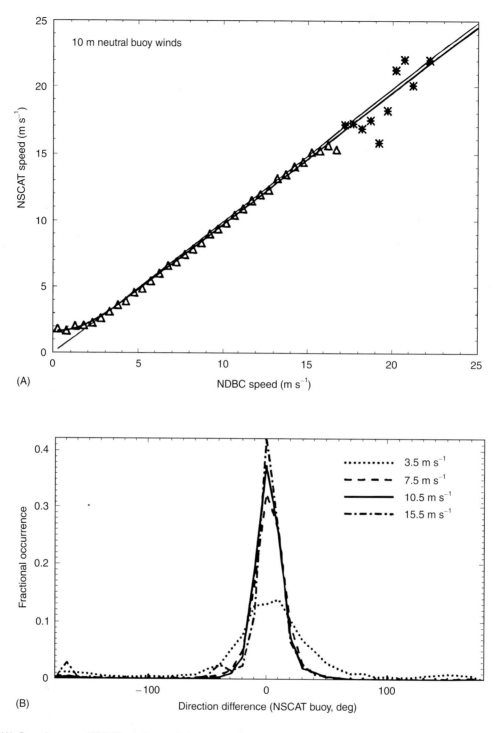

Figure 5 (A) Sample mean NSCAT wind speeds in 0.5 m s⁻¹ buoy wind speed bins. All NSCAT measurements were located within 50 km and 30 min of buoy measurements. Triangles denote > 100 NSCAT measurements in the mean; asterisks denote 5–99 samples. (B) Distributions of directional differences (NSCAT buoy) for 1 m s⁻¹ buoy wind speed bins centered on the indicated wind speeds. (Reproduced with permission from Freilich and Dunbar, 1999.)

application of satellite scatterometry to low-frequency, large-scale events. Interpolation of the raw data can reduce this problem somewhat but may introduce false signals. The obvious way to alleviate the problem is to put fleets of satellite scatterometers into orbit, an idea that may not be completely impossible in this day of smaller, cheaper satellites.

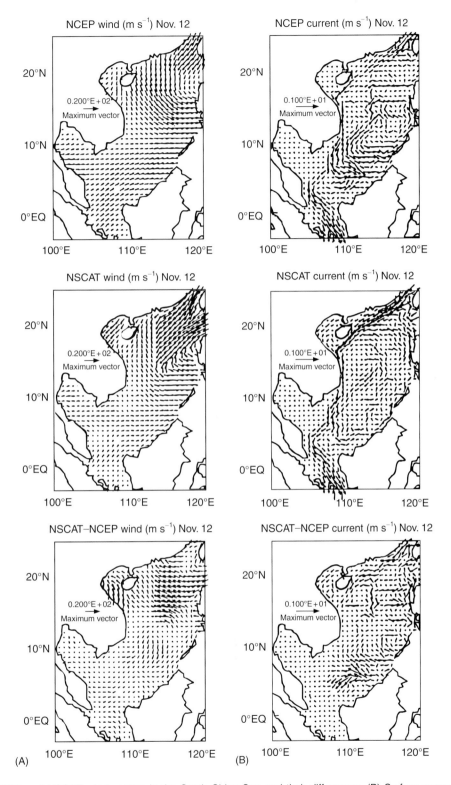

Figure 6 (A) NCEP and NSCAT wind vectors in the South China Sea and their differences. (B) Surface currents in the South China Sea from NCEP model under NCEP and NSCAT wind forcing. (Reproduced with permission from Chu *et al.*, 1999.)

Another limitation on satellite scatterometry that has been observed many times is the mismeasurement of surface wind fields in the presence of rain. Without an accompanying microwave radiometer, it is difficult to ascertain from scatterometer measurements alone when rain is occurring on the surface.

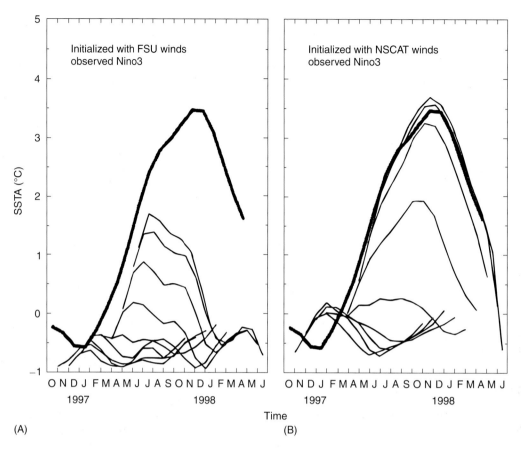

Figure 7 Forecasts of sea surface temperature anomalies in the eastern tropical Pacific for the 1997–98 El Niño by the Lamont-Doherty Earth Observatory model using (A) Florida State University winds and (B) NSCAT winds. Heavy curves are observations; other lines are predictions made at various times (Reproduced with permission from Chen et al., 1999.)

Since heavy rainfall can attenuate a Ku-band scatterometer signal by more than a factor of 10 every kilometer, serious mismeasurement of surface wind fields can occur when rain is present. Furthermore, microwave return from the sea surface is also changed by rainfall, and this change is not well understood. Measurements indicate that σ_o becomes more isotropic in the presence of rain, and this effect is being utilized in an attempt to develop a rain flag from scatterometer data alone. Such a flag would allow users to determine when rain effects might be present in scatterometer data, but not what the proper wind vectors are.

Winds below $3\,\mathrm{m\,s^{-1}}$ are difficult to measure accurately using satellite scatterometry. Present geophysical model functions are smoothly decreasing functions of wind speed that are nonzero even at zero wind speeds. The Europeans have chosen not to provide wind speed estimates below $3\,\mathrm{m\,s^{-1}}$ using their ERS scatterometers. Recent studies have shown that, in addition to the directional variability inherent in low winds, microwave backscatter is virtually zero below some threshold wind speed in the

neighborhood of $2\,\mathrm{m\,s^{-1}}$ to $4.5\,\mathrm{m\,s^{-1}}$, depending on incidence angle and water temperature. Variability of the wind over the surface footprint of the scatterometer, however, obscures this threshold in most satellite scatterometry measurements. The geophysical model function at these low wind speeds therefore appears to depend on both the mean wind vector over the footprint and its variability. Suggestions for improving low wind speed measurements have included specifying a variability-dependent model function that depends on geographic region and season, as well as using details of the probability distribution on σ_o. Implementation of such suggestions in future retrieval schemes offers hope of improving low wind speed measurements.

Finally, high wind speeds have also proven to be a problem for scatterometry. Recent studies indicate that σ_o increases less rapidly with wind speed above about $25\,\mathrm{m\,s^{-1}}$ than current model functions predict. The result is that wind speeds above this value tend to be underestimated in scatterometry wind retrievals. Proposals have been offered for improved high wind speed model functions based on simultaneous scatterometry/radiometry measurements from aircraft flying through hurricanes. Implementation of such model functions promises to yield better wind retrievals at high wind speeds.

Conclusion

Satellite-based microwave scatterometry is a mature technology that has proven itself capable of yielding global oceanic wind speeds of unprecedented accuracy and spatial coverage. Satellites currently in orbit and planned for future missions promise a continuous long-term series of global wind measurements that can aid in climate studies. Based on the time-series presently available, satellite scatterometer measurements have proven themself capable of improving present oceanographic and atmospheric models. Future improvements in scatterometry promise to make this technology even more valuable in studying the dynamics of the atmosphere and oceans.

See also

Aircraft Remote Sensing. Air–Sea Gas Exchange. Heat and Momentum Fluxes at the Sea Surface. History of Satellite Oceanography and Introductory Concepts. Satellite Altimetry. Satellite Measurements of Salinity. Satellite Passive Microwave Measurements of Sea Ice. Satellite Remote Sensing SAR. Satellite Remote Sensing of Sea Surface Temperatures. Sensors for Mean Meteorology. Sensors for Micrometeorological Flux Measurements. Surface, Gravity and Capillary Waves. Wave Generation by Wind. Wind and Buoyancy-forced Upper Ocean. Wind Driven Circulation.

Further Reading

Brown RA and Zeng L (1994) Estimating central pressures of oceanic midlatitude cyclones. *Journal of Applied Meteorology* 33: 1088–1095.

Chen D, Cane MA and Zebiak SE (1999) The impact of NSCAT winds on predicting the 1997/1998 El Niño: A case study with the Lamont-Doherty Earth Observatory model. *Journal of Geophysical Research* 104: 11322–11327.

Chu PC, Lu S and Liu WT (1999) Uncertainty of South China Sea prediction using NSCAT and National Centers for Environmental Prediction winds during tropical storm Ernie, 1996. *Journal of Geophysical Research* 104: 11273–11289.

Donelan MA and Pierson WJ (1987) Radar scattering and equilibrium ranges in wind-generated waves with application to scatterometry. *Journal of Geophysical Research* 92: 4971–5029.

Fisher RE (1972) Standard deviation of scatterometer measurements from space. *IEEE Transactions Geoscience Electronics* GE-10: 106–113.

Freilich MH (1997) Validation of vector magnitude datasets: effects of random component errors. *Journal of Atmospheric and Oceanic Technology* 14: 695–703.

Freilich MH and Dunbar RS (1999) The accuracy of the NSCAT 1 vector winds: Comparisons with National Data Buoy Center buoys. *Journal of Geophysical Research* 104: 11231–11246.

Fu Lee-Lueng and Yi Chao (1997) The sensitivity of a global ocean model to wind forcing. A test using sea level and wind observations from satellites and operational wind analysis. *Geophysical Research Letters* 24: 1783–1786.

Graf J, Sasaki C, Winn C *et al.* (1998) NASA scatterometer experiment. *Acta Astronautica* 43: 397–407.

Jones WL, Schroeder LC and Mitchell JL (1977) Aircraft measurements of the microwave scattering signature of the ocean. *IEEE Journal of Oceanic Engineering* OE-2: 52–61.

Kelly KA, Dickinson S and Yu Z (1999) NSCAT tropical wind stress maps: implications for improving ocean modeling. *Journal of Geophysical Research* 104: 11291–11310.

Moore RK and Pierson WJ (1996) *Measuring Sea State and Estimating Surface Winds from a Polar Orbiting Satellite*. In: Proceedings of the International Symposium on Electromagnetic Sensing of Earth from Satellites, Miami Beach, FL, 1966.

Moore RK and Fung AK (1979) Radar determination of winds at sea. *Proceedings of the IEEE* 67: 1504–1521.

Naderi FM, Freilich MH and Long DG (1991) Spaceborne radar measurement of wind velocity over the ocean – An overview of the NSCAT Scatterometer system. *Proceedings of the IEEE* 79: 850–866.

Pierson WJ Jr (1983) The measurement of the synoptic scale wind over the ocean. *Journal of Geophysical Research* 88: 1682–1780.

SATELLITE REMOTE SENSING OF SEA SURFACE TEMPERATURES

P. J. Minnett, University of Miami, Miami, FL, USA

doi:10.1006/rwos.2001.0343

Introduction

The ocean surface is the interface between the two dominant, fluid components of the Earth's climate system: the oceans and atmosphere. The heat moved around the planet by the oceans and atmosphere helps make much of the Earth's surface habitable, and the interactions between the two, that take place through the interface, are important in shaping the climate system. The exchange between the ocean and atmosphere of heat, moisture, and gases (such as CO_2) are determined, at least in part, by the sea surface temperature (SST). Unlike many other critical variables of the climate system, such as cloud cover, temperature is a well-defined physical variable that can be measured with relative ease. It can also be measured to useful accuracy by instruments on observation satellites.

The major advantage of satellite remote sensing of SST is the high-resolution global coverage provided by a single sensor, or suite of sensors on similar satellites, that produces a consistent data set. By the use of onboard calibration, the accuracy of the time-series of measurements can be maintained over years, even decades, to provide data sets of relevance to research into the global climate system. The rapid processing of satellite data permits the use of the global-scale SST fields in applications where the immediacy of the data is of prime importance, such as weather forecasting – particularly the prediction of the intensification of tropical storms and hurricanes.

Measurement Principle

The determination of the SST from space is based on measuring the thermal emission of electromagnetic radiation from the sea surface. The instruments, called radiometers, determine the radiant energy flux, B_λ, within distinct intervals of the electromagnetic spectrum. From these the brightness temperature (the temperature of a perfectly emitting 'black-body' source that would emit the same radiant flux) can be calculated by the Planck equation:

$$B_\lambda(T) = 2hc^2\lambda^{-5}(e^{hc/(\lambda kT)} - 1)^{-1} \qquad [1]$$

where h is Planck's constant, c is the speed of light in a vacuum, k is Boltzmann's constant, λ is the wavelength and T is the temperature. The spectral intervals (wavelengths) are chosen where three conditions are met: (1) the sea emits a measurable amount of radiant energy, (2) the atmosphere is sufficiently transparent to allow the energy to propagate to the spacecraft, and (3) current technology exists to build radiometers that can measure the energy to the required level of accuracy within the bounds of size, weight, and power consumption imposed by the spacecraft. In reality these constrain the instruments to two relatively narrow regions of the infrared part of the spectrum and to low-frequency microwaves. The infrared regions, the so-called atmospheric windows, are situated between wavelengths of 3.5–4.1 μm and 8–12 μm (**Figure 1**); the microwave measurements are made at frequencies of 6–12 GHz.

As the electromagnetic radiation propagates through the atmosphere, some of it is absorbed and scattered out of the field of view of the radiometer, thereby attenuating the original signal. If the attenuation is sufficiently strong none of the radiation from the sea reaches the height of the satellite, and such is the case when clouds are present in the field of view of infrared radiometers. Even in clear-sky conditions a significant fraction of the sea surface emission is absorbed in the infrared windows. This energy is re-emitted, but at a temperature characteristic of that height in the atmosphere. Consequently the brightness temperatures measured through the clear atmosphere by a spacecraft radiometer are cooler than would be measured by a similar device just above the surface. This atmospheric effect, frequently referred to as the temperature deficit, must be corrected accurately if the derived sea surface temperatures are to be used quantitatively.

Infrared Atmospheric Correction Algorithms

The peak of the Planck function for temperatures typical of the sea surface is close to the longer wavelength infrared window, which is therefore well suited to SST measurement (**Figure 1**). However, the main atmospheric constituent in this spectral interval that contributes to the temperature

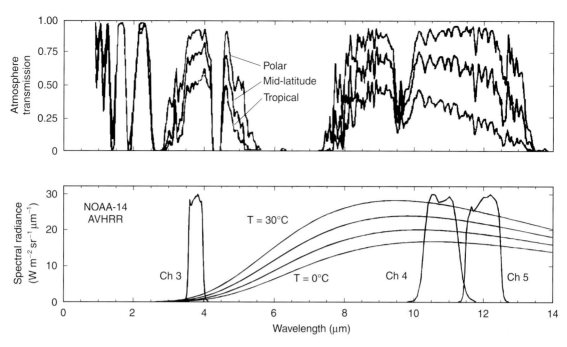

Figure 1 Spectra of atmospheric transmission in the infrared (wavelengths 1–14 µm) calculated for three typical atmospheres from diverse parts of the ocean; polar, mid-latitude and tropical with integrated water vapor content of 7 kg m^{-2} (polar), 29 kg m^{-2} (mid-latitude) and 54 kg m^{-2} (tropical). Regions where the transmission is high are well suited to satellite remote sensing of SST. The lower panel shows the electromagnetic radiative flux for four sea surface temperatures (0, 10, 20, and 30°C) with the relative spectral response functions for channels 3, 4, and 5 of the AVHRR on the NOAA-14 satellite. The so-called 'split-window' channels, 4 and 5, are situated where the sea surface emission is high, and where the atmosphere is comparatively clear but exhibits a strong dependence on atmospheric water vapor content.

deficit is water vapor, which is very variable both in space and time. Other molecular species that contribute to the temperature deficit are quite well mixed throughout the atmosphere, and therefore inflict a relatively constant temperature deficit that is simple to correct.

The variability of water vapor requires an atmospheric correction algorithm based on the information contained in the measurements themselves. This is achieved by making measurements at distinct spectral intervals in the windows when the water vapor attenuation is different. These spectral intervals are defined by the characteristics of the radiometer and are usually referred to as bands or channels (**Figure 1**). By invoking the hypothesis that the difference in the brightness temperatures measured in two channels, i and j, is related to the temperature deficit in one of them, the atmospheric correction algorithm can be formulated thus:

$$SST_{ij} - T_i = f(T_i - T_j) \qquad [2]$$

where SST_{ij} is the derived SST and T_i, T_j are the brightness temperatures in channels i, j.

Further, by assuming that the atmospheric attenuation is small in these channels, so that the radiative transfer can be linearized, and that the channels are spectrally close so that Planck's function can be linearized, the algorithm can be expressed in the very simple form:

$$SST_{ij} = a_o + a_i T_i + a_j T_j \qquad [3]$$

where are a_o, a_i, and a_j are coefficients. These are determined by regression analysis of either coincident satellite and *in situ* measurements, mainly from buoys, or of simulated satellite measurements derived by radiative transfer modeling of the propagation of the infrared radiation from the sea surface through a representative set of atmospheric profiles.

The simple algorithm has been applied for many years in the operational derivation of the sea surface from measurements of the Advanced Very High Resolution Radiometer (AVHRR, see below), the product of which is called the multi-channel SST (MCSST), where i refers to channel 4 and j to channel 5.

More complex forms of the algorithms have been developed to compensate for some of the shortcomings of the linearization. One such widely applied algorithm takes the form:

$$SST_{ij} = b_o + b_1 T_i + b_2(T_i - T_j)SST_r$$
$$+ b_3(T_i - T_j)(\sec \theta - 1) \qquad [4]$$

where SST_r is a reference SST (or first-guess temperature), and θ is the zenith angle to the satellite radiometer measured at the sea surface. When applied to AVHRR data, with i and j referring to channels 4 and 5 derived SST is called the nonlinear SST (NLSST). A refinement is called the Pathfinder SST (PFSST) in the research program designed to post-process AVHRR data over a decade or so to provide a consistent data set for climate research. In the PFSST, the coefficients are derived on a monthly basis for two different atmospheric regimes, distinguished by the value of the T_4–T_5 differences being above or below 0.7 K, by comparison with measurements from buoys.

The atmospheric correction algorithms work effectively only in the clear atmosphere. The presence of clouds in the field of view of the infrared radiometer contaminates the measurement so that such pixels must be identified and removed from the SST retrieval process. It is not necessary for the entire pixel to be obscured, even a portion as small as 3–5%, dependent on cloud type and height, can produce unacceptable errors in the SST measurement. Thin, semi-transparent cloud, such as cirrus, can have similar effects to subpixel geometric obscuration by optically thick clouds. Consequently, great attention must be paid in the SST derivation to the identification of measurements contaminated by even a small amount of clouds. This is the principle disadvantage to SST measurement by spaceborne infrared radiometry. Since there are large areas of cloud cover over the open ocean, it may be necessary to composite the cloud-free parts of many images to obtain a complete picture of the SST over an ocean basin.

Similarly, aerosols in the atmosphere can introduce significant errors in SST measurement. Volcanic aerosols injected into the cold stratosphere by violent eruptions produce unmistakable signals that can bias the SST too cold by several degrees. A more insidious problem is caused by less readily identified aerosols at lower, warmer levels of the atmosphere that can introduce systematic errors of a much smaller amplitude.

Microwave Measurements

Microwave radiometers use a similar measurement principle to infrared radiometers, having several spectral channels to provide the information to correct for extraneous effects, and black-body calibration targets to ensure the accuracy of the measurements. The suite of channels is selected to include sensitivity to the parameters interfering with the SST measurements, such as cloud droplets and surface wind speed, which occurs with microwaves at higher frequencies. A simple combination of the

Table 1 Relative merits of infrared and microwave radiometers for sea surface temperature measurement

Infrared	Microwave
Good spatial resolution (~ 1 km)	Poor spatial resolution (~ 50 km)
Surface obscured by clouds	Clouds largely transparent, but measurement perturbed by heavy rain
No side-lobe contamination	Side-lobe contamination prevents measurements close to coasts or ice
Aperture is reasonably small; instrument can be compact for spacecraft use	Antenna is large to achieve spatial resolution from polar orbit heights (~ 800 km above the sea surface)
4 km resolution possible from geosynchronous orbit; can provide rapid sampling data	Distance to geosynchronous orbit too large to permit useful spatial resolution with current antenna sizes

brightness temperature, such as eqn [2], can retrieve the SST.

The relative merits of infrared and microwave radiometers for measuring SST are summarized in **Table 1**.

Characteristics of Satellite-derived SST

Because of the very limited penetration depth of infrared and microwave electromagnetic radiation in sea water the temperature measurements are limited to the sea surface. Indeed, the penetration depth is typically less than in the infrared, so that temperature derived from infrared measurements is characteristic of the so-called skin of the ocean. The skin temperature is generally several tenths of a degree cooler than the temperature measured just below, as a result of heat loss from the ocean to atmosphere. On days of high insolation and low wind speed, the absorption of sunlight causes an increase in near surface temperature so that the water just below the skin layer is up to a few degrees warmer than that measured a few meters deeper, beyond the influence of the diurnal heating. For those people interested in a temperature characteristic of a depth of a few meters or more, the decoupling of the skin and deeper, bulk temperatures is perceived as a disadvantage of using satellite SST. However, algorithms generated by comparisons between satellite and *in situ* measurements from buoys include a mean skin effect masquerading as part of the atmospheric effect, and so the application of these results in an estimate of bulk temperatures.

The greatest advantage offered by satellite remote sensing is, of course, coverage. A single, broad-swath,

Table 2 Spectral characteristics of current and planned satellite-borne infrared radiometers

AVHRR		ATSR		MODIS		OCTS		GLI	
λ (μm)	NEΔT (K)	λ (μm)	NEΔT (K)	λ (μm)	NEΔT (K)	λ (μm)	NEΔT (K)	λ (μm)	NEΔT (K)
3.75	0.12	3.7	0.019	3.75	0.05	3.7	0.15	3.715	< 0.15
				3.96	0.05				
				4.05	0.05				
				8.55	0.05	8.52	0.15	8.3	< 0.1
10.5	0.12	10.8	0.028	11.03	0.04	10.8	0.15	10.8	< 0.1
11.5	0.12	12.0	0.025	12.02	0.04	11.9	0.15	12	< 0.1

imaging radiometer on a polar-orbiting satellite can provide global coverage twice per day. An imaging radiometer on a geosynchronous satellite can sample much more frequently, once per half-hour for the Earth's disk, or smaller segments every few minutes, but the spatial extent of the data is limited to that part of the globe visible from the satellite.

The satellite measurements of SST are also reasonably accurate. Current estimates for routine measurements show absolute accuracies of ± 0.3 to ± 0.5 K when compared to similar measurements from ships, aircraft, and buoys.

Spacecraft Instruments

All successful instruments have several attributes in common: a mechanism for scanning the Earth's surface to generate imagery, good detectors, and a mechanism for real-time, in-flight calibration. Calibration involves the use of one or more black-body calibration targets, the temperatures of which are accurately measured and telemetered along with the imagery. If only one black-body is available a measurement of cold dark space provides the equivalent of a very cold calibration target. Two calibration points are needed to provide in-flight calibration; nonlinear behavior of the detectors is accounted for by means of pre-launch instrument characterization measurements.

The detectors themselves inject noise into the data stream, at a level that is strongly dependent on their temperature. Therefore, infrared radiometers require cooled detectors, typically operating from 78 K (− 195°C) to 105 K (− 168°C) to reduce the noise equivalent temperature difference (NEΔT) to the levels shown in **Table 2**.

The Advanced Very High Resolution Radiometer (AVHRR)

The satellite instrument that has contributed the most to the study of the temperature of the ocean surface is the AVHRR that first flew on TIROS-N

launched in late 1978. AVHRRs have flown on successive satellites of the NOAA series from NOAA-6 to NOAA-14, with generally two operational at any given time. The NOAA satellites are in a near-polar, sun-synchronous orbit at a height of about 780 km above the Earth's surface and with an orbital period of about 100 min. The overpass times of the two NOAA satellites are about 2.30 a.m. and p.m. and about 7.30 a.m. and p.m. local time. The AVHRR has five channels: 1 and 2 at ~ 0.65 and ~ 0.85 μm are responsive to reflected sunlight and are used to detect clouds and identify coastlines in the images from the daytime part of each orbit. Channels 4 and 5 (**Table 2** and **Figure 1**) are in the atmospheric window close to the peak of the thermal emission from the sea surface and are used primarily for the measurement of sea surface temperature. Channel 3, positioned at the shorter wavelength atmospheric window, is responsive to both surface emission and reflected sunlight. During the nighttime part of each orbit, measurements of channel 3 brightness temperatures can be used with those from channels 4 and 5 in variants of the atmospheric correction algorithm to determine SST. The presence of reflected sunlight during the daytime part of the orbit prevents much of these data from being used for SST measurement. Because of the tilting of the sea surface by waves, the area contaminated by reflected sunlight (sun glitter) can be quite extensive, and is dependent on the local surface wind speed. It is limited to the point of specular reflection only in very calm seas.

The images in each channel are constructed by scanning the field of view of the AVHRR across the Earth's surface by a mirror inclined at 45° to the direction of flight (**Figure 2A**). The rate of rotation, 6.67 Hz, is such that successive scan lines are contiguous at the surface directly below the satellite. The width of the swath (~ 2700 km) means that the swaths from successive orbits overlap so that the whole Earth's surface is covered without gaps each day.

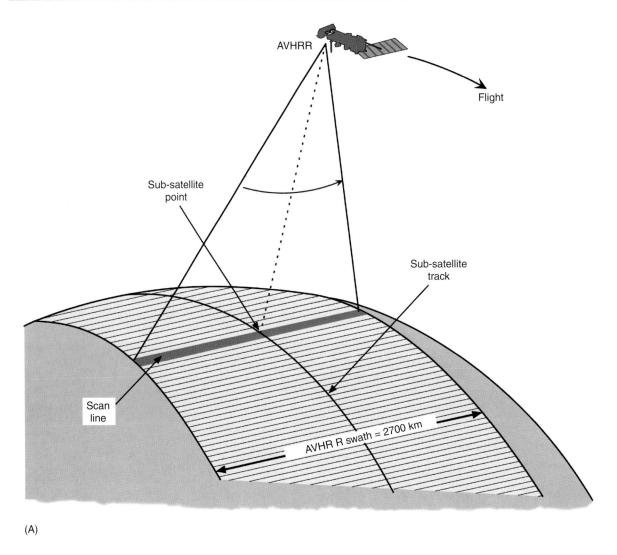

(A)

Figure 2 Scan geometries of AVHRR (A) and ATSR (B). The continuous wide swath of the AVHRR is constructed by linear scan lines aligned across the direction of motion of the subsatellite point. The swaths of the ATSR are generated by an inclined conical scan, which covers the same swath through two different atmospheric path lengths. The swath is limited to 512 km by geometrical constraints. Both radiometers are on sun-synchronous, polar-orbiting satellites.

The Along-Track Scanning Radiometer (ATSR)

An alternative approach to correcting the effects of the intervening atmosphere is to make a brightness temperature measurement of the same area of sea surface through two different atmospheric path lengths. The pairs of such measurements must be made in quick succession, so that the SST and atmospheric conditions do not change in the time interval. This approach is that used by the ATSR, two of which have flown on the European satellites ERS-1 and ERS-2.

The ATSR has infrared channels in the atmospheric windows comparable to those of AVHRR, but the rotating scan mirror sweeps out a cone inclined from the vertical by its half-angle (**Figure 2B**).

The field of view of the ATSR sweeps out a curved path on the sea surface, beginning at the point directly below the satellite, moving out sideways and forwards. Half a mirror revolution later, the field of view is about 900 km ahead of the subsatellite track in the center of the 'forward view'. The path of the field of view returns to the subsatellite point, which, during the period of the mirror rotation, has moved 1 km ahead of the starting point. Thus the pixels forming the successive swaths through the nadir point are contiguous. The orbital motion of the satellite means that the nadir point overlays the center of the forward view after about 2 min. The atmospheric path length of the measurement at nadir is simply the thickness of the atmosphere, whereas the slant path to the center of the

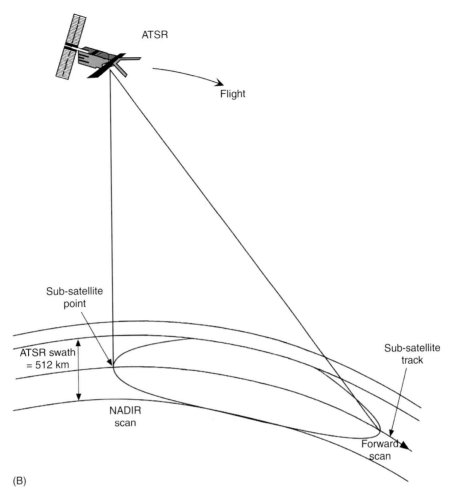

(B)

Figure 2 *Continued*

forward view is almost double that, resulting in colder brightness temperatures. The differences in the brightness temperatures between the forward and nadir swaths are a direct measurement of the effect of the atmosphere and permit a more accurate determination of the sea surface temperature. The atmospheric correction algorithm takes the form:

$$SST = c_o + \sum_i c_{n,i} T_{n,i} + \sum_i c_{f,i} T_{f,i} \qquad [5]$$

where the subscripts n and f refer to measurements from the nadir and forward views, i indicates two or three atmospheric window channels and the set of c are coefficients. The coefficients, derived by radiative transfer simulations, have an explicit latitudinal dependence.

Accurate calibration of the brightness temperatures is achieved by using two onboard black-body cavities, situated between the apertures for the nadir and forward views such that they are scanned each rotation of the mirror. One calibration target is at the spacecraft ambient temperature while the other

is heated, so that the measured brightness temperatures of the sea surface are straddled by the calibration temperatures.

The limitation of the simple scanning geometry of the ATSR is a relatively narrow swath width of 512 km. The ERS satellites have at various times in their missions been placed in orbits with repeat patterns of 3, 35, and 168 days, and given the narrow ATSR swath, complete coverage of the globe has been possible only for the 35 and 186 day cycles. This disadvantage is offset by the intended improvement in absolute accuracy of the atmospheric correction, and of its better insensitivity to aerosol effects.

The Moderate Resolution Imaging Spectroradiometer (MODIS)

The MODIS is a 36-band imaging radiometer on the NASA Earth Observing System (EOS) satellites *Terra*, launched in December 1999, and *Aqua*, planned for launch by late 2001. MODIS is much more complex than other radiometers used for SST measurement, but uses the same atmospheric windows. In addition to the usual two bands in the 10–12 µm interval, MODIS has three narrow bands in the 3.7–4.1 µm windows, which, although limited by sun-glitter effects during the day, hold the potential for much more accurate measurement of SST during the night. Several of the other 31 bands of MODIS contribute to the SST measurement by better identification of residual cloud and aerosol contamination.

The swath width of MODIS, at 2330 km, is somewhat narrower than that of AVHRR, with the result that a single day's coverage is not entire, but the gaps from one day are filled in on the next. The spatial resolution of the infrared window bands is 1 km at nadir.

The GOES Imager

SST measurements from geosynchronous orbit are made using the infrared window channels of the GOES Imager. This is a five-channel instrument that remains above a given point on the Equator. The image of the Earth's disk is constructed by scanning the field of view along horizontal lines by an oscillating mirror. The latitudinal increments of the scan line are done by tilting the axis of the scan mirror. The spatial resolution of the infrared channels is 2.3 km (east–west) by 4 km (north–south) at the subsatellite point. There are two imagers in orbit at the same time on the two GOES satellites, covering the western Atlantic Ocean (GOES-East) and the eastern Pacific Ocean (GOES-West). The other parts of the global oceans visible from geosynchronous orbit are covered by three other satellites operated by Japan, India, and the European Meteorological Satellite organization (Eumetsat). Each carries an infrared imager, but with lesser capabilities than the GOES Imager.

TRMM Microwave Imager (TMI)

The TMI is a nine-channel microwave radiometer on the Tropical Rainfall Measuring Mission satellite, launched in 1997. The nine channels are centered at five frequencies: 10.65, 19.35, 21.3, 37.0, and 85.5 GHz, with four of them being measured at two polarizations. The 10.65 GHz channels confer a sensitivity to SST, at least in the higher SST range found in the tropics, that has been absent in microwave radiometers since the SMMR (Scanning Multifrequency Microwave Radiometer) that flew on the short-lived Seasat in 1978 and on Nimbus-7 from 1978 to 1987. Although SSTs were derived from SMMR measurements, these lacked the spatial

resolution and absolute accuracy to compete with those of the AVHHR. The TMI complements AVHRR data by providing SSTs in the tropics where persistent clouds can be a problem for infrared retrievals. Instead of a rotating mirror, TMI, like other microwave imagers, uses an oscillating parabolic antenna to direct the radiation through a feed-horn into the radiometer.

The swath width of TMI is 759 km and the orbit of TRMM restricts SST measurements to within 38.5° of the equator. The beam width of the 10.65 GHz channels produces a footprint of 37 × 63 km, but the data are over-sampled to produce 104 pixels across the swath.

Applications

With absolute accuracies of satellite-derived SST fields of ~ 0.5 K or better, and even smaller relative uncertainties, many oceanographic features are resolved. These can be studied in a way that was hitherto impossible. They range from basin-scale perturbations to frontal instabilities on the scales of tens of kilometers. SST images have revealed the great complexity of ocean surface currents; this complexity was suspected from shipboard and aircraft measurements, and by acoustically tacking neutrally buoyant floats. However, before the advent of infrared imagery the synoptic view of oceanic variability was elusive, if not impossible.

El Niño

The El Niño Southern Oscillation (ENSO) phenomenon has become a well-known feature of the coupled ocean–atmosphere system in terms of perturbations that have a direct influence on people's lives, mainly by altering the normal rainfall patterns causing draughts or deluges – both of which imperil lives, livestock, and property.

The normal SST distribution in the topical Pacific Ocean is a region of very warm surface waters in the west, with a zonal gradient to cooler water in the east; superimposed on this is a tongue of cool surface water extending westward along the Equator. This situation is associated with heavy rainfall over the western tropical Pacific, which is in turn associated with lower level atmospheric convergence and deep atmospheric convection. The atmospheric convergence and convection are part of the large-scale global circulation. The warm area of surface water, enclosed by the 28°C isotherm, is commonly referred to as the 'Warm Pool' and in the normal situation is confined to the western part of the tropical Pacific. During an El Niño event the warm surface water, and associated convection and rainfall, migrate eastward perturbing the global atmospheric circulation. El Niño events occur up to a few times per decade and are of very variable intensity. Detailed knowledge of the shape, area, position, and movement of the Warm Pool can be provided from satellite-derived SST to help study the phenomenon and forecast its consequences.

Figure 3 shows part of the global SST fields derived from the Pathfinder SST algorithm applied to AVHRR measurements. The tropical Pacific SST field in the normal situation (December 1993) is shown in the upper panel, while the lower panel shows the anomalous field during the El Niño event of 1997–98. This was one of the strongest El Niños on record, but also the best documented and forecast. Seasonal predictions of disturbed patterns of winds and rainfall had an unprecedented level of accuracy and provided improved useful forecasts for agriculture in many affected areas. Milder than usual hurricane and tropical cyclone seasons were successfully forecast, as were much wetter winters and severe coastal erosion on the Pacific coasts of the Americas.

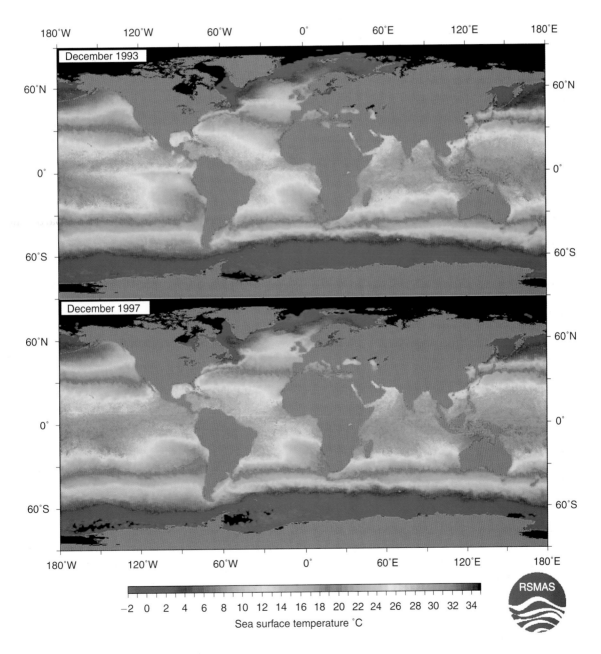

Figure 3 Global maps of SST derived from the AVHRR Pathfinder data sets. These are monthly composites of cloud-free pixels and show the normal situation in the tropical Pacific Ocean (above) and the perturbed state during an El Niño event (below).

Hurricane Intensification

The Atlantic hurricane season in 1999 was one of the most damaging on record in terms of land-falling storms in the eastern USA, Caribbean, and Central America. Much of the damage was not a result of high winds, but of torrential rainfall. Accurate forecasting of the path and intensity of these land-falling storms is very important, and a vital component of this forecasting is detailed knowledge of SST patterns in the path of the hurricanes. The SST is indicative of the heat stored in the upper ocean that is available to intensify the storms, and SSTs of $> 26°C$ appear to be necessary to trigger the intensification of the hurricanes. Satellite-derived SST maps are used in the prediction of the development of storm propagation across the Atlantic Ocean from the area off Cape Verde where atmospheric disturbances spawn the nascent storms. Closer to the USA and Caribbean, the SST field is important in determining the sudden intensification that can occur just before landfall. After the hurricane has passed, they sometimes leave a wake of cooler water in the surface that is readily identifiable in the satellite-derived SST fields.

Frontal Positions

One of the earliest features identified in infrared images of SST were the positions of ocean fronts, which delineate the boundaries between dissimilar surface water masses. Obvious examples are western boundary currents, such as the Gulf Stream in the Atlantic Ocean (**Figure 4**) and the Kuroshio in the Pacific Ocean, both of which transport warm surface water poleward and away from the western coastlines. In the Atlantic, the path of the warm surface water of the Gulf Stream can be followed in SST images across the ocean, into the Norwegian Sea, and into the Arctic Ocean. The surface water loses heat to the atmosphere, and to adjacent cooler waters on this path from the Gulf of Mexico to the Arctic, producing a marked zonal difference in the climates of the opposite coasts of the Atlantic and Greenland-Norwegian Seas. Instabilities in the fronts at the sides of the currents have been revealed in great detail in the SST images. Some of the large-scale instabilities can lead to loops on scales of a few tens to hundreds of kilometers that can become 'pinched off' from the flow and evolve as independent features, migrating away from the currents. When these occur on the equator side of the current these are called 'Warm Core Rings' and can exist for many months; in the case of the Gulf Stream these can propagate into the Sargasso Sea.

Figure 5 shows a series of instabilities along the boundaries of the Equatorial current system in the Pacific Ocean. The extent and structure of these features were first described by analysis of satellite SST images.

Coral Bleaching

Elevated SSTs in the tropics have adverse influences on living coral reefs. When the temperatures exceed the local average summertime maximum for several days, the symbiotic relationship between the coral polyps and their algae breaks down and the reef-building animals die. The result is extensive areas where the coral reef is reduced to the skeletal structure without the living and growing tissue, giving the reef a white appearance. Time-series of AVHRR-derived SST have been shown to be valuable predictors of reef areas around the globe that are threatened by warmer than usual water temperatures. Although it is not possible to alter the outcome, SST maps have been useful in determining the scale of the problem and identifying threatened, or vulnerable reefs.

The 'Global Thermometer'

Some of the most pressing problems facing environmental scientists are associated with the issue of global climate change: whether such changes are natural or anthropogenic, whether they can be forecast accurately on regional scales over decades, and whether undesirable consequences can be avoided. The past decade has seen many air temperature records being surpassed and indeed the planet appears to be warming on a global scale. However, the air temperature record is rather patchy in its distribution, with most weather stations clustered on Northern Hemisphere continents.

Global SST maps derived from satellites provide an alternative approach to measuring the Earth's temperature in a more consistent fashion. However, because of the very large thermal inertia of the ocean (it takes as much heat to raise the temperature of only the top meter of the ocean through one degree as it does for the whole atmosphere), the SST changes indicative of global warming are small. Climate change forecast models indicate a rate of temperature increase of only a few tenths of a degree per decade, and this is far from certain because of our incomplete understanding of how the climate system functions, especially in terms of various feedback factors such as those involving changes in cloud and aerosol properties. Such a rate of temperature increase will require SST records of several decades length before the signal, if present, can be unequivocally identified above the

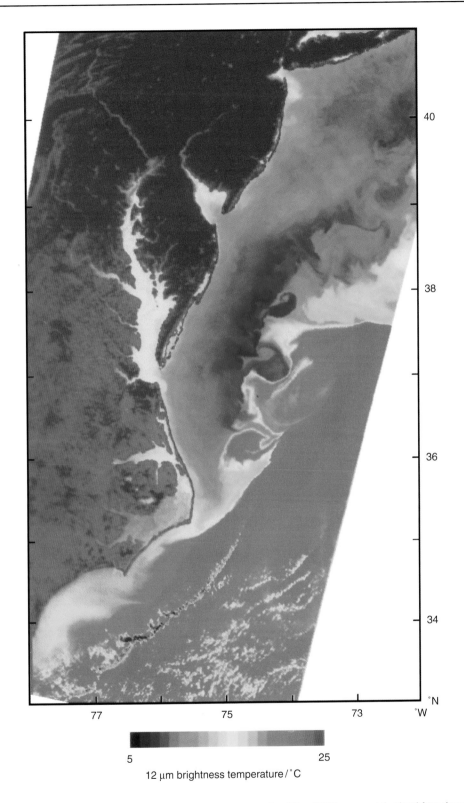

Figure 4 Brightness temperature image derived from the measurements of the ATSR on a nearly cloud-free day over the eastern coast of the USA. The warm core of the Gulf Stream is very apparent; it departs from the coast at Cape Hatteras. The cool, shelf water from the north entrains the warmer outflows from the Chesapeake and Delaware Bays. The north wall of the Gulf Stream reveals very complex structure associated with frontal instabilities that lead to exchanges between the Gulf Stream and inshore waters. The small-scale multicolored patterns over the warm Gulf Stream waters to the south indicate the presence of cloud. This image was taken at 15.18 UTC on 21 May 1992, and is derived from nadir view data from the 12 μm channel. (Generated from data © NERC/ESA/RAL/BNSC, 1992.)

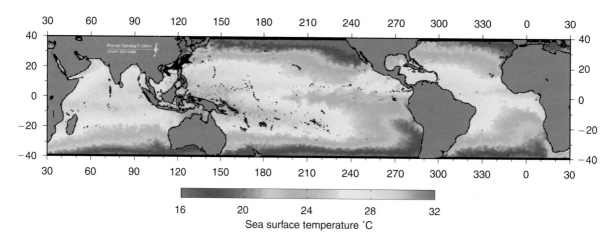

Figure 5 Tropical SSTs produced by microwave radiometer measurements from the TRMM (Tropical Rainfall Measuring Mission) Microwave Imager (TMI). This is a composite generated from data taken during the week ending December 22, 1999. The latitudinal extent of the data is limited by the orbital geometry of the TRMM satellite. The measurement is much less influenced by clouds than those in the infrared, but the black pixels in parts of the oceans where there are no islands indicate areas of heavy rainfall. The image reveals the cold tongue of surface water along the Equator in the Pacific Ocean and cold water off the Pacific coast of South America, indicating a non-El Niño situation. Note that the color scale is different from that used in **Figure 3**. The image was produced by Remote Sensing Systems, sponsored in part by NASA'S Earth Science Information Partnerships (ESIP) (a federation of information sites for Earth science); and by the NOAA/NASA Pathfinder Program for early EOS products; principal investigator: Frank Wentz.

uncertainties in the accuracy of the satellite-derived SSTs. Furthermore, the inherent natural variability of the global SST fields tends to mask any small, slow changes. Difficult though this task may be, global satellite-derived SSTs are an important component in climate change research.

Air-sea Exchanges

The SST fields play further indirect roles in the climate system in terms of modulating the exchanges of heat and greenhouse gases between the ocean and atmosphere. Although SST is only one of several variables that control these exchanges, the SST distributions, and their evolution on seasonal timescales can help provide insight into the global patterns of the air–sea exchanges. An example of this is the study of tropical cloud formation over the ocean, a consequence of air–sea heat and moisture exchange, in terms of SST distributions.

Future Developments

Over the next several years continuing improvement of the atmospheric correction algorithms can be anticipated to achieve better accuracies in the derived SST fields, particularly in the presence of atmospheric aerosols. This will involve the incorporation of information from additional spectral channels, such as those on MODIS or other EOS era satellite instruments. Improvements in SST coverage, at least in the tropics, can be expected in areas of heavy, persistent cloud cover by melding SST

retrievals from high-resolution infrared sensors with those from microwave radiometers, such as the TMI.

Continuing improvements in methods of validating the SST retrieval algorithms will improve our understanding of the error characteristics of the SST fields, guiding both the appropriate applications of the data and also improvements to the algorithms.

On the hardware front, a new generation of infrared radiometers designed for SST measurements will be introduced on the new operational satellite series, the National Polar-Orbiting Environmental Satellite Environments System (NPOESS) that will replace both the civilian (NOAA-n) and military (DMSP, Defense Meteorological Satellite Program) meteorological satellites. The new radiometer, called VIIRS (the Visible and Infrared Imaging Radiometer Suite), will replace the AVHRR and MODIS. The prototype VIIRS will fly on the NPP (NPOESS Preparatory Program) satellite scheduled for launch in late 2005. At present, the design details of the VIIRS are not finalized, but the physics of the measurement constrains the instrument to use the same atmospheric window channels as previous and current instruments, and have comparable, or better, measurement accuracies.

The ATSR series will continue with at least one more model, called the Advanced ATSR (AATSR) to fly on Envisat to be launched in 2001. The SST capability of this will be comparable to that of its predecessors.

Thus, the time-series of global SSTs that now extends for two decades will continue into the future to provide invaluable information for climate and oceanographic research.

See also

Air–Sea Gas Exchange. Carbon Dioxide (CO₂) Cycle. Coral Reef and other Tropical Fisheries. Current Systems in the Indian Ocean. Current Systems in the Southern Ocean. Dispersion in Shallow Seas. Electrical Properties of Sea Water. El Niño Southern Oscillation (ENSO). Evaporation and Humidity. Heat and Momentum Fluxes at the Sea Surface. IR Radiometers. Ocean Circulation. Ocean Color from Satellites. Penetrating Short-wave Radiation. Radiative Transfer in the Ocean. Satellite Altimetry. Satellite Measurements of Salinity. Satellite Oceanography, History and Introductory Concepts. Satellite Remote Sensing SAR. Shelf-sea and Slope Fronts. Thermohaline Circulation. Upper Ocean Time and Space Variability.

Further Reading

Barton IJ (1995) Satellite-derived sea surface temperatures: Current status. *Journal of Geophysical Research* 100: 8777–8790.

Gurney RJ, Foster JL and Parkinson CL (eds) (1993) *Atlas of Satellite Observations Related to Global Change.* Cambridge: Cambridge University Press.

Ikeda M and Dobson FW (1995) *Oceanographic Applications of Remote Sensing.* London: CRC Press.

Kearns EJ, Hanafin JA, Evans RH, Minnett PJ and Brown OB (2000) An independent assessment of Pathfinder AVHRR sea surface temperature accuracy using the Marine-Atmosphere Emitted Radiance Interferometer (M-AERI). Bulletin of the American Meteorological Society. 81: 1525–1536.

Kidder SQ and Vonder Haar TH (1995) *Satellite Meteorology: An Introduction.* London: Academic Press.

Legeckis R and Zhu T (1997) Sea surface temperature from the GEOS-8 geostationary satellite. *Bulletin of the American Meteorological Society* 78: 1971–1983.

May DA, Parmeter MM, Olszewski DS and Mckenzie BD (1998) Operational processing of satellite sea surface temperature retrievals at the Naval Oceanographic Office. *Bulletin of the American Meteorological Society,* 79: 397–407.

Robinson IS (1985) *Satellite Oceanography: An Introduction for Oceanographers and Remote-sensing Scientists.* Chichester: Ellis Horwood.

Stewart RH (1985) *Methods of Satellite Oceanography.* Berkeley, CA: University of California Press.

Victorov S (1996) *Regional Satellite Oceanography.* London: Taylor and Francis.

SATELLITE REMOTE SENSING SAR

A. K. Liu and S. Y. Wu, NASA Goddard Space Flight Center, Greenbelt, MD, USA

doi:10.1006/rwos.2001.0339

Introduction

Synthetic aperture radar (SAR) is a side-looking imaging radar usually operating on either an aircraft or a spacecraft. The radar transmits a series of short, coherent pulses to the ground producing a footprint whose size is inversely proportional to the antenna size, its aperture. Because the antenna size is generally small, the footprint is large and any particular target is illuminated by several hundred radar pulses. Intensive signal processing involving the detection of small Doppler shifts in the reflected signals from targets to the moving radar produces a high resolution image that is equivalent to one that would have been collected by a radar with a much larger aperture. The resulting larger aperture is the 'synthetic aperture' and is equal to the distance traveled by the spacecraft while the radar antenna is collecting information about the target. SAR techniques depend on precise determination of the relative position and velocity of the radar with respect to the target, and on how well the return signal is processed.

SAR instruments transmit radar signals, thus providing their own illumination, and then measure the strength and phase of the signals scattered back to the instrument. Radar waves have much longer wavelengths compared with light, allowing them to penetrate clouds with little distortion. In effect, radar's longer wavelengths average the properties of air with the properties and shapes of many individual water droplets, and are only affected while entering and exiting the cloud. Therefore, microwave radar can 'see' through clouds.

SAR images of the ocean surface are used to detect a variety of ocean features, such as refracting surface gravity waves, oceanic internal waves, wind fields, oceanic fronts, coastal eddies, and intense low pressure systems (i.e. hurricanes and polar lows), since they all influence the short wind waves responsible for radar backscatter. In addition, SAR is the only sensor that provides measurements of the directional wave spectrum from space. Reliable coastal wind vectors may be estimated from calibrated SAR images using the radar cross-section. The ability of a SAR to provide valuable information on the type, condition, and motion of the sea ice and surface signatures of swells, wind fronts, and eddies near the ice edge has also been amply demonstrated.

With all-weather, day/night imaging capability, SAR penetrates clouds, smoke, haze, and darkness to acquire high quality images of the Earth's surface. This makes SAR the frequent sensor of choice for cloudy coastal regions. Space agencies from the USA, Canada, and Europe use SAR imagery on an operational basis for sea ice monitoring, and for the detection of icebergs, ships, and oil spills. However, there can be considerable ambiguity in the interpretation of physical processes responsible for the observed ocean features. Therefore, the SAR imaging mechanisms of ocean features are briefly described here to illustrate how SAR imaging is used operationally in applications such as environmental monitoring, fishery support, and marine surveillance.

History

The first spaceborne SAR was flown on the US satellite Seasat in 1978. Although Seasat only lasted 3 months, analysis of its data confirmed the sensitivity of SAR to the geometry of surface features. On March 31, 1991 the Soviet Union became the next country to operate an earth-orbiting SAR with the launch of Almaz-1. Almaz-1 returned to earth in 1992 after operating for about 18 months. The European Space Agency (ESA) launched its first remote sensing satellite, ERS-1, with a C-band SAR on July 17, 1991. Shortly thereafter, the JERS-1 satellite, developed by the National Space Development Agency of Japan (NASDA), was launched on February 11, 1992 with an L-band SAR. This was followed a few years later by Radarsat-1, the first Canadian remote sensing satellite, launched in November 1995. Radarsat-1 has a ScanSAR mode with a 500 km swath and a 100 m resolution, an innovative variation of the conventional SAR (with a swath of 100 km and a resolution of 25 m). ERS-2 was launched in April 1995 by ESA, and Envisat-1 with an Advanced SAR is underway with a scheduled launch date in July 2001. The Canadian Space Agency (CSA) has Radarsat-2 planned for 2002, and NASDA has Advanced Land Observing Satellite (ALOS) approved for 2003. **Table 1** shows all major ocean-oriented spaceborne SAR missions worldwide from 1978 to 2003.

Table 1   Major ocean-oriented spaceborne SAR missions

Platform	Nation	Launch	Band[a]	Status
Seasat	USA	1978	L	Ended
Almaz-1	USSR	1991	S	Ended
ERS-1	Europe	1991	C	Standby
JERS-1	Japan	1992	L	Ended
ERS-2	Europe	1995	C	Operational
Radarsat-1	Canada	1995	C	Operational
Envisat-1	Europe	2001	C	Launch scheduled
Radarsat-2	Canada	2002	C	Approved
ALOS	Japan	2003	L	Approved

[a]Some frequently used radar wavelengths are: 3.1 cm for X-band, 5.66 cm for C-band, 10.0 cm for S-band, and 23.5 cm for L-band.

Aside from these free-flying missions, a number of early spaceborne SAR experiments in the USA were conducted using shuttle imaging radar (SIR) systems flown on NASA's Space Shuttle. The SIR-A and SIR-B experiments, in November 1981 and October 1984, respectively, were designed to study radar system performance and obtain sample data of the land using various incidence angles. The SIR-B experiment provided a unique opportunity for studying ocean wave spectra due to the relatively lower orbit of the Shuttle as compared with satellites. The low orbital altitude increases the frequency range of ocean waves that could be reliably imaged, because blurring of the detected waves caused by the motion of ocean surface during the imaging process is reduced. The final SIR mission, SIR-C in April and October 1994, simultaneously recorded SAR data at three wavelengths (L-, C-, and X-bands). These multiple-frequency data from SIR-C improved our understanding of the radar scattering properties of the ocean surface.

Imaging Mechanism of Ocean Features

For a radar with an incidence angle of $20°$–$50°$, such as all spaceborne SARs, backscatter from the ocean surface is produced primarily by the Bragg resonant scattering mechanism. That is, surface waves traveling in the radar range (across-track) direction with a wavelength of $\lambda/(2 \sin \theta)$, called the Bragg resonant waves, account for most of the backscattering. In this formula, λ is the radar wavelength, and θ is the incidence angle. In general, the Bragg resonant waves are short gravity waves with wavelengths in the range of 3–30 cm, depending on the radar wavelength or band, as shown in **Table 1**. Because SAR is most sensitive to waves of this wavelength, or roughness of this scale, any

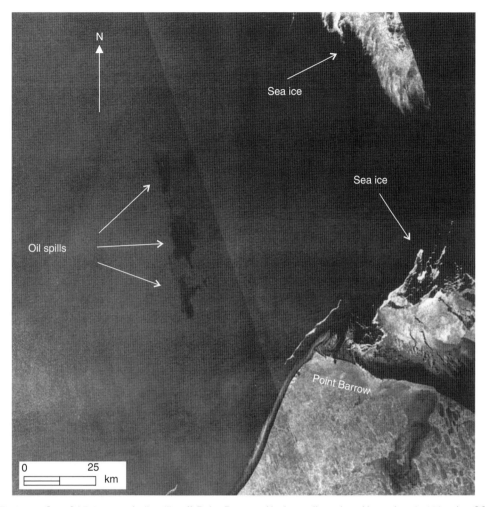

Figure 1 Radarsat ScanSAR image of oil spills off Point Barrow, Alaska, collected on November 2, 1997. (© CSA 1997.)

ocean phenomenon or process that produces modulation in these particular wavelengths is theoretically detectable by SAR. The radar cross-section of the ocean surface is affected by any geophysical variable, such as wind stress, current shear, or surface slicks, that can modulate the ocean surface roughness at Bragg-scattering scales. Thus, SARs have proven to be an excellent means of mapping ocean features.

For ocean current features, the essential element of the surface manifestation is the interaction between the current field and the wind-driven ocean surface waves. The effect of the surface current is to alter the short-wave spectrum from its equilibrium value, while the natural processes of wave energy input from the wind restores the ambient equilibrium spectrum. A linear SAR system is one for which the variation of the SAR image intensity is proportional to the gradient of the surface velocity. The proportionality depends on radar wavelength, radar incidence angle, angle between the radar look direction and the current direction, azimuth angle,

and the wind velocity. Under high wind condition, large wind waves may overwhelm the weaker current feature. When current flows in the cross-wind direction, the wave–current interaction is relatively weak, causing a weak radar backscattering signal.

For ocean frontal features, the change in surface brightness across a front in a SAR image is caused by the change in wind stress exerted onto the ocean surface. The wind stress in turn depends on wind speed and direction, air–sea temperature difference, and surface contamination. The effects of wind stress upon surface ripples and therefore upon radar cross-section, have been modeled and demonstrated as shown in the example below. In the high wind stress area, the ocean surface is rougher and appears as a brighter area in a SAR image. On the other side of the front where the wind is lower, the surface is smoother and appears as a darker area in a SAR image.

The reason why surface films are detectable on radar images is that oil films have a dampening effect on short surface waves. Radar is remarkably

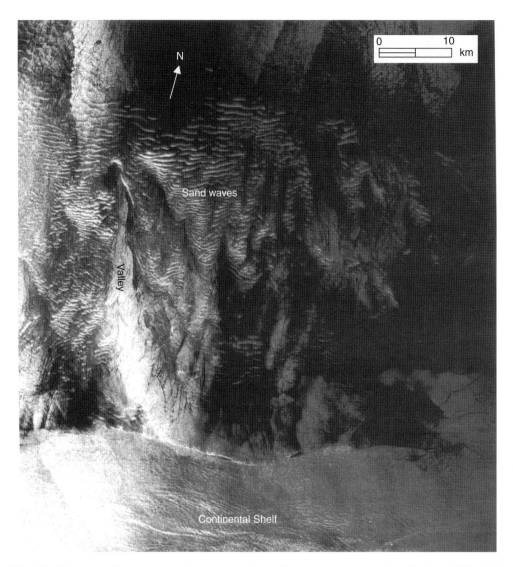

Figure 2 ERS-1 SAR image of shallow water bathymetry at Taiwan Tan acquired on July 27, 1994. (© ESA 1994.)

sensitive to small changes in the roughness of sea surface. Oil slicks also have a dark appearance in radar images and are thus similar to the appearance of areas of low winds. The distinctive shape and sharp boundary of localized surface films allows them to be distinguished from the relatively large regions of low wind.

Examples of Ocean Features from SAR Applications

A number of important SAR applications have emerged recently, particularly since ERS-1/2, and Radarsat-1 data became available and the ability to process SAR data has improved. In the USA, the National Oceanic and Atmospheric Administration (NOAA) and the National Ice Center use SAR im-

agery on an operational basis for sea ice monitoring, iceberg detection, fishing enforcement, oil spill detection, wind and storm information. In Canada, sea ice surveillance is now a proven near-real-time operation, and new marine and coastal applications for SAR imagery are still emerging. In Europe, research on SAR imaging of ocean waves has received great attention in the past 10 years, and has contributed to better global ocean wave forecasts. However, the role of SAR in the coastal observing system still remains at the research and development stage. For reference, examples of some typical ocean features from SAR applications are provided below.

For marine environmental monitoring, features such as oil spills, bathymetry, and polar lows are important for tracking and can often be identified easily with SAR. In early November 1997,

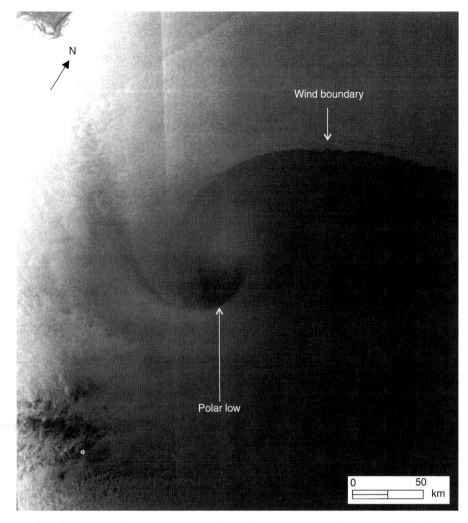

N

Wind boundary

Polar low

0   50
          km

Figure 3 Radarsat ScanSAR image of a polar low in the Bering Sea collected on February 5, 1998. (© CSA 1998.)

Radarsat's SAR sensor captured an oil spill off Point Barrow, Alaska. The oil slicks showed up clearly on the ScanSAR imagery on November 2, 3 and 9. The oil spill is suspected to be associated with the Alaskan Oil Pipeline. **Figure 1** shows a scene containing the oil slicks cropped out from the original ScanSAR image for a closer look. Tracking oil spills using SAR is useful for planning clean-up activities. Early detection, monitoring, containment, and clean up of oil spills are crucial to the protection of the environment.

Under favorable wind conditions with strong tidal current, the surface signature of bottom topography in shallow water has often been observed in SAR images. **Figure 2**, showing an ERS-1 SAR image of the shallow water bathymetry of Taiwan Tan collected on July 27, 1994, is such an example. Taiwan Tan is located south west of Taiwan in the Taiwan Strait. Typical water depth there is around 30 m with a valley in the middle and the continental shelf

break to the south. An extensive sand wave field (**Figure 2**) is developed at Taiwan Tan regularly by wind, tidal current, and surface waves. Monitoring the changes of bathymetry is critical to ship navigation, especially in the areas where water is shallow and ship traffic is heavy.

Intense low pressure systems in polar regions, often referred to as polar lows, may develop above regions between colder ice/land and warmer ocean during cold air outbreaks. These intense polar lows are formed off major jet streams in cold air masses. Since they usually occur near polar regions where data are sparse, SAR images have been a useful tool for studying these phenomena. **Figure 3** shows a Radarsat ScanSAR image of a polar low in the Bering Sea (centered at 58.0°N, 174.9°E) collected on February 5, 1998. It has a wind boundary to the north spiraling all the way to the center of the storm that separates the high wind (bright) area from the low wind (dark) area. The rippled character along

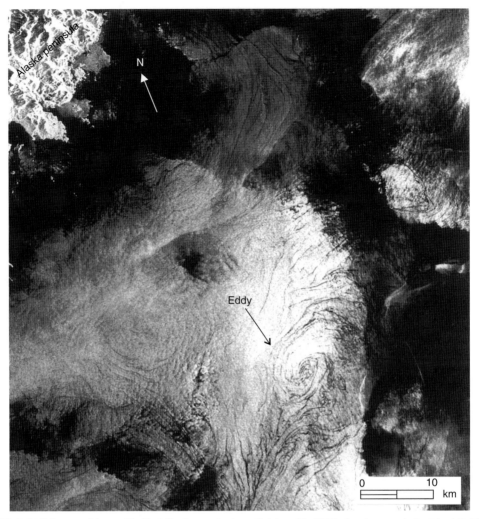

Figure 4 ERS-1 SAR image of lower Shelikof Strait, acquired on October 23, 1991 showing a spiral eddy. (© ESA 1991.)

the wind boundary indicates the presence of an instability disturbance induced by the shear flow, which in turn is caused by the substantial difference in wind speed across the boundary.

Ocean features such as eddies, fronts, and ice edges can result in changes in water temperature, turbulence, or transport and may be the primary determinant of recruitment to fisheries. The survival of larvae is enhanced if they remain on the continental shelf and ultimately recruit to nearshore nursery areas. Features such as fronts and eddies can retain larval patches within the shelf zone. **Figure 4** shows an ERS-1 SAR image acquired on October 23, 1991 (centered at 56.69°N, 156.07°W) in lower Shelikof Strait, the Gulf of Alaska. In this image, an eddy with a diameter of approximately 20 km is visible due to low wind conditions. The eddy is characterized by spiraling curvilinear lines which are most likely associated with current shears, surface films, and to a lesser extent temperature contrasts. SAR

has the potential to locate these eddies over extensive areas in coastal oceans.

A Radarsat ScanSAR image over the Gulf of Mexico taken on November 23, 1997 (**Figure 5**) shows a distinct, nearly straight front stretching at least 300 km in length. The center of the front in this scene is in the Gulf of Mexico some 400 km south west of New Orleans. The frontal orientation is about 76° east of north. Closer inspection reveals that there are many surface film-like filaments on the south side of the front. Concurrent wind data suggest that surface currents converge along the front. Therefore, the formation of this front is probably caused by the accumulation of natural surface films brought about by the convergence around the front. This example highlights SAR's sensitivity to the changes of wind speed and the presence of surface films.

The edge of the sea ice has been found to be highly productive for plankton spring bloom and

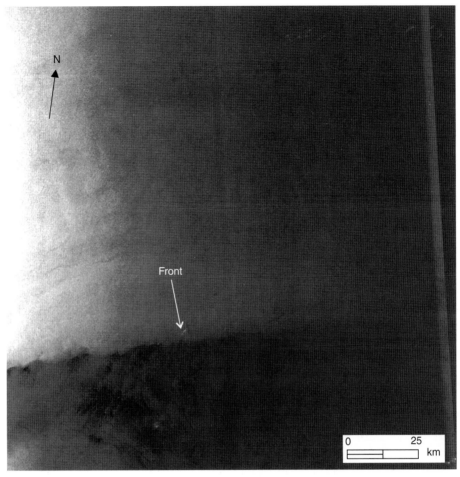

Figure 5 Radarsat ScanSAR image collected over the Gulf of Mexico on November 23, 1997 showing a frontal boundary. (© CSA 1997.)

fishery feeding. In the Bering Sea, fish abundance is highly correlated with yearly ice extent because for their survival many species of fish prefer the cold pools left behind after ice retreat. SAR images are very useful for tracking the movement of the ice edge. **Figure 6** shows a Radarsat ScanSAR image near the ice edge in the Bering Sea collected on February 29, 2000 (centered at 59.6°N and 177.3°W). The sea ice pack with ice bands extending from the ice edge can be clearly seen as the bright area because sea ice surface is rougher than ocean surface (dark area). In the same image, a front is also visible and may or may not be associated with the ice edge to the north. The cold water near the ice edge dampens wave action and appears as a darker area compared with the other side of the front, where it shows up as brighter area due to higher wind and higher sea states.

Information on surface and internal waves, as well as ship wakes, are very important and valuable for marine surveillance and ship navigation. The principal use of SAR for oceanographic studies has

been for the detection of ocean waves. The wave direction and height derived from SAR data can be incorporated into models of wind–wave forecast and other applications such as wave–current interaction. **Figure 7** shows an ERS-1 SAR image of long surface waves (or swells) in the lower Shelikof Strait collected on October 17, 1991 (centered at 56.11°N, 156.36°W). The location is close to that of the spiral eddy shown in **Figure 4**. Because of the higher winds and higher sea states at the time the image was acquired, the eddy is less conspicuous in this SAR image taken 6 days earlier. Although the direct surface signature of the eddy cannot be discerned clearly in this image, the wave refraction in the eddy area can still be observed. The rays of the wave field can be traced out directly from the SAR image. The ray pattern provides information on the wave refraction pattern and on the relative variation of wave energy along a ray through wave–current interaction.

Tidal currents flowing over submarine topographic features such as a sill or continental shelf in

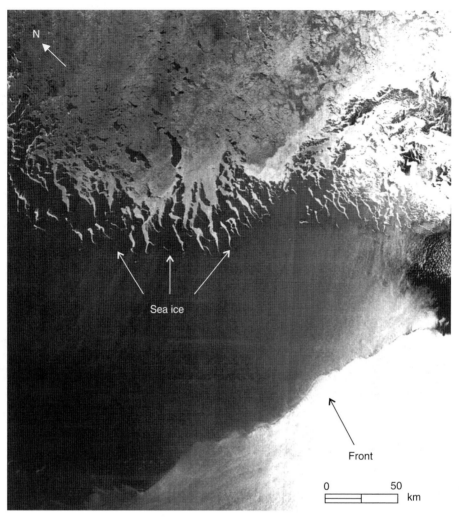

Figure 6 Radarsat ScanSAR image collected over the Bering Sea on February 29, 2000 showing a front near the ice edge. (© CSA 2000.)

a stratified ocean can generate nonlinear internal waves of tidal frequency. This phenomenon has been studied by many investigators. Direct observations have lent valuable insight into the internal wave generation process and explained the role they play in the transfer of energy from tides to ocean mixing. These nonlinear internal waves are apparently generated by internal mixing as tidal currents flow over bottom features and propagate in the open ocean. In the South China Sea near DongSha Island, enormous westward propagating internal waves from the open ocean are often confronted by coral reefs on the continental shelf. As a result, the waves are diffracted upon passing the reefs. **Figure 8** shows a Radarsat ScanSAR image collected over the northern South China Sea on April 26, 1998, showing at least three packets of internal waves. Each packet consists of a series of internal waves, and the pattern of each wave is characterized by

a bright band followed immediately by a dark band. The bright/dark bands indicate the contrast in ocean rough/smooth surfaces caused by convergence/divergence areas induced by the internal waves. At times, the wave 'crest' as observed by SAR from the length of bands can be over 200 km long. After passing the DongSha coral reefs, the waves regroup themselves into two separate packets of internal waves. Later, they interact with each other and emerge as a single wave packet again. SAR can be a very useful tool for studying these shelf processes and the effect of the internal waves on oil drilling platforms, nutrient mixing, and sediment transport.

Ships and their wakes are commonly observable in high-resolution satellite SAR imagery. Detection of ships and ship wakes by means of remote sensing can be useful in the areas of national defense intelligence, shipping traffic, and fishing enforcement. **Figure 9** is an ERS-1 SAR image collected on May 31,

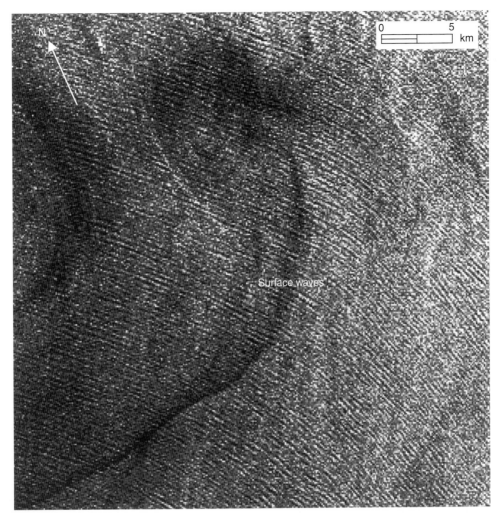

Figure 7 ERS-1 SAR image of lower Shelikof Strait, obtained on October 17, 1991 showing long surface gravity waves refracted by current. (© ESA 1991.)

1995 near the northern coast of Taiwan. The image is centered at 25.62°N and 121.15°E, approximately 30 km offshore in the East China Sea. A surface ship heading north east, represented by a bright spot, can be easily identified. Behind this ship, a long dark turbulence wake is clearly visible. The turbulent wake dampens any short waves, resulting in an area with low backscattering as indicated by the arrow A in **Figure 9**. Near the ship, the dark wake is accompanied by a bright line which may be caused by the vortex shed by the ship into its wake. The ship track follows the busy shipping lane between Hong Kong, Taiwan, and Japan. The ambient dark slicks are natural surface films induced by upwelling on the continental shelf. In the lower part of the image, another ship turbulent wake (long and dark linear feature oriented east–west) can be identified near the location B in **Figure 9**. A faint bright line connects to the end of this turbulent wake, forming a V-shaped wake in the box B. The faintness of this second ship may be caused by very low backscattering of the ship configuration or the wake could have been formed by a submarine. In the latter case, it must have been operating very close to the ocean surface, since the surface wake is observable. The ship wake is pointing to the east, indicating that the faint ship was moving from mainland China toward the open ocean.

Discussion

As mentioned earlier, SAR has the unique capability of operating during the day or night and under all weather conditions. With repeated coverage, spaceborne SAR instruments provide the most efficient means to monitor and study the changes in important elements of the marine environment. As demonstrated by the above examples, the use of SAR-

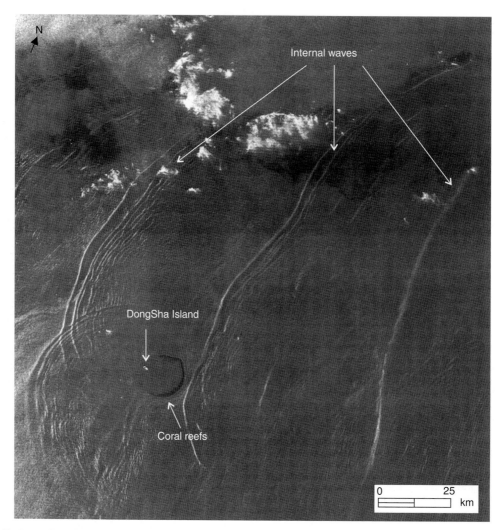

Figure 8 Radarsat ScanSAR image collected over the South China Sea on April 26, 1998 showing three internal wave packets. (© CSA 1998.)

derived observations to track eddies, fronts, ice edges, and oil slicks can supply valuable information and can aid in the management of the fishing industry and the protection of the environment. In overcast coastal areas at high latitudes, the uniformly cold sea surface temperature and persistent cloud cover preclude optical and infrared measurement of surface temperature features, and obscure ocean color observations. The mapping of ocean features by SAR in these challenging coastal regions is, therefore, a potentially major application for satellite-based SAR, particularly for the wider swath Scan-SAR mode. Furthermore, SAR data provide unique information for studying the health of the Earth system, as well as critical data for natural hazards and resource assessments.

The prospect of SAR data collection extending well into the twenty-first century gives impetus to current research in SAR applications in ocean science and opens the doors to change detection studies on decadal timescales. The next step is to move into the operational use of SAR data to complement ground measurements. The challenge is to increase cooperation in the scheduling, processing, dissemination, and pricing of SAR data from all SAR satellites between international space agencies. Such cooperation might permit near-real-time high-resolution coastal SAR measurements of sufficient temporal and spatial coverage to impact weather forecasting for selected heavily populated coastal regions. It is necessary to bear in mind that each satellite image is a snapshot and can be complemented with buoy and ship measurements. Ultimately, these data sets should be integrated by numerical models. Such validated and calibrated models will prove extremely useful in understanding a wide variety of oceanic processes.

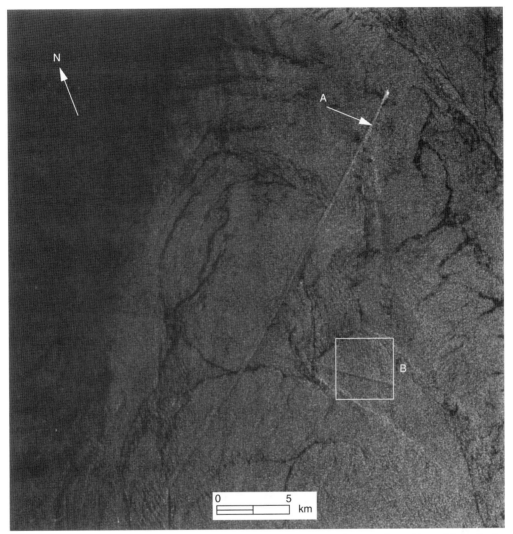

Figure 9 ERS-1 SAR image of East China Sea, obtained on May 31, 1995 showing a surface ship and its wake (arrow A) and a V-shaped wake in box B. (© ESA 1995.)

See also

Aircraft Remote Sensing. Beaches, Physical Processes affecting. Ice–Ocean Interaction. Satellite Altimetry. Satellite Oceanography, History and Introductory Concepts. Satellite Passive Microwave Measurements of Sea Ice. Satellite Remote Sensing Microwave Scatterometers. Satellite Remote Sensing of Sea Surface Temperatures. Surface Films. East Australian Current. Wave Generation by Wind.

Further Reading

Alaska SAR Facility User Working Group (1999) *The Critical Role of SAR in Earth System Science.* (http://www.asf.alaska.edu/)

Beal RC and Pichel WG (eds) (2000) *Coastal and Marine Applications of Wide Swath SAR.* Johns Hopkins APL Technical Digest, 21.

European Space Agency (1995) *Scientific Achievements of ERS-1.* ESA SP-1176/I.

Fu L and Holt B (1982) *Seasat Views Oceans and Sea Ice with Synthetic Aperture Radar,* JPL Publication, pp. 81–120. Pasadena, CA: NASA, JPL/CIT.

Hsu MK, Liu AK and Liu C (2000) An internal wave study in the China Seas and Yellow Sea by SAR. *Continental Shelf Research* 20: 389–410.

Liu AK, Peng CY and Schumacher JD (1994) Wave-current interaction study in the Gulf of Alaska for detection of eddies by SAR. *Journal of Geophysical Research* 99: 10075–10085.

Liu AK, Peng CY and Weingartner TJ (1994) Ocean–ice interaction in the marginal ice zone using SAR. *Journal of Geophysical Research* 99: 22391–22400.

Liu AK, Peng CY and Chang YS (1996) Mystery ship detected in SAR image. *EOS, Transactions, American Geophysical Union* 77: 17–18.

Liu AK, Peng CY and Chang YS (1997) Wavelet analysis of satellite images for coastal watch. *IEEE Journal of Oceanic Engineering* 22: 9–17.

Tsatsoulis C and Kwok R (1998) *Analysis of SAR Data of the Polar Oceans.* Berlin: Springer-Verlag.

SCHOOLING

See **FISH SCHOOLING**

SEA ICE

Overview

W. F. Weeks, Portland, OR, USA

doi:10.1006/rwos.2001.0001

Introduction

Sea ice, any form of ice found at sea that originated from the freezing of sea water, has historically been among the least-studied of all the phenomena that have a significant effect on the surface heat balance of the Earth. Fortunately, this neglect has recently lessened as the result of improvements in observational and operational capabilities in the polar ocean areas. As a result, considerable information is now available on the nature and behavior of sea ice as well as on its role in affecting the weather, the climate, and the oceanography of the polar regions and possibly of the planet as a whole.

Extent

Although the majority of Earth's population has never seen sea ice, in area it is extremely extensive: 7% of the surface of the Earth is covered by this material during some time of the year. In the northern hemisphere the area covered by sea ice varies between 8×10^6 and $15 \times 10^6 \, km^2$, with the smaller number representing the area of multiyear (MY) ice remaining at the end of summer. In summer this corresponds roughly to the contiguous area of the United States and to twice that area in winter, or to between 5% and 10% of the surface of the northern hemisphere ocean. At maximum extent, the ice extends down the western side of the major ocean basins, following the pattern of cold currents and reaching the Gulf of St. Lawrence (Atlantic) and the Okhotsk Sea off the north coast of Japan (Pacific). The most southerly site in the northern hemisphere where an extensive sea ice cover forms is the Gulf of Bo Hai, located off the east coast of China at 40°N. At the end of the summer the perennial MY ice pack

of the Arctic is primarily confined to the central Arctic Ocean with minor extensions into the Canadian Arctic Archipelago and along the east coast of Greenland.

In the southern hemisphere the sea ice area varies between 3×10^6 and $20 \times 10^6 \, km^2$, covering between 1.5% and 10% of the ocean surface. The amount of MY ice in the Antarctic is appreciably less than in the Arctic, even though the total area affected by sea ice in the Antarctic is approximately a third larger than in the Arctic. These differences are largely caused by differences in the spatial distributions of land and ocean. The Arctic Ocean is effectively land-locked to the south, with only one major exit located between Greenland and Svalbard. The Southern Ocean, on the other hand, is essentially completely unbounded to the north, allowing unrestricted drift of the ice in that direction, resulting in the melting of nearly all of the previous season's growth.

Geophysical Importance

In addition to its considerable extent, there are good reasons to be concerned with the health and behavior of the world's sea ice covers. Sea ice serves as an insulative lid on the surface of the polar oceans. This suppresses the exchange of heat between the cold polar air above the ice and the relatively warm sea water below the ice. Not only is the ice itself a good insulator, but it provides a surface that supports a snow cover that is also an excellent insulator. In addition, when the sea ice forms with its attendant snow cover, it changes the surface albedo, α (i.e., the reflection coefficient for visible radiation) of the sea from that of open water ($\alpha = 0.15$) to that of newly formed snow ($\alpha = 0.85$), leading to a 70% decrease in the amount of incoming short-wave solar radiation that is absorbed. As a result, there are inherent positive feedbacks associated with the existence of a sea ice cover. For instance, a climatic warming will presumably reduce both the extent and the thickness of the sea ice. Both of these changes will, in turn, result in increases in the temperature of the atmosphere and of the sea, which will further reduce ice thickness and extent. It is this positive feedback that is a major factor in producing the unusually large increases in arctic temperatures that are forecast by numerical models simulating the effect of the accumulation of greenhouse gases.

The presence of an ice cover limits not only the flux of heat into the atmosphere but also the flux of moisture. This effect is revealed by the common presence of linear, local clouds associated with individual leads (cracks in the sea ice that are covered with either open water or thinner ice). In fact, sea ice exerts a significant influence on the radiative energy balance of the complete atmosphere–sea ice–ocean system. For instance, as the ice thickness increases in the range between 0 and 70 cm, there is an increase in the radiation absorption in the ice and a decrease in the ocean. There is also a decrease in the radiation adsorption by the total atmosphere–ice–ocean system. It is also known that the upper 10 cm of the ice can absorb over 50% of the total solar radiation, and that decreases in ice extent produce increases in atmospheric moisture or cloudiness, in turn altering the surface radiation budget and increasing the amount of precipitation. Furthermore, all the ultraviolet and infrared radiation is absorbed in the upper 50 cm of the ice; only visible radiation penetrates into the lower portions of thicker ice and into the upper ocean beneath the ice. Significant changes in the extent and/or thickness of sea ice would result in major changes in the climatology of the polar regions. For instance, recent computer simulations in which the ice extent in the southern hemisphere was held constant and the amount of open water (leads) within the pack was varied showed significant changes in storm frequencies, intensities and tracks, precipitation, cloudiness, and air temperature.

However, there are even less obvious but perhaps equally important air–ice and ice–ocean interactions. Sea ice drastically reduces wave-induced mixing in the upper ocean, thereby favoring the existence of a 25–50 m thick, low-salinity surface layer in the Arctic Ocean that forms as the result of desalination processes associated with ice formation and the influx of fresh water from the great rivers of northern Siberia. This stable, low-density surface layer prevents the heat contained in the comparatively warm (temperatures of up to $+3°C$) but more saline denser water beneath the surface layer from affecting the ice cover. As sea ice rejects roughly two-thirds of the salt initially present in the sea water from which the ice forms, the freezing process is equivalent to distillation, producing both a low-salinity component (the ice layer itself) and a high-salinity component (the rejected brine). Both of these components play important geophysical roles. Over shallow shelf seas, the rejected brine, which is dense, cold, and rich in CO_2, sinks to the bottom, ultimately feeding the deep-water and the bottom-water layers of the world ocean. Such processes are particularly effective in regions where large polynyas exist (semipermanent open water and thin-ice areas at sites where climatically much thicker ice would be anticipated).

The 'fresh' sea ice layer also has an important geophysical role to play in that its exodus from the Arctic Basin via the East Greenland Drift Stream represents a fresh water transport of $2366 \text{ km}^3 \text{ y}^{-1}$ (c. 0.075 Sv). This is a discharge equivalent to roughly twice that of North America's four largest rivers combined (the Mississippi, St. Lawrence, Columbia, and Mackenzie) and in the world is second only to the Amazon. This fresh surface water layer is transported with little dispersion at least as far as the Denmark Strait and in all probability can be followed completely around the subpolar gyre of the North Atlantic. Even more interesting is the speculation that during the last few decades this fresh water flux has been sufficient to alter or even stop the convective regimes of the Greenland, Iceland and Norwegian Seas and perhaps also of the Labrador Sea. This is a sea ice-driven, small-scale analogue of the so-called halocline catastrophe that has been proposed for past deglaciations, when it has been argued that large fresh water runoff from melting glaciers severely limited convective regimes in portions of the world ocean. The difference is that, in the present instance, the increase in the fresh water flux that is required is not dramatic because at near-freezing temperatures the salinity of the sea water is appreciably more important than the water temperature in controlling its density. It has been proposed that this process has contributed to the low near-surface salinities and heavy winter ice conditions observed north of Iceland between 1965 and 1971, to the decrease in convection described for the Labrador Sea during 1968–1971, and perhaps to the so-called 'great salinity anomaly' that freshened much of the upper North Atlantic during the last 25 years of the twentieth century. In the Antarctic, comparable phenomena may be associated with freezing in the southern Weddell Sea and ice transport northward along the Antarctic Peninsula.

Sea ice also has important biological effects at both ends of the marine food chain. It provides a substrate for a special category of marine life, the ice biota, consisting primarily of diatoms. These form a significant portion of the total primary production and, in turn, support specialized grazers and species at higher trophic levels, including amphipods, copepods, worms, fish, and birds. At the upper end of the food chain, seals and walruses use ice extensively as a platform on which to haul out and give birth to young. Polar bears use the ice as a platform while hunting. Also important is the fact that in shelf seas such as the Bering and Chukchi, which are well mixed in the winter, the melting of the ice cover in the spring lowers the surface salinity, increasing the stability of the water column. The reduced mixing concentrates phytoplankton in the near-surface photic zone, thereby enhancing the overall intensity of the spring bloom. Finally, there are the direct effects of sea ice on human activities. The most important of these are its barrier action in limiting the use of otherwise highly advantageous ocean routes between the northern Pacific regions and Europe and its contribution to the numerous operational difficulties that must be overcome to achieve the safe extraction of the presumed oil and gas resources of the polar shelf seas.

Properties

Because ice is a thermal insulator, the thicker the ice, the slower it grows, other conditions being equal. As sea ice either ablates or stops growing during the summer, there is a maximum thickness of first-year (FY) ice that can form during a specific year. The exact value is, of course, dependent upon the local climate and oceanographic conditions, reaching values of slightly over 2 m in the Arctic and as much as almost 3 m at certain Antarctic sites. It is also clear that during the winter the heat flux from areas of open water into the polar atmosphere is significantly greater than the flux through even thin ice and is as much as 200 times greater than the flux through MY ice. This means that, even if open water and thin ice areas comprise less than 1–2% of the winter ice pack, lead areas must still be considered in order to obtain realistic estimates of ocean–atmosphere thermal interactions.

If an ice floe survives a summer, during the second winter the thickness of the additional ice that is added is less than the thickness of nearby FY ice for two reasons: it starts to freeze later and it grows slower. Nevertheless, by the end of the winter, the second-year ice will be thicker than the nearby FY ice. Assuming that the above process is repeated in subsequent years, an amount of ice is ablated away each summer (largely from the upper ice surface) and an amount is added each winter (largely on the lower ice surface). As the year pass, the ice melted on top each summer remains the same (assuming no change in the climate over the ice), while the ice forming on the bottom becomes less and less as

a result of the increased insulating effect of the thickening overlying ice. Ultimately, a rough equilibrium is reached, with the thickness of the ice added in the winter becoming equal to the ice ablated in the summer. Such steady-state MY ice floes can be layer cakes of ten or more annual layers with total thicknesses in the range 3.5–4.5 m. Much of the uncertainty in estimating the equilibrium thickness of such floes is the result of uncertainties in the oceanic heat flux. However, in sheltered fiord sites in the Arctic where the oceanic heat flux is presumed to be near zero, MY fast ice with thicknesses up to roughly 15–20 m is known to occur. Another important factor affecting MY ice thickness is the formation of melt ponds on the upper ice surface during the summer in that the thicknesses and areal extent of these shallow-water bodies is important in controlling the total amount of short-wave radiation that is absorbed. For instance, a melt pond with a depth of only 5 cm can absorb nearly half the total energy absorbed by the whole system. The problem here is that good regional descriptive characterizations of these features are lacking as the result of the characteristic low clouds and fog that occur over the Arctic ice packs in the summer. Particularly lacking are field observations on melt pond depths as a function of environmental variables. Also needed are assessments of how much of the meltwater remains ponded on the surface of the ice as contrasted with draining into the underlying sea water. Thermodynamically these are very different situations.

Conditions in the Antarctic are, surprisingly, rather different. There, surface melt rates within the pack are small compared to the rates at the northern boundary of the pack. The stronger winds and lower humidities encountered over the pack also favor evaporation and minimize surface melting. The limited ablation that occurs appears to be controlled by heat transfer processes at the ice–water interface. As a result, the ice remains relatively cold throughout the summer. In any case, as most of the Antarctic pack is advected rapidly to the north, where it encounters warmer water at the Antarctic convergence and melts rapidly, only small amounts of MY ice remain at the end of summer.

Sea ice properties are very different from those of lake or river ice. The reason for the difference is that when sea water freezes, roughly one-third of the salt in the sea water is initially entrapped within the ice in the form of brine inclusions. As a result, initial ice salinities are typically in the range 10–12‰. At low temperatures (below $-8.7°C$), solid hydrated salts also form within the ice. The composition of the brine in sea ice is a unique function of the temperature, with the brine composition becoming more saline as the temperature decreases. Therefore, the brine volume (the volumetric amount of liquid brine in the ice) is determined by the ice temperature and the bulk ice salinity. Not only is the temperature of the ice different at different levels in the ice sheet but the salinity of the ice decreases further as the ice ages ultimately reaching a value of $\sim 3\%_{0}$ in MY ice. Brine volumes are usually lower in the colder upper portions of the ice and higher in the warmer, lower portions. They are particularly low in the above-sea-level part of MY ice as the result of the salt having drained almost completely from this ice. In fact, the upper layers of thick MY ice and of aged pressure ridges produce excellent drinking water when melted. As brine volume is the single most important parameter controlling the thermal, electrical, and mechanical properties of sea ice, these properties show associated large changes both vertically in the same ice sheet and between ice sheets of differing ages and histories. To add complexity to this situation, exactly how the brine is distributed within the sea ice also affects ice properties.

There are several different structural types of sea ice, each with characteristic crystal sizes and preferred crystal orientations and property variations. The two most common structural types are called congelation and frazil. In congelation ice, large elongated crystals extend completely through the ice sheet, producing a structure that is similar to that found in directionally solidified metals. In the Arctic, large areas of congelation ice show crystal orientations that are so similar as to cause the ice to have directionally dependent properties in the horizontal plane as if the ice were a giant single crystal. Frazil, on the other hand, is composed of small, randomly oriented equiaxed crystals that are not vertically elongated. Congelation is more common in the Arctic, while frazil is more common in the Antarctic, reflecting the more turbulent conditions characteristically found in the Southern Ocean.

Two of the more unusual sea ice types are both subsets of so-called 'underwater ice.' The first of these is referred to as platelet ice and is particularly common around margins of the Antarctic continent at locations where ice shelves exist. Such shelves not only comprise 30% of the coastline of Antarctica, they also can be up to 250 m thick. Platelet ice is composed of a loose open mesh of large platelets that are roughly triangular in shape with dimensions of 4–5 cm. In the few locations that have been studied, platelet ice does not start to develop until the fast ice has reached a thickness of several tens of centimeters. Then the platelets develop beneath the fast ice, forming a layer that can be several meters thick. The fast ice appears to serve as a superstrate that facilitates the initial nucleation of the platelets. Ultimately, as the fast ice thickens, it incorporates some of the upper platelets. In the McMurdo Sound region, platelets have been observed forming on fish traps at a depth of 70 m. At locations near the Filchner Ice Shelf, platelets have been found in trawls taken at 250 m. This ice type appears to be the result of crystal growth into water that has been supercooled a fraction of a degree. The mechanism appears to be as follows. There is evidence that melting is occurring on the bottom of some of the deeper portions of the Antarctic ice shelves. This results in a water layer at the ice–water interface that is not only less saline and therefore less dense than the underlying seawater, but also is exactly at its freezing point at that depth because it is in direct contact with the shelf ice. When this water starts to flow outward and upward along the base of the shelf, supercooling develops as a result of adiabatic decompression. This in turn drives the formation of the platelet ice.

The second unusual ice type is a special type of frazil that results from what has been termed suspension freezing. The conditions necessary for its formation include strong winds, intense turbulence in an open water area of a shallow sea and extreme sub-freezing temperatures. Such conditions are characteristically found either during the initial formation of an ice cover in the fall or in regions where polynya formation is occurring, typically by newly formed ice being blown off of a coast or a fast ice area leaving in its wake an area of open water. When such conditions occur, the water column can become supercooled, allowing large quantities of frazil crystals to form and be swept downward by turbulence throughout the whole water column. Studies of benthic microfossils included in sea ice during such events suggest that supercooling commonly reaches depths of 20–25 m and occasionally to as much as 50 m. The frazil ice crystals that form occur in the form of 1–3 mm diameter discoids that are extremely sticky. As a result, they are not only effective in scavenging particulate matter from the water column but they also adhere to material on the bottom, where they continue to grow fed by the supercooled water. Such so-called anchor ice appears to form selectively on coarser material. The resulting spongy ice masses that develop can be quite large and, when the turbulence subsides, are quite buoyant and capable of floating appreciable quantities of attached sediment to the surface. There it commonly becomes incorporated in the overlying sea ice. In rivers, rocks weighing as much as 30 kg have been observed to be incorporated into an ice cover by this mechanism. Recent interest in this subject has been the result of the possibility that this mechanism has been effective in incorporating hazardous material into sea ice sheets, which can then serve as a long-distance transport mechanism.

Drift and Deformation

If sea ice were motionless, ice thickness would be controlled completely by the thermal characteristics of the lower atmosphere and the upper ocean. Such ice sheets would presumably have thicknesses and physical properties that change slowly and continuously from region to region. However, even a casual examination of an area of pack ice reveals striking local lateral changes in ice thicknesses and characteristics. These changes are invariably caused by ice movements produced by the forces exerted on the ice by winds and currents. Such motions are rarely uniform and lead to the build-up of stresses within ice sheets. If these stresses become large enough, cracks may form and widen, resulting in the formation of leads. Such features can vary in width from a few meters to several kilometers and in length from a few hundred meters to several hundred kilometers. As mentioned earlier, during much of the year in the polar regions, once a lead forms it is immediately covered with a thin skim of ice that thickens with time. This is an ever-changing process associated with the movement of weather systems as one lead system becomes inactive and is replaced by another system oriented in a different direction. As lead formation occurs at varied intervals throughout the ice growth season, the end result is an ice cover composed of a variety of thicknesses of uniform sheet ice.

However, when real pack ice thickness distributions are examined (**Figure 1**), one finds that there is a significant amount of ice thicker than the 4.5–5.0 m maximum that might be expected for steady-state MY ice floes. This thicker ice forms by the closing of leads, a process that commonly results in the piling of broken ice fragments into long, irregular features referred to as pressure ridges. There are many small ridges and large ridges are rare. Nevertheless, the large ridges are very impressive, the largest free-floating sail height and keel depth reported to date in the Arctic being 13 and 47 m, respectively (values not from the same ridge). Particularly heavily deformed ice commonly occurs in a band of ~ 150 km running between the north coast of Greenland and the Canadian Arctic Islands and the south coast of the Beaufort Sea. The limited data available on Antarctic ridges suggest that they are generally smaller and less frequent than ridges in

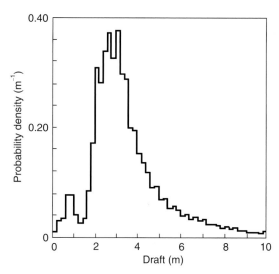

Figure 1 The distribution of sea ice drafts expressed as probability density as determined via the use of upward-looking sonar along a 1400 km track taken in April 1976 in the Beaufort Sea. All ice thicker than ~ 4 m is believed to be the result of deformation. The peak probabilities that occur in the range between 2.4 and 3.8 m represent the thicknesses of undeformed MY ice, while the values less than 1.2 m come from ice that ice that recently formed in newly formed leads.

the Arctic Ocean. The general pattern of the ridging is also different in that the long sinuous ridges characteristic of the Arctic Ocean are not observed. Instead, the deformation can be better described as irregular hummocking accompanied by the extensive rafting of one floe over another. Floe sizes are also smaller as the result of the passage of large-amplitude swells through the ice. These swells, which are generated by the intense Southern Ocean storms that move to the north of the ice edge, result in the fracturing of the larger floes while the large vertical motions facilitate the rafting process.

Pressure ridges are of considerable importance for a variety of reasons. First, they change the surface roughness at the air–ice and water–ice interfaces, thereby altering the effective surface tractions exerted by winds and currents. Second, they act as plows, forming gouges in the sea floor up to 8 m deep when they ground and are pushed along by the ungrounded pack as it drifts over the shallower (< 60 m) regions of the polar continental shelves. Third, as the thickest sea ice masses, they are a major hazard that must be considered in the design of offshore structures. Finally, and most importantly, the ridging process provides a mechanical procedure for transferring the thinner ice in the leads directly and rapidly into the thickest ice categories.

Considerable information on the drift and deformation of sea ice has recently become available through the combined use of data buoy and satellite observations. This information shows that, on the average, there are commonly two primary ice motion features in the Arctic Basin. These are the Beaufort Gyre, a large clockwise circulation located in the Beaufort Sea, and the Trans-Polar Drift Stream, which transports ice formed on the Siberian Shelf over the Pole to Fram Strait between Greenland and Svalbard. The time required for the ice to complete one circuit of the gyre averages 5 years, while the transit time for the Drift Stream is roughly 3 years, with about 9% of the sea ice of the Arctic Basin (919 000 km^2) moving south through Fram Strait and out of the basin each year. There are many interesting features of the ice drift that exist over shorter time intervals. For instance, recent observations show that the Beaufort Gyre may run backward (counterclockwise) over appreciable periods of time, particularly in the summer and fall. There have even been suggestions that such reversals can occur on decadal timescales. Typical pack ice velocities range from 0 to 20 cm s^{-1}, although extreme velocities of up to 220 cm s^{-1} (4.3 knots) have been recorded during storms. During winter, periods of zero ice motion are not rare. During summers, when considerable open water is present in the pack, the ice appears to be in continuous motion. The highest drift velocities are invariably observed near the edge of the pack. Not only are such locations commonly windy, but the floes are able to move toward the free edge with minimal inter-floe interference. Ice drift near the Antarctic continent is generally westerly, becoming easterly further to the north, but in all cases showing a consistent northerly diverging drift toward the free ice edge.

Trends

Considering the anticipated geophysical consequences of changes in the extent of sea ice, it is not surprising that there is considerable scientific interest in the subject. Is sea ice expanding and thickening, heralding a new glacial age, or retreating and thinning before the onslaught of a greenhouse-gas-induced heatwave? One thing that should be clear from the preceding discussion is that the ice is surprisingly thin and variable. Small changes in meteorological and oceanographic forcing should result in significant changes in the extent and state of the ice cover. They could also produce feedbacks that might have significant and complex climatic consequences.

Before we examine what is known about sea ice variations, let us first examine other related observations that have a direct bearing on the question of

sea ice trends. Land station records for 1966–1996 show that the air temperatures have increased, with the largest increases occurring winter and spring over both north-west North America and Eurasia, a conclusion that is supported by increasing permafrost temperatures. In addition, meteorological observations collected on Russian NP drifting stations deployed in the Arctic Basin show significant warming trends for the spring and summer periods. It has also recently been suggested that when proxy temperature sources are considered, they indicate that the late twentieth-century Arctic temperatures were the highest in the past 400 years.

Recent oceanographic observations also relate to the above questions. In the late 1980s the balance between the Atlantic water entering the Arctic Basin and the Pacific water appears to have changed, resulting in an increase in the areal extent of the more saline, warmer Atlantic water. In addition, the Atlantic water is shallower than in the past, resulting in temperature increases of as much as 2°C and salinity increases of up to 2.5‰ at depths of 200 m. The halocline, which isolates the cold near-surface layer and the overlying sea ice cover from the underlying warmer water, also appears to be thinning, a fact that could profoundly affect the state of the sea ice cover and the surface energy budget in the Arctic. Changes revealed by the motions of data buoys placed on the ice show that there has been a weakening of the Beaufort Sea Gyre and an associated increased divergence of the ice peak. There are also indications that the MY ice in the center of the Beaufort Gyre is less prevalent and thinner than in the past and that the amount of surface melt increased from ∼ 0.8 m in the mid-1970s to ∼ 2 m in 1997. This conclusion is supported by the operational difficulties encountered by recent field programs such as SHEBA that attempted to maintain on-ice measurements. The increased melt is also in agreement with observed decreases in the salinity of the near-surface water layer.

It is currently believed that these changes appear to be related to atmospheric changes in the Polar Basin where the mean atmospheric surface pressure is decreasing and has been below the 1979–95 mean every year since 1988. Before about 1988–99 the Beaufort High was usually centered over 180° longitude. After this time the high was both weaker and typically confined to more western longitudes, a fact that may account for lighter ice conditions in the western Arctic. There also has been a recent pronounced increase in the frequency of cyclonic storms in the Arctic Basin.

So are there also direct measurements indicating decreases in ice extent and thickness? Historical data based on direct observations of sea ice extent are rare, although significant long-term records do exist for a few regions such as Iceland where sea ice has an important effect on both fishing and transportation. In monitoring the health of the world's sea ice covers the use of satellite remote sensing is essential because of the vast remote areas that must be surveyed. Unfortunately, the satellite record is very short. If data from only microwave remote sensing systems are considered, because of their all-weather capabilities, the record is even shorter, starting in 1973. As there was a 2-year data gap between 1976 and 1978, only 25 years of data are available to date. The imagery shows that there are definitely large seasonal, interannual and regional variations in ice extent. For instance, a decrease in ice extent in the Kara and Barents Seas contrasts with an increase in the Baffin Bay/Davis Strait region and out-of-phase fluctuations occur between the Bering and the Okhotsk Seas. The most recent study, which examined passive microwave data up to December 1996, concludes that the areal extent of Arctic sea ice has decreased by 2.9% ± 0.4% per decade. In addition, record or near-record minimum areas of Arctic sea ice have been observed in 1990, 1991, 1993, 1995, and 1997. A particularly extreme recession of the ice along the Beauford coast was also noted in the fall of 1998. Russian ice reconnaissance maps also show that a significant reduction in ice extent and concentration has occurred over much of the Russian Arctic Shelf since 1987.

Has a systematic variation also been observed in ice thickness? Unfortunately there is, at present, no satellite-borne remote sensing technique that can measure sea ice thicknesses effectively from above. There is also little optimism about the possibilities of developing such techniques because the extremely lossy nature of sea ice limits penetration of electromagnetic signals. Current ice thickness information comes from two very different techniques: *in situ* drilling and upward-looking, submarine-mounted sonar. Although drilling is an impractical technique for regional studies, upward-looking sonar is an extremely effective procedure. The submarine passes under the ice at a known depth and the sonar determines the distance to the underside of the ice by measuring the travel times of the sound waves. The result is an accurate, well-resolved under-ice profile from which ice draft distributions can be determined and ice thickness distributions can be estimated based on the assumption of isostacy. Although there have been a large number of under-ice cruises starting with the USS *Nautilus* in 1958, to date only a few studies have been published that examine temporal variations in ice thickness in the

Arctic. The first compared the results of two nearly identical cruises: that of the USS *Nautilus* in 1958 with that of the USS *Queenfish* in 1970. Decreases in mean ice thickness were observed in the Canadian Basin (3.08–2.39 m) and in the Eurasian Basin (4.06–3.57 m). The second study has compared the results of two Royal Navy cruises made in 1976 and 1987, and obtained a 15% decrease in mean ice thickness for a 300 000 km^2 area north of Greenland. Although these studies showed similar trends, the fact that they each only utilized two years' data caused many scientists to feel that a conclusive trend had not been established. However, a recent study has been able to examine this problem in more detail by comparing data from three submarine cruises made in the 1990s (1993, 1996, 1997) with the results of similar cruises made between 1958 and 1976. The area examined was the deep Arctic Basin and the comparisons used only data from the late summer and fall periods. It was found that the mean ice draft decreased by about 1.3 m from 3.1 m in 1958–76 to 1.8 m in the 1990s, with a larger decrease occurring in the central and eastern Arctic than in the Beaufort and Chukchi Seas. This is a very large difference, indicating that the volume of ice in the region surveyed is down by some 40%. Furthermore, an examination of the data from the 1990s suggests that the decrease in thickness is continuing at a rate of about 0.1 m y^{-1}.

Off the Antarctic the situation is not as clear. One study has suggested a major retreat in maximum sea ice extent over the last century based on comparisons of current satellite data with the earlier positions of whaling ships reportedly operating along the ice edge. As it is very difficult to access exactly where the ice edge is located on the basis of only ship-board observations, this claim has met with some skepticism. An examination of the satellite observations indicates a very slight increase in areal extent since 1973. As there are no upward-looking sonar data for the Antarctic Seas, the thickness database there is far smaller than in the Arctic. However, limited drilling and airborne laser profiles of the upper surface of the ice indicate that in many areas the undeformed ice is very thin (60–80 cm) and that the amount of deformed ice is not only significantly less than in the Arctic but adds roughly only 10 cm to the mean ice thickness (**Figure 2**).

What is one to make of all of this? It is obvious that, at least in the Arctic, a change appears to be under way that extends from the top of the atmosphere to depths below 100 m in the ocean. In the middle of this is the sea ice cover, which, as has been shown, is extremely sensitive to environmental

Figure 2   (A) Ice gouging along the coast of the Beaufort Sea. (B) Aerial photograph of an area of pack ice in the Arctic Ocean showing a recently refrozen large lead that has developed in the first year. The thinner newly formed ice is probably less than 10 cm thick. (C) A representative pressure ridge in the Arctic Ocean. (D) A rubble field of highly deformed first-year sea ice developed along the Alaskan coast of the Beaufort Sea. The tower in the far distance is located at a small research station on one of the numerous off-shore islands located along this coast. (E) Deformed sea ice along the NW Passage, Canada. (F) Aerial photograph of pack ice in the Arctic Ocean.

changes. What is not known is whether these changes are part of some cycle or represent a climatic regime change in which the positive feedbacks associated with the presence of a sea ice cover play an important role. Also not understood are the interconnections between what is happening in the Arctic and other changes both inside and outside

the Arctic. For instance, could changes in the Arctic system drive significant lower-latitude atmospheric and oceanographic changes or are the Arctic changes driven by more dynamic lower-latitude processes? In the Antarctic the picture is even less clear, although changes are known to be underway, as evidenced by the recent breakup of ice shelves along the eastern coast of the Antarctic Peninsula. Not surprisingly, the scientific community is currently devoting considerable energy to attempting to answer these questions. One could say that a cold subject is heating up.

See also

Antarctic Circumpolar Current. Arctic Basin Circulation. Icebergs. Sea Ice: Variations in Extent and Thickness.

Further Reading

Cavelieri DJ, Gloersen P, Parkinson CL, Comiso JC and Zwally HJ (1997) Observed hemispheric asymmetry in global sea ice changes. *Science* 278(5340): 1104–1106.

Dyer I and Chryssostomidis C (eds) (1993) *Arctic Technology and Policy*. New York: Hemisphere.

Leppäranta M (ed.) (1998) *Physics of Ice-covered Seas*, 2 vols. Helsinki. Helsinki University Printing House.

McLaren AS (1989) The underice thickness distribution of the Arctic basin as recorded in 1958 and 1970. *Journal of Geophysical Research* 94(C4): 4971–4983.

Morison JH, Aagaard K and Steele M (1998) *Study of the Arctic Change Workshop*. (Report on the Study of the Arctic Change Workshop held 10–12 November 1997, University of Washington, Seattle, WA). Arctic System Science Ocean–Atmosphere–Ice Interactions Report No. 8 (August 1998).

Rothrock DA, Yu Y and Maykut G (1999) Thinning of the Arctic sea ice cover. *Geophysical Research Letters* 26.

Untersteiner N (ed.) (1986) *The Geophysics of Sea Ice*. NATO Advanced Science Institutes Series B, Physics, vol. 146. New York: Plenum Press.

Variations in Extent and Thickness

P. Wadhams, University of Cambridge, Cambridge, UK

Copyright © 2001 Academic Press

doi:10.1006/rwos.2001.0004

This review considers the seasonal and interannual variability of sea ice extent and thickness in the Arctic and Antarctic, and the downward trends which have recently been shown to exist in Arctic thickness and extent. There is no evidence at present for any thinning or retreat of the Antarctic sea ice cover.

Sea Ice Extent

Arctic

Seasonal variability   The best way of surveying sea ice extent and its variability is by the use of satellite imagery, and the most useful imagery on the large scale is passive microwave, which identifies types of surface through their natural microwave emissions, a function of surface temperature and emissivity. **Figure 1** shows ice extent and concentration maps for the Arctic for each month, averaged over the period 1979–87, derived from the multifrequency SMMR (scanning multichannel microwave radiometer) sensor aboard the Nimbus-7 satellite. This instrument gives ice concentration and, through comparison of emissions at different frequencies, the percentage of the ice cover that is multiyear ice, i.e. ice which has survived at least one

summer of melt. The ice concentrations are estimated to be accurate to $\pm 7\%$.

At the time of maximum advance, in February and March (**Figure 1A**), an ice cover fills the entire Arctic Ocean. The Siberian shelf seas are also ice-covered to the coast, although the warm inflow from the Norwegian Atlantic Current keeps the western part of the Barents Sea open. There is also a bight of open water to the west of Svalbard, kept open by the warm West Spitsbergen Current and formerly known as Whalers' Bay because it allowed sailing whalers to reach high latitudes. It is here that open sea is found closest to the Pole in winter – beyond 81° in some years. The east coast of Greenland has a sea ice cover along its entire length (although in mild winters the ice fails to reach Cape Farewell); this is transported out of Fram Strait by the Trans Polar Drift Stream and advected southward in the East Greenland Current, the strongest part of the current (and so the fastest ice drift) being concentrated at the shelf break. The Odden ice tongue at 72–75°N can be seen in these averaged maps as a distinct bulge in the ice edge, visible from January until April with an ice concentration of 20–50%. During any given year the Odden feature usually develops in the shape of a tongue, covering the region influenced by the Jan Mayen Current (a cold eastward offshoot of the East Greenland Current) and composed mainly of locally formed pancake ice.

Moving round Cape Farewell there is a thin band of ice off West Greenland (called the 'Storis'), the

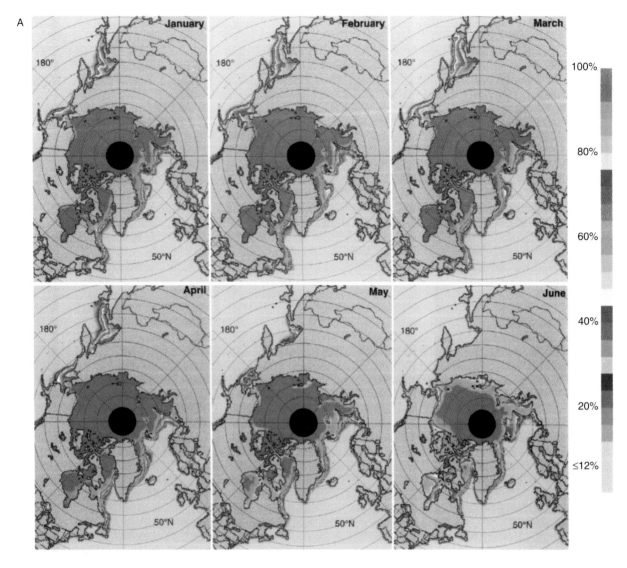

Figure 1 Ice extent and concentration maps for the Arctic for (A) winter, (B) summer months, averaged over the period 1979–87, derived from the multifrequency SMMR sensor aboard the Nimbus 7 satellite. (Reproduced with permission from Gloersen et al., 1992.)

limit of ice transported out of the Arctic Basin, which often merges with the dense locally formed ice cover of Baffin Bay and Davis Strait. The whole of the Canadian Arctic Archipelago, Hudson Bay, and Hudson Strait are ice-covered, and on the western side of Davis Strait the ice stream of the Labrador Current carries ice out of Baffin Bay southward towards Newfoundland. The southernmost ice limit of this drift stream is usually the north coast of Newfoundland, where the ice is separated by the bulk of the island from an independently formed ice cover filling the Gulf of St Lawrence, with the ice-filled St Lawrence River and Great Lakes behind. Further to the west a complete ice cover extends across the Arctic coasts of north-west Canada and

Alaska and fills the Bering Sea, at somewhat lower concentration, as far as the shelf break. Sea ice also fills the Sea of Okhotsk and the northern end of the Sea of Japan, with the north coast of Hokkaido experiencing the lowest latitude sea ice (44°) in the Northern Hemisphere.

In April the ice begins to retreat from its low latitude extremes. By May the Gulf of St Lawrence is clear, as is most of the Sea of Okhotsk and some of the Bering Sea. The Odden ice tongue has disappeared and the ice edge is retreating up the east coast of Greenland. By June the Pacific south of Bering Strait is ice-free, with the ice concentration reducing in Hudson Bay and several Arctic coastal locations. August and September (**Figure 1B**) are the

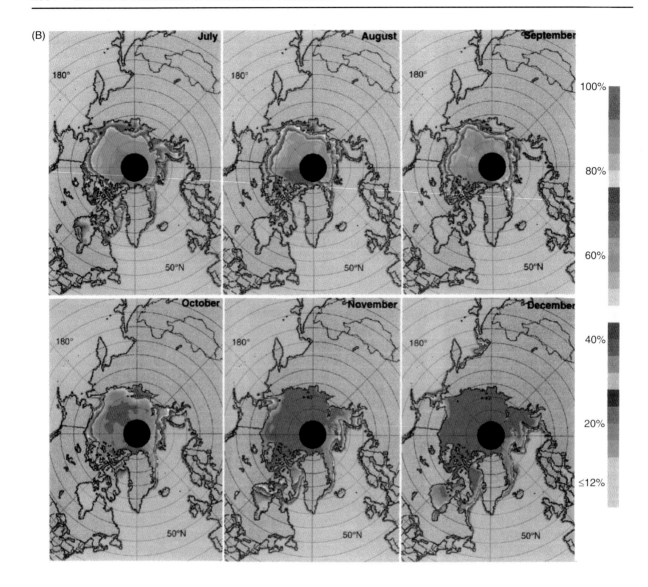

Figure 1 *Continued*

months of greatest retreat, constituting the brief Arctic summer. During these months the Barents and Kara Seas are ice-free as far as the shelf break, with the Arctic pack retreating to (and occasionally beyond) northern Svalbard and Franz Josef Land. The Laptev and East Siberian Sea are generally ice-free, with ice in some years remaining to block choke points such as the Vilkitsky Strait south of Severnaya Zemlya. This allows marine transport through a Northern Sea Route across the top of Russia, but with a need for icebreaker escort through the central ice-choked region. In East Greenland the ice has retreated northwards to about 72–73° (a latitude which varies greatly from year to year), while the whole system of Baffin Bay, Hudson Bay, and Labrador is ice-free. Occasionally a small mass of ice, called the 'Middle Ice', remains at the northern end of Baffin Bay. In the Canadian Arctic Archipelago the winter fast ice which filled the channels usually breaks up and partly melts or moves out, but in some years ice remains to clog vital channels, and the Northwest Passage is not such a dependably navigable seaway as the Northern Sea Route. There is usually a slot of open water across the north of Alaska, but again in some years the main Arctic ice edge moves south to touch the Alaskan coast, making navigation very difficult for anything but a full icebreaker.

By October new ice has formed in many of the areas which were open in summer, especially around the Arctic Ocean coasts, and in November–January there is steady advance everywhere towards the winter peak. The Sea of Okhotsk acquires its first ice cover in December, and the Odden starts

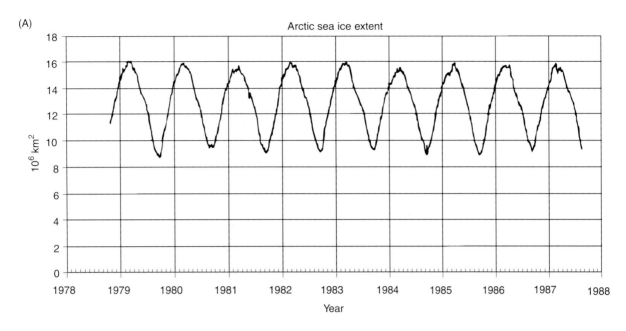

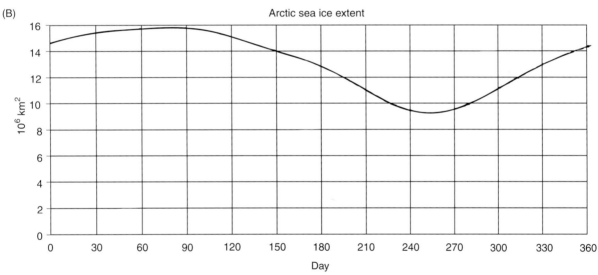

Figure 2 (A) The cycle of Arctic sea ice extent for the 1979–87 period; (B) the averaged seasonal cycle. (Reproduced with permission from Gloersen *et al.*, 1992.).

to appear; Baffin Bay and Hudson Bay are already fully ice-covered.

The averaged seasonal cycle for the 1979–87 era (**Figure 2B**) gives a maximum extent – 'extent' here is defined as the total area of sea within the 15% ice concentration contour – of $15.7 \times 10^6 \, \text{km}^2$ in late March, and a minimum of $9.3 \times 10^6 \, \text{km}^2$ in early September. For sea ice area, derived as extent multiplied by concentration, the figures are 13.9 and $6.2 \times 10^6 \, \text{km}^2$ in winter and summer.

Results from SMMR multiyear ice retrievals show that in the Arctic multiyear ice is found in the highest concentrations within the central Arctic Ocean, in the area controlled by the Beaufort Gyre.

This is not surprising, since the area is permanently ice-covered and floes circulate on closed paths which take 7–10 years for a complete circuit. It was found that multiyear fractions of 50–60% are typical for the Gyre region, rising to 80% in the very centre. Multiyear fractions of 30–40% are found in the part of the Trans Polar Drift Stream fringing the Beaufort Gyre, while in the rest of this current and in peripheral areas of the Arctic the multi-year fraction is $\leqslant 20\%$.

Interannual variability The seasonal cycle described above varies in detail from year to year, and there is evidence from an extension of the record to

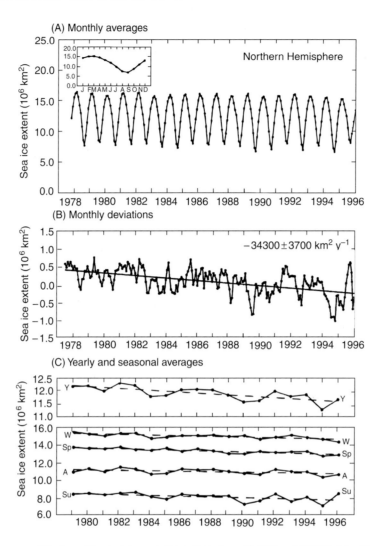

Figure 3 (A) Monthly averaged Northern Hemisphere sea ice extents from SMMR and SSM/I data, November 1978–December 1996. Inset shows average seasonal cycle. (B) Monthly deviations of the extents from the 18-year average, with linear trend shown. (C) Yearly and seasonally averaged ice extents: W = January–March, Sp = April–June, Su = July–September, A = October–December. (Reproduced with permission from Parkinson *et al.* (1999) *Journal of Geophysical Research*, 104: 20837–20856.

the present day that a steady decrease of the overall ice extent in the Arctic has been taking place. Analyses where datasets from the SMMR and newer SSM/I sensors have been combined and reconciled show that the sea ice extent in the Arctic has declined at a decadal rate of some 2.8–3% since 1978, with a more rapid recent decline of 4.3% between 1987 and 1994. **Figure 3** shows how an apparently fairly stable annual cycle of large amplitude (**Figure 3A**) reveals a distinct downward trend of area (**Figure 3B**) when anomalies from interannual monthly means are considered. (**Figure 3C**) shows that the downward trend occurs for every season of the year; the estimated mean annual loss of ice area is $(34\,300 \pm 3700)$ km^2.

The steady hemispheric decline in sea ice extent masks more violent regional changes. In the Bering

Sea there was a sudden downward shift of sea ice area in 1976, indicating a regime shift in the wind stress field as the Aleutian Low moved its position. In the Arctic Basin a passive microwave analysis of the length of the ice-covered season during 1979–86 showed a see-saw effect, with amelioration in the Russian Arctic, Greenland, Barents, and Okhotsk Seas and a worsening in the Labrador Sea, Hudson Bay, and the Beaufort Sea. A further analysis extended the coverage to 1978–96 and confirmed these results: the Kara/Barents Sea region had the highest rate of decline in area, of 10.5% per decade, followed by the seas of Okhotsk and Japan and the central Arctic Basin at 9.7 and 8.7% respectively. Lesser declines were experienced by the Greenland Sea (4.5%), Hudson Bay (1.4%), and the Canadian Arctic Archipelago (0.6%). Increases were registered in the Bering Sea (1% – the starting date being later than the 1976 collapse), Gulf of St Lawrence (2%), and Baffin Bay/Labrador Sea (3.1%). Taking the more modern data into account it is clear that the see-saw effect discovered over 1979–86 has been largely subsumed into a general retreat.

A particularly important area of sea ice retreat has been the central part of the Greenland Sea gyre, in the vicinity of 75°N 0–5°W. This is normally the site of strong wintertime convection, driven by salt fluxes from local ice growth over the cold Jan Mayen Current, which produces the tongue-like Odden feature. Cold off-ice winds move newly formed ice eastward within the tongue so that the net salt flux in the western part of the feature is strongly positive; this is where convection occurs. Tracer experiments have shown that deep convection has failed to reach the bottom since about 1971 and in recent years has been greatly reduced in volume and confined to the uppermost 1000 m, while ice production has also been reduced, with no Odden forming at all in 1994, 1995, and 2000. The salt flux produces the causal link between ice retreat and convection shut-off.

What is the reason for these changes? Climatic simulations by GCMs predict that the global warming effect due to increased atmospheric CO_2 should be amplified in the polar regions, particularly the Arctic, mainly through the ice–albedo feedback effect. However, a more immediate cause of many of the observed changes can be identified as a changed pattern of atmospheric circulation in high latitudes. In the North Atlantic sector of the Northern Hemisphere this can be represented by the North Atlantic Oscillation (NAO) index, the wintertime difference in pressure between Iceland and Portugal, which was low or negative through most of the 1950s–1970s but has been rising since the 1980s and which has been highly positive throughout the current decade. A high positive NAO index is associated with an anomalous low pressure center over Iceland which involves enhanced west and north-west winds over the Labrador Sea (cold winds which cause increased cooling hence increased convection), enhanced east winds over the Greenland Sea in the 72–75° latitude range (causing a reduction in local ice growth in the Odden ice tongue, and a reduced separation between growth and decay regions, hence a reduced rate of convection); enhanced north-east winds in the Fram Strait area, causing an increased area flux of ice through the Strait (although the ice may have a reduced thickness, so the volume flux is not necessarily increased); and an enhanced wind-driven flow of the North Atlantic Current, allowing more warm water to enter the Atlantic layer of the Arctic Ocean.

Within the Arctic Basin this pattern is incorporated into a large-scale wintertime pattern (with associated index) called the Arctic Oscillation (AO), which involves a see-saw of sea level pressure (SLP) between the Arctic Basin and the surrounding zonal ring. The anomaly appears to extend into the upper atmosphere, and so represents an oscillation in the strength of the whole polar vortex. **Figure 4** shows the results of an analysis of the differences in SLP over the Arctic Ocean between high and low NAO years (corresponding also to different phases of the AO index). What had been thought of as the Arctic 'norm', i.e. a high over the Beaufort Sea leading to the familiar Beaufort Gyre and Trans Polar Drift Stream as the resulting free drift (along the isobars) ice circulation pattern, is actually the result of a low NAO (situation B). With a high NAO (situation A) the Beaufort High is suppressed and squeezed towards the Alaskan coast, causing a reduction in the area and strength of the Beaufort Gyre, and a tendency for ice produced on the Siberian shelves to turn east and perform a longer circuit within the basin before emerging from the Fram Strait (this applies also to the trajectory of fresh water from Siberian rivers). The weakening of the Beaufort Gyre may well explain anomalously low summer sea ice extents observed in the Beaufort Sea in 1996–98 (**Figure 5**), since locally melting ice is not replaced by new inputs of ice from the north east. The difference field (situation C) aptly demonstrates these changes if one considers the differential ice drift vectors as occurring along the isobars shown. Situations A and B represent two distinct patterns of Arctic Ocean circulation, which may be called 'cyclonic' and 'anticyclonic' respectively.

Antarctic

Seasonal variability The sea ice cover in the Antarctic is one of the most climatically important features of the Southern Hemisphere. Its enormous seasonal variation in extent greatly outstrips that of Arctic sea ice, and makes it second only to Northern Hemisphere snow extent as a varying cryospheric feature on the Earth's surface. **Figure 6** shows monthly averaged sea ice extent and concentration maps for the Antarctic, derived in the same way as **Figure 1** from SMMR passive microwave data, and covering the same period, 1979–87.

With the seasons reversed, the maximum ice extent occurs in August and September. At its maximum (**Figure 6A**) the ice cover is circumpolar in extent. Moving clockwise, the ice limit reaches 55°S in the Indian Ocean sector at about 15°E, but lies at

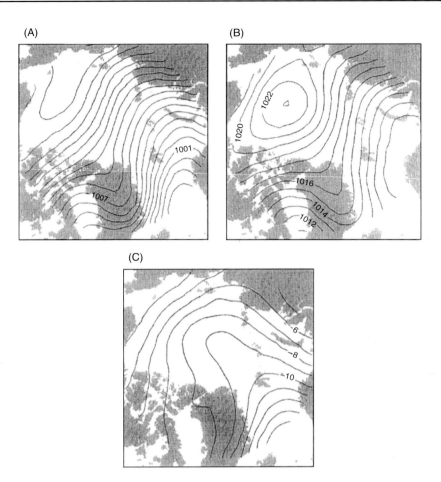

Figure 4 Pressure fields over the Arctic Ocean corresponding to (A) high North Atlantic Oscillation index, (B) low NAO index, (C) the difference field. (Reproduced with permission from Kwok and Rothrock (1999) *Journal of Geophysical Research,* 104: 5177–5189.

about 60°S around most of the rest of east Antarctica, then slips even further south to 65°S off the Ross Sea. The edge moves slightly north again to 62°S at 150°W, then again shifts southward to 66°S off the Amundsen Sea before moving north again to engulf the South Shetland and South Orkney Islands off the Antarctic Peninsula and complete the circle. The zonal variation in latitude of this winter maximum therefore amounts to some 11°. It has been found that the winter advance of the ice edge follows closely the advance of the 271.2°C isotherm in surface air temperature (freezing point of sea water) and almost coincides with this isotherm at the time of maximum advance. The ice limit is therefore mainly determined thermodynamically, with the gross zonal variations in the winter ice limit matching zonal variations in the freezing isotherm (due to the distribution of continents in the Southern Hemisphere). Smaller-scale variations in the maximum ice limit may be related to deflections in the Antarctic Circumpolar Current as it crosses submarine ridges. Note that within the ice limit the ice concentration

is generally less than the almost 100% concentration found in the Arctic Ocean in winter. Even in the areas of greatest concentration, the central Weddell and Ross Seas, it is only in the range 92–96%, while there is a broad marginal ice zone facing the open Southern Ocean over which the concentration steadily diminishes over an outer band of width 200–300 km. This has been found to be a zone over which the advancing winter ice edge is composed of pancake ice, maintained as small cakes by the turbulent effect of the strong wave field.

Ice retreat begins in October and is rapid in November and December. Again the retreat is circumpolar but has interesting regional features. In the sector off Enderby Land at 0–20°E a large gulf opens up in December to join a coastal region of reduced ice concentration which opens in November. This is a much attenuated version of a winter polynya which was detected in the middle of the pack ice in this sector during 1974–76, but which has only recurred very occasionally as an open-water feature since that date, e.g. in 1994. It was

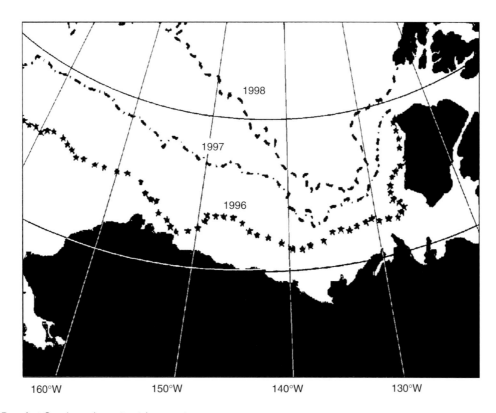

Figure 5 Beaufort Sea ice edge retreat in recent summers.

known as the Weddell Polynya and lay over the Maud Rise, a plateau of reduced water depth. The area was investigated in winter 1986 by the Winter Weddell Sea Project (WWSP) cruise of FS 'Polarstern', and it was found that the region is already part of the Antarctic Divergence, where upwelling of warmer water can occur, and that additional circulating currents and the doming of isopycnals over the rise could allow enough heat to reach the surface to keep the region ice-free in winter. Since the occurrence is irregular, the region is presumably balanced on the edge of instability. The 1986 winter cover was of high concentration but was very thin. The December distribution also shows an open-water region appearing in the Ross Sea, the so-called Ross Sea Polynya, with ice still present to the north. In November and December a series of small coastal polynyas can be seen to be actively opening along the east Antarctica coast.

By January (**Figure 6B**) further retreat has occurred. The Ross Sea is now completely open, east Antarctica has only a narrow fringe of ice around it, and large ice expanses are confined to the eastern Ross Sea, the Amundsen-Bellingshausen Sea sector (60–140°W), and the western half of the Weddell Sea. The month of furthest retreat is February. Ice remains in these three regions, but most of the east

Antarctic coastline is almost ice-free, as is the tip of the Antarctic Peninsula. This is the season when supply ships can reach Antarctic bases, when tourist ships visit Antarctica, and when most oceanographic research cruises are carried out. It can be seen that the ice concentration in the center of the western Weddell Sea massif is still 92–96%. This is the region which bears the most resemblance to the central Arctic Ocean; it is the only part of the Antarctic to contain significant amounts of multi-year ice, it is very difficult to navigate, and consequently even its bathymetry is not as well known as that of other parts of the Antarctic Ocean.

By March the very short Antarctic summer is over and ice advance begins. The first advances take place within the Ross and Weddell Seas, then circumpolar advance begins in April. During May and June the Weddell Sea ice swells out to the north east, while around the whole of Antarctica the ice edge continues to advance until the August peak.

Figure 7 is the Antarctic equivalent of **Figure 2**. The annual cycle of ice extent can be seen to have a much higher amplitude than in the Arctic, and the year-to-year variability of the peaks and troughs is also somewhat greater. During the 8.8-year record, the February average extent varies from 3.4 to $4.3 \times 10^6 \, km^2$, while the September average extent

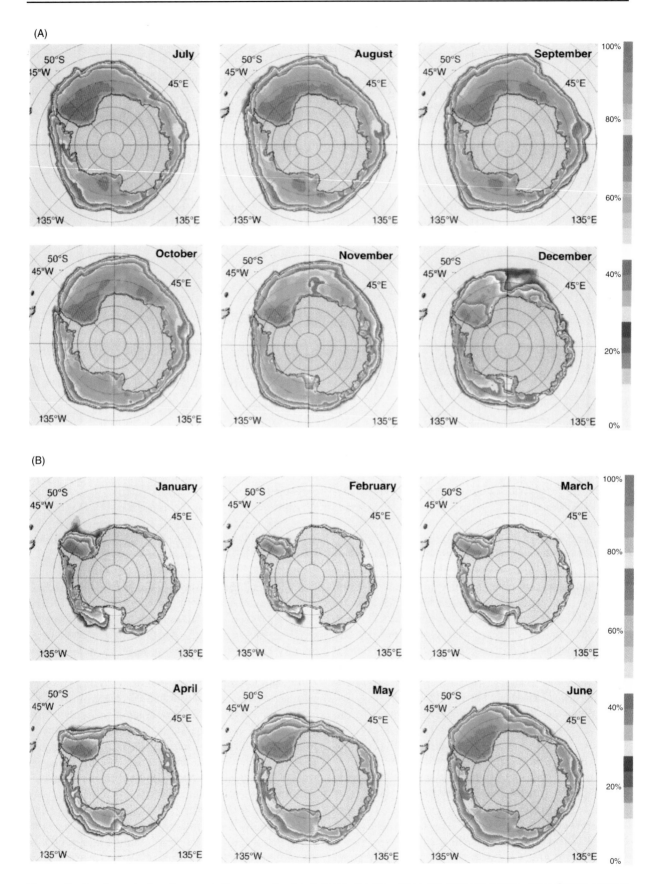

Figure 6 Sea ice extent and concentration maps for the Antarctic for (A) winter, (B) summer months, from SMMR passive microwave data averaged over the period 1979–87. (Reproduced with permission from Gloersen *et al.*, 1992.)

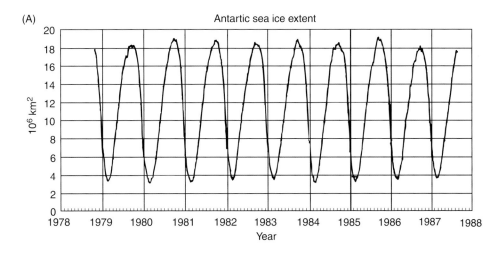

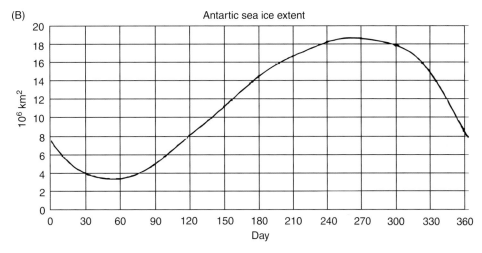

Figure 7 (A) The cycle of Antarctic sea ice extent for the 1979–87 period; (B) the averaged seasonal cycle. (Reproduced with permission from Gloersen *et al.*, 1992.)

varies from 15.5 to $19.1 \times 10^6 \, \text{km}^2$, both covering ranges of ± 10–12%. The overall average cycle (**Figure 7b**) shows a retreat which is steeper than the advance. The mean minimum extent, at the end of February, is $3.6 \times 10^6 \, \text{km}^2$, while the mean maximum extent, in the middle of September, is $18.8 \times 10^6 \, \text{km}^2$. Because of low average ice concentrations within the pack, the corresponding minimum and maximum ice areas are 2.1 and $15.0 \times 10^6 \, \text{km}^2$.

The winter ice extent in the Antarctic exceeds that of the Arctic winter, while the summer minima are very much lower. This implies that the combined Arctic and Antarctic sea ice extent should be greatest during the Arctic summer. **Figure 8** shows that in fact the peak occurs in October after a plateau during the summer and that the global minimum occurs in late February. The range is approximately 19–$29 \times 10^6 \, \text{km}^2$, with a high interannual variability for both maxima and minima.

Interannual variability Observational data on Antarctic sea ice extent show no significant trend. Passive microwave data for the 1988–94 period show no evidence of an overall trend in extent, but some evidence suggesting that anomaly patterns propagate eastward, offering support for the idea of an Antarctic circumpolar wave in surface pressure, wind, temperature, and sea ice extent. During the 7 years of the study, ice seasons shortened in the east Ross Sea, Amundsen Sea, west Weddell Sea, offshore eastern Weddell Sea, and east Antarctica between 40° and 80°E. Ice seasons lengthened in the west Ross Sea, Bellingshausen Sea, central Weddell Sea, and the region 80°E–135°E. Earlier evidence of a major ice retreat in the Bellingshausen Sea in the summer of 1988–91 was shown to be a short-lived phenomenon. A longer-term statistical analysis of passive microwave data from 1978 onwards gave a small and not statistically significant upward trend in overall Antarctic ice extent, of some 1.3% per

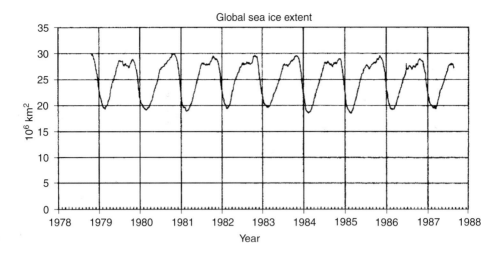

Figure 8 Combined global cycle of sea ice extent, 1979–87 (Reproduced with permission from Gloersen *et al.*, 1992.)

decade. The conclusion is that sea ice extent and GCM predictions agree in showing no strong trend.

Sea Ice Thickness

Arctic

Knowledge of the regional and temporal variability of ice thickness in the Arctic comes mainly from upward sonar profiling by submarines. Therefore the level of knowledge depends on whether submarines have been able to operate in the area concerned. Until now, data have been obtained mainly from British submarines operating in the Greenland Sea and Eurasian Basin since 1971, and from US submarines operating in the Canada and Eurasian Basins since 1958.

Results show that the ice in Baffin Bay is largely thin first-year ice with a modal thickness of 0.5–1.5 m. In the southern Greenland Sea too, the ice, although composed largely of partly melted multiyear ice, also has a modal thickness of about 1 m, with the decline in mean thickness from Fram Strait giving a measure of the freshwater input to the Greenland Sea at different latitudes. Over the Arctic Basin itself there is a gradation in mean ice thickness from the Soviet Arctic, across the Pole and towards the coasts of north Greenland and the Canadian Arctic Archipelago, where the highest mean thicknesses of some 7–8 m are observed. These overall variations are in accord with the predictions of numerical models which take account of ice dynamics and deformation as well as ice thermodynamics. The overall basin mean is about 5 m in winter and 4 m in summer.

In order to assess whether significant changes are occurring in a region of the Arctic it is necessary to obtain area-averaged observations of mean ice thickness over the same region using the same equipment at different seasons or in different years. Ideally the region should be as large as possible, to allow assessment of whether changes are basin-wide or simply regional. Also the measurements should be repeated annually in order to distinguish between a fluctuation and a trend. Because of the unsystematic nature of Arctic submarine deployments this goal has not yet been achieved, but a number of comparisons have been carried out which strongly suggest that a significant thinning has been occurring. Some of these have been made possible by very large new datasets which have been obtained from the US SCICEX civilian submarine program during 1993–99.

SCICEX data obtained in September–October of 1993, 1996, and 1997 have been compared with data obtained during six summer cruises during the period 1958–76. Twenty-nine crossing places were identified, where a submarine track from the recent period crossed one from the early period, and the corresponding tracks (of average length 160 km) were compared in thickness. In each case the mean thicknesses obtained were adjusted to a standard date of September 15 using an ice–ocean model to account for seasonal variability. The 29 matched datasets were divided into six geographical regions (**Figure 9**). The decline in mean ice draft was significant for every region and increased across the Arctic from the Canada Basin towards Europe – it was 0.9 m in the Chukchi Cap and Beaufort Sea, 1.3 m in the Canada Basin, 1.4 m near the North Pole, 1.7 m in the Nansen Basin, and 1.8 m in the eastern

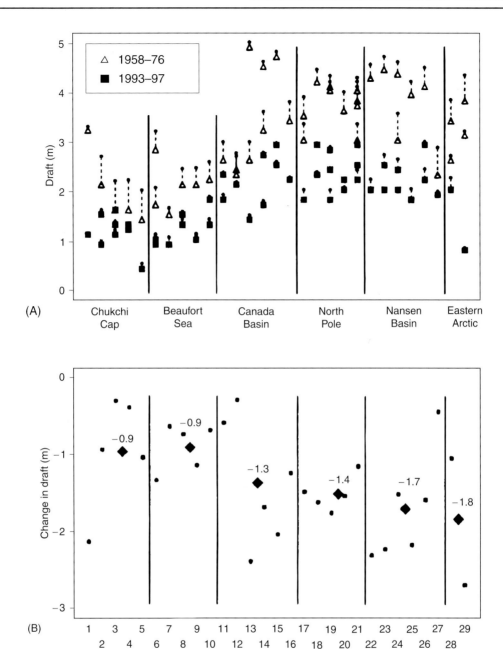

Figure 9 (A) Mean ice drafts at crossings of early cruises with cruises in the 1990s. Early data (1958–76) are shown by open triangles and those from the 1990s by solid squares, both seasonally adjusted to September 15. The small dots show the original data before the seasonal adjustment. The crossings are grouped into six regions separated by the solid lines and named appropriately. (B) Changes in mean ice draft at cruise crossings (dots) from the early data to the 1990s. The change in the mean ice draft for all crossings in each region is shown by a large diamond. (Reproduced with permission from Rothrock *et al.* (1999) *Geophysical Research Letters* (1999) 26: 3469–3472.)

Arctic. Overall, the mean change in draft was from 3.1 m in the early period to 1.8 m in the recent period, a decline of 42%.

The authors of the study commented that the decline in mean draft could arise thermodynamically from any of the following flux increases:

1. A $4 \, \text{W m}^{-2}$ increase in ocean heat flux,

2. A $13 \, \text{W m}^{-2}$ increase in poleward atmospheric heat transport, or

3. A $23 \, \text{W m}^{-2}$ increase in downwelling shortwave radiation during summer.

Clearly a change in ice dynamics can also produce a change in mean ice draft, although it is not known what change in wind forcing would be needed to

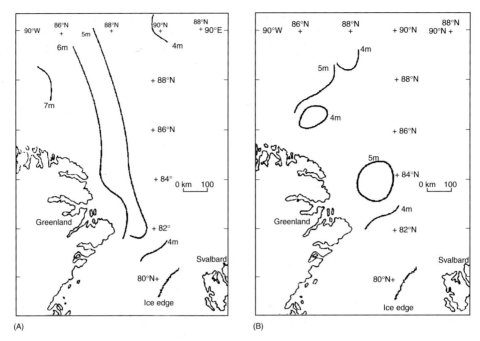

Figure 10 Contour maps of mean ice drafts from the Eurasian Basin measured from British submarines, October 1976 and May 1987. (Reproduced with permission from Wadhams (1990) *Nature* 345: 795–797.)

account for the magnitude and distribution of the observed draft decrease.

This is the most extensive comparison so far, but it should be noted that all datasets involved are from summer, mostly late summer, so that the reported decline refers to only one season of the year, and that most track comparisons occur over the North Pole region and Canada Basin, with few in the Eurasian Arctic and none south of 84° in the Eurasian Basin.

Complementary to this study are comparisons from the Eurasian Basin and Greenland Sea made using data from British submarine cruises. One comparison (**Figure 10**) involved datasets from a triangular region extending from Fram Strait and the north of Greenland to the North Pole, recorded in October 1976 and May 1987. Mean drafts were computed over 50 km sections, and each value was positioned at the centroid of the section concerned; the results were contoured to give the maps shown in **Figure 10**. There was a decrease of 15% in mean draft averaged over the whole area (300 000 km²), from 5.34 m in 1976 to 4.55 m in 1987. Profiles along individual matching track lines showed that the decrease was concentrated in the region south of 88°N and between 30° and 50°W. By comparison of the entire shape of the probability density functions of ice draft, the conclusion was that the main contribution to the loss of volume was the replacement of multiyear and ridged ice by young and first-year ice.

For instance, taking ice of 2–5 m thickness as an indicator of undeformed multiyear ice fraction, this declined from 47.6% in 1976 to 39.1% in 1987, a relative decline of 18%. This is in agreement with a recent finding that multiyear ice fraction in the Arctic (estimated from passive microwave data) suffered a 14% decrease during the period 1978–98. This multiyear ice variability correlated well with mean ice thickness variability from the eastern Arctic as estimated from surface oscillation measurements made from Russian drifting stations (note that the oscillation technique, an inference based on the peak period of swell propagating through the ice, has not been validated against direct measurements).

The British study did not correct for seasonal variability between the 1976 measurements, made in October, and those of 1987, made in April–May. If this is done using a model, the decrease in mean ice draft (standardized to September 15) becomes much greater at 42%, since April–May is the time of greatest ice thickness. This is in excellent agreement with results for the entire overall US dataset, yet occurred within a period of only 11 years. This indicates either that thinning occurs faster in the Eurasian Basin than elsewhere in the Arctic or that it is invalid to compare datasets from different times of year simply by standardizing to 'summer' through use of a model.

The latter problem is largely overcome in an analysis of the most recent British dataset, obtained in

Table 1 Mean drafts in 1° bins of latitude in 1976 and 1996

Latitude range	Mean draft m in 1996	Mean draft m in 1976	1996 as % of 1976
81–82	1.57	5.84	26.9
82–83	2.15	5.87	36.6
83–84	2.88	4.90	58.7
84–85	3.09	4.64	66.6
85–86	3.54	4.57	77.4
86–87	3.64	4.64	78.5
87–88	2.36	4.60	51.2
88–89	3.24	4.41	73.4
89–90	2.19	3.94	55.5
Overall	2.74	4.82	56.8

September 1996 by HMS 'Trafalgar'. These data can be compared directly with results from October 1976. The two submarines followed similar courses between 81°N and 90°N on about the 0° meridian, and it was found that about 2100 km of track from each submarine, when divided into 100 km sections, were close enough in correspondence to count as 'crossing tracks'. The overall decline in mean ice thickness between 1976 and 1996 was 43%, in remarkably close agreement with US results. The mean drafts in 1° bins of latitude were as shown in **Table 1**.

It can be seen that there was a significant decrease of mean draft at every latitude, but that the decline is largest just north of Fram Strait and near the Pole itself. A characteristic of the ice cover observed from below was the large amount of completely open water present at all latitudes. A seasonality correction to the 1976 data for the slight difference in mean draft between October and September brings the ratio to 59.0% for September, a decline of 41%. Thus the British and the US data are in remarkably good agreement in describing a very significant decrease in the thickness of Arctic Ocean sea ice.

A cautionary note must be sounded in that these significant decreases in thickness derived from spatially averaged data conceal large random variabilities at given locations. Time-series of ice draft at fixed locations have been obtained from moored upward sonar systems, of which the most comprehensive set spans Fram Strait. An analysis of data from 1991 to 1998 showed that interseasonal and interannual variability in thickness far exceed any trend, although of course the length of the dataset is only 7 years.

A possible direct cause of the observed thinning is the recent discovery that the Atlantic sublayer in the Arctic Ocean, which lies beneath the polar surface water and which derives from the North Atlantic Current, has warmed substantially (by 1–2°C at 200 m depth) and increased its range of influence relative to water of Pacific origin. The front separating the two water types has now shifted from the

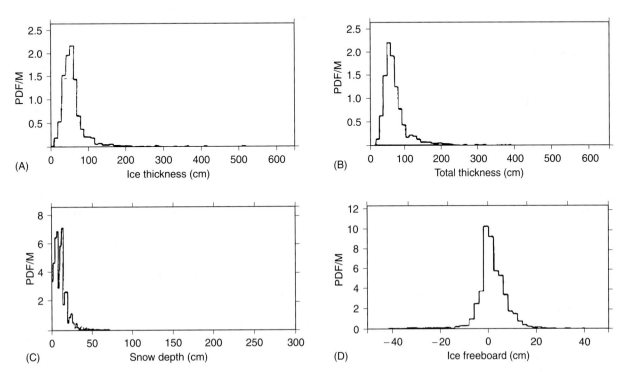

Figure 11 Thickness distributions of Antarctic first-year sea ice (A), ice plus snow (B), and snow alone (C). The distribution of ice freeboards is shown in (D); a negative value permits water infiltration. (Reproduced with permission from Wadhams *et al.* (1987) *Journal of Geophysical Research* 92: 14535–14552.)

Lomonosov to the Alpha-Mendeleyev Ridge. This warmer and shallower sublayer should increase the ocean heat flux into the bottom of the ice. This is enhanced by the fact that the structure of the polar surface layer has itself changed. In the Eurasian Basin there was formerly a cold halocline layer in the 100–200 m depth range, where temperature stayed cold with increasing depth despite salinity rising. Its existence was due to riverine input from Siberia, which has recently diverted eastward due to a changed atmospheric circulation, causing a retreat of the cold halocline and possibly associated with the recently observed summertime retreat of sea ice in the Beaufort Sea sector (**Figure 5**).

Antarctic

Knowledge of ice thickness in the Antarctic is much less extensive than in the Arctic, since systematic data have been obtained mainly by repetitive drilling except for the use of moored upward sonar at certain sites in the Weddell Sea. The winter pack ice in the Antarctic is of global importance because of its vast extent and large seasonal cycle, so the deter-

mination of winter ice thickness remains a high research priority.

In 1986 the first deep penetration into the circumpolar Antarctic pack during early winter, the time of ice edge advance, was accomplished by the WWSP cruise of FS 'Polarstern', during which systematic ice thickness measurements (at 1 m intervals along lines of about 100 holes) were made throughout the eastern part of the Weddell-Enderby Basin, from the ice edge to the coast, covering Maud Rise and representing a typical cross-section of the first-year circumpolar Antarctic pack during the season of advance. After a spring cruise in 1988 a second winter cruise was carried out in 1989: the Winter Weddell Gyre Study (WWGS) involved a crossing of the Weddell Sea from the tip of the Antarctic Peninsula to Kap Norvegia in the east during September–October, and thus allowed the multiyear ice regime of the western Weddell Sea to be studied in midwinter.

In the advancing Antarctic pack, composed of first-year consolidated pancake ice, the ice thickness distribution was as shown in **Figure 11**. Note the

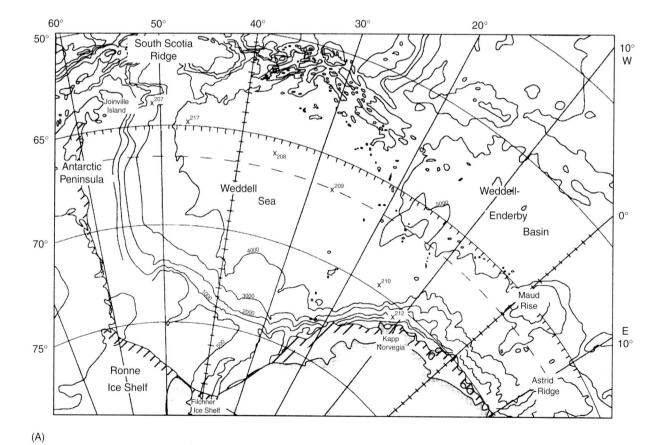

(A)

Figure 12 (A) Locations of six moored upward-looking echo sounders in the Weddell Sea. (B) Mean ice drafts December 1990–December 1992 from these sounders. (Reproduced with permission from Jeffries, 1998.)

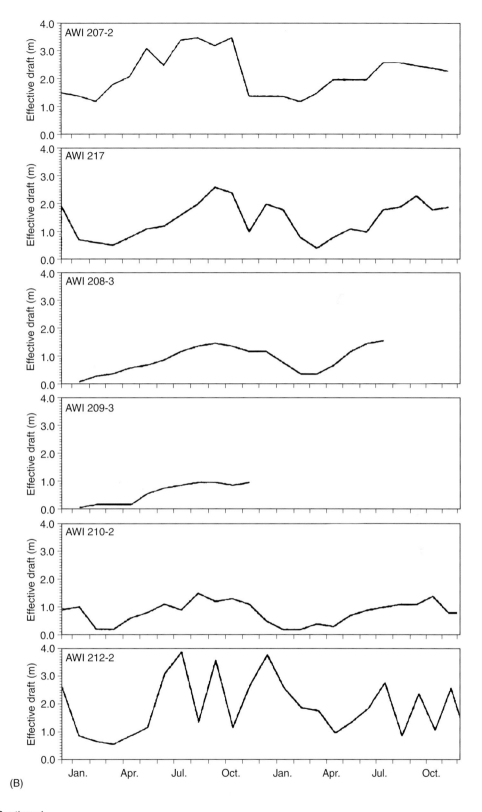

Figure 12 Continued

peak at the very low value of 50–60 cm, with a peak in snow cover thickness at 14–16 cm. The snow cover was sufficient to push the ice surface below water level in some 17% of holes drilled, and this leads to water infiltration into the snow layer and the formation of a new type of ice, snow-ice, at the boundary between ice and overlying snow. It is reasonable to suppose that the pancake ice-forming mechanism is typical of the entire circumpolar advancing ice edge in winter (neglecting embayments such as the Ross and Weddell Seas).

Multiyear ice was measured in 1989 in the western Weddell Sea; the Weddell Gyre carries ice from the eastern Weddell Sea deep into high southern latitudes in the southern Weddell Sea off the Filchner-Ronne Ice Shelf, and then northward up the eastern side of the Peninsula to the north-west Weddell Sea. This journey takes about 18 months, and so permits much of the ice to mature into multiyear (strictly, second-year) ice. Multiyear ice can be identified by its structure in cores, and by the very thick snow cover which it acquires, which is almost always sufficient to depress the ice surface below the waterline. Ice drilling showed that the mean thickness of undeformed multiyear ice (1.17 m), was about double that of first-year ice (0.60 m). The presence of ridging roughly doubles the mean draft of the 100 m floe sections in which it occurs (0.60–1.03 m in first-year; 1.17–2.51 m in multiyear). In addition, snow is very much deeper on multiyear ice (0.63–0.79 m) than on first-year ice (0.16–0.23 m).

Drilling from ships has the three advantages that data can be obtained from many locations, that first-year ice and multiyear ice can be clearly discriminated, and that the detailed structure of ridges can be measured. In other respects, however, moored upward-looking echo sounders (ULES) give far more information. Data were collected from six such ULES systems moored across the Weddell Sea (**Figure 12A**) during a 2 year period from December 1990 to December 1992. The results for mean drafts (**Figure 12B**) are in good agreement with drilling data for winter, but also reveal the annual cycle ('effective draft' in this figure is true mean draft, i.e. including the open water component). It can be seen that the westernmost ULES, in the Weddell Sea outflow (207), has a cycle ranging from just over 1 m in summer to about 3 m in winter, in good agreement with ridged multiyear ice sampled from drilling. The thickness diminishes considerably over the central Weddell Sea (208, 209) but then rises again near the Enderby Land coast (212) to a very variable mean value. This last ULES, very close to the coast, is in a shear zone where much ridging can occur as well as deformation around grounded icebergs.

In summary, Antarctic sea ice of a given age is much thinner on average than Arctic sea ice, but the overlying snow cover can be thicker. The reasons for the great snow thickness in multiyear Antarctic ice are that the snow does not necessarily melt

during the first summer, while during its second year it enters the inner part of the Weddell Sea where precipitation is greater. In the central Arctic snow depth may reach 40 cm by the end of the first winter, but the snow melts in summer so that snow thickness on multiyear ice is a function only of time of year. In the Antarctic the snow thickness is sufficient to push the ice–snow interface below sea level in 17% of cases sampled for first-year ice and up to 53% for second- and multiyear ice. It has been estimated that the resulting snow-ice (or so-called 'meteoric ice') makes up 16% of the ice mass in the Weddell Sea. Finally, in the Antarctic much of the ice has a fine-grained structure of randomly oriented crystals, formed from the freezing of a frazil ice suspension to form pancakes, then the freezing together of pancakes to form consolidated pancake ice, the typical first-year ice type forming in the advancing winter ice edge region. In the Arctic most ice has formed by congelation growth and so shows a crystal fabric of columnar-grained ice with horizontal c-axes. A mechanical difference is that in the Antarctic most ridges appear to be formed by buckling and crushing of the material of the floes themselves, and so are composed of a small number of fairly thick blocks extending to modest depths – typically 6 m or less. In the Arctic ridges tend to be formed by the crushing of thin ice in refrozen leads between floes, and so are composed of a large mass of small blocks, extending to greater depths – typically 10–20 m, with significant numbers extending to 30 m or more and even to 40–50 m in extreme cases.

See also

Antarctic Circumpolar Current. Arctic Basin Circulation. Coupled Sea Ice-Ocean Models. Current Systems in the Atlantic Ocean. Current Systems in the Southern Ocean. General Circulation Models. Icebergs. Ice–Ocean Interaction. Ice-shelf Stability. Polynyas. Sea Ice: Overview. Weddell Sea Circulation.

Further Reading

Ackley SF and Weeks WF (eds). (1990). *Sea Ice Properties and Processes*, CRREL Monograph 90-1. Hanover, NH: US Army Cold Regions Research and Engineering Laboratory.

Gloersen P, Campbell WJ, Cavalieri DJ *et al.* (1992) *Arctic and Antarctic Sea Ice, 1978–1987: Satellite Passive-microwave Observations and Analysis*. National Aeronautics and Space Administration, Report NASA SP-511.

Jeffries MO (ed.) (1998) *Antarctic Sea Ice: Physical Processes, Interactions and Variability*, Antarctic

Research Series 74. Washington: American Geophysical Union.

Leppäranta M (ed.) (1998) *Physics of Ice-Covered Seas*, vols 1 and 2. University of Helsinki Press.

Wadhams P (2000) *Ice in the Ocean*. London: Gordon and Breach.

Wadhams P, Dowdeswell JA and Schofield AN (eds) (1996) *The Arctic and Environmental Change*. London: Gordon and Breach Publishers.

Wadhams P, Gascard J-C and Miller L (eds) (1999) The European Subpolar Ocean Programme: ESOP. *Deep-Sea Research II* 46: 1011–1530 (special issue).

Wheeler PA (ed.) (1997) 1994 Arctic Ocean Section. *Deep-Sea Research II* 44: (Special issue)

Zwally HJ, Comiso JC, Parkinson CL *et al.* (1983). *Antarctic Sea Ice 1973–1976: Satellite Passive Microwave Observations*. Washington, DC: NASA, Report. SP-459.

SEA LEVEL CHANGE

J. A. Church, Antarctic CRC and CSIRO Marine Research, Tasmania, Australia
J. M. Gregory, Hadley Centre, Berkshire, UK

doi:10.1006/rwos.2001.0268

Introduction

Sea-level changes on a wide range of time and space scales. Here we consider changes in mean sea level, that is, sea level averaged over a sufficient period of time to remove fluctuations associated with surface waves, tides, and individual storm surge events. We focus principally on changes in sea level over the last hundred years or so and on how it might change over the next one hundred years. However, to understand these changes we need to consider what has happened since the last glacial maximum 20 000 years ago. We also consider the longer-term implications of changes in the earth's climate arising from changes in atmospheric greenhouse gas concentrations.

Changes in mean sea level can be measured with respect to the nearby land (relative sea level) or a fixed reference frame. Relative sea level, which changes as either the height of the ocean surface or the height of the land changes, can be measured by a coastal tide gauge.

The world ocean, which has an average depth of about 3800 m, contains over 97% of the earth's water. The Antarctic ice sheet, the Greenland ice sheet, and the hundred thousand nonpolar glaciers/ice caps, presently contain water sufficient to raise sea level by 61 m, 7 m, and 0.5 m respectively if they were entirely melted. Ground water stored shallower than 4000 m depth is equivalent to about 25 m (12 m stored shallower than 750 m) of sea-level change. Lakes and rivers hold the equivalent of less than 1 m, while the atmosphere accounts for only about 0.04 m.

On the time-scales of millions of years, continental drift and sedimentation change the volume of the ocean basins, and hence affect sea level. A major influence is the volume of mid-ocean ridges, which is related to the arrangement of the continental plates and the rate of sea floor spreading.

Sea level also changes when mass is exchanged between any of the terrestrial, ice, or atmospheric reservoirs and the ocean. During glacial times (ice ages), water is removed from the ocean and stored in large ice sheets in high-latitude regions. Variations in the surface loading of the earth's crust by water and ice change the shape of the earth as a result of the elastic response of the lithosphere and viscous flow of material in the earth's mantle and thus change the level of the land and relative sea level. These changes in the distribution of mass alter the gravitational field of the earth, thus changing sea level. Relative sea level can also be affected by local tectonic activities as well as by the land sinking when ground water is extracted or sedimentation increases. Sea water density is a function of temperature. As a result, sea level will change if the ocean's temperature varies (as a result of thermal expansion) without any change in mass.

Sea-Level Changes Since the Last Glacial Maximum

On timescales of thousands to hundreds of thousands of years, the most important processes affecting sea level are those associated with the growth and decay of the ice sheets through glacial–interglacial cycles. These are also relevant to current and future sea-level rise because they are the cause of ongoing land movements (as a result of changing surface loads and the resultant small changes in the shape of the earth – postglacial rebound) and ongoing changes in the ice sheets.

Sea-level variations during a glacial cycle exceed 100 m in amplitude, with rates of up to tens of millimetres per year during periods of rapid decay of the ice sheets (**Figure 1**). At the last glacial

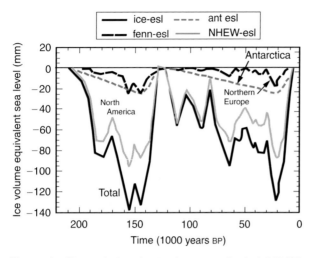

Figure 1 Change in ice sheet volume over the last 200 000 years. Fenn, Fennoscandian; ant, Antarctic; NHEW; North America; esl, equivalent sea level. (Reproduced from Lambeck, 1998.)

maximum (about 21 000 years ago), sea level was more than 120 m below current levels. The largest contribution to this sea-level lowering was the additional ice that formed the North American (Laurentide) and European (Fennoscandian) ice sheets. In addition, the Antarctic ice sheet was larger than at present and there were smaller ice sheets in presently ice-free areas.

Observed Recent Sea-Level Change

Long-term relative sea-level changes have been inferred from the geological records, such as radiocarbon dates of shorelines displaced from present day sea level, and information from corals and sediment cores. Today, the most common method of measuring sea level relative to a local datum is by tide gauges at coastal and island sites. A global data set is maintained by the Permanent Service for Mean Sea Level (PSMSL). During the 1990s, sea level has been measured globally with satellites.

Tide-gauge Observations

Unfortunately, determination of global-averaged sea-level rise is severely limited by the small number of gauges (mostly in Europe and North America) with long records (up to several hundred years, **Figure 2**). To correct for vertical land motions, some sea-level change estimates have used geological data, whereas others have used rates of present-day vertical land movement calculated from models of postglacial rebound.

A widely accepted estimate of the current rate of global-average sea-level rise is about 1.8 mm y^{-1}. This estimate is based on a set of 24 long tide-gauge records, corrected for land movements resulting from deglaciation. However, other analyses produce different results. For example, recent analyses suggest that sea-level change in the British Isles, the North Sea region and Fennoscandia has been about 1 mm y^{-1} during the past century. The various assessments of the global-average rate of sea-level change over the past century are not all consistent within stated uncertainties, indicating further sources of error. The treatment of vertical land movements remains a source of potential inconsistency, perhaps amounting to 0.5 mm y^{-1}. Other sources of

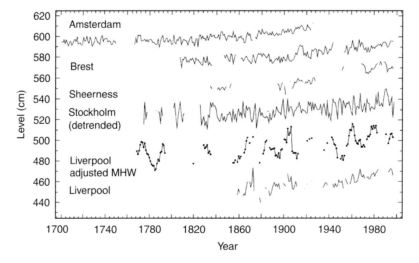

Figure 2 Time series of relative sea level over the last 300 years from several European coastal locations. For the Stockholm record, the trend over the period 1774 to 1873 has been removed from the entire data set. For Liverpool two series are given. These are the mean sea level and, for a longer period, the mean high water (MHW) level. (Reproduced with permission from Woodworth, 1999.)

error include variability over periods of years and longer and any spatial distribution in regional sea level rise (perhaps several tenths of a millimeter per year).

Comparison of the rates of sea-level rise over the last 100 years ($1.0-2.0 \, \text{mm y}^{-1}$) and over the last two millennia ($0.1-0.2 \, \text{mm y}^{-1}$) suggests the rate has accelerated fairly recently. From the few very long tide-gauge records (**Figure 2**), it appears that an acceleration of about $0.3-0.9 \, \text{mm y}^{-1}$ per century occurred over the nineteenth and twentieth century. However, there is little indication that sea-level rise accelerated during the twentieth century.

Altimeter Observations

Following the advent of high-quality satellite radar altimeter missions in the 1990s, near-global and homogeneous measurement of sea level is possible, thereby overcoming the inhomogeneous spatial sampling from coastal and island tide gauges. However, clarifying rates of global sea-level change requires continuous satellite operations over many years and careful control of biases within and between missions.

To date, the TOPEX/POSEIDON satellite-altimeter mission, with its (near) global coverage from 66°N to 66°S (almost all of the ice-free oceans) from late 1992 to the present, has proved to be of most value in producing direct estimates of sea-level change. The present data allow global-average sea level to be estimated to a precision of several millimeters every 10 days, with the absolute accuracy limited by systematic errors. The most recent estimates of global-average sea level rise based on the short (since 1992) TOPEX/POSEIDON time series range from $2.1 \, \text{mm y}^{-1}$ to $3.1 \, \text{mm y}^{-1}$.

The altimeter record for the 1990s indicates a rate of sea-level rise above the average for the twentieth century. It is not yet clear if this is a result of an increase in the rate of sea-level rise, systematic differences between the tide-gauge and altimeter data sets or the shortness of the record.

Processes Determining Present Rates of Sea-Level Change

The major factors determining sea-level change during the twentieth and twenty-first century are ocean thermal expansion, the melting of nonpolar glaciers and ice caps, variation in the mass of the Antarctic and Greenland ice sheets, and changes in terrestrial storage.

Projections of climate change caused by human activity rely principally on detailed computer models referred to as atmosphere–ocean general circulation models (AOGCMs). These simulate the global three-dimensional behavior of the ocean and atmosphere by numerical solution of equations representing the underlying physics. For simulations of the next hundred years, future atmospheric concentrations of gases that may affect the climate (especially carbon dioxide from combustion of fossil fuels) are estimated on the basis of assumptions about future population growth, economic growth, and technological change. AOGCM experiments indicate that the global-average temperature may rise by 1.4–5.8°C between 1990 and 2100, but there is a great deal of regional and seasonal variation in the predicted changes in temperature, sea level, precipitation, winds, and other parameters.

Ocean Thermal Expansion

The broad pattern of sea level is maintained by surface winds, air–sea fluxes of heat and fresh water (precipitation, evaporation, and fresh water runoff from the land), and internal ocean dynamics. Mean sea level varies on seasonal and longer timescales. A particularly striking example of local sea-level variations occurs in the Pacific Ocean during El Niño events. When the trade winds abate, warm water moves eastward along the equator, rapidly raising sea level in the east and lowering it in the west by about 20 cm.

As the ocean warms, its density decreases. Thus, even at constant mass, the volume of the ocean increases. This thermal expansion is larger at higher temperatures and is one of the main contributors to recent and future sea-level change. Salinity changes within the ocean also have a significant impact on the local density, and thus on local sea level, but have little effect on the global-average sea level.

The rate of global temperature rise depends strongly on the rate at which heat is moved from the ocean surface layers into the deep ocean; if the ocean absorbs heat more readily, climate change is retarded but sea level rises more rapidly. Therefore, time-dependent climate change simulation requires a model that represents the sequestration of heat in the ocean and the evolution of temperature as a function of depth. The large heat capacity of the ocean means that there will be considerable delay before the full effects of surface warming are felt throughout the depth of the ocean. As a result, the ocean will not be in equilibrium and global-average sea level will continue to rise for centuries after atmospheric greenhouse gas concentrations have stabilized. The geographical distribution of sea-level change may take many decades to arrive at its final state.

While the evidence is still somewhat fragmentary, and in some cases contradictory, observations indicate ocean warming and thus thermal expansion, particularly in the subtropical gyres, at rates resulting in sea-level rise of order $1\,\mathrm{mm\,y^{-1}}$. The observations are mostly over the last few decades, but some observations date back to early in the twentieth century. The evidence is most convincing for the subtropical gyre of the North Atlantic, for which the longest temperature records (up to 73 years) and most complete oceanographic data sets exist. However, the pattern also extends into the South Atlantic and the Pacific and Indian oceans. The only areas of substantial ocean cooling are the subpolar gyres of the North Atlantic and perhaps the North Pacific. To date, the only estimate of a global average rate of sea-level rise from thermal expansion is $0.55\,\mathrm{mm\,y^{-1}}$.

The warming in the Pacific and Indian Oceans is confined to the main thermocline (mostly the upper $1\,\mathrm{km}$) of the subtropical gyres. This contrasts with the North Atlantic, where the warming is also seen at greater depths.

AOGCM simulations of sea level suggest that during the twentieth century the average rate of change due to thermal expansion was of the order of 0.3–$0.8\,\mathrm{mm\,y^{-1}}$ (**Figure 3**). The rate rises to 0.6–$1.1\,\mathrm{mm\,y^{-1}}$ in recent decades, similar to the observational estimates of ocean thermal expansion.

Nonpolar Glaciers and Ice Caps

Nonpolar glaciers and ice caps are rather sensitive to climate change, and rapid changes in their mass contribute significantly to sea-level change. Glaciers gain mass by accumulating snow, and lose mass (ablation) by melting at the surface or base. Net accumulation occurs at higher altitude, net ablation at lower altitude. Ice may also be removed by discharge into a floating ice shelf and/or by direct calving of icebergs into the sea.

In the past decade, estimates of the regional totals of the area and volume of glaciers have been improved. However, there are continuous mass balance records longer than 20 years for only about 40 glaciers worldwide. Owing to the paucity of measurements, the changes in mass balance are estimated as a function of climate.

On the global average, increased precipitation during the twenty-first century is estimated to offset only 5% of the increased ablation resulting from warmer temperatures, although it might be significant in particular localities. (For instance, while glaciers in most parts of the world have had negative mass balance in the past 20 years, southern Scandinavian glaciers have been advancing, largely because of increases in precipitation.) A detailed computa-

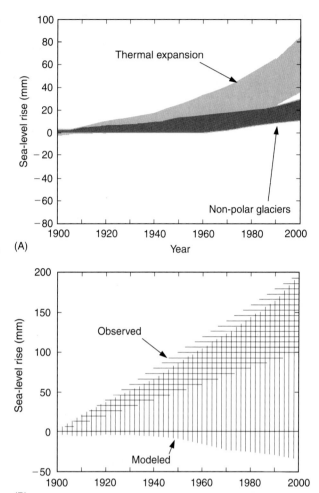

Figure 3 Computed sea-level rise from 1900 to 2000 AD. (A) The estimated thermal expansion is shown by the light stippling; the estimated nonpolar glacial contribution is shown by the medium-density stippling. (B) The computed total sea level change during the twentieth century is shown by the vertical hatching and the observed sea-level change is shown by the horizontal hatching.

tion of transient response also requires allowance for the contracting area of glaciers.

Recent estimates of glacier mass balance, based on both observations and model studies, indicate a contribution to global-average sea level of 0.2 to $0.4\,\mathrm{mm\,y^{-1}}$ during the twentieth century. The model results shown in **Figure 3** indicate an average rate of 0.1 to $0.3\,\mathrm{mm\,y^{-1}}$.

Greenland and Antarctic Ice Sheets

A small fractional change in the volume of the Greenland and Antarctic ice sheets would have a significant effect on sea level. The average annual solid precipitation falling onto the ice sheets is equivalent to $6.5\,\mathrm{mm}$ of sea level, but this input is approximately balanced by loss from melting and iceberg calving. In the Antarctic, temperatures are so low that surface melting is negligible, and the ice sheet loses mass mainly by ice discharge into floating ice shelves, which melt at their underside and eventually break up to form icebergs. In Greenland, summer temperatures are high enough to cause widespread surface melting, which accounts for about half of the ice loss, the remainder being discharged as icebergs or into small ice shelves.

The surface mass balance plays the dominant role in sea-level changes on a century timescale, because changes in ice discharge generally involve response times of the order of 10^2 to 10^4 years. In view of these long timescales, it is unlikely that the ice sheets have completely adjusted to the transition from the previous glacial conditions. Their present contribution to sea-level change may therefore include a term related to this ongoing adjustment, in addition to the effects of climate change over the last hundred years. The current rate of change of volume of the polar ice sheets can be assessed by estimating the individual mass balance terms or by monitoring surface elevation changes directly (such as by airborne and satellite altimetry during the 1990s). However, these techniques give results with large uncertainties. Indirect methods (including numerical modeling of ice-sheets, observed sea-level changes over the last few millennia, and changes in the earth's rotation parameters) give narrower bounds, suggesting that the present contribution of the ice sheets to sea level is a few tenths of a millimeter per year at most.

Calculations suggest that, over the next hundred years, surface melting is likely to remain negligible in Antarctica. However, projected increases in precipitation would result in a net negative sea-level contribution from the Antarctic ice sheet. On the other hand, in Greenland, surface melting is projected to increase at a rate more than enough to offset changes in precipitation, resulting in a positive contribution to sea-level rise.

Changes in Terrestrial Storage

Changes in terrestrial storage include reductions in the volumes of some of the world's lakes (e.g., the Caspian and Aral seas), ground water extraction in excess of natural recharge, more water being impounded in reservoirs (with some seeping into aquifers), and possibly changes in surface runoff. Order-of-magnitude evaluations of these terms are uncertain but suggest that each of the contributions could be several tenths of millimeter per year, with a small net effect (**Figure 3**). If dam building continues at the same rate as in the last 50 years of the twentieth century, there may be a tendency to reduce sea-level rise. Changes in volumes of lakes and rivers will make only a negligible contribution.

Permafrost currently occupies about 25% of land area in the northern hemisphere. Climate warming leads to some thawing of permafrost, with partial runoff into the ocean. The contribution to sea level in the twentieth century is probably less than 5 mm.

Projected Sea-Level Changes for the Twenty-first Century

Detailed projections of changes in sea level derived from AOCGM results are given in material listed as Further Reading. The major components are thermal expansion of the ocean (a few tens of centimeters), melting of nonpolar glaciers (about 10–20 cm), melting of Greenland ice sheet (several centimeters), and increased storage in the Antarctic (several centimeters).

After allowance for the continuing changes in the ice sheets since the last glacial maximum and the melting of permafrost (but not including changes in terrestrial storage), total projected sea-level rise during the twenty-first century is currently estimated to be between about 9 and 88 cm (**Figure 4**).

Regional Sea-Level Change

Estimates of the regional distribution of sea-level rise are available from several AOGCMs. Our confidence in these distributions is low because there is little similarity between model results. However, models agree on the qualitative conclusion that the range of regional variation is substantial compared with the global-average sea-level rise. One common

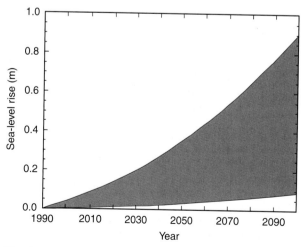

Figure 4 The estimated range of future global-average sea-level rise from 1990 to 2100 AD for a range of plausible projections in atmospheric greenhouse gas concentrations.

feature is that nearly all models predict less than average sea-level rise in the Southern Ocean.

The most serious impacts of sea-level change on coastal communities and ecosystems will occur during the exceptionally high water levels known as storm surges produced by low air pressure or driving winds. As well as changing mean sea level, climate change could also affect the frequency and severity of these meteorological conditions, making storm surges more or less severe at particular locations.

Longer-term Changes

Even if greenhouse gas concentrations were to be stabilized, sea level would continue to rise for several hundred years. After 500 years, sea-level rise from thermal expansion could be about 0.3–2 m but may be only half of its eventual level. Glaciers presently contain the equivalent of 0.5 m of sea level. If the CO_2 levels projected for 2100 AD were sustained, there would be further reductions in glacier mass.

Ice sheets will continue to react to climatic change during the present millennium. The Greenland ice sheet is particularly vulnerable. Some models suggest that with a warming of several degrees no ice sheet could be sustained on Greenland. Complete melting of the ice sheet would take at least a thousand years and probably longer.

Most of the ice in Antarctica forms the East Antarctic Ice Sheet, which would disintegrate only if extreme warming took place, beyond what is currently thought possible. The West Antarctic Ice Sheet (WAIS) has attracted special attention because it contains enough ice to raise sea level by 6 m and because of suggestions that instabilities associated with its being grounded below sea level may result in rapid ice discharge when the surrounding ice shelves are weakened. However, there is now general agreement that major loss of grounded ice, and accelerated sea-level rise, is very unlikely during the twenty-first century. The contribution of this ice sheet to sea level change will probably not exceed 3 m over the next millennium.

Summary

On timescales of decades to tens of thousands of years, sea-level change results from exchanges of mass between the polar ice sheets, the nonpolar glaciers, terrestrial water storage, and the ocean. Sea level also changes if the density of the ocean changes (as a result of changing temperature) even though there is no change in mass. During the last century, sea level is estimated to have risen by 10–20 cm, as a result of combination of thermal expansion of the ocean as its temperature rose and increased mass of the ocean from melting glaciers and ice sheets. Over the twenty-first century, sea level is expected to rise as a result of anthropogenic climate change. The main contributors to this rise are expected to be thermal expansion of the ocean and the partial melting of nonpolar glaciers and the Greenland ice sheet. Increased precipitation in Antarctica is expected to offset some of the rise from other contributions. Changes in terrestrial storage are uncertain but may also partially offset rises from other contributions. After allowance for the continuing changes in the ice sheets since the last glacial maximum, the total projected sea-level rise over the twenty-first century is currently estimated to be between about 9 and 88 cm.

See also

Abrupt Climate Change. Elemental Distribution: Overview. El Niño Southern Oscillation (ENSO). El Niño Southern Oscillation (ENSO) Models. Forward Problem in Numerical Models. Glacial Crustal Rebound, Sea levels and Shorelines. Ice-shelf Stability. Icebergs. International Organizations. Ocean Circulation. Ocean Subduction. Past Climate From Corals. Satellite Altimetry. Sea Level Variations Over Geologic Time.

Further Reading

Church JA, Gregory JM, Huybrechts P *et al.* (2001) Changes in sea level. In: Houghton JT (ed.) *Climate Change 2001*; *The Scientific Basis*. Cambridge: Cambridge University Press.

Douglas BC, Keaney M and Leatherman SP (eds) (2000) *Sea Level Rise: History and Consequences*, 232 pp. San Diego: Academic Press.

Fleming K, Johnston P, Zwartz D *et al.* (1998) Refining the eustatic sea-level curve since the Last Glacial Maximum using far- and intermediate-field sites. *Earth and Planetary Science Letters* 163: 327–342.

Lambeck K (1998) Northern European Stage 3 ice sheet and shoreline reconstructions: Preliminary results. News 5, Stage 3 Project, Godwin Institute for Quaternary Research, 9 pp.

Peltier WR (1998) Postglacial variations in the level of the sea: implications for climate dynamics and solid-earth geophysics. *Review of Geophysics* 36: 603–689.

Summerfield MA (1991) *Global Geomorphology*. Harlowe: Longman.

Warrick RA, Barrow EM and Wigley TML (1993) *Climate and Sea Level Change: Observations, Projections and Implications*. Cambridge: Cambridge University Press.

WWW Pages of the Permanent Service for Mean Sea Level, http://www.pol.ac.uk/psmsl/

SEA LEVEL VARIATIONS OVER GEOLOGIC TIME

M. A. Kominz, Western Michigan University, Kalamazoo, MI, USA

doi:10.1006/rwos.2001.0255

Introduction

Sea level changes have occurred throughout Earth history. The magnitudes and timing of sea level changes are extremely variable. They provide considerable insight into the tectonic and climatic history of the Earth, but remain difficult to determine with accuracy.

Sea level, where the world oceans intersect the continents, is hardly fixed, as anyone who has stood on the shore for 6 hours or more can attest. But the ever-changing tidal flows are small compared with longer-term fluctuations that have occurred in Earth history. How much has sea level changed? How long did it take? How do we know? What does it tell us about the history of the Earth?

In order to answer these questions, we need to consider a basic question: what causes sea level to change? Locally, sea level may change if tectonic forces cause the land to move up or down. However, this article will focus on global changes in sea level. Thus, the variations in sea level must be due to one of two possibilities: (1) changes in the volume of water in the oceans or (2) changes in the volume of the ocean basins.

Sea Level Change due to Volume of Water in the Ocean Basin

The two main reservoirs of water on Earth are the oceans (currently about 97% of all water) and glaciers (currently about 2.7%). Not surprisingly, for at least the last three billion years, the main variable controlling the volume of water filling the ocean basins has been the amount of water present in glaciers on the continents. For example, about 20 000 years ago, great ice sheets covered northern North America and Europe. The volume of ice in these glaciers removed enough water from the oceans to expose most continental shelves. Since then there has been a sea level rise (actually from about 20 000 to about 11 000 years ago) of about 120 m (**Figure 1A**).

A number of methods have been used to establish the magnitude and timing of this sea level change.

Dredging on the continental shelves reveals human activity near the present shelf-slope boundary. These data suggest that sea level was much lower a relatively short time ago. Study of ancient corals shows that coral species which today live only in very shallow water are now over 100 m deep. The carbonate skeletons of the coral, which once thrived in the shallow waters of the tropics, yield a detailed picture of the timing of sea level rise, and, thus, the melting of the glaciers. Carbon-14, a radioactive isotope formed by carbon-12 interacting with high-energy solar radiation in Earth's atmosphere (*see* **Cosmogenic Isotopes**) allows us to determine the age of Earth materials, which are about 30 thousand years old.

This is just the most recent of many, large changes in sea level caused by glaciers, (**Figure 1B**). These variations in climate and subsequent sea level changes have been tied to quasi-periodic variations in the Earth's orbit and the tilt of the Earth's spin axis. The record of sea level change can be estimated by observing the stable isotope, oxygen-18 in the tests (shells) of dead organisms (*see* **Cenozoic Climate − Oxygen Isotope Evidence**). When marine microorganisms build their tests from the calcium, carbon, and oxygen present in sea water they incorporate both the abundant oxygen-16 and the rare oxygen-18 isotopes. Water in the atmosphere generally has a lower oxygen-18 to oxygen-16 ratio because evaporation of the lighter isotope requires less energy. As a result, the snow that eventually forms the glaciers is depleted in oxygen-18, leaving the ocean proportionately oxygen-18-enriched. When the microorganisms die, their tests sink to the seafloor to become part of the deep marine sedimentary record. The oxygen-18 to oxygen-16 ratio present in the fossil tests has been calibrated to the sea level change, which occurred from 20 000 to 11 000 years ago, allowing the magnitude of sea level change from older times to be estimated. This technique does have uncertainties. Unfortunately, the amount of oxygen-18 which organisms incorporate in their tests is affected not only by the amount of oxygen-18 present but also by the temperature and salinity of the water. For example, the organisms take up less oxygen-18 in warmer waters. Thus, during glacial times, the tests are even more enriched in oxygen-18, and any oxygen isotope record reveals a combined record of changing local temperature and salinity in addition to the record of global glaciation.

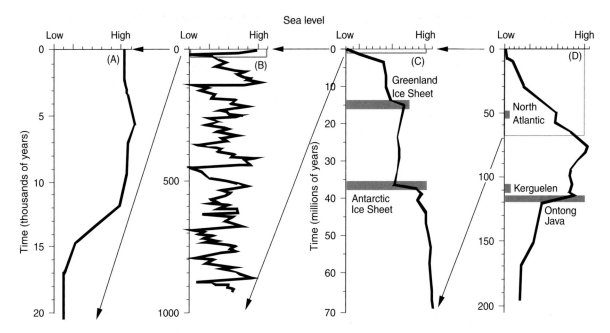

Figure 1 (A) Estimates of sea level change over the last 20 000 years. Amplitude is about 120 m. (B) Northern Hemisphere glaciers over the last million years or so generated major sea level fluctuations, with amplitudes as high as 125 m. (C) The long-term oxygen isotope record reveals rapid growth of the Antarctic and Greenland ice sheets (indicated by gray bars) as Earth's climate cooled. (D) Long-term sea level change as indicated from variations in deep-ocean volume. Dominant effects include spreading rates and lengths of mid-ocean ridges, emplacement of large igneous provinces (the largest, marine LIPs are indicated by gray bars), breakup of supercontinents, and subduction of the Indian continent. The Berggren *et al.* (1995) chronostratigraphic timescale was used in (C) and (D).

Moving back in time through the Cenozoic (zero to 65 Ma), paleoceanographic data remain excellent due to relatively continuous sedimentation on the ocean floor (as compared with shallow marine and terrestrial sedimentation). Oxygen-18 in the fossil shells suggests a general cooling for about the last 50 million years. Two rapid increases in the oxygen-18 to oxygen-16 ratio about 12.5 Ma and about 28 Ma are observed (**Figure 1C**). The formation of the Greenland Ice Sheet and the Antarctic Ice Sheet are assumed to be the cause of these rapid isotope shifts. Where oxygen-18 data have been collected with a resolution finer than 20 000 years, high-frequency variations are seen which are presumed to correspond to a combination of temperature change and glacial growth and decay. We hypothesize that the magnitudes of these high-frequency sea level changes were considerably less in the earlier part of the Cenozoic than those observed over the last million years. This is because considerably less ice was involved.

Although large continental glaciers are not common in Earth history they are known to have been present during a number of extended periods ('ice house' climate, in contrast to 'greenhouse' or warm climate conditions). Ample evidence of glaciation is

found in the continental sedimentary record. In particular, there is evidence of glaciation from about 2.7 to 2.1 billion years ago. Additionally, a long period of glaciation occurred shortly before the first fossils of multicellular organisms, from about 1 billion to 540 million years ago. Some scientists now believe that during this glaciation, the entire Earth froze over, generating a 'snowball earth'. Such conditions would have caused a large sea level fall. Evidence of large continental glaciers are also seen in Ordovician to Silurian rocks (~ 420 to 450 Ma), in Devonian rocks (~ 380 to 390 Ma), and in Carboniferous to Permian rocks (~ 350 to 270 Ma).

If these glaciations were caused by similar mechanisms to those envisioned for the Plio-Pleistocene (**Figure 1B**), then predictable, high-frequency, periodic growth and retreat of the glaciers should be observed in strata which form the geologic record. This is certainly the case for the Carboniferous through Permian glaciation. In the central United States, UK, and Europe, the sedimentary rocks have a distinctly cyclic character. They cycle in repetitive vertical successions of marine deposits, near-shore deposits, often including coals, into fluvial sedimentary rocks. The deposition of marine rocks over large areas, which had only recently been nonmarine, suggests very large-scale sea level changes. When the duration of the entire record is taken into account, periodicities of about 100 and 400 thousand years are suggested for these large sea level changes. This is consistent with an origin due to a response to changes in the eccentricity of the Earth's orbit. Higher-frequency cyclicity associated with the tilt of the spin axis and precession of the equinox is more difficult to prove, but has been suggested by detailed observations.

It is fair to say that large-scale (10 to > 100 m), relatively high-frequency (20 000–400 000 years; often termed 'Milankovitch scale') variations in sea level occurred during intervals of time when continental glaciers were present on Earth (ice house climate). This indicates that the variations of Earth's orbit and the tilt of its spin axis played a major role in controlling the climate. During the rest of Earth history, when glaciation was not a dominant climatic force (greenhouse climate), sea level changes corresponding to Earth's orbit did occur. In this case, the mechanism for changing the volume of water in the ocean basins is much less clear.

There is no geological record of continental ice sheets in many portions of Earth history. These time periods are generally called 'greenhouse' climates. However, there is ample evidence of Milankovitch scale variations during these periods. In shallow marine sediments, evidence of orbitally driven sea level changes has been observed in Cambrian and Cretaceous age sediments. The magnitudes of sea level change required (perhaps 5–20 m) are far less than have been observed during glacial climates. A possible source for these variations could be variations in average ocean-water temperature. Water expands as it is heated. If ocean bottom-water sources were equatorial rather than polar, as they are today, bottom-water temperatures of about 2°C today might have been about 16°C in the past. This would generate a sea level change of about 11 m. Other causes of sea level change during greenhouse periods have been postulated to be a result of variations in the magnitude of water trapped in inland lakes and seas, and variations in volumes of alpine glaciers. Deep marine sediments of Cretaceous age also show fluctuations between oxygenated and anoxic conditions. It is possible that these variations were generated when global sea level change restricted flow from the rest of the world's ocean to a young ocean basin. In a more recent example, tectonics caused a restriction at the Straits of Gibraltar. In that case, evaporation generated extreme sea level changes and restricted their entrance into the Mediterranean region.

Sea Level Change due to Changing Volume of the Ocean Basin

Tectonics is thought to be the main driving force of long-term (≥ 50 million years) sea level change. Plate tectonics changes the shape and/or the areal extent of the ocean basins.

Plate tectonics is constantly reshaping surface features of the Earth while the amount of water present has been stable for about the last four billion years. The reshaping changes the total area taken up by oceans over time. When a supercontinent forms, subduction of one continent beneath another decreases Earth's ratio of continental to oceanic area, generating a sea level fall. In a current example, the continental plate including India is diving under Asia to generate the Tibetan Plateau and the Himalayan Mountains. This has probably generated a sea level fall of about 70 m over the last 50 million years. The process of continental breakup has the opposite effect. The continents are stretched, generating passive margins and increasing the ratio of continental to oceanic area on a global scale (**Figure 2A**). This results in a sea level rise. Increments of sea level rise resulting from continental breakup over the last 200 million years amount to about 100 m of sea level rise.

Some bathymetric features within the oceans are large enough to generate significant changes in sea level as they change size and shape. The largest physiographic feature on Earth is the mid-ocean ridge system, with a length of about 60 000 km and a width of 500–2000 km. New ocean crust and lithosphere are generated along rifts in the center of these ridges. The ocean crust is increasingly old, cold, and dense away from the rift. It is the heat of ocean lithosphere formation that actually generates this feature. Thus, rifting of continents forms new ridges, increasing the proportionate area of young, shallow, ocean floor to older, deeper ocean floor (**Figure 2B**). Additionally, the width of the ridge is a function of the rates at which the plates are moving apart. Fast spreading ridges (e.g. the East Pacific Rise) are very broad while slow spreading ridges (e.g. the North Atlantic Ridge) are quite narrow. If the average spreading rates for all ridges decreases, the average volume taken up by ocean ridges would decrease. In this case, the volume of the ocean basin available for water would increase and a sea level fall would occur. Finally, entire ridges may be removed in the process of subduction, generating fairly rapid sea level fall.

Scientists have made quantitative estimates of sea level change due to changing ocean ridge volumes. Since ridge volume is dependent on the age of the

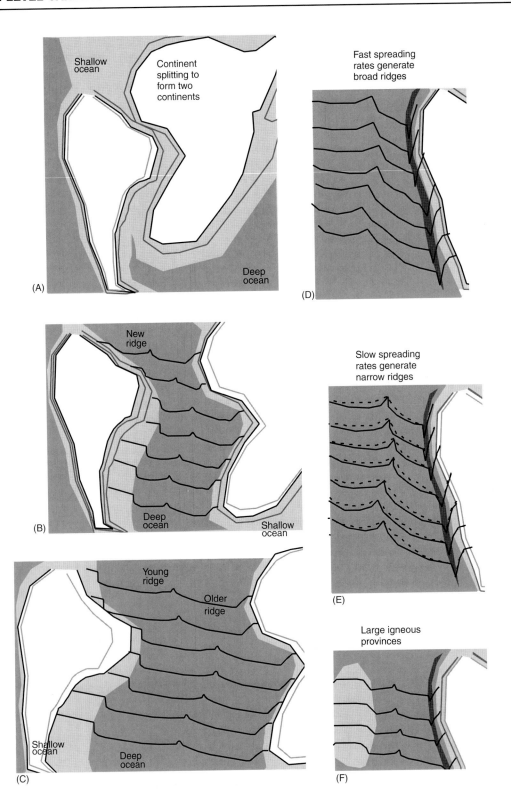

Figure 2 Diagrams showing a few of the factors which affect the ocean volume. (A) Early breakup of a large continent increases the area of continental crust by generating passive margins, causing sea level to rise. (B) Shortly after breakup a new ocean is formed with very young ocean crust. This young crust must be replacing relatively old crust via subduction, generating additional sea level rise. (C) The average age of the ocean between the continents becomes older so that young, shallow ocean crust is replaced with older, deeper crust so that sea level falls. (D) Fast spreading rates are associated with relatively high sea level. (E) Relatively slow spreading ridges (solid lines in ocean) take up less volume in the oceans than high spreading rate ridges (dashed lines in ocean), resulting in relatively low sea level. (F) Emplacement of large igneous provinces generates oceanic plateaus, displaces ocean water, and causes a sea level rise.

ocean floor, where the age of the ocean floor is known, ridge volumes can be estimated. Seafloor magnetic anomalies are used to estimate the age of the ocean floor, and thus, spreading histories of the oceans (*see* **Magnetics**). The oldest ocean crust is about 200 million years old. Older oceanic crust has been subducted. Thus, it is not surprising that quantitative estimates of sea level change due to ridge volumes are increasingly uncertain and cannot be calculated before about 90 million years. Sea level is estimated to have fallen about 230 m (± 120 m) due to ridge volume changes in the last 80 million years.

Large igneous provinces (LIPs) are occasionally intruded into the oceans, forming large oceanic plateaus (*see* **Igneous Provinces**). The volcanism associated with LIPs tends to occur over a relatively short period of time, causing a rapid sea level rise. However, these features subside slowly as the lithosphere cools, generating a slow increase in ocean volume, and a long-term sea level fall. The largest marine LIP, the Ontong Java Plateau, was emplaced in the Pacific Ocean between about 120 and 115 Ma (**Figure 1D**). Over that interval it may have generated a sea level rise of around 50 m.

In summary, over the last 200 million years, long-term sea level change (**Figure 1D**) can be largely attributed to tectonics. Continental crust expanded by extension as the supercontinents Gondwana and Laurasia split to form the continents we see today. This process began about 200 Ma when North America separated from Africa and continues with the East African Rift system and the formation of the Red Sea. The generation of large oceans occurred early in this period and there was an overall rise in sea level from about 200 to about 90 million years. New continental crust, new mid-ocean ridges, and very fast spreading rates were responsible for the long-term rise (**Figure 1D**). Subsequently, a significant decrease in spreading rates, a reduction in the total length of mid-ocean ridges, and continent–continent collision coupled with an increase in glacial ice (**Figure 1C**) have resulted in a large-scale sea level fall (**Figure 1D**). Late Cretaceous volcanism associated with the Ontong Java Plateau, a large igneous province (*see* **Igneous Provinces**), generated a significant sea level rise, while subsequent cooling has enhanced the 90 million year sea level fall. Estimates of sea level change from changing ocean shape remain quite uncertain. Magnitudes and timing of stretching associated with continental breakup, estimates of shortening during continental assembly, volumes of large igneous provinces, and volumes of mid-ocean ridges improve as data are gathered. However, the exact configuration of past continents and oceans can only be a mystery due to the recycling character of plate tectonics.

Sea Level Change Estimated from Observations on the Continents

Long-term Sea Level Change

Estimates of sea level change are also made from sedimentary strata deposited on the continents. This is actually an excellent place to obtain observations of sea level change not only because past sea level has been much higher than it is now, but also because in many places the continents have subsequently uplifted. That is, in the past they were below sea level, but now they are well above it. For example, studies of 500–400 million year old sedimentary rocks which are now uplifted in the Rocky Mountains and the Appalachian Mountains indicate that there was a rise and fall of sea level with an estimated magnitude of 200–400 m. This example also exemplifies the main problem with using the continental sedimentary record to estimate sea level change. The continents are not fixed and move vertically in response to tectonic driving forces. Thus, any indicator of sea level change on the continents is an indicator of relative sea level change. Obtaining a global signature from these observations remains extremely problematic. Additionally, the continental sedimentary record contains long periods of non-deposition, which results in a spotty record of Earth history. Nonetheless a great deal of information about sea level change has been obtained and is summarized here.

The most straightforward source of information about past sea level change is the location of the strand line (the beach) on a stable continental craton (a part of the continent, which was not involved in local tectonics). Ideally, its present height is that of sea level at the time of deposition. There are two problems encountered with this approach. Unfortunately, the nature of land–ocean interaction at their point of contact is such that those sediments are rarely preserved. Where they can be observed, there is considerable controversy over which elements have moved, the continents or sea level. However, data from the past 100 million years tend to be consistent with calculations derived from estimates of ocean volume change. This is not saying a lot since uncertainties are very large (see above).

Continental hypsography (cumulative area versus height) coupled with the areal extent of preserved marine sediments has been used to estimate past sea

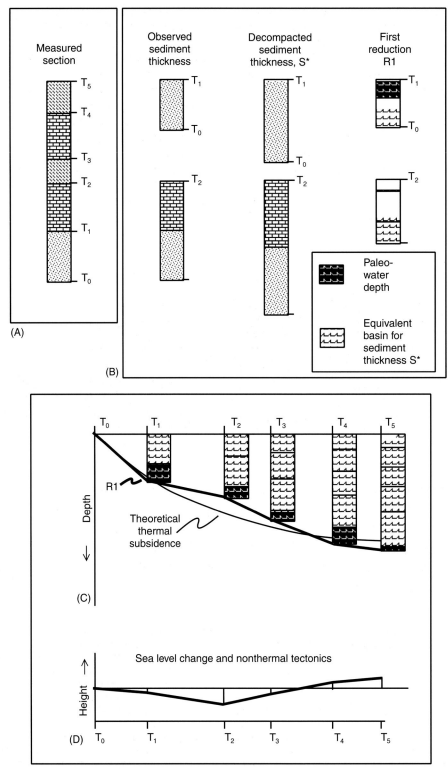

Figure 3 Diagrams depicting the backstripping method for obtaining sea level estimates in a thermally subsiding basin. (A) A stratigraphic section is measured either from exposed sedimentary rocks or from drilling. These data include lithologies, ages, and porosity. Note that the oldest strata are always at the base of the section (T0). (B) Porosity data are used to estimate the thickness that each sediment section would have had at the time of deposition (S*). They are also used to obtain sediment density so that the sediments can be unloaded to determine how deep the basin would have been in the absence of the sediment load (R1). This calculation also requires an estimate of the paleo-water depth (the water depth at the time of deposition). (C) A plot of R1 versus time is compared (by least-squares fit) to theoretical tectonic subsidence in a thermal setting. (D) The difference between R1 and thermal subsidence yields a quantitative estimate of sea level change if other, nonthermal tectonics, did not occur at this location.

level. In this case only an average result can be obtained, because marine sediments spanning a time interval (generally 5–10 million years) have been used. Again, uncertainties are large, but results are consistent with calculations derived from estimates of ocean volume change.

Backstripping is an analytical tool, which has been used to estimate sea level change. In this technique, the vertical succession of sedimentary layers is progressively decompacted and unloaded (**Figure 3A, B**). The resulting hole is a combination of the subsidence generated by tectonics and by sea level change (**Figure 3B, C**). If the tectonic portion can be established then an estimate of sea level change can

be determined (**Figure 3C, D**). This method is generally used in basins generated by the cooling of a thermal anomaly (e.g. passive margins). In these basins, the tectonic signature is predictable (exponential decay) and can be calibrated to the well-known subsidence of the mid-ocean ridge.

The backstripping method has been applied to sedimentary strata drilled from passive continental margins of both the east coast of North America and the west coast of Africa. Again, estimates of sea level suggest a rise of about 100–300 m from about 200–110 Ma followed by a fall to the present level (**Figure 1D**). Young interior basins, such as the Paris Basin, yield similar results. Older, thermally driven

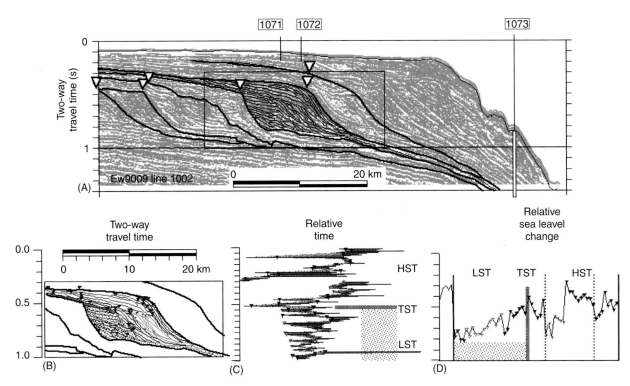

Figure 4 Example of the sequence stratigraphic approach to estimates of sea level change. (A) Multichannel seismic data (gray) from the Baltimore Canyon Trough, offshore New Jersey, USA (Miller *et al.*, 1998). Black lines are interpretations traced on the seismic data. Thick dark lines indicate third-order Miocene-aged (5–23 Ma) sequence boundaries. They are identified by truncation of the finer black lines. Upside-down deltas indicate a significant break in slope associated with each identified sequence boundary. Labeled vertical lines (1071–1073) show the locations of Ocean Drilling Project wells (**Deep-sea Drilling Methodology**), used to help date the sequences. The rectangle in the center is analyzed in greater detail. (B) Detailed interpretation of a single third-order sequence from (A). Upside-down deltas indicate a significant break in slope associated with each of the detailed sediment packages. Stippled fill indicates the low stand systems tract (LST) associated with this sequence. The gray packages are the transgressive systems tract (TST), and the overlying sediments are the high stand systems tract (HST). (C) Relationships between detailed sediment packages (in B) are used to establish a chronostratigraphy (time framework). Youngest sediment is at the top. Each observed seismic reflection is interpreted as a time horizon, and each is assigned equal duration. Horizontal distance is the same as in (A) and (B). A change in sediment type is indicated at the break in slope from coarser near-shore sediments (stippled pattern) to finer, offshore sediments (parallel, sloping lines). Sedimentation may be present offshore but at very low rates. LST, TST, and HST as in (B). (D) Relative sea level change is obtained by assuming a consistent depth relation at the change in slope indicated in (B). Age control is from the chronostratigraphy indicated in (C). Time gets younger to the right. The vertical scale is in two-way travel time, and would require conversion to depth for a final estimate of the magnitude of sea level change. LST, TST, and HST as in (B). Note that in (B), (C), and (D), higher frequency cycles (probably fourth-order) are present within this (third-order) sequence. Tracing and interpretations are from the author's graduate level quantitative stratigraphy class project (1998, Western Michigan University).

basins have also been analyzed. This was the method used to determine the (approximately 200 m) sea level rise and fall associated with the breakup of a Pre-Cambrian supercontinent in earliest Phanerozoic time.

Million Year Scale Sea Level Change

In addition to long-term changes in sea level there is evidence of fluctuations that are considerably shorter than the 50–100 million year variations discussed above, but longer than those caused by orbital variations (≤ 0.4 million years). These variations appear to be dominated by durations which last either tens of millions of years or a half to three million years. These sea level variations are sometimes termed second- and third-order sea level change, respectively. There is considerable debate concerning the source of these sea level fluctuations. They have been attributed to tectonics and changing ocean basin volumes, to the growth and decay of glaciers, or to continental uplift and subsidence, which is independent of global sea level change. As noted above, the tectonic record of subsidence and uplift is intertwined with the stratigraphic record of global sea level change on the continents. Synchronicity of observations of sea level change on

a global scale would lead most geoscientists to suggest that these signals were caused by global sea level change. However, at present, it is nearly impossible to globally determine the age equivalency of events which occur during intervals as short as a half to two million years. These data limitations are the main reason for the heated controversy over third-order sea level.

Quantitative estimates of second-order sea level variations are equally difficult to obtain. Although the debate is not as heated, these somewhat longer-term variations are not much larger than the third-order variations so that the interference of the two signals makes definition of the beginning, ending and/or magnitude of second-order sea level change equally problematic. Recognizing that our understanding of second- and third-order (million year scale) sea level fluctuations is limited, a brief review of that limited knowledge follows.

Sequence stratigraphy is an analytical method of interpreting sedimentary strata that has been used to investigate second- and third-order relative sea level changes. This paradigm requires a vertical succession of sedimentary strata which is analyzed in at least a two-dimensional, basinal setting. Packages of sedimentary strata, separated by unconformities, are observed and interpreted mainly in terms of their

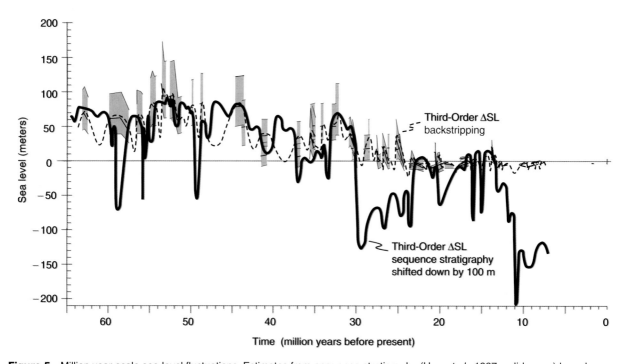

Figure 5 Million year scale sea level fluctuations. Estimates from sequence stratigraphy (Haq *et al.*, 1987; solid curve) have been shifted down by 100 m to allow comparison with estimates of sea level from backstripping (Kominz *et al.*, 1998; dashed curve). Where sediments are present, the backstripping results, with uncertainty ranges, are indicated by gray fill. Between backstrip observations, lack of preserved sediment is presumed to have been a result of sea level fall. The Berggren *et al.* (1995) biostratigraphic timescale was used.

internal geometries (e.g. **Figure 4**). The unconformities are assumed to have been generated by relative sea level fall, and thus, reflect either global sea level or local or regional tectonics. This method of stratigraphic analysis has been instrumental in hydrocarbon exploration since its introduction in the late 1970s. One of the bulwarks of this approach is the 'global sea level curve' most recently published by Haq *et al.* (1987). This curve is a compilation of relative sea level curves generated from sequence stratigraphic analysis in basins around the world. While sequence stratigraphy is capable of estimating relative heights of relative sea level, it does not estimate absolute magnitudes. Absolute dating requires isotope data or correlation via fossil data into the chronostratigraphic timescale (*see* **Geomagnetic Polarity Timescale**). However, the two-dimensional nature of the data allows for good to excellent relative age control.

Backstripping has been used, on a considerably more limited basis, in an attempt to determine million year scale sea level change. This approach is rarely applied because it requires very detailed, quantitative, estimates of sediment ages, paleoenvironments and compaction in a thermal tectonic setting. A promising area of research is the application of this method to coastal plain boreholes from the mid-Atlantic coast of North America. Here an intensive Ocean Drilling Project survey is underway which is providing sufficiently detailed data for this type of analysis (*see* **Deep-sea Drilling Methodology**). Initial results suggest that magnitudes of million year scale sea level change are roughly one-half to one-third that reported by Haq *et al*. However, in glacial times, the timing of the cycles was quite consistent with those of this 'global sea level curve' derived by application of sequence stratigraphy (**Figure 5**). Thus, it seems reasonable to conclude that, at least during glacial times, global, third-order sea level changes did occur.

Summary

Sea level changes are either a response to changing ocean volume or to changes in the volume of water contained in the ocean. The timing of sea level change ranges from tens of thousands of years to over 100 million years. Magnitudes also vary significantly but may have been as great as 200 m or more. Estimates of sea level change currently suffer from significant ranges of uncertainty, both in magnitude and in timing. However, scientists are converging on consistent estimates of sea level changes by using very different data and analytical approaches.

See also

Cosmogenic Isotopes. Deep-sea Drilling Methodology. Geomagnetic Polarity Timescale. Igneous Provinces. Magnetics.

Further Reading

Allen PA and Allen JR (1990) *Basin Analysis: Principles & Applications.* Oxford: Blackwell Scientific Publications.

Berggren WA, Kent DV, Swisher CC, Aubry MP (1995) A revised Cenozoic geochronology and chronostratigraphy. In: Berggren WA, Kent DV and Hardenbol J (eds) *Geochronology, Time Scales and Global Stratigraphic Correlations: A Unified Temporal Framework for an Historical Geology.* SEPM Special Publication No. 54, pp. 131–212.

Bond GC (1979) Evidence of some uplifts of large magnitude in continental platforms. *Tectonophysics* 61: 285–305.

Coffin MF and Eldholm O (1994) Large igneous provinces: crustal structure, dimensions, and external consequences. *Reviews of Geophysics* 32: 1–36.

Crowley TJ and North GR (1991) *Paleoclimatology.* Oxford Monographs on Geology and Geophysics, no. 18.

Fairbanks RG (1989) A 17,000-year glacio-eustatic sea level record: influence of glacial melting rates on the Younger Dryas event and deep-ocean circulation. *Nature* 6250: 637–642.

Hallam A (1992) *Phanerozoic Sea-Level Changes.* NY: Columbia University Press.

Haq BU, Hardenbol J and Vail PR (1987) Chronology of fluctuating sea levels since the Triassic (250 million years ago to present). *Science* 235: 1156–1167.

Harrison CGA (1990) Long term eustasy and epeirogeny in continents. In: *Sea-Level Change*, pp. 141–158. Washington, DC: US National Research Council Geophysics Study Committee.

Hauffman PF and Schrag DP (2000) Snowball Earth. *Scientific American* 282: 68–75.

Kominz MA, Miller KG and Browning JV (1998) Long-term and short term global Cenozoic sea-level estimates. *Geology* 26: 311–314.

Miall AD (1997) *The Geology of Stratigraphic Sequences.* Berlin: Springer-Verlag.

Miller KG, Fairbanks RG, Mountain GS (1987) Tertiary oxygen isotope synthesis, sea level history, and continental margin erosion. *Paleoceanography* 2: 1–19.

Miller KG, Mountain GS, Browning J et al. (1998) Cenozoic global sea level, sequences, and the New Jersey transect; results from coastal plain and continental slope drilling. *Reviews of Geophysics* 36: 569–601.

Sahagian DL (1988) Ocean temperature-induced change in lithospheric thermal structure: a mechanism for long-term eustatic sea level change. *Journal of Geology* 96: 254–261.

Wilgus CK, Hastings BS, Kendall CG St C et al. (1988) *Sea Level Changes: An Integrated Approach.* Special Publication no. 42. Society of Economic Paleontologists and Mineralogists.

SEA LIONS

See **SEALS**

SEA OTTERS

J. L. Bodkin, US Geological Survey, Alaska, USA

doi:10.1006/rwos.2001.0434

Introduction

A century ago sea otters were on the verge of extinction. Reduced from several hundred thousand individuals, the cause of their decline was simply the human harvest of what is arguably the finest fur in the animal kingdom. They persisted only because they became so rare that, despite intensive efforts, they could no longer be found. Probably less than a few dozen individuals remained in each of 13 remote populations scattered between California and Russia. By about 1950 it was clear that several of those isolated populations were recovering. Today more than 100 000 sea otters occur throughout much of their historic range, between Baja California, Mexico, and Japan, although suitable unoccupied habitat remains (**Figure 1**). As previous habitat is reoccupied, either through natural dispersal or translocation, sea otter populations and the coastal marine ecosystem they occupy, can be studied before, during, and following population recovery. Because of concern for the conservation of the species, as well as conflicts between humans and sea otters over coastal marine resources, nearly continuous research programs studying this process have been supported during the past 50 years. Because sea otters occur near coastlines, bring their prey to the surface to consume, and can be easily observed and handled, they may be more amenable to study than most marine mammals. Both the accessibility of the species and the serendipitous 'experimental' situation provided by their widespread removal and subsequent recovery have provided a depth of understanding into the ecological role of sea otters in coastal communities and the response of sea otters to changing ecological conditions that may be unprecedented among the large mammals.

Sea otters are unique, both among the other otters, to which they are most closely related, and the other marine mammals, with which they share the oceans as a common habitat. They are the only fully marine species of Lutrinae, or otter subfamily of mustelids, having evolved relatively recently from their terrestrial and freshwater ancestors. Thus, the natural history of sea otters is a result of both their phylogenetic history and the adaptations required by a life at sea. As a result of these sometimes opposing pressures, sea otters display characteristics of their recent ancestors, the other otters, and also exhibit attributes that result from those adaptations that are common among the other mammals of the sea.

Description

The sea otter is well known as the smallest of the marine mammals but also the largest of the Lutrinae. The sexes are moderately dimorphic with males attaining maximum weights up to 45 kg and 158 cm total lengths. Adult females attain weights up to 36 kg and total lengths to 140 cm. Newborn pups weigh about 2 kg and are about 60 cm in length. Dentition is highly modified with broad, flattened molars for crushing hard-shelled invertebrates, rounded, blunt canines for puncturing and prying prey, and spade-shaped protruding incisors for scraping tissues out of shelled prey (**Figure 2**). The body is elongated and the tail is relatively short (less than one-third of total length) and flattened compared to other otters. The fore legs are short and powerful with sensitive paws used to locate and manipulate prey and in grooming, traits held in common with the clawless otters (*Aonyx* spp. or crab-eating otters) also known to forage principally on invertebrates. The fore legs are not used in aquatic locomotion. The extruding claws present in sea otters are unique among the mustelids and are useful in digging for prey in soft sediment habitats. The hind feet are enlarged and flattened relative to other otters and are the primary source of underwater locomotion. The external ear is small and similar to the ear of the otariid pinniped. Fur color ranges from brown to nearly black and a general lightening from the head downward may occur with aging.

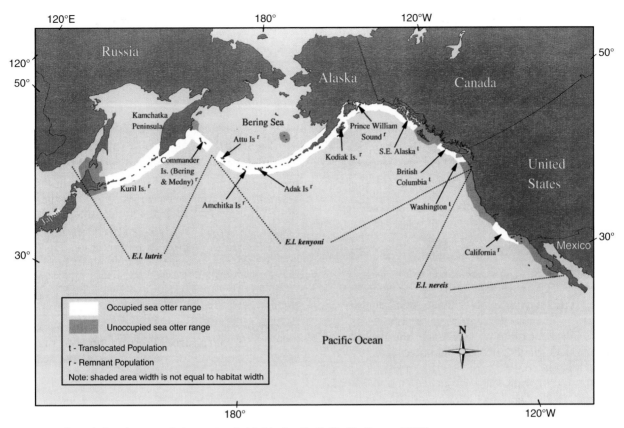

Figure 1 Occupied and unoccupied sea otter habitat in the North Pacific Ocean (2000).

The pelage is composed of bundles made up of a single guard hair and numerous underfur hairs, at a ratio of about 1 : 70. Hair density ranges to 165 000 cm^{-2} and is the densest hair among mammals. Most marine mammals insulate with a blubber layer, but sea otters are the only one to rely exclusively on an air layer trapped in the fur for insulation. Although air is a superior insulator, it requires constant grooming to maintain insulating quality. It is this means of insulation (fur) that made the sea otter so valuable to humans and makes it susceptible to oil spills and other similar contaminants, that can reduce insulation. Additionally, fur and the air it contains, allow the sea otter to be positively buoyant, a trait shared with some marine mammals but with none of the other otters.

Range and Habitat

Primitive sea otters (the extinct *Enhydridon* and *Enhydritherium*) are recognized from Africa, Eurasia, and North America. Specimens of *Enhydra* date to the late Pliocene/early Pleistocene about 1–3 million years ago. The modern sea otter occurs only in the north Pacific ocean, from central Baja California, Mexico to the northern islands of Japan (**Figure 1**). The northern distribution is limited by the southern extent of winter sea ice that can limit access to foraging habitat. Southern range limits are less well understood, but are likely to be related to increasing water temperatures and reduced productivity at lower latitudes and constraints imposed by the dense fur the sea otter uses to retain heat.

Although sea otters occupy and utilize all coastal marine habitats, from protected bays and estuaries

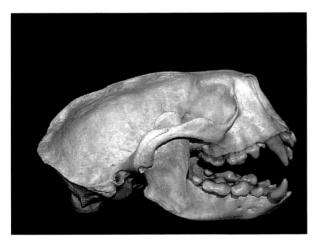

Figure 2 Photo of sea otter skull and dentition.

to exposed outer coasts and offshore islands, their habitat requirements are defined by their ability to dive to the seafloor to forage. Although they may haul out on intertidal or supratidal shores, their habitat is limited landward by the sea/land interface and no aspect of their life history requires leaving the ocean. The seaward limit to their distribution is defined by their diving ability and is approximated by the 100 m depth contour. Although sea otters may be found at the surface in water deeper than 100 m, either resting or swimming, they must maintain relatively frequent access to the seafloor to obtain food. Sea otters forage in diverse bottom types, from fine mud and sand to rocky reefs. In soft sediment communities they prey largely on burrowing infauna, whereas in consolidated rocky habitats epifauna are the primary prey.

Life History

Following a gestation of about six months, the female sea otter gives birth to a single, relatively large pup. In contrast, all other otters have multiple offspring, whereas all other marine mammals have single offspring. Pupping can occur during any month, but appears to become more seasonal with increasing latitude with most pups born at high latitudes arriving in late spring. The average length of pup dependency is about six months, resulting in an average reproductive interval of about one year. Females breed within a few days of weaning a pup, and thus may be either pregnant or with a dependent young throughout most of their adult life. The juvenile female attains sexual maturity at between two and four years and the male at about age three years, although social maturity, particularly among males may be delayed for several more years. The sea otter is polygynous; male sea otters gain access to females through territories where other males are excluded. The reproductive system results in a general segregation of the sexes; most adult females and few territorial males occupy most of the habitat and nonterritorial males reside in dense aggregations. Similar reproductive systems are common among the pinnipeds, whereas most otter species are organized around family units consisting of a mother, her offspring and one or more males.

Survival in sea otters is largely age dependent. Survival of dependent pups is variable ranging from about 0.20 to 0.85, and is likely dependent on maternal experience, food availability, and environmental conditions. Among dependent pups, survival is lowest in the first few weeks after parturition, then increasing and remaining high through dependency. A second period of reduced survival follows weaning and may result in annual mortality rates up to 0.60. Once a sea otter attains adulthood, survival rates are high, around 0.90, but decline later in life.

Maximum longevity is about 20 years in females and 15 years in males. Sources of mortality include a number of predators, most notably the white shark (*Carcharadon charcharias*) and the killer whale (*Orca orcinus*). Bald eagles (*Haliaeetus leucocephalus*) may be a significant cause of very young pup mortality. Terrestrial predators, including wolves (*Canis lupus*), bears (*Ursos arctos*) and wolverine (*Gulo gulo*) may kill sea otters when they come ashore, although such instances are likely rare. Pathological disorders related to enteritis and pneumonia are common among beach-cast carcasses and may be related to inadequate food resources, although such causes of mortality generally coincide with late winter periods of inclement weather. Nonlethal gastrointestinal parasites are common and lethal infestations are occasionally observed. Among older animals, tooth wear can lead to abscesses and systemic infection, eventually contributing to death.

Adaptations to Life at Sea

Locomotion

Adaptations seen in the sea otter reflect a transition away from a terrestrial, and toward an aquatic existence. The sea otter has lost the terrestrial running ability present in other otters, and although walking and bounding remain, they are awkward. The reduced tail length decreases balance while on land. The sea otter is similar to other otters in utilizing the tail as a supplementary means of propulsion, but more similar to the phocid seal in primarily relying on the hind flippers for aquatic locomotion. The hind feet of the sea otter are more highly adapted compared to other otters, but less modified than the hind legs of both pinnipeds and cetaceans. The hind feet are enlarged through elongation of the digits, flattened and flipper like and provide the primary source of underwater locomotion. While swimming, the extended hind flippers of the sea otter approximate the lunate pattern and undulating movement of the fluke of cetaceans.

The primary method of aquatic locomotion in sea otters while submersed is accomplished by craniocaudal thrusts of the pelvic limbs, including bending of the lumbar, sacral, and caudal regions for increased speed. The sea otter loses swimming efficiency in the resistance and turbulence during the recovery stroke and through spaces between the flippers and tail. Travel velocities over distances less than 3 km are in the range of about 0.5–0.7 m s^{-1}, with a maximum of about $2.5\,\mathrm{m\,s^{-1}}$. Estimated sustainable rates of travel over longer distances are in the range of 0.16–$1.5\,\mathrm{m\,s^{-1}}$. Rates of travel during foraging dives average about $1.0\,\mathrm{m\,s^{-1}}$.

A unique paddling motion, consisting of alternating vertical thrusts and recovery of the hind flippers is a common means of surface locomotion. The tail is capable of propelling the sea otter slowly, usually during either resting or feeding.

Thermoregulation

Sea otters, similar to all homeotherms, must strike a balance between conserving body heat in cold environments, and dissipating heat when internal production exceeds need. This process is particularly sensitive in the high latitudes where water is cold and heat loss potential is high and when a relatively inflexible insulator such as air is used. The small size of the sea otter magnifies the heat loss problem because of a high surface area to volume ratio. One way to offset high heat loss is through the generation of additional internal heat. The sea otter accomplishes this through a metabolic rate 2.4–3.2 times higher than predicted in a terrestrial mammal of similar size. This elevated metabolic rate requires elevated levels of energy intake. As a result of this increased energy requirement, sea otters consume 20–33% of their body mass per day in food. In northern latitudes, sea otters may haul-out more frequently during the winter as a means to conserve heat.

Although air is a more efficient insulator than blubber (10 mm of sea otter fur is approximately equal to 70 mm of blubber), the fur requires high maintenance costs and does not readily allow heat dissipation that may be required following physical exertion. To maintain the integrity of the fur, up to several hours per day are spent grooming, primarily before and after foraging, but also during foraging and resting activities. To dissipate excess internal heat, and possibly absorb solar radiation the sea otter's hind flippers are highly vascularized and with relatively sparse hair. Heat can be conserved by closing the digits and placing the flippers against the abdomen, or dissipated by expanding the digits and exposing the interdigital webbing to the environment. Because air is compressible, sea otters likely lose insulation while diving, and may undergo unregulated heat loss during deep dives, however the heat loss may be offset to some extent by the elevated metabolic heat produced during diving.

Diet and Diving

Sea otters prey principally on sessile or slow-moving benthic invertebrates in nearshore habitats through-out their range, from protected inshore waters to exposed outer coast habitats. In contrast, most other otters and most pinnipeds and odontocete cetaceans rely largely on a fish-based diet. Although capable of using vision to forage, the primary sensory modality used to locate and acquire prey appears to be tactile, since otters feed in highly turbid water, and at night. Foraging in rocky habitats and kelp forests consists of hunting prey on the substrate or in crevices. Foraging in soft sediment habitats often requires excavating large quantities of sediments to extract infauna such as clams.

Although the number of species preyed on by sea otters exceeds 150, only a few of these predominate in the diet, depending on latitude, habitat type, season, and length of occupation by sea otters. Generally, otters foraging over rocky substrates and in kelp forests mainly consume decapod crustaceans (primarily species of *Cancer*, *Pugettia*, and *Telmessus*), mollusks (including gastropods, bivalves, and cephalopods) and echinoderms (species of *Strongylocentrotus*). In protected bays with soft sediments, otters mainly consume infaunal bivalves (species of *Saxidomus*, *Protothaca*, *Macoma*, *Mya*, and *Tresus*) whereas along exposed coasts of soft sediments, *Tivela stultorum*, *Siliqua* spp. are common prey. Mussels (species of *Mytilus* and *Modiolis*) are a common prey in most habitats where they occur and may be particularly important in providing nourishment for juvenile sea otters foraging in shallow water. Sea urchins are relatively minor components of the sea otter's diet in Prince William Sound and the Kodiak archipelago, but are a principal component of the diet in the Aleutian Islands. In the Aleutian, Commander and Kuril Islands, a variety of fin fish (including hexagrammids, gaddids, cottids, perciformes, cyclopterids, and scorpaenids) are present in the diet. For unknown reasons, fish are rarely consumed by sea otters in regions to the east of the Aleutian Islands.

Sea otters also exploit normally rare, but episodically abundant prey. Examples include squid (*Loligo* sp.) and pelagic red crabs (*Pleuroncodes planipes*) in California and smooth lumpsuckers (*Aptocyclus ventricosus*) in the Aleutian Islands. The presence of abundant episodic prey may allow temporary release from food limitation and result in increased survival. Sea otters, on occasion, attack and consume sea birds, including teal, scoters, loons, gulls, grebes, and cormorants, although this behavior is apparently rare.

The sea otter is the most proficient diver among the otters, but one of the least proficient among the other marine mammals. Diving occurs during foraging but also while traveling, grooming, and during

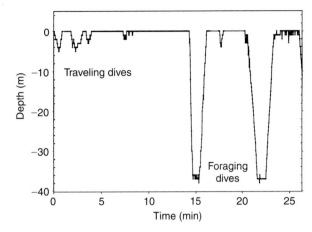

Figure 3 Examples of foraging and traveling sea otter dive profiles. Ascent/descent rates and percentage of dive time at bottom vary between dive types.

social interactions. Foraging dives are characterized by rapid ascent and descent rates with relatively long bottom times. Traveling dives are characterized by slow ascent and descent rates and relatively short times at depth (**Figure 3**). The maximum recorded dive depth in the sea otter is 101 m and most diving is to depths less than 60 m. Mean and maximum dive depths appear to vary between sexes and among individuals. Generally, males appear to have average dive depths greater than females, although some females regularly dive to depths greater than some males. Maximum reported dive duration is 260 s. Average dive durations are in the range 60–120 s and differ among areas and are likely influenced by both dive depth and prey availability. Foraging success rates are generally high, from 0.70 to > 0.90 although the caloric return per unit effort probably varies relative to prey availability.

Reproduction

In contrast to the pinnipeds and polar bears, as well as the other otters, sea otter reproduction has no obligatory terrestrial component. In this regard, the sea otter is more similar to the cetacea or sirenia. Sea otters throughout their range are capable of reproducing throughout the year. Alternatively, most pinnipeds and cetaceans display strong seasonal, synchronous reproductive cycles. Likely the most conspicuous reproductive adaptation that separates sea otters from the other lutrines and aligns them with the other marine mammals is reduction in litter size. All other lutrines routinely give birth to and successfully raise multiple offspring from a single litter. Although sea otters infrequently conceive twin fetuses, there are no known records of a mother successfully raising more than one pup.

This trait of single large offspring appears to be strongly selected for in marine mammals (**Figure 4**) with the exception of polar bears.

The Role of Sea Otters in Structuring Coastal Marine Communities

A keystone predator is one whose effects on community structure and function are disproportionately large, relative to their own abundance. Sea otters provide one of the best documented examples of the keystone predator concept in the ecological literature. Although other factors can influence rocky nearshore marine communities, particularly when sea otters are absent, the generality of the sea otter effect is supported with empirical data from many sites throughout the North Pacific rim.

There are several factors that have contributed to our understanding of the role sea otters play in structuring coastal communities. First, sea otters are distributed along a narrow band of relatively shallow habitat where they forage almost exclusively on generally large and conspicuous benthic invertebrates. Because water depths are shallow, scientists, aided with scuba have been able to rigorously characterize and quantify coastal marine communities where sea otters occur. Because sea otters bring their prey to the surface prior to consumption, the type, number, and sizes of prey they consume are straightforward to determine. Furthermore, many of the preferred prey of sea otters are either ecologically (e.g., grazers) or economically (e.g., support fisheries) important, prompting a long-standing interest in the sea otter and its effect on communities. And finally, because sea otters were removed from

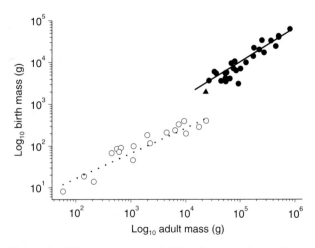

Figure 4 Birth mass versus adult female mass of sea otters relative to the other mustelids and marine mammals. ●, Pinnipeds; ○, Mustelids; ▲, *Enhydra lutris*; —, Pinniped, regression, ···, mustelid regression.

most of their range prior to 1900, but have recovered most of that range during the last half of the twentieth century, we have been afforded the opportunity to repeatedly observe coastal communities both with and without sea otters present, as well as before and after sea otters recolonize an area. This combination of factors may be nearly impossible to duplicate in other communities that support large carnivores.

Kelp Forests

The majority of studies on sea otter ecosystem effects has taken place in kelp forest communities that occur over rocky reefs and has led to a generalized sea otter paradigm. This scenario describes the rocky reef community in the absence of sea otters as being dominated by the effects of the herbivorous sea urchins (*Strongylocentrotus* spp.) that can effectively eliminate kelp populations, resulting in what have come to be termed 'urchin barrens' (**Figure 5A**). The urchin barren is characterized by low species diversity, low algal biomass, and abundant and large urchin populations. In areas of the Aleutian Islands dominated by urchin barrens the primary source of organic carbon in nearshore food webs results from fixation of carbon by phytoplankton or microalgae.

Sea urchins may be the most preferred of the many species preyed on by sea otters. As a result, when sea otters recolonize urchin-dominated habitat, urchin populations are soon reduced in abundance with few if any large individuals remaining. Following reduced urchin abundance, herbivory on kelps by urchins is reduced. In response to reduced herbivory, kelp, and other algal populations increase in abundance often forming a multiple-layer forest culminating in a surface canopy forming kelp forest (**Figure 5**). The kelp forest in turn supports a high diversity of associated taxa, including gastropods, crustaceans, and fishes. The effects of the kelp forests may extend to birds and other mammals that may benefit from kelp forest-associated prey populations. In areas of the Aleutian Islands dominated by kelp forests, the primary source of organic carbon in nearshore food webs results from fixation of carbon by kelps, or macroalgae.

Soft-Sediment Subtidal

Studies throughout the North Pacific indicate that sea otters can have predictable and measurable effects on invertebrate populations in sedimentary communities. The pattern typically involves a predation-related reduction in prey abundance and a shift in the size-class composition toward smaller

(A)

(B)

Figure 5 Photos of (A) urchin barrens and (B) kelp forest.

individuals, similar to patterns seen in rocky reefs. This pattern has been observed for several species of bivalve clams and Dungeness crabs (*Cancer magister*). The green sea urchin (*Strongylocentrotus droebachiensis*) can be common and abundant in sedimentary habitats in Alaska, particularly where sea otters have been absent for extended periods of time. In such areas, as sea otters recover prior habitat, the green urchin may be a preferred prey. The effect of sea otter foraging on the green urchin in sedimentary habitats is similar to the otter effect observed on urchins in rocky reefs. Urchin abundance is reduced and surviving individuals are small and may achieve refuge by occupying small interstitial spaces created by larger sediment sizes. Although sea urchins may provide an abundant initial resource during sea otter recolonization, clams appear to be the primary dietary item in soft sediment habitats occupied for extended periods.

Cascading trophic effects of sea otter foraging are less well described in sedimentary habitats, when contrasted to rocky reefs. It is likely that the reduction of urchins has a positive effect on algal productivity, especially where the larger sediment sizes (e.g., rocks and boulders) required to support macroalgae are present. In the process of foraging on clams, sea otters discard large numbers of shells after removing the live prey. These shell remains provide additional hard substrate that can result in increased rates of recruitment of some species such as anemones and kelps.

The effects of otter foraging in sedimentary habitats include disturbance to the community in the form of sediment excavation and the creation of foraging pits. These pits are rapidly occupied by the sea star (*Pycnopodia helianthoides*). *Pycnopodia* densities are higher near the otter pits where they may prey on small clams exposed, but not consumed, by sea otters.

Responses to Changing Ecological Conditions

Contrasts between sea otter populations at recently (below equilibrium) and long-occupied sites (equilibrium) as well as contrasts at sites over time provide evidence of how sea otters can modify life history characteristics in response to changing population densities and the resulting changes in ecological conditions. The data summarized below were collected at sites sampled over time, both during and following recolonization, and also at different locations where sea otters were either present for long or short periods of time.

Diet

In populations colonizing unoccupied habitat, sea otters feed largely on the most abundant and energetically profitable prey. Preferred prey species likely differ between areas but include the largest individuals of taxa such as gastropods, bivalves, echinoids, and crustaceans. Over time, as populations approach carrying capacity and the availability of unoccupied habitat diminishes, preferred prey of the largest sizes become scarce and smaller individuals and less preferred prey are consumed with increasing frequency. Several consequences of this pattern of events are evident, and have been repeatedly observed. One is an increase in dietary diversity over time as otter populations recolonize new habitats and grow toward resource limitation. A relatively few species are replaced by more species, of smaller size, and at least in the Aleutian Islands, a new prey type (e.g., fish) may become prevalent in the diet. This in turn may eventually lead to a new and elevated equilibrium density. Another result of declining prey is an increase in the quantity of time spent foraging as equilibrium density is approached. In below-equilibrium populations, sea otters may spend as little as 5 h each day foraging, whereas in equilibrium populations 12 h per day or more may be required. Finally, declining body conditions and total weights have been seen in sea otters as equilibrium density is attained. Mass/length ratios are consistently greater in below-equilibrium populations compared to those at or near equilibrium. At Bering Island as the population reached and exceeded carrying capacity over a 10 year period (**Figure 6**), average weights of adult males declined from 32 kg to 25 kg.

Reproduction

Studies of age specific reproductive rates have produced generally consistent results in populations both below and at equilibrium densities. Although a small proportion of females may attain sexual maturity at age two most are mature at age four, regardless of population status. Annual adult female reproductive rates are uniformly high, at around 0.85–0.95, among all populations despite differences in availability of food resources. Under the range of ecological conditions studied, sea otter reproductive output does not appear to play a major role in population regulation, although it seems likely that as ecological or environmental conditions deteriorate, at some point reproductive output should be adversely affected.

Survival

Greater food availability, and the resulting improved body condition can result in significantly higher juvenile sea otter survival rates in below-equilibrium populations. Coupled with high reproductive output, high survival has resulted in sea otter population growth rates in several translocated or recolonizing populations that have approached the theoretical maximum for the species of about 0.24 per year. Declining prey availability does not appear to affect the ages and sexes equally. Survival of dependent pups was nearly twice as high (0.83) in a population below equilibrium, compared to one at equilibrium (0.47). Postweaning survival appears similarly affected, increasing during periods of increasing prey availability. Adult survival appears high and uniform at about 0.90, among both equilibrium and below-equilibrium populations. Survival appears to be greater among females of all age classes, compared to males. At Bering Island in

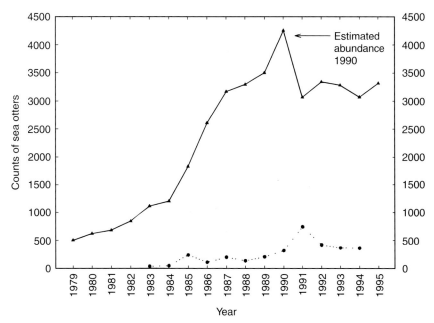

Figure 6 Sea otter abundance (▲, average annual growth 1979–1989 = 0.24) and carcass recovery rates (●) during a period of recolonization and equilibration with prey resources, Bering Island, Russia.

Russia (**Figure 6**) during a year when about 0.28 of the population was recovered as beach cast carcasses, 0.80 of the 742 carcasses were male. Higher rates of male mortality are associated with higher densities of sea otters in male groups and increased competition for food. Thus sex ratios in populations of sea otters are generally skewed toward females. Available data suggest that survival, particularly among juveniles, is the primary life history variable responsible for regulating sea otter populations.

Populations

Trends in sea otter populations today vary widely from rapidly increasing in Canada, Washington and south east Alaska, to stable or changing slightly in Prince William Sound, the Commander Islands, and California, to declining rapidly throughout the entire Aleutian Archipelago. Rapidly increasing population sizes are easily understood by abundant food and space resources and increases should continue to be seen until those resources become limiting. Relatively stable populations can be generally characterized by food limitation and birth rates that approximate death rates. The recent large-scale declines in the Aleutian Archipelago are unprecedented in recent times. Our view of sea otter populations has been largely influenced by events in the past century when food and space were generally unlimited. However, as resources become limiting, it is likely that other mechanisms, such as predation or disease will play increasingly important roles in structuring sea otter populations.

See also

Marine Mammals, History of Exploitation.

Further Reading

Estes JA (1989) Adaptations for aquatic living in carnivores. In: Gittleman JL (ed.) *Carnivore Behavior Ecology and Evolution*, pp. 242–282. New York: Cornell University Press.

Estes JA and Duggins DO (1995) Sea otters and kelp forests in Alaska: generality and variation in a community ecology paradigm. *Ecological Monographs* 65: 75–100.

Kenyon KW (1969) *The Sea Otter in the Eastern Pacific Ocean. North American Fauna* no. 68. Washington, DC: Bureau of Sport Fisheries and Wildlife, Dept. of Interior.

Kruuk H (1995) *Wild Otters, Predation and Populations.* Oxford: Oxford University Press.

Reynolds JE III and Rommel SE (eds.) (1999) *Biology of Marine Mammals.* Washington and London: Smithsonian Institute Press.

Riedman ML and Estes JA (1990) The sea otter (*Enhydra lutris*): behavior, ecology, and natural history. US Fish and Wildlife Service Biological Report 90(14).

VanBlaricom GR and Estes JA (eds) (1988) The community ecology of sea otters. *Ecological Studies*, vol. 65. New York: Springer-Verlag.

SEA SURFACE TEMPERATURES

See **SATELLITE REMOTE SENSING OF SEA SURFACE TEMPERATURES**

SEA TURTLES

F. V. Paladino, Indiana-Purdue University at
Fort Wayne, Fort Wayne, IN, USA
S. J. Morreale, Cornell University, Ithaca,
NY, USA

doi:10.1006/rwos.2001.0443

Introduction

There are seven living species of sea turtles that
include six representatives from the family
Cheloniidae and one from the family Dermochelyidae. These, along with two other extinct families, Toxochelyidae and Prostegidae, had all evolved
by the early Cretaceous, more than 100 Ma (million
years ago). Today the cheloniids are represented by
the loggerhead turtle (*Caretta caretta*), the green
turtle *Chelonia mydas*, the hawksbill turtle (*Eretmochelys imbricata*), the flatback turtle (*Natator
depressus*), and two congeneric turtles, the Kemp's
ridley turtle (*Lepidochelys kempii*) and the olive
ridley turtle *(Lepidochelys olivacea)*. The only remaining member of the dermochelyid family is the
largest of all the living turtles, the leatherback turtle
(*Dermochelys coriacea*).

Sea turtles evolved from a terrestrial ancestor, and
like all reptiles, use lungs to breathe air. Nevertheless, a sea turtle spends virtually its entire life in the
water and, not surprisingly, has several extreme
adaptations for an aquatic existence. The rear limbs
of all sea turtles are relatively short and broadly
flattened flippers. In the water, these are used as
paddles or rudders to steer the turtle's movements
whereas on land they are used to push the turtle
forward and to scoop out the cup shaped nesting
chamber in the sand (**Figure 1**). The front limbs are
highly modified structures that have taken the appearance of a wing. Internally the front limbs have
a shortened radius and ulna (forearm bones) and
greatly elongated digits that provide the support for
the flattened, blade-like wing structure. Functionally
the front flippers are nearly identical to bird wings,
providing a lift-based propulsion system that is not
seen in other turtles.

Other special adaptations help sea turtles live in
the marine environment. All marine turtles are well
suited for diving, with specialized features in their
blood, lungs, and heart that enable them to stay
submerged comfortably for periods from 20 min to
more than an hour. Since everything they eat and
drink comes from the ocean, their kidneys are designed to minimize salt uptake and conserve water
and they have highly developed glands in their
heads that concentrate and excrete salt in the form
of tears.

Despite all their adaptations and highly specialized mechanisms for life in the ocean, sea turtles are
inescapably tied to land at some stage of their life.
Turtles, like all other reptiles, have amniotic eggs
which contain protective membranes that allow for
complete embryonic development within the protected environment of the egg. This great advance-
ment in evolution provided many advantages for
reptiles over their predecessors: fish and amphibians.
However, turtle eggs must develop in a terrestrial
environment. Thus, in order to reproduce, female
sea turtles need to emerge from the water and come
ashore to lay eggs. All seven species lay their eggs
on sandy beaches in warmer tropical and subtropical regions of the world.

Eggs are deposited on the beach in nest chambers,
which can be as deep as 1 m below the sand surface.
The adult female crawls on land, excavates a chamber into which it lays 50–130 eggs, and returns to
the water after covering the new nest. Eggs usually
take 50–70 days to fully develop and produce
hatchlings that clamber to the beach surface at
night. The length of the incubation primarily is
influenced by the prevailing temperature of the nest
during development; warmer temperatures hasten
the hatchlings' development rates.

Nest temperature also plays a key role in determining the sex of hatchlings. Sea turtles share with
other reptiles a phenomenon known as temper-
ature-dependent sex determination (TSD). In sea
turtles, warmer nest temperatures produce females,
whereas cooler temperatures generate males. More
specifically, it appears that there is a crucial period
in the middle trimester of incubation during which

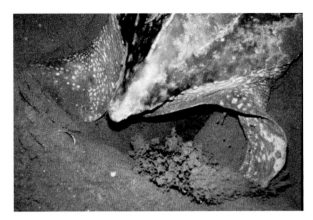

Figure 1 Leatherback digging nest.

temperature acts on the sex-determining mechanism in the developing embryo. Temperatures of > 30°C generate female sea turtles, and temperatures cooler than 29°C produce males. There is a pivotal range between these temperatures that can produce either sex. During rainy periods it is very common to have nests that are exclusively male, whereas a sunny dry climate tends to produce many nests of all females. Extended periods of extreme weather can produce an extremely skewed sex ratio for an entire beach over an entire nesting season. It is of much concern that global warming trends could have drastic effects on reptiles for which a balanced sex ratio totally depends on temperature.

During a single nesting season an individual female nests several different times, emerging from the water to lay eggs at roughly two-week intervals. Thus a female's reproductive output can total sometimes more than 1000 eggs by the end of the nesting season. However, once finished, most individuals do not return to nest again for several years. There is much variation in the measured intervals of return, between 1 and 9 years, before subsequent nesting bouts. Alternatively, males may mate at a much more frequent rate. Mating occurs in the water, and has been observed in some areas, usually near the nesting beaches. However, timing and location of mating is not known for many nesting populations, and is virtually unknown for some species.

Sea turtles have evolved a life history strategy that includes long-lived adults, a high reproductive potential, and high mortality as hatchlings or juveniles. As adults they have few predators with the exception of some of the larger sharks and crocodilians. Age to reproductive maturity is estimated at 8–20 years depending on the species and feeding conditions. Their eating habits include herbivores, like the green and black turtle, carnivores like the ridley turtles that eat crustaceans, and even spongivores,

like the hawksbill. Almost all sea turtle hatchlings have a pelagic dispersal phase which is poorly documented and understood. Juveniles and adults congregate in feeding areas after the hatchling has attained a certain size. All sea turtles are currently listed in Appendix I of the Convention on International Trade in Endangered Species of Flora or Fauna (CITES Convention) and all species except the flatback (*Natator depressus*) from Australia, are also listed as threatened or endangered by the World Conservation Union (IUCN). Historically populations of sea turtles have declined worldwide over the past 20 years due to loss of nesting beach habitat, harvest of their eggs due to natural predation or human consumption, and the extensive adult mortality brought about by fishing pressure from the expansion of shrimp and drift net or longline ocean fisheries.

General Sea Turtle Biology

The first fossil turtles appear in the Upper Triassic (210–223 Ma) and the sea turtles of today are descendants of that lineage. The current species of the family Cheloniidae evolved about 2–6 Ma and the sole representative of the Dermochelyidae family is estimated to have evolved about 20 Ma. All sea turtles are characterized by an anapsid skull, in which the region behind the eye socket is completely covered by bone without any openings, and the presence of a bony upper shell (carapace) and a similar bony lower shell (plastron). In all of the cheloniids the carapace and plastron consist of paired bony plates, in contrast to the sole Dermochelyid which has no evident bony plates in the leathery carapace and plastron, but instead interlocking cartilaginous osteoderms.

Like all reptiles, sea turtles have indeterminate growth, i.e., they continue to grow throughout their entire lifetime, which is estimated to be 40–70 years. Growth rates are often rapid in early life stages depending on amount of food available and environmental conditions, but diminish greatly after sexual maturity. As turtles get very old, growth probably becomes negligible. Sea turtles can grow to be quite large; and the leatherback is by far the largest, with most adults ranging between 300 and 500 kg with lengths of over 2 m. The cheloniid turtles are all smaller ranging from the smallest, the Kemp's and olive ridleys, to the flatback, which may weigh as much as 350 kg.

Sea turtles can not retract their necks or limbs into their shell like other turtle species and have evolved highly specialized paddle-like, hind limbs and wing-like front limbs with reduced nails on the

Figure 2 Single nail on fore-flipper of a juvenile loggerhead.

forelimbs limited to one or two claw-like growths (**Figure 2**) that are designed to facilitate clasping of the carapace on the female by the male during copulation. Fertilization is internal and sea turtles lay terrestrial nests on sandy beaches in the tropics and subtropics. Adults migrate from feeding areas to nesting beaches at intervals of 1–9 years, distances ranging from hundreds to thousands of nautical kilometers, to lay nests on land. Nest behavior can be solitary or aggregate in phenomena called an 'Arribada' where up to 75 000–350 000 female turtles will emerge to nest on one beach over a period of three to five consecutive evenings (**Figure 3**). Egg clutches of 50–200 eggs are buried in the sand at 20–40 cm below the surface. The nests are unattended and hatch 45–70 days later into hatchlings that average 15–30 g in weight. The sex of a hatchling sea turtle is determined by the temperature at which the nest is incubated (temperature dependent sex determination) during the middle trimester of development (critical period). For most sea turtles, temperatures above 29.5°C result in the development of a female whereas those eggs incubated at temperatures below 29°C become males. Sea turtles lay cleidoic eggs which are 2–6 cm in diameter and have a typical leathery inorganic shell constructed by the shell gland in the oviduct of the laying females. Nesting is seasonal and controlled by photoperiod cues integrated by the pineal gland and also influenced by the level of nutrition and fat stores available.

Sea turtle physiology is well adapted for deep and shallow diving that can average from 15 min to 2 hours in length. Their tissues contain high levels of respiratory pigments like myoglobin as a reserve oxygen store during the breath hold/dive. Respiratory tidal volumes are quite large (2–6 liters per breath) and allow for rapid washout of carbon dioxide and oxygen uptake during the brief periods spent on the surface breathing (**Figure 4**). Thus in a normal one hour period a sea turtle may spend 50 min under the water and only 10 min at the surface breathing and operating entirely aerobically and accrue no oxygen debt. Cardiovascular adaptations include counter current heat exchangers in the flippers to reduce or enhance heat exchange with the surrounding water. Control of the vascular tree is

Figure 3 Olive ridley arribada at Nancite Costa Rica.

much higher than the level of arterioles and can permit almost complete restriction of blood flow to all tissues but the heart, brain, central nervous system, and kidney during the deepest dives. They also have evolved temperature and pressure adapted enzymes that will operate well at both the surface and at extreme depths. Leatherbacks are the deepest diving sea turtles; dives of over 750 m in depth have been recorded.

All sea turtles have lacrimal salt glands (**Figure 5**) and reptilian kidneys with short loops of Henle to regulate ion and water levels. Lacrimal salt glands allow sea turtles to drink sea water from birth and maintain water balance despite the high levels of inorganic salts in both their marine diet and the salty ocean water they drink. Diets range from: plant materials like sea grass, *thallasia* and algae for herbivores like the green and black sea turtles; crabs and crustaceans for the ridley turtles; sponges for the loggerhead; and jellyfish and other Cniderians are the sole diet of leatherbacks.

Dermochelyidae (1 genera, 1 species)

The genus *Dermochelys*: *Dermochelys coriacea* (**Linne'**) **the leatherback turtle** Leatherbacks are

Figure 4 Leatherback breathing at surface.

Figure 5 Clear salt gland secretion from base of eye.

Figure 6 Leatherback osteoderm with ridges on carapace.

the largest (up to 600 kg as adults) and most ancient of the sea turtles diverging from the other turtle families in the Cretaceous. About 20 Ma *Dermochelys* evolved to a body form very similar to that seen today. Their shell consists of cartilaginous osteoderms and they do not have the characteristic laminae and plates found in the plastron and carapace of other sea turtles. The leatherback shell is streamlined with seven cartilaginous narrow ridges on the carapace and five ridges on the plastron that direct water flow in a laminar manner over their entire body (**Figure 6**). The appearance is a distinctive black with white spots, a smooth, scaleless carapace skin and a white smooth-skinned plastron. The head has two saber tooth like projections on the upper beak that overlap the front of the lower beak and serve to pierce the air bladder of floating cniderians that are a large component of their diet when available. Their nesting distribution is worldwide with colonies in Africa, Islands across the Caribbean, Florida, Pacific Mexico, the Pacific and Caribbean coastlines of Central America, South America, Malaysia, New Guinea, and Sri Lanka. Remarkably they are found in subpolar oceans at surface water temperatures of 7–10°C while maintaining a core body temperature above 20°C.

Leatherbacks are pelagic and remain in the open ocean throughout their life feeding on soft-bodied invertebrates such as cniderians, ctenophores, and salps. Average adult females weigh about 300 kg and may take only 8–10 years to attain that size after emerging from their nests at about 24 g. Sex is determined by TSD with a pivotal temperature of 29.5°C. They build a simple cup shaped body pit and lay the largest eggs, about the size of a tennis ball, in clutches of 60–110 yolked eggs. Unlike other sea turtles leatherbacks also deposit 30–60 smaller yolkless eggs that are infertile and only contain

albumins. These yolkless eggs tend to be laid late in the nesting process and are primarily on the top of the clutch and may serve as a reservoir for air when hatchlings emerge and congregate prior to the frenzied digging to emerge from their nest chamber.

Unlike other sea turtles leatherbacks crawl on their wrists while on land, rotating both front flippers simultaneously and pushing with both rear flippers in unison. This contrasts with the alternating right to left front and rear flipper crawls of the Cheloniidae. In the ocean their enlarged paddle-like front flippers generate enormous thrust providing excellent propulsion and allowing leatherbacks to migrate an average of 70 km per day when leaving the nesting beaches. Leatherback migrations from Central American beaches are along narrow 'corridors' that are about 100 km wide (**Figure 7**). These oceanic migratory corridors provide important insights into the complex reproductive behavior of these animals. Genetic studies have demonstrated

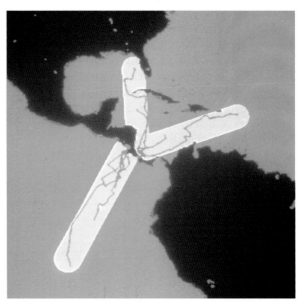

Figure 7 Leatherback migration corridors.

Figure 8 Green turtle.

that there is excellent gene flow across all the ocean basins with a strong natal homing and distinct genetic haplotypes in different nesting populations.

Cheloniidae (5 genera, 6 species, 1 race)

The genus *Chelonia*: *Chelonia mydas* (Linne') the green turtle and *Chelonia mydas agassizi* (Bocourt) the black turtle The genus *Chelonia* contains only one living species the green turtle *Chelonia mydas mydas* (Linne') that also has a Pacific–Mexican population that is considered by some researchers as a subspecies or race, the Pacific green turtle or black turtle, *Chelonia mydas agassizi* (Bocourt). The green turtle (**Figure 8**) actually has a brownish colored carapace and scales on the legs, with a yellowish plastron and has one pair of prefrontal head scales with four pairs of lateral laminae. The name 'green turtle' comes from the large greenish fat deposit found under the carapace which is highly desired for the cooking of turtle soup. The black turtle (**Figure 9**) is distinguished by the dark black color of both

Figure 9 Black turtle (with transmitter).

the carapace and plastron in the adult form. This subspecies or race is found only along the Pacific coastline of Mexico and extends in smaller populations along the Pacific Central American coastline to Panama. *Chelonia* are herbivorous; the green turtle eats marine algae and other marine plants of the genera *Zostera*, *Thallassina*, *Enhaus*, *Posidonia*, and *Halodule* and is readily found in the Caribbean in eel-grass (*Zostera*) beds, whereas black turtles rely heavily on red algae and other submerged vegetation. These turtles live in the shallow shoals along the equatorial coastlines as far North as New England in the Atlantic and San Diego in the Pacific and as far south as the Cape of Good Hope in the Atlantic and Chile in the Pacific. The main breeding rookeries include the Coast of Central America, many islands in the West Indies, Ascension Island, Bermuda, the Florida Coast, islands off Sarawak (Malaysia), Vera Cruz coastline in Mexico (center for black turtles), and islands off the coast of Australia. Adult females emerge on sandy beaches at night to construct nests in which they lay 75–200 golf ball-sized leathery eggs. The eggs in these covered nests develop unattended and the young emerge 45–60 days latter. The sex of the hatchlings is determined by the temperature at which the eggs are incubated during the third trimester of development called TSD. Temperatures above 29°C tend to produce all female hatchlings whereas those below 29°C produce males. The hatchlings have a high mortality in the first year and spend a pelagic period before reappearing in juvenile/adult feeding grounds, such as the reefs off Bermuda, the Azores, and Heron Island in the Pacific. As adults they average 150–300 kg with straight line carapace lengths of 65–90 cm. Black turtles tend to be smaller than green turtles. Genetic studies have shown a strong natal homing of the females to the beaches where they were hatched with significant gene flow between rookeries probably due to mating with males from different natal beaches found on common feeding grounds.

The genus *Lepidochelys*: *Lepidochelys kempii* (Garman) the Kemp's ridley, *Lepidochelys olivacea* (Eschscholtz) the olive ridley The Kemp's or Atlantic Ridley and the olive or Pacific Ridley turtles are the two distinct species of this genus. The Kemp's ridley (**Figure 10**) is the rarest of the sea turtles whereas the olive ridley is the most abundant. Anatomically ridleys are the smallest sea turtles. Adult Kemp's usually have five pairs of grey-colored coastal scutes and olive ridleys have 5–9 olive-colored pairs. The adults have a straight line carapace length of 55–70 cm and weigh on average

Figure 10 Juvenile Kemp's ridley.

100–200 kg. Both species have an interesting reproductive behavior in that they have communal nesting called 'arribadas' (Spanish for arrival). There are a number of arribada beaches where females come out by the thousands on sandy beaches 2–6 km long on three or four successive evenings once a month during the nesting seasons. A percentage of the total population are solitary nesters that emerge on other nearby beaches or on arribada beaches on nights other than the arribadas. These solitary nesters have a very high hatching success of their individual nests (about 80%). It is unknown what proportion of the total population of females nest in this solitary manner and what role they play in the contribution of new recruits into the reproductive adult numbers. Arribadas in Gahirmatha, India have been described with 100 000 turtles emerging in one night to nest communally only inches apart. Many nests are dug up by successive nesting females in the same or subsequent nights of the arribada. The hatching success of these arribada nests is about 4–8% which is the lowest among all sea turtles. Kemp's ridley turtles have arribadas on only one nesting beach in the world, Rancho Nuevo (Caribbean), Mexico. Historically Kemp's arribadas were estimated at 20 000–40 000 female turtles per night in the 1950s, but recent arribadas average only 300–400 individuals per night. This dramatic decline has been attributed to the numbers of adults and juveniles killed in the shrimping nets of the fisheries in the Gulf of Mexico and the American Atlantic coastline. These genera feed primarily on crustaceans, crabs and shrimp, and as a result have been in direct competition with these well-developed fisheries for many years and have suffered dire consequences. Olive ridleys have arribada beaches in Pacific Mexico, Nicaragua, Costa Rica, and along the Bay of Bengal in India. Very little is known about the ocean life of these sea turtle species and it is believed that after hatching they also spend 1–4 years in a pelagic phase associated with floating *Sargassum* and then reappear in the coastal estuaries and neretic zones worldwide at a straight carapace length of about 20–30 cm. Although Kemp's ridley turtles feed primarily on crabs in inshore areas, olive ridley turtles have a more varied diet that includes salps (*Mettcalfina*), jellyfish, fish, benthic invertebrates, mollusks, crabs, shrimp and bryozoans. This more varied diet may account for the different deep ocean and nearshore habitats in which the olive ridleys are found.

The genus *Eretmochelys*: *Eretmochelys imbricata* (Linne'), the hawksbill This species is the most sought after sea turtle for the beauty of the carapace. Historically eyeglass frames and hair combs were made from the carapace of hawksbills. The head is distinguished by a narrow, elongated snout-like mouth and jaw that resembles the beak of a raptor (**Figure 11**). They have four pairs of thick laminae and 11 peripheral bones in the carapace. They tend to be more solitary in their nesting behavior than the other sea turtles but this may be due to their reduced populations. Other than the kemp's ridley turtles they tend to be the smallest turtles averaging 75–150 kg as adults. They are common residents of coral reefs worldwide and appear as juveniles at about 20–25 cm straight length carapace after a pelagic developmental phase in the floating *Sargassum*. What is impressive about the adults is that their diet is 90% sponges. This is one of the few spongivorous animals and electron micrographs of their gastrointestinal tract has shown the microvilli with millions of silica spicules imbedded in the tissue. Tunicates, sea anemone, bryozoans, coelenterates, mollusks, and marine plants have also been found to be important components of the hawksbill diet.

Figure 11 Raptor-like beak of hawksbill.

Hawksbills have the largest clutch size of any of the chelonids averaging 130 eggs per nest. They also lay the smallest eggs with a mean diameter of 37.8 mm and mean average weight of 26.6 g. This results in the smallest hatchling of all sea turtles with an average weight of 14.8 g. Other than possibly the olive ridley turtle they have the lowest clutch frequency of only 2.74 clutches per nesting season. Hawksbills also tend to be the quickest to construct their nest when out on a nesting beach; they spend only about 45 min to complete the process whereas other turtles take between one to two hours.

Habitats include shallow costal waters and they are readily found on muddy bottoms or on coral reefs. The genetic structure of Atlantic and Pacific populations indicate that hawksbills like other sea turtles have strong natal homing and form distinct nesting populations. They also have TSD and hatchlings also have a pelagic phase before appearing in coastal waters and feeding grounds at about the size of a dinner plate.

Cuba has requested that these turtles be upgraded to CITES Appendix II status which would allow farming of 'local resident populations.' Genetic studies, however, have conclusively demonstrated that these Cuban populations are not local and include turtles from other regions of the Caribbean. These kinds of controversies will increase as human pressures for turtle products, competition for the same food resources such as shrimp, incidental capture due to longline fishing practices, and beach alteration and use by humans increases.

The genus *Caretta*: *Caretta caretta* (Linne'): the loggerhead turtle The loggerhead, like olive ridley turtles, are distinguished by two pairs of prefrontal scales, three enlarged poreless inframarginal laminae, and more than four pairs of lateral laminae. They have a beak like snout that is very broad and not narrow like the hawksbill and they have the largest and broadest head and jaw of the chelonids (**Figure 12**). They nest throughout the Atlantic, Caribbean, Central America, South America, Mediterranean, West Africa, South Africa, through the Indian Ocean, Australia, Eastern and Western Pacific Coastlines.

Loggerheads are primarily carnivorous but have a varied diet that includes crabs (including horseshoe crabs), mollusks, tube worms, sea pens, fish, vegetation, sea pansies, whip corals, sea anenomies, and barnacles and shrimp. Their habitats appear to be quite diverse and they will shift between deeper continental shelf areas and up into shallow river estuaries and lagoons. This sea turtle is the only turtle that has large resident and nesting populations

Figure 12 Broad head of juvenile loggerhead.

across the North American coastline from Virginia to Florida and the nesting rookeries in Georgia and Florida are currently doing well despite severe human pressures.

Loggerheads that nest on islands near Japan have been shown to develop and grow to adult size along the Mexican coastline and then migrate 10 000 km across the Pacific to nest. Genetic studies have confirmed that the Baja Mexico haplotypes are the same as the female adults nesting on these Japanese islands confirming strong natal homing in this genus. Loggerheads make a simple nest at night and lay about 100–120 golf ball-sized eggs. The hatchlings spend a pelagic phase and reappear at 25–30 cm straight carapace length in coastal bays, estuaries and lagoons as well as along the continental shelf and open oceans (**Figure 13**). The orientation and homing of loggerhead hatchlings has been extensively studied and it has been demonstrated that these turtles can detect the direction and intensity of magnetic fields and magnetic inclination angles. This ability together with the use of olfactory cues and chemical imprinting on olfactory cues from natal beaches may account for the natal homing ability of all sea turtles.

The genus *Natator*: *Natator depressus* the flatback This genus has a very limited distribution

Figure 13 Juvenile loggerhead turtle.

and is only found in the waters off Australia yet it appears that populations are not endangered at this time. Their nesting beaches are primarily in northern and south-central Queensland. Most of the nesting beaches are quite remote which has protected this species from severe impact by humans. Apart from the Kemp's ridley this is the only sea turtle to nest in significant numbers during the daytime. It is thought that nighttime nesting has evolved as a behavior to reduce predation and detection but there may also be thermoregulatory considerations. During the daytime in the tropics and subtropics both radiant heat loads from the sun and thermal heat loads from the hot sands may significantly heat the turtles past their critical thermal tolerance. In fact, a number of daytime-nesting flatbacks may die due to overheating if the females do not time their emergence and return to the water to coincide closely with the high tides. If individuals are stranded on the nesting beach when the tides are low and the females must traverse long expanses of beach to emerge or return to the sea during hot and sunny daylight hours, there is a higher potential to overheat and die. On the other hand green turtles in the French Frigate Shoals area of the Pacific are known to emerge on beaches or remain exposed during daylight in shallow tidal lagoons during nonnesting periods and appear to heat up to either aid in digestion or destroy ectoparasites.

Flatback turtles are characterized by a compressed appearance and profile of the carapace with fairly thin and oily scutes. Flatbacks are also distinguished by four pairs of laminae on the carapace and the rim of the shell tends to coil upwards toward the rear. Flatbacks have a head that is very similar to the Kemp's ridley with the exception of a pair of preocular scales between the maxilla and prefrontal scales on the head. These turtles tend to be the largest chelonid, weighing up to 400 kg as adults,

and lay the second largest egg with diameters of about 51.5 mm weighing about 51.5 g, smaller only than the leatherback which has a mean egg diameter of 53.4 mm and mean egg weights of 75.9 g. Flatbacks have the smallest clutch size of the chelonids, laying a mean of 53 eggs per clutch, and will lay about three clutches per nesting season. There is only one readily accessible nesting beach for this species at Mon Repos in Queensland, Australia. Other isolated rookeries like Crab Island are found along the Gulf of Carpentaria and Great Barrier Reef.

It is believed that flatbacks may be the only sea turtle that does not have an extended pelagic period in the open ocean. Hatchlings and juveniles spend the early posthatchling stage in shallow, protected coastal waters on the north-eastern Australian continental shelf and Gulf of Carpentaria. Their juvenile and adult diet is poorly known but appears to include snails, soft corals, mollusks, bryozoans, and sea pens.

See also

International Organizations. Sandy Beaches, Biology of.

Further Reading

Bustard R (1973) *Sea Turtles: Natural History and Conservation*. New York: Taplinger Publishing.

Carr A (1986) *The Sea Turtle: So Excellent a Fishe*. Austin: University of Texas Press.

Carr A (1991) *Handbook of Turtles*. Ithaca, NY: Comstock Publishing Associates of Cornell University Press.

Gibbons JW (1987) Why do turtles live so long? *BioScience* 37(4): 262–269.

Lutz PL and Musick J (eds) (1997) *The Biology of Sea Turtles*. Boca Raton: CRC Press.

Rieppel O (2000) Turtle origins. *Science* 283: 945–946.

SEABIRD CONSERVATION

J. Burger, Rutgers University, Piscataway, NJ, USA

Copyright © 2001 Academic Press

doi:10.1006/rwos.2001.0240

Introduction

Conservation is the preservation and protection of plants, animals, communities, or ecosystems; for marine birds, this means preserving and protecting them in all of their diverse habitats, at all times

of the year. Conservation implies some form of management, even if the management is limited to leaving the system alone, or monitoring it, without intervention or human disturbance. The appropriate degree of management is often controversial. Some argue that we should merely protect seabirds and their nesting habitats from further human influences, leaving them alone to survive or to perish. For many species, however, this solution is not practical because they do not live on remote islands, in inaccessible sites, or places that could be totally ignored by people. For other species, their nesting and foraging habitats have been so invaded by human activities that they must adapt to new, less suitable conditions. For some species, their declines have been so severe that only aggressive intervention will save them. Even species that appear to be unaffected by people have suffered from exotic feral animals and diseases that have come ashore, brought by early seafarers in dugout canoes or later by mariners in larger boats with more places for invading species to hide.

For marine birds there is compelling evidence that the activities of man over centuries have changed their habitats, their nesting biology, and their foraging ecology. Thus, we have a responsibility to conserve the world's marine birds. For conservation to be effective, the breeding biology, natural history, foraging ecology, and interactions with humans must be well understood, and the factors that contribute to their overall reproductive success and survival known.

Marine Bird Biology and Conservation

In this article, seabirds, or marine birds, include both the traditional seabird species (penguins, petrels and shearwaters, albatrosses, tropicbirds, gannets and boobies, frigate-birds, auks, gulls and terns, and pelicans) and closely related species that spend less time at sea (cormorants, skimmers), and also other species that spend a great deal of their life cycle along coasts or at sea, but may nest inland, such as shore birds. Seabirds are distributed worldwide, and nest in a variety of habitats, from remote oceanic islands that are little more than coral atolls, to massive rocky cliffs, saltmarsh islands, sandy beaches or grassy meadows, and even rooftops. While most species of seabirds nest colonially, some breed in loose colonies of scattered nests, and still others nest solitarily. Understanding the nesting pattern and habitat preferences of marine birds is essential to understanding the options for conservation of these birds on their nesting colonies. Without such information, appropriate habitats might not be preserved.

The attention of conservationists is normally directed to protecting seabirds while they are breeding, but marine birds spend most of their lives away from the breeding colonies. Although some terns and gulls first breed at 2 or 3 years of age, other large gulls and other seabirds breed when they are much older. Some albatrosses do not breed until they are 10 years old. Nonbreeders often wander the oceans, bays, and estuaries, and do not return to the nesting colonies until they are ready to breed. They face a wide range of threats during this period, and these often pose more difficult conservation issues because the birds are dispersed, and are not easy to protect. In some cases, such as roseate terns (*Sterna dougallii*), we do not even know where the vast majority of overwintering adults spend their time. Since many species forage over the vast oceans during the nonbreeding season, or before they reach adulthood, conditions at sea are critical for their long-term survival. While landscape ecology has dominated thought for terrestrial systems, few of its tenets have been applied to oceanic ecosystems, or to the conservation needs of marine birds.

A brief description of the factors that affect the success of marine bird populations will be enumerated before discussion of conservation strategies and management options for the protection and preservation of marine bird populations.

Threats to Marine Birds

Marine bird conservation can be thought of as the relationship between hazards or threats, marine bird vulnerabilities, and management. The schematic in **Figure 1** illustrates the major kinds of hazards faced by marine birds, and indeed all birds, and the different kinds of vulnerabilities they face. The outcomes shown in **Figure 1** are the major ones; however, there are many others that contribute to the overall decline in population levels (**Figure 2**). Conservation involves some form of intervention or management for each of these hazards, to preserve and conserve the species.

Marine Bird Vulnerabilities

Factors that affect marine bird vulnerability include the stage in the life cycle, their activity patterns, and

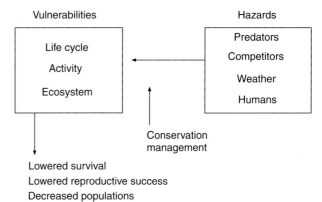

Figure 1 Marine avian conservation is the relationship between the hazards marine birds must face, along with their vulnerabilities.

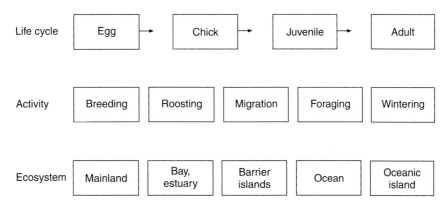

Figure 2 The primary vulnerabilities marine birds face deal with aspects of their life cycles, activity patterns and the ecosystems they inhabit.

their ecosystem (**Figure 3**). Marine birds are differentially vulnerable during different life stages. Many of the life cycle vulnerabilities are reduced by nesting in remote oceanic islands (albatrosses, many petrels, many penguins) or in inaccessible locations, such as cliffs (many alcids, kittiwakes, *Rissa tridactyla*) or tall trees. However, not all marine birds nest in such inaccessible sites, and some sites that were inaccessible for centuries are now inhabited by people and their commensal animals.

For many species, the egg stage is the most vulnerable to predators, since eggs are sufficiently small that a wide range of predators can eat them. Eggs are placed in one location, and are entirely dependent upon parents for protection from inclement weather, accidents, predators, and people. In many cultures, bird eggs, particularly seabird eggs, play a key role, either as a source of protein or as part of cultural traditions. In some cultures, the eggs of particular species are considered aphrodisiacs and are highly prized and sought after.

Egging is still practiced by humans in many places in the world, usually without any legal restrictions. Even where egging is illegal, either the authorities overlook the practice or it is impossible to enforce,

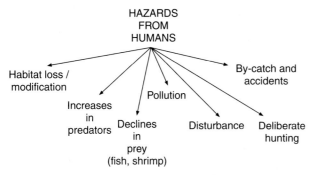

Figure 3 Humans provide a wide range of hazards, including direct and indirect effects.

or it is sufficiently clandestine to be difficult to apprehend the eggers.

Chicks are nearly as vulnerable as eggs, although many seabirds are semiprecocial at birth and are able to move about somewhat within a few days of hatching. The more precocial, the more likely the chick can move about to hide from predators or people, or seek protection from inclement weather. Nonetheless, chicks are unable to fly, and thus cannot avoid most ground predators and many aerial predators if they cannot hide sufficiently. The prefledging period can last for weeks (small species such as terns) to 6 months for albatrosses.

The relative vulnerability of juveniles and adults is usually the same, at least with respect to body size. Most juveniles are as large as adults and can fly, and are thus able to avoid predators. Juveniles, however, are less experienced with predators and with foraging, and so are less adept at avoiding predators and at foraging efficiently. The relative vulnerability of juveniles and adults depends on their size, habitat, type of predator, and antipredator behavior. For example, species that nest high in trees (Bonaparte's gull, *Larus philadelphia*) or on cliffs (e.g., kittiwakes, some murres, some alcids) are not exposed to ground predators that cannot reach them. Species that nest on islands far removed from the mainland are less vulnerable to ground predators (e.g., rats, foxes), unless these have been introduced or have unintentionally reached the islands. Marine birds that are especially aggressive in their defense of their nests (such as most terns) can sometimes successfully defend their eggs and young from small predators by mobbing or attacks.

Activity patterns also influence their vulnerability. When marine birds are breeding, they are tied to a particular nest site, and must either abandon their eggs or chicks or stay to protect them from inclement weather, predators, or people. At other times of the year, sea birds are not as tied to one location, and can move to avoid these threats. Foraging birds are vulnerable not only to predators and humans, but to accidents from being caught in fishing gill nets, drift nets, or longlines, from which mortality can be massive, especially to petrels and albatrosses.

The choice of habitats or ecosystems also determines their relative vulnerability to different types of hazards. Marine birds nesting on oceanic islands are generally removed from ground predators and most aerial predators but face devastation when such predators reach these islands. Cats and rats have proven to be the most serious threat to seabirds nesting on oceanic and barrier islands. The threat from predators increases the closer nesting islands are to the mainland, a usual source of predators and people. Similarly, the threat from storm and hurricane tides is greater near-shore, particularly for ground-nesting seabirds.

Marine Bird Hazards

The major hazards and challenges to survival of marine birds are from competitors (for mates, food, nesting sites), predators, inclement weather, and humans (**Figure 1**). Of these, humans are the greatest problem for the conservation of marine birds and strongly influence the other three types of hazards. Humans affect marine birds in a wide range of ways, by changing the environment around seabirds (**Figure 4**), ultimately causing population declines. While other hazards, such as competition for food and inclement weather are widespread, marine birds have always faced these challenges.

Habitat loss and modification are the greatest threats to marine birds that nest in coastal regions, and for birds nesting on near-shore islands. Direct loss of habitat is often less severe on remote oceanic islands, although recent losses of habitat on the Galapagos and other islands are causes for concern. Habitat loss can also include a decrease in available foraging habitat, either directly through its loss or through increased activities that decrease prey abundance or their ability to forage within that habitat.

Humans cause a wide range of other problems:

- Introducing predators to remote islands, and increasing the number of predators on islands and coastal habitats. For example, rats and cats have

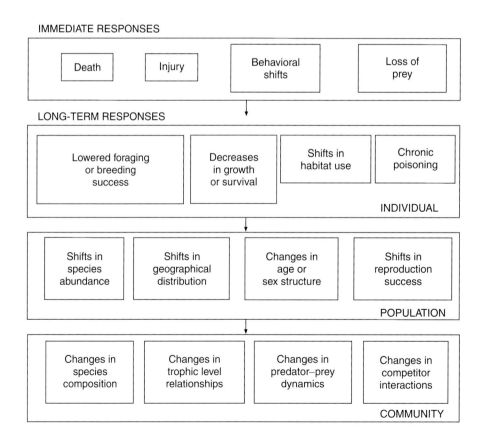

Figure 4 Human activities can affect marine birds in a variety of ways, including immediate and long-term effects.

been introduced to many remote islands, either deliberately or accidentally. Further, because of the construction of bridges and the presence of human foods (garbage), foxes, raccoons and other predators have reached many coastal islands.

- Decreasing available prey through overfishing, habitat loss, or pollution. Coastal habitat loss can decrease fish production because of loss of nursery areas, and pollution can further decrease reproduction of prey fish used by seabirds.
- Decreasing reproductive success, causing behavioral deficits, or direct mortality because of pollution. Contaminants, such as lead and mercury, can reduce locomotion, feeding behavior, and parental recognition in young, leading to decreased reproductive success.
- Decreasing survival or causing injuries because birds are inadvertently caught in fishing lines, gillnets, or ropes attached to longlines.
- Decreasing reproductive success or foraging success because of deliberate or accidental disturbance of nesting, foraging, roosting, or migrating marine birds. For some marine birds the presence of recreational fishing boats, personal watercraft, and commercial fishing boats reduces the area in which they can forage.
- Deliberate collection of eggs, and killing of chicks or adults for food, medicine, or other purposes. On many seabird nesting islands in the Caribbean, and elsewhere, the eggs of terns and other species are still collected for food. Egging of murres and other species is also practiced by some native peoples in the Arctic.

Conservation and Management of Marine Birds

Conservation of marine birds is a global problem, and global solutions are needed. This is particularly true for problems that occur at sea, where the birds roost, migrate, and forage. No one governmental jurisdiction controls the world's population of most species. Education, active protection and management, international treaties and agreements, and international enforcement may be required to solve some of the major threats to marine birds. However, the conservation and management of marine birds also involves intervention in each of the above hazards, and this can often be accomplished locally or regionally with positive results.

Habitat loss can be partly mitigated by providing other suitable habitats or nesting spaces nearby; predators can be eliminated or controlled; fishing can be managed so that stocks are not depleted

to a point where there are no longer sufficient resources for marine birds; contamination can be reduced by legal enforcement; human disturbance can be reduced by laws, wardening, and voluntary action; by-catch can be reduced by redesigning fishing gear or changing the spatial or temporal patterns of fishing; and deliberate or illegal hunting can be reduced by education and legal enforcement. Each will be discussed below.

Habitat creation and modification is one of the most useful conservation tools because it can be practiced locally to protect a particular marine bird nesting or foraging area. In many coastal regions, nesting habitat has been created for beach-nesting terns, shore birds, and other species by placing sand on islands that are otherwise unsuitable, extending sandy spits to create suitable habitat, and removing vegetation to keep the appropriate successional stage. In some places grassy meadows are preserved for nesting birds, while in others sand cliffs have been modified, and concrete slabs have been provided for cormorants, gannets, and boobies (albeit to make it easier to collect the guano for fertilizer). Habitat modification can also include creation of nest sites. Burrows have been constructed for petrels; chick shelters have been created for terns; and platforms have been built for cormorants, anhingas, and terns.

The increase in the diversity and number of introduced and exotic predators on oceanic and other islands is a major problem for many marine birds. One of the largest problems marine birds face worldwide is the introduction of cats and rats to remote nesting islands. Since most marine birds on remote islands nest on the ground, their eggs and chicks are vulnerable to cats and rats. Some governments, such as that of New Zealand, have devoted considerable time and resources to removal of these two predators from remote nesting islands, but the effort is enormous.

Many marine birds evolved on remote islands where there were no mammalian predators. These species lack antipredator behaviors that allow them to defend themselves, as seen in albatrosses, which do not leave their nests while rats gnaw them. Most sea birds on remote islands nest on, or under the ground, where they are vulnerable to ground predators, and they do not leave their nests when approached. Rats and cats have proven to be the most significant threat to sea birds worldwide, and their eradication is essential if some marine birds are to survive. New Zealand has invested heavily in eradicating invasive species on some of its offshore islands, allowing sea birds and other endemic species to survive. Cats, however, are extremely difficult to remove, even from small offshore islands, and up to three years were required to remove them completely from some New Zealand islands. Such a program involves a major commitment of time, money, and personnel by local or federal governments.

Simply removing predators, however, does not always result in immediate increases in seabird populations. Sometimes unusual management practices are required, such as use of decoys and play-back of vocalizations to attract birds to former colony sites. Steve Kress of National Audubon reestablished Atlantic puffins (*Fratercula arctica*) on nesting colonies in Maine by a long-term program of predator removal, decoys, and the playback of puffin vocalizations.

Increasing observations of chick mortality from starvation or other breeding failures have focused attention on food availability, and the declines in fish stocks. Declines in prey can be caused by sea level changes, water temperature changes, increases in predators and competitors, and other natural factors. However, they can also be caused by overfishing that depletes the breeding stocks, and reducing the production of small fish that serve as prey for seabirds. There are two mechanisms at work: in some cases fishermen take the larger fish, thereby removing the breeding stock, with a resultant decline in small fish for prey. This may have happened in the northern Atlantic. In other cases, fishermen take small fish, thereby competing directly with the seabirds, as partially happened off the coast of Peru.

Overfishing is a complicated problem that often requires not only local fisheries management but national and international treaties and laws. Even then, fisheries biologists, politicians, importers/exporters, and lawyers see no reasons to maintain the levels of fish stocks necessary to provide for the foraging needs of sea birds. Nonetheless, the involvement of conservationists interested in preserving marine bird populations must extend to fisheries issues, for this is one of the major conservation challenges that seabirds face.

By-catch in gill nets, drift nets, and longlines is also a fisheries problem. With the advent of long-lines, millions of seabirds of other species are caught annually in the miles of baited lines behind fishing vessels. The control and reduction in the number of such fishing gear is critical to reducing seabird mortality. Longlines are major problems for seabirds in the oceans of the Southern Hemisphere, although Australia and New Zealand are requiring bird-deterrents on longline boats.

Pollution is another threat to sea birds: Pollutants include heavy metals, organics, pesticides, plastics,

and oil, among others. Oil spills have often received the most attention because there are often massive and conspicuous die-offs of sea birds following major oil spills. Usually, however, the carcass counts underestimate the actual mortality because the spills happen at sea or in bad conditions where the carcasses are never found or do not reach land before they are scavenged or decay. Although direct mortality is severe from oil spills, one of the greatest problems following an oil spill is the decline of local breeding populations, as happened following the *Exxon Valdez* in Alaska. Ten years after the spill some seabird species had still not recovered to pre-spill levels. Partially this resulted from a lack of excess reproduction on nearby islands, where predators such as foxes kept reproduction low.

While major oil spills have the potential to cause massive die-offs of birds that are foraging and breeding nearby, or migrating through the area, chronic oil pollution is also a serious threat. Many coastal areas, particularly near major ports, experience chronic oil spillage that accounts for far more oil than the massive oil spills that receive national attention. Chronic pollution can cause more subtle effects such as changes in foraging behavior, deficits in begging, weight loss, and internal lesions.

When there are highly localized population declines as a result of pollution, such as oil spills, or of inclement weather, predators, or other causes, the management options are limited. However, one method to encourage rapid recovery is to manage the breeding colonies outside of the affected area, allowing them to serve as sources for the depleted colonies. In the case of the *Exxon Valdez*, for example, there were numerous active colonies immediately outside of the spill impact zone. However, reproduction on many of these islands was suboptimal owing to the presence of predators (foxes). Fox removal would no doubt increase reproductive success on those islands, providing surplus birds that could colonize the depleted colonies within the spill zone itself.

Management for reductions in marine pollutants, including oil, can be accomplished by education, negotiations with companies, laws and treaties, and sanctions. For example, following the *Exxon Valdez*, the U.S. government passed the Oil Pollution Act that ensured that by 2020 all ships entering U.S. waters would have double hulls and many other safety measures to reduce the possibility of large oil spills.

Another threat to marine birds is through atmospheric deposition of mercury, cadmium, lead, and other contaminants. At present, mercury and other contaminants have been found in the tissues of birds throughout the world, including the Arctic and Antarctic. While atmospheric deposition is greatest in the Northern Hemisphere, contaminants from the north are reaching the Southern Hemisphere. The problem of atmospheric deposition of mercury, and oxides of nitrogen and sulfur, can be managed only by regional, national, and international laws that control emissions from industrial and other sources, although some regional negotiations can be successful.

Marine birds have been very useful as indicators of coastal and marine pollution because they integrate over time and space. While monitoring of sediment and water is costly and time-consuming, monitoring of the tissues of birds (especially feathers) can be used to indicate where there may be a problem. Declines in marine bird populations such as occurred with DDT, were instrumental in regulating contaminants. Marine birds have been especially useful as bioindicators in the Great Lakes for polychlorinated biphenyls (PCBs), on the East Coast of North America and in northern Europe for mercury, and in the Everglades for mercury.

Human disturbance is a major threat to seabirds, both in coastal habitats and on oceanic islands. While the level and kinds of human disturbance to marine birds in coastal regions is much higher than for oceanic islands, the birds that nest on oceanic islands did not evolve with human disturbance and are far less equipped to deal with it. Disturbance to breeding and feeding assemblages can be deliberate or accidental, when people come close without even realizing they are doing so, and fail to notice or be concerned. Sometimes colonial birds mob people who enter their colony, but the people do not see any eggs or chicks (because they are cryptic), and so are unaware they are causing any damage. Chicks and eggs, however, can be exposed to heat or cold stress during these disturbances.

Human disturbance can be managed by education, monitoring (by volunteers, paid wardens, or law enforcement officers), physical barriers (signs, strings, fences, barricades), laws, and treaties. In most cases, however, it is worth meeting with affected parties to figure out how to reduce the disturbance to the birds while still allowing for the human activities. This can be done by limiting access temporally and keeping people away during the breeding season, or by posting the sensitive location but allowing human activities in other regions. Compliance will be far higher if the interested parties are included in the development of the conservation strategy, rather than merely being informed at a later point. Moreover, such people often have creative solutions that are successful.

The deliberate collecting of eggs and marine birds themselves can be managed by education, negotiations, laws and treaties, and enforcement. In places where the collection of eggs or adult birds is needed as a source of protein or for cultural reasons, mutual education by the affected people and managers will be far more successful. In many cases, indigenous peoples have maintained a sustainable harvest of seabird eggs and adults for centuries without ill effect to the seabird populations. However, if the populations of these people increase, the pressure on seabird populations may exceed their reproductive capacity. People normally took only the first eggs, and allowed the birds to re-lay and raise young. Conservation was often accomplished because individuals 'owned' a particular section of the colony, and their 'section' was passed down from generation to generation. There was thus strong incentive to preserve the population and not to overuse the resource, particularly since most seabirds show nest site tenacity and will return to the same place to nest year after year. When 'governments' took over the protection of seabird colonies, no one owned them any longer, and they suffered the fate of many 'commons' resources: they were exploited to the full with devastating results. Whereas subsistence hunting of seabirds and their eggs was successfully managed for centuries, the populations suffered overnight with the advent of government control and the availability of new technologies (snowmobiles, guns). More recently, the use of personal watercraft has increased in some coastal areas, destroying nurseries for fish and shellfish, disturbing foraging activities, and disrupting the nesting activities of terns and other species.

Hunting by nontraditional hunters can also be managed by education, persuasion, laws, and treaties. However, both types of hunting can be managed only when there are sufficient data to provide understanding of the breeding biology, population dynamics, and population levels. Without such information on each species of marine bird, it is impossible to determine the level of hunting that the populations can withstand. Extensive egging and hunting of marine birds by native peoples still occurs in some regions, such as that of the murres in Newfoundland and Greenland.

On a few islands, some seabird populations have both suffered and benefitted at the hands of the military. Some species nested on islands that were used as bombing ranges (Culebra, Puerto Rico) or were cleared for air transport (Midway) or were directly bombed (Midway, during the Second World War). In these cases, conservation could only involve governmental agreements to stop these activities, and of course, in the case of war, it is no doubt out of the hands of conservationists. However, military occupancy may protect colonies by excluding those who would exploit the birds.

Conclusions

Conservation of marine birds is a function of understanding the hazards that a given species or group of species face, understanding the species vulnerabilities and possible outcomes, and devising methods to reduce or eliminate these threats so that the species can flourish. Methods range from preserving habitat and preventing any form of disturbance (including egging and hunting), to more complicated and costly procedures such as wardening, and attracting birds back to former nesting colonies.

The conservation methods that are generally available include education, creation of nesting habitat and nest sites, the elimination of predators, the cessation of overfishing, building of barriers, use of wardens and guards, use of decoys and vocalization, creation of laws and treaties, and the enforcement of these laws. In most cases, the creation of coalitions of people with differing interests in seabirds, to reach mutually agreeable solutions, will be the most effective and long-lasting. Although ecotourism may pose the threat of increased disturbance or beach development, it can be managed as a source of revenue to sustain conservation efforts.

It is necessary to bear in mind that conservation of seabirds is not merely a matter of protecting and preserving nesting assemblages, but of protecting their migratory and wintering habitat and assuring an adequate food supply. Assuring a sufficient food supply can place marine birds in direct conflict with commercial and recreational fishermen, and with other marine activities, such as transportation of oil and other industrial products, use of personal watercraft and boats, and development of shoreline industries and communities. Conservation of marine birds, like many other conservation problems, is a matter of involving all interested parties in solving a 'commons' issue.

See also

Alcidae. Ecosystem Effects of Fishing. Laridae, Sternidae and Rynchopidae. Oil Pollution. Pelecaniformes. Procellariiformes. Seabird Conservation. Seabird Foraging Ecology. Seabird Migration. Seabird Population Dynamics. Seabird Reproductive Ecology. Seabirds and Fisheries Interactions. Seabirds as Indicators of Ocean Pollution. Sphenisciformes.

Further Reading

Burger J (2001) Tourism and ecosystems. *Encyclopedia of Global Environmental Change*. Chichester: Wiley.

Burger J and Gochfeld M (1994) Predation and effects of humans on island-nesting seabirds. In: Nettleship DN, Burger J and Gochfeld M (eds) *Threats to Seabirds on Islands*, pp. 39–67. Cambridge: International Council for Bird Preservation Technical Publication.

Croxall JP, Evans PGH and Schreiber RW (1984) *Status and Conservation of the World's Seabirds*. Cambridge: International Council for Bird Preservation Technical Publication No. 2.

Kress S (1982) The return of the Atlantic Puffin to Eastern Egg Rock, Maine. *Living Bird Quarterly* 1: 11–14.

Moors PJ (1985) *Conservation of Island Birds*. Cambridge: International Council for Bird Preservation Technical Publication.

Nettleship DN, Burger J and Gochfeld M (eds) (1994) *Threats to Seabirds on Islands*. Cambridge: International Council for Bird Preservation Technical Publication.

Vermeer K, Briggs KT, Morgan KH and Siegel-Causey D (1993) *The Status, Ecology, and Conservation of Marine Birds of the North Pacific*. Ottawa: Canadian Wildlife Service Special Publication.

SEABIRD FORAGING ECOLOGY

L. T. Balance, NOAA-NMFS, La Jolla, CA, USA
D. G. Ainley, H.T. Harvey & Associates, San Jose, CA, USA
G. L. Hunt, Jr., University of California, Irvine, CA, USA

doi:10.1006/rwos.2001.0233

Introduction

Though bound to the land for reproduction, most seabirds spend 90% of their life at sea where they forage over hundreds to thousands of kilometers in a matter of days, or dive to depths from the surface to several hundred meters. Although many details of seabird reproductive biology have been successfully elucidated, much of their life at sea remains a mystery owing to logistical constraints placed on research at sea. Even so, we now know a considerable amount about seabird foraging ecology in terms of foraging habitat, behavior, and strategy, as well as the ways in which seabirds associate with or partition prey resources.

Foraging Habitat

Seabirds predictably associate with a wide spectrum of physical marine features. Most studies implicitly assume that these features serve to increase prey abundance or availability. In some cases, a physical feature is found to correlate directly with an increase in prey; in others, the causal mechanisms are postulated. To date, the general conclusion with respect to seabird distribution as related to oceanographic features is that seabirds associate with large-scale currents and regimes that affect physiological temperature limits and/or the general level of prey abundance (through primary production), and with small-scale oceanographic features that affect prey dispersion and availability.

Water Masses

In practically every ocean, a strong relation between sea bird distribution and water masses has been reported, mostly identified through temperature and/or salinity profiles (**Figure 1**). These correlations occur at macroscales (1000–3000 km, e.g. associations with currents or ocean regimes), as well as

mesoscales (100–1000 km, e.g. associations with warm- or cold-core rings within current systems). The question of why species associate with different water types has not been adequately resolved. At issue are questions of whether a seabird responds directly to habitat features that differ with water mass (and may affect, for instance, thermoregulation), or directly to prey, assumed to change with water mass or current system.

Environmental Gradients

Physical gradients, including boundaries between currents, eddies, and water masses, in both the horizontal and vertical plane, are often sites of elevated seabird abundance. Seabirds respond to the strength of gradients more than the presence of them. In shelf ecosystems, e.g. the eastern Bering Sea shelf and off the California coast, cross-shelf gradients are stronger than along-shelf gradients, and sea-bird distribution and abundance shows a corresponding strong gradient across, as opposed to along the shelf. At larger scales the same pattern is evident, e.g. crossing as opposed to moving parallel with boundary currents.

Physical gradients can affect prey abundance and availability to seabirds in several ways. First, they can affect nutrient levels and therefore primary production, as in eastern boundary currents. Second, they can passively concentrate prey by carrying planktonic organisms through upwelling, downwelling, and convergence. Finally, they can maintain property gradients (fronts, see below) to which prey actively respond. In the open ocean, where currents

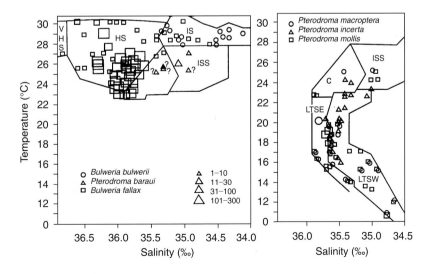

Figure 1 Distribution of gadfly petrels in the Indian Ocean corresponding to various regimes of surface temperature and salinity: (A) warm-water species, and (B) cool-water species. The relative size of symbols is proportional to the number of sightings. Symbols for water masses as follows: VHS, very high salinity; HS, high salinity; IS, intermediate salinity; ISS, intermediate salinity south; C, common water; LTSE, low temperature southeast; LTSW, low temperature southwest. (Redrawn from Pocklington R (1979) *Marine Biology* 51: 9–21.)

and dynamic processes are less active, prey behavior should be the principal mechanism responsible for seabird aggregation. In these cases, locations of aggregations are unpredictable, and this has important consequences for the adaptations necessary for seabirds to locate and exploit them. In contrast, in continental shelf systems, currents impinge upon topographically fixed features, such as reefs, creating physical gradients predictable in space and time to which seabirds can go directly. Thus, the first and second mechanisms are the more important in shelf systems, and aggregations are so predictable that seabirds learn where and when to be in order to eat.

Fronts

Much effort has been devoted successfully to identifying correlations between seabird abundance and fronts, or those gradients exhibiting dramatic change in temperature, density, or current velocity (**Figure 2**). Results indicate a considerable range of variation in the strength of seabird responses to fronts. This may be due to the fact that fronts influence sea-bird distribution only on a small scale. The factors behind the range of response is of interest in itself.

Nevertheless, fronts are important determinants of prey capture. Two hypotheses have been proposed to account for this: (1) that frontal zones enhance primary production, which in turn increases prey supply, e.g. boundaries of cold- or warm-core rings in the Gulf Stream; and (2) that frontal zones serve to concentrate prey directly into exploitable patches, e.g. current rips among the Aleutian Island passages.

Topographic Features

Topographic features serve to deflect currents, and can be sites of strong horizontal and vertical changes in current velocity, thus, concentrating prey through a variety of mechanisms (**Figure 2**). For example, seamounts are often sites of seabird aggregation, likely related to the fact that they are also sites of increased density and heightened migratory activity for organisms comprising the deep scattering layer. A second example are topographic features in relatively shallow water, including depressions in the tops of reefs and ridges across the slope of marine escarpments, which may physically trap euphausiids as they attempt to migrate downward in the morning. A third example is the downstream, eddy effect of islands that occur in strong current systems.

Depth gradient itself is sometimes correlated with increased abundance of seabirds, and water depth in

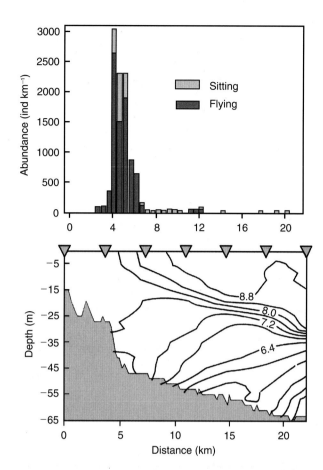

Figure 2 The aggregation of foraging shearwaters at fronts, in this case the area where an elevation in bottom topography forces a transition between stratified and well-mixed water. The bottom panel shows isotherms. (Redrawn from Hunt GL *et al.*, *Marine Ecology Progress Series* 141: 1–11.)

general has long been related to seabird abundance and species composition. Depth-related differences in primary productivity explain large-scale patterns between shelf and oceanic waters. Within shelf systems themselves, several hypotheses explain changes in species composition and abundance with depth. First, primary production may be diverted into one of two food webs, benthic or pelagic, and this may explain differences in organisms of upper trophic levels in inner versus outer shelf systems, e.g. the eastern Bering Sea. Alternatively, the fact that interactions between flow patterns in the upper water column and bottom topography will be strong in inner shelf areas but decoupled in outer shelf areas, may result in differences in predictability of prey and consequently, differences in the species that exploit them, e.g. most coastal shelf systems. Finally, depending on diving ability and depth, certain seabirds may be able to exploit bottom substrate, whereas others may not.

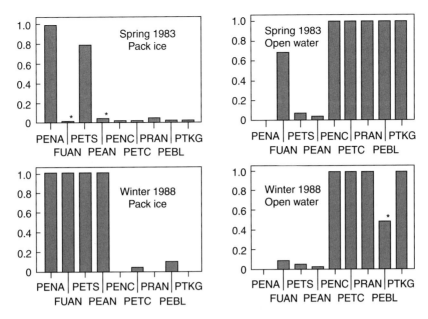

Figure 3 The correspondence of various seabird species either to pack ice or open water immediately offshore of the ice; Scotia-Weddell Confluence, Southern Ocean. Abbreviations: PENA, Adélie penguin; FUAN, Antarctic fulmar; PETS, snow petrel; PEAN, Antarctic petrel; PENC, chinstrap penguin; PETC, pintado petrel; PRAN, Antarctic prion; PEBL, blue petrel; PTKG, Kerguelen petrel. (Redrawn from Ainley *et al.* (1994) *Journal of Animal Ecology* 63: 347–364.)

Sea Ice

A strong association of individual species and characteristic assemblages exists with sea ice features. On one hand, certain species are obligate associates with sea ice; on the other, sea ice can limit access to the water column, and in some cases, aggregation of seabirds near the ice margin is a simple response to this barrier (**Figure 3**).

Sea ice often enhances foraging opportunities. First, there is an increased abundance of mesopelagic organisms beneath the ice, believed to be a phototactic response of these organisms to reduced light levels. Second, primary production is enhanced beneath the ice or at its edge, and in turn leads to increased abundance of primary and secondary consumers. The abundance and degree of concentration of this sympagic fauna varies with ice age, ice structure, and depth to the bottom. As a result, Arctic ice, often multiyear in nature, has a speciose sympagic fauna as compared to Antarctic ice, which is often annual. The ice zone can be divided into at least three habitats, a region of leads within the ice itself, the ice edge, and the zone seaward of the ice. Each zone is exploited by different seabird species, and the relative importance of zones appears to differ between Arctic and Antarctic oceans, with the seaward zone being particularly important in Antarctic systems.

Foraging Behavior

Most seabird species take prey within a half meter of the sea surface. This they accomplish by capturing prey that either are airborne themselves (as a result of escaping subsurface predators, see below), or are shallow enough that, to grasp prey, the bird dips its beak below the surface (dipping) or crashes into the surface and extends its head, neck and upper body downward (surface plunging, contact dipping). Other species feed on dead prey floating at the surface. Another group takes prey within about 20 m of the surface either by flying into the water to continue flight-like wing movements below the surface (pursuit plunging) or by using the momentum of an aerial dive (plunging).

Finally, a number of seabirds can dive using either their wings or feet for propulsion. Dive depth is related to body size, which in turn relates to physiological capabilities of diving. The deepest dives recorded by a bird (penguin) reach 535 m, but most species are confined to the upper 100 m. A highly specialized group of species capture prey by stealing them from other birds.

Foraging Strategies

The issue of how seabirds locate their prey is far from completely understood. Below is a summary of known strategies, most of which depend on sea-birds locating some feature which itself serves to reliably concentrate prey.

Physical Features

Physical features are important to foraging sea-birds, because they serve to concentrate prey in space and time, and because they often occur under predictable circumstances (**Figure 2**). This issue was addressed above.

Subsurface Predators

Subsurface predators commonly drive prey to the surface because the air–water interface acts as a boundary beyond which most prey cannot escape. Under these circumstances, seabirds can access these same prey from the air. A wide array of subsurface predators are important to seabirds: large predatory fishes (e.g. tuna), particularly in relatively barren waters of the tropics; marine mammals, both cetaceans and pinnipeds; and marine birds themselves.

Subsurface predators increase the prey available to birds in at least three ways. First, they drive prey to the surface. Second, they injure or disorient prey, which then drift to the surface and are accessible to surface foragers. Third, they leave scraps on which seabirds forage, particularly when the prey themselves are large.

Feeding subsurface predator schools can be highly visible and the degree of association of birds with these prey patches is often great. One investigation found 79% of the variability in seabird density could be explained by the number of gray whale mud plumes; this visibility may have been responsible for the higher correlation than typically reported for other studies attempting to relate seabird and prey abundance (see below).

Feeding Flocks

Seabirds in most of the world's oceans exploit clumped prey by feeding in multispecies flocks. Studies in all latitudes report that seabirds in flocks, often in a very few disproportionately large aggregations, account for the majority of all individuals seen feeding. Although flocks can result from passive aggregation at a shared resource, evidence indicates that seabirds benefit in some way from the presence of other individuals. First, as noted above, some seabird species act as subsurface predators, making prey available to surface-feeding seabirds, e.g. alcids driving prey to the surface for larids in coastal Alaska. Second, a small number of seabird species are kleptoparasitic, obtaining their prey from other seabirds, e.g. jaegers, skuas and a few other species.

Third, vulnerability of individual fish in a school may increase with the number of birds feeding in the flock. Finally, flocks are highly visible signals of the location of a prey patch, e.g. species keying on frigate birds circling high over tuna schools.

Certain species are disproportionately responsible for these signals, simply through their highly visual flight characteristics. Such species are termed 'catalysts'. There is strong evidence that seabirds follow these visual signals, in some cases distinguishing between searching and feeding catalyst species, and between those feeding on a single prey item and those feeding on a clumped patch.

Nocturnal Feeding

Many fishes and invertebrates remain at depth during the day and migrate to the surface after dark. During crepuscular or dark periods, surface densities of prey can be 1000 times greater than during the day. This migration is more significant in low than high latitudes, and in oceanic than neritic waters. Many studies indirectly infer nocturnal feeding from the presence of vertically migrating species in seabird diets or from circadian activity patterns. Little direct evidence exists.

Among the indirect data on how seabirds might locate prey at night is a considerable body of information on olfaction. In particular, members of the avian order Procellariiformes possess olfactory lobes disproportionately large compared to the total brain size and compared to most vertebrates. Experiments show a marked ability of these birds to differentiate among odors and, especially, to find sources of odors that are trophically meaningful.

Maximization of Search Area

Feeding opportunities in the open ocean are often patchily distributed requiring seabirds to travel over large areas in search for prey. This ability to search large areas is tightly linked to adaptations for flight proficiency. Seabirds capable of wide-area search exhibit morphological adaptations of the wing and tail that enhance energy-efficient flight. They also modify their flight behavior to take advantage of wind as an energy source. Penguins, loons, grebes, cormorants and alcids, the pre-eminent seabird divers (see above), sacrifice wide-area search capabilities for what is needed for diving: high density and small wings. Therefore, divers are limited to areas of high prey availability.

An investigation into the relationship between wind and foraging behavior has revealed taxon-specific preferences in flight direction dependent on wind direction, and in turn, to wing morphology and presumed prey distribution. Procellariiformes, primarily oceanic foragers, preferentially fly across the wind, whereas Pelecaniformes and Charadriiformes, the majority of which forage over shelf and slope waters, preferentially fly into and across headwinds. Because prey on a global scale are more patchy in oceanic than shelf and slope waters, across-wind flight may allow Procellariiformes to cover more area at lower energetic cost, whereas headwind flight allows slower ground speeds, possibly increasing the probability of detecting prey and decreasing response time once a prey item is detected. Flying up- or across-wind among procellariids also maximizes probabilities of finding prey using olfaction.

Associations With Prey

Positive or significant correlations have been identified between seabird and prey abundance, in the Bering Sea, eastern North Atlantic, Barents Sea, and in locations throughout the Antarctic, although rarely at scales smaller than about 2–3 km. In some studies, however, the correlation is weak to nonexistent, or a negative correlation is reported. From these results, several general principles have arisen. First, the strength of the correlation increases with the spatial scale at which measurements are made. Second, correlations between planktonic-feeding seabirds and their prey are lower than correlations between piscivorous seabirds and their prey. The reasons for this pattern may relate to differences in patch characteristics dependent on prey species, or to differences in search mode of various predators. Third, correlations are not as strong as expected and in many cases, a correlation is found only with repeated surveys. Many hypotheses have been proposed to explain the latter, including: (1) seabirds are unable consistently to locate large prey patches; (2) prey are sufficiently abundant that seabirds do not need to locate largest prey patches; (3) prey are actively avoiding seabirds; (4) prey patches are continuously moving so that a time lag exists between patch formation, discovery by the seabird, and measurement by the researcher; (5) extremely large prey patches are disproportionately important to seabirds so that they spend much of their time searching for or in transit to and from such patches; and (6) our means to measure prey patches (usually hydroacoustically) is a mismatch to the biology and attributes of the predators. Finally, different seabird species may respond on the basis of threshold levels of prey abundance, and these thresholds vary seasonally as well as with reproductive status of the bird (breeders require more food than nonbreeders, which comprise at least half the typical sea-bird population).

Resource Partitioning

Food is considered to be an important resource regulating seabird populations. Accordingly, much research has focused on identifying differences in the way coexisting species exploit prey. At sea, the fact that different oceanic regimes or currents support different prey communities as well as different sea-bird communities has been used to support the idea that the geographic range of seabird species is a response to the presence of specific prey. Contrasting these patterns, however, several colony-based studies report high diet overlap between species, leading to speculation that dietary differences may reflect differences in foraging habitat as opposed to prey selection. Evidence from at-sea research indicates broad overlap in the species and/or sizes of prey taken by coexisting seabird species, often despite species-specific feeding methods, body size or habitat segregation.

Prey Selection

Under certain circumstances seabirds do make choices as to what prey they will attempt to capture (Figure 4). Among breeding species of the western North Atlantic and North Sea, in cases where a prey stock, such as capelin or sandlance, are being exploited by a large array of species, birds key in on fish that provide the highest energy package, in this case fish in reproductive condition. In the Bering Sea, breeding auklets have been observed to ignore smaller zooplankton to take the most energy-dense copepod species available. Finally, in the Antarctic, during winter with almost constant darkness or near-darkness when the mesopelagic community is near the surface most of the time, seabirds have been documented to avoid smaller prey (euphausiids) to take larger and more energy-rich prey (myctophids; Figure 5). However, although this is evidence for active prey selection, in these cases there is broad dietary overlap among seabird predators.

Prey/Predator Size

Body-size differences among coexisting sea bird species have been used to imply diet segregation by prey size (Figure 5). In general, discounting penguins but realizing there are a number of exceptions, the larger seabirds (i.e. those > 1500 g) tend to take fish and squid; the smaller species tend to take juvenile or larval fish and squid, along with zoo-

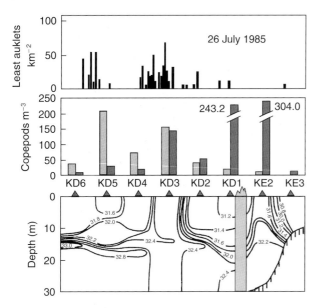

Figure 4 Aggregation of least auklets over concentrations of the copepods *Eucalanus bungii* and *Neocalanus* spp. (light bars), and *Calanus marshallae* (dark bars). All three are confirmed prey of least auklets. Birds virtually ignored huge concentrations of *C. marshallae*, which were much closer to the breeding colony on King Island to preferentially feed on larger and presumably more energy-rich *Neocalanus* spp. (Redrawn from Hunt GL Jr and Harrison NM (1990) *Marine Ecology Progress Series* 65: 141–150.)

planktonic invertebrates. It comes down to energetic cost-efficiency of foraging and the morphology of the seabird foraging apparatus: the bill, which picks one prey item at a time (the lone exception, perhaps, being one or two species groups, e.g. prions, that may filter-feed). Some degree of dietary size segregation is apparent among the few studies that have investigated all species breeding at single sites, e.g. a tropical oceanic island. Other studies at sea report little, if any, dietary separation, even though a 1000-fold difference in seabird size can exist. The implication is that seabirds often forage opportunistically depending on the availability of prey in their preferred habitat, and that differences in habitat are more important than differences in prey selection in facilitating predator coexistence.

Habitat Type/Time

Species or assemblages often segregate according to habitat with little evidence of interactions among seabirds that significantly influence their pelagic distributions. Instead, the implication is that species respond to physical and biological characteristics of environments according to their individual needs and flight or diving capabilities. Spatial segregation can occur with respect to simple habitat features. For example, the Antarctic avifauna is divided into

one assemblage associated with pack-ice covered waters and the other with ice-free waters (**Figure 3**). The species composition of these assemblages changes little over time; assemblage distribution tracks the distribution of ice features in the absence of differences in prey communities between the two habitat types, and there is little spatial overlap between the two assemblages. Spatial segregation, with respect to species or assemblages, can also occur along environmental gradients with respect to physical, chemical, and biological features of a seabird's habitat, in both the horizontal and vertical dimension. This has been well documented especially in shelf waters, where differences in foraging habitat, particularly as determined by depth, lead to differences in diet; it is also evident in the pelagic tropical waters along productivity gradients.

A recurrent theme is that seabird species sort out along prey density gradients, regardless of prey identity. Such segregation has been recorded even within the same prey patch, with certain species exploiting the center and others the periphery. The idea that oligotrophic waters, having reduced prey availability, can only be exploited by highly aerial species with efficient locomotion, whereas productive waters are necessary for diving species is one that occurs in a wide variety of studies conducted in tropical, temperate, and polar systems.

Finally, segregation can occur with respect to time, specifically with respect to those species that feed at night versus during the day. A few seabird species are adapted to feed only at night.

Mutualism and Kleptoparasitism

The sea bird flocking community in the North Pacific comprises species having complementary foraging behaviors, thus, indicating a degree of integration within the community. In particular, the feeding behavior of catalyst and diving species could be interpreted as mutualistic, catalysts signaling the location of a prey patch, and divers increasing or maintaining prey concentration. Certain authors have speculated that this relationship could have resulted from coevolution of behavior designed to increase the mutualistic benefit of the association. Kleptoparasitism was also proposed to stabilize these feeding flocks by forcing alcids to forage at the edges of a prey patch where they are less vulnerable to piracy, thus, maintaining patch density and ultimately, increasing prey availability to all flock members.

Morphological or Physiological Factors

Differential resource use is sometimes ascribed to species-specific morphological or physiological

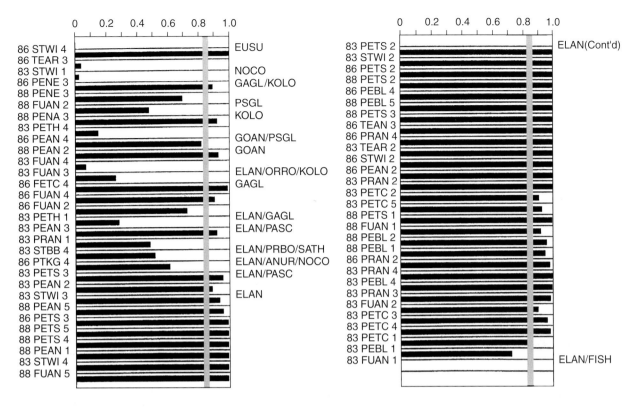

Figure 5 Diet overlap among Antarctic seabirds. Even though a 1000-fold difference in predator size existed, there was no appreciable separation of diet for many sea-bird species. The vertical stippled line indicates the level at which diet is considered to be similar on the basis of prey species overlap. Bird species to the left of bars, prey species to the right. Each bird species name is preceded by the year in which collection was made, and followed by a number that denotes habitat (1, open water; 2, sparse ice; 3, heavy ice). Bird species: PENE, emperor penguin; PENA, Adélie penguin; FUAN, Antarctic fulmar; PETS, snow petrel; PEAN, Antarctic petrel; PENC, chinstrap penguin; PETC, pintado petrel; PRAN Antarctic prion; PEBL, blue petrel; PTKG, kerguelen petrel; STWI, Wilson's storm-petrel; STBB, black-bellied storm-petrel; TEAR, Arctic tern; TEAN, Antarctic tern. Prey names: ANUR, *Anuropis* spp.; EUSU *Euphausia superba*; ELAN, *Electrona antarctica*; GAGL, *Galiteuthis glacialis*; GOAN, *Gonatus antarcticus*; KOLO, *Kondakovia longimana*; NOCO, *Notolepis coatsi*; ORRO, *Orchomene rossi*; PASC, *Pasiphaea scotia*; PRBO, *Protomyctophum bolini*; PSGL, *Psychroteuthis glacialis*; SATH, *Salpa thompsoni*. (Redrawn from Ainley DG *et al.* (1992) *Marine Ecology Progress Series* 90: 207–221.)

factors affecting flight or diving capabilities. Several examples exist. First, terns differ in their ability to feed successfully in dense flocks over predatory fishes as a function of a given species' ability to hover for prolonged periods of time. Second, differential metabolic demands may be responsible for species-specific differences in the threshold prey density to which alcids respond. Finally, differential flight costs correlate with species-specific patterns in resource use, e.g. along productivity gradients in the tropics, or the amount of foraging habitat that can be exploited (near-shore vs. offshore).

Ultimately, many of these morphological and physiological adaptations are driven by body size. Body size influences depth of dive capabilities, cost of transport, and basal metabolic rate. Additionally, body size can frequently be used to predict the outcome of interference competition (below).

Competition

Interference competition apparently does occur between seabirds at sea. It is referred to most often in the context of feeding flocks, taking the form of aggressive encounters, and collisions between feeding birds. The proximate limiting resource identified in many of these cases is access to prey, i.e. space over the prey patch. In another situation, shearwaters in the North Pacific feed by pursuit plunging in large groups, by which they disperse, decimate, or drive prey deeper into the water column thereby reducing the availability of prey to surface-feeding species. This same mechanism has been proposed for tropical boobies, which by plunge diving may also drive prey beyond the reach of surface feeders.

Despite widespread discussion of trophic competition, supporting data are sparse and some evidence indicates it to be not important in structuring some sea-bird communities. For example, one study in the Antarctic found no habitat expansion of the pack-ice assemblage into adjacent open waters seasonally vacated by another community (**Figure 3**), a shift that might be expected if competition affected community structure and habitat selection. In that study, sufficient epipelagic prey were available in the ice-free waters to be exploited successfully by sea birds (**Figure 5**).

Competition with Fisheries

Many of the forage species sought by sea birds are the same sought by industrial fisheries. The result is conflict, particularly in eastern boundary currents, where clupeid fishes are dominant and are of ideal size and shape to be consumed by sea birds. The tracking of bird populations with fish stocks has been especially well documented in the Benguela and Peru currents, where not only have fish stocks been heavily exploited but so have guano deposits accumulated by the sea birds. The bird populations have responded closely to geographic, temporal and numerical variation in the fish stocks. Well documented, also, have been fish stocks and avian predator populations in the North Sea. There, commercial depletion of predatory fish benefited sea-bird populations by reducing competition for forage fish; when fisheries turned to the forage fish themselves, seabird populations declined. In some areas, it has been proposed to use statistical models of predator populations as an indicator of fish-stock status independent of fishery data, for instance, the Convention for the Conservation of Antarctic Living Marine Resources. Much information is needed to calibrate seabird responses to prey populations before seabirds can be used reliably to estimate prey stocks.

See also

Benguela Current. Canary and Portugal Currents. Seabird Migration. Seabird Responses to Climate Change. Seabirds and Fisheries Interactions. Sea Ice: Variations in Extent and Thickness.

Further Reading

Ainley DG, O'Connor EF and Boekelheide RJ (1984) *The Marine Ecology of Birds in the Ross Sea, Antarctica.* American Ornithologists' Union, Monograph 32. Washington, DC.

Ainley DG and Boekelheide RJ (eds) (1990) *Seabirds of the Farallon Islands: Ecology, Dynamics and Structure of an Upwelling-system Community.* Stanford, CA: Stanford University Press.

Ashmole NP (1971) Seabird ecology and the marine environment, In Farner DS, King JR and Parkes KC (eds) *Avian Biology*, vol. 1, pp. 223–286. New York: Academic Press.

Briggs KT, Tyler WB, Lewis DB and Carlson DR (1987) *Bird Communities at Sea off California: 1975–1983.* Studies in Avian Biology, No. 11. Berkeley, CA: Cooper Ornithological Society.

Burger J, Olla BL and Winn WE (eds) (1980) *Behavior of Marine Animals*, vol. 4: *Birds*. New York: Plenum Press.

Cooper J (ed.) (1981) *Proceedings of the Symposium on Birds of Sea and Shore.* African Seabird Group. Cape Town, Republic of South Africa.

Croxall JP (ed.) (1987) *Seabirds: Feeding Ecology and Role in Marine Ecosystems.* London: Cambridge University Press.

Furness RW and Greenwood JJD (1983) Birds as monitors of environmental change. London and New York: Chapman & Hall.

Furness RW and Monaghan P (1987) *Seabird Ecology.* London: Blackie.

Montevecchi WA and Gaston AJ (eds) (1991) *Studies of high latitude seabirds* 1: *Behavioural, Energetic and Oceanographic Aspects of Seabird Feeding Ecology.* Canadian Wildlife Service, Occasional Papers 68. Ottawa, Canada.

Nettleship DN, Sanger GA and Springer PF (1982) *Marine Birds: Their Feeding Ecology and Commercial Fisheries Relationships.* Canadian Wildlife Service, Special Publication. Ottawa, Canada.

Vermeer K, Briggs KT and Siegel Causey D (eds) (1992) Ecology and conservation of marine birds of the temperate North Pacific. Canadian Wildlife Service, Special Publication. Ottawa, Canada.

Whittow GC and Rahn H (eds) (1984) *Seabird Energetics.* New York: Plenum Press.

SEABIRD MIGRATION

L. B. Spear, H. T. Harvey & Associates, San Jose, CA, USA

Copyright © 2001 Academic Press

doi:10.1006/rwos.2001.0238

Introduction

Bird migration is one of the most fascinating phenomena in our living environment, and accordingly has been studied since ancient times, particularly among nonmarine species. Studies of migration and navigation of nonmarine species have become quite sophisticated, examining in detail subjects including orientation and navigation, and physiological and morphological adaptations. In contrast, studies of migration among marine birds have been fewer and more simplistic. Indeed, until recently, much of the information on migration of seabirds had come from the recovery of individuals ringed (i.e., metal rings are attached to the legs, each stamped with a unique set of numbers) at their breeding sites. Although ringing has revealed considerable information about the migrations of species that stay close to coasts (thus, facilitating recoveries), it has provided little insight into the movements of species that stay far at sea during the nonbreeding period. In addition, studies at sea have been few, owing to the immense size of the world's oceans and the inherent logistical difficulties. Fortunately, however, an upsurge in pelagic investigations has occurred in the past 20 years, owing to the advent of ground-position satellite (GPS) telemeters that can be attached to larger avian species (e.g., albatrosses), and ship-board studies of the flight direction and flight behavior of birds on the high seas.

Background

Migration among seabirds ultimately is a response to the seasonally changing altitude of the sun's position, causing changes in environmental conditions to which the birds must adapt to survive and reproduce. Individual seabirds have the ability to go to a precise wintering location and then return to a precise breeding location. The duration between trips to and from wintering and breeding sites can be annual (as in adults), or last several years in the case of subadults of some species (notably the Procellariiformes, see below). In the latter case, fledgelings go to sea and do not return to land until reaching sexual maturity at ages of up to 10 years or more, such as in the wandering and royal albatrosses (*Diomedea exulans* and *D. epomophora*). On reaching breeding age, many seabirds return to the same colony from which they originated, in fact, they often nest on or adjacent to the exact location where they were hatched. After first breeding, a large proportion also return each year to the same nest site to breed. Furthermore, individual adults have the ability to return with pinpoint precision to the same wintering location each year following breeding. These locations are usually those where these individuals foraged during their subadult years.

Seabirds can home in on their breeding sites during all types of weather, during darkness (e.g., some species return to nesting burrows under dense vegetation only during the night, even during dense fog), and can navigate distances at sea approaching a global scale. The latter was demonstrated in two experiments in which the Manx shearwater (*Puffinus puffinus*) and 18 Laysan albatrosses (*Phoebastria immutibilis*) were taken from their breeding

sites (where they were attending eggs or young) and air-freighted to locations thousands of kilometers away. Many were released at locations where they surely had not been before, and in environments for which they are not adapted. The shearwater returned to its nest site in Wales in 12.5 days, covering the 5200 km (shortest) distance from its release site at Boston, Massachusetts, at a rate of 415 km/day. Fourteen of the 18 albatrosses returned to their nest sites on Sand Island, Hawaii, with median trip duration and distance flown of 12 days, and 275 km/day (straight-line), respectively. One bird, released in Washington, took 10.1 days to cover the 5200 km distance back to Sand Island, although it probably flew a longer, tacking course because of the headwinds that it would have encountered if it flew directly to the island.

Sensory Mechanisms Used for Orientation and Navigation

As noted above, studies among terrestrial species have provided many insights into the sensory mechanisms by which birds navigate over long distances. Given the ability of seabirds to navigate long distances across the open ocean, they most likely use one or a combination of the sensory mechanisms indicated for terrestrial species. These mechanisms include endogenous (genetically transmitted) vector navigation, olfactory and time-compensated sun orientation; star, magnetic, UV, and polarized light orientation.

Endogenous vector navigation, in which birds are genetically programmed to follow the correct course and to start their migrations at the correct time, is thought to explain how young birds that have never migrated before, and that frequently do not accompany experienced individuals (as in the case for many seabirds), find their wintering areas. In this regard, spatial and temporal orientation appear to be coded to both celestial rotation and the geomagnetic field, this information being contained within a heritable, endogenous program.

Yet, migrants encounter many uncertainties due to unpredictable weather which can disrupt vector navigation. Through a series of experimental studies examining hypotheses addressing this problem, it has become the general consensus that birds have a compass sense as well as a very extensive grid/mosaic map sense. Birds can integrate combinations of time-compensated sun inclination (particularly at sunset), star, magnetic, UV, and polarized light cues, although the possibility that these capabilities vary among species remain open. Nevertheless, the evidence indicates that for short-term orientation, magnetic cues take precedence over celestial, that visual cues at sunset over-ride both the latter, and that polarized sky light is used during dusk orientation. The basis for the map aspect employed for navigation is not well studied and consists of two hypotheses: perceptions of a magnetic grid and/or an olfactory mosaic/gradient.

Physiological and Behavioral Adaptations

Many species of terrestrial birds have major shifts in their physiology just preceding the migration period, including dramatic increases in food intake and body fat, hypertrophy of breast muscles, increased hematocrit levels, and increases in body protein. Although few physiological studies of seabirds exist, most indicate a lack of, or smaller, physiological changes then occurs in land birds. During the post-breeding phase of migration, seabirds are generally lighter, with lower fat reserves, than when returning to their colonies at the beginning of the next breeding season. For example, adult sooty shearwaters (*P. griseus*) weigh 20–25% less during the post-breeding migration than when returning during the prenuptial period. Similar differences in the pre- and post-breeding body mass also occur in several Charadriformes (e.g., gulls, terns, auks).

Several factors could be responsible for the lack of more obvious physiological adaptations for migration among seabirds. First, most of the terrestrial species in which major physiological changes occur have small body masses (<75 g) and, thus, lower flight efficiency than larger species such as most seabirds. Second, and probably most importantly, many of the terrestrial species perform long-distance nonstop flights, often at high elevations, over obstacles where they cannot feed (e.g., large bodies of water, or deserts). In contrast, such migrations are rare among seabirds, because seabirds nearly always migrate at low elevations over the ocean, facilitating frequent feeding along the migration route. Hence, seabirds are seldom far from a habitat offering feeding opportunities, even during transequatorial migrations by species that feed mostly in temperate or boreal latitudes.

The post-breeding migration of seabirds is generally more leisurely than the pre-breeding migration. One reason is that seabirds, especially those moving longer distances, feed more during the post-breeding than the pre-breeding movement. This behavior is probably related to the poorer body condition of seabirds just after breeding. Two examples include the sooty shearwater and the Arctic tern. Both perform transequatorial migrations, although the shear-water is a southern hemisphere breeder, with its post-breeding migration during the boreal spring, whereas the tern is a northern hemisphere breeder that leaves its breeding grounds in the boreal autumn. Thus, the post-breeding movements by the two species are 6 months out of phase, indicating that seasonal differences in ocean productivity in equatorial waters (highest in the boreal autumn) are unrelated to the low feeding rate during the pre-breeding migration. Faster prenuptial migration probably occurs because ample fat reserves have been obtained on the winter grounds (i.e., feeding is not required). In addition, higher wing loading (from higher fat reserves) facilitates faster flight. Thus, the seasonal differences in fat reserves among species of seabirds are not likely adaptations for migration *per se*, but instead, are important in the life histories of many because early arrival on the breeding grounds facilitates the acquisition of a higher quality nesting territory favorable for successful breeding and because the amount of time available for foraging after arrival at the colony is greatly reduced.

Morphological Adaptations

The distance that seabirds migrate is strongly related to morphology. The most important morphological feature is the shape of the wings, measured as the aspect ratio, a dimensionless value calculated as the wing span^2 divided by total wing area. Hence, birds with high aspect ratios have narrower wings. They also have less profile drag (i.e., less friction with the air), lower air turbulence, and generally migrate longer distances compared with birds with lower aspect ratios. For instance, in the Laridae many species of terns, and smaller gulls and skuas with long narrow wings are transequatorial migrants, whereas larger gulls and skuas with lower aspect ratios usually move much shorter distances or are even sedentary.

Wing loading (wing area divided by body mass) is also related to migration patterns in seabirds, although this relationship differs with flight styles used by different seabird taxa, and is also confounded with aspect ratio. Higher wing loading requires swifter flight for the birds to remain airborne. Within taxa of seabirds that typically use gliding or flap-gliding flight (e.g., albatrosses, shearwaters, and petrels) those with higher wing loading tend to have longer migrations. The gliding species with higher wing loading also tend to have higher aspect ratios (**Figure 1**), which increases their flight efficiency. Swift, energy-efficient flight equates to longer distances travelled; however, the gliding species are heavily dependent on wind energy for flight because of their flight mode and high wing loading. As a result, those with higher wing loadings are confined to higher latitudes where the wind is usually stronger, although some species with moderate to high wing loading do make transequatorial crossings.

In contrast to gliders, seabirds that use flapping flight, such as Larids, have an inverse relationship between wing loading and migration distance (**Figure 1**). This is undoubtedly related to the lower flight efficiency of flapping species with higher wing loading compared with those with lower wing loading, all other factors being equal. Another factor is the inverse relationship between wing loading and aspect ratio when comparing gliding versus flapping species (with the exception of Alcids – see below).

That is, aspect ratio decreases (and drag increases) with increases in wing loading among most seabird taxa that typically use flapping flight.

As noted above, Alcids (auks, auklets, murres, murrelets, and guillemots) are an exception to the wing loading versus aspect ratio relationship for species using flapping flight. Aspect ratios of Alcids increase with wing loading (**Figure 1**), i.e., a relationship similar to that of Procellariiformes (primarily gliders). However, unlike the Procellariiformes and Larids, there does not appear to be a relationship between wing loading, or aspect ratio, and migration tendency among Alcids. This may be because Alcid wings are highly adapted for underwater propulsion (i.e., similar to penguins) resulting in small wing sizes and the highest wing loading among avian species. Thus, their foraging ranges

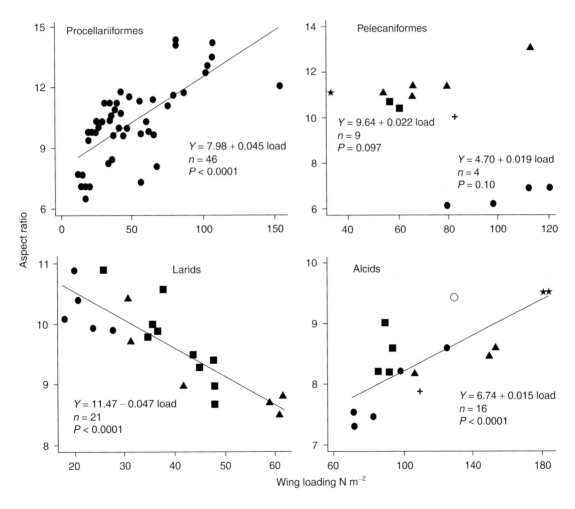

Figure 1 Relationship between wing loading and aspect ratio of four groups of seabirds, as indicated by data presented in 1997 by Spear and Ainley. Each point denotes the average for a given species. Lines indicate the best regression fit. Values of *n* are number of species, and *P* values indicate the level of significance by which slopes deviated from zero. Among the Pelecaniformes, the cormorants (●) were considered to be distinct from six species representing pelicans (+), tropic birds (■), boobies (▲), and frigate birds (★). Larids included terns (●), skuas (▲), and gulls (■). Alcids included auklets (●) murrelets (■), puffins (▲), pigeon guillemot (+), razorbill (○), and murres (★).

and nonbreeding movements are short. Indeed, many Alcid species are flightless during part of the dispersal period due to primary molt, indicating that a significant part of the dispersal is traversed by swimming.

Types of Migration

Several types of 'migration' are recognized. These include 'true migration' in which all members of a population move from a breeding area to a wintering area disjunct from the former; and 'partial migration', in which some members of a population migrate and others do not. Yet another type of movement, 'dispersal', is found in populations in which individuals move various distances after the breeding season, such that they occur at all distances within a given radius of the breeding site. Although a large proportion of the terrestrial avifauna exhibits true annual migration, this is rare among seabirds, probably because the marine environment is more stable in that it does not have the extreme seasonal warming and cooling of land masses. In addition, seabirds usually expand their at-sea ranges without having to cross as many barriers as are encountered during movements by terrestrial species. Even so, seabirds have the notoriety of having the longest distance migrants among the animal kingdom. The following is a review of the movement patterns of seabirds.

Sphenisciformes

Penguins (Spheniscidae) The pelagic ranges of the penguins are confined to the southern hemisphere. Relatively little is known of the post-breeding movements of these 17 flightless species. The four species of the genus *Spheniscus* have the lowest latitude distributions among penguins and probably move the shortest distances. Two, the Galapagos and Humboldt penguins (*S. mendiculus* and *S. humboldti*), are relatively sedentary along the coasts of the Galapagos Islands, and Peru to northern Chile, respectively. These species rarely are found more than 10 km from shore. The two other *Spheniscus* species usually stay within 50 km of the coasts of Africa (African penguin, *S. demerus*) and southern South America (Magellanic penguin, *S. magallenicus*), although there have been sightings of the latter to 250 km offshore. Satellite telemetry has shown that the Magellanic penguins breeding on Punta Tombo, Argentina, disperse for up to 4000 km northeastward to wintering areas off Brazil, although other population members winter along the coast of Argentina, closer to the colony. Three of the four telemetered Magellanic

penguins which moved to coastal Brazil traveled 20 km/day.

Dispersal by some of the species breeding in higher, temperature to subpolar, latitudes (king, *Aptenodytes patagonicus*; rockhopper, *Eudyptes chrysocome*; Snares Island, *E. robustus*; Fiordland crested, *E. pachyrhynchus*; erect-crested, *E. sclateri*; royal, *E. schlegeli*; macaroni, *E. chytsolophus*; Gentoo, *Pygoscelis papua*; yellow-eyed, *Megadyptes antipodes*; and little penguins, *Eudyptula minor*) is probably more extensive than is true for *Spheniscus*, as they are often seen hundreds of kilometers from shore. Finally, two of the three penguin species that breed only on the Antarctic continent (Adelie and chinstrap penguins, *Pygoscelis adeliae* and *P. antarctica*) migrate north to areas near or inside of the ice pack (Adelies; e.g., polynyas, leads) or open water (chinstraps) after waters near the continent become covered with solid ice. Movements of the third species, the Emperor penguin (*Aptenodytes forsteri*), are not well know; however, there are sighting records for this species from the coasts of Argentina and New Zealand.

Procellariiformes

Albatrosses (Diomedeidae) The albatrosses are under extensive taxonomic reclassification, and number between 14 and 24 species. Post-breeding movements of albatrosses are mostly longitudinal; no species undertakes transequatorial movements. These birds range long distances from their colonies during the nonbreeding period as well as during foraging trips undertaken while they are breeding. For example, six wandering albatrosses equipped with satellite transmitters flew 3660–15 200 km during single foraging trips after being relieved by their mates from incubation duties at the nest. Even breeding waved albatrosses (*Phoebastria irrorata*), with the smallest pelagic range among the group, have a round-trip commute of no less than 2000 km between the breeding colony on the Galapagos Islands and the nearest edge of their foraging area along the coast of Ecuador and Peru.

Most albatross species range farthest from their colonies during the nonbreeding period. In fact, many species are partially migratory (as opposed to being dispersers). As explained above, these species are considered as partially migratory because individuals are found in both the wintering and breeding areas during winter, but do not winter (or winter in small numbers) between the two locations. For example, Buller's (*Thalassarche bulleri*), Chatham (*T. eremita*), Salvin's (*T. salvini*), and shy (*T. cauta*) albatrosses breed on islands near New Zealand (the first three) and Australia (shy).

Although some birds stay near the colonies throughout the year, large proportions of the Buller's, Chatham, and Salvin's albatrosses migrate at least 8500 km eastward across the South Pacific to the coast of South America, and many shy albatrosses migrate westward across the Indian Ocean to the coast of South Africa.

Three other species, the wandering, black-browed (*T. melanophris*), and gray-headed (*T. crysostoma*) albatrosses, have breeding colonies located circumpolarly across southern latitudes near 50°S. The South Georgia populations may be partially migratory, although the distinction from dispersive is not clear. South Georgian wandering albatross fly north to waters off Argentina, and then eastward to important wintering areas off South Africa. Some continue to Australian waters and may even circumnavigate the Southern Ocean. One of several of these birds equipped with a satellite transmitter averaged 690 km/day. A large proportion (*c.* 85%) of South Georgian black-browed albatrosses also winter off South Africa, and many South Georgian gray-headed albatrosses are thought to fly westward to waters off the Pacific coast of Chile, and then to New Zealand.

The waved albatross is unique among the albatross group. Besides having the smallest pelagic range, it breeds near the Equator, i.e., at a latitude > 25° lower than any of the other species. Furthermore, the foraging area used while breeding is the same, or nearly the same, as that used post-breeding. Thus, the classification of this species as a 'disperser', or even as 'partially migratory', is appropriate only in that it leaves the colony post-breeding (and does not occupy the 900 km stretch between the Galapagos and the mainland), although the size of the foraging area changes little.

The post-breeding movements of yellow-nosed (*T. chlororhynchos*), sooty (*Phoebetria fusca*), light-mantled sooty (*P. palpebrata*), royal, black-footed (*Phoebastria nigripes*), short-tailed (*P. albatrus*), and Laysan albatrosses are less clear, although each apparently disperses post-breeding to seas adjacent to their colonies.

Fulmars, Shearwaters, Petrels, Prions, and Diving Petrels (Procellariidae) The migration tendencies of this family of 78 species is not well known, although those that have been studied are either partially migratory or dispersers. The pelagic range of 38 species (49%) is confined to the southern hemisphere, including the 18 species of fulmarine petrels and prions, three of the four species of diving petrels, seven (33%) of the shearwaters, and 10 (29%) of the gadfly petrels (**Table 1**). Another 19 (25%) of the 78 species perform extensive transequatorial

migrations. These include nine species (43%) of the shearwaters and 10 species (29%) of gadfly petrels. Twenty (26%) others are primarily tropical, including one species of diving petrel, five species (24%) of shearwaters, and 14 (40%) gadfly petrels. Many of the 'tropical' species disperse across the Equator, but like species having nontransequatorial movements, movements of these species usually have a greater longitudinal than latitudinal component, and are usually of shorter distances than those of transequatorial migrants.

The detailed migration/dispersal routes of most Procellariids are poorly known. Two exceptions are the sooty shearwater and Juan Fernandez petrels, which are very abundant and appear to be partial migrants. Two populations of sooty shearwaters exist, one breeding in New Zealand and the other in Chile. Many individuals from each population migrate to and from the North Pacific each year, and none winter in the equatorial Pacific. Observations of flight directions in the equatorial Pacific indicate that many complete a figure of eight route (*c.* 40 500 km). The route apparently involves easterly flight from New Zealand to the Peru Current in winter, northwesterly flight to the western North Pacific in spring, eastward movement to the eastern North Pacific during summer, and southwest flight to New Zealand during autumn (**Figure 2**). Most are probably nonbreeders, possibly from both the New Zealand and Chilean populations. Many, probably breeders from both populations, likely use shorter routes to and from the North Pacific (*c.* 28 000–29 000 km). Other (nonmigratory) individuals from both populations apparently stay in the southern hemisphere. Migration routes are coordinated with wind regimes in the Pacific, such that the usual flight direction utilizes quartering tail winds (**Figure 2**).

Juan Fernandez petrels breed in the Juan Fernandez archipelago off Chile. Many migrate into the North Pacific where they winter mostly between 5°N and 20°N. Another large component of the population stays in the South Pacific, mostly between 12°S and 35°S. Collections of specimens indicate that the great majority found in the North Pacific are subadults, whereas those in the South Pacific are predominantly adults.

It is likely that many other Procellariids also perform partial migration (e.g., Cook's, white-winged, black-winged, and mottled petrels), or even complete migrations (e.g., Cook's petrel, Hutton's shearwater, Magenta and Bonin petrels.

Storm Petrels (Oceanitidae) The 19 species of storm petrels include eight with nontransequatorial

Table 1 Migration tendencies (transequatorial, nontransequatorial, and tropical) of the 78 species of Procellariids and 19 species of Oceanitids[a]

PROCELLARIIDAE
FULMARINE PETRELS Nontransequatorial (12)
Northern giant petrel *Macronectes halli*
Southern giant petrel *M. giganteus*
Northern fulmar *Fulmarus glacialis*
Southern fulmar *F. glacialoides*
Antarctic petrel *Thalassoica antarctica*
Cape petrel *Daption capnse*

Snow petrel *Pagodroma nivea*
White-chinned petrel *Procellaria aequinoctialis*
Parkinson's petrel *P. parkinsoni*
Westland petrel *P. westlandica*
Grey petrel *P. cinerea*

GADFLY PETRELS
Nontransequatorial (10)
Great-winged petrel *Pterodroma macoptera*
Atlantic petrel *P. incerta*
Kerguelen petrel *P. brevirostris*
Magenta petrel *P. magentae*
Soft-plumaged petrel *P. mollis*
Barau's petrel *P. baraui*
White-headed petrel *P. lessoni*
Bonin petrel *P. hypoleuca*
Chatham petrel *P. axillaris*
Defilippe's petrel *P. defilippiana*

Tropical (14)
Phoenix petrel *P. alba*
Trinidad petrel *P. arminjoniana*
Herald petrel *P. heraldica*
Hendersons's petrel *P. atrata*
Tahiti petrel *P. rostrata*
Mascarene petrel *P. aterrima*
Bermuda petrel *P. cahow*
Black-capped petrel *P. hasitata*
Hawiian petrel *P. sandwichensis*
Galapagos petrel *P. phaeopygia*
White-winged petrel *P. leucoptera*
Collared petrel *P. brevipes*
Bulwer's petrel *Bulweria bulwerii*
Jouanin's petrel *B. fallax*

Transequatorial (10)
Mottled petrel *P. inexpectata*
Murphy's petrel *P. ultima*
Solander's petrel *P. solandri*
Kermadec petrel *P. neglecta*
Juan Fernandez petrel *P. externa*
White-necked petrel *P. cervicalis*
Cook's petrel *P. cooki*
Black-winged petrel *P. nigripennis*
Pycroft's petrel *P. pycrofti*
Stejneger's petrel *P. longirostris*

Unknown (1)
Macgillivray's petrel *P. macgillivrayi*

PRIONS Nontransequatorial (7)
Blue petrel *Halobaena caerulea*
Broad-billed prion *Pachyptila vittata*
Antarctic prion *P. desolata*
Salvin's prion *P. salvini*
Fairy prion *P. turtur*
Fulmar prion *P. crassirostris*
Slender-billed prion *P. belcheri*

DIVING PETRELS
Nontransequatorial (3)
Georgian diving petrel *Pelecanoides georgicus*
Common diving petrel *P. urinatrix*
Magellan diving petrel *P. magellani*

Tropical (1)
Peruvian diving petrel *P. garnoti*

SHEARWATERS (21 species)
Nontransequatorial (7)
Little shearwater *Puffinus assimilis*
Black-vented shearwater *P. opisthomelas*
Fluttering shearwater *P. gavia*
Hutton's shearwater *P. huttoni*
Heinroth's shearwater *P. heinrothi*
Balearic shearwater *P. mauretanicus*
Levantine shearwater *P. yelkouan*

Transequatorial (9)
Streaked shearwater *Calonectris leucomelas*
Cory's shearwater *C. diomedea*
Pink-footed shearwater *Puffinus creatopus*
Flesh-footed shearwater *P. carneipes*
Great shearwater *P. gravis*
Buller's shearwater *P. bulleri*
Sooty shearwater *P. griseus*
Short-tailed shearwater *P. tenuirostris*
Manx shearwater *P. puffinus*

Tropical (5)
Wedge-tailed shearwater *P. pacificus*
Christmas shearwater *P. nativitatis*
Newell's shearwater *P. newelli*
Townsend's shearwater *P. auricularis*
Audubon's shearwater *P. lherminieri*

Table 1 *Continued*

OCEANITIDAE
STORM PETRELS

Nontransequatorial (8)	Transequatorial (5)

Nontransequatorial (8)
Gray-backed storm petrel *Garrodia nereis*
White-faced storm petrel *Pelagodroma marina*
Black-bellied storm petrel *Fregetta tropica*
White-bellied storm petrel *F. grallaria*
Tristram's storm petrel *O. tristrami*
Swinhoe's storm petrel *O. monorhis*
Ashy storm petrel *O. homochroa*
Least storm petrel *O. microsoma*

Transequatorial (5)
Wilson's storm petrel *Oceanites oceanicus*
British storm petrel *Hydrobates pelagicus*
Leach's storm petrel *Oceanodroma leucorhoa*
Black storm petrel *O. melania*
Matsudaira's storm petrel *O. matsudairae*

Tropical (6)
Elliot's storm petrel *Oceanites gracilis*
White-throated storm petrol *Nesofregetta fuliginosa*
Wedge-rumped storm petrel *Oceanodroma tethys*
Harcourt's storm petrel *O. castro*
Markham's storm petrel *O. markhami*
Hornby's storm petrel *O. hornbyi*

[a]Tropical species are those in which most individuals stay between the tropic of Cancer/Capricorn.

movements, five transequatorial, and six that stay primarily in tropical latitudes. Storm petrels are mostly dispersive. Leach's and Wilson's storm petrels disperse the farthest and also are the most abundant, with circumpolar distributions. The former breeds mostly between 25°N and 50°N, and winters as far south as waters near New Zealand (about 35°S). Similarly, the Wilson's storm petrel breeds from about 45°S to 60°S, and winters north to about 40°N.

Two species of storm petrels that may be migratory, or partially migratory, include the white-faced and white-bellied storm petrels. Many white-faced storm petrels (race *maoriana*) that breed adjacent to New Zealand apparently migrate east to waters off Chile and Peru. In warm-water (El Niño) years, some birds continue westward from the Peru Current, out along the Equator, in association with waters of the South Equatorial Current. The *grallaria* race of white-bellied storm petrel is represented by a very small population breeding on islands north of New Zealand and Australia. This population is particularly interesting in that new information from the equatorial Pacific indicates that many or all of these birds migrate about 2500 km to a relatively small (1 million km²) section of waters between 135°W and 145°W and between 5°S and about 12°S, adjacent to the Marquesas.

Pelecaniformes

Pelicans (Pelecanidae, 4 marine species), boobies and gannets (Sulidae, 9 species), cormorants (Phalacrocoracidae; 29 marine species), frigate birds (Fregatidae; 5 species), and tropic birds (Phaethontidae, 3 species) Distributions of the marine species of pelicans and cormorants are coastal, whereas those of boobies, frigate birds, and tropic birds are pelagic. Movements of most Pelecaniformes are dispersive, and none of the nontropical species performs transequatorial migrations. Many species, including all boobies, frigate birds, and tropic birds, are tropical; most of the cormorants, pelicans, and gannets prefer temperate to polar latitudes. Movements of the tropical species are primarily nondirectional (i.e., direction can include combinations of north, south, east, and west orientation), whereas those of temperate to polar species usually have a stronger latitudinal component.

Charadriiformes

Phalaropes (Scolopacidae) Two of the three phalarope species breed in the northern hemisphere, either in the continental interiors (red-necked phalarope, *Phalaropus lobatus*) or Arctic slopes (red phalarope, *P. fulicarius*), and perform extensive migrations to marine habitat. Movements of a large proportion of both populations, particularly those of the red phalarope, are transequatorial, with many individuals wintering along the west coasts of South America and Africa. A major concentration also winters in the Panama Bight.

Skuas (Stercoracidae) Like the two phalaropes, the three smaller skuas (pomarine, *Stercorarius pomarinus*; Arctic, *S. parasiticus*; and long-tailed, *S. longicaudus*) breed on the Arctic slope and winter in oceanic waters. A large proportion of these birds also perform extensive transequatorial migrations, with large numbers wintering off the west coasts of South America and South Africa. Individuals are also occasionally seen off Australia, New Zealand, and in the Indian Ocean. Another large percentage winters off Mexico, Central America, and northern

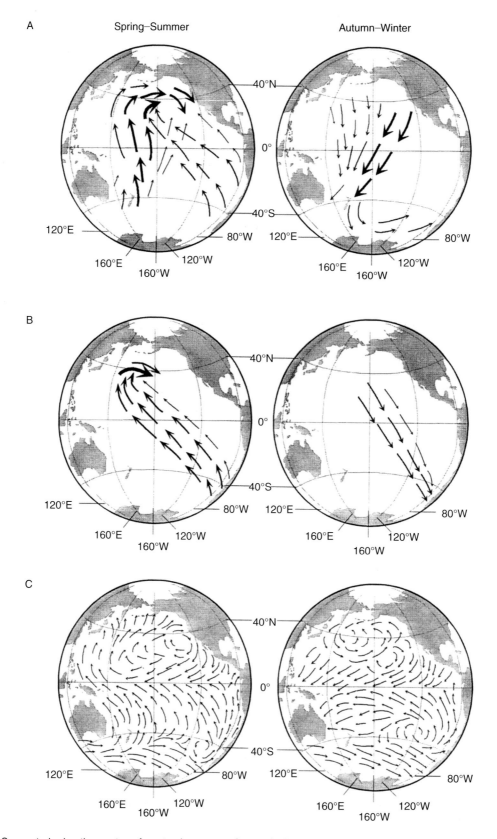

Figure 2 Suggested migration routes of sooty shearwaters from colonies near (A) New Zealand and Australia, and (B) Chile. Sizes of vectors reflect differences in the number of shearwaters, as suggested in 1974 by Shuntov from observations in the South and North Pacific, and in 1999 by Spear and Ainley, from observations in the equatorial Pacific. (C) Wind regimes of the Pacific Ocean during two seasonal periods as depicted in 1966 by Gentilli.

Table 2 Migration tendencies of the Larids (transequatorial, nontransequatorial, and tropical), including 48 species of gulls and 44 species of terns[a]

GULLS

Nontransequatorial (40)

Great black-backed gull *Larus marinus*
Western gull *L. occidentalis*
Yellow-footed gull *L. livens*
Herring gull *L. argentatus*
Kelp gull *L. dominicanus*
Glaucous-winged gull *L. glaucescens*
Glaucous gull *L. hyperboreus*
Slaty-backed gull *L. shistisagus*
Great black-headed gull *L. ichthyaetus*
Indian black-headed gull *L. brunnicephalus*
Chinese black-headed gull *L. saundersi*
Pacific gull *L. pacificus*
Band-tailed gull *L. belcheri*
Iceland gull *L. glaucoides*
Kumlien's gull *L. kumlieni*
Thayer's gull *L. thayeri*
California gull *L. californicus*
Black-tailed gull *L. crassirostris*
Sooty gull *L. hemprichii*
White-eyed gull *L. leucophthalmus*
Dolphin gull *L. scoresbii*
Common gull *L. canus*
Ring-billed gull *L. delewarensis*
Black-headed gull *L. ridibundus*
Laughing gull *L. atricilla*
Bonaparte's gull *L. philadelphia*

Relict gull *L. relictus*
Hartlaub's gull *L. hartlaubii*
Heermann's gull *L. heermanni*
Brown-hooded gull *L. maculipennis*
Silver gull *L. novaehollandiae*
Black-billed gull *L. bulleri*
Little gull *L. minutus*
Audouin's gull *L. audouinii*
Mediterranean gull *L. melanocephalus*
Slender-billed gull *L. genei*
Black-legged kittiwake *Rissa tridactyla*
Red-legged kittiwake *R. brevirostris*
Ross's gull *Rhodostethia rosea*
Ivory gull *Pagophila eburnean*

Transequatorial (3)

Lesser black-backed gull *L. fuscus*
Franklin's gull *L. pipixcan*
Sabine's gull *L. sabini*

Tropical (5)

Lava gull *L. fuliginosis*
Gray gull *L. modestus*
Gray-headed gull *L. cirrocephalus*
Andean gull *L. serranus*
Swallow-tailed gull *L. furcatus*

TERNS

Nontransequatorial (14)

Caspian tern *Sterna caspia*
South American tern *S. hirundinacea*
Antarctic tern *S. vittata*
Kerguelen tern *S. virgata*
Forster's tern *S. forsteri*
Trudeau's tern *S. trudeaui*
Roseate tern *S. dougalii*
White-fronted tern *S. striata*
Aleutian tern *S. aleutica*
Fairy tern *S. nereis*
Black-fronted tern *S. albostriata*
Damara tern *S. balaenarum*
Little tern *S. albifrons*
Least tern *S. antillarum*

Transequatorial (8)

Sandwich tern *S. sandvicensis*
Common tern *S. hirundo*
Arctic tern *S. paradisaea*
Royal tern *S. maxima*
Elegant tern *S. elegans*
Black tern *Childonias niger*
Whiskered tern *C. hybridus*
White-winged tern *C. leucopterus*

Tropical (22)

Large-billed tern *S. simplex*
Gull-billed tern *S. nilotica*
Indian River tern *S. aurantia*
White-cheeked tern *S. repressa*
Black-napped tern *S. sumatrana*
Black-billed tern *S. meganogastra*
Gray-backed tern *S. lunata*
Bridled tern *S. anaethetus*
Sooty tern *S. furcata*
Amazon tern *S. superciliaris*
Peruvian tern *S. lorata*
Crested tern *S. bergii*
Lesser-crested tern *S. bengalensis*
Chinese crested tern *S. bernsteini*
Cayenne tern *S. eurygnatha*
Saunder's little tern *S. saundersi*
Blue-gray noddy *procelsterna cerulea*
Brown noddy *Anos stolidus*
Black noddy *A. minutus*
Lesser noddy *A. tenuirostris*
Inca tern *Larosterna inca*
White tern *Gygis alba*

[a]Tropical species are those in which most individuals stay between tropic of Cancer/Capricorn.

Africa. Finally, a minority, primarily adults, stay in temperate to subpolar latitudes of the northern hemisphere during winter.

The four larger skuas (great, *Catharacta skua*; brown, *C. lonnbergi*; South Polar, *C. maccormicki*; and Chilean, *C. chilensis*) generally perform only short, dispersive post-breeding movements. An exception is the South Polar skua, many of which disperse widely from their Antarctic breeding sites, even into the more northern latitudes (e.g., Alaska) of the northern hemisphere.

Gulls and Terns (Laridae) The 48 gull species (subfamily, Larinae) are mostly dispersive, although for some species the post-breeding dispersal of some birds can extend for thousands of kilometres. As noted above, the larger species, with higher wing loading, tend to move shorter distances post-breeding than do smaller species. The five migratory species include the Franklin's, Sabine's, Thayer's, lesser black-backed, and California gulls. Of these, the movements of three (6% of the 48 species) are transequatorial (**Table 2**). In fact, these three species are the only ones with regular transequatorial movements among the gulls that have nontropical breeding grounds. Of the remaining species, the range of 40 (83%) is confined to one hemisphere or the other, and that of five others (10%) is confined within tropical latitudes.

Of the five migrants, only the Sabine's and Thayer's gulls breed in Arctic latitudes. Interestingly, other species that also breed in the Arctic, including the Ivory, Ross's, glaucous, and Iceland gulls remain there, or disperse relatively short distances to subArctic latitudes, during winter. It is likely that different foraging habitats or prey requirements are responsible for these differences in movement patterns.

The post-breeding, dispersive movements of larger gull species have been studied in detail. These studies indicate that breeding adults leave the colony and fly quickly to specific locations, such as a particular bay or fishing port. The locations, 'vacation spots', are those with which they have become familiar during their subadult years (first 4 years of life), such that the birds know the foraging logistics and availability of a predictable food supply. This is important. After reaching the vacation location the adults must molt and replace the primary wing feathers (making them less mobile for about a month) and replenish body reserves in preparation for the next breeding season.

The 44 species of terns (subfamily, Sterninae) are mostly smaller than gulls and have higher aspect ratios (i.e., longer, narrower wings). Not surprisingly, this group is represented by a higher proportion of species that migrate, including eight species (18%) whose migrations are transequatorial, compared with 14 species (32%) whose movements are mostly confined to one hemisphere, and 21 species (50%) that remain primarily within tropical latitudes.

Arctic terns are unique because they have the longest migrations known among the world's animal species. These terns breed in the Arctic to 80°N, and winter in the Southern Ocean to 75°S. Based on band returns and observations at sea, some Arctic

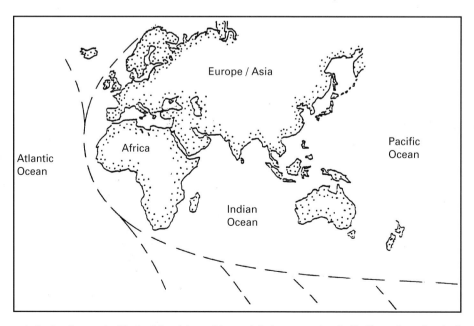

Figure 3 Suggested migration route (dashed lines) to-and-from wintering areas by Arctic Terns breeding in Scandanavia and eastern Canada, from band returns summarized in 1983 by Mead, and from at-sea observations in the Indian Ocean reported in 1996 by Stahl et al.

terns from Scandinavia and eastern Canada fly to and from waters off Australia, New Zealand, and the Pacific (**Figure 3**). The shortest, round-trip flight to the Pacific exceeds 50 000 km. These birds can live up to 25 years, indicating that the lifetime migration distance could exceed 1 million km.

Murres, murrelets, auks, auklets, puffins, and guillemots (Alcidae) The 22 species of Alcids, confined to the northern hemisphere, have dispersive post-breeding movements. Compared with other seabirds, their dispersal distances are short (see Morphological Adaptations and Flight Behavior). The primary reasons are: (1) their very high wing loading and, thus, inefficient flight; (2) they are highly adapted pursuit divers that can exploit a range of subsurface habitats; and (3) they occur in waters of the Arctic and boundary currents where prey are abundant. In summary, long distance flights by Alcids are impractical, and are not required. This life history trait is like that of penguins, another group highly adapted for pursuit diving, but is in marked contrast to the movements of other seabirds with poorer diving abilities. Alcids with the longest distance dispersal are the little auk (*Alle alle*), tufted puffin (*Fratercula cirrhata*), horned puffin (*F. corniculata*), Atlantic puffin (*F. arctica*), and parakeet auklet (*Cyclorrhynchus psittacula*). Some individuals representing these species disperse up to 1000 km or more into the pelagic waters of the North Atlantic and North Pacific.

See also

Alcidae. Laridae, Sternidae and Rynchopidae. Pelecaniformes. Phalaropes. Procellariiformes. Seabird Conservation. Seabird Foraging Ecology.

Seabird Overview. Seabird Population Dynamics. Seabird Reproductive Ecology. Seabird Responses to Climate Change. Seabirds and Fisheries Interactions. Seabirds as Indicators of Ocean Pollution. Sphenisciformes.

Further Reading

Able KP (1995) Orientation and navigation: A perspective of fifty years of research. *Condor* 97: 592–604.

Berthold P (1993) *Bird Migration: A General Survey.* Oxford: Oxford University Press.

Gentilli J (1966) Wind principles. *In*: Fairbridge RW (ed) *The Encyclopedia of Oceanography*, pp. 989–993, New York: Reinhold.

Harrison P (1983) *Seabirds, an Identification Guide.* Boston, MA: Houghton-Mifflin.

Mead C (1983) *Bird Migration.* New York: Facts On File, Inc.

Pennycuick CJ (1989) *Bird Flight Performance.* New York: Oxford University Press.

Prince PA, Croxall JP, Trathan PN and Wood AG (1998) The pelagic distribution of albatrosses and their relationships with fisheries. In: Robertson G and Gales R (eds) *Albatross Biology and Conservation*, pp. 137–167. Chipping Norton, Australia: Surrey Beatty & Sons.

Shuntov VP (1974) *Seabirds and the Biological Structure of the Ocean.* [Translated from Russian.] Springfield, VA: US Department of Commerce.

Spear LB and Ainley DG (1997) Flight behaviour of seabirds in relation to wind direction and wing morphology. *Ibis* 139: 221–233.

Spear LB and Ainley DG (1999) Migration routes of Sooty Shearwaters in the Pacific Ocean. *Condor* 101: 205–218.

Stahl JC, Bartle JA, Jouventin, P, Roux JP and Weimerskirch H (1996) *Atlas of Seabird Distribution in the South-west Indian Ocean.* Villiers en Bois, France: Centre National de la Recherche Scientifique.

SEABIRD OVERVIEW

G. L. Hunt, University of California, Irvine, CA, USA

doi:10.1006/rwos.2001.0226

Seabirds or marine birds are species that make their living from the ocean. Of the approximately 9700 species of birds in the world, about 300–350 are considered seabirds. The definition as to what constitutes a seabird differs among authors, but generally includes the penguins (Sphenisciformes), petrels and albatrosses (Procellariiformes), pelicans, boobies and cormorants (Pelecaniformes), and the gulls, terns and auks (Lariformes) (**Table 1**). Sometimes included are loons (Gaviiformes), grebes (Podicipediformes), and those ducks that forage at sea throughout the year or during the winter (Anseriformes). Bird species that are restricted to obtaining their prey by wading along the margins of the sea, such as herons or sandpipers, are not included.

The distribution of types of seabirds shows striking differences between the northern and southern hemispheres, particularly at high latitudes. Best known are the restrictions of the auks (Alcidae) to

Table 1 Distribution and species richness of families of seabirds

Common name of family	Family	Total number of species	Number of species nesting south of 30°S	Number of species nesting between 30°S and 30°N	Number of species nesting north of 30°N	Period of sea use	Flight type (flapping, flap-gliding or soaring)	Foraging region (primarily neritic or oceanic)
Penguins	Spheniscidae	18	16	2	0	Year-round	Wing-propelled swimming	Mostly neritic
Loons or divers	Gaviidae	5	0	0	5	Migration and winter	Flapping	Neritic
Grebes	Podicipedidae	21	12	13	8	Migration and winter	Flapping	Neritic
Albatrosses	Diomedeidae	17	14	3	3	Year-round	Soaring	Oceanic
Petrels and shearwaters	Procellariidae	66	44	20	15	Year-round	Flap-gliding	Mostly oceanic
Storm petrels	Hydrobatidae	20	9	11	10	Year-round	Flap-gliding	Mostly oceanic
Diving petrels	Pelecanoididae	4	4	1	0	Year-round	Flapping	Mostly neritic
Frigate birds	Fregatidae	5	2	5	2	Year-round	Soaring	Mostly oceanic
Tropic birds	Phaethontidae	3	2	3	3	Year-round	Flapping, with some soaring	Oceanic
Gannets and boobies	Sulidae	9	6	7	5	Year-round	Flapping, with some gliding	Both neritic and oceanic
Pelicans	Pelecanidae	8	4	7	5	Most species fresh water	Flapping, and some gliding	Neritic
Cormorants	Phalacrocoracidae	28	20	17	10	Most species year-round	Flapping	Neritic
Sea ducks	Anatidae (part)	19	5	0	11	Migration and winter	Flapping	Neritic
Phalaropes	Scolopacidae	3	0	0	3	Migration and winter	Flapping	Neritic
Skuas, gulls, terns, noddies and skimmers	Laridae	100	36	49	59	Most species year-round	Flapping	Mostly neritic; some tropical terns and noddies oceanic
Auks	Alcidae	23	0	2	22	Year-round	Flapping	Neritic

(Adapted with permission from Harrison 1983.)

the northern hemisphere, and the penguins (Spheniscidae), diving petrels (Pelecanoididae), and most species of albatrosses (Diomedeidae) to the southern oceans. There are also many more species of shearwaters and petrels (Procellariidae) that nest in the southern hemisphere than in the north.

The cost of flight varies greatly among seabird species. Some groups, such as diving petrels and auks, have heavy wing loading and require flapping flight that is relatively expensive. Others have low wing-loading and are able to make use of wind energy and soar extensively, flapping only occasionally. These species, including albatrosses and many of the petrels, can travel relatively long distances with minimal energy expenditure. Between these extremes are birds that alternate flapping flight with soaring or gliding. These species-specific differences in cost of flight probably had profound effects on the types of birds that were able to survive in differ-

ent climate domains of the oceans. Seabirds in the southern hemisphere tend to be efficient fliers that cover large areas in search of food by using the wind to enhance their soaring flight. These birds may range up to 5000 km from their colonies in search of prey. In contrast, in the northern hemisphere, most species of seabirds depend on relatively expensive flapping flight, and forage within 100 km or less of their colonies. Costs of flight have also clearly had a strong effect on the types of foraging techniques and chick-provisioning routines used by different groups of seabirds.

During the breeding season, most seabirds nest in dense colonies on predator-free islands or on inaccessible cliffs. A few nest at low densities or as scattered individuals. All seabirds provision their young at the nest, and feed them at intervals from once every few hours to up to 15 days, depending upon the species. Because of their need to periodically return to their nests with food, seabirds are good examples of central place foragers.

The life history patterns of seabirds include long life spans, delayed reproduction, and small numbers of young produced in any one year. Interannual survival for adult birds, where estimated, is generally thought to be in the order of 90–95% or better. In contrast, survival through the first winter after fledging may be below 50%. The age of first reproduction varies between species as a function of expected life span. Most seabirds do not commence breeding until they are at least 3 or 4 years old, and some of the albatrosses delay the year of first breeding until they are 15 years of age. Clutch size is small. Most oceanic foragers lay only one egg per year. Some species of neritic foragers lay clutches of two or three eggs, with cormorants laying clutches of six or more eggs. Clutch sizes for species nesting in productive upwelling zones may be larger than is typical for their taxonomic relatives in less productive areas.

For most species of seabirds, reproductive rates are low and, except for those species adapted to upwelling regions, there is little opportunity to increase reproductive output in years of high prey abundance. In some ocean regions, years with successful reproduction are the exception. In those cases, it is likely that a few good years and a small percentage of particularly able parents may account for the majority of the young produced. Reproductive success improves with experience, and for many species mate retention from one year to the next is high, with diminished reproductive success associated with the changing of mates. Thus, high survival rates for adult birds are critical to maintain stable populations.

Seabirds obtain their prey by a wide variety of methods including pursuing fish and plankton at depths as great as 300 m or more, seizing prey on the wing from the surface of the ocean, and stealing prey from others (**Figure 1**). Prey most commonly taken by seabirds includes small fish and squid, and large zooplankton such as euphausiids and amphipods. A few seabird species specialize on copepods. The use of gelatinous zooplankton is known for albatrosses, petrels, and auks. Gelatinous

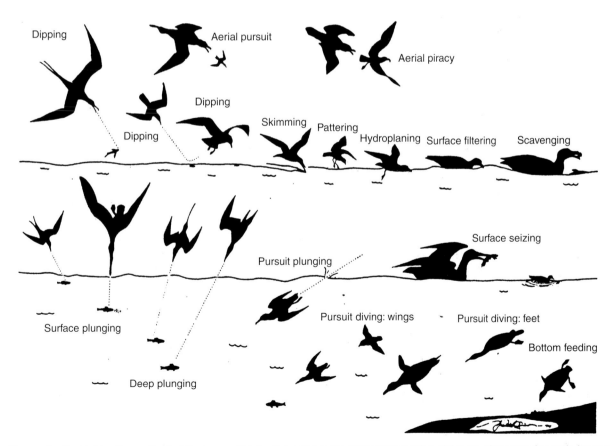

Figure 1 Seabird feeding methods. The types of birds silhouetted in the drawing from left to right: *top*, skua pursuing a phalarope; skua pursuing a gull; *second row*, frigate bird, noddy, gull, skimmer, storm petrel, prion, Cape petrel, giant petrel; *third row*, tern, pelican, tropic bird, gannet, albatross, phalarope; *underwater*, shearwater, murre, diving petrel, penguin, cormorant, a sea duck. (Reproduced with permission from Ashmole, 1971.)

zooplankton may be more commonly consumed by other species than is appreciated, as remains of gelatinous species are difficult to detect in stomachs or preserved food samples.

Foraging methods used by seabirds vary with the density of prey and may also vary with water clarity. Plunge diving for prey and surface seizing of prey are particularly common in clear tropical waters. In more turbid and prey-rich polar and subpolar waters, many seabirds pursue their prey beneath the sea. In the Southern Ocean, where continental shelf regions are deep and limited in area, most seabirds other than penguins are efficient fliers that can cover vast areas in search of patchy prey. In the northern hemisphere, most species of seabirds forage over broad, shallow continental shelves where interactions between currents and bathymetry result in predictable concentrations of prey. These prime foraging locations are often close to shore, where tidal currents create convergence fronts and other physical features that force aggregations of zooplankton upon which small fish forage. Frontal systems associated with shelf edges are important in both hemispheres.

Marine birds are assumed to play only an insignificant role in oceanic carbon cycling. However, where their consumption has been examined, they have been found to take significant amounts of prey. In the southeastern Bering Sea, seabirds have been estimated to consume between $0.12\,g\,C\,m^{-2}\,a^{-1}$ and $0.29\,g\,C\,m^{-2}\,a^{-1}$ depending on the region of the shelf and the year. Since the seabirds are foraging between the second and third trophic levels, this consumption accounts for between 12 and $29\,g\,C\,m^{-2}\,a^{-1}$ of the $200–400\,g\,C\,m^{-2}\,a^{-1}$ of primary production over the shelf. Where estimates have been made of the proportion of secondary production consumed, seabirds have been found to take between 5 and 30% of local secondary production (**Table 2**). In the North Sea, seabird consumption of sand eels is in the order of 197 000 tonnes. Over half of this consumption is concentrated near the Shetland Islands. However, seabird consumption of sand eels in the North Sea as a whole is small compared with the overall production of sand eels or the take by the industrial fishery which focuses on offshore banks. Recent calculations of the fraction of total exploitable stocks in the eastern Bering Sea that are consumed by seabirds suggest that about 3% of walleye pollock and 1% of herring are taken by birds. Recent shifts in the diets of shearwaters in the eastern Bering Sea could mean that the estimate for pollock is low by a factor of two or more. In the North Pacific, depending on the region of concern, estimates of prey consumption by marine birds vary between $0.01\,t\,km^{-2}$ and $1.72\,t\,km^{-2}$ for the summer months of June, July and August.

The effects of fisheries on seabirds are almost always greater than the effect of birds on fisheries. Seabirds have been used by fishers as indicators of the presence of large predatory fish. Commercial fishing activity frequently provides offal and discards to seabirds, and a number of seabird populations have benefited from this supplemental source of food. Indeed, juvenile albatrosses in the North Pacific may be experiencing increased winter survival rates thanks to the availability of discards and offal from commercial fisheries. However, seabirds attracted to fishing vessels do not distinguish between discarded fish and baited hooks. Between 1990 and 1994, > 30 000 Laysan albatrosses and

Table 2 Community energetics models of fish consumption by seabirds (modified from Hunt et al., 1996[a])

Location	Estimated % pelagic fish production consumed	Major consumers	Major prey species
Oregon Coast	22	Shearwaters, storm petrels, cormorants, murre	Northern anchovy, juvenile hake
Foula, Shetland Islands	29	Fulmar, murre, shag, puffin	Sand eels
North Sea	5–8	Fulmar, gulls, terns, murre, puffin	Sand eels
North Sea	5–10	Fulmar, gannet, shag, gulls, kittiwake, terns, razorbill, murre, puffin	Sand eels
Saldanha Bay, South Africa	29	Penguin, gannet, cormorant	Pilchard
Benguela Region	6	Gannet, cormorant	Pilchard
Vancouver Island, British Columbia[b]	11, 17, and 21	Shearwater, murre	Juvenile herring

[a]Hunt GL, Barrett RT, Joiris C and Montevecchi W (1996) Seabird/fish interactions: an introduction. *ICES Coop. Res. Rep.* 216: 2–5.

[b]Logerwell EA and Hargraves NB (1997) *Seabird Impacts on Forage Fish: Population and Behavioral Interactions.* Proceedings of Forage Fishes in Marine Ecosystems. Alaska Sea Grant College Program AK-SG-97-01.

20 000 black-footed albatrosses were killed on long-lines off Hawaii. In the southern hemisphere, up to 20 000 shy albatrosses and 10 000 wandering albatrosses are entangled annually in the blue-fin tuna long-line fishery. Approximately 10% of the world population of wandering albatrosses is killed annually. A loss of this magnitude cannot be sustained in a species that has a life history strategy that depends on extraordinarily high annual rates of adult survival. Not surprisingly, populations of 6 of 17 species of albatross are now declining rapidly.

Seabirds have proven to be useful indicators of changes in marine ecosystems. Because of their position at the top of marine food chains, they tend to accumulate a number of pollutants. For example, DDE, which concentrates in lipids, can result in eggshell thinning in some species. Eggshell thinning in various pelican and cormorant populations was one of the first indicators that DDE was present in certain coastal waters. Likewise, recent work has shown seabirds bio-accumulate mercury and other heavy metals. Seabirds also provide information about the status of populations of prey on which they depend. Changes in prey abundance are reflected by changes in annual rates of production of young, or in extreme cases, changes in seabird population size. Climate-driven changes in prey abundance, at scales from years to hundreds of years, are reflected by changes in seabird distribution, abundance, and reproductive output. Although it has been a goal of a number of studies, the ability to estimate the standing stocks of prey populations based on indices derived from seabirds has yet to be accomplished. Calibration of the responses of the seabirds to shifts in prey abundance has been difficult to achieve.

The most serious threats to the conservation of seabirds are those that result in the deaths of adult birds, particularly those individuals that have been successful breeders. Seabird populations can usually recover from a single instance of mortality, but chronic elevated rates of mortality are devastating. Seabirds are extremely vulnerable to predation in their colonies. Thus, the presence of foxes, cats, rats, and other introduced predators on islands where seabirds nest is of great concern. Past introductions of predators have resulted in the extirpation of nesting birds, and the removal of predators has resulted in their return. Also of great concern is the annual drowning of high numbers of seabirds in gill nets and on long-lines. The situation with respect to albatrosses and other large procellariiform birds is of particular concern, as these birds require long adult life spans to insure the production of sufficient young to maintain stable populations. Oil spills can have devastating local effects, but if the resulting pollution is not long-lasting, populations are likely to recover. Chronic pollution by oil or other chemicals may have both lethal and sublethal effects, and can damage populations of seabirds over time. Competition for resources such as nesting space or food is occasionally of concern. Development of islands and beaches affects a few populations of seabirds. Likewise there are a few instances where fisheries may be competing with seabirds for particular size-classes or species of prey. However, since most fisheries target large predatory fish and discard offal and small fishes, many of which would otherwise not have been available to seabirds, it is unclear how widespread competitive interactions with fisheries may be.

See also

Alcidae. Laridae, Sternidae and Rynchopidae. Pelecaniformes. Phalaropes. Procellariiformes. Seabird Conservation. Seabird Migration. Seabird Reproductive Ecology. Seabird Responses to Climate Change. Seabirds and Fisheries Interactions. Seabirds as Indicators of Ocean Pollution. Sphenisciformes.

Further Reading

Ashmole NP (1971) Seabird ecology and the marine environment. In: Farner DS and King JR (eds) *Avian Biology*, vol. 1, pp. 224–286. New York: Academic Press.

Croxall JP (1987) *Seabirds: Feeding Ecology and Role in Marine Ecosystems*. Cambridge: Cambridge University Press.

Enticott J and Tipling D (1997) *Seabirds of the World*. Mechanicsburg, PA: Stackpole Books.

Furness RW and Monaghan P (1987) *Seabird Ecology*. London: Blackie Books.

Harrison P (1983) *Seabirds: An Identification Guide*. Boston: Houghton Mifflin.

Hunt GL and Nettleship DN (1988) *Seabirds of High-latitude Northern and Southern Environments*. In: Oulette H (ed.) Proceedings of the XIX International Ornithological Congress 1986, pp. 1143–1155. Ottawa: University of Ottawa Press.

Hunt GL and Schneider D (1987) Scale dependent processes in the physical and biological environment of marine birds. In: Croxall JC (ed.) *Seabirds: Feeding Biology and Role in Marine Ecosystems*, pp. 7–41. Cambridge: Cambridge University Press.

Hunt GL, Mehlum F, Russell RW *et al.* (1999) *Physical Processes, Prey Abundance, and the Foraging Ecology of Seabirds*. In: Adams NJ and Slotow R (eds) Proceedings of the 22nd International Ornithological Congress, Durban, pp. 2040–2056. Johannesburg: BirdLife South Africa.

Murphy RC (1936) *Oceanic Birds of South America*, vols 1 and 2. Macmillan, New York: American Museum of Natural History.

Nelson JB (1978) *The Sulidae: Gannets and Boobies.* Oxford: Oxford University Press.

Pennycuick CJ (1989) *Bird Flight Performance. A Practical Calculation Manual.* Oxford: Oxford University Press.

Schneider DC, Hunt GL, Jr and Harrison NM (1986) Mass and energy transfer to seabirds in the southeastern Bering Sea. *Continental Shelf Research* 5: 241–257.

Warham J (1990) *The Petrels: Their Ecology and Breeding Systems.* London: Academic Press.

Warham J (1996) *The Behaviour, Population Biology and Physiology of the Petrels.* London: Academic Press.

Whittow GC and Rahn H (1984) *Seabird Energetics.* New York: Plenum Press.

SEABIRD POPULATION DYNAMICS

G. L. Hunt, University of California, Irvine, CA, USA

doi:10.1006/rwos.2001.0234

The population biology of seabirds is characterized by delayed breeding, low reproductive rates, and long life spans. During the breeding season, the distribution of seabirds is clumped around breeding colonies, whereas when not breeding, birds are more dispersed. These population traits have important consequences for interactions between people and seabirds. The aggregation of large portions of the adult population of a species in colonies means that a single catastrophic event, such as an oil spill, can kill a large segment of the local breeding population. Although seabird populations can withstand the failure to produce young in one or even a few years without suffering severe population-level consequences, the loss of adults has an immediate and long-lasting impact on population dynamics. Even a small decrease in adult survival rates may cause population decline.

Seabirds breed in colonies on islands, cliffs, and other places where they are protected from attacks by terrestrial predators. Species that forage at large distances from their colonies usually choose locations for their colonies that are less vulnerable to incursions by predators than are the colonies of species that forage in the immediate vicinity of the colony. For the offshore species, the cost of increased travel to a more protected site may be minor compared with the benefit of freedom from unwanted visitors. In contrast, for species that need to forage close to their colonies, even a short increase in the distance traveled between colony and foraging site may mean that it is uneconomical to occupy a particular breeding site.

Colony size tends to vary with the distance that a species travels in search of food. Inshore-foraging seabirds may nest singly or in small groups, whereas species that forage far at sea may have colonies that are comprised of hundreds of thousands of pairs. Two hypotheses have been offered to explain this trend. One hypothesis focuses on the issue of food availability. If birds forage far from their colonies, there is a much greater area in which food may be encountered than if foraging is restricted to a small radius around the colony, and thus a larger size colony can be supported. This hypothesis assumes that seabird colony size is limited by food availability. For species that forage near their colonies, there is evidence that reproductive parameters sensitive to prey availability, such as chick growth rates and fledging success, vary negatively with colony size. Likewise, there is evidence that colony size and location may be sensitive to the size and location of neighboring colonies. Evidence that seabirds depress prey populations near their colonies is limited. The second hypothesis focuses on the role that colonies may play in the process of information acquisition by birds seeking prey. When birds forage far from their colony, there may be a need for large numbers of birds so that those flying out from the colony are able to observe successful returning foragers and thereby work their way to productive foraging areas using the stream of birds returning to the colony for guidance. The longer the distance from the colony, the greater the number of birds that are required to provide an unbroken stream of birds to guide the out-bound individuals. The evolution of a system of this sort is possible because each individual will benefit from information on food resources gained by being part of a large colony. Selection for large colony size will continue so long as the colony is not so large that food supplies are severely depressed.

Seabirds show considerable philopatry, with individuals often returning to the same colony, or even the same part of the colony from which they fledged. Once a nest site or territory is established, individuals and pairs may use the same site in subsequent years. Pairing tends to be for multiple seasons ('for life'), particularly when pairs are successful at raising young. Divorce does occur, and may be most frequent after failure to raise young. However, experience is important, including experience with a particular partner, so changing partners may result in a period of adjustment to a new partner and consequent lower reproductive success.

The philopatry of seabirds and their tendency to remain in the same part of a colony once they are established may have important implications for the genetic structure of seabird populations. Few data are presently available, but there is some evidence for closer genetic ties for individuals nesting in close proximity. If this is proven to be generally true, there would be important ramifications for understanding a genetic basis for the evolution of coloniality based on inclusive fitness and the reciprocal aid of relatives.

Thus far, there is scant information regarding the genetic distance between individuals in different colonies. The issue of whether colonies are discrete populations (stocks in fisheries parlance) or whether there is considerable exchange between colonies has important implications not only for the evolution of local variation, but also for the conservation and management of seabirds. Little information is available as to whether some colonies are net exporters of recruits (sources) and others net importers (sinks), or if when a colony is decreased in size by a catastrophe unrelated to food resources, it will be quickly replenished by recruits from other colonies.

Delay in the age at which breeding commences is a striking characteristic of seabird population biology. Most species do not begin breeding until they are in their third or fourth year, and some groups, such as the albatrosses, delay breeding until they are 10 or more years of age. Birds that commence breeding at a younger age than is usual for their species have reduced reproductive success in their first year of breeding when compared with individuals that wait to breed at an older age. Additionally, birds that commence breeding at an early age tend to have an elevated mortality rate in the first year of breeding. The delay in the commencement of breeding may be a reflection of the long period needed for young birds to acquire the skills for efficient foraging. Several studies have shown that sub-adult birds are less efficient foragers, and have a lower success rate in capturing prey than do adult birds foraging in the same area. Additionally, sub-adult birds may visit the colony where they will breed for one or more years before they make their first attempt at breeding. This time may be necessary for learning where prey is to be found in the vicinity of the colony. The delay of breeding is

particularly great in the Procellariiform birds, and these birds forage over vast areas of the ocean. It may be particularly challenging for them to learn where prey may be most predictably located within the potentially huge foraging arena available to them. The delay in the commencement of breeding may also provide time for birds in newly forming pairs to learn each others behavioral rhythms so that they will be more effective parents, although evidence to test this hypothesis is lacking. Certainly, coordination between the members of a pair is an essential ingredient of successful reproduction.

Most species of marine birds lay small clutches of eggs, with typical clutch size being one or two eggs. There is a tendency for species that forage far from their colonies to have smaller clutches (one egg) than those which forage inshore near the colony (two or three eggs). Likewise, species that live and forage in areas of strong upwelling tend to have considerably larger clutch sizes (three or four eggs) than closely related species that forage in mid-ocean regions. Two factors may come into play here. First, birds that forage at great distances from their colonies may be unable to transport sufficient prey sufficiently quickly to raise more than one offspring. In many of these species, individual parents may visit the colony to provision their young at intervals from 1 to 10 days or more. In some species of Procellariiformes, adults alternate long and short foraging trips while provisioning their young. After short trips, chicks gain mass, but adults lose body mass, whereas, after long trips, adults have gained mass, although the chicks will have not benefited as greatly as after short trips. These results point to the constraints on the ability of these birds to raise larger broods.

Many species that forage close to their colonies (inshore-foraging species) raise larger broods than offshore-foraging species. Parent birds of the inshore-foraging species usually make multiple provisioning visits to the colony during a day. Although they may not be able to carry more food per trip than similar-sized species that forage offshore, the possibility of multiple trips allows sufficiently high rates of food delivery to permit supplying the needs of more than one offspring. Multiple-chick broods are common in pelicans, cormorants, gulls, and some species of terns, and are also found in the most-inshore-feeding alcids.

Species breeding in upwelling systems and subject to the boom or bust economy of an ecosystem with extreme interannual variation in productivity may be a special case. In these species, periodic declines in ecosystem production, as may occur in El Niño years off Peru or California, can result not only in reproductive failures, but also considerable mortality of adults. In these cases, the potential for high reproductive output in the years when prey is plentiful may be necessary to offset adult mortality rates that may be higher than is typical for most seabird species.

For the offshore-foraging species, annual fluctuations in reproductive output may be small, particularly when compared with inshore-foraging species. This low variability may be a reflection of the wide expanse of ocean over which they can search for food. For the inshore species, there may be many years in which no young are fledged and only a few years in which they are successful at fledging young. This interannual variation may reflect the dependence of inshore-foraging species on localized upwelling and other forms of physical forcing of prey patches, which may show considerable interannual variation in the amount of prey present. Thus, the 'good' years become disproportionately important for the success of the local population. Additionally, there is limited evidence that suggests that a few individuals within a seabird population may account for a large proportion of the young produced. The implications of these findings, if further work shows them to be generally true, will be considerable for management efforts toward the conservation of seabirds. Loss of the most productive individuals or disruption of breeding in one of the rare years in which food resources are sufficient to lead to strong production of young will have a disproportionate impact on population stability. However, identifying the most productive individuals, or predicting which years will be critical for good reproductive output will be a challenge.

The third component of seabird life histories that must accompany delayed commencement of reproduction and low rates of reproduction is an extended period of survival during which reproduction is possible. The life spans of seabirds are among the longest of any birds. Survival to the age of 20 years or more is probably common in most species, and the larger species of Procellariids are known to live in excess of 40 years. Knowledge of the actual life spans of seabirds is difficult to obtain, as most of the marking devices used until recently have had much shorter life spans than the birds whose life spans were being measured. The result is a considerable underestimate of the expected life spans of seabirds, and thus of their possible life time reproductive potential. Despite these problems, estimates of adult survival rates of 92–95% are the norm. Thus, even a small decrease in the rate of adult survival will have a proportionally large impact on mortality rates.

Survival of recently fledged young and juveniles is considerably lower than that of adults. The highest rates of mortality are most likely to occur in the first few months of independence, when young are learning to fend for themselves. Mortality is also high during the first winter, again probably due to experience. Possibly as many as 50% of fledglings fail to survive to their first spring. Survival rates for juveniles are higher, but there may be an increase in mortality as birds begin to enter the breeding population and encounter the aggression of neighbors and the increased energetic demands of caring for young.

There are conflicting hypotheses about what limits the size of seabird populations, although in most cases it is believed that food rather than predation or disease is the most likely limiting factor. Phillip Ashmole has argued that seabirds are likely to be most stressed for food during the breeding season, when large numbers of individuals are concentrated in the vicinity of breeding colonies. In contrast, during winter, seabirds are spread over vast reaches of ocean and are not tied to specific colonies and their associated foraging areas. During winter, seabirds should be free to move about and take advantage of prey, wherever it may be found. As discussed above, there is a modest suite of data that suggests that prey availability may limit colony size and reproductive output of breeding seabirds. There are even fewer data to test the hypothesis favored by David Lack, that seabird populations are limited by wintertime food supplies and survival rates. Few studies of seabirds have focused on winter ecology during winter conditions, although this is the time when 'wrecks' (masses of dead birds driven ashore) are most common. Marine birds are sensitive to ocean conditions, with foraging success being reduced during periods of high winds. It is likely that in winter, even if there is no change in the amount of prey present, prey may be harder to obtain. Additional information on winter food stress comes from species that perform transequatorial migrations, thus allowing them to winter in the summer of the opposite hemisphere. Even under these presumably benign conditions, Southern Hemisphere shearwaters wintering in the North Pacific Ocean and Bering Sea experience occasional episodes of mass mortality, apparently from starvation. There is increasing evidence that decreases in the productivity of the California Current system are causing declines in the numbers of both migrant shearwaters and locally breeding species. The question as to which is the most stressful season for seabirds remains to be resolved.

Because of the sensitivity of seabird population dynamics to changes in adult survival rates, any factor that increases adult mortality is potentially detrimental to the conservation of seabirds. Thus, the loss of adult birds in fishery bycatch is of great concern. Breeding adults caught in gill nets and/or on long lines result in the loss not only of the adult, but also the chick for which it was caring. The loss of a breeding adult may also result in lower subsequent reproductive output by the surviving parent because it will likely have lower reproductive success during the first year with a new partner. The group most vulnerable to bycatch on long lines appears to be the Procellariiformes, which make shallow dives to grab baited hooks as they enter or leave the water. These are amongst the longest-lived of seabirds, and the curtailment of their breeding lives has a severe impact on their populations. Indeed, the populations of many species of albatross are declining at an alarming rate. A second major threat to seabirds is the presence of introduced predators on the islands where the birds breed. Rats, cats, foxes, and even snakes kill both chicks and attending adults. Again, loss of adult breeding birds has the most potentially serious impact on the future stability of the population. Reduction of anthropogenic sources of adult mortality in seabirds must be one of the most urgent imperatives for conservation biologists and managers.

See also

Alcidae. Procellariiformes. Seabird Conservation. Seabird Reproductive Ecology. Seabird Responses to Climate Change. Seabirds and Fisheries Interactions.

Further Reading

Ainley DG and Boekelheide RJ (1990) *Seabirds of the Farallon Islands: Ecology, Dynamics, and Structure of an Upwelling-system Community*. Stanford, California: Stanford University Press.

Ashmole NP (1963) The regulation of numbers of tropical oceanic birds. *Ibis* 103b: 458–473.

Ashmole NP (1971) Seabird ecology and the marine environment. In: Farner DS and King JR (eds) *Avian Biology*, Vol. 1, pp. 224–286. New York: Academic Press.

Furness RW and Monaghan P (1987) *Seabird Ecology*. London: Blackie Books.

Gaston AJ and Jones IL (1998). *The Auks: Family Alcidae*. Oxford: Oxford University Press.

Hunt GL Jr (1980) Mate selection and mating systems in seabirds. In: Burger J, Olla BL and Winn HE (eds) *Behavior of Marine Animals*, Vol. 4, pp. 113–151. New York: Plenum Press.

Lack D (1966) *Population Studies of Birds*. Oxford: Clarendon Press.

Lack D (1968) *Ecological Adaptations for Breeding in Birds*. London: Chapman and Hall.

Nelson JB (1978) *The Sulidae: Gannets and Boobies*. Oxford: Oxford University Press.

Warham J (1990) *The Petrels: Their Ecology and Breeding Systems*. London: Academic Press.

Warham J (1996) *The Behaviour, Population Biology and Physiology of the Petrels*. London: Academic Press.

Williams TD (1995) *The Penguins*. Oxford: Oxford University Press.

Wittenberger JF and Hunt GL Jr (1985) The adaptive significance of coloniality in birds. In: Farner DS, King JF and Parkes KC. *Avian Biology*, Vol. VIII. Orlando, FL: Academic Press.

Wooller RD, Bradley JS, Skira IJ and Serventy DL (1989) Short-tailed shearwater. In: Newton I (ed.) *Lifetime Reproduction in Birds*. London: Academic Press.

SEABIRD REPRODUCTIVE ECOLOGY

L. S. Davis and R. J. Cuthbert, University of Otago, Dunedin, New Zealand

doi:10.1006/rwos.2001.0239

Introduction

Finding the food necessary to produce and raise offspring is a fundamental problem that animals face. The oceans, despite being a productive and rich environment, rarely provide a steady or reliable source of food; instead, feeding opportunities are patchily distributed in both time and space. As a consequence, seabirds, from 20 g least petrels (*Halocyptena microsoma*) up to 37 kg emperor penguins (*Aptenodytes patagonicus*), have had their life history strategies shaped by the need to cope with this ebb and flow of resources.

The British ornithologist David Lack published two landmark books in 1954 and 1968 that influenced the way that we think about reproduction in birds. Lack hypothesized that the clutch size of birds has evolved so that it represents the maximum number of chicks that can be reared. His logic suggested that food is most constraining when parents have chicks (i.e., the period when they need to both feed themselves and provide enough food for their chicks to grow) and this must inevitably limit how many chicks parents can care for adequately. It was an attractive idea, linking reproductive effort to food supply. For marine birds, which by definition must feed in a different place from where they breed, this seemed particularly relevant.

However, more recent theoretical developments suggest that it is not so simple. Most seabirds are long-lived and, in the face of many potential breeding opportunities over a lifetime, there are likely to be trade-offs between reproductive effort and adult survival. The advantages of investing in a reproductive attempt must be weighed against the costs in terms of the breeder's survival and future reproductive potential. In essence, natural selection acts as the scales. The birds do not need to make conscious choices: over time, those animals with the best strategies for balancing survival and reproduction – known as life history strategies – will leave more offspring and the genes controlling their behavior will become more prevalent.

Inshore and Offshore Strategies

Seabirds, by the nature of their prey and their requirements for a solid substrate upon which to lay their eggs, are forced to have a dichotomy between the place where they breed and the place where they feed. As such, the distance of their feeding grounds from the colony will potentially have major implications for their reproductive ecology. David Lack noted that seabirds can be divided into two broad categories: inshore foragers and offshore foragers. The amount of food that can be brought back to the nest by those birds that must travel a long way to reach their prey will be limited by transport costs and logistics. As a consequence, Lack observed that offshore foragers tend to have lower rates of reproduction and take longer to fledge their chicks (**Table 1**). To offset this, offshore foragers tend to be longer-lived, with higher rates of annual survival.

Lack saw inshore and offshore foraging as strategies to deal with a suite of issues that seabirds must face with respect to breeding: at what age to begin breeding, the best time to breed, whether to breed or not, how much to invest in breeding, and whether to abandon a breeding attempt. One way or another, the solutions all involve a relationship

between the reproductive behavior of the birds and their feeding ecology.

At What Age to Begin Breeding?

To maximize lifetime reproductive output in a Darwinian world it might seem obvious that animals should begin breeding as soon as they are able to do so. Paradoxically, this is not always the case, and especially for seabirds. In some species, at least, breeding exacts a toll on the breeder and this cost tends to be disproportionately high for young, inexperienced breeders. Relatively few (17%) Adelie penguins (*Pygoscelis adeliae*) that begin breeding as 4-year-olds survive to see their 12th birthday, whereas those that do not start breeding until 7 years of age almost invariably get to 12 years and beyond. Female fulmars (*Fulmarus glacialis*) that put great effort into breeding in their first few years lead shorter lives and overall produce fewer offspring than those that do not do so.

In addition, young breeders also tend to be less successful in their breeding attempts. Natural selection, rather than simply maximizing the number of offspring produced, gives a competitive edge to those strategies that maximize the number of offspring produced over a lifetime that go on to breed successfully themselves. Just partaking in a breeding attempt will not be advantageous unless the offspring produced are likely to survive and breed. Fulmars that begin breeding later than the modal age for first-time breeders are much more likely to fledge offspring at their first attempt.

Potential young breeders, then, are faced with a double dose of reality: they are less likely to be successful and they are likely to reduce the number of future breeding opportunities available to them. Consequently, many sea birds delay the onset of breeding until well after they are physiologically capable of breeding. For example, fulmars reach adult size at one year of age, but usually they do not begin breeding until males are 8 years old and females are 12 years old. Gray-headed albatross (*Diomedea chrysostoma*) wait even longer and on average begin breeding when they are 13 years old.

Of course, such wait-and-see strategies only make sense in long-lived birds where they can be reasonably certain that they will live to breed another day. In situations where life spans are short and/or interannual survival rates are low (e.g., owing to annual migrations), birds should be inclined to breed at younger ages when the opportunity arises. Common diving-petrels (*Pelecanoides urinatrix*), the most pelagic of all the diving-petrels, begin breeding as 2-year-olds. Mean age of first breeding of

Table 1 Life-history characteristics of inshore and offshore foraging seabirds

Characteristics	Inshore	Offshore
Clutch size	2–3	1
Breeding attempts in season	1–2	1
Relay eggs	Yes	No
Annual survival	High	Very high

penguins (excluding crested penguins) is correlated with their annual survival rates: those species with the shortest expected life spans begin breeding earlier. (The crested penguins are a law unto themselves, not initiating breeding until 7 or 8 years of age, even though their annual survival rates are not high by penguin standards, suggesting that there are other costs to breeding in these species that mitigate the potential advantage to be gained from breeding earlier.)

When Is the Best Time to Breed?

As well as the age at which birds begin breeding, natural selection also influences the timing of breeding within each season. In an environment where resources fluctuate greatly, it will be highly advantageous to breed when resources are abundant. David Lack suggested that, as the chick-rearing period is the period of greatest food demand, breeding should be initiated so that the chick-rearing period coincides with the period when food is most abundant. Evidence for this comes from many seabird studies, which show that those chicks that hatch late in the breeding season are less likely to fledge, or, if they do, they will be smaller and in poorer condition (which will detrimentally affect their survival prospects). (While prey availability is one factor that has shaped the timing of breeding, other factors are also likely to be important. Amongst alcids, for example, late breeders are more susceptible to predation.)

Selection for the timing of breeding to coincide with peaks in seasonal food abundance has resulted in some species of sea birds being highly synchronous in their laying. In roseate terns (*Sterna dougallii*) and short-tailed shearwaters (*Puffinus tenuirostris*) almost all eggs in a colony are laid within the space of 2–3 days. In the short-tailed shearwater, not only is the laying date highly synchronous, but it is also very constant. In this species, breeding off Tasmania, the mean laying date of 25–26 November has not altered in over 100 years, a fact that is well known by the local people who harvest the eggs for food.

The timing of breeding is most likely to be important for those sea birds breeding in high latitudes, where food supply is more variable owing to the more pronounced seasonality and where other environmental factors – such as the brief break-up of pack ice in the polar summer – are often critical for successful breeding. In contrast, sea birds breeding in the tropics experience a more stable environment with a more constant and steady food supply. The lack of clearly defined seasons in the tropics

has enabled some species to forsake the seasonal concept altogether. Audubon's shearwater (*Puffinus lherminieri*) and the Christmas shearwater (*Puffinus nativitatis*) have no defined breeding season and there is little synchrony in laying. One breeding cycle takes about 6 months and, after a 2-month postnuptial molt, pairs begin breeding again.

Whether to Breed or Not?

Individuals face an important decision at the start of every season: whether to breed or not. If a bird is in poor condition or if feeding resources are poor, then a strategy of skipping a breeding attempt, rather than risking its survival and prospects of breeding in subsequent years, may be more profitable.

The occurrence of nonbreeding or 'sabbatical' years was first recognized in the 1930s, and has since been found to occur commonly in many groups of sea birds, including cormorants, gulls, terns, petrels, and penguins. In a detailed 15-year study on Cory's shearwater (*Calonectris diomedea*), it was found that on average around 10% of the breeding population failed to breed in any given year and, while most birds were absent for just a single year, some birds took sabbaticals of up to 7 years' duration. While there are presently few comparative data on the rate at which seabirds skip breeding attempts, the strategy of taking sabbaticals is likely to be most beneficial in long-lived birds. The exact mechanism by which birds 'decide' whether to breed is unknown, but it is likely to involve a physiological response to the body condition of the bird reaching some minimum threshold. As body condition will be influenced by the availability of food, it can essentially function as an indicator for resource abundance.

Other seabirds are constrained from breeding in every year simply because of the duration of their breeding cycle. Among these are some of the largest representatives of both the flying and flightless sea birds; the wandering and royal albatrosses and the king penguin. (The largest of the penguin species, the emperor penguin, gets around the difficulty of managing to fledge a very large chick and still maintain an annual breeding cycle, by initiating breeding during the heart of the Antarctic winter.) Breeding in these species – from the arrival of adults at the very start of the season to the fledging of the chick – takes around 14 months. Hence, successful breeders (those that fledge chicks) are able to breed only in every other year (albatross) or twice in every three years (king penguins).

How Much to Invest in Breeding?

As there will be trade-offs between reproductive effort and adult survival in their effects on lifetime reproductive success, it behoves long-lived species not to invest too much in any given breeding attempt. Resources should be invested in ways that maximize lifetime reproductive success.

For many species of sea bird, evidence suggests that the body condition of individuals (usually measured as the mass of a bird after scaling for body size) is the proximate factor that regulates the level of investment of resources. Experimental and observational studies have shown that individuals will continue to feed a chick or incubate an egg until they reach a low threshold of body condition. At this point, birds will then either desert the egg or feed the chick at a lower rate so as to maintain their own body condition and enhance their survival prospects. Evidence for this comes from the Antarctic petrel (*Thalassoica antarctica*), a long-lived offshore forager. In an experiment where the costs of provisioning the chick were raised, the extra costs were passed onto the chick (by feeding the chick smaller meals and less often) rather than onto the parents (whose body condition remained constant). Relatively few comparable studies have been performed on inshore foraging seabirds but, because of their lower life expectancy and faster rate of reproduction, they could be expected to more readily deplete their body condition (at the risk of their own survival) to ensure the success of a breeding attempt.

While supplying chicks with enough food may be a constraint for many species of seabird, increasing evidence suggests that the availability of food at other periods of the season – courtship, egg production, and incubation – may also affect the level of resources invested into a breeding attempt. In lesser black-backed gulls (*Larus fuscus*), the body condition of females during the prelaying period is an important factor determining both the number of eggs laid and the size of eggs that are produced (egg size is related to the hatching and survival prospects of chicks). A similar relationship between body condition and egg size is found in Hutton's shearwater (*Puffinus huttoni*) (**Figure 1**), which breeds at altitudes of 1200–1800 m in the mountains of New Zealand. In this species the number of available breeding burrows is limited and adults must regularly fly into the breeding colonies to compete for burrows. As a result of this activity, parents lose body condition during the courtship period, and this subsequently affects the amount of resources that females can invest into the egg.

Figure 1 Hutton's shearwaters (*Puffinus huttoni*) exemplify the dichotomy that seabirds face in their need to breed on land and forage on distant marine resources. In this species the breeding colonies are located inland at 1200–1800 m in the Seaward Kaikoura Mountains, South Island, New Zealand. (Photo: R. Cuthbert.)

Whether to Abandon a Breeding Attempt?

Having embarked on a breeding attempt, it need not pay a parent to persist with it come hell or high water. Natural selection is not a romantic. Reproductive strategies that enhance lifetime reproductive output will be favored, even if they involve the abandonment of eggs or cute and fluffy chicks.

Internal development of offspring, similar to that in mammals, was never going to be a likely evolutionary option for flying birds, where there were always advantages to be gained from eschewing weight. But the laying of eggs gives birds the opportunity to adjust their reproductive investment early in a breeding attempt, before the costs incurred are too high. When will it pay parents, then, to terminate investment?

When Resources are Low

Parental investment theory predicts that parents should be more inclined to terminate investment when resources are low, especially if the survival or future reproductive potential of the parent is threatened by persisting with a breeding opportunity in the face of inadequate resources. Seabirds are typically monogamous and have eggs that require a long period of incubation. This necessitates both parents sharing the incubation duties, usually with one bird sitting on the nest while the other is at sea feeding. The sitting bird is unable to eat and, inevitably, there must be a limit to how long it can sustain itself from its fat reserves. Eventually an incubating bird will abandon the nest and eggs.

It is difficult to quantify the proportion of nests lost because of desertion by an unrelieved incubating bird. Frequent surveillance is required to establish causes of loss (very often researchers are faced with just an empty nest, making it hard to ascribe a cause of loss with any certainty) and, compounding this, disturbance caused by frequent surveillance can itself precipitate desertions. Nevertheless, it is clear from those detailed studies that have been carried out on Procellariiformes (petrels and albatross) and Sphenisciformes (penguins), that desertions by unrelieved birds can constitute a major form of egg loss. This is particularly so in poor food years, when increases in the duration of foraging trips are associated with a greater likelihood that the foraging bird's partner will have abandoned the nest by the time it returns.

Early in a Breeding Attempt

Parental investment theory also predicts that parents should be more inclined to abandon their investment early in a breeding attempt (when reproductive costs are low). It is not at all clear how well this prediction is supported, or even how relevant it is for seabirds. In a broad sense, abandonment of the nest may be more likely to occur during incubation rather than during chick rearing in some seabirds; but drawing any definitive support for the prediction from this is complicated because a chick offers a very different stimulus from that of an egg and, as chicks need to be fed, nest reliefs are typically more frequent during chick rearing, reducing the probability that the attending bird will exhaust its energy reserves. Further, although parental investment theory predicts that the sex that has invested least in a breeding attempt should be the one most inclined to terminate investment (because it has the least to lose), this is not always the case (e.g., often at the time of desertion in Adelie penguins (*Pygoscelis adeliae*), it is the abandoning bird that has invested the most). When there is only one breeding opportunity each year, the costs of termination are potentially very high and should not be undertaken

lightly, so seabirds often tend to stick at it until the point where their own survival comes under threat. Hence, ecological constraints, reflected in food supply and foraging times, dictate when seabirds are likely to abandon breeding opportunities rather than levels of relative expenditure.

When Young

Theoretically, young breeders have less to lose from terminating investment in a breeding attempt, because they have more future reproductive opportunities ahead of them than do older breeders. While hazard functions, which display the risk to eggs of being abandoned by parents, do show a heightened risk of abandonment from young breeders in species such as Adelie penguins, it is difficult to separate out the effects of experience from this. That is: Are young breeders more inclined to abandon the breeding attempt because of a lack of experience or because of a strategy to enhance lifetime reproductive output?

There is also controversy about whether the level of investment by seabirds increases with age. One classic study of California gulls (*Larus californicus*) has revealed that the reproductive success of older breeders is greater than for other breeders. The authors put this down to increased effort on the part of the older parents as they had fewer future breeding opportunities. While this explanation sits nicely with parental investment theory, it is also possible that the higher than average breeding success of older breeders simply reflects the effect of experience.

In kittiwakes (*Rissa tridactyla*), reproductive success improves with experience and it has been argued that, even from a theoretical perspective, differences in age-related investment should only occur if annual survival decreases with age. The problem with this last argument is twofold: (1) given a finite maximum life span, even if annual survival rates do not change dramatically until very near the end, birds will have fewer potential breeding opportunities the closer they get to that maximum, and (2) it is very difficult to detect changes in annual survival rates in old birds because there are relatively few studies with the longitudinal data needed to follow annual survival of long-lived seabirds and, even where they exist, sample sizes involving the very oldest birds tend to be very low.

In fulmars, breeding success increases with experience, but only until their 10th year of breeding. Thereafter it remains constant. Data from a 20-year study of little penguins (*Eudyptula minor*) reveals that breeding performance increases with breeding experience up to the 7th year of breeding, but declines from then on. What is emerging from many of these long-term studies of seabirds is that the quality of individual breeders may override theoretical arguments about age or experience. That is, certain 'high-quality' breeders consistently rear more offspring successfully, they are able to do so more readily in the face of environmental perturbations, and they have better survival. If this is generally the case, as evidenced by kittiwakes, yellow-eyed penguins (*Megadyptes antipodes*) and shags (*Phalacrocorax aristotelis*), then it could also explain the apparent higher productivity of older breeders: that is, those surviving to an older age may be the most successful breeders anyway. Certainly, the one factor that consistently emerges from these studies as explaining most variation in lifetime reproductive output is breeding life span: you have to live to breed.

Inshore and Offshore Strategies Revisited

While it is useful to think of sea bird species as either inshore or offshore foragers, it is important to note that there is a continuum between these extremes. Moreover, studies on several albatross and petrels have shown that some species adopt a strategy of mixing long and short foraging trips that enables them to exploit both inshore and offshore resources. For example, sooty shearwaters (*Puffinus griseus*) breeding on the Snares Islands south of New Zealand make several short and frequent feeding trips, which are good for enhancing chick growth. However, parents use up body reserves during short trips and, when their condition reaches a certain threshold, they embark on a long foraging trip to Antarctic waters that enables them to replenish their own body condition.

Rearing only one chick is a characteristic of offshore foragers, but some researchers have claimed that they are not working to full capacity and could potentially feed more than one chick. Twinning experiments carried out on gannets (*Sula* sp.) and Manx shearwaters (*Puffinus puffinus*) show that, at least in some cases, offshore foragers are capable of providing enough food to rear two chicks. However, these experiments fail to take into account limits at other stages of the breeding cycle and, in particular, impacts on the future reproductive opportunities of the parents.

It is also important to note that the concept of being an inshore or offshore forager has little relevance for some species of seabirds. For example, some skuas (Stercorariidae) and frigate-birds (Fregatidae) obtain much of their food through

Table 2　Reproductive tendencies of inshore and offshore foraging seabirds

Breeding decision	Inshore	Offshore
Age at which to begin breeding	Younger	Older
Best time to breed	Not so critical	Critical
Whether to breed	More inclined	Less inclined
How much to invest	More	Less
Whether to abandon	Less inclined	More inclined

chasing and harassing individuals of other seabird species until they drop or regurgitate their prey (a behavior known as kleptoparasitism), and giant petrels (*Macronectes giganteus*) specialize in scavenging on carcasses.

These species are the exceptions, however, and for most sea birds the concept of an inshore–offshore foraging continuum provides a useful framework for assessing their reproductive tendencies (Table 2).

Coloniality

A discussion of reproductive ecology in seabirds would not be complete without mention of coloniality. Ninety-eight percent of seabirds nest in colonies: groups of nests that can range from a few to a few hundred thousand or more. It has been suggested that seabirds nest together because of habitat constraints, but there are many species where areas suitable for breeding are not limiting and yet the birds are still colonial. Other hypotheses suggest that the colony benefits reproductive behavior through social stimulation, enhanced opportunities for extra-pair matings, allowing assessment of the suitability of a habitat for breeding, and reducing losses due to predation. However, many authors maintain that the main advantage of coloniality is the enhanced foraging efficiency that it promotes. A colony may act as an information centre, whereby birds in the colony learn the location of food by observing or following successful foragers. Alternatively, it may be more efficient to forage in groups and recruitment of members to a foraging group may be facilitated by living together.

Whatever the advantages of colonial living, given the widespread occurrence of colonial breeding in seabirds, they would appear to apply equally well to both inshore and offshore foragers.

Global Effects on Reproductive Strategies

Finally, when considering reproductive strategies of seabirds, large-scale effects need to be borne in mind. Marine resources vary spatially and temporally. While seasonal patterns occur with a certain level of predictability, environmental stochasticity means that from the breeding seabird's perspective there will be good years and poor years. Population viability analyses reveal that populations of sea birds can be particularly susceptible to catastrophic events. One such cause of reduced breeding performance can be El Niño and La Niña events, which often result in a reduction of available prey or necessitate a switch to poorer-quality alternative prey.

Such environmental uncertainty could be expected to impact upon reproductive life histories. Increases in foraging distances and conservation of breeding life span may be expected. That is, selection should favor offshore foraging strategies that enlarge the potential feeding zone and maximize life span to take advantage of as many good years as possible. Hence, in determining the efficacy of offshore foragers in particular, data from single-season studies are unlikely to be sufficient. Conversely, inshore foragers, which tend to rely on a locally abundant and predictable food supply, are the species most likely to be affected by human activities (e.g., emission of greenhouse gases) that increase environmental stochasticity.

Conclusions

The breeding of seabirds is dictated by life history strategies that have evolved principally in response to food supply. In that sense, David Lack was correct. However, the situation is more complex than he envisaged. At opposite ends of a continuum there are two broad strategies adopted by seabirds: the James Bond strategy of the inshore foragers (breed fast, die young, and have a good-looking corpse) and the more conservative strategy of the offshore foragers that seeks to maximize lifetime reproductive output by withholding and adjusting investment as necessary to maximize breeding life span. As we learn more about seabirds, we discover that their reproductive behavior is not fixed but can be relatively plastic, adjusting just where they are on that continuum in response to environmental conditions and food availability.

See also

Laridae, Sternidae and Rynchopidae. Seabird Foraging Ecology. Seabird Overview.

Further Reading

Furness RW and Monaghan P (1987) *Seabird Ecology*. Glasgow: Blackie.

Lack D (1954) *The Natural Regulation of Animal Numbers*. Oxford: Clarendon Press.

Lack D (1968) *Ecological Adaptations for Breeding in Birds*. Oxford: Clarendon Press.

Stearns SC (1992) *The Evolution of Life Histories*. Oxford: Oxford University Press.

Trivers RL (1972) Parental investment and sexual selection. In: Campbell B (ed.) *Sexual Selection and the Descent of Man 1871–1971*, pp. 136–179. Chicago: Aldine.

SEABIRD RESPONSES TO CLIMATE CHANGE

David G. Ainley, H.T. Harvey and Associates, San Jose, CA, USA

G. J. Divoky, University of Alaska, Fairbanks, AK, USA

Copyright © 2001 Academic Press

doi:10.1006/rwos.2001.0237

Introduction

This article reviews examples showing how seabirds have responded to changes in atmospheric and marine climate. Direct and indirect responses take the form of expansions or contractions of range; increases or decreases in populations or densities within existing ranges; and changes in annual cycle, i.e., timing of reproduction. Direct responses are those related to environmental factors that affect the physical suitability of a habitat, e.g., warmer or colder temperatures exceeding the physiological tolerances of a given species. Other factors that can affect seabirds directly include: presence/absence of sea ice, temperature, rain and snowfall rates, wind, and sea level. Indirect responses are those mediated through the availability or abundance of resources such as food or nest sites, both of which are also affected by climate change.

Seabird response to climate change may be most apparent in polar regions and eastern boundary currents, where cooler waters exist in the place of the warm waters that otherwise would be present. In analyses of terrestrial systems, where data are in much greater supply than marine systems, it has been found that range expansion to higher (cooler but warming) latitudes has been far more common than retraction from lower latitudes, leading to speculation that cool margins might be more immediately responsive to thermal variation than warm margins. This pattern is evident among sea birds, too. During periods of changing climate, alteration of air temperatures is most immediate and rapid at high latitudes due to patterns of atmospheric circulation. Additionally, the seasonal ice and snow cover characteristic of polar regions responds with great sensitivity to changes in air temperatures. Changes in atmospheric circulation also affect eastern boundary currents because such currents exist only because of wind-induced upwelling.

Seabird response to climate change, especially in eastern boundary currents but true elsewhere, appears to be mediated often by El Niño or La Niña. In other words, change is expressed stepwise, each step coinciding with one of these major, short-term climatic perturbations. Intensive studies of seabird populations have been conducted, with a few exceptions, only since the 1970s; and studies of seabird responses to El Niño and La Niña, although having a longer history in the Peruvian Current upwelling system, have become commonplace elsewhere only since the 1980s. Therefore, our knowledge of sea-bird responses to climate change, a long-term process, is in its infancy. The problem is exacerbated by the long generation time of seabirds, which is 15–70 years depending on species.

Evidence of Sea-bird Response to Prehistoric Climate Change

Reviewed here are well-documented cases in which currently extant seabird species have responded to climate change during the Pleistocene and Holocene (last 3 million years, i.e., the period during which humans have existed).

Southern Ocean

Presently, 98% of Antarctica is ice covered, and only 5% of the coastline is ice free. During the Last Glacial Maximum (LGM: 19 000 BP), marking the end of the Pleistocene and beginning of the Holocene, even less ice-free terrain existed as the ice sheets grew outward to the continental shelf break and their mass pushed the continent downward. Most likely, land-nesting penguins (Antarctic genus *Pygoscelis*) could not have nested on the Antarctic continent, or at best at just a few localities (e.g., Cape Adare, northernmost Victoria Land). With warming, loss of mass and subsequent retreat of the ice, the continent emerged.

The marine-based West Antarctic Ice Sheet (WAIS) may have begun to retreat first, followed by the land-based East Antarctic Ice Sheet (EAIS). Many Adélie penguin colonies now exist on the raised beaches remaining from successive periods of rapid ice retreat. Carbon-dated bones from the oldest beaches indicate that Adélie penguins colonized sites soon after they were exposed. In the Ross Sea, the WAIS receded south-eastward from the shelf break to its present position near Ross Island approximately 6200 BP. Penguin remains from Cape Bird, Ross Island (southwestern Ross Sea), date back to 7070 ± 180 BP; those from the adjacent southern Victoria Land coast (Terra Nova Bay) date to 7505 ± 230 BP. Adélie penguin remains at capes Royds and Barne (Ross Island), which are closest to the ice-sheet front, date back to 500 BP and 375 BP, respectively. The near-coast Windmill Islands, Indian Ocean sector of Antarctica, were covered by the EAIS during the LGM. The first islands were deglaciated about 8000 BP, and the last about 5500 BP. Penguin material from the latter was dated back to 4280–4530 BP, with evidence for occupation 500–1000 years earlier. Therefore, as in Victoria Land, soon after the sea and land were free from glaciers, Adélie penguins established colonies.

The study of raised beaches at Terra Nova Bay also investigated colony extinction. In that area several colonies were occupied 3905–4930 BP, but not since. The period of occupancy, called 'the penguin optimum' by geologists, corresponds to one of a warmer climate than at present. Currently, this section of Victoria Land is mostly sea-ice bound and penguins nest only at Terra Nova Bay owing to a small, persistent polynya (open-water area in the sea ice).

A study that investigated four extinct colonies of chinstrap penguin in the northern part of the Antarctic Peninsula confirmed the rapidity with which colonies can be founded or deserted due to fluctuations in environmental conditions. The colonies were dated at about 240–440 BP. The chinstrap penguin, an open-water loving species, occupied these former colonies during infrequent warmer periods indicated in glacial ice cores from the region. Sea ice is now too extensive for this species offshore of these colonies. Likewise, abandoned Adélie penguin nesting areas in this region were occupied during the Little Ice Age (AD 1500–1850), but since have been abandoned as sea ice has dissipated in recent years (see below).

South-east Atlantic

A well-documented avifaunal change from the Pleistocene to Recent times is based on bone deposits at Ascension and St Helena Islands. During glacial maxima, winds were stronger and upwelling of cold water was more pronounced in the region. This pushed the 23°C surface isotherm north of St Helena, thus accounting for the cool-water seabird avifauna that was present then. Included were some extinct species, as well as the still extant sooty shearwater and white-throated storm petrel. As the glacial period passed, the waters around St Helena became warmer, thereby encouraging a warm-water avifauna similar to that which exists today at Ascension Island; the cool-water group died out or decreased substantially in representation. Now, a tropical avifauna resides at St Helena including boobies, a frigatebird not present earlier, and Audubon's shearwater. Most recently these have been extirpated by introduced mammals.

North-west Atlantic/Gulf of Mexico

Another well-documented change in the marine bird fauna from Plio-Pleistocene to Recent times is available for Florida. The region experienced several major fluctuations in sea level during glacial and interglacial periods. When sea level decreased markedly to expose the Isthmus of Panama, thus, changing circulation, there was a cessation of upwelling and cool, productive conditions. As a result, a resident cool-water avifauna became extinct. Subsequently, during periods of glacial advance and cooler conditions, more northerly species visited the area; and, conversely, during warmer, interglacial periods, these species disappeared.

Direct Responses to Recent Climate Change

A general warming, especially obvious after the mid-1970s, has occurred in ocean temperatures especially west of the American continents. Reviewed here are sea-bird responses to this change.

Chukchi Sea

The Arctic lacks the extensive water–ice boundaries of the Antarctic and as a result fewer seabird species will be directly affected by the climate-related changes in ice edges. Reconstructions of northern Alaska climatology based on tree rings show that temperatures in northern Alaska are now the warmest within the last 400 years with the last century seeing the most rapid rise in temperatures following the end of the Little Ice Age (AD 1500–1850). Decreases in ice cover in the western Arctic in the last 40 years have been documented, but the recent beginnings of regional ornithological research

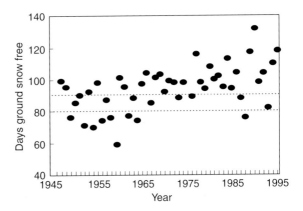

Figure 1 Changes in the length of the annual snow-free period at Barrow, Alaska, 1947–1995. Dashed lines show the number of days that black guillemots and horned puffins require a snow-free cavity (80 and 90 days, respectively). Black guillemots first bred near Barrow in 1966 and horned puffins in 1986. (Redrawn from Divoky, 1998.)

precludes examining the response of birds to these changes.

Changes in distribution and abundance related to snow cover have been found for certain cavity-nesting members of the auk family (Alcidae). Black guillemots and horned puffins require snow-free cavities for a minimum of 80 and 90 days, respectively, to successfully pair, lay and incubate eggs, and raise their chicks(s). Until the mid-1960s the snow-free period in the northern Chukchi Sea was usually shorter than 80 days but with increasing spring air temperatures, the annual date of spring snow-melt has advanced more than 5 days per decade over the last five decades (**Figure 1**). The annual snow-free period now regularly exceeds 90 days, which reduces the likelihood of chicks being trapped in their nest sites before fledging. This has allowed black guillemots and horned puffins to establish colonies (range expansion) and increase population size in the northern Chukchi.

California Current

Avifaunal changes in this region are particularly telling because the central portion of the California Current marks a transitional area where subtropical and subarctic marine faunas meet, and where north–south faunal shifts have been documented at a variety of temporal scales. Of interest here is the invasion of brown pelicans northward from their 'usual' wintering grounds in central California to the Columbia River mouth and Puget Sound, and the invasion of various terns and black skimmers northward from Mexico into California. The pelican and terns are tropical and subtropical species.

During the last 30 years, air and sea temperatures have increased noticeably in the California Current region. The response of seabirds may be mediated by thermoregulation as evidenced by differences in the amount of subcutaneous fat held by polar and tropical seabird species, and by the behavioral responses of seabirds to inclement air temperatures.

The brown pelican story is particularly well documented and may offer clues to the mechanism by which similar invasions come about. Only during the very intense El Niño of 1982–83 did an unusual number of pelicans move northward to the Columbia River mouth. They had done this prior to 1900, but then came a several-decade long period of cooler temperatures. Initially, the recent invasion involved juveniles. In subsequent years, these same birds returned to the area, and young-of-the-year birds followed. Most recently, large numbers of adult pelicans have become a usual feature feeding on anchovies that have been present all along. This is an example of how tradition, or the lack thereof, may facilitate the establishment (or demise) of expanded range, in this case, compatible with climate change.

The ranges of skimmers and terns have also expanded in pulses coinciding with El Niño. This pattern is especially clear in the case of the black skimmer, a species whose summer range on the east coast of North America retracts southward in winter. On the west coast, almost every step in a northward expansion of range from Mexico has coincided with ocean warming and, in most cases, El Niño: first California record, 1962 (not connected to ocean warming); first invasion *en masse*, 1968 (El Niño); first nesting at Salton Sea, 1972 (El Niño); first nesting on coast (San Diego), 1976 (El Niño); first nesting farther north at Newport Bay and inland, 1986 (El Niño). Thereafter, for Southern California as a whole, any tie to El Niño became obscure, as (1) average sea-surface temperatures off California became equivalent to those reached during the intense 1982–83 El Niño, and (2) population increase became propelled not just by birds dispersing north from Mexico, but also by recruits raised locally. By 1995, breeding had expanded north to Central California. In California, with warm temperatures year round, skimmers winter near where they breed.

The invasion northward by tropical/subtropical terns also relates to El Niño or other warm-water incursions. The first US colony (and second colony in the world) of elegant tern, a species usually present off California as post-breeders (July through October), was established at San Diego in 1959, during the strongest El Niño event in modern times.

A third colony, farther north, was established in 1987 (warm-water year). The colony grew rapidly, and in 1992–93 (El Niño) numbers increased 300% (to 3000 pairs). The tie to El Niño for elegant terns is confused by the strong correlation, initially, between numbers breeding in San Diego and the biomass of certain prey (anchovies), which had also increased. Recently, however, anchovies have decreased. During the intense 1997–98 El Niño, hundreds of elegant terns were observed in courtship during spring even farther north (central California). No colony formed.

Climate change, and El Niño as well, may be involved in the invasion of Laysan albatross to breed on Isla de Guadalupe, in the California Current off northern Mexico. No historical precedent exists for the breeding by this species anywhere near this region. First nesting occurred in 1983 (El Niño) and by 1988, 35–40 adults were present, including 12 pairs. Ocean temperatures off northern Mexico are now similar to those in the Hawaiian Islands, where nesting of this species was confined until recently. In the California Current, sea temperatures are the warmest during the autumn and winter, which, formerly, were the only seasons when these albatross occurred there. With rising temperatures in the California Current, more and more Laysan albatross have been remaining longer into the summer each year. Related, too, may be the strengthening of winds off the North American west coast to rival more closely the trade winds that buffet the Hawaiian Islands. Albatross depend on persistent winds for efficient flight, and such winds may limit where albatrosses occur, at least at sea.

Several other warm-water species have become more prevalent in the California Current. During recent years, dark-rumped petrel, a species unknown in the California Current region previously, has occurred regularly, and other tropical species, such as Parkinson's petrel and swallow-tailed gull have been observed for the first time in the region.

In response to warmer temperatures coincident with these northward invasions of species from tropical and subtropical regions, a northward retraction of subarctic species appears to be underway, perhaps related indirectly to effects of prey availability. Nowadays, there are markedly fewer black-footed albatross and sooty and pink-footed shearwaters present in the California Current system than occurred just 20 years ago (**Figure 2**). Cassin's auklet is becoming much less common at sea in central California, and its breeding populations have also been declining. Similarly, the southern edge of the breeding range of common murres has retreated north. The species no longer breeds in Southern

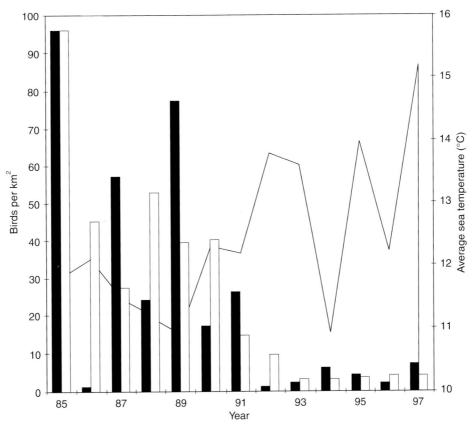

Figure 2 The density (■) (plus 3-point moving average, □) of a cool-water species, the sooty shearwater, in the central portion of the California Current, in conjunction with changes in marine climate (sea surface temperature, (—)), 1985–1997.

California (Channel Islands) and numbers have become greatly reduced in Central California (**Figure 3**). Moreover, California Current breeding populations have lost much of their capacity, demonstrated amply as late as the 1970s, to recover from catastrophic losses. The latter changes may or may not be more involved with alterations of the food web (see below).

Northern Bellingshausen Sea

Ocean and air temperatures have been increasing and the extent of pack ice has been decreasing for the past few decades in waters west of the Antarctic peninsula. In response, populations of the Adélie penguin, a pack-ice species, have been declining, while those of its congener, the open-water dwelling chinstrap penguin have been increasing (**Figure 4**). The pattern contrasts markedly with the stability of populations on the east side of the Antarctic Peninsula, which is much colder and sea-ice extent has changed little.

The reduction in Adélie penguin populations has been exacerbated by an increase in snowfall coincident with the increased temperatures. Deeper snow drifts do not dissipate early enough for eggs to be laid on dry land (causing a loss of nesting habitat and eggs), thus also delaying the breeding season so that fledging occurs too late in the summer to accommodate this species' normal breeding cycle. This pattern is the reverse of that described above for black guillemots in the Arctic. The penguin reduction has affected mostly the smaller, outlying subcolonies, with major decreases being stepwise and occurring during El Niño.

Similar to the chinstrap penguin, some other species, more typical of the Subantarctic, have been expanding southward along the west side of the Antarctic Peninsula. These include the brown skua, blue-eyed shag, southern elephant seal and Antarctic fur seal.

Ross Sea

A large portion (32%) of the world's Adélie penguin population occurs in the Ross Sea (South Pacific sector), the southernmost incursion of ocean on the planet (to 78°S). This species is an obligate inhabitant of pack ice, but heavy pack ice negatively affects reproductive success and population growth. Pack-ice extent decreased noticeably in the late 1970s and early 1980s and air temperatures have also been

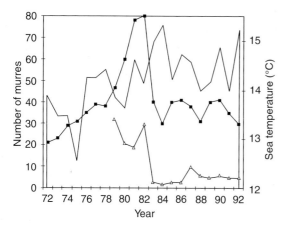

Figure 3 Changes in the number of breeding common murres in central (California, ■) and northern (Washington, △) portions of the California Current during recent decades. Sea surface temperature (—) from central California shown for comparison. During the 1970s, populations had the capacity to recover from reductions due to anthropogenic factors (e.g., oil spills). Since the 1982–83 El Niño event and continued higher temperatures, however, the species' capacity for population recovery has been lost (cf. **Figure 5**).

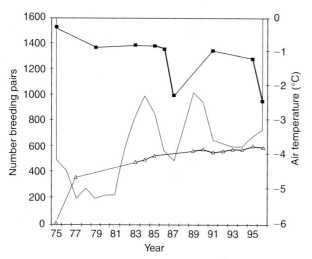

Figure 4 Changes in the number of breeding pairs of two species of penguins at Arthur Harbor, Anvers Island, Antarctica (64°S, 64°W), 1975–1996. The zoogeographic range of Adélie penguins (■) is centered well to the south of this site; the range of chinstrap penguins (△) is centered well to the north. Arthur Harbor is located within a narrow zone of overlap (200 km) between the two species and is at the northern periphery of the Adélie penguins' range.

rising. The increasing trends in population size of Adélie penguins in the Ross Sea are opposite to those in the Bellingshausen Sea (see above; **Figure 5**). The patterns, however, are consistent with the direction of climate change: warmer temperatures, less extensive pack ice. As pack ice has become more dispersed in the far south, the penguin has benefited.

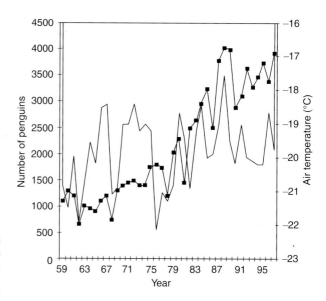

Figure 5 Changes in numbers of breeding pairs of Adélie penguins (■) at Cape Royds, Ross Island, Antarctica (77°S, 166°E), during the past four decades. This is the southernmost breeding site for any penguin species. Although changes in sea ice (less extensive now) is the major direct factor involved, average air temperatures (—) (indirect factor) of the region are shown.

As with the Antarctic Peninsula region, sub-antarctic species are invading southward. The first brown skua was reported in the southern Ross Sea in 1966; the first known breeding attempt occurred in 1982; and the first known successful nesting occurred in 1996. The first elephant seal in the Ross Sea was reported in 1974; at present, several individuals occur there every year.

Indirect Responses to Recent Climate Change

California Current

The volume of zooplankton has declined over the past few decades, coincident with reduced upwelling and increased sea-surface temperatures. In response, numbers of sooty shearwaters, formerly, the most abundant avian species in the California Current avifauna, have declined 90% since the 1980s (**Figure 2**). The shearwater feeds heavily on euphausiids in the California Current during spring. The decline, however, has occurred in a stepwise fashion with each El Niño or warm-water event. Sooty shearwaters are now ignoring the Peru and California currents as wintering areas, and favoring instead those waters of the central North Pacific transition zone, which have been cooling and increasing in productivity (see below).

The appearance of the elegant and royal terns as nesting species in California (see above) may in part be linked to the surge in abundance in northern anchovy, which replaced the sardine in the 1960s–1980s. More recently, the sardine has rebounded and the anchovy has declined, but the tern populations continue to grow (see above). Similarly, the former breeding by the brown pelican as far north as Central California was linked to the former presence of sardines. However, the pelicans recently invaded northward (see above) long before the sardine resurgence began. Farthest north the pelicans feed on anchovies.

Central Pacific

In the central North Pacific gyre, the standing crop of chlorophyll-containing organisms increased gradually between 1965 and 1985, especially after the mid-1970s. This was related to an increase in storminess (winds), which in turn caused deeper mixing and the infusion of nutrients into surface waters. The phenomenon reached a maximum during the early 1980s, after which the algal standing crop subsided. As ocean production increased, so did the reproductive success of red-billed tropicbirds and red-footed boobies nesting in the Leeward Hawaiian Islands (southern part of the gyre). When production subsided, so did the breeding success of these and other species (lobsters, seals) in the region. Allowing for lags of several years as dictated by demographic characteristics, the increased breeding success presumably led to larger populations of these seabird species.

Significant changes in the species composition of seabirds in the central Pacific (south of Hawaii) occurred between the mid-1960s and late 1980s. Densities of Juan Fernandez and black-winged petrels and short-tailed shearwaters were lower in the 1980s, but densities of Stejneger's and mottled petrels and sooty shearwaters were higher. In the case of the latter, the apparent shift in migration route (and possibly destination) is consistent with the decrease in sooty shearwaters in the California Current (see above).

Peru Current

The Peruvian guano birds – Peruvian pelican, piquero, and guanay – provide the best-documented example of changes in seabird populations due to changes in prey availability. Since the time of the Incas, the numbers of guano birds have been strongly correlated with biomass of the seabirds' primary prey, the anchoveta. El Niño 1957 (and earlier episodes) caused crashes in anchoveta and guano bird populations, but these were followed by full recovery. Then, with the disappearance of the anchoveta beginning in the 1960s (due to over-fishing and other factors), each subsequent El Niño (1965, 1972, etc.) brought weaker recovery of the seabird populations.

Apparently, the carrying capacity of the guano birds' marine habitat had changed, but population decreases occurred stepwise, coinciding with mortality caused by each El Niño. However, more than just fishing caused anchoveta to decrease; without fishing pressure, the anchoveta recovered quickly (to its lower level) following El Niño 1982–83, and trends in the sardine were contrary to those of the anchoveta beginning in the late 1960s. The seabirds that remain have shifted their breeding sites southward to southern Peru and northern Chile in response to the southward, compensatory increase in sardines. A coincident shift has occurred in the zooplankton and mesopelagic fish fauna. All may be related to an atmospherically driven change in ocean circulation, bringing more subtropical or oceanic water onshore. It is not just breeding seabird species that have been affected, nonbreeding species of the region, such as sooty shearwater, have been wintering elsewhere than the Peru Current (see above).

Trends in penguin populations on the Galapagos confirm that a system-wide change has occurred off western South America (**Figure 6**). Galapagos penguins respond positively to cool water and negatively to warm-water conditions; until recently they recovered after each El Niño, just like the Peruvian guano birds. Then came El Niño 1982–83. The population declined threefold, followed by just a slight recovery. Apparently, the carrying capacity of the habitat of this seabird, too, is much different now than a few decades ago. Like the diving, cool-water species of the California Current (see above), due to climate change, the penguin has lost its capacity for population growth and recovery.

Gulf of Alaska and Bering Sea

A major 'regime' shift involving the physical and biological make-up of the Gulf of Alaska and Bering Sea is illustrated amply by oceanographic and fisheries data. Widespread changes and switches in populations of ecologically competing fish and invertebrate populations have been underway since the mid-1970s. Ironically, seabird populations in the region show few geographically consistent patterns that can be linked to the biological oceanographic trends. There have been no range expansions or retractions, no doubt because this region, in spite of its great size, does not constitute

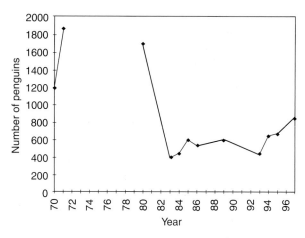

Figure 6 Changes in numbers of breeding Galápagos penguins, 1970–1997. With the 1982–83 El Niño event, the species lost the capacity to recover from periodic events leading to increased mortality. Compare to **Figure 3**, which represents another cool-water species having a similar life-history strategy, but which resides in the other eastern Pacific boundary current (i.e., the two species are ecological complements in the Northern and Southern Hemispheres).

a faunal transition; and from within the region, species to species, some colonies have shown increases, others decreases, and others stability. Unfortunately, the picture has been muddled by the introduction of exotic terrestrial mammals to many seabird nesting islands. Such introductions have caused disappearance and serious declines in seabird numbers. In turn, the subsequent eradication of mammals has allowed recolonization and increases in seabird numbers.

In the Bering Sea, changes in the population biology of seabirds have been linked to decadal shifts in the intensity and location of the Aleutian Low Pressure System (the Pacific Decadal Oscillation), which affects sea surface temperatures among other things. For periods of 15–30 years the pressure (North Pacific Index) shifts from values that are above (high pressure state) or below (low pressure state) the long-term average. Kittiwakes in the central Bering Sea (but not necessarily the Gulf of Alaska) do better with warmer sea temperatures; in addition, the relationship of kittiwake productivity to sea surface temperature changes sign with switches from the high to low pressure state. Similarly, the dominance among congeneric species of murres at various sympatric breeding localities may flip-flop depending on pressure state. Although these links to climate have been identified, cause–effect relationships remain obscure. At the Pribilof Islands in the Bering Sea, declines in seabird numbers, particularly of kittiwakes, coincided with the regime shift that began in 1976. Accompanying these declines has

been a shift in diets, in which lipid-poor juvenile walleye pollock have been substituted for lipid-rich forage fishes such as sand lance and capelin. Thus, the regime shifts may have altered trophic pathways to forage fishes and in turn the availability of these fish to seabirds. Analogous patterns are seen among the seabirds of Prince William Sound.

Chukchi Sea

Decrease of pack ice extent in response to recent global temperature increases has been more pronounced in the Arctic than the Antarctic. This decrease has resulted in changes in the availability of under-ice fauna, fish, and zooplankton that are important to certain arctic seabirds. The decline of black guillemot populations in northern Alaska in the last decade may be associated with this pack ice decrease.

North Atlantic

The North Atlantic is geographically confined and has been subject to intense human fishery pressure for centuries. Nevertheless, patterns linked to climate change have emerged. In the North Sea, between 1955 and 1987, direct, positive links exist between the frequency of westerly winds and such biological variables as zooplankton volumes, stock size of young herring (a seabird prey), and the following parameters of black-legged kittiwake biology: date of laying, number of eggs laid per nest, and breeding success. As westerly winds subsided through to about 1980, zooplankton and herring decreased, kittiwake laying date retarded and breeding declined. Then, with a switch to increased westerly winds, the biological parameters reversed.

In the western Atlantic, changes in the fish and seabird fauna correlate to warming sea surface temperatures near Newfoundland. Since the late 1800s, mackerel have moved in, as has one of their predators, the Atlantic gannet. Consequently, the latter has been establishing breeding colonies farther and farther north along the coast.

South-east Atlantic, Benguela Current

A record of changes in sea-bird populations relative to the abundance and distribution of prey in the Benguela Current is equivalent to that of the Peruvian upwelling system. As in other eastern boundary currents, Benguela stocks of anchovy and sardine have flip-flopped on 20–30 year cycles for as long as records exist (to the early 1900s). Like other eastern boundary currents, the region of concentration of prey fish, anchovy versus sardine, changes with sardines being relatively more abundant poleward in the region compared to anchovies. Thus, similar to the Peruvian situation, seabird populations have shifted, but patterns also are apparent at smaller timescales depending on interannual changes in spawning areas of the fish. As with all eastern boundary currents, the role of climate in changing pelagic fish populations is being intensively debated.

See also

Benguela Current. California and Alaska Currents. Canary and Portugal Currents. El Niño Southern Oscillation (ENSO). Polynyas. Seabird Foraging Ecology. Sea Ice: Overview. Sea Level Change. Upwelling Ecosystems.

Further Reading

Aebischer NJ, Coulson JC and Colebrook JM (1990) Parallel long term trends across four marine trophic levels and weather. *Nature* 347: 753–755.

Crawford JM and Shelton PA (1978) Pelagic fish and seabird interrelationships off the coasts of South West and South Africa. *Biological Conservation* 14: 85–109.

Decker MB, Hunt Jr GL and Byrd Jr GV (1996) The relationship between sea surface temperature, the abundance of juvenile walleye pollock (*Theragra chalcogramma*), and the reproductive performance and diets of seabirds at the Pribilof islands, southeastern Bering Sea. *Canadian Journal of Fish and Aquatic Science* 121: 425–437.

Divoky GJ (1998) *Factors Affecting Growth of a Black Guillemot Colony in Northern Alaska.* PhD Dissertation, University of Alaska, Fairbanks.

Emslie SD (1998) *Avian Community, Climate, and Sea-level Changes in the Plio-Pleistocene of the Florida Peninsula.* Ornithological Monograph No. 50. Washington, DC: American Ornithological Union.

Furness RW and Greenwood JJD (eds) (1992) *Birds as Monitors of Environmental Change.* London, New York: Chapman and Hall.

Olson SL (1975) *Paleornithology of St Helena Island, South Atlantic Ocean.* Smithsonian Contributions to Paleobiology, No. 23. Washington, DC.

Smith RC, Ainley D, Baker K *et al.* (1999) Marine ecosystem sensitivity to climate change. *BioScience* 49(5): 393–404.

Springer AM (1998) Is it all climate change? Why marine bird and mammal populations fluctuate in the North Pacific. In: Holloway G, Muller P and Henderson D (eds) *Biotic Impacts of Extratropical Climate Variability in the Pacific.* SOEST Special Publication. Honolulu: University of Hawaii.

Stuiver M, Denton GH, Hughes T and Fastook JL (1981) History of the marine ice sheet in West Antarctica during the last glaciation: a working hypothesis. In: Denton GH and Hughes T (eds) *The Last Great Ice Sheets.* New York: Wiley.

SEABIRDS AND FISHERIES INTERACTIONS

C. J. Camphuysen, Netherlands Institute for Sea Research, Texel, The Netherlands

doi:10.1006/rwos.2001.0235

Introduction

Of the ~ 9000 species of birds in the world, only about 350 may be classified as seabirds (**Table 1**). The exact number of species depends on the definition of a 'seabird' and on recent taxonomic conventions. Here, in addition to the four orders that are generally considered as seabirds (all Sphenisciformes, all Pelecaniformes, all Procellariiformes, and some Charadriiformes), also divers or loons (Gaviiformes) and seaducks (members of the much larger order of Anseriformes) are included. Following this selection, it is apparent that only 4% of all species of birds utilize the greater part of the world's surface: the oceans. Seabirds prey upon fish, squid, and other marine organisms and the productivity of the oceans and regional variations in food availability determine the distribution, breeding success, and ultimately the numbers of seabirds on our planet. Most seabirds are largely piscivorous, but, particularly in the penguins, petrels and storm petrels, and northern auks, many species are planktivorous. Some seabirds exploit marine resources only outside the breeding season and shift to terrestrial feeding during nesting. Several species of seaduck exploit shellfish resources in shallow seas and these may shift to fish prey only under exceptional conditions.

Seabirds vary substantially in size and morphology. From the smallest of the storm-petrels (the least petrel *Halocyptena microsoma*, ~ 20 g) to the largest of the albatrosses (wandering albatross *Diomedea exulans*, up to 11 kg with a wingspan of up to 3.5 m) or penguins (emperor penguin *Aptenodytes forsteri*, 19–46 kg flightless), there is a great variability in capacity for flight, time spent at sea, preferred types of prey, and foraging techniques. Seabirds are wide-ranging organisms and they can be both predators and scavengers of marine

Table 1 Orders and families selected for this review as 'sea birds'[a]

Order	Selected families	Genera	Species	Taxa
Sphenisciformes	Penguins	6	17	26
Gaviiformes	Divers (or loons)	1	4	6
Procellariiformes	Albatrosses, petrels, shearwaters, storm-petrels, diving petrels	23	108	175
Pelecaniformes	Tropicbirds, pelicans, gannets and boobies, cormorants, darters, frigatebirds	6	65	117
Anseriformes	Steamer ducks, seaduck, mergansers, and allies	1	22	33
Charadriiformes	Skuas, gulls, terns, skimmers, auks	31	127	261
		68	343	618

[a]The taxonomy follows Del Hogo J, Elliott A and Sargatal J (1992, 1996) *Handbook of the Birds of the World*, vols 1 and 3. Barcelona: Lynx Edition.

prey. Most are essentially surface foragers, obtaining their prey from the top meters of the water column. Penguins, some shearwaters, cormorants, and alcids are exceptions that are known to dive several tens or even hundreds of meters deep. Several species, most notably some gulls, have successfully adapted to humans and exploit new, artificial sources of food such as fishery waste at trawlers, rubbish tips, litter at holiday resorts, etc. Other species, such as the great auk *Penguinis impennis*, the only recent flightless alcid, became overexploited and were driven to extinction.

Seabirds can either be harmed by or benefit from the fishing activities of humans. Direct effects of fisheries involve, for example, the killing of seabirds in fishing gear (by-catch). Indirect effects mostly operate through changes in food supplies of birds. Surface feeders are most vulnerable to baited hooks set in long-line fisheries; pursuit diving seabirds are most likely to suffer from gillnets set on the bottom of coastal seas. Surface feeders, particularly the larger, scavenging species, usually profit most from discarded undersized fish and offal in commercial fisheries. In recent decades, there has been tremendous progress in our quantitative understanding of the processes in marine ecosystems, including the interactions of seabirds and fisheries. Some of the most profound effects of fisheries on seabirds are discussed and illustrated here.

Seabird Feeding Techniques

Seabirds exploit the marine environment in a number of ways: in and from the air, at the water surface, by plunge diving up to a few meters deep and by wing- or foot-propelled pursuit diving, reaching over a 100 m in depth. Several coastal seabirds, such as divers, grebes, cormorants, and seaducks rarely exceed 20 m in depth while foraging over the bottom, while several pelagic seabirds, in-

cluding, perhaps unexpectedly, some of the generally more aerial shearwaters, reach many tens of meters deep and sometimes well over 100 m (e.g., 180 m by the common guillemot *Uria aalge*). The feeding technique largely determines the sort of conflict or contact that seabirds may have with commercial fisheries. Scavenging at fishing vessels, for example, is typical for large surface feeders. Birds that get caught by baited hooks on long-lines are also mainly large surface feeders, particularly those that scavenge at dead fish, squids or corpses of marine mammals during normal life. Most birds that drown in fish nets (behind trawlers or in gill nets set adrift or set on the bottom) are mainly pursuit diving seabirds, such as auks, cormorants, and seaduck.

Many seabirds are gape-limited, piscivorous species that prey on small shoaling fish in areas where high concentrations of prey occur. Obviously, restricted areas in which large schools of fish are formed are often equally attractive to fishermen. Diving seabirds and fishermen may target the same prey (although possibly different size classes of fish), while commercial fisheries may provide an additional source of food for surface feeders, simply by discarding unwanted material and undersized fish. In practice, it appears the most undersized fish are suitable at least for the largest of scavengers (mollymawk albatrosses and gannets). For the remaining part, the smaller fraction, there is often intense competition among the scavengers, leading to the establishment of complex dominance hierarchies and species-specific feeding success rates. Scavenging seabirds tend to select slightly smaller prey than they can normally handle, thus facilitating a rapid process of picking up and swallowing the food and minimizing the risk of being robbed by competitors.

Diving for prey or flying long distances for food is an energetically expensive way of foraging, requiring very high intake rates. Such high intake rates are

SEABIRDS AND FISHERIES INTERACTIONS

impossible under most conditions, so that concentrations of prey are either sought out, or deliberately formed by concerted action of (a group of) predators. Seabirds are famous for exploiting prey concentrated and herded to the surface by cetaceans or large predatory fish such as tuna. Several diving seabirds are known to concentrate the herd prey to the surface, which is in turn, and often through a mechanism of commensalism, exploited by surface feeding birds that are incapable of reaching the prey by other means. High concentrations of prey may also be found 'more naturally' along shelf breaks, in oceanic or coastal fronts, in upwelling areas, and along pack ice or glaciers. The need for concentrated resources of food and the availability of prey in terms of being within reach of the predator, and of suitable size and fat content, is more important than the actual stock. Hence, fishery statistics and stock assessments are usually inadequate to describe or estimate the food resource of individual species of seabirds or other marine predators. A complicating factor is that the availability of prey may be largely determined by the presence and (foraging) activities of another marine predator, so that shifts in the abundance of one species may lead to food shortages and starvation of another, even if the food resource itself is apparently unchanged and sufficiently large.

History

When humans gradually developed and improved fishing techniques through the centuries and finally set out in boats to fish at sea, seabirds were immediately met with in large numbers. It was soon learned that seabirds could be used for navigation at sea and early fishermen successfully followed foraging seabirds as indicators of the presence of fish that were otherwise difficult to find. The peaceful coexistence of seabirds and fishermen did not last long. This was partly because seabirds themselves, or their eggs and offspring, were considered attractive sources of protein, perfectly suitable for human consumption. In addition, seabirds were increasingly considered to be competitors or even pests and were persecuted for that reason. Until the beginning of the twentieth century, many species of seabirds were heavily exploited or persecuted in much of their range and some were nearly driven to extinction.

Human fisheries developed further, and particularly when sailing vessels were gradually replaced by modern purse seiners and powerful trawlers with engines in the nineteenth century, overfishing increasingly caused problems. The interactions between fisheries and marine top-predators such as seabirds now became more complex. In the twentieth century, many species of seabirds have greatly increased in numbers, and have successfully expanded their breeding range. Over the last few decades, the growth in some of these populations has ceased, but seabird numbers are generally at a historically high level. The drastic increase in numbers of seabirds has led to speculation about what might have caused these trends. While the impact of direct exploitation of seabirds in the old days was largely beyond doubt, the effects of overfishing were much less clear-cut. Some seabirds may have benefited from rich, new food supplies made available, first, by offal from whaling and later from commercial fishing vessels. However, even more complex trophic interactions may have occurred, because fisheries are generally directed at much larger prey (marketable fish) than seabirds eat. By cropping large piscivorous predators and cannibals, such fisheries benefit seabirds by increasing the abundance of small fish. As such, even commercial 'overfishing' might have been beneficial for seabirds.

The Exploitation and Culling of Seabirds

Seabirds have been exploited as food for as long as humans have established coastal communities. At some archaeological sites it appeared that such communities were often entirely reliant on seabirds and fish for their protein intake. Fishermen killed seabirds at sea or visited seabird colonies to obtain flesh with which to bait their hooks on a large scale, at least until the late nineteenth century. As an example, the harvest of the northern gannets in the Gulf of St. Lawrence is likely to have reduced the local population from over 100 000 pairs in the early 1800s to below 750 pairs by the turn of the century, when bird protection laws were enacted. Today, seabirds are still harvested by local fishermen in areas like Indonesia, and boobies are probably still killed to be used as bait in lobster traps off Brazil.

Fish-eating birds, particularly those capable of eating relatively large fish, have always been blamed by fishermen for depleting local fish stocks and were often held responsible for declining catches. This has led to mass killing or culls, even if no evidence was provided to support the impact of the birds or to show that the birds had negatively affected the fish populations concerned. For that reason, both in Europe (great cormorant *Phalacrocorax carbo*) and in the New World (double-crested cormorant *Phalacrocorax auritus*), cormorants were at least regionally nearly driven to extinction. So far there is no scientific evidence that a cull of any marine predator has enhanced any commercial fishery.

Recognition of the need to conserve nature rather than to relentlessly exploit its resources developed largely during the twentieth century. As a result, the mass slaughter of birds, even of species that were generally considered as pests, largely came to a halt. Yet, for example, the rise and fall of herring gull populations in The Netherlands during the twentieth century closely followed trends in levels of persecution. When breeding colonies became protected, around the 1920s, all populations increased. This increase ceased when culling was introduced to 'manage' gull numbers for the protection of other birds and to maintain the population at 'acceptable levels.' Culling in the late 1960s was not so intense that populations declined, but when full protection measures were taken in the 1970s, populations exploded to unprecedented levels and peaked in the 1980s.

By-catches of Seabirds in Commercial Fisheries

Seabird populations are negatively affected by the extra mortality induced by commercial fisheries all over the world: any net or line set with baited hooks carries the risk of catching seabirds as a bycatch.

Many surface-foraging seabirds commonly scavenge on dead or moribund prey and such birds are likely to try to steal bait from longline hooks during line setting. Large numbers of seabirds become hooked and subsequently drown as the longline sinks below the sea surface, a problem only fully appreciated in the last few decades. Pelagic long-lining in tropical and temperate seas concentrates mainly on tuna, swordfish, and sharks. Demersal long-lining in cold temperate waters of the continental shelves of the Atlantic and the Pacific, and in the Southern Ocean mainly concentrates on bottom-dwelling fish such as large gadoids and flatfish. Both types of longline fisheries use baited hooks. Procellariiform birds (mainly albatrosses and petrels) are the principal by-catch in longline fisheries all over the globe (**Table 2**). The by-catch of long-liners operating in the north-east Atlantic is known to have killed as many as 1.75 birds per 1000 hooks (95% of which were northern fulmars). At night, mortality rates are substantially lower (0.02 birds/1000 hooks). When these figures are multiplied by the nearly 500 million hooks set in one year by the Norwegian auto-line fleet alone, the annual mortality of seabirds must be very large. In the Pacific demersal and pelagic longline fisheries, an estimated 'several thousand' black-footed and Laysan albatrosses are killed annually. Catch rates of seabirds by

Table 2 Review of longline fisheries in the world and the principal sea bird bycatch[a]

Region	Countries mainly involved	Target fish	Principal by-catch
North-east Atlantic longline fisheries	Norway, Iceland, Faeroes	Demersal fish (e.g., cod)	Northern fulmar
North Pacific longline fisheries; Bering Sea, Sea of Okhotsk and the Gulf of Alaska	USA, Canada, Russia	Demersal fish (e.g., cod, halibut)	Albatrosses, northern fulmar
North Pacific longline fisheries; international waters	USA, Japan, Korea, Taiwan	Pelagic fish (e.g., swordfish)	Albatrosses
Southern continental shelf demersal longline fisheries; Pacific and Atlantic coasts of South America	Chile, Argentina, Uruguay and Brazil	Hakes, ling	Albatrosses and petrels
Southern continental shelf demersal longline fisheries	Atlantic coast of southern Africa (South Africa, Namibia)	Hakes, ling	Albatrosses and petrels
Southern continental shelf demersal longline fisheries	Australia, New Zealand	Hakes, ling	Albatrosses and petrels
Southern Ocean Patagonian toothfish longline fishery; sub-Antarctic islands and seamounts of the Southern Ocean		Patagonian toothfish	Mollymawk albatrosses, wandering albatrosses, white-chinned petrel, giant petrels
Southern Ocean bluefin tuna longline fishery	Australia, Japan, New Zealand, Korea, Taiwan	Bluefin tuna	Albatrosses, petrels

[a]From Tasker *et al.* (1999).

Patagonian toothfish long-liners were estimated at 145 000 birds killed during the 1996/97 season alone, mainly mollymawk albatrosses and the white-chinned petrel, with smaller numbers of wandering albatrosses and giant petrels. For some of these species, population decreases have been recorded at their breeding grounds that are thought to be due to longline-induced mortality.

Perhaps even larger numbers of seabirds drown in gill nets (set nets and drift nets combined), whether these are still in commercial exploitation and set by fishermen or have been lost and have established themselves on the seafloor as 'ghost nets' (continuing their catches). Surface gill nets are mainly used for squid, salmon, and small tunalike bonitos; bottom nets are used for demersal fish such as cod and various species of flatfish (Table 3). Much of the mortality in these nets will go unnoticed, as fishermen are known to 'hide' the casualties by sinking them, and most certainly are very reluctant to report any seabird kills. Serious problems with gill nets are reported from Newfoundland (inshore capelin fisheries, offshore salmon fisheries), Greenland (offshore salmon fisheries), the North Pacific (offshore, salmon), northern Norway (offshore and nearshore, cod and salmon), the Baltic (shallow waters, demersal fish such as flatfish and cod), and around Britain and Ireland (mainly nearshore, salmon and bass). It is mainly pursuit diving alcids, shearwaters, and seaducks (Baltic) that are known to drown in vast numbers (Table 3). For example, prior to the 1992 moratorium on high-seas drift nets in the North Pacific, ~ 500 000 seabirds were drowned annually (mainly shearwaters). More recent data showed that approximately 50 000 seaducks drown annually in an area as small as the IJsselmeer area in The Netherlands, including 1100, or 4.5% of the world population, of smew *Mergus albellus*.

A less serious but unnecessary threat to seabirds is that seabirds pick up floating debris, including netting, nylon line and ropes, from the sea surface to use as nesting material instead of seaweed, or simply by accident. As a result, the birds become entangled and die from starvation. The northern gannet *Morus bassanus* and great cormorant ranked highest among 90 species of stranded marine birds of which 140 000 corpses were checked for entanglements as cause of death in the southern North Sea. Some 5% of all beached northern gannets checked ($n = 1395$) were entangled in ropes or fishing gear, while 2% of all great cormorants ($n = 310$) had suffered a similar fate. Inspection of northern gannet nests on the Shetland Islands in Britain revealed that 92% of all nests contained at least some plastics, while 50% contained virtually nothing else. Both chicks and adult birds are seen to become entangled in the colony and most of these casualties die from starvation. While the mortality associated with this type of pollution may be quite low, the amount of debris floating around at sea has increased substantially in recent decades and sightings of gannets carrying around nets, plastics, and ropes are now very common, particularly at the main fishing grounds.

Discards and Offal as Food for Scavenging Seabirds

Albatrosses, petrels, shearwaters, gannets, and most gulls are common scavengers at fishing vessels. Away from fishing vessels, most of these birds are

Table 3 Review of gill-net fisheries in the northern hemisphere and principal sea bird by-catch[a]

Region	Countries mainly involved	Area, target fish	Principal by-catch
NW Atlantic	Newfoundland (Canada)	Inshore, spawning capelin Offshore, salmon	Common guillemot, Atlantic puffin, black guillemot, Brünnich's guillemot, great shearwater, northern gannet
NW Atlantic	Greenland	Offshore, salmon	Brünnich's guillemot
N Pacific	Japan	High-seas drift-netting, salmon	Sooty and short-tailed shearwater, black-footed albatross, tufted puffin, horned puffin
		Offshore, squid	Japanese murrelet
		Inshore, salmon	Ancient murrelet, marbled murrelet
Pacific	California	Near-shore	Common guillemot
NE Atlantic	Norway	Near-shore, cod, salmon	Common and Brünnichs guillemot
Baltic	Poland, Sweden, Germany	Inshore, flatfish, cod	Common eider, common scoter, common guillemot, razorbill, long-tailed duck
North Sea	IJsselmeer, The Netherlands	Inshore, eel, freshwater fish	Red-breasted merganser, goosander, smew, scaup, tufted duck
NE Atlantic	Britain and Ireland	Inshore, bass, salmon	Common guillemot, razorbill

[a]From Tasker *et al.* (1999).

surface feeders, specialized in catching zooplankton, squid, or small fish or as scavengers on carrion. Some of the gulls also have terrestrial feeding modes; the gannets and boobies are deep plunge diving seabirds; while most shearwaters naturally feed by pursuit plunging (reaching considerable depth). Pursuit diving piscivorous seabirds, such as cormorants and larger auks, are occasionally observed in association with fishing vessels, but, even if these birds benefit from human fisheries, they have very different feeding techniques from the 'ordinary' scavengers at trawlers.

Consumption rates of scavenging seabirds have been measured onboard fishing vessels during sessions of experimental discarding. Overall consumption rates vary greatly in different parts of the world but are generally highest for offal (liver and guts of gutted fish), and discarded roundfish (usually undersize roundfish and nontarget species), and considerably lower for discarded flatfish and benthic invertebrates. In the North Sea, consumption rates (proportion of prey items taken by seabirds of all discards and fishery waste produced) of the most common forms of fishery waste ranged from < 10% in benthic invertebrates to 25% of the flatfish, over 80% of all roundfish, and 92% of all offal (**Table 4**). In several studies it was shown that consumption rates in winter were particularly high. Seabirds tend to select prey that is easy to handle and to swallow, which appeared to be more important than its energetic equivalent. Several species demonstrated strong preferences for certain types of prey (whether or not forced by competitors). For example, most spiny grey gurnards were picked up by lesser black-backed gulls *Larus graellsii*, while smooth and slender whitings, offered simultaneously, were often ignored. The consumption of both flatfish and benthic invertebrates, less preferred food for most scavenging seabirds, increases when competition is high, but is often negligible when the number of scavengers was low in proportion to the amount of discards supplied.

Although discards and offal are an important additional source of food for scavenging seabirds, the reproductive output of individuals that take nearly only fishery waste is not usually very high. For example, the chick growth index of great skuas *Stercorarius skua* breeding on the Shetland Islands declined considerably when more than 50% of the prey delivered by the parents comprised of discards and offal. Similarly, the reproductive output of lesser black-backed gulls in the southern North Sea was high in years when clupeoid fish dominated chick diets, but low when chicks were mainly provisioned with fishery waste. In the northern fulmar

Fulmarus glacialis, one of the commoner scavengers behind fishing vessels in the North Sea, no correlation was found between numbers of fishing vessels or the amount of discards produced per km^2 in a given area and the numbers of fulmars at sea, suggesting that commercial fisheries were not the prime determinant of their distribution at sea.

Seaduck, Aquacultures and Fisheries for Clams

Several seaduck, such as common eider *Somateria mollissima* and common scoter *Melanitta nigra*, feed on bivalves during most of the winter. In aquacultures in coastal, shallow seas, these duck are often considered and treated as pests and are disturbed, shot, or otherwise scared away. In so-called mussel farms, where mussels are manipulated to grow on ropes or other structures, a single eider duck can do considerable damage by stripping off hundreds of shells to obtain a single specimen or perhaps a few to eat. In areas as the Wadden Sea in The Netherlands, where musselseed is harvested in one place, to be dumped to grow to marketable size in certain licensed, cultivated parts of that sea area, eiders tend to be attracted in vast numbers, and fishermen scare these birds away by establishing small teams of people with speed boats that enter the mussel cultures at regular intervals. In the presence of alternative feeding sites, such activities cause little or no problems to the seaduck. However, most (natural) shellfish occur in dense banks that grow for a number of years and die off at times. For example, in the southern North Sea, large concentrations of up to 160 000 common scoters along with smaller numbers of velvet scoters *Melanitta fusca* winter over banks holding stocks of the bivalve *Spisula subtruncata* within approximately 10 km of the coast. These seaduck are easily disturbed and the appearance of fishing vessels in the recently established *Spisula* fishery has led to both disturbance and local depletion of food stocks. With these banks of shellfish now being exploited, and usually overexploited, and although the future of the bivalves itself is not set at risk, the local feeding conditions of seaduck deteriorate to such levels that mass mortality due to starvation or departure are the only alternatives. To complicate the interactions between seaduck and fisheries in the Wadden Sea even further, the natural resources of eider duck within the Wadden Sea (old, established mussel banks on the mudflats) were removed in the early 1990s and have still not recovered, so that the wintering population of nearly 200 000 eiders, seeking alternative prey in the coastal waters, is now in direct competition with

Table 4 Consumption rates of common forms of discards and other forms of fishery waste (fraction consumed by sea birds of all prey offered) by North Sea seabirds from sessions of experimental discarding at sea and the estimated energetic equivalents (kJ g^{-1}) of discarded organisms (where known). Shown are total numbers offered, numbers swallowed (consumed) or pecked on (usually to reach and feed on intestines) and the number that sink, the fraction consumed (swallowed or pecked on), and the general body shape of fish

	Consumed	Pecked on	Sunk	Total	Percentage consumed	Body shape	Energetic equivalent (kJ g^{-1})
Swimming crab	46		160	206	22.3		3.5
Common starfish	265		1558	1823	14.5		2.0
Hermit crab	7		158	165	4.2		3.5
Sea-mouse	1		124	125	0.8		2.0
Unidentified starfish	2		662	664	0.3		2.0
Brittlestar			450	450	0.0		2.0
Masked crab			246	246	0.0		3.5
All benthic invertebrates	375	4	3779	4158	9.1		
Shrimps	34	1	84	119	29.4		3.5
Cephalopods	62	1	72	135	46.7		
Sole	54	1	104	159	34.6	Slender, supple	
Long rough dab	242	17	527	786	33.0	Rough, slender	
Dab	1102	11	3170	4283	26.0	Stiff, rather wide	
Lemon sole	43	7	162	212	23.6	Smooth, slender	
Plaice	46	1	681	728	6.5	Stiff, rather wide	
All flatfish	1578	41	4850	6469	25.0		4.0
Offal	7533		650	8183	92.1		9.0
Bib	1038		42	1080	96.1	Smooth, slender	4.0
Poor cod	403		22	425	94.8	Smooth, slender	4.0
Blue whiting	98	3	6	107	94.4	Smooth, slender	4.0
Norway pout	5571	21	336	5928	94.3	Smooth, slender	4.0
Lesser Argentine	311		26	337	92.3	Smooth, slender	4.0
Herring or sprat	125		19	144	86.8	Smooth, slender	6.5
Whiting	8420	383	1610	10413	84.5	Smooth, slender	4.0
Herring	6595	97	1512	8204	81.6	Smooth, slender	6.5
Haddock	4213	194	1080	5487	80.3	Smooth, slender	4.0
Sand eel	1145		345	1490	76.8	Smooth, slender	5.0
Sprat	2824	1	917	3742	75.5	Smooth, slender	6.5
Greater sand eel	203		69	272	74.6	Smooth, slender	5.0
Tub gurnard	126	2	47	175	73.1	Spiny, hooked	4.0
Grey gurnard	1242	92	521	1855	71.9	Spiny, hooked	4.0
Cod	1473	63	653	2189	70.2	Smooth, slender	4.0
Mackerel	583	21	318	922	65.5	Smooth, slender	6.5
Hooknose	109		103	212	51.4	Hooked	
Scad	263	12	304	579	47.5	Rather smooth, slender	4.0
Dragonet	145		186	331	43.8	Spiny	
All roundfish	35273	895	8258	44426	81.4		4.0

over 100 000 scoters. Subsequent overfishing of these coastal stocks in 1999 resulted in unprecedented mass mortality of eiders in winter 1999/2000.

The Effects of Overfishing

Stock Depletion

Most piscivorous seabirds specialize on small shoaling fish, such as herring, sardines, anchovy, capelin, or sand eels. Small fatty fish are particularly important prey in the breeding season, when the energetic requirements of the chick(s) are to be met. Competition between fisheries and seabirds for prey resources has been documented for several areas, including the North Pacific (Mexico–Oregon, anchovy), Californian waters (sardine), Peru (anchovy), the North Sea (sand eels), northern Norway (herring), and the Barents Sea (capelin). Major fish stock collapses occurred in each of these areas and, although in most cases recruitment failures, El Niño events, severe winters, or other natural factors were suggested to have caused the crashes, there were poorly managed industrial fisheries running that were not stopped in time to avoid havoc. In all these cases, mass mortality of seabirds, or at least major disruptions in breeding success, occurred. In one of the more famous examples, Atlantic puffins, facing structural food shortages after the depletion of herring stocks off the Lofoten Islands in Norway, suffered from 22 consecutive years with virtually total chick mortality due to insufficient provisioning rates. A recent collapse in sand eel stocks around Shetland and Orkney had the most severe effects on surface feeding seabirds, while deep pursuit diving species maintained high reproductive success. In this event, the foraging mechanisms rather than the fish stocks themselves were apparently damaged (surface swarming sand eels disappeared), but also the local industrial fisheries for sand eels had very low catch rates. The collapse of the Peruvian anchoveta in the early 1970s, coinciding with an El Niño event but in which the stocks were heavily overfished at first, led to exceptionally high mortality rates in local breeding seabirds (boobies and cormorants, or guano birds). The capelin crash in the Barents Sea led to mass mortality of common guillemots (highly specialized on this type of prey), but only to minor reductions in stocks of Brünnich's guillemots (with a more diverse prey spectrum).

The Seabird Paradox

The paradox is, however, that overfishing can also be beneficial for seabirds. There is a general belief that stocks of immature fish in the North Sea have increased, largely thanks to the overfishing of large (predatory) fish. Similarly, congruent shifts in sand eel abundance in the North Atlantic ecosystems were explained by the relative scarcity of large predatory fish as a result of overfishing. At present, the heavily exploited North Sea is a pool of young (immature) fish rather than a balanced ecosystem. A much larger proportion of young fish survives in the absence of mature conspecifics and other (fish) predators than previously, many of which become available to seabirds as discards during commercial trawling operations. The drastic overfishing of mackerel has led to a major increase of sand eels stocks, but the fraction previously taken by this predatory fish (a summer visitor) is now removed by rapidly developing Danish and Norwegian industrial fisheries. The increased food supply (in terms of a greater proportion of fish of suitable size for seabirds) is thought to have been beneficial for seabirds and is suspected to have caused the dramatic

expansion and growth of most populations of seabirds in the NE Atlantic over the last hundred years or so. Substantial increases have been recorded in breeding populations of a variety of seabirds, including surface feeders, scavengers, plunge divers, and pursuit diving birds (**Table 5**). Although it is now generally accepted that most of these birds have profited from the overfishing of large predatory fish (as piscivorous competitors), part of the increase will have been caused by the relaxation of persecution and exploitation of these species. As far as known, there has not been a large increase in populations of molluskivorous seaduck, nor in the more coastal piscivorous divers wintering in these waters, while the increase in the breeding population of the equally coastal and nonmigratory European shags *Stictocarbo aristotelis* has been modest.

Discussion and Conclusions

Fisheries and seabirds compete for the same resources and, although some aspects of human fisheries are beneficial for seabirds, several of the side-effects do great harm to birds. Some piscivorous seabirds are persecuted by fishermen because they are thought to deplete local fish stocks. However, fisheries probably always have greater effects on seabirds than vice versa and there are no examples of fish stock recoveries after an avian predator has been removed from an ecosystem. The annual losses of seabirds, most notably albatrosses, petrels, and several species of auks in longline fisheries and in drifting or bottom-set gill nets are immense. It has been suggested that the gross overfishing of large predatory fish over the last century has led to increases in the survival and stocks of young fish. There is circumstantial evidence, though there are few factual data, that seabirds have profited from this newly established and abundant food resource. There is little doubt that the production of discards (unwanted by-catch of small fish, unmarketable species of fish and benthic invertebrates) and offal (discarded waste of gutted marketable fish) in commercial fisheries is of great significance for some species of seabirds. Discards and offal benefit a group of 'scavenging' seabirds, by 'offering' prey that would otherwise be unavailable and out of reach for these birds. Catches by industrial fisheries, usually targeting the staple foods of marine predators such as seabirds, cetaceans, and seals, have increased dramatically over the last 40 years. Major crashes in (local) fish stocks are not usually attributed to industrial fisheries with certainty, but few of these fisheries are adequately managed such that havoc can actually be prevented. Several case

Table 5 General trends in sea bird populations in the North Sea

	Population	Principal prey	Habitat	Trend
Divers (or loons)	Winter	Fish	Coastal	Stable or decline
	Breeding	Fish	Inland	Decline
Northern fulmar	Breeding	Plankton, squid, fish, offal	Pelagic	Increase
Manx shearwater	Breeding	Fish, squid	Pelagic	Stable
European storm-petrel	Breeding	Plankton, nekton, small fish	Pelagic	Stable
Leach's storm-petrel	Breeding	Plankton, nekton, small fish	Pelagic	Stable or increase
Northern gannet	Breeding	Shoaling fish	Offshore	Increase
Great cormorant	Breeding	Fish	Coastal	Increase
European shag	Breeding	Fish	Coastal	Increase
Common eider	Breeding/winter	Mollusks	Coastal	Stable/slow increase
Common scoter	Winter	Mollusks	Coastal	Stable?
Velvet scoter	Winter	Mollusks	Coastal	Stable?
Arctic skua	Breeding	Robbing birds (fish)	Coastal	Stable or Increase
Great skua	Breeding	Fish, birds, discards	Offshore	Increase
Common gull	Breeding	Largely terrestrial	Coastal/land	Stable or increase
Herring gull	Breeding	Fish, mollusks, benthic inv., discards	Coastal	Increase, recent decline
Lesser black-backed gull	Breeding	Fish, discards	Offshore	Increase
Great black-backed gull	Breeding	Fish, birds, discards	Offshore	Stable
Black-legged kittiwake	Breeding	Fish	Pelagic	Increase
Sandwich tern	Breeding	Fish	Coastal	Slow increase
Common tern	Breeding	Fish	Coastal/land	Stable
Arctic tern	Breeding	Fish	Coastal/offshore	Increase
Common guillemot	Breeding	Fish	Offshore	Increase
Razorbill	Breeding	Fish	Offshore	Increase
Black guillemot	Breeding	Fish	Coastal	Stable?
Atlantic puffin	Breeding	Fish	Offshore	Gradual increase

studies have indicated that poor reproductive output in seabirds, mass mortalities due to starvation, or complete breeding failures in some breeding seasons could be attributed to fish stock depletion and probably to overfishing.

See also

El Niño Southern Oscillation (ENSO). El Niño Southern Oscillation (ENSO) Models. Fisheries: Multispecies Dynamics. Seabird Foraging Ecology.

Further Reading

Anderson DW, Gress F, Mais KF and Kelly PR (1980) Brown pelicans as anchovy stock indicators and their relationships to commercial fishing. *California Cooperative Oceanic Fisheries Investigations Reports* 21: 54–61.

Anker-Nilssen T (1987) The breeding performance of Puffins *Fratercula arctica* on Røst, northern Norway in 1979–1985. *Fauna Norvegica Ser. C., Cinclus* 10: 21–38.

Au DW and Pitman RL (1988) Seabird relationships with tropical tunas and dolphins. In: Burger J (ed.) *Seabirds and Other Marine Vertebrates: Competition, Predation and Other Interactions*, pp. 174–212. New York: Columbia University Press.

Camphuysen CJ and Garthe S (1999) Sea birds and commercial fisheries: population trends of piscivorous sea birds explained? In: Kaiser MJ and Groot SJ de (eds) *Effects of Fishing on Non-target Species and Habitats: Biological, Conservation and Socio-Economic Issues*, pp. 163–184. Oxford: Blackwell Science.

Camphuysen CJ and Garthe S (1997) Distribution and scavenging habits of Northern Fulmars in the North Sea. *ICES Journal of Marine Science* 54: 654–683.

Camphuysen CJ and Webb A (1999) Multi-species feeding associations in North Sea seabirds: jointly exploiting a patch environment. *Ardea* 87(2): 177–198.

Crawford RJM and Shelton PA (1978) Pelagic fish and seabird interrelationships off the coasts of South West Africa. *Biological Conservation* 14: 85–109.

Croxall JP (ed.) (1987) *Seabirds: Feeding Ecology and Role in Marine Ecosystems*. Cambridge: Cambridge University Press.

Duffy DC and Schneider DC (1994) Seabird–fishery interactions: a manager's guide. In: Nettleship DN, Burger J and Gochfeld M (eds) *Seabirds on Islands – Threats, Case Studies and Action Plans*, pp. 26–38. Birdlife Conservation Series No. 1. Cambridge: Birdlife International.

Dunnet GM, Furness RW, Tasker ML and Becker PH (1990) Seabird ecology in the North Sea. *Netherlands Journal of Sea Research* 26: 387–425.

Furness RW (1982) Competition between fisheries and seabird communities. *Advances in Marine Biology* 20: 225–307.

Garthe S, Camphuysen CJ and Furness RW (1996) Amounts of discards in commercial fisheries and their significance as food for seabirds in the North Sea. *Marine Ecology Progress Series* 136, 1–11.

Hunt GL and Furness RW (eds) (1996) *Seabird/Fish Interactions, with Particular Reference to Seabirds in the North Sea.* ICES Cooperative Research Report No. 216. Copenhagen: International Council for Exploration of the Sea.

Jones LL and DeGange AR (1988) Interactions between seabirds and fisheries in the North Pacific Ocean. In: Burger J (ed.) *Seabirds and Other Marine Vertebrates: Competition, Predation and Other Interactions*, pp. 269–291. New York: Columbia University Press.

Montevecchi WA, Birt VL and Cairns DK (1988) Dietary changes of seabirds associated with local fisheries failures. *Biology of the Ocean 5*: 153–161.

Robertson G and Gales R (eds) (1998) *Albatross Biology and Conservation.* Chipping Norton: Surrey Beatty & Sons.

Springer AM, Roseneau DG, Lloyd DS, McRoy CP and Murphy EC (1986) Seabird responses to fluctuating prey availability in the eastern Bering Sea. *Marine Ecology Progress Series* 32: 1–12.

Tasker ML, Camphuysen CJ, Cooper J, *et al.* (1999) The impacts of fishing on marine birds. *ICES Journal of Marine Science* 57: 531–547.

SEABIRDS AS INDICATORS OF OCEAN POLLUTION

W. A. Montevecchi, Memorial University of Newfoundland, Newfoundland, Canada

doi:10.1006/rwos.2001.0236

Background

As wide-ranging, upper and multi-trophic level consumers, marine birds can provide useful indication of ocean pollutants. Seabirds are the most visible marine animals, and individuals, chicks, and eggs are relatively easily sampled, often nonlethally, over wide oceanographic regions. Birds also appeal to the general public who often go to great lengths to protect them. Hence, there is opportunity to help preserve marine ecosystems by monitoring and protecting sea birds and the habitats and prey on which they depend.

Pollutants are assayed to measure levels or rates of change of environmental pollution and to assess biological effects including those on humans. Both nominal and ordinal (qualitative) and interval and ratio (quantitative) measurements are possible. However, physiological, behavioral, taxomonic, and seasonal variations can limit the usefulness of different avian assays in reflecting variation in environmental levels of ocean pollution. Quantitative assays can be problematic because pollutants and other environmental stresses frequently occur in combination in indicator organisms, so it is often difficult or impossible to delineate the effects of a specific pollutant. The problem is complicated when different pollutants have synergistic or additive effects. Hence, determining the most appropriate assay for a pollutant to be monitored and then calibrating the assay are critical problems in all bio-monitoring programs.

Pelagic seabirds such as albatrosses and petrels can provide information on oceanic food webs, whereas coastal and littoral species such as auks and terns can provide information on inshore trophic interactions. Birds that feed at different trophic levels, such as gannets on large pelagic fishes, cormorants on benthic fishes, and sea ducks on bivalves, can be targeted to address different monitoring questions.

Many problems associated with pollution in the ocean are the result of nontarget organisms being affected by chemical management tools. Agricultural and forestry practices have been major sources of organochlorine and of other pesticide and herbicide treatments that affect birds and other nontarget organisms. Assays using marine birds also yield information about industrial chemicals, heavy metals and radionuclides. Pollutant levels reflect toxin sources in regional as well as local environments and are frequently high in estuaries and adjacent waters. Moreover, many chemical and metal pollutants are transported atmospherically, as well as aquatically, over great distances from contact zones – often to pristine polar regions. The movements of contaminated animals can also carry pollutants from source interactions to distant sites. Marine oil pollution is a global problem that results from both highly publicized spills and more extensively from long-term chronic low levels of illegal discharges. In both of these situations, research with seabirds has provided scientists with a means of studying and quantifying biological effects and of raising public awareness and concern about ocean health. Discarded and lost fishing gear and plastics are relatively recent and highly persistent sources of marine pollution that are increasing with expanding global use.

History

Widespread uses of synthetic chemicals following World War II rapidly created environmental problems. Extensive application of organophosphate pesticides poisoned many nontarget animals. Some of the first indications of their harmful effects came to light during the 1960s, many from studies of birds. Resultant public outcries were largely responsible for the banning of DDT and other organochlorine pesticides in North America and Europe. As organochlorines were phased out, background environmental levels soon decreased, and the reproductive success of brown pelicans and ospreys increased to pre-pesticide levels.

Organochlorines have low solubility in water and high environmental persistence. They were replaced largely by water-soluble organophosphates, carbomates, and other compounds that are less environmentally persistent and rapidly metabolized, but that may still be highly toxic to nontarget organisms. Many organochlorines are still used in South America, Asia, and Africa and affect the avifauna that migrate to and from these areas from other regions where these pesticides have been outlawed. Organochlorines accumulate in lipid (i.e. are lipophilic) and can bio-accumulate throughout an animal's life, as well as bio-amplify across trophic levels. Hence, effects of organochlorines and other lipophilic pollutants (e.g. methyl-mercury) are often most evident and can be best monitored through effects on top predators especially birds. However, lipophilic pollutants tend to covary, and an animal's lipid levels change seasonally as well as with food stress. In this respect, chicks can yield useful relatively immediate local assays of environmental toxins. Eggs can be useful assays, although their pollutant loads can also be influenced by food conditions and clutch sequence. For pollutants that are water soluble, it is normally more useful to use bio-monitors at lower trophic levels.

Bio-assay Calibration

The sampling of different species, life stages, age classes, and tissues have different utilities with respect to assaying different pollutants over different time and space scales. Dry-weight analyses are important for comparative purposes to control for variation in the water contents of tissues. Different classes of pollutants are addressed in separate sections below.

Organochlorines

DDT (dichloro-diphenyl-trichloroethane), its primary and stable metabolite DDE, and cyclodienes (dieldrin, aldrin, heptachlor) were used in insecticide applications. Many birds of prey and fish-eating birds accumulated organochlorines up to orders of magnitude above those in their prey and up to a million times greater than background environmental levels. Females often shunt some of their toxic burdens into eggs, and some organochlorines can decrease avian egg viability even at very low concentrations. DDT via DDE was identified as the agent responsible for the shell thinning of brown pelican eggs in the western USA and of osprey eggs that resulted in the species' precipitous decline in the eastern USA during the 1960s. DDE inhibits enzymatic (ATPase) activity in the shell gland preventing calcium transport, causing shell thinning and hence breakage during incubation. Relatedly, dieldrin, a powerful neurotoxin, was deemed responsible for the population crash of peregrine falcons in the UK. Organochlorines, PCBs, and other toxins have been detected in birds and other animals in polar regions as a result of atmospheric transportation.

PCBs

PCBs (polychlorinated biphenyls) are industrial compounds that were used in paints and as fluids in electrical and mechanical equipment. They have long-term persistence and wide dispersal in the marine environment, including polar regions. They too are lipophilic and hence bio-accumulate and magnify in higher levels of food webs. They seem to be highly toxic to seals and marine mammals, although not to birds. PCB production was greatly decreased during the 1970s, and assays with avian tissues or eggs could prove informative for levels of contamination in marine food webs. Because most of the PCBs produced are either at refuse sites or still in use, it is important to continue to monitor their presence, effects and environmental dispersion. Terns are the most sensitive seabird species known to the toxic effects of PCBs.

Heavy Metals

Copper, mercury, lead, and cadmium produce the most serious forms of heavy metal pollution in marine environments. Pollutant levels generally parallel those of regional environments with highest levels being found in the Mediterranean, intermediate levels around the British Isles, and lowest levels in northern Norway and eastern Canada. Many metals, particularly copper, mercury, and lead, and the chemical tributyltin (TBT) have been incorporated into marine paints as anti-fouling agents. These biocides are lethal to bivalves and have resulted in their elimination from many benthic communities.

Metals tend to accumulate in very specific body tissues (e.g. cadmium in kidneys) and assays are usually targeted precisely. Avian eggs also provide useful assays of some metals and have been found to exhibit oceanographic trends in mercury contamination.

Interestingly, birds shunt body burdens of mercury and perhaps tributyltin to growing feathers that are molted annually. Assays of mercury levels in feathers permit the assessment of both spatial and temporal fluctuations in contamination. Feather assays of methyl-mercury are attractive for many reasons: (i) removing selective feathers is harmless to the animal, (ii) feathers can be easily sampled and stored without freezing or other preparation, (iii) feathers can be simultaneously collected over large oceanographic regions, (iv) the metal burdens of the same individuals can be compared in successive years, (iv) historical trends in pollution can be obtained by analyzing feathers from museum specimens.

Mercurial relationships with feathers are particularly interesting in that inorganic mercury that is deposited atmospherically, like other heavy metals, adheres to feather surfaces and can be measured. In comparison, methyl-mercury that is derived from food sources is incorporated into the keratin structures of feathers. The use of small body feathers from circumscribed plumage sites, such as the scapulars, is proving most amenable for comparative analyses. Mercury levels in the feathers of adults tend to be more variable than those of nestlings that are accumulated during a brief period from food obtained in the vicinity of the nest. The same holds for eggs.

Mercury levels in the feathers of puffins, shearwaters, and skuas have increased during the past century in the UK. Levels in the Baltic sea birds also increased from the beginning of the twentieth century, and this trend was attributed to the extensive use of alkyl-mercury as a seed treatment in Scandinavia. Mercury levels in the feathers of auks in the North Atlantic during the 1970s and earlier were much lower than those of auks in the Baltic Sea, reflecting the higher pollutant levels there.

An important caution about the use of feathers for historical analysis is that if the species' diet has changed trophic levels (as occurs at times) over the study period, this could influence the levels of metals in the feathers. Hence it is most conservative in historical reconstructions to target specialist species with narrow dietary breadths. However, stable isotopic determinations of trophic level can also be derived from feathers to assess possible dietary changes.

Fish absorb mercury directly from water through their gills and can also be used to assess contamination by heavy metals. However, unlike fish, birds do not bio-accumulate mercury over their lifetimes and so offer different assaying possibilities. Evidence suggests that mesopelagic and deep-water fishes accumulate higher levels of mercury than epipelagic and coastal fishes. These findings have led to the hypothesis that inorganic mercury is converted to methyl-mercury in low oxygen environments in deep oceans and that this may facilitate uptake in these fishes and then hence by the birds that prey on them. Research is ongoing.

Lead weights used by fishers and lead shot used by hunters are major sources of contamination. Lead concentrations are highest in estuarine and inshore areas, and lead toxicity has been responsible for mortality among swans, marine waterfowl, and seabirds. This mortality has attracted considerable public attention that is resulting in (albeit too slowly) the replacement of lead with nontoxic materials. Blood and enzyme analyses can be used to assess environmental lead levels. Eiders often exhibit high levels of copper in livers and appear to be the best potential avian species for assaying this metal in marine environments. Cadmium and other metals that are not lipid soluble are not assayed in feathers, so there is no advantage in assaying birds compared to other taxa. All atmospherically deposited metals can be evaluated by accumulations on feather surfaces.

Oil Pollution

Chronic illegal discharges of unsegregated ballast and bilge water and tank flushes at sea pose long-term environmental problems for birds and other marine organisms. Unlike highly publicized situations involving tanker spills, oiled birds found on beaches are often the first and sometimes the only evidence that a pollution event, likely illegal, has occurred. Standardized beach bird surveys have generated robust intra- and inter-regionally comparable databases for decades. However, these surveys appear to be possible only in regions inhabited by large numbers of pursuit-diving birds (auks, penguins) that are highly vulnerable to oil at sea (**Figure 1**). Surveys have shown divergent trends in oil pollution in different regions: decreasing in the north-east Atlantic and increasing in the north-west Atlantic. Oil pollution is extensive in the Mediterranean where, owing to an absence of auks, beach bird surveys have not proven tractable.

Once oiled, birds often swim to shore to get out of the cold water at high latitudes and to attempt to

Figure 1 A murre entangled in a small piece of net (on left) and an oiled murre, both recovered on a beach on the south coast of Newfoundland.

clean their plumage on land. Carcass trajectories are influenced by winds and currents. The numbers of oiled birds that are recovered on shore are fractions of those oiled at sea, and estimates of these proportions are being assessed by experiments involving the release of drift blocks at sea. Most oiled seabirds come ashore on beaches exposed to dominant wind directions. Such geographic features are essential considerations in selecting beach bird survey sites.

Species composition can at times be indicative of the distance of a pollution event from the coast. For example, murres tend to be oiled farther from shore than dovekies and sea ducks, and when the composition of oiled species changes from the former to the latter it can reflect the shoreward movement of an oil slick. Gas chromatographic and mass spectrometric techniques have been used to 'fingerprint' the compositions of hydrocarbons removed from oiled seabirds and have been used in efforts to prosecute specific ships alleged to have polluted. There is also the possibility of 'marking' oil shipments so that they could be identified in the case of discharge or spillage. Separators have to be installed as mandatory equipment on ships to promote oil recycling, in much the same way as anti-pollution devices have been legislated for automobiles. However, until binding, enforceable international agreements require the recycling and discharging of used oil at land-based flushing facilities, hydrocarbon pollution in the world's oceans will be a fact of life and death.

Radionuclides

Radionuclides (e.g. cesium) are released from weapon testing and use and from industrial accidents, such as those associated with power-generat-

ing facilities. As is the case with many other pollutants, specific radionuclides are taken up in specific tissues. These chemicals are monitored in shellfish with concerns for human consumption. Hence, shellfish predators including shorebirds, gulls, and sea ducks are likely the best avian species to assay with reference to this source of pollution. Seabirds could potentially provide useful indications of the bioavailability of certain radionuclides over global oceanographic regions.

Plastic

Plastic biodegrades slowly, is very environmentally persistent and occurs in all of the world's oceans. The replacement of twine nets with synthetic monofilament nets has extended the existence of lost and discarded nets adrift at sea that entangle and kill many fish, birds, and mammals (**Figure 1**). The levels of mortality associated with these by-catches can be very high and produce large-scale negative population effects.

Thousands of plastic pellets litter each square kilometer of ocean surface and often accumulate at fronts. Many marine animals ingest plastic, especially small particles of industrial plastics such as styrofoam. Ingestion can create gastrointestinal problems and result in mortality. Petrels appear most vulnerable to small bits of plastic. PCBs and other toxic chemicals often adhere to the surfaces of plastic debris and hence can also increase the contaminant burdens of animals that ingest plastic. Some seabird species incorporate plastic strapping, netting, line, and other solid objects collected from the ocean surface into nests. Cormorant and gannet nests with plastic built into them have increased in recent decades, reflecting increased plastic pollution at sea.

Acid Rain

Acid precipitation of industrial releases of sulfur dioxide, ammonia, and nitrogen oxides has generated many environmental problems. Most of these have been evidenced in terrestrial and freshwater environments. However, as the generation of this pollution increases, effects are to be expected in marine environments.

Conclusion

As pollutant inputs to the environment continue to accelerate, diversify, and combine in novel ways, new effects on marine birds will be detected and new avian bio-assays will continue to be developed and needed. A healthy and diverse avifauna supported by a natural diverse prey base is indicative of a well functioning and healthy marine environment. Decreases in avian diversity are evident in regions with increased pollution levels. Nature abhors vacuum and life proliferates, but biodiversity creates the fabric of life that sustains the natural functioning of large-scale ecosystem processes. For the sake of the oceans and for our own benefit, it is essential to do everything possible to understand human-induced threats to the world's oceans and with or without that understanding to protect and preserve them.

See also

Anti-fouling Materials. Atmospheric Input of Pollutants. Chlorinated Hydrocarbons. Metal Pollution.

Oil Pollution. Pollution Control. Pollution: Effects on Marine Communities. Pollution, Solids. Radioactive Wastes.

Further Reading

Furness RW and Greenwood JJD (eds) (1993) *Birds as Monitors of Environmental Change*. London: Chapman and Hall.

Furness RW and Rainbow PS (eds) (1990) *Heavy Metals in the Marine Environment*. New York: CRC Press.

Poole A (1987) *The Osprey: A Natural and Unnatural History*. Cambridge: Cambridge University Press.

Ratcliffe DA (1967) Decrease in eggshell weight in certain birds of prey. *Nature* 215: 208–210.

Spitzer PR, Riseborough RW, Walker W *et al.* (1978) Productivity of ospreys in Connecticut – Long Island increase as DDE residues decline. *Science* 202: 203–205.

SEALS

I. L. Boyd, Natural Environment Research Council, Cambridge, UK

doi:10.1006/rwos.2001.0432

Taxonomy

The seals, or Pinnipedia, are the suborder of the Carnivora that includes the Phocidae (earless or 'true' seals), Otariidae (eared seals, including fur seals and sea lions) and the Odobenidae (walrus). They are related to the bears, based on a common ancestry with terrestrial arctoid carnivores (**Figure 1**). The otariids retain more of the ancestral characteristics than the other two groups but all have a more or less aquatic lifestyle and display highly developed morphological and physiological adaptations to an aquatic existence.

The Pinnipedia are made up of 34 species and 48 species/subspecies groupings (**Table 1**). However, with the advent of new methods based on DNA analysis for examining phylogeny and also because of new methods used to track animals at sea many of these groupings are questionable. Several groups that were thought to have been different species have overlapping ranges and are likely to interbreed. It seems most probable that the southern fur seals (*Arctocephalus* sp., Table 1) are not distinct species. Conversely, some of the North Atlantic phocid pinnipeds that are classified as single species are likely to be better represented as a group of subspecies.

The gray seal is a particular example in which three genetically distinct populations (NW Atlantic, NE Atlantic and Baltic) are recognized.

Distribution and Abundance

The greatest diversity and absolute abundances of pinnipeds occurs at temperate and polar latitudes (Table 1). Only three phocid seal species, the monk seals, are truly tropical species and all of these are either highly endangered or, in one case, may be extinct. Among the otariids, fur seals and sea lions extend their distributions into the tropics but their absolute abundance in these locations is low compared with the populations at higher latitudes.

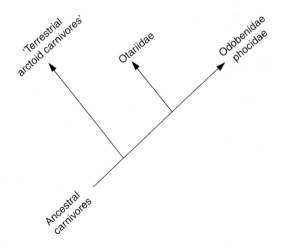

Figure 1 Pinniped phylogeny from a cladogram based on postcranial morphology. (Reproduced from Berta *et al.*, 1989.)

Probably > 50% of the biomass of pinnipeds in the world is derived from one species, the crabeater seal. Some estimates of the abundance of this species have undoubtedly been exaggerated but it is nevertheless the dominant species. This is partly because its main habitat is the vast Antarctic pack ice. The ringed seal has a comparable distribution in the Arctic and it also has abundances numbered in the millions. The relative numbers of the different species are shown in Table 1.

Although the status of the populations of some species/subspecies groups is unknown, only seven (22%) groupings are in decline whereas seventeen (55%) of these groups are increasing in abundance. However, twenty-three groups (48%) are classified as being in need of some form of active conservation management. Threats to pinnipeds can be represented as: (1) *direct threats* from harvesting and international trade, incidental catch by commercial fisheries and direct killing by fishermen; (2) *intermediate threats* from episodic mass mortalities, habitat degeneration (including environmental pollution, competition for food with humans, disturbance and changes to the physical environment); (3) *longer-term threats* from climate change and reduction of genetical diversity.

Morphological and Physiological Adaptations to Aquatic Life

Morphological adaptations include the modification of limbs to form flippers for swimming, the development of a streamlined fusiform shape, the presence of insulation in the form of fur and/or a subcutaneous layer of blubber and increased visual acuity for foraging at extremely low light levels. Unlike cetaceans, an ability to echolocate has not been confirmed in pinnipeds, although some studies have purported to show that seals are capable of behavior that is consistent with echolocation abilities.

Physiological adaptations include a highly developed dive response. On submergence this involves the rapid reduction of heart rate, reduced peripheral circulation and the sequestration of large amounts of oxygen bound to myoglobin in the muscles. Pinnipeds also have high concentrations of red blood cells in the blood and changed morphology of the red blood cells themselves, which take on a 'cocked-hat' shape. The architecture of the venous system is also modified, especially amongst phocids, to allow a larger volume of blood to be stored. Included within the dive response is the ability, when at the surface between dives, to increase heart rate rapidly to clear and reprocess metabolic waste products and to re-oxygenate the tissue in preparation for the next dive.

Thermal Constraints

Perhaps the greatest single constraint on the evolution of pinnipeds as aquatic animals is the problem associated with thermoregulating in cold water that is 25 times more conductive to heat than air. Since pinnipeds are endothermic homeotherms that normally maintain a body temperature of 36–38°C, they are presented with a significant thermal challenge when they are immersed in water at or close to freezing. The observation that the greatest number and species diversity of pinnipeds is found in temperate and polar regions suggests that they have adapted well to this challenge. However, the cost associated with this seems to be that pinnipeds have retained a non-aquatic phase in their life histories. By virtue of their relatively small body size, newborn pinnipeds cannot survive for long in cold water and so they are born on land or ice and remain there until they have built up sufficient insulation to allow them to go to sea. Unlike cetaceans, pinnipeds do not appear to have solved the problem of giving birth to young with insulation already developed but many cetaceans have a much larger body size than pinnipeds (thus reducing the thermal challenge to the newborn) and many species also migrate to warmer waters to give birth. Cetaceans appear to have developed a different strategy to deal with the cold.

Pinniped Life Histories: The Constraints of Aquatic Life

An important consequence of the necessity for nonaquatic births in pinnipeds is that mothers are more or less separated from their foraging grounds by the need to occupy land or ice during the period of offspring dependency. Food abundance in the marine environment is not evenly distributed and has a degree of unpredictability in space and time. Therefore, pinniped mothers have had to trade-off the necessity to find a location to give birth which is safe from predation, since pinnipeds are vulnerable when on land, with the need to feed herself and her pup throughout the period of offspring dependency. This has apparently led to two different types of maternal behavior.

In small pinnipeds including all the fur seals, most of the sea lions and small phocids, it is not possible, by virtue of body size alone, for mothers to carry sufficient energy reserves at birth to support both her and her offspring until the offspring is able to be independent. This means that mothers must

Table 1 Species and common names, abundances, trends in abundance and conservation status for the Pinnipedia

Taxonomic classification		Common name	Distribution	Abundance[a] ($\log_{10}$ scale)	Trend in abundance	Conservation status
Family Otariidae						
Subfamily Otariinae						
Eumetopias jubatus		Steller sea lion	North Pacific	***	Declining	Threatened
Zalophus californianus subspecies:	*californianus*	California sea lion	Western USA	****	Increasing	No threat
	wollebaeki	Galapagos sea lion	Galapagos Is.	***	Stable (?)	Rare
	japonicus	Japanese sea lion	Japan	*	Unknown	Possibly extinct
Ontaria byronia		Southern sea lion	South America	***	Declining	Threatened
Neophoca cinerea		Australian sea lion	Southern Australia	**	Stable	Rare
Phocarctos hookeri		Hooker's sea lion	Southern New Zealand	**	Declining	Endangered
Subfamily Arctocephalinae						
Arctocephalus townsendi		Guadalupe fur seal	Guadalupe Is.	**	Increasing	Rare
Arctocephalus galapagoensis		Galapagos fur seal	Galapagos Is.	***	Stable (?)	Rare
Arctocephalus philipii		Juan Fernandez fur seal	Juan Fernandez Is.	**	Increasing	Rare
Arctocephalus australis subspecies:	*australis*	Falkland fur seal	Falkland Is.	***	Increasing (?)	Locally vulnerable
	gracilis	South American fur seal	South America	****	Increasing	No threat
Arctocephalus tropicalis		Sub-Antarctic fur seal	Sub-Antarctic	*****	Increasing	No threat
Arctocephalus gazella		Antarctic fur seal	Antarctic	******	Increasing	No threat
Arctocephalus pusillus subspecies:	*pusillus*	South African fur seal	Southern Africa	******	Increasing	No threat
	doriferus	Australian fur seal	Southern Australia	***	Increasing	No threat
Arctocephalus forsteri		New Zealand fur seal	Australia and New Zealand	***	Increasing	No threat
Callorhinus ursinus		Northern fur seal	North Pacific	******	Declining (?)	No threat
Family Odobenidae						
Subfamily Odobeninae						
Odobenus rosmarus subspecies:	*rosmarus*	Atlantic walrus	Arctic	****	Unknown	Commercially threatened
	divergens	Pacific walrus	Arctic	**	Unknown	Commercially threatened
Family Phocidae						
Subfamily Phocinae						
Halichoerus grypus		Grey seal	North Atlantic	****	Increasing	No threat
Phoca vitulina subspecies:	*vitulina*	Eastern Atlantic harbor seal	NE Atlantic	***	Increasing	No threat
	concolor	Western Atlantic harbor seal	NW Atlantic	***	Increasing	No threat
	stejnegeri	Western Pacific harbor seal	NW Pacific	**	Unknown	No threat
	richardsi	Eastern Pacific harbor seal	NE Pacific	****	Increasing	Rare
	mellonae	Ungava seal	Quebec, Canada	*	Unknown	Rare

Table 1 Continued

Taxonomic classification		Distribution	Abundance[a] (log_{10} scale)	Trend in abundance	Conservation status
Phoca largha		Spotted seal	Bering Sea	****	Unknown	No threat
Phoca hispida subspecies:		Arctic ringed seal	Arctic	*****	Unknown	No threat
	ochotensis	Okhotsk sea ringed seal	Sea of Okhotsk	****	Unknown	No threat
	botnica	Baltic ringed seal	Baltic Sea	**	Increasing (?)	Threatened
	Saimensis	Saimaa seal	Lake Saimaa, Finland	*	Unknown	Endangered
	ladogensis	Ladoga seal	Lake Ladoga, Russia	***	Stable – increasing	Vulnerable
Phoca caspica		Caspian seal	Caspian Sea	****	Stable	Vulnerable
Phoca sibirica		Baikal seal	Lake Baikal	***	Unknown	Vulnerable
Phoca groenlandica		Harp seal	NW Atlantic & Arctic	*****	Increasing	No threat
Phoca fasciata		Ribbon seal	Bering Sea & Arctic	****	Unknown	No threat
Cystophora cristata		Hooded seal	N Atlantic & Arctic	****	Unknown	No threat
Erignathus barbatus subspecies:	*barbatus*	Atlantic bearded seal	Arctic	****	Unknown	No threat
	nauticus	Pacific bearded seal	Arctic	****	Unknown	No threat
Subfamily Monachinae						
Monachus monachus		Mediterranean monk seal	Mediterranean	*	Declining	Endangered
Monachus tropicalis		West Indian monk seal	Caribbean Sea	*	Unknown	Possibly extinct
Monachus schauinslandi		Hawaiian monk seal	Hawaiian Islands	*	Declining	Endangered
Leptonychotes weddellii		Weddell seal	Artarctica	****	Stable	No threat
Ommatophoca rossii		Ross seal	Antarctica	***	Unknown	No threat
Lobodon carcinophagus		Crabeater seal	Antarctica	*******	Stable	No threat
Hydrurga leptonyx		Leopard seal	Antarctica	***	Stable	No threat
Mirounga leonina		Southern elephant seal	Sub-Antarctic	****	Declining (?)	No threat
Mirounga angustirostris		Northern elephant seal	Sub-Antarctic	****	Increasing	No threat

[a]The number of asterisks denote the range in the size of the world populations: *, 0–1000; **, 1000–10 000; ***, 10 000–100 000; ****, 100 000–1 000 000; *****, 1 000 000–10 000 000; ******, 10 000 000–100 000 000.

supplement their energy reserves by feeding during lactation. In the case of the smallest pinnipeds, the fur seals, mothers rely almost entirely on the energy from foraging and have very few reserves. Therefore, these small pinnipeds are restricted to breeding at sites which are close enough to food for the mothers to be able to make foraging trips on time scales that are less than the time it would take their pup to starve. Consequently, lactation in these small pinnipeds tends to be extended over several months and, in a few cases, can last over a year.

In contrast, pinnipeds of large body size (the transition in this case between large and small appears to occur at a maternal body mass of about 100 kg) are able to carry sufficient energy reserves to allow mothers to feed both themselves and their offspring while they are ashore. In these species, the tendency is for mothers to make only a single visit ashore and for her not to feed during lactation. As a result, these mothers have a short lactation and, in the case of the hooded seals, this is reduced to only four days, but 15–30 days is more normal.

These types of behaviors, which stem directly from the combined physical restrictions of thermoregulation in newborn pups and maternal body size, have had two further important consequences. The first of these is that larger pinnipeds are better able to exploit food resources at greater distance from the birth site and they have been shown to range over whole ocean basins in search of food. This is a necessary consequence of their larger size because, in contrast to small pinnipeds, they must exploit richer food sources because of their greater absolute food requirement. Since richer food sources are also rarer food sources, the large pinnipeds have fewer options as to where they can forage profitably. Thus, with some exceptions the large pinnipeds occupy much larger ranges than the small pinnipeds.

By mammalian standards all pinnipeds are of large body size. Even though pups are well developed at birth, this means that it takes several years for most pinnipeds to grow to a body size large enough for them to become sexually mature. The minimum duration to reach sexual maturity is about 3 years and the maximum, in species such as the grey seal, is 5–6 years. Thereafter, they only produce a single pup each year and the individuals of most species will fail to produce a pup about one year in four. However, since females may live for 20 to >40 years, they are relatively long-lived animals.

Mating Systems

The second consequence of the physical restriction that thermoregulation in newborn pups places on pinnipeds is the mating system. This feature of pinniped biology has been the subject of intensive investigations, largely because it is much the most dramatic and obvious part of pinniped life-histories. There has been much speculation as to why pinnipeds should mostly have developed mating behavior involving dense aggregations and apparent extreme polygyny but it is likely to be a consequence of the necessity for mothers to give birth out of the water. Restrictions in the availability of appropriate breeding habitat (defined in terms of both its proximity to food and its protection from predation), together with reduced risk of predation that individuals have when they are in groups, probably combined to increase the fitness of those mothers that had a tendency to give birth in groups. It is also considerably more efficient, in energetic terms, to have to return to land only once during each reproductive cycle. Females have made use of an ancestral characteristic involving the existence of a postpartum estrus at which most females are mated and become pregnant. Without this, females would have been required to seek a mate at a time when, for many species, the population would be highly dispersed over a wide area while foraging. This means that appropriate mates would have been more difficult to find than at a time of year when the population is highly aggregated.

Males that are present when there is the greatest chance of mating will be most likely to gain greatest genetical fitness. Thus, in almost all pinnipeds, a competitive mating system has developed around the rookeries of females with their pups. Moreover, this has led to selection for male morphological and behavioral characteristics that confer greater ability to dominate matings. The male hooded seal has developed a deep red septum between his nostrils that can be blown out like a balloon as a display organ; male elephant seals have developed loud vocalizations which, together with their enlarged rostrum, make a formidable display; male harbor and Weddell seals have complex and loud underwater vocalizations that are almost certainly part of a competitive mating system; and in most species there is a marked sexual dimorphism of body size in which males can be six to eight times the mass of females. This sexual dimorphism in mass may serve a double function: increased mass leads to increased muscle power and the ability to fight off rivals for matings and increased mass also confers increased staying power allowing individual males to fast while they are on the breeding grounds and maintaining their presence amongst receptive females for as long as possible. However, this larger body mass also has a cost in that, because of their greater absolute energy expenditure, the larger males must find richer food sources to be able to feed profitably and recover their condition between breeding seasons. A consequence of this is that males have lower survival rates than the smaller females.

An exception to much of this is found in many of the seals that breed on ice. In these cases, mothers often have the option to give birth in close proximity to food and, at least in the Antarctic where there are no polar bears, they are relatively safe from predation. There is no sexual dimorphism in the crabeater seal, a species which gives birth in the Antarctic pack ice without any detectable aggregation of mothers. In the Arctic, the harp seal is the rough ecological equivalent of the crabeater seal and in this species there are large aggregations, known in Canada as whelping patches. It would appear that one of the main contrasts between these species is that harp seals are exposed to polar bear predation whereas crabeater seals are not. In neither case is there marked sexual dimorphism of body size despite evidence for competitive mating. As in the case of the Weddell and harbor seals, which have only small sexual dimorphism of body size, these species mainly mate in the water rather than on land. Therefore, it may be that large body size in male pinnipeds is mainly a characteristic that is an advantage to those that have terrestrial mating.

Diet

Seals are mostly fish-eating although the majority of species have a broad diet that also includes squid, molluscs, crustaceans, polychaete worms and, in certain cases other vertebrates including seabirds and other seals. Even those that prey mainly on fish take a broad range of species although there is a tendency for specialization on oil-bearing species such as herring, capelin, sand eels/lance, sardines and anchovies. This is because these species have a high energy content and they are often in shoals so that they may be an energetically more profitable form of prey than many other species.

Perhaps the most specialized pinniped in terms of diet is the walrus which forages mainly on benthic molluscs, crustaceans and polychaetes. Its dentition is adapted to crushing the shells of molluscs and their tusks are used to stir up the sediment on the sea bed to disturb the prey within. In the Arctic, bearded seals have a similar feeding habit and several other species feed regularly on benthic invertebrates. Among gray seals there is evidence that some individuals specialize in different types of prey. For most species, feeding occurs mainly within

the water column and may be associated with particular oceanic features, such as fronts or upwellings of deep water that are likely to contain higher concentrations of prey. Seals may migrate distances of up to several thousand kilometers to find these relatively rich veins of food.

The crabeater seal feeds almost entirely on Antarctic krill (a small shrimp-like crustacean) that it gathers mainly from the underside of ice floes where the krill themselves feed on the single-celled algae that grow within the brine channels within the ice. Antarctic fur seals also feed on krill to the north of the Antarctic pack ice edge and many of the Antarctic seals rely to varying degrees on krill as a source of food. In fact Antarctic krill probably sustain more than half of the world's biomass of seals and also sustain a substantial proportion of the biomass of the world's sea birds and whales. The dentition of crabeater seals is modified to help strain these small shrimps from the water.

Many species of seals will, on occasions, eat sea birds. Male sea lions of several species have been recorded as snacking on sea birds and male Antarctic fur seals regularly feed on penguins. However, the most specialized predator of sea birds and other seals is the leopard seal. This powerful predator is found mainly in the Antarctic pack ice and is credited with being the most significant cause of death amongst juvenile crabeater seals even though few cases of direct predation have been observed. Individual leopard seals may specialize on specific types of prey because the same individuals have been observed preying on Antarctic fur seals at one location in successive years and one of these has been seen at two locations over 1000 km apart where young Antarctic fur seals can be found.

Diving for Food

The development over the past decade of microelectronic instruments for measuring the behavior of pinnipeds has put the adaptations for aquatic life in these animals into a new perspective. Some pinnipeds are capable of very long and very deep dives in search of their prey. The result of this diving ability is that pinnipeds are able to exploit on a regular basis any food that is in the upper 500 m of the water column. In general, larger body size confers greater diving ability mainly because the rate at which animals of large body size use their oxygen store is less than that for small individuals. Ultimately, it is the amount of oxygen carried in the tissues that determines how long a pinniped can stay submerged and time submerged limits the depth to which pinnipeds can dive. Consequently the largest pinnipeds, elephant seals, dive longer and deeper than any others.

On average adult elephant seals dive to what seems to us as a punishing schedule. Average dive durations can exceed 30 minutes with about 2 minutes between dives and elephant seals maintain this pattern of diving for months on end, only stopping every few days to 'rest' at the surface for a slightly longer interval than normal but usually much less than an hour. Technically, elephant seals are more correctly seen as surfacers rather than divers.

Occasionally elephant seals dive to depths of 1500 m and dives can last up to 2 hours with no apparent effect on the time spent at the surface between dives. It is still a mystery to physiologists how elephant seals, and many other species including hooded seals and Weddell seals, manage to have such extended dives. Many physiologists believe that free-ranging seals like elephant seals are able to reduce their metabolic rate while submerged to such an extent that they can conserve precious oxygen stores and they can then rely on aerobic metabolism throughout the dives. This strategy may allow these animals to access food resources that the majority of air-breathing animals cannot reach. As described above, this is likely to be of critical importance to these large-bodied animals because of their need to find rich food sources.

See also

Krill. Marine Mammal Evolution and Taxonomy. Polar Ecosystems.

Further Reading

Berta A, Ray CE and Wyss AR (1989) Skeleton of the oldest known pinniped, *Enaliarctos mealsi. Science* 244: 60–62.

Laws RM (ed.) (1993) *Antarctic Seals*. Cambridge: Cambridge University Press.

Reijnders P, Brasseur S, van der Toorn J *et al.* (1993) *Seals, Fur Seals, Sea Lions, and Walrus. Status and Conservation Action Plan*. Gland, Switzerland: IUCN.

Reynolds JE III and Rommel SA (eds.) (1999) *Biology of Marine Mammals*. Washington and London: Smithsonian Institution Press.

SEAMOUNTS AND OFF-RIDGE VOLCANISM

R. Batiza, Ocean Sciences, National Science Foundation, VA, USA

doi:10.1006/rwos.2001.0097

Summary

There are three major types of off-axis volcanism forming the abundant seamounts, islands, ridges, plateaus, and other volcanic landforms in the world's oceans. (1) The generally small seamounts that form near the axes of medium and fast-spreading ridges but less so at slow-spreading ones. These are most likely a result of mantle upwelling and melting in a wide zone below mid-ocean ridges, although off-axis 'mini plumes' cannot be ruled out. (2) The huge oceanic plateaus and linear volcanic chains that form from starting plumes and trailing plume conduits respectively. It is widely believed that mantle plumes originate in the lower mantle, perhaps near the core–mantle boundary. (3) Off-ridge volcanism that is not due to plumes, but which chemically and isotopically resembles plume volcanism. Emerging data indicate that much off-axis volcanism previously ascribed to mantle plumes is not plume-related. Several distinct types of activity seem to be the result of various forms of intraplate mantle upwelling, or pervasively available asthenosphere melt rising in conduits opened by intraplate stresses, or both. Seamounts, ridges, and plateaus produced by off-axis volcanism play important roles in ocean circulation, as biological habitats, and in biogeochemical cycles involving the ocean crust.

Introduction

The seafloor that is produced at mid-ocean ridges is ideally quite uniform, and except for the regular abyssal hills and the rugged linear traces of ridge offsets, it is essentially featureless. In strong contrast, real ocean crust in the main ocean basins and the marginal basins and seas is decorated with volcanic islands, seamounts, ridges, and platforms that range in size from tiny lava piles only tens of meters high to vast volcanic outpourings covering huge areas of seafloor. Volcanoes that are active close to mid-ocean ridges are related to the ridge processes that build the ocean crust, whereas those erupting farther away, so-called off-axis, off-ridge, or intraplate volcanic features, are the result of processes that are unrelated to mid-ocean ridges. The largest oceanic volcanic features, oceanic plateaus and linear chains of islands and seamounts (known as large igneous provinces or LIPs) (*see* **Large Igneous Provinces**), are considered to be the result of rising plumes of hot material that may originate as deep as the core-mantle boundary. Laboratory models suggest that when first initiated, plumes consist of large buoyant 'heads' (so-called starting plumes) trailed by much narrower cylindrical conduits that continue to feed material upward. In these models, the massive starting plume experiences decompression melting and eruption of this melt produces large oceanic plateaus. When starting plumes rise below continents, they produce huge volcanic outpourings called flood basalts. After passage of the starting plume, further melting of the rising cylindrical conduit can build linear island and seamount chains on the moving, over-riding plate. Plumes are found within plate interiors and also at and near mid-ocean ridges, with which they interact.

While volumetrically plumes may be responsible for most off-ridge volcanism, there are other forms of off-ridge volcanism that do not appear to be related to mantle plumes, although chemically and isotopically their magmas are very similar to those of supposed plumes. These diverse and less voluminous volcanic features include individual isolated seamounts, *en echelon* volcanic ridges, clustered seamounts, and lava fields. The distinct tectonic settings in which they occur suggest that their origin is related to stresses induced in moving lithospheric plates; however, the manner in which melt is produced is uncertain.

This article describes near-ridge seamounts, plume-related volcanism, and off-axis volcanism that is not related to mantle plumes. For each of these three major types, their characteristics are reviewed briefly and the evidence for their origin and evolution is discussed. A common theme is the question of whether volcanism is principally controlled by the availability of mantle-derived melt, or alternatively, the extent to which the thermomechanical properties of ocean lithosphere variably influence the eruption of this melt. It is clearly more difficult for magma to penetrate and erupt through thick, cold, and fast-moving lithosphere than through thin, hot, and slow-moving lithosphere.

Another common thread is the extent to which different kinds of off-axis volcanism can be linked with patterns of mantle flow occurring at various levels within the Earth's mantle: flow which is linked in fundamental ways to the Earth's heat loss and the dominant plate tectonic processes that control the dynamic outer layer of the Earth. Finally, diverse oceanic volcanic features interact in important ways with ocean currents and biological organisms. Because of volcanic degassing, hydro-

thermal activity, and slow weathering processes, these volcanic features also affect the chemistry of sea water and influence patterns of sedimentation. the oceanographic effects of seamounts and other off-axis volcanic features are briefly discussed.

Near-ridge Seamounts

The most abundant seamounts on Earth, probably numbering in the millions, are the relatively small, mostly submerged volcanoes that occur on the flanks of mid-ocean ridges. They originate at and grow fairly close to the active mid-ocean ridges, so despite their huge numbers only a small percentage are active at any given time, and because of their small size they contribute only a few percent of material to the ocean crust. Although the existence of abundant seamounts on the ocean floor has been known since the earliest exploration of the ocean, the availability in the early 1980s of multibeam swath mapping sonar systems (*see* **Ships**) has made it possible to study large numbers of these seamounts. Several dozen have been studied in detail with deep-sea research submersibles (*see* **Manned Submersibles, Deep Water**).

Individual volcanic seamounts vary in size from small dome-shaped lava piles only tens of meters high, to large volcanic edifices several kilometers in height. Commonly, they have steep outer slopes, flat or nearly flat circular summit areas, and collapse features such as calderas and pit craters (**Figure 1**). In general, the smallest volcanoes tend to have the most diverse shapes. They occur as both individual volcanoes and as linear groups consisting of a few to several dozen individual volcanoes (**Figure 2**). In general, those volcanoes comprising chains tend to be larger than the isolated individual ones. Large numbers of seamounts, mostly occurring as linear chains, have been mapped on the flanks of the Juan de Fuca ridge and both the northern (**Figure 3**) and southern (**Figure 4**) East Pacific Rise (EPR).

At the Juan de Fuca ridge and along the southern East Pacific Rise (**Figure 4**), there is a marked asymmetry to the distribution of seamount chains, with most chains present on the Pacific plate. This asymmetry is absent or much less marked along the Pacific-Cocos portion of the northern EPR, and occurs, but with the opposite sense, along the Pacific-Rivera boundary. In contrast with seamount chains, isolated small seamounts near the southern EPR are symmetrically distributed on both flanks of the EPR axis.

Studies at Santa Barbara indicate that near-axis seamounts, whether isolated or in chains, form close

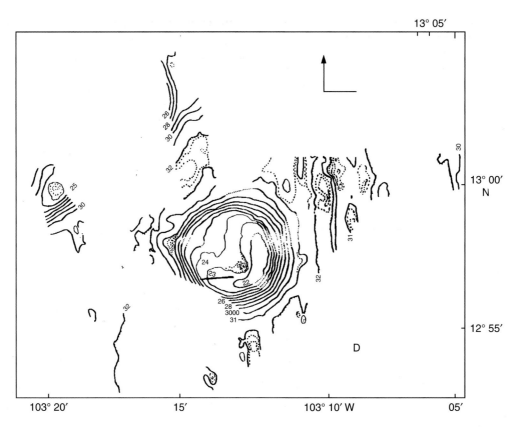

Figure 1 Seabeam map of Seamount 'D' in the eastern equatorial Pacific. Depth contours are in meters (four digits) or hundreds of meters (two digits). The arrow shows the direction of ridge-parallel abyssal hills. Note the relatively flat summit region and the caldera that is breached to the northwest. (Reproduced with permission from Elsevier from Batiza R and Vanko D (1983) Volcanic development of small oceanic central volcanoes on the flanks of the East Pacific Rise inferred from narrow-beam echo-sounder surveys. *Marine Geology* 54: 53–90.)

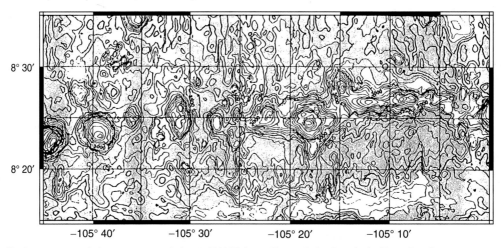

Figure 2 Seabeam map of the seamount chain at 8°20'N (see Figure 3 for location). Note the diverse seamount shapes. (Reproduced with permission from Schierer DS and Macdonald KC (1995) Near-axis seamounts on the flanks of the East Pacific Rise, 8°N to 17°N. *Journal of Geophysical Research* 100: 2239–2259.)

to the axis in a zone that is about 0.2–0.3 million years wide and is independent of spreading rate. Many may continue to grow within a wider zone and a much smaller number may continue to be active at even great distances (several hundred kilo- meters) from the axis. Far from the axis it is difficult to distinguish near-axis seamounts that remain ac- tive for very long periods, from near-axis seamounts that are volcanically reactivated, from true intra- plate volcanism that was initiated far from the axis.

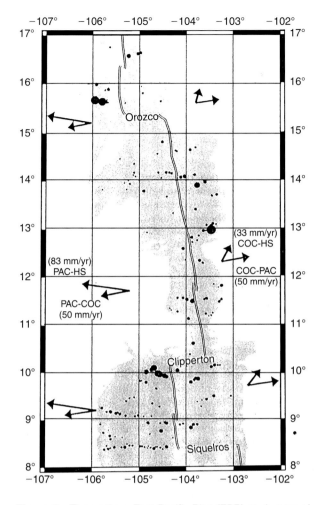

Figure 3 The northern East Pacific Rise (EPR) study area of Schierer and Macdonald (1995). Seamounts > 200 m in height are shown as dots, and the double line is the axis of the EPR. The arrows show the magnitude and direction of relative and absolute plate motions. (Reproduced with permission from Schierer DS and Macdonald KC (1995) Near-axis seamounts on the flanks of the East Pacific Rise, 8°N to 17°N. *Journal of Geophysical Research* 100: 2239–2259.)

In such cases, the distinction between ridge-related volcanism and true intraplate volcanism can be somewhat blurred.

Near-axis seamounts occur most commonly on the flanks of inflated ridges with large cross-sectional areas and abundant melt supply, and the abundance of large ones (> 400 m high) is strongly correlated with spreading rate (**Figure 5**). Abundant seamounts characterize not only modern fast-spreading ridge flanks, but also crust produced at fast spreading rates in the past, for example, in the Indian Ocean before the collision of India with Asia.

While near-axis seamounts form preferentially at inflated portions of fast-spreading ridges, they also occur near offsets (*see* **Mid-Ocean Ridge Tectonics**,

Volcanism and Geomorphology). such as overlapping spreading centers (OSCs), where they form closer to the ridge axis. They may also occur on fracture zones, although this is much more common on old versus young ocean crust. At the slow-spreading Mid-Atlantic Ridge (MAR), studies show that small seamounts are very common within the floor of the axial valley. Many or most of these appear to be a manifestation of ridge axis volcanism from both primary volcanic vents as well as off-axis eruptions fed by lava tubes.

Exactly how and why seamounts form near mid-ocean ridge axes is not known, although the composition of their lavas suggests strongly that they have the same mantle sources as volcanics erupted at the axis. Since the zone of melting that feeds the axes of fast-spreading ridges is very wide, extending several hundred kilometers on both sides of the axis, it is possible that near-axis seamounts are simply due to rising axial melt that was ineffectively focused at the axis. This idea explains their chemistry but not why they so commonly form chains. An appealing idea to explain chains is that they are due to mantle heterogeneities akin to 'mini' mantle plumes, which would help explain the occurrence of chains trending in the direction of absolute plate motion and possibly the observed asymmetry of distribution on the flanks of some ridges, as seen at the Juan de Fuca ridge. However, on the Cocos plate, where the absolute and relative motion are very different, most chains are parallel to relative motion, suggesting that perhaps the movement of the lithosphere or convection rolls parallel to relative motion might trigger seamount formation.

However, not all near-axis seamount chains trend parallel or subparallel to relative or absolute plate motion. Lonsdale showed that the oblique trend of the Larson seamounts near the EPR at ~ 21°N is consistent with its being fed by an asthenospheric melt diapir rising beneath the ridge axis, as envisioned in the model of Schouten and others. A problem with testing this idea further is that, along most of the EPR, the relative and absolute plate motions are quite similar and in this case the Schouten *et al.* trend is not distinct enough from the absolute and relative motion directions to be recognized. In summary, a widely applicable, self-consistent hypothesis to explain all the observations of near-axis seamounts and seamount chains is not yet available.

Finally, on the flanks of the southern EPR (**Figure 4**), numerous chains of near-axis seamounts show an inverse correlation of seamount volume between adjacent chains, suggesting that the magma might originate in plume-like sources in the upper

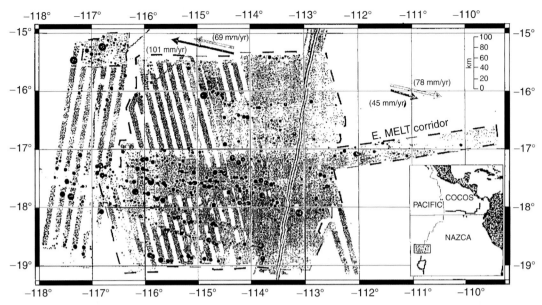

Figure 4 Study area along the southern East Pacific Rise showing the EPR axis (double line) and seamounts as dots. Arrows show the relative (gray) and absolute (black) plate motion vectors. Later studies show much more complete mapping on the Nazca plate to the east of the EPR. Note the very abundant seamount chains present especially on the Pacific plate. (Reproduced with permission from Klewer from Scheirer DS, Macdonald KC, Forsyth DW and Shen Y (1996) Abundant seamounts of the Rano Rahi seamount field near the southern East Pacific Rise, 15° to 19°S. *Marine Geophysical Researches* 18: 14–52.)

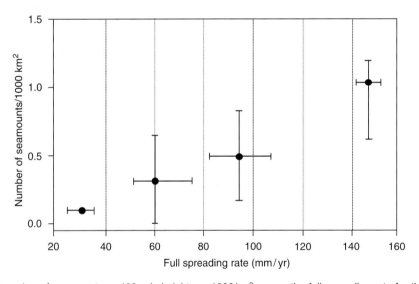

Figure 5 Plot of number of seamounts > 400 m in height per 1000 km² versus the full spreading rate for the Mid-Atlantic Ridge, various medium-spreading ridges, and the northern and southern EPR. Faster spreading ridges produce more near-axis seamounts and inflated ridge segments produce more and larger seamounts than segments with smaller cross-sectional area. (Reproduced with permission from Schierer DS and Macdonald KC (1995) Near-axis seamounts on the flanks of the East Pacific Rise, 8°N to 17°N. *Journal of Geophysical Research* 100: 2239–2259.)

mantle. This is an interesting observation, suggesting that near-axis seamounts are controlled by melt availability. However, the fact that seamounts form in a narrow zone near the axis that corresponds to a lithosphere thickness of 4–8 km and is independent of spreading rate suggests that the lithosphere plays an important role in the origin of near-axis seamounts. In general, the extent to which near-axis seamounts are controlled by magma availability

(and whether their sources are distinct from those feeding the axis), or lithospheric vulnerability or both, is presently unknown.

Off-ridge Plume-related Volcanism

At the opposite end of size spectrum from the small near-axis seamounts are the huge oceanic plateaus

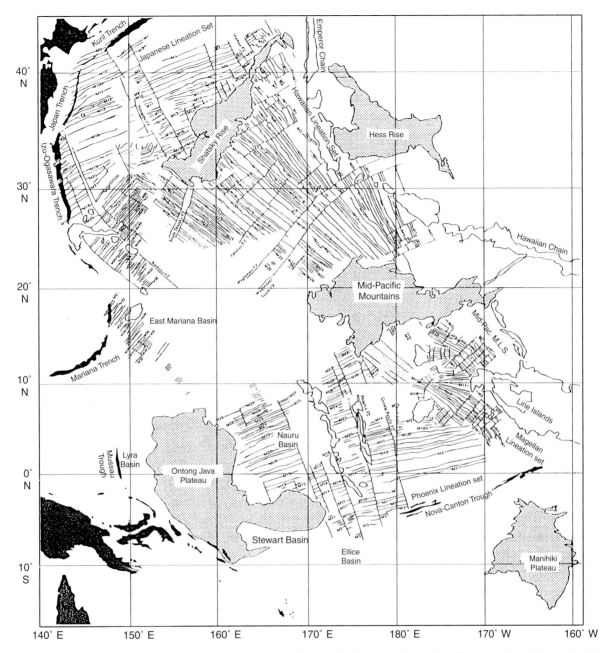

Figure 6 Map of the west and central Pacific showing the major oceanic plateaus of the region. Note also the outline north of the Mid-Pacific Mountains of the Hawaii-Emperor seamount chain with its dogleg just south of the Hess Rise. (Reproduced with permission from Neal CR, Mahoney JJ, Kroenke LW, Duncan RA and Petterson MG (1997) The Ontong Java Plateau. In: *Large Igneous Provinces*, Geophysical Monograph 100, pp. 183–216. Washington, DC: AGU.)

(**Figure 6**) that occur in all the major ocean basins. As previously discussed, these large igneous provinces (LIPs) are thought to be the result of melting of starting plumes, and it has been proposed that the Pacific plateaus were produced by an immense superplume or group of plumes in Cretaceous time. In addition to these huge plateaus, mantle plumes are thought to produce the long linear island and seamount chains that are so common in the ocean basins (**Figure 7**). Widely held corollaries of the

plume hypothesis are that plumes are nearly fixed relative to one another and that they originate in the lower mantle, possibly at the core–mantle boundary. Further, the conventional wisdom is that most of the intraplate volcanism on the planet is due to plumes. The Hawaii-Emperor chain of islands, atolls, seamounts, and drowned islands (guyots) is the classic example of a 'well-behaved' plume, with an orderly and predictable age progression of eruptive ages and bend in direction (**Figure 6**) at 43 Ma

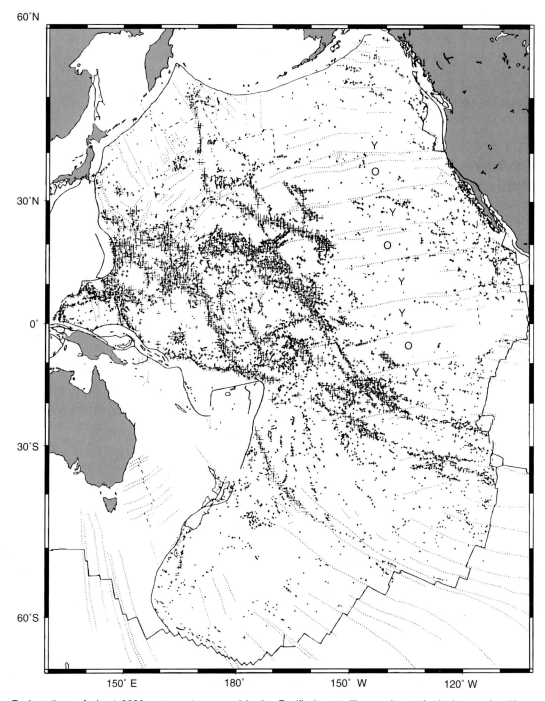

Figure 7 Locations of about 9000 seamounts mapped in the Pacific by satellite gravity methods (crosses), with cross size proportional to the maximum vertical gravity gradient. Note that the western and central Pacific have the most numerous large seamounts. Note that while many seamounts are clustered into linear chains and equant clusters, some are relatively isolated. (Reproduced with permission from Wessel P and Lyons S (1997) Distribution of large Pacific seamounts from Geosat/ERS-1: Implications for the history of intraplate volcanism. *Journal of Geophysical Research* 102: 22 459–22 475.)

when the Pacific plate motion changed from NNW to WNW. Finally, the composition of Hawaiian lavas and those of many other suspected mantle plumes are distinct from the sources that supply mid-ocean ridges, consistent with the hypothesis

that plumes sample a different and perhaps deeper region of the Earth's mantle.

Interestingly, in many cases mantle plumes are close to or centered on active mid-ocean ridges, in which case the plume and ridge interact and mixing

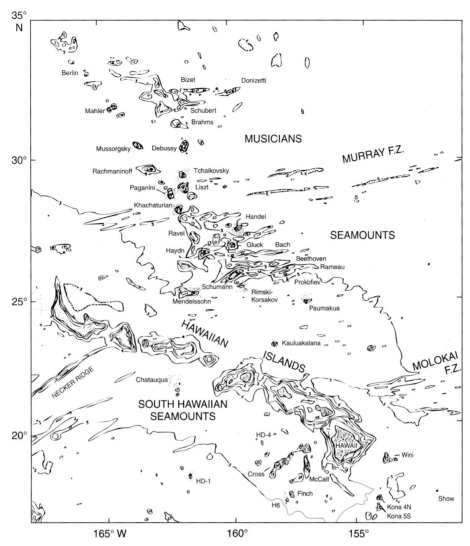

Figure 8 Generalized map of the central Pacific (contours in km) showing a portion of the Hawaiian island chain and the Musicians seamounts. Note that the group comprises a chain of NW trending seamounts including Mahler, Berlin, and Paganini and also E–W trending ridges such as those including Bizet and Donizetti to the north and Bach and Beethoven to the south. (Reproduced with permission from Sager WW and Pringle MS (1987) Paleomagnetic constraints on the origin and evolution of the Musicians and south Hawaiian seamounts, central Pacific Ocean. In: *Seamounts, Islands, and Atolls*, Geophysical Monograph 43, pp. 133–162 Washington, DC: AGU.)

of mantle sources is observed. Iceland is the classic example of a ridge-centered plume; whereas Galapagos is a good example of plume–ridge interaction. In addition to mixing between plume and ridge mantle sources, plume–ridge interaction can lead to the formation of the second type of hot spot island chain discussed by Morgan, in which case the orientation of the linear chain is not parallel to the absolute plate motion (as for normal plumes), but rather has a trend intermediate between the absolute and relative plate motions. An example of this type of seamount chain may be the Musicians seamounts (**Figure 8**), which consists of a western chain of seamounts oriented NW, with roughly E–W trend-

ing ridges progressing eastward. The NW trending chain has the proper orientation for a normal hot spot chain, whereas the E–W ridges appear to have been produced by plume–ridge interaction and are intermediate in trend between the absolute and relative plate motions in the Cretaceous when the Musicians plume interacted with the Pacific-Farallon spreading center.

Off-axis Volcanism not Related to Plumes

There is increasing evidence that plumes may not be the only, or even the most abundant form of intra-plate volcanism within the ocean basins. While studies of non-plume intraplate volcanism are just beginning, at least several distinct types of occurrences have been documented. A considerable obstacle to non-plume hypotheses of intraplate volcanism has been the general belief that mantle upwelling is required for melting, as at ridges and plumes, combined with the fact that most models of mantle convection show no upwelling in intraplate regions. One way around this problem is to show that secondary upwelling can occur in intraplate regions, as with Richter and Parson's longitudinal upper mantle convective rolls (called Richter rolls). Another way is to invoke localized upward mantle flow into depressions or recesses in the base of the lithosphere. A final possibility is to invoke diffuse regional mantle upwelling, as might be generated by a weak mantle plume. A completely different way around the problem, discussed by Green and others, is to cause melting not by decompression, but rather by an influx of volatiles, as is thought to occur at convergent margins. In mid-plate settings, volatiles could perhaps migrate upward from the low velocity zone of the asthenosphere. If this occurs, then magmas may generally be present and available below most ocean lithosphere, and would need only an appropriate pathway for eruption.

Recent surveys have documented the presence on older Pacific seafloor, of long, *en echelon*, linear ridges whose trend is distinct from that of plume traces on the Pacific plate. For example, the Puka Puka ridges (**Figure 9**), stretch for at least several thousand kilometers and their morphology suggests that they are due to eruptions accompanying tensional cracking of the Pacific plate. Interestingly, the lavas of the Puka Puka ridges are chemically similar to those of supposed plumes on the Pacific plate; however, the trend of the ridges and the measured age progression of volcanism indicate that the ridges could not be due to a mantle plume.

Another form of intraplate volcanism not involving mantle plumes is the very common formation of large volcanoes of alkali basalt within the axes of inactive or fossil spreading ridges. Lonsdale has documented their very common existence in fossil ridges of the extinct Pacific-Farallon ridge system, the Mathematician fossil ridge, and the fossil Galapagos Rise (**Figure 10**). In some cases, these volcanoes are large enough to form islands, for example, Guadalupe Island off the coast of Baja California. Samples from these islands, and rarer samples from submerged seamounts, indicate that these lavas also are indistinguishable from supposed plume lavas on the Pacific plate.

A third form of non-plume volcanism that appears to be fairly widespread is associated with flexure caused by loading of the ocean lithosphere. Examples of this type of volcanism include lava fields found on the flexural arch associated with the Hawaiian island chain. The so-called North Arch and South Arch lava fields contain lavas not unlike those of the Hawaiian plume, but they erupted several hundred kilometers from the presumed location of the plume. An example on a smaller scale is Jasper seamount (**Figure 11**), which is surrounded by a ring of seamounts built on its flexural arch. Finally, there is the example of the southern Austral islands and seamounts, which recent studies suggest cannot be explained by a mantle plume, as previously proposed. Instead, it appears that these volcanoes are the result of available melts erupting in response to flexural loading by nearby edifices. In all these cases where samples are available, the lavas are chemically and isotopically similar to supposed plume lavas.

The most incompletely documented occurrences of non-plume volcanism, but potentially important in terms of volumes, are isolated large seamounts and groups of seamounts forming clusters rather than linear chains. About a dozen examples of iso-

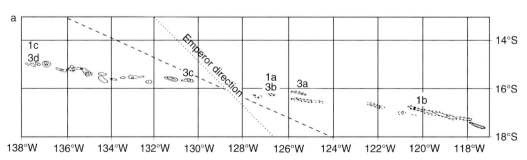

Figure 9 Bathymetric map of part of the Puka Puka Ridges showing only the ridges for clarity. Note their *en echelon* geometry and their trend which is distinct from the Hawaiian chain direction (dashed) and the Emperor chain direction (dotted). (Reproduced with permission from Lynch MA (1999) Linear ridge groups: evidence for tensional cracking in the Pacific Plate. *Journal of Geophysical Research* 104: 29 321–29 333.)

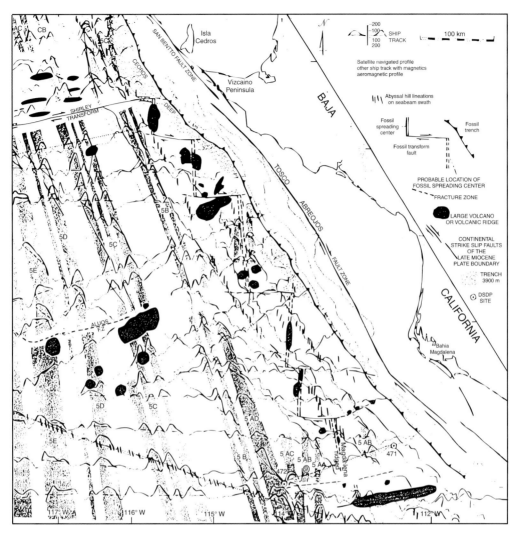

Figure 10 Interpretation of magnetic anomalies off the coast of Baja California showing the locations of probable fossil spreading centers (double dashed lines). Note that many of the fossil spreading centres have large volcanoes or volcanic ridges built in their axes. (Reproduced with permission of the American Association of Petroleum Geologists from Lonsdale P (1991) Structural patterns of the Pacific floor offshore of Peninsular California. In: *The Gulf and Peninsular Province of the Californias*, AAPG Memoir 47, pp. 87–125.Tulsa, OK: AAPG.)

lated large intraplate seamounts not due to mantle plumes have been documented in several studies, for example Vesteris seamount (**Figure 12**). However, there are potentially many hundreds or even thousands of such volcanoes in the ocean basins. It is possible that every large volcano that is not clearly associated with a linear chain is a member of this group. Additional examples of large isolated intraplate seamounts are Shimada seamount and Henderson seamount in the eastern Pacific. As shown by the recent studies of Wessel and Kroenke, there are many large seamounts in the Pacific that are not members of linear chains. Further, Chapel and Small have shown that while many of the very largest volcanoes in the Pacific are associated with

linear chains, many large volcanoes are clustered in nonlinear groups.

In addition to these forms of non-plume volcanism, recent studies have questioned the plume origin of several linear island chains. Wessel and Kroenke propose that many short island chains without clear age progressions, such as the Cook-Australs, the Marqueses, and the Society Islands, are 'crackspots': sites of extensional volcanism along reactivated zones of weakness induced by intraplate stresses. While these ideas are controversial, they suggest that the Pacific may contain only about five mantle plumes, instead of the several dozen that have previously been proposed. If these suggestions prove to be correct, then much, perhaps even most intraplate

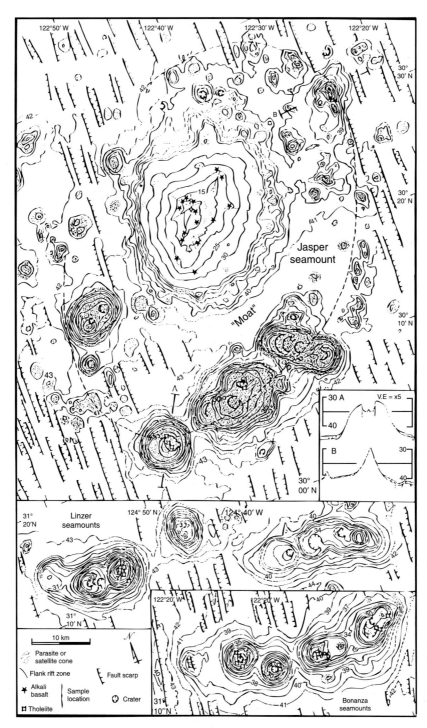

Figure 11 Bathymetric maps of several seamounts (two-digit contours are hundreds of meters). Note the elliptical group of seamounts surrounding Jasper seamount. These are presumed to have erupted on Jasper's flexural arch, similar to seamounts and lava beds elsewhere in the Pacific basin. The Linzer and Bonanza Seamounts are additional examples of near-ridge seamounts built on older Pacific crust, although Linzer, like Jasper, may be part of the Fieberling hot spot chain. (Reproduced with permission of the American Association of Petroleum Geologists from Lonsdale P (1991) Structural patterns of the Pacific floor offshore of Peninsular California. In: *The Gulf and Peninsular Province of the Californias*, AAPG Memoir 47, pp. 87–125. Tulsa, OK: AAPG.)

volcanism in the oceans will be of the non-plume type, with important implications for mantle convection and mechanisms of advective heat loss. Haase has shown that chemically and isotopically, various types of plume and non-plume intraplate volcanism seem to define a single population that

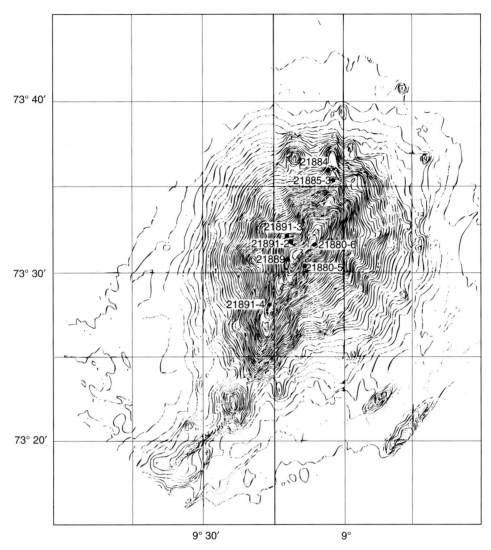

Figure 12 Bathymetric map of Vesteris seamount, an isolated intraplate seamount in the north Atlantic ocean. Note the volcanic rift zones shown with heavy dark lines. Sample locations and numbers are shown. (Reproduced with permission from Oxford University Press from Haase KM and Devey CW (1994) The petrology and geochemistry of Vesteris seamount, Greenland basin – an intraplate alkaline volcano of non-plume origin. *Journal of Petrology* 35: 295–328.)

exhibits chemical systematics as a function of the age of the lithosphere affected by intraplate volcanism. This chemical coherence, along with emerging evidence for the volumetric importance of non-plume volcanism, suggests that much oceanic intraplate volcanism originates in the upper mantle, not in the lower mantle as suggested by the plume hypothesis.

Oceanographic Effects

Seamounts of all types, including large plateaus and platforms, may contribute about 10% or more to the mass of oceanic crust. Since seamounts are volcanic and host active hydrothermal convective systems, seamounts should have a significant effect

on element cycles involving sea water and its dynamic interaction with the ocean crust. Likewise, because normal abyssal sedimentation patterns are severely disturbed in the vicinity of seamounts, they exert a significant influence on the average composition of oceanic sediments, including their hydrothermal and biogenic components (*see* **Authigenic Deposits; Calcium Carbonates**).

Seamounts may also have a significant influence on global ocean circulation patterns (*see* **Ocean Circulation; Water Types and Water Masses**) because their presence induces much greater mixing than is measured in areas with smooth bottom topography. At a more local scale, seamounts have a great effect on circulation patterns and currents, which in turn have very important effects on seamount biota, including populations of fishes (*see* **Pelagic Fish; Deep-Sea Fauna**). In general, seamounts host very diverse and abundant faunas, with important effects on oceanic biology. Thus, while seamounts and off-axis volcanism are interesting on their own, seamounts are also of great interest as obstacles to current flow, biological habitats, and for biogeochemical cycles involving the ocean crust.

See also

Authigenic Deposits. Calcium Carbonates. Deep-sea Fauna. Manned Submersibles, Deep Water. Ocean Circulation. Igneous Provinces. Mid-ocean Ridge Tectonics, Volcanism and Geomorphology. Pelagic Fishes. Ships. Water Types and Water Masses.

Further Reading

Batiza R, Smith T and Niu Y (1989) Geologic and petrologic evolution of seamounts near the EPR based on submersible and camera study. *Marine Geophysical Researches* 11: 169–236.

Floyd PA (ed.) (1991) *Oceanic Basalts*. Glasgow: Blackie and Son.

Green DH and Falloon TJ (1998) Pyrolite: A Ringwood concept and its current expression. In: Jackson I (ed.) *The Earth's Mantle*, pp. 311–378. Cambridge: Cambridge University Press.

Haase KM (1996) The relationship between the age of the lithosphere and the composition of oceanic magmas: constraints on partial melting, mantle sources and the thermal structure of the plates. *Earth and Planetary Science Letters* 144: 75–92.

Larson RL (1991) Latest pulse of Earth: Evidence for a mid-Cretaceous superplume. *Geology* 19: 547–550.

Lueck RG and Mudge TD (1997) Topographically induced mixing around a shallow seamount. *Science* 276: 1831–1833.

McNutt MK, Caress DW, Reynolds J, Johrdahl KA and Duncan RA (1997) Failure of plume theory to explain midplate volcanism in the southern Austral Island. *Nature* 389: 479–482.

Morgan WJ (1978) Rodriquez, Darwin, Amsterdam, ..., A second type of hotspot island. *Journal of Geophysical Research* 83: 5355–5360.

Richter FM and Parsons B (1975) On the interaction of two scales of convection in the mantle. *Journal of Geophysical Research* 80: 2529–2541.

Schilling J-G (1991) Fluxes and excess temperatures of mantle plumes inferred from their interaction with migrating mid-ocean ridges. *Nature* 352: 397–403.

Schmidt R and Schmincke H-U (2000) Seamounts and island building In: *Encyclopedia of Volcanoes*, pp. 383–402. London: Academic Press.

Schouten H, Klitgord KD and Whitehead JA (1985) Segmentation of mid-ocean ridges. *Nature* 317: 225–229.

Smith DK and Cann JR (1999) Constructing the upper crust of the Mid-Atlantic Ridge: A reinterpretation based on the Puna Ridge, Kilauea Volcano. *Journal of Geophysical Research* 104: 25 379–25 399.

Wessel P and Kroenke LW (2000) The Ontong Java Plateau and late Neogene changes in Pacific Plate motion. *Journal of Geophysical Research* 105: 28 255–28 277.

White SM, Macdonald KC, Scheirer DS and Cormier M-H (1998) Distribution of isolated volcanoes on the flanks of the East Pacific Rise, 15.3°–20°S. *Journal of Geophysical Research* 103: 30 371–30 384.

SEDIMENT CHRONOLOGIES

J. K. Cochran, State University of New York, Stony Brook, NY, USA

doi:10.1006/rwos.2001.0182

Introduction

Although the stratigraphic record preserved in deep-sea sediments can span up to 200 Ma, techniques of isotopic dating commonly used to extract sediment accumulation time scales are useful for only a fraction of this range. In addition, the temporal record is blurred by the mixing activities of the benthic fauna living in the upper centimeters of the sediment column. Radionuclide distributions in the sediments provide the most straightforward way of resolving mixing and accumulation rates in deep-sea sediment over the past $\sim$ 5–7 Ma. The basis for these techniques is the supply of radionuclides to the oceanic water column, followed by their scavenging onto sinking particles and transport to the sediment–water interface. Decay of the radionuclides following burial provides chronometers with which mixing and accumulation rates can be determined.

Radionuclide Supply to the Sediment–Water Interface

Table 1 lists the most frequently used radionuclides for determining chronologies of deep-sea sediments. Many of these are members of the naturally occurring ^{238}U and ^{235}U decay series. Both ^{238}U and ^{235}U, as well as ^{234}U, are supplied to the oceans by rivers

Table 1 Radionuclides useful in determining chronologies of deep-sea sediments

Radionuclide	Half-life	Source	Use	Useful time range
^{234}Th	24 days	Dissolved ^{238}U	Particle mixing	100 days
^{210}Pb	22 years	Dissolved ^{226}Ra, atmospheric deposition	Particle mixing	100 years
^{14}C^a	5730 years	Cosmogenic production	Sediment accumulation	35 000 years
^{231}Pa	32 000 years	Dissolved ^{235}U	Sediment accumulation	150 000 years
^{230}Th	75 000 years	Dissolved ^{234}U	Sediment accumulation	400 000 years
^{10}Be	1.5×10^6 years	Cosmogenic production		7×10^6 years
^{239,240}Pu	6600, 24 000 years	Anthropogenic: atomic weapons testing	Particle mixing	Since input (1954)
^{137}Cs	30 years	Anthropogenic: atomic weapons testing	Particle mixing	Since input (1954)

^{a} ^{14}C also has an anthropogenic source from atmospheric testing of atomic weapons.

and are stably dissolved in sea water as the uranyl tricarbonate species $[UO_2(CO_3)_3]^{-4}$. In sea water these three U isotopes decay to ^{234}Th, ^{231}Pa, and ^{230}Th, respectively. The extent of removal of these radionuclides from the oceanic water column is a function of the rate of scavenging relative to the rate of decay. ^{234}Th has a relatively short half-life and can be effectively scavenged in near-surface and near-bottom waters of the open ocean and in the nearshore. ^{230}Th and ^{231}Pa, on the other hand, both have long half-lives and are efficiently scavenged. (While removal of ^{230}Th is nearly quantitative in the open ocean water column, ^{231}Pa shows some spatial variations in the extent of scavenging, with more effective removal at ocean margins.)

^{210}Pb is another ^{238}U decay series radionuclide that has been applied to deep-sea sediment chronologies. ^{210}Pb is produced from dissolved ^{226}Ra in sea water but is also added to the surface ocean from the atmosphere, where it is produced from decay of ^{222}Rn. Like thorium and protactinium, ^{210}Pb is scavenged from sea water and carried to the sediments in association with sinking particles. Owing to its short half-life, ^{210}Pb has been used principally to determine the rate at which the surface sediments are mixed by organisms.

Two other radionuclides that are supplied to the oceans from the atmosphere are ^{14}C and ^{10}Be. Both are produced in the atmosphere from the interaction of cosmic rays with atmospheric gases. (^{14}C also has been produced from atmospheric testing of atomic weapons.) ^{14}C is transferred from the dissolved inorganic carbon pool to calcium carbonate tests and to organic matter and is carried to the sea floor with sinking biogenic particles. ^{10}Be is scavenged onto particle surfaces, much like thorium and protactinium.

Radionuclides produced in association with atmospheric testing of atomic weapons provide pulse-input tracers to the oceans. Both ^{137}Cs and ^{239,240}Pu have been introduced to the oceans in this fashion,

and their input peaked in 1963–64 as a consequence of the imposition of the ban on atmospheric weapons testing. Fractions of the oceanic inventories of both cesium and plutonium have been transferred to deep-sea sediments via scavenging onto sinking particles. In deep-sea sediments, the distributions of plutonium and ^{137}Cs are useful for constraining rates of particle mixing.

Radionuclides are commonly measured by detection of the α, β or γ emissions given off when they decay. This approach takes advantage of the fact that the radioactivity (defined as λN, the product of the decay constant λ and the number of atoms N) is often more readily measurable than the number of atoms (i.e., the concentration). As radiation interacts with matter, ions are produced and radiation detection involves measuring the electric currents that result. Both gas-filled and solid-state detectors are used. Measurement of radioactivity often involves chemical separation and purification of the element of interest, followed by preparation of an appropriate source for counting. Recent advances in mass spectrometry permit direct determination of atom concentrations for uranium, plutonium, and long-lived thorium isotopes by thermal ionization mass spectrometry (TIMS), as well as radiocarbon and ^{10}Be using tandem accelerators as mass spectrometers.

Principles of Determination of Chronologies

Once deposited at the sediment–water interface, particle–reactive radionuclides are subject to decay as well as downward transport by burial and particle mixing. These processes are represented by the general diagenetic equation applied to radionuclides:

$$\frac{\partial A}{\partial t} = D_B \frac{\partial^2 A}{\partial x^2} - S\frac{\partial A}{\partial x} - \lambda A \qquad [1]$$

where A is the nuclide radioactivity (dpm/cm^3 sediment), D_B is the particle mixing coefficient (cm^2/x), S is the accumulation rate (cm/y), λ is the decay constant (y), x is depth in the sediment column (with $x = 0$ taken to be the sediment–water interface), and t is time. Certain underlying assumptions are made in the formulation of eqn [1]. These include no chemical mobilization of the radionuclide in the sediment column and constant sediment porosity.

Particle mixing of deep-sea sediments by benthic organisms is often parametrized as an eddy diffusion-like process, although nonlocal models invoking mixing at discrete depths also have been applied. Except in sediments deposited in anoxic basins, mixing of deep-sea sediments by organisms is commonly active in the upper 2–10 cm of the sediment column, possibly because this near-interface zone contains the most recently deposited organic material. Evidence from multiple profiles of long-lived radionuclides in deep-sea sediments suggests that particle mixing by organisms generally does not extend below the surficial mixed zone. This pattern is in contrast to that observed in estuarine and coastal sediments, which can be mixed to depths in excess of 1 m by organisms. Such deep mixing perturbs radionuclide profiles and makes extraction of sediment chronologies difficult in coastal sediments.

For the uranium and thorium decay series radionuclides, the assumption is usually made that the depth profiles are in steady state (i.e., invariant with time) because production and supply from the overlying water column are continuous. The solution to eqn [1] can be written as

$$A(x) = C \exp(\alpha x) + F \exp(\beta x) \quad [2]$$

If sediments are mixed to a depth L (cm), the constants in eqn [2] can be evaluated with the boundary conditions $A = A_0$ at $x = 0$ and D_B $(\delta A/\delta x) = 0$ at $x = L$.

$$F = \frac{-A_0 \alpha \exp(\alpha L)}{\beta \exp(\beta L) - \alpha \exp(\alpha L)} \quad [3]$$

$$C = A_0 - F \quad [4]$$

$$\alpha = \frac{S + \sqrt{S^2 + 4\lambda D_B}}{2D_B} \quad [5]$$

$$\beta = \frac{S - \sqrt{S^2 + 4\lambda D_B}}{2D_B} \quad [6]$$

If the depth of the mixed layer is greater than the penetration depth of the tracers, an approximation

to the solution of eqn [1] is given by

$$A(x) = A_0 \exp\left[\left(\frac{S - \sqrt{S^2 + 4\lambda D_B}}{2D_B}\right)(x)\right] \quad [7]$$

If particle mixing is negligible $(D_B = 0)$ below this mixed zone, eqn [1] reduces to

$$\frac{\partial A}{\partial t} = S \frac{\partial A}{\partial x} - \lambda A \quad [8]$$

For the condition, $A = A_0$ at $x = L$, eqn [8] is solved as

$$A(x) = A_0 \exp\left(-\frac{\lambda(x - L)}{S}\right) \quad [9]$$

Values of the sediment accumulation rate are determined from eqn [9] by plotting ln A versus depth (x), such that

$$\ln A = \ln A_L - \lambda(x - L)/S \quad [10]$$

where A_L is the activity at the base of the mixed layer and $(x - L)$ is depth below the mixed layer. The sediment accumulation rate S is thus determined from the slope of the ln A–x plot.

Deep-sea sediments often contain detrital minerals supplied to the oceans by riverine and atmospheric transport. For the radionuclides in the uranium decay series, these minerals contain small amounts of the parent radionuclides ^{238}U, ^{234}U, ^{235}U, and ^{226}Ra. Activities of ^{234}Th, ^{230}Th, ^{231}Pa, and ^{210}Pb will decrease to equilibrium with the parent activity. For chronometric purposes, the parent activity is subtracted from the measured daughter activity to obtain a quantity that can be used in eqns [1] and [8]. This quantity, termed the 'excess' activity, corresponds to that scavenged from sea water, and given sufficient time ($\sim$ 5 half-lives) approaches zero with depth in the sediment column. Indeed, the useful time range of the uranium series radionuclides, as well as the cosmogenic chronometers, is approximately 5 half-lives, after which the activity is only $\sim 3\%$ of the initial value. **Table 1** gives the useful time ranges of the radionuclides commonly measured in deep-sea sediments.

Application of eqn [1] to anthropogenic radionuclides such as ^{137}Cs or ^{239,240}Pu does not permit the assumption of steady state because these nuclides were supplied at varying rates with time. Non-steady-state solutions have been formulated, but require the assumption of an input function for the radionuclides to the sediment–water interface.

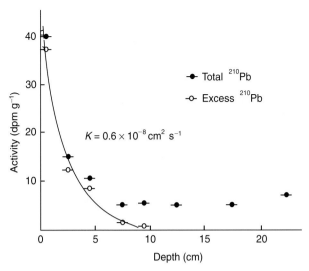

Figure 1 Excess ^{210}Pb activity versus depth in a sediment core from the Mid-Atlantic Ridge. The activity is mixed to ~ 8 cm by the benthic fauna. The rate of mixing (D_B in eqn [7]) is 0.6×10^{-8} cm^2 s^{-1} (~ 0.2 cm^2 a^{-1}). (Reprinted from Nozaki Y, Cochran JK, Turekian KK and Keller G. Radiocarbon and ^{210}Pb distribution in submersible-taken deep-sea cores from Project FAMOUS. *Earth and Planetary Science Letters*, vol. 34, pp. 167–173, copyright 1977, with permission from Elsevier Science.)

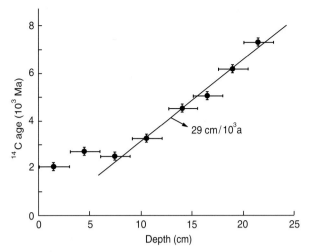

Figure 2 Radiocarbon age versus depth in a sediment core from the Mid-Atlantic Ridge. The age is homogenized in the upper 8 cm owing to mixing by the benthic fauna (see **Figure 1**). Below 8 cm, a sediment accumulation rate of 2.9 cm ka^{-1} is calculated (eqn [9]). (Reprinted from Nozaki Y, Cochran JK, Turekian KK and Keller G, Radiocarbon and ^{210}Pb distribution in submersible-taken deep-sea cores from Project FAMOUS. *Earth and Planetary Science Letters*, vol. 34, pp. 167–173, copyright 1977, with permission from Elsevier Science.)

Models including both a constant input since the peak introduction of ^{137}Cs and plutonium to the ocean or a pulse input (maximum in 1963–64) have been used. The validity of these input scenarios is a significant limitation on the use of anthropogenic radionuclides in sediment mixing studies.

Examples of Radionuclide Profiles in Deep-sea Sediments

Figure 1 shows a profile of excess ^{210}Pb in a sediment core taken in a sediment pond in the Mid-Atlantic Ridge. ^{210}Pb in this core is mixed to a depth of 8 cm. Such a depth of mixing is quite typical of deep-sea sediments, and below this depth the sediment is undisturbed by mixing by the benthic fauna. **Figure 2** shows the radiocarbon profile in the same core. The rate of sediment accumulation may be calculated from the gradient in radiocarbon ages with depth. Indeed, radiocarbon is unique among the chronometers considered here in providing absolute ages for a given depth in the sediment column. This is possible because radiocarbon can be related to the activity in pre-industrial, pre-bomb carbon to provide an absolute age for the carbon fraction being analyzed. All the other chronometers discussed herein provide relative ages by relating the activity at depth to that at the sediment–water interface or the base of the mixed zone (eqn [9]).

For a long-lived radionuclide such as ^{230}Th, mixing will tend to homogenize the activity in the mixed zone. Below that depth, ^{230}Th will decrease consistently with its decay constant and the sediment accumulation rate (eqn [9]). **Figure 3** shows excess ^{230}Th profiles in three deep-sea cores from the Pacific Ocean. The mixing of the surficial layers is quite clear from the profile. The gradient in activity with depth below the mixed zone yields sediment accumulation rates of 0.14 to 0.30 cm per 1000 years. Sediment accumulation rates of deep-sea sediments determined by the excess ^{230}Th and ^{231}Pa methods typically range from millimeters to centimeters per 1000 years.

Profiles of the anthropogenic radionuclides ^{137}Cs and ^{239,240}Pu in a sediment core of the deep Pacific Ocean are shown in **Figure 4**. A non-steady-state solution to eqn [1] must be applied to these profiles because the radionuclides have been added to the oceans only since 1945 and the profiles are evolving in time as the radionuclides are removed from the overlying water column and added to the sediment–water interface. Particle mixing rates determined from these profiles are 0.36 or 1.4 cm^2 a^{-1} depending on the input function chosen. A pulse input of the radionuclides at the time of maximum fallout to the earth's surface (1963) provides mixing rates that are most similar to that obtained from ^{210}Pb in the same core. Mixing rates

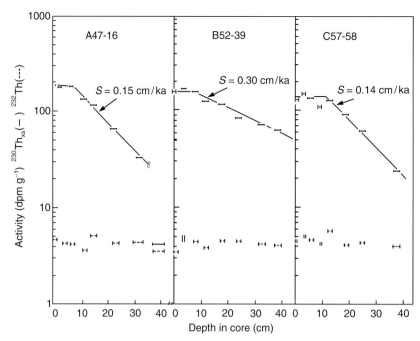

Figure 3 Excess ^{230}Th activities versus depth in three cores from the north equatorial Pacific. The activities are homogenized in the upper ~ 10 cm as a consequence of particle mixing by the benthic fauna. Accumulation rates calculated from the decreasing portions of the profiles are 0.14–0.3 cm ka^{-1}. (Reprinted from Cochran JK and Krishnaswami S. Radium, thorium, uranium and ^{210}Pb in deep-sea sediment and sediment pore water from the North Equator Pacific. *American Journal of Science*, vol. 280, pp. 847–889, copyright 1980, with permission from American Journal of Science.)

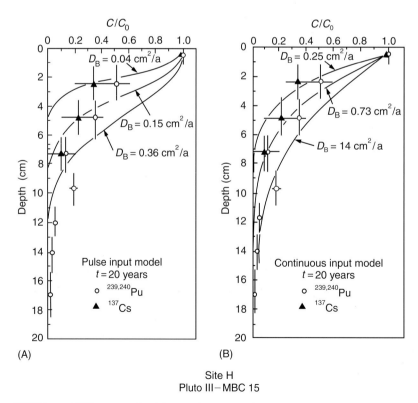

Figure 4 Activities of ^{239,240}Pu and ^{137}Cs versus depth in a sediment core from the equatorial Pacific. The activities are normalized to the value in the surficial depth interval and are modeled using pulse and continuous inputs of these anthropogenic radionuclides to the sediment–water interface. The profiles are the result of mixing by the benthic fauna. (Reprinted from Cochran JK Particle mixing rates in sediments of the eastern Equatorial Pacific: Evidence from ^{210}Pb, ^{239,240}Pu and ^{137}Cs distributions at MANOP sites. *Geochimica et Cosmochimica Acta*, vol. 49, pp. 1195–1210, copyright 1985, with permission from Elsevier Science.)

of deep-sea sediments determined from short-lived and recently input radionuclides are generally $< 1\,cm^2\,y^{-1}$. (In shallow water sediments, mixing rates can be two orders of magnitude greater than observed in the deep sea.) The rate and depth of mixing of sediments determines the extent to which changes in paleoceanographic indicators (e.g., oxygen isotopes) can be resolved.

Long-lived radionuclides such as [10]Be offer the opportunity to extend radionuclide chronologies of deep-sea sediments to several million years. Recent advances in the measurement of [10]Be by accelerator mass spectrometry (AMS) permit analysis of small samples and high-quality chronologies to be determined using this radionuclide. Longer chronologies are especially useful in interpreting the record of parameters such as oxygen or carbon isotopes that are linked to paleoceanographic changes. Indeed it has become common to use the now well-established stratigraphy of oxygen isotopes to 'date' depth horizons of deep-sea sediments, yet it is important to recognize that the oxygen isotope stratigraphy was first established through the use of uranium series radionuclides (principally excess [230]Th).

Final mention must be made of the dating of horizons preserved in deep-sea sediments via the potassium–argon method. The method is based on the decay of [40]K (half-life $= 1.2 \times 10^9$ y) to stable [40]Ar, a noble gas. The method is useful only for materials whose initial argon was lost when the rock was formed. Subsequent production of [40]Ar in the rock is from [40]K decay and the [40]Ar/[40]K ratio serves as an indicator of the rock's age. The method can be used to date volcanic materials that are deposited at the sediment–water interface, for example, as volcanic dust or ash associated with a volcanic eruption. Because of the long half-life of [40]K, this method has potential for dating sediments on long timescales, but because of the particular

requirements (volcanic material deposited at the sediment–water interface), it is not often possible to use it.

See also

Cosmogenic Isotopes. Ocean Margin Sediments. Radiocarbon. Stable Carbon Isotope Variations in the Ocean. Temporal Variability of Particle Flux. Uranium–Thorium Decay Series in the Water Column. Uranium–Thorium Series Isotopes in the Ocean.

Further Reading

Berner RA (1980) *Early Diagenesis: A Theoretical Approach.* Princeton: Princeton University Press: 241pp.

Boudreau BP (1997) *Diagenetic Models and Their Implementation,* 414 pp. Heidelberg; Germany; Springer-Verlag.

Cochran JK (1992) The oceanic chemistry of the uranium- and thorium-series nuclides. In: Ivanovich M and Harmon RS (eds) *Uranium Series Disequilibrium: Applications to Earth, Marine, and Environmental Sciences,* 2nd edn, pp. 334–395. Oxford: Oxford University Press.

Huh C-A and Kadko DC (1992) Marine sediments and sedimentation processes. In: Ivanovich M and Harmon RS (eds) *Uranium Series Disequilibrium: Applications to Earth, Marine, and Environmental Sciences,* 2nd edn, pp. 460–486. Oxford: Oxford University Press.

Libes SM (1992) *Marine Biogeochemistry,* pp. 517–556. New York: Wiley.

Turekian KK (1996) *Global Environmental Change.* Englewood Cliffs, NJ: Prentice-Hall.

Turekian KK and Cochran JK (1978) Determination of marine chronologies using natural radionuclides. In: Riley JP and Chester R (eds) *Chemical Oceanography,* vol. 7, pp. 313–360. New York: Academic Press.

Turekian KK, Cochran JK and DeMaster DJ (1978) Bioturbation in deep-sea deposits: rates and consequences. *Oceanus* 21: 34–41.

SEDIMENTARY RECORD, RECONSTRUCTION OF PRODUCTIVITY FROM THE

G. Wefer, Universität Bremen, Bremen, Germany
W. H. Berger, Scripps Institution of Oceanography, La Jolla, CA, USA

doi:10.1006/rwos.2001.0251

Introduction

Reconstruction of productivity patterns is of great interest because of important links of productivity to current patterns, mixing of water masses, wind stress, the global carbon cycle, hydrocarbon resources, and biogeography. The history of productivity is reflected in the flux of organic carbon into the sediment. There are a number of fluxes other than organic carbon that can be useful in assessing productivity fluctuations through time. Among others, fluxes of opal and of carbonate have been used, as well as the flux of particulate barite. In addition, microfossil assemblages contain clues to the intensity of production, as some species occur preferentially in high-productivity regions whereas others avoid these.

One marker for the fertility of subsurface waters (that is, for nutrient availability) is the carbon isotope ratio of totally dissolved inorganic carbon within that water ($^{13}C/^{12}C$, expressed as $\delta^{13}C$). In today's ocean, values of $\delta^{13}C$ of totally dissolved inorganic carbon are negatively correlated with nitrate and phosphate contents. Another useful tracer of phosphate content in subsurface waters is the Cd/Ca ratio. The correlation between this ratio and phosphate concentrations is quite well documented. A rather new development in the search for clues to ocean fertility is the analysis of the $^{15}N/^{14}N$ ratio in organic matter, which tracks nitrate utilization. The fractionation dynamics in the environment of growth are analogous to those of carbon isotopes. These various markers are captured within the organisms growing within the water tagged by the isotopic or elemental ratios.

Today's high production areas are in the temperate to high latitudes where wind-driven mixing is strong and where days are long in summer, and in equatorial and coastal upwelling regions (**Figures 1** and **2**). Favorable sites for burial of organic carbon are on the upper continental slope, not far off the coast where upwelling occurs, but far enough to reach depths of reasonably quiet water, where organic matter can settle and stay, embedded

within silty sediment. The Pleistocene record in sediments on the continental slope shows large fluctuations in the burial rates of organic carbon which are generally interpreted as productivity fluctuations (unless redeposition from terrigenous sources is responsible).

Productivity Proxies

Organic Matter (and Oxygen Demand)

Generally speaking, there is a relationship between productivity in surface waters and organic carbon accumulation in underlying sediments. Below the central gyres (the deserts of the ocean) organic carbon content in sediments is extremely low. In upwelling areas, the organic carbon content is high, and in many cases sufficient for sulfate reduction and pyrite formation. From this observation, it may be expected that at any one place a change in the content of organic carbon indicates a change in productivity through time.

The range of variation in productivity of surface water spans roughly a factor of 10 (excluding estuaries and inner shelf), while the content in sediments varies by, for example, a factor of 40. The transfer of carbon from the surface waters to the sediments depends on the leakage of carbon out of the pelagic food web and the associated downward transport of organic matter, both in particulate and in dissolved

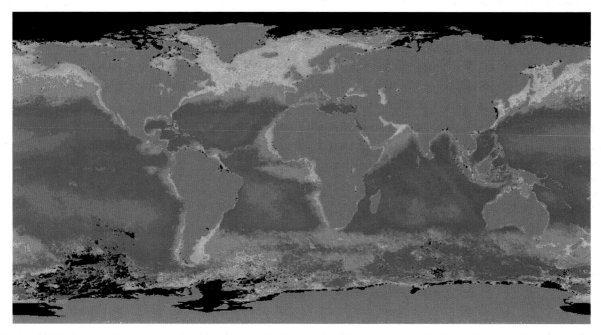

Figure 1 Pigment distribution in surface waters, inferred from color scanning data aboard C7CS satellite (November 1978–June 1986). Orange, high productivity ($> 150\,g\,C\,m^{-2}\,y^{-1}$); deep blue: low productivity ($50\,g\,C\,m^{-2}\,y^{-1}$). Sources: NASA/Goddard Space Flight Center, MD, USA, compiled by B. Davenport, Bremen.)

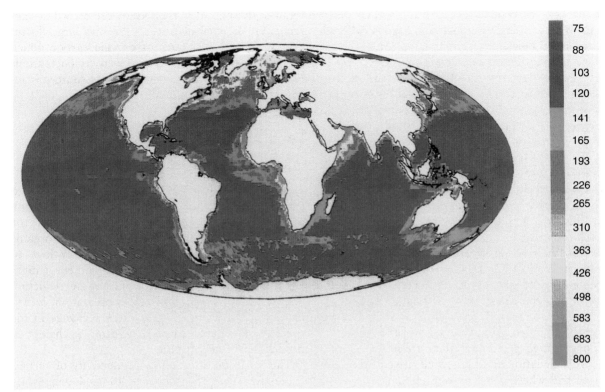

Figure 2 Primary production in grams carbon per m^2 and year, estimated from satellite radiometer data. (Reproduced with permission from Longhurst *et al.*, 1995.)

form. Favourable sites for burial of organic carbon are on the upper continental slope, for several reasons. The coastal setting results in high productivity, the shelf provides additional carbon, and setting and decomposition in the water column are of short duration.

The equation relating productivity (PP) to C_{org} content is of the form:

$$PP(a)/PP(b) = [C_{org}(a)/C_{org}(b)]^q$$

where q is usually between 0.6 and 0.8. Taking $C_{org}(b)$ as the Holocene standard value, the ratios downcore, after exponentiation to 0.7, are a reasonable estimate of the stratigraphic sequence of the factors of change of productivity. The great precision suggested by more complicated formulations must be largely doubted, especially since the influx of organic material redeposited from the shelves in regions close to continental margins can materially influence results.

The quantitative reconstruction of productivity from organic matter content was introduced just over 20 years ago, by P. Müller and collaborators. This work established that glacial periods showed higher productivity in the eastern North Atlantic than interglacial ones (**Figure 3**). Similar findings

were subsequently made for the eastern and western equatorial Pacific and for the upwelling areas which depend on trade wind stress to power them.

Within the sediment, C_{org} is constantly being destroyed by bacteria. This is especially true close to the seafloor, but it is also evident several meters below. The destruction first proceeds by using free oxygen, but subsequently occurs by the reduction of nitrate, manganese oxide, iron oxide, and dissolved sulfate. The latter leads to precipitation of iron sulfide. Thus, an oxygen debt is built up within the sediment. Instead of using C_{org} as a productivity indicator (which results in a general trend for lower estimates with increasing age of sediment) it is reasonable to substitute oxygen demand (reducing power) as a proxy.

Opal (Mobile Silica)

A number of biologically derived substances other than organic carbon can also be useful in assessing productivity fluctuations. There is a good correlation in the flux of organic carbon and of carbonate in the open ocean, as seen in sediment traps. However, carbonate is commonly readily dissolved in sediments accumulating below areas of high production, during early diagenesis. Unlike carbonate, opal content is high in sediments below high productivity

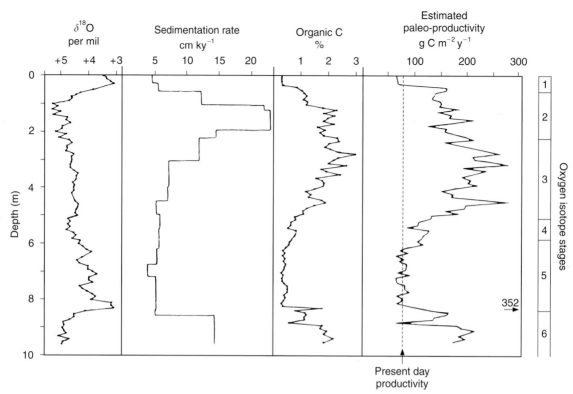

Figure 3   Paleo-productivity estimates for Meteor Core 12392–1 on the continental rise off the Spanish Sahara, north-west Africa. Sedimentation rate and organic carbon content are input variables from which productivity is calculated. Even numbers represent glacials and uneven numbers represent interglacials. (Reproduced with permission from Müller and Suess 1979 and Müller et al., 1983.)

regions (e.g. in coastal upwelling regions), and low elsewhere (e.g. in oligotrophic areas). Its flux (as seen in traps) is well-correlated with that of C_{org}, the ratio of opal to C_{org} being especially high around the Antarctic.

Opal is originally brought into the sediment by the deposition of the skeletons of diatoms, radiolarians, and silicoflagellates, but has a tendency to migrate, especially in older sediments. Well over one half of the ocean's opal flux is concentrated in the Southern Ocean, where diatomaceous ooze accumulates over extensive regions. This reflects above all the high diatom productivity resulting from very deep mixing (which brings silicate to the photic zone) and from seasonal contrast (diatom blooms during onset of warming and stratification). The great efficiency with which silica is extracted from the ocean around the Antarctic continent implies that the overall concentration of silicate in the world ocean will depend to a large extent on changes in this efficiency through time. The state of the North Atlantic (which exports silica at present, North Atlantic Deep Water production being vigorous) also is important in setting the background level, as is the intensity of coastal upwelling and

equatorial upwelling. Focused upwelling, by providing a smaller region from which to redissolve silica on the seafloor, has the effect of decreasing overall silicate concentrations. This effect has increased, overall, since the end of the Eocene period (Auversian Facies Shift, about 40 Ma) and has led to an overall removal of silica from deep-sea deposits (in favor of deposits in the ocean margins and in eastern equatorial regions).

Along the equator in the Pacific (and also to some extent in the Atlantic) the content of opal in sediments is considerably greater in the east than in the west, because of the higher supply of diatoms in the east, which in turn depends on the supply of nutrients and silicate, brought east by subsurface currents. In contrast, the western regions have rather thick warm-water layers which are depleted in nutrients. Quaternary changes in opal deposition in east and west are not in phase, presumably because of this strong element of asymmetry. The east–west contrast in opal sedimentation in the equatorial Pacific is much greater than the contrast in productivity. Thus, opal flux as a productivity index greatly amplifies the primary signal in this setting. Presumably, the cause of this amplification is that much of the silica is redissolved, not as a proportion of what is coming down, but as a background loss, which is largely independent of the amount of material delivered. This process has the effect of greatly increasing the initial differences in deposition. In the end, residuals which are chiefly composed of shells and shell fragments resistant to dissolution are compared. Such shells (that is, well-silicified large shells) are generated disproportionately in the more productive areas.

Because of the diatom-limiting nature of silicate concentrations opal is not a reliable proxy for productivity as such, but can only proxy for diatom production (as well as production of radiolarians and silicoflagellates). This is brought out well when studying the productivity record of the western equatorial Pacific. Here productivity increased by a factor near 1.5 over the present, during glacial time as judged from other productivity proxies. Instead of a higher rate of deposition of opal, however, we see a decreased rate is seen. Out-of-phase relationships between opal deposition and other productivity proxies are also known from deposits off south-western Africa, and from sediments in the Santa Barbara Basin and from off Peru. Clues to possible causes for these discrepancies may be found in the contrast of diatom sedimentation between equatorial Pacific and equatorial Atlantic. At the same level of productivity, opaline sediments are notoriously less in evidence in the Atlantic than in the Pacific Ocean, presumably due to a weaker silicification of diatom tests in the Atlantic. The cause of the asymmetry is taken to be the lower concentration of silicate in subsurface waters of the Atlantic, compared with those in the Pacific. Apparently, the glacial northern and central Pacific was more like the northern and central Atlantic today, i.e. there was much less silicate dissolved in its intermediate waters. This proposition agrees well with the other observations suggesting more 'Atlantic' conditions in the glacial Pacific, including the carbonate record (lowered carbonate compensation depth).

Isotope Ratios in Carbon and Nitrogen

In the reconstruction of productivity it is important to distinguish between flux proxies (which tag the export of biogenic materials from the surface to the seafloor) and nutrient proxies (which contain information about the nutrient content of the water wherein the production takes place). Isotopic ratios of carbon and nitrogen (expressed as deviation from a standard ratio, $\delta^{13}C$, $\delta^{15}N$) are nutrient proxies. The classical marker for the fertility of subsurface waters (i.e. the nutrients potentially available through mixing) is the difference in $\delta^{13}C$ values of surface waters and subsurface waters (it is usually given as $\Delta\delta^{13}C$), as seen in shallow-living and deep-living planktonic foraminifers. In the deeper waters, $\delta^{13}C$ decreases as nutrient contents rise, because the oxidation of organic matter within the thermocline sets free both carbon (with an excess content of ^{12}C) and the nutrients nitrate and phosphate. When the nutrient-rich deeper waters are brought to the surface, the $\delta^{13}C$ increases, because ^{12}C is extracted preferentially into organic matter and sinks with the excess production. The process continues until the nutrients are used up. Thus, the nutrient-free water retains a memory of former nutrient content through the anomalous enrichment in ^{13}C, over background. Unfortunately, the background is not usually well known, since the water is reset by exchange with the atmosphere and, on long time-scales, by exchange with large carbon reservoirs. Hence the use of differences in $\delta^{13}C$: the larger the difference, the higher the nutrient content of the thermocline water.

In analogy to the difference in $\delta^{13}C$ between surface water and subsurface water the seasonal difference can also be used, providing that productivity is strongly pulsed, and thus samples the differing conditions of shallow and deep waters. Other differences are that between planktonic and benthic species (comparing surface water conditions with the deep ocean in general), and differences between benthic species living on or in the sediment, which provide clues to the $\delta^{13}C$ gradient between seafloor and interstitial waters within the uppermost sediment. This latter difference is expected to increase as the interstitial waters lose oxygen to the bacterial combustion of the organic matter.

A new tool in the reconstruction of productivity-related conditions is the use of the stable isotopes of nitrogen in organic matter, that is, the ratio of $^{15}N/^{14}N$ (expressed as $\delta^{15}N$). Assimilation of nitrate by phytoplankton is accompanied by nitrogen-isotope fractionation and produces a strong gradient in $\delta^{15}N$ as the source nitrate is consumed (with ^{15}N preferentially left behind) and organic matter is exported from the photic zone (with nitrate set free at depth, now enriched in ^{14}N). As a result of these processes, temporal and spatial changes in the balance between nitrate advection and consumption in the upper water layers will lead to corresponding changes in the $\delta^{15}N$ values of particulate nitrogen settling out of the productive zone and moving to the seafloor.

Elemental Ratios (Cd/Ca) and Trace Elements (Ba)

Many rare metals act as nutrients, that is, they are depleted in surface waters and enriched within

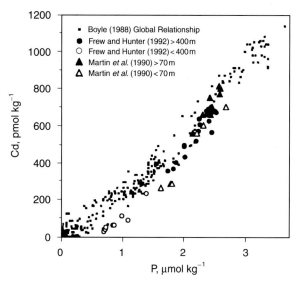

Figure 4   Correlations between Cd and phosphate in the open ocean below the mixed layer. (Adapted with permission from Boyle 1994.)

thermocline waters and at depth. If such elements are incorporated into organic matter or shells, their abundance can provide clues to the intensity of upwelling or the general level of concentration in upwelled waters, or both. Prime examples are the cadmium/calcium ratio in calcareous shells and the deposition of barium, which persists as barite (barium sulfate) within the sediment.

The Cd/Ca ratio was introduced by E. Boyle, in the 1980s, as a tracer of phosphate content in ocean waters. The correlation between the metal ratio and the nutrient content is remarkably good (**Figure 4**). It rests on the fact that cadmium is rare in sea water and tends to be extracted from surface water together with phosphate, being liberated again in deeper waters, upon oxidation of the organic matter binding it. Thus, cadmium and phosphorus are precipitated and redissolved together. Difficulties arise from the requirements of unusually demanding chemical procedures when extracting cadmium from shells in the seafloor (to avoid contamination) and from the possibility that correlations may not be stable through geologic time. The sources of sinks of cadmium, calcium, and phosphorus are all different. For example, cadmium tends to be precipitated as sulfide in anaerobic conditions. Thus, whenever anaerobic conditions expand (as perhaps during certain phases of the glacial–interglacial cycle) there will be a tendency for increased extraction of this element. When used as a proxy for phosphate, then, indications for phosphate would be lowered during times of poor deep-water ventilation, independently of the true phosphate content. Nevertheless, the

method has proved useful as a tracer of deep-water phosphate content, as recorded in the shells of benthic foraminifers, for any one time period.

Barite has long been recognized as an indicator of productivity. For example, E. Goldberg and G. Arrhenius, in the 1950s, documented a distinct peak of barite accumulation below the equatorial upwelling region in the eastern Pacific, in a north-to-south transect. Compared with other paleo-productivity proxies such as organic matter, barite is quite refractory. It is not clear exactly how barium enters the export flux. In low latitudes, it may be surmised that incorporation into skeletons by acantharians (Radiolaria), and subsequent precipitation as microcrystals of barite within sinking aggregates, is important. Elsewhere, diatom flux and organic matter flux presumably bring the barium with them. In any case, changing barite abundances agree well with productivity fluctuations on glacial–interglacial timescales.

Barite, as a flux proxy, is subject to variations in availability, much like opal. Therefore it may not be a 'pure' indicator of productivity. Problems arise especially when the record is contaminated by terrigenous barite particles. In addition, small changes in oxygen concentration, at sensitive levels, can influence the abundance of sulfate ions in interstitial waters and hence the stability of microscopic barite crystals.

Microfossil Assemblages

Within each group of planktonic organisms, some species occur preferentially in high-productivity regions while others avoid these, or cannot compete in bloom situations. Thus, among the shelled plankton, the relative abundances in the sediment contain clues to the intensity of production at the time of sedimentation.

Microfossil assemblages will have aspects of flux proxies but also reflect ecologically important conditions such as sequences of mixing and stratification, or seasonality in general. Among the proxies for flux are the diatom species directly involved in upwelling blooms, such as *Chaetoceros*. In addition, there are more subtle changes in species composition that reflect differences in the style and intensity of mixing and production. In the eastern equatorial Atlantic, for example, changes have been found in the diatom flora that indicate increased productivity during glacial time.

Among planktonic foraminifers, a number of species have been identified as indicators of high productivity. In low latitudes, for example, these include the forms *Globigerina bulloides*, *Neogloboquadrina dutertrei*, and *Globorotalia tumida*. In temperate latitudes, *Globigerina quinqueloba* is

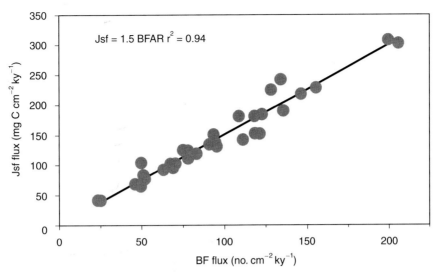

Figure 5 Benthic foraminifera accumulation rate (BFAR) and flux of organic matter to the seafloor (Jsf). Data based on box-core data from eastern and western equatorial Pacific, and from South Atlantic, as given in Table 1, Herguera and Berger, 1991. The graph compares core-top data with productivity estimates (PP) based on the map in Berger *et al.* (1989). Jsf is taken as $0.25 \cdot PP^2/Z + 0.5 \cdot PP/Z^{0.5}$ and from this relationship PP can be estimated from BFAR, by solving for PP. Z is water depth.

useful as a productivity indicator, while *N. pachyderma* (sin.) indicates very cold upwelling water. Thus, the ratio of these species to their more warm-loving contemporaries should provide good indicators of upwelling. Multivariate statistics that use relative abundance of species yield interesting results. In some cases, results from different reconstructions (e.g. organic carbon versus microfauna) differ considerably. The reasons are poorly understood as yet.

Benthic foraminifers live on the organic material falling to the seafloor. Thus, their abundance should vary with the food supply from above. This is indeed the case, as shown by comparing the accumulation of benthic foraminifers with the overlying productivity for various regions (see **Figure 5**). As the food supply changes, the chemical conditions on the seafloor also change, and therefore the bacterial flora is affected. In turn, this must influence the composition of the benthic fauna. Indeed, the open-ocean faunas (where food may be presumed to be in short supply) are dominated by *Cibicidoides, Eponides, Melonis, Oridorsalis,* and others, while the faunas below the productive upwelling regions along the continental margins are dominated by *Uvigerina, Bolivina, Bulimina,* and associated forms. Various observations suggest that it will be difficult to use species composition in benthic foraminifers as productivity indicators, within high production regions. Other factors besides productivity (oxygen deficiency, sulfide abundance) may be more important at high levels of organic matter flux. However, at low to intermediate levels of

productivity faunal changes should be readily detectable.

Examples for Productivity Reconstructions

Equatorial Upwelling

One of the earliest attempts at quantitative reconstruction of equatorial upwelling was carried out by G.O.S. Arrhenius, working with sediments from the eastern equatorial Pacific, raised by the Swedish Deep-Sea Expedition. He used changes in the size distribution of diatoms to argue for greatly increased productivity during glacial periods, which he related to increased upwelling from increased trade wind stress. These suggestions have been fully confirmed, most recently by the study of barite deposition see **Figure 6A**). The results show a strong precessional cycle, as expected if wind is the driving force. (Winds are influenced both by the overall planetary temperature gradient, which greatly increases during glacial periods as the northern polar front migrates toward the equator, and by the intensity of the summer monsoon which varies on a precessional cycle.)

Interestingly, the western equatorial Pacific shows a quite similar pattern, as seen in the oxygen demand data (see **Figure 6B**). Again, the precessional signal is quite strong. These results likewise confirm earlier studies showing increased glacial productivity in this region. (For the eastern part, the factor of change is thought to be somewhat greater than in

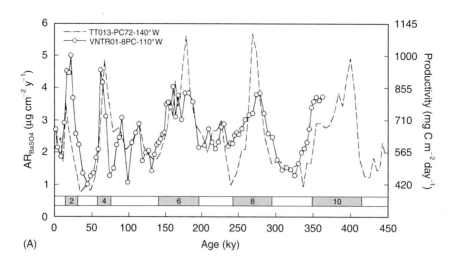

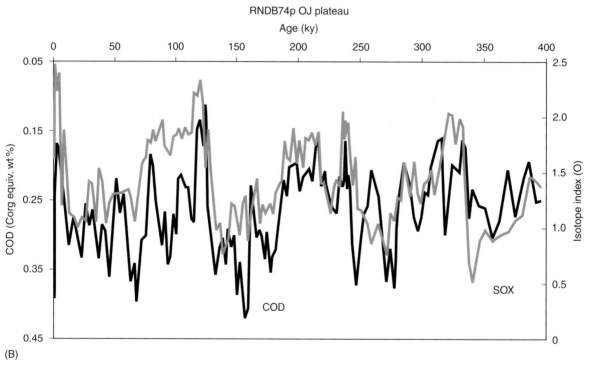

Figure 6 Productivity history in the equatorial Pacific. (A) Accumulation rate of barite (AR_{BaSO_4}) and productivity in the central and eastern equatorial Pacific (Reproduced with permission from Kastner, 1999.) Even numbers represent interglacials and uneven numbers represent glacials. (B) Oxygen demand in the western equatorial Pacific Ontong Java Plateau (Oxygen isotope data core RNDB 74P reproduced with permission from Perks, 1999. COD, organic carbon demand; SOX, oxygen isotope index. Data from Ontong Java Plateau 0°, 20.48°N, 159°22.49′E, 2547 m depth, reproduced with permission from Yasuda and Berger (unpublished observation).

the west, for example, about 2 in the east, and around 1.5 in the west, for the glacial-to-interglacial contrast.) As mentioned, the opal deposition does not reflect this general pattern. While varying in phase with glacial–interglacial conditions in the east (similar to organic matter and barite) it varies at counterpoint phase in the west. Therefore a strongly decreased silicate content in thermocline waters in the west is indicated. In the present ocean, low

silicate values go parallel with decreased phosphate values (with the silicate decreasing much faster than the phosphate). Therefore it is likely that the phosphate content of intermediate waters was decreased over large parts of the central Pacific, during glacial time. Thus the increased productivity in the western equatorial Pacific has to be the result of very vigorous mixing, and not of increased nutrient concentrations.

In the Atlantic, also, equatorial upwelling increased during glacial times, especially in the eastern regions (see Further Reading section: particularly, Berger and Herguera, 1992 and Wefer *et al.*, 1996.

Coastal Upwelling Centers

The increase in glacial productivity off north-west Africa has been mentioned (**Figure 3**); it is well established and has been confirmed many times. Much information has recently become available from off the coast of central and southern Africa. Off the mouth of the Congo River, productivity was high during glacials and low during interglacial stages, as a rule; the maximum range suggested by the data exceeds a factor of three (**Figure 7A**), certainly for the contrast between the last glacial and the late Holocene stages. A precessional influence seems quite strong. The enormous accumulation of organic matter and the correspondingly high accumulation in opal during the last glacial maximum are noteworthy. The lack of response to the Stage 4 glacial is puzzling. A comparison core is needed to exclude the possibility of missing material.

Off Angola the fluctuations in the rates of carbon accumulation are comparable to those off the Congo River (**Figure 7B**). However, the opal accumulation is rather insensitive to the variation in

productivity seen in the C_{org}. As for the western equatorial Pacific, relatively low silicate values must be postulated for glacial periods, which in essence compensate for the increase in upwelling. By the above argument (phosphate decreases with silicate but at a lesser rate) the nutrient content of water welling up during glacial periods was less than during interglacials, but strong winds overcame the effect by producing much more vigorous mixing and upwelling than today.

Matuyama Opal Maximum off Southwestern Africa

The various caveats regarding productivity reconstruction which have been pointed out and illustrated make it extremely difficult to take quantitative reconstruction far into the past. Indeed for the more distant time periods (e.g. pre-Neogene) the sign of the change is not clear (mainly because lack of oxygen simulates many of the indicators of high production, and because of the unreliability of the opal record).

Recently, a concerted effort has been made by Leg 175 of the Ocean Drilling Program to attempt a comprehensive reconstruction of productivity of the Benguela upwelling system, for the late Neogene period. Results so far are both puzzling and enlightening. When contemplating opal deposition

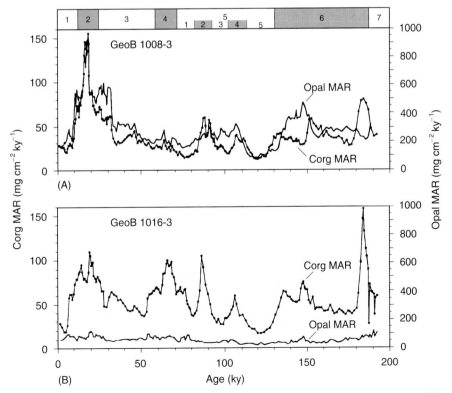

Figure 7 Accumulation rates of organic carbon and of opal in the South Atlantic. (A) Congo fan area. (B) Off Angola. Even numbers are glacial and uneven numbers are interglacial isotope stages. (Reproduced with permission from Schneider, 1991).

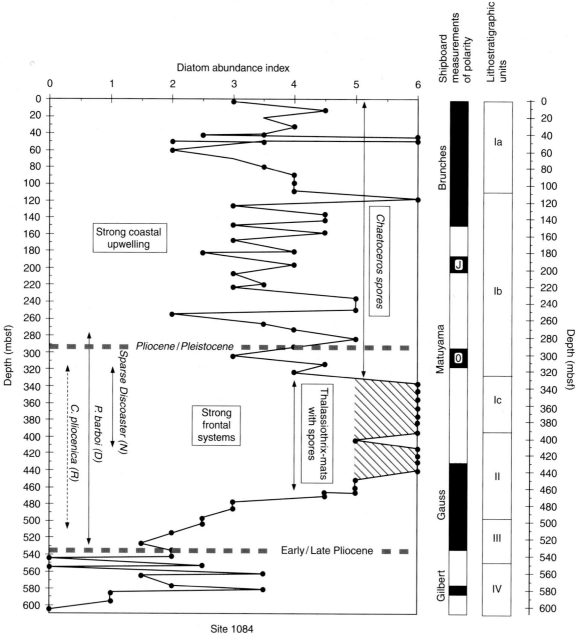

Figure 8   The Matuyama Diatom Maximum off south-western Africa, as seen in Hole 1084A. Shipboard measurements of polarity and lithographic units are given on the right-hand side. The time interval of subantarctic influence in the late Pliocene is exemplified by the presence of *Proboscia barboi* (diatom) and *Cycladophora pliocenica* and by the sparse occurrence of *Discoaster* nannofossil species. Strong coastal upwelling dominated over Site 1084 in the Pleistocene, whereas strong frontal systems (among the BCC (Benguela Coastal Current), the upwelling filaments, and the BOC (Benguela Oceanic Current) dominated the late Pliocene, as exemplified by *Chaetoceros* resting spores and Thalassiothrix-rich sediments, respectively. (Reproduced with permission from Wefer *et al.* (1998), as drawn by C.B. Lange.)

of the last 3 My, a rapid increase in deposition in the late Pliocene is found, and then an overall decrease to lower values as the system moves into the Quaternary and increased overall ice buildup. Parallel to the trend of decreasing opal deposition, there is, paradoxically, an increase in spores of the diatom *Chaetoceros*, indicating increased upwelling. The

maximum itself, between 2.2 and 2.0 My, is characterized by a rich mixture of various diatom floras, including Antarctic, central ocean, and upwelling. Dense deposits of almost pure diatom ooze ('diatom mats') are also found in this portion, suggesting self-sedimenting blooms along major ocean frontal systems (**Figure 8**).

The peculiar stratigraphy of the Matuyama Diatom Maximum – maximum deposition of opal along a long-term trend of temperature decrease – emphasizes the presence of several competing factors as the climate moved into the northern ice ages. These include winds and eddy formation off southwestern Africa, access of Antarctic waters and waters from the Agulhas Current from the Indian Ocean, as well as intensity of coastal upwelling and the quality of the upwelling water. The overall message in terms of geochemistry is that the quality of the upwelling water suffered (i.e. nutrients decrease) as the world moved into the Quaternary ice ages. In principle, this agrees with the reverse opal deposition (with respect to other productivity indicators) observed within the glacial–interglacial themselves.

Conclusions

Various sediment properties deliver useful information for reconstructing past productivity of the ocean. Major improvements have been made in the attempt to understand productivity conditions in the past. Nevertheless, proxies require further calibration and testing with time-series and time-slices. Water samples, plankton nets, and sediment trap investigations, in conjunction with laboratory experiments, are of great importance in making these calibrations. It has to be taken into account that a single proxy is not sufficient to document productivity conditions, so a number of proxies should always be used (multiproxy approach).

See also

Conservative Elements. Nitrogen Isotopes in the Ocean.

Further Reading

Antoine D, André JM and Morel A (1996) Oceanic primary production. 2. Estimation at global scale from satellite (coastal zone color scanner) chlorophyll. *Global Biogeochemical Cycles* 10: 43–55.

Berger WH and Herguera JC (1992) Reading the sedimentary record of the ocean's productivity. In: Falkowski PG and Woodhead AD (eds) *Primary Productivity and Biogeochemical Cycles in the Sea*, pp. 455–486. New York: Plenum Press.

Berger WH and Wefer G (1996) Expeditions into the past: paleoceanographic studies in the South Atlantic. In: Wefer G, Berger WH, Siedler G and Webb D (eds) *The South Atlantic: Present and Past Circulation*, pp. 363–410. Berlin, Heidelberg, New York: Springer-Verlag.

Berger WH, Smetacek VS and Wefer G (eds) (1989) *Productivity of the Ocean: Present and Past Circulation*. Chichester: J Wiley & Sons.

Berger WH, Herguera JC, Lange CB and Schneider R (1994) Paleoproductivity: flux proxies versus nutrient proxies and other problems concerning the Quaternary productivity record. In: Zahn R, Kaminski M, Labeyrie LD and Pederson TF (eds) *Carbon Cycling in the Glacial Ocean: Constraints on the Ocean's Role in Global Change*, pp. 385–412. Berlin: Springer-Verlag.

Boyle EA (1988) Cadmium: chemical tracer of deepwater paleoceanography. *Paleoceanography* 3: 471–489.

Boyle EA (1994) A comparison of carbon isotopes and cadmium in the modern and glacial maximum ocean: can we account for the discrepancies? In: Zahn R, Pedersen TF, Kaminski MA and Labeyrie L (eds) *Carbon Cycling in the Glacial Ocean: Constraints on the Ocean's Role in Global Change*, NATO ASI Series, vol. 17, pp. 167–194. Berlin: Springer-Verlag.

Fischer G and Wefer G (eds) (1999) *Use of Proxies in Paleoceanography. Examples from the South Atlantic.* Berlin: Springer-Verlag.

Frew RD and Hunter KA (1992) Influence of Southern Ocean Waters on the cadmium-phosphate properties of the global ocean. *Nature* 360: 144–146.

Herguera JC and Berger WH (1991) Paleoproductivity: glacial to postglacial change in the western equatorial Pacific, from benthic foraminifera. *Geology* 19: 1173–1176.

Kastner M (1999) Oceanic minerals: Their origin, nature of their environment, and significance. *Proceedings of the National Academy of Science USA* 96: 3380–3387.

Lange CB and Berger WH (1993) *Diatom Productivity and Preservation in the Western Equatorial Pacific: The Quaternary Record*. Proceedings of the Ocean Drilling Program, Scientific Results 130: 509–523.

Lange CB, Berger WH, Lin H-L and Wefer G (2000) The early Matuyama Diatom Maximum off SW Africa, Benguela Current System (ODP Leg 175). *Marine Geology* 161: 93–114.

Lisitzin AP (1996) *Oceanic Sedimentation, Lithology and Geochemistry.* (Translated from Russian, originally published in 1978). Washington, DC: American Geophysical Union.

Longhurst AL, Sathyendranath S, Platt T and Caverhill C (1995) An estimate of global primary production from satellite radiometer data. *Journal of Plankton Research* 17: 1245–1271.

Martin JH, Gordon RM and Fitzwater SE (1990) Iron in Antarctic Waters. *Nature* 345: 156–158.

Müller PJ and Suess E (1979) Productivity, sedimentation rate, and sedimentary organic matter in the oceans – I. Organic carbon preservation. *Deep-Sea Research* 26: 1347–1362.

Müller PJ, Erlenkeuser H and von Grafenstein R (1983) Glacial–interglacial cycles in oceanic productivity inferred from organic carbon contents in eastern North

Atlantic sediment cores. In: Suess E and Thiede J (eds) *Coastal Upwelling: 1st Sediment Record B*, pp. 365–398. New York: Plenum Press.

Paytan A and Kastner M (1996) Benthic Ba fluxes in the central equatorial Pacific, implications for the oceanic Ba cycle. *Earth and Planetary Science Letters* 142: 439–450.

Perks HM (1999) *Climatic and Oceanographic Controls on the Burial and Preservation of Organic Matter in Equatorial Pacific Deep-sea Sediments*. PhD thesis, University of California, San Diego.

Perks HM and Keeling RF (1998) A 400 kyr record of combustion oxygen demand in the western equatorial Pacific: Evidence for a precessionally forced climate response. *Paleoceanography* 13: 63–69.

Sarnthein M, Pflaumann U, Ross R, Tiedemann R and Winn K (1992) Transfer functions to reconstruct ocean paleoproductivity, a comparison. In: Summerhayes CP, Prell WL and Emeis KC (eds) *Upwelling Systems: Evolution Since the Early Miocene*. Geology Society Special Publication 64: 411–427.

Schneider RR (1991) *Spätquartäre Produktivitätsänderungen im östlichen Angola-Becken: Reaktion auf Variationen im Passat-Monsun-Windsystem und in der Advektion des Benguela-Küstenstroms*. Reports, Fachbereich Geowissenschaften, Universität Bremen.

Wefer G and Fischer G (1993) Seasonal patterns of vertical particle flux in equatorial and coastal upwelling areas of the eastern Atlantic. *Deep-Sea Research* 40: 1613–1645.

Wefer G, Berger WH, Siedler G and Webb D (eds) (1996) *The South Atlantic. Present and Past Circulation*. Berlin: Springer-Verlag.

Wefer G, Berger WH, Richter C *et al.* (1998) Proceedings Ocean Drilling Program Initial Reports Leg 175. College Station, TX (Ocean Drilling Program).

Wefer G, Berger WH, Bijma J and Fischer G (1999) Clues to ocean history: a brief overview of proxies. In: Fischer G and Wefer G (eds) *Use of Proxies in Paleoceanography. Examples from the South Atlantic*, pp. 1–68. Berlin: Springer-Verlag.

SEICHES

D. C. Chapman, Woods Hole Oceanographic Institution, Woods Hole, MA, USA
G. S. Giese, Woods Hole Oceanographic Institution, Woods Hole, MA, USA

doi:10.1006/rwos.2001.0128

Introduction

Seiches are resonant oscillations, or 'normal modes', of lakes and coastal waters; that is, they are standing waves with unique frequencies, 'eigenfrequencies', imposed by the dimensions of the basins in which they occur. For example, the basic behavior of a seiche in a rectangular basin is depicted in **Figure 1**. Each panel shows a snapshot of sea level and currents every quarter-period through one seiche cycle. Water moves back and forth across the basin in a periodic oscillation, alternately raising and lowering sea level at the basin sides. Sea level pivots about a 'node' in the middle of the basin at which the sea level never changes. Currents are maximum at the center (beneath the node) when the sea level is horizontal, and they vanish when the sea level is at its extremes.

Seiches can be excited by many diverse environmental phenomena such as seismic disturbances, internal and surface gravity waves (including other normal modes of adjoining basins), winds, and atmospheric pressure disturbances. Once excited, seiches are noticeable under ordinary conditions because of the periodic changes in water level or currents associated with them (**Figure 1**). At some locations and times, such sea-level oscillations and currents produce hazardous or even destructive conditions. Notable examples are the catastrophic seiches of Nagasaki Harbor in Japan that are locally known as 'abiki', and those of Ciutadella Harbor on Menorca Island in Spain, called 'rissaga'. At both locations extreme seiche-produced sea-level oscillations greater than 3 m have been reported. Although seiches in most harbors do not reach such heights, the currents associated with them can still be dangerous, and for this reason the study of coastal and harbor seiches and their causes is of practical significance to harbor management and design. In this article we place emphasis on marine seiches, especially those in coastal and harbor waters.

History

In 1781, J. L. Lagrange found that the propagation velocity of a 'long' water wave (one whose wavelength is long compared to the water depth h) is given by $(gh)^{1/2}$, where g is gravitational acceleration. Merian showed in 1828 that such a wave, reflecting back and forth from the ends of a closed rectangular basin of length L, produces a standing wave with a period T, given by

$$T = \frac{2L}{n(gh)^{1/2}}$$ [1]

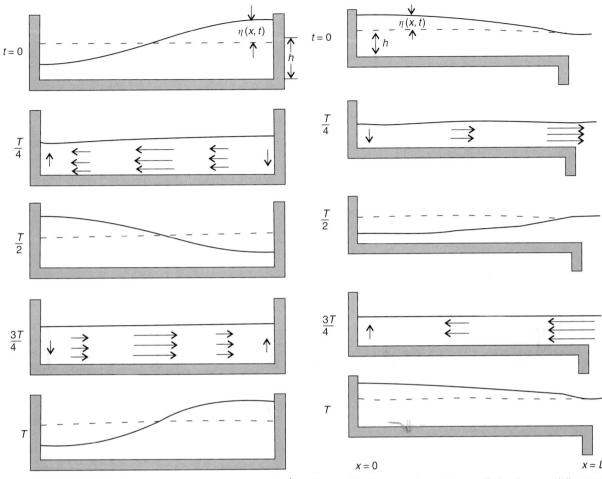

Figure 1 Diagram of a mode-one seiche oscillation in a closed basin through one period T. Panels show the sea surface and currents (arrows) each quarter-period. The basin length is L. The undisturbed water depth is h, and the deviation from this depth is denoted by η. Note the node at $x = L/2$ where the sea surface never moves (i.e. $\eta = 0$).

Figure 2 Diagram of a seiche oscillation in a partially open basin through one period T. Panels show the sea surface and currents (arrows) each quarter-period. The basin length is L with the open end at $x = L$. The undisturbed water depth is h, and the deviation from this depth is denoted by η. Note that the node (where $\eta = 0$) is located at the open end.

where $n = 1, 2, 3, \dots$ is the number of nodes of the wave (**Figure 1**) and designates the 'harmonic mode' of the oscillation. Eqn [1] is known as Merian's formula.

F. A. Forel, between 1869 and 1895, applied Merian's formula with much success to Swiss lakes, in particular Lake Geneva, the oscillations of which had long been recognized by local inhabitants who referred to them as 'seiches', apparently from the Latin word 'siccus' meaning 'dry'. Forel's seiche studies were of great interest to scientists around the world and by the turn of the century many were contributing to descriptive and theoretical aspects of the phenomenon. Perhaps most noteworthy was G. Chrystal who, in 1904 and 1905, developed a comprehensive analytical theory of free oscillations in closed basins of complex form.

By the end of the nineteenth century, it was widely recognized that seiches also occurred in open basins, such as harbors and coastal bays, either as lateral oscillations reflecting from side-to-side across the basin, or more frequently, as longitudinal oscillations between the basin head and mouth. Longitudinal harbor oscillations are dynamically equivalent to lake seiches with a node at the open mouth (**Figure 2**), and a modified version of Merian's formula for such open basins gives the period as

$$T = \frac{4L}{(2n - 1)(gh)^{1/2}} \qquad [2]$$

where $n = 1, 2, 3, \dots$. The dynamics leading to both eqns [1] and [2] are discussed below.

Interest in coastal seiches was fanned by the development of highly accurate mechanical tide recorders (*see* **Tides**) which frequently revealed surprisingly regular higher-frequency or 'secondary' oscillations in addition to ordinary tides. Even greater motivation was provided by F. Omori's observation, reported in 1900, that the periods of destructive sea waves (*see* **Tsunamis**) in harbors were often the same as those of the ordinary 'secondary waves' in those same harbors. This led directly to a major field, laboratory, and theoretical study that was carried out in Japan from 1903 through 1906 by K. Honda, T. Terada, Y. Yoshida, and D. Isitani, who concluded that coastal bays can be likened to a series of resonators, all excited by the same sea with its many frequencies of motion, but each oscillating at its own particular frequencies – the seiche frequencies.

Most twentieth century seiche research can be traced back to problems or processes recognized in the important work of Honda and his colleagues. In particular, it became widely accepted that most coastal and harbor seiching is forced by open sea processes. A. Defant developed numerical modeling techniques which modern computers have made very efficient, and B. W. Wilson applied the theory to ocean engineering problems. Many others have made major contributions during the twentieth century.

Dynamics

The dynamics of seiches are easiest to understand by considering several idealized situations with simplified physics. More complex geometries and physics have been considered in seiche studies, but the basic features developed here apply qualitatively to those studies.

In a basin in which the water depth is much smaller than the basin length, fluid motions may be described by the depth-averaged velocity u and the deviation of the sea surface from its resting position η. Changes in these quantities are related through momentum and mass conservation equations:

$$\frac{\partial u}{\partial t} = -g\frac{\partial \eta}{\partial x} - ru \qquad [3]$$

$$\frac{\partial \eta}{\partial t} + h\frac{\partial u}{\partial x} = 0 \qquad [4]$$

in which h is the fluid depth at rest, r is a coefficient of frictional damping, g is gravitational acceleration, x is the horizontal distance and t is time (see **Figure 1**). Nonlinear and rotation effects have been neglected, and only motions in the x direction are

considered. Eqn [3] states that the fluid velocity changes in response to the pressure gradient introduced by the tilting of the sea surface, and is retarded by frictional processes. Eqn [4] states that the sea-surface changes in response to convergences and divergences in the horizontal velocity field; that is, where fluid accumulates ($\partial u/\partial x < 0$) the sea surface must rise, and vice versa. Eqns [3] and [4] can be combined to form a single equation for either η or u, each having the same form. For example,

$$\frac{\partial^2 u}{\partial t^2} + r\frac{\partial u}{\partial t} - gh\frac{\partial^2 u}{\partial x^2} = 0 \qquad [5]$$

Closed Basins

The simplest seiche occurs in a closed basin with no connection to a larger body of water, such as a lake or even a soup bowl. **Figure 1** shows a closed basin with constant depth h and vertical sidewalls. At the sides of the basin, the velocity must vanish because fluid cannot flow through the walls, so $u = 0$ at $x = 0$ and L. Solutions of eqn [5] that satisfy these conditions and oscillate in time with frequency ω are

$$u = u_0 e^{-rt/2}\cos(\omega t)\sin\left(\frac{n\pi}{L}x\right) \qquad [6]$$

where u_0 is the maximum current, $n = 1, 2, 3, \ldots$, and $\omega = [gh(n\pi/L)^2 - r^2/4]^{1/2}$. The corresponding sea-surface elevation is

$$\eta = -\frac{u_0\omega L}{gn\pi}e^{-rt/2}\left[\sin(\omega t) - \frac{r}{2\omega}\cos(\omega t)\right]\cos\left(\frac{n\pi}{L}x\right) \qquad [7]$$

Eqns [6] and [7] represent the normal modes or seiches of the basin. The integer n defines the harmonic mode of the seiche and corresponds to the number of velocity maxima and sea-level nodes (locations where sea level does not change) which occur where $\cos(n\pi x/L) = 0$.

The spatial structure of the lowest or fundamental mode seiche ($n = 1$) is shown schematically in **Figure 1** through one period and was described above. Sea level rises and falls at each sidewall, pivoting about the node at $x = L/2$. The velocity vanishes at the sidewalls and reaches a maximum at the node. Sea level and velocity are almost 90° out of phase; the velocity is zero everywhere when the sea level has its maximum displacement, whereas the velocity is maximum when the sea level is horizontal.

The effect of friction is to cause a gradual exponential decay or damping of the oscillations and a slight decrease in seiche frequency with a shift in phase between u and η. If friction is weak (small r), the seiche may oscillate through many periods before fully dissipating. In this case, the frequency is close to the undamped value, $\omega \approx (gh)^{1/2}n\pi/L$ with period ($T = 2\pi/\omega$) given by Merian's formula, eqn [1]. If friction is sufficiently strong (very large r), the seiche may fully dissipate without oscillating at all. This occurs when $r > 2(gh)^{1/2}n\pi/L$, for which the frequency ω becomes imaginary.

The speed of a surface gravity wave in this basin is $(gh)^{1/2}$, so the period of the fundamental seiche ($n = 1$) is equivalent to the time it takes a surface gravity wave to travel across the basin and back. Thus, the seiche may be thought of as a surface gravity wave that repeatedly travels back and forth across the basin, perfectly reflecting off the sidewalls and creating a standing wave pattern.

Partially Open Basins

Seiches may also occur in basins that are connected to larger bodies of water at some part of the basin boundary (**Figure 2**). For example, harbors and inlets are open to the continental shelf at their mouths. The continental shelf itself can also be considered a partially open basin in that the shallow shelf is connected to the deep ocean at the shelf edge. The effect of the opening can be understood by considering the seiche in terms of surface gravity waves. A gravity wave propagates from the opening to the solid boundary where it reflects perfectly and travels back toward the opening. However, on reaching the opening it is not totally reflected. Some of the wave energy escapes from the basin into the larger body of water, thereby reducing the amplitude of the reflected wave. The reflected portion of the wave again propagates toward the closed side-wall and reflects back toward the open side. Each reflection from the open side reduces the energy in the oscillation, essentially acting like the frictional effects described above. This loss of energy due to the radiation of waves into the deep basin is called 'radiation damping.' Its effect is to produce a decaying response in the partially open basin, similar to frictional decay.

In general, a wider basin mouth produces greater radiation damping, and hence a weaker resonant seiche response to any forcing. Conversely, a narrower mouth reduces radiation damping and hence increases the amplification of fundamental mode seiches, theoretically becoming infinite as the basin mouth vanishes. However, seiches are typically forced through the basin mouth (see below), so a narrow mouth is expected to limit the forcing and yield a decreased seiche response. J. Miles and W. Munk pointed out this apparent contradiction in 1961 and referred to it as the 'harbor paradox.' Later reports raised a number of questions concerning the validity of the harbor paradox, among them the fact that frictional damping, which would increase with a narrowing of the mouth, was not included in its formulation.

The seiche modes of an idealized partially open basin (**Figure 2**) can be found by solving eqn [5] subject to a prescribed periodic sea-level oscillation at the open side; $\eta = \eta_0 \cos(\sigma t)$ at $x = L$ where σ is the frequency of oscillation. For simplicity, friction is neglected by setting $r = 0$. The response in the basin is

$$\eta = \eta_0 \cos(\sigma t)\frac{\cos(kx)}{\cos(kL)} \qquad [8]$$

$$u = \eta_0 (g/h)^{1/2}\sin(\sigma t)\frac{\sin(kx)}{\cos(kL)} \qquad [9]$$

where the wavenumber k is related to the frequency by $\sigma = (gh)^{1/2}k$. The response is similar to that in the closed basin, with the velocity and sea level again 90° out of phase. The spatial structure (k) is now determined by the forcing frequency. Notice that both the velocity and sea-level amplitudes are inversely proportional to $\cos(kL)$, which implies that the response will approach infinity (resonance) when $\cos(kL) = 0$. This occurs when $k = (n\pi - \pi/2)/L$, or equivalently when $\sigma = (gh)^{1/2}(n\pi - \pi/2)/L$ where $n = 1, 2, 3 \ldots$ is any integer.

These resonances correspond to the fundamental seiche modes for the partially open basin. They are sometimes called 'quarter-wave resonances' because their spatial structure consists of odd multiples of quarter wavelengths with a node at the open side of the basin ($x = L$). The first mode ($n = 1$) contains one-quarter wavelength inside the basin (as in **Figure 2**), so its total wavelength is equal to four times the basin width L, and its period is $T = 2\pi/\sigma = 4L/(gh)^{1/2}$. Other modes have periods given by eqn [2].

Despite the fact that these modes decay in time owing to radiation damping, they are expected to be the dominant motions in the basin because their amplitudes are potentially so large. That is, if the forcing consists of many frequencies simultaneously, those closest to the seiche frequencies will cause the largest response and will remain after the response at other frequencies has decayed. Furthermore, higher modes ($n \geq 2$) have shorter length scales and higher frequencies, so they are more likely to be dissipated by frictional forces, leaving the first mode to dominate the response. This is much like the ringing of a bell. A single strike of the hammer excites vibrations at many frequencies, yet the fundamental resonant frequency is the one that is heard. Finally, the enormous amplification of the resonant response means that a small-amplitude forcing of the basin can excite a much larger response in the basin.

Observations in the laboratory as well as nature reveal that seiches have somewhat longer periods than those calculated for the equivalent idealized open basins discussed above. This increase is similar to that which would be produced by an extension in the basin length, L, and it results from the fact that the water at the basin mouth has inertia and therefore is disturbed by, and participates in, the oscillation. This 'mouth correction', which was described by Lord Rayleigh in 1878 with respect to air vibrations, increases with the ratio of mouth width to basin length. In the case of a fully open square harbor, the actual period is approximately one-third greater than in the idealized case.

In nature, forcing often consists of multiple frequencies within a narrow range or 'band'. In this case, the response depends on the relative strength of the forcing in the narrow band and the resonant response at the seiche frequency closest to the dominant band. If the response at the dominant forcing frequency is stronger than the response at the resonant frequency, then oscillations will occur primarily at the forcing frequency. For example, the forcing frequency σ in eqns [8] and [9] may be different from any seiche frequency, and the energy in the forcing at the resonant seiche frequency may be so small that it is not amplified enough to overwhelm the response at the dominant forcing frequency. In this case, the observed oscillations, sometimes referred to as 'forced seiches', will have frequencies different from the 'free seiche' frequencies discussed above.

Generating Mechanisms and Observations

Seiches in harbors and coastal regions may be directly generated by a variety of forces, some of which are depicted in **Figure 3**: (1) atmospheric pressure fluctuations; (2) surface wind stress (*see* **Storm Surges**); (3) surface gravity waves caused by seismic activity (*see* **Tsunamis**); (4) surface gravity waves formed by wind (*see* **Wave Generation by Wind**); and (5) internal gravity waves (*see* **Internal Waves and Internal Tides**). It should be kept in mind that each of these forcing mechanisms can also generate or enhance other forcing mechanisms, thereby indirectly causing seiches. Thus, precise identification of the cause of seiching at any particular harbor or coast can be difficult.

To be effective the forcing must cause a change in the volume of water in the basin, and hence the sea level, which is usually accomplished by a change in the inflow or outflow at the open side of the basin. The amplitude of the seiche response depends on

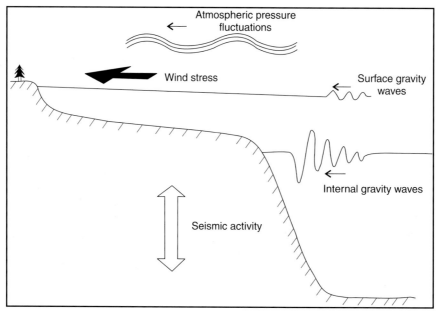

Figure 3 Sketch of various forcing mechanisms that are known to excite harbor and coastal seiches. Arrows for atmospheric pressure fluctuations, surface gravity waves, and internal gravity waves indicate propagation. Arrows for wind stress and seismic activity indicate direction of forced motions.

both the form and the time dependence of the forcing. The first mode seiche typically has a period somewhere between a few minutes and an hour or so, so the forcing must have some energy near this period to generate large seiches.

As an example, a sudden increase in atmospheric pressure over a harbor could force an outflow of water, thus lowering the harbor sea level. When the atmospheric pressure returns to normal, the harbor rapidly refills, initiating harbor seiching. However, although atmospheric pressure may change rapidly enough to match seiche frequencies, the magnitude of such high frequency fluctuations typically produces sea-level changes of only a few centimeters, so direct forcing is unlikely though not unknown. Several examples of direct forcing of fundamental and higher-mode free oscillations of shelves, bays, and harbors by atmospheric pressure fluctuations have been described. In the case of the observations at Table Bay Harbor, in Cape Town, South Africa, it was noted that 'a necessary ingredient ... was found to be that the pressure waves approach ... from the direction of the open sea.'

In lakes, seiches are frequently generated by relaxation of direct wind stress, and since wind stress acting on a harbor can easily produce an outflow of water, this might seem to be a significant generation mechanism in coastal waters as well. However, strong winds rarely change rapidly enough to initiate harbor seiching directly. That is, the typical timescales for changes in strong winds are too long to match the seiche mode periods. Nevertheless, wind relaxation seiches have been observed in fiords and long bays such as Buzzards Bay in Massachusetts, USA.

Tsunamis are rare, but they consist of large surface gravity waves that can generate enormous inflows into coastal regions, causing strong seiches, especially in large harbors. The resulting seiches, which may be a mix of free and forced oscillations, were a major motivation for early harbor seiche research as noted earlier (see **Tsunamis**). Direct generation of seiches by local seismic disturbances (as distinct from forcing by seismically generated tsunamis) is well established but very unusual. For example, the great Alaskan earthquake of 1964 produced remarkable seiches in the bayous along the Gulf of Mexico coast of Louisiana, USA. Similar phenomena are the sometimes very destructive oscillations excited by sudden slides of earth and glacier debris into high-latitude fiords and bays.

Most wind-generated surface gravity waves tend to occur at higher frequencies than seiches, so they are not effective as direct forcing mechanisms. However, in some exposed coastal locations, wind-generated swells combine to form oscillations, called 'infragravity' waves, with periods of minutes. These low-frequency surface waves are a well-known agent for excitation of seiches in small basins with periods less than about 10 minutes. Noteworthy are observations of 2–6 min seiches in Duncan Basin of Table Bay Harbor, Cape Town, South Africa, that occur at times of stormy weather. In 1993, similar short period seiches were reported in Barbers Point Harbor at Oahu, Hawaii, and their relationship to local swell and infragravity waves was demonstrated (see **Surface, Gravity and Capillary Waves**).

Perhaps the most effective way of directly exciting harbor and coastal seiches is by internal gravity waves. These internal waves can have large amplitudes and their frequency content often includes seiche frequencies. Furthermore, internal gravity waves are capable of traveling long distances in the ocean before delivering their energy to a harbor or coastline. In recent years this mechanism has been suggested as an explanation for the frequently reported and sometimes hazardous harbor seiches with periods in the range of 10–100 min. There is little or no evidence of a seismic origin for these seiches, and their frequency does not match that of ordinary ocean wind-generated surface waves. Their forcing has often been ascribed to meteorologically produced long surface waves. For example, it has been suggested that the 'abiki' of Nagasaki Harbor and the 'Marrobbio' in the Strait of Sicily may be forced by the passage of large low-pressure atmospheric fronts. In 1996, evidence was found that the hazardous 10-minute 'rissaga' of Ciutadella Harbor, Spain, and offshore normal modes are similarly excited by surface waves generated by atmospheric pressure oscillations, and it was proposed that the term 'meteorological tsunamis' be applied to all such seiche events.

However, attributing the cause of remotely generated harbor seiches to meteorologically forced surface waves does not account for observations that such seiches are frequently associated with ocean tides. In 1908 it was noted that in many cases harbor seiche activity occurs at specific tidal phases. In the 1980s, a clear association was found between tidal and seiche amplitudes and it was suggested that tide-generated internal waves could be a significant agent for excitation of coastal and harbor seiches. In 1990, a study of the fundamental theoretical questions concerning transfer of momentum from internal waves to seiche modes and the wide frequency gap between tides and harbor seiches indicated that the high-frequency energy content of tide-generated internal solitary waves is sufficient to account for the energy of the recorded seiches, and

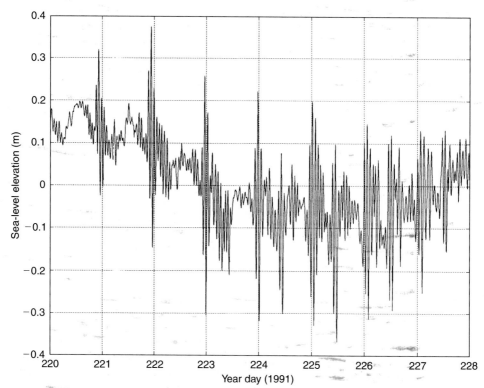

Figure 4 An example of harbor seiches from Puerto Princesa at Palawan Island in the Philippines. The tidal signal has been removed from this sea-level record to accentuate the bursts of 75-min harbor seiches. The seiches are excited by the arrival at the harbor mouth of internal wave packets produced by strong tidal current flow across a shallow sill some 450 km away.

a dynamical model for the generating process was published.

Observations at Palawan Island in the Philippines have demonstrated that harbor seiches can be forced by tide-generated internal waves and, as might be expected, there was also a strong dependency between seiche activity and water column density stratification. Periods of maximum seiche activity are associated with periods of strong tides, an example of which is given in **Figure 4**, which shows an 8-day sea-level record from Puerto Princesa at Palawan Island with the tidal signal removed. Bursts of 75-min harbor seiches are excited by the arrival at the harbor mouth of internal wave packets produced by strong tidal current flow across a shallow sill some 450 km away. The internal wave packets require 2.5 days to reach the harbor, producing a similar delay between tidal and seiche patterns. As an illustration, note the change in seiche activity from a diurnal to a semidiurnal pattern that is evident in **Figure 4**. A similar shift in tidal current patterns occurred at the internal wave generation site several days earlier.

More recent observations at Ciutadella Harbor in Spain point to a second process producing internal wave-generated seiches. Often the largest seiche events at that harbor occur under a specific set of conditions – seasonal warming of the sea surface and extremely small tides – which combine to produce very stable conditions in the upper water column. It has been suggested that under those conditions, meteorological processes can produce internal waves by inducing flow over shallow topography and that these meteorologically produced internal waves are responsible for the observed seiche activity.

See also

Internal Tides. Internal Waves. Storm Surges. Surface, Gravity and Capillary Waves. Tides. Tsunamis. Wave Generation by Wind.

Further Reading

Chapman DC and Giese GS (1990) A model for the generation of coastal seiches by deep-sea internal waves. *Journal of Physical Oceanography* 20: 1459–1467.

Chrystal G (1905) On the hydrodynamic theory of seiches. *Transactions of the Royal Society of Edinburgh* 41: 599–649.

Defant A (1961) *Physical Oceanography*, vol 2. New York: Pergamon Press.

Forel FA (1892) *Le Leman (Collected Papers)*, 2 vols. Lausanne, Switzerland: Rouge.

Giese GS, Chapman DC, Collins MG, Encarnacion R and Jacinto G (1998) The coupling between harbor seiches at Palawan Island and Sulu Sea internal solitons. *Journal of Physical Oceanography* 28: 2418–2426.

Honda K, Terada T, Yoshida Y and Isitani D (1908) Secondary undulations of oceanic tides. *Journal of the College of Science, Imperial University, Tokyo* 24: 1–113.

Korgen BJ (1995) Seiches. *American Scientist* 83: 330–341.

Miles JW (1974) Harbor seiching. *Annual Review of Fluid Mechanics* 6: 17–35.

Okihiro M, Guza RT and Seymour RT (1993) Excitation of seiche observed in a small harbor. *Journal of Geophysical Research* 98: 18 201–18 211.

Rabinovich AB and Monserrat S (1996) Meteorological tsunamis near the Balearic and Kuril islands: descriptive and statistical analysis. *Natural Hazards* 13: 55–90.

Wilson BW (1972) Seiches. *Advances in Hydroscience* 8: 1–94.

SEISMIC STRUCTURE

A. Harding, University of California, San Diego, CA, USA

doi:10.1006/rwos.2001.0258

Introduction

Seismic exploration of the oceans began in earnest in the 1950s. The early seismic experiments were refraction in nature using explosives as sources. The principal data were first arrival, P-wave travel times, which were analyzed to produce primarily one-dimensional models of compressional velocity as a function of depth. Within a decade, the results of these experiments had convincingly demonstrated that the crust beneath the ocean crust was much thinner than continental crust. Moreover, the structure of the deep ocean was unexpectedly uniform, particularly when compared with the continents. In light of this uniformity it made sense to talk of average or 'normal' oceanic crust. The first compilations described the average seismic structure in terms of constant velocity layers, with the igneous crust being divided into an upper layer 2 and an underlying layer 3.

Today, the scale and scope of seismic experiments is much greater, routinely resulting in two- and three-dimensional images of the oceanic crust. Experiments can use arrays of ocean bottom seismographs and/or multichannel streamers to record a wide range of reflection and refraction signals. The source is typically an airgun array, which is much more repeatable than explosives and produces much more densely sampled seismic sections. Seismic models of the oceanic crust are now typically continuous functions of both the horizontal and vertical position, but are still principally P-wave or compressional models, because S-waves can only be produced indirectly through mode conversion in active source experiments.

In spite of their greater resolving power, modern experiments are still too limited in their geographic scope to act as a general database for looking at many of the questions concerning oceanic seismic structure. The main vehicle for looking at the general seismic structure of the oceans is still the catalog of one-dimensional P-wave velocity models built up over approximately 40 years of experiments. The original simple layer terminology, with slight elaboration, is by now firmly entrenched as the means of describing the principal seismic features of the oceanic crust; despite the fact that the representation of the underlying velocity structure has changed significantly over time. The next section discusses the evolution of the velocity model and the layer description. Subsequent sections discuss the interpretation of seismic structure in terms of geologic structure; the seismic structure of anomalous crust; and the relationship of seismic structure to such influences as spreading rate and age.

Normal Oceanic Crust

Table 1 reproduces one of the first definitions of average or 'normal' oceanic crust by Raitt (1963). Even in this era before plate tectonics, Raitt excluded from consideration any areas such as oceanic plateaus that he thought atypical of the deep ocean. Today, compilations count as normal crust formed at midocean ridges away from fracture zones. The early refraction experiments typically consisted of a small set of widely spaced instruments. They were analyzed using the slope-intercept method in which a set of straight lines was fitted to first arrival travel times. This type of analysis naturally leads to stair-step or 'layer-cake' models consisting of a stack of uniform velocity layers separated by steps in velocity. Although their limitations as a description of the earth were recognized, these models provided a simple and convenient means of comparing geographically diverse data sets. Raitt divided the oceanic crust into three layers and included a fourth to represent the upper mantle. The top layer (layer 1), was a variable thickness sedimentary layer. Below this came the two layers that together comprised the igneous oceanic crust, a thinner more variable velocity layer (layer 2), and a thicker, more uniform velocity layer (layer 3). Layer 3 is the most characteristically oceanic of the layers. Arrivals from this layer are the most prominent arrivals in typical refraction profiles. The uniformity of high velocities within this layer mark layer 3 as being compositionally distinct from continental crust. At the base of layer 3 is the Mohorovic discontinuity or Moho, identifiable as such because the velocities of layer 4 were comparable to those seen in the upper mantle beneath continents.

As refraction data sets with better spatial sampling became available, the systematic errors inherent in fitting a few straight lines to the first arrival travel times became more noticeable. This was especially true for layer 2 first arrivals, which appear over a relatively short-range window, but have noticeable curvature because of the wide range of layer 2 velocities. The initial resolution of this problem was to fit more lines to the data and divide layer 2 into smaller, constant velocity sublayers termed 2A, 2B, and 2C. However, there was a more fundamental problem: the layer-cake models were not consistent with the waveform and amplitude behavior of the data. This flaw became apparent when, instead of just using travel times, the entire recorded wavefield began to be modeled using synthetic seismograms. Waveform modeling led to a recasting of the one-dimensional model in terms of smoothly varying velocities, constant velocity gradients, or finely layered stair-steps. Large velocity steps or interfaces are now included in the models only if they are consistent with the amplitude behavior. The stair-step representation is a tacit admission that there is a limit to the resolution of finite bandwidth data. A stair-step model is indistinguishable from a continuous gradient provided the layering is finer than the vertical resolution of data, which for refraction data is some significant fraction of a wavelength. Today, purely travel time analysis based upon densely sampled primary and secondary arrival times and accumulated knowledge can yield accurate models, but seismogram modeling is still required to achieve the best resolution.

The change in the style of the velocity models is illustrated in **Figure 1**, which shows models for a recent Pacific data set. A change in gradient rather than a jump in velocity marks the boundary between layer 2 and 3 in most modern models.

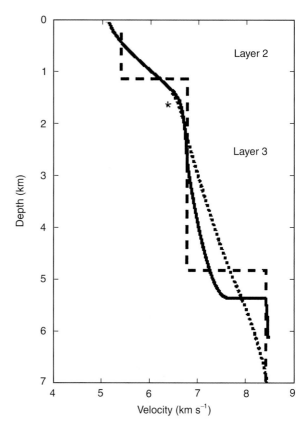

Figure 1 Velocity models for a recent Pacific data set.

Paradoxically, the jump in velocity at the Moho – present in the traditional layer-cake model – is not required by the first arrival times, but is required to fit the secondary arrival times and amplitude behavior of the data. The example also illustrates another general problem with layer-cake models, which is that they systematically underestimate both layer and total crustal thickness.

Table 1 also presents a more modern summary of average oceanic crustal structure. All models included in this compilation were the result of synthetic seismogram analysis. A range of velocities and typical gradients now characterizes layers 2 and 3. Layer 2 is a region of rapidly increasing velocity at the top of the crust, with typical gradients from $1–3 s^{-1}$, while layer 3 is a thicker region of more uniform velocity with gradients between 0 and $0.2 s^{-1}$. The layer thicknesses are little changed from the Raitt compilation but are systematically thicker and thus total crustal thickness is also larger.

A virtue of the layer description is that it captures the main features of the seismic data and results without being too precise. So, while the style of velocity model has moved away from one that is strictly layered, the description in terms of layers persists both for historical continuity and linguistic

Table 1 Traditional and modern summaries of average oceanic crystal structure

Parameter	Traditional[a]		Modern[b]		
	Velocity (km s^{-1})	Thickness (km)	Velocity (km s^{-1})	Thickness (km)	Representative gradient (s^{-1})
Layer 2 (igneous crust)	5.07 ± 0.63	1.71 ± 0.75	2.5–6.6	2.11 ± 0.55	1–3
Layer 3 (igneous crust)	6.69 ± 0.26	4.86 ± 1.42	6.6–7.6	4.97 ± 0.90	0.1–0.2
Layer 4 (upper mantle)	8.13 ± 0.24		> 7.6		
Total igneous crust		6.57 ± 1.61		7.08 ± 0.78	

[a]From Raitt (1963).
[b]Modified from White et al. (1992).

convenience. Without a simple layer-cake model, defining the layer 2/3 transition can be somewhat problematic. Ideally the velocity function will show a small velocity jump or resolvable inflection at the boundary, but often the layer 2/3 boundary is taken as being either at a change in the general velocity gradient of the model or at a particular velocity just below representative layer 3 velocities, sometimes both. The chosen velocity is typically somewhere between 6.5 and 6.7 km s^{-1}.

The use of layers 2 and 3 is almost universally and consistently applied when summarizing the basic features of both seismic data sets and models. More problematic is the use of layers 2A, 2B, and 2C to describe subintervals layer 2. While the use of these layers is widespread in the literature, their application is more variable and has evolved in conjunction with changes in model style and resolution. As a result caution is needed when comparing models from disparate experiments, particularly across tectonic and geographic regions. Today, many authors subdivide layer 2 into 2A and 2B only. Layer 2A is widely recognized as a well defined surficial layer in young oceanic crust near ridges, being associated with velocities < 3 km s^{-1} and a transition to velocities > 4 km s^{-1} at its base. In this division, layer 2B is simply the lower part of layer 2.

Interpretation of Seismic Structure

The nature of the relationship between the seismic and geologic structure of the oceanic crust is the subject of ongoing debate. The adoption of plate tectonics and seafloor spreading provided a framework for understanding the initially surprising uniformity and simplicity of ocean seismic structure. It also gave rise to the hope that there would be a correspondingly simple and universal interpretation of the seismic layering in terms of geologic structure. However, this expectation has receded as

the complexity of the seismic models has increased and our understanding of the diverse magmatic, tectonic, and hydrothermal processes shaping crustal structure, both spatially and temporally, has improved. Fundamentally, there is no unique, unambiguous interpretation of seismic velocity in terms of rock type or geologic structure. Having coincident P- and S-wave velocity models plus other geophysical data can considerably reduce this ambiguity but ultimately cannot eliminate it. Ideally, reference drill holes through the full oceanic crustal section would be used to calibrate seismic and other geophysical results. These would allow the dominant processes, controlling, for example, the layer 2/3 boundary or the nature of layer 2A to be identified in different tectonic settings. Unfortunately, to date only a limited number of drill holes have penetrated a significant depth into the oceanic crust, and none have penetrated a full crustal section. As a result there has been only limited opportunity for direct comparison between seismic and *in situ* structure. When interpreting seismic results, we must still rely heavily on inferences that draw upon a number of less direct sources including seafloor observations, analogy with ophiolites, and laboratory measurements on dredged rock samples.

The simplest, most straightforward interpretation of seismic velocity is to assume that velocity is dependent on composition and that different velocities indicate different rock types. This reasoning has guided the traditional interpretation of the Moho boundary. In seismic models, the characteristic signature of the Moho is an increase in velocity of between ~0.5 and 1.0 km s^{-1} to velocities > 7.6 km s^{-1}. The increase may occur either as a simple interface (at the resolution of the data) or as a transition region up to a kilometer thick. In reflection data, the Moho is often observed as a low frequency, ~10 Hz, quasi-continuous event. In general, layer 3 is considered to be predominantly gabbros, and the most common interpretation of the

Moho is as the boundary between a mafic gabbroic crust and an ultramafic upper mantle. The observed Moho structures reflect either a simple contact or a transition zone of interleaved mafic and ultramafic material. These interpretations are supported by observations within ophiolites, reference sections of oceanic lithosphere obducted onto land. However, partially serpentinized, ultramafic peridotites can have P- and S-wave velocities that are for practical purposes indistinguishable from gabbros, when the degree of alteration is between 20 and 40%. For lesser degrees of alteration, the serpentinized rocks will have velocities that are intermediate between gabbros and peridotite and are thus distinguishable. Although widespread serpentinization is unlikely, particularly at fast and intermediate spreading rates, it may be locally important at segment boundaries at slow spreading rates where faulting and extension expose peridotites to pervasive alteration by deeply circulating sea water.

A further complication in interpreting the Moho is the fact that the igneous crust may contain cumulate ultramafic rocks such as dunite, which crystallized as sills from a melt. Seismically, these are indistinguishable from the residual upper mantle harburgites that yielded the crustal melts. Thus a distinction is sometimes made between Moho defined seismically, and a petrologic Moho separating cumulate rocks from source rocks.

There is very little intrinsic difference in composition or velocity to the basaltic rocks – pillow lavas and sheet flows, sheeted dikes, and gabbros – that typically constitute the upper part of the oceanic crust. Certainly not enough to explain the large range of layer 2 velocities or the difference between layers 2 and 3. Instead the seismic character is attributed primarily to the cracks, fractures, and fissures that permeate the upper crust. These reduce the effective stiffness of the rock matrix, and hence the velocity of the upper crust at seismic wavelengths. The relationship between velocity and the size, shape, orientation, and distribution of cracks is a complex nonlinear one that affects P- and S-wave velocities differently. However, at least informally, it is often crack volume or porosity that is taken as the primary control. The large velocity gradients within layer 2 are then seen as being the result of a progressive closure of cracks or reduction in porosity with depth, as confining pressure increases. Once most cracks are closed, velocities are only weakly pressure-dependent and can have the low velocity gradients characteristic of layer 3. An analogous velocity behavior, but with a smaller velocity range is observed in individual rock samples subject to increasing confining pressure.

From this perspective, there can only be a structural or lithologic interpretation of the seismic layers if there is a structural dependence to the crack distribution. Support for such an interpretation comes from composite velocity profiles through ophiolites, constructed using laboratory measurements on hand samples at suitable confining pressures. These profiles showed a broad agreement with oceanic results and led to the standard ophiolite interpretation of seismic structure in which layer 2 is equated with the extrusive section of pillow lavas and sheeted dikes and layer 3 with the intrusive gabbroic section. Moreover, the relative and absolute thicknesses of the extrusive and intrusive sections in ophiolites are comparable to those of the oceanic seismic layers. At a more detailed level, measurements on ophiolites often show a reduction in porosity at the transition from pillow basalts to sheeted dikes: an observation used to bolster the inference that layer 2A is equivalent to pillow lava section in young oceanic crust.

As the only available complete exposures of oceanic crustal sections, ophiolites have had a historically influential role in guiding the interpretation of seismic layering. However, a number of cautions are in order. First, there are inherent uncertainties in extrapolating seismic velocities measured on hand samples to the larger scales and lower frequencies characteristic of seismic experiments. Second, the seismic velocities of ophiolites could have been modified during the obduction process. Finally, most ophiolites are thought to have been produced in back arcs or marginal basin settings and thus while valuable structural analogs may not be representative of the ocean basins as a whole. Ultimately, the ophiolite model can only be used as a guide, albeit an important one, for interpreting oceanic structure, seismic or otherwise and conclusions drawn from it must be weighed against other constraints.

The traditional ophiolite model is a convenient and widely used shorthand for describing seismic structure, that is useful provided that too much is not asked or expected of it. The porosity interpretation of upper crustal velocities gives the basic layer 2/layer 3 division of seismic models a sort of universality that transcends, within bounds, changes in the underlying lithologic structure, and emphasizes the need for taking tectonic setting into consideration when interpreting seismic structure. Any process that either resets or significantly modifies this crack/porosity distribution of the upper crust will imprint itself on the seismic structure. For example, near fracture zones, fracturing and faulting are usually inferred to be dominant controls on layer

2 structure. At Deep Sea Drilling Project Hole 504B, a deep penetration hole in 5.9 million year old crust formed at intermediate spreading rates, the base of seismic layer 2 is found to lie within the sheeted dike section. It is thought that progressive filling of cracks by hydrothermal alteration processes has, over time, raised the depth of the layer 2/3 boundary.

Anomalous Crust

Definitions of normal seismic structure focus on oceanic crust formed at midocean ridges and specifically exclude structures such as fracture zones or oceanic plateaus as anomalous. Particularly at slow spreading rates, fracture zones and the traces of segment boundaries are part of the warp and weft of ocean fabric, making up about 20% of the seafloor. These have been most extensively studied in the northern Atlantic, where the ridge is segmented on scales of 20–100 km, and most segment boundaries are associated with attenuated crust. The degree of crustal thinning shows no simple dependence of segment offset, although the large-offset fracture zones may indeed contain the most extreme structure. Fracture zones and segment traces typically exhibit an inner and outer region of influence. In the outer region, there is gradual thinning of the crust towards the trace extending over a distance of perhaps 20 km. This region is marked by a deepening of the seafloor and a simultaneous shoaling of the Moho. Within the outer region, seismic structure is a thinned but recognizable version of normal crust, and the most extreme structure is associated with the inner region. Here, within a ~ 10 km wide zone, the crust may be < 3 km thick and in one-dimension may appear to be all layer 2 down to the Moho. Looked at in cross-section, there is a coalescence of a gradually thickening layer 2 with a Moho transition region, and the elimination of layer 3. The structure at segment boundaries can be explained as a combination of reduced magma supply, much of it feeding laterally from the segment centers, and pervasive faulting, possibly low angle in nature.

Velocities intermediate between crust and mantle values, 7.1–7.4 km s^{-1}, are observed beneath small offset transforms and nontransform offsets below about 4 km depth, most likely indicating partial serpentinization of the mantle by deeply circulating water. The upper limit of serpentinization is difficult to determine seismically because of the overlap in velocity between gabbros and altered peridotites at high degrees of alteration. But, from seafloor observations, at least some serpentinites must lie at shallower depths.

In the fast spread Pacific, fracture zones – which are spaced at intervals of a few hundred kilometers – affect only a relatively small fraction of the total crust. In addition, seismic studies suggest that the crustal structure of fracture zones is essentially a slightly thinner version of normal crust with a well-defined layer 3. In addition there can be some thickening and slowing of layer 2 in the vicinity of the transform associated with upper crustal faulting.

The term oceanic large igneous provinces (LIPs) provides a convenient umbrella under which to group such features as oceanic plateaus, aseismic ridges, seamount groups, and volcanic passive margins. They are massive emplacements of mostly mafic extrusive and intrusive material whose origin lies outside the basic framework of seafloor spreading. Together they account for much of the anomalous structure apparent in maps of seafloor bathymetry. At present, the rate of LIP emplacement, including the continents, is estimated to be equal to about 5–10% of midocean ridge production. However, during the formation of the largest LIPs, such as the Ontong Java plateau, off-axis volcanism was a significant fraction of midocean ridge rates. Many LIPs can trace their origin to either transient or persistent (hot-spot) mantle plumes; as such they provide a valuable window into the dynamics of the mantle. Where plumes interact with ridges, they can significantly affect the resulting crustal structure. The most notable example of this, at present, is the influence of the Iceland hot-spot on spreading along the Reykjanes ridge, where the crust is about 10 km thick and includes an approximately 7 km thick layer 3.

Two general features of LIPs seismic structure are a thickened crust, up to 25 km thick, and a high velocity, lower crustal body, reaching up to 7.6 km s^{-1}. However, in detail, the seismic structure depends on the style and setting of their emplacement, including whether the emplacement was submarine or subareal, intraplate or plate boundary. For example, the Kerguelen-Heard Plateau (a province of the larger Kerguelen LIP) is estimated to have 19–21 km thick igneous crust, the majority of which is a 17 km thick layer 3 with velocities between 6.6 and 7.4 km s^{-1}. The plateau is inferred to have formed by seafloor spreading in the vicinity of a hot-spot similar to Iceland. If this is the case, the greater thickness and higher velocities of layer 3 can be attributed to greater than normal extents of partial melting within the upwelling mantle. An example of an intraplate setting is the formation of the Marquesas Island hot-spot. Seismic data reveal that in addition to the extrusive volcanism responsible for the islands, significant intrusive emplacement has created a crustal root beneath the previously existing oceanic crust. Combined, the total crust is up to 17 km thick. The crustal root, with velocities between 7.3 and 7.75 km s^{-1}, may be purely intrusive or a mixture of intrusive rocks with preexisting mantle peridotites.

Systematic Features of the Oceanic Crust

For the most part, seismic investigations of the oceanic crust tend to focus on specific geologic problems. As a consequence, the catalog of published seismic results has sampling biases that make it less than ideal for looking at certain more general questions. There are for example a relatively large number of good measurements of young Pacific crust and old Atlantic crust, but fewer on old Pacific crust and only a handful of measurements on crust formed at ultra-slow spreading rates. Older data sets analyzed by the slope-intercept method are often discounted unless they are the only data available for a particular region. Nevertheless there are a number of systematic features of oceanic crust that can be discerned from compilations of seismic results.

Spreading Rate Dependence of Average Crustal Thickness

Although the style of crustal accretion varies considerably between slow and fast spreading ridges, the average thickness of the crust produced including fracture zones is remarkably uniform at 7 ± 1 km for full spreading rates between 20 and 150 mm a^{-1}. This result indicates that the rate of crustal production is linearly related to spreading rate over this range. Crustal thicknesses are more variable at slower spreading rates, reflecting the more focused magma supply and greater tectonic extension. At ultra-slow spreading rates below 20 mm a^{-1}, there is a measurable and rapid decrease in average crustal thickness. This reduction is expected theoretically, as conductive heat loss inhibits melt production in the upwelling mantle.

Age Dependence of Crustal Structure

The clearest and strongest aging signal in the oceanic crust is the approximate doubling of surficial velocities with age from about 2.5 km s^{-1} at the ridge axis to 5 km s^{-1} off-axis. This increase in velocity was first reported in the mid-1970s based on compilations of surface sonobuoy data. Originally, the velocity signal was interpreted as being asso-

ciated with a thinning of layer 2A over a period of 20–40 Ma. However, the same data can equally well be explained as simply the increase in velocity of a constant thickness layer, and a compilation of modern seismic data sets indicates that layer 2A velocities increase much more rapidly, almost doubling in < 10 Ma. While both of these inferences are supported by individual flowline profiles extending out from the ridge axis, the distribution bias of modern seismic data sets to the ridge axes makes it hard to assess the robustness of this result.

The increase in layer 2A velocity with age is due to hydrothermal alteration sealing cracks within the upper crust. There need not be a correspondingly large decrease in porosity, as alteration that preferentially seals the small aspect ratio cracks will produce a large velocity increase for a small porosity reduction. Given this mechanism, similar, albeit smaller increases in layer 2B velocities might be expected. Such an increase is not apparent in present compilations, although a small systematic change would be masked by the intrinsic variability of layer 2B and the variability induced by different analysis methods. There is though some indication of systematic change with layer 2B from analysis of ratios of P- and S-wave velocity and as noted in the previous section alteration is thought to have raised the layer 2/3 boundary at Hole 504B.

Anisotropic Structure

Two types of anisotropic structure are frequently reported for the oceanic crust and upper mantle. The P-wave velocities of the upper mantle are found to be faster in the fossil spreading direction, than in the original ridge parallel direction, with the difference being around 7%. This is due to the preferential alignment of the fast a-axis of olivine crystals in the direction of spreading as mantle upwells beneath the midocean ridge.

The other region of the crust that exhibits anisotropy is the extrusive upper crust, which has a fast P-wave propagation direction parallel to ridge axis at all spreading rates. The peak-to-peak magnitude of the anisotropy averages ~ 10%. Like the velocity structure, this shallow anisotropy is generally ascribed to the crack distribution within the upper crust. Extensional forces in the spreading direction are thought to produce thin cracks and fissures that preferentially align parallel to the ridge axis.

See also

Mid-Ocean Ridge Tectonics, Volcanism and Geomorphology. Seismology Sensors.

Further Reading

Carlson RL (1998) Seismic velocities in the uppermost oceanic crust: Age dependence and the fate of layer 2A. *Journal of Geophysical Research* 103: 7069–7077.

Fowler CMR (1990) *The Solid Earth*. Cambridge: Cambridge University Press.

Horen H, Zamora M and Dubuisson G (1996) Seismic waves velocities and anisotropy in serpentinized peridotites from Xigaze ophiolite: Abundance of serpentine in slow spreading ridges. *Geophysical Research Letters* 23: 9–12.

Raitt RW (1963) The crustal rocks. In: Hill MN (ed.) *The Sea*, vol. 3. New York: Interscience.

Spudich P and Orcutt J (1980) A new look at the seismic velocity structure of the oceanic crust. *Reviews in Geophysics* 18: 627–645.

White RS, McKenzie D and O'Nions RK (1992) Oceanic crustal thickness from seismic measurements and rare earth element inversions. *Journal of Geophysical Research* 97: 19 683–19 715.

SEISMOLOGY SENSORS

L. M. Dorman, University of California, San Diego, La Jolla, CA. USA

doi:10.1006/rwos.2001.0334

Introduction

A glance at the globe shows that the Earth's surface is largely water-covered. The logical consequence of this is that seismic studies based on land seismic stations alone will be severely biased because of two factors. The existence of large expanses of ocean distant from land means that many small earthquakes underneath the ocean will remain unobserved. The difference in seismic velocity structure between continent and ocean intruduces a bias in locations, with oceanic earthquakes which are located using only stations on one side of the event being pulled tens of kilometers landward. Additionally, the depths of shallow subduction zone events, which are covered by water, will be very poorly determined. Thus seafloor seismic stations are necessary both for completeness of coverage as well as for precise location of events which are tectonically important. This paper summarizes the status of seafloor seismic instrumentation.

The alternative methods for providing coverage are temporary (pop-up) instruments and permanently connected systems. The high costs of seafloor cabling has thus far precluded dedicated cables of significant length for seismic purposes, although efforts have been made to use existing, disused wires. Accordingly, the main emphasis of this report will be temporary instruments.

Large ongoing programs to investigate oceanic spreading centers (RIDGE) and subductions (MARGINS) have provided impetus for the upgrading of seismic capabilities in oceanic areas.

The past few years has seen a blossoming of ocean bottom seismograph (OBS) instrumentation, both in number and in their capabilities. Active experimental programs are in place in the USA, Europe, and Japan. Increases in the reliability of electronics and in the capacity of storage devices has allowed the development of instruments which are much more reliable and useful. Major construction programs in Japan and the USA are producing hundreds of instruments, a number which allows imaging experiments which have been heretofore associated with the petroleum exploration industry. This contrasts sharply with the severely underdetermined experiments which have characterized earthquake

Figure 1 The UTIG OBS, a particularly 'clean' mechanical design, which has been in use for many years, with evolving electronics. The anchor is 1.2 m on each side. (Photograph by Gail Christeson, UTIG.)

Table 1 Characteristics of short period ocean bottom seismometers

Parameter	UTIG	WHOI-SP	SIO-IGPP-SP	GEOMAR-SP	JAMSTEC-ORI[a]
Contact/website	Nakamura/http://www.ig.utexas.edu/research/projects/obs/obs.html	Detrick/Collins http://www.obsip.org	Orcutt/Babcock http://www.obsip.org	Flüh/Bialas http://www.geomar.de	–
Seismic sensor(s)	3-component 4.5 Hz Mark Products L-15B or Oyo GS-11D	2 Hz Mark Products L-22 4.5 Hz or L-28 4.5 Hz, vertical component	2 Hz Mark Products L-22, vertical component	optionally uses Webb BB sensor	4.5 Hz vertical
Frequency response	4.5–100 Hz	2–X Hz	2–X Hz	0.05–30 Hz/0.01–X Hz	4.5–100 Hz
Nominal sensitivity	2.5 nm s^{-1}	–	–	–	–
Hydrophone	OAS E-2PD crystal	Hightech crystal	Hightech HYI-90-U	OAS E4SD or Cox-Webb DPG	–
Frequency Response	3–100 Hz	5–X Hz	50 mHz-X	0.05–30 Hz	–
Digitizer type, dynamic range, sample rates	14 bits + gain-ranging, 126 dB,112 dB re electronic noise	Quanterra Q330 24-bit,126 dB, 1–200 Hz	Cirrus/Crystal CS5321-CS5322 24-bit, 124–130 dB	oversampling, 120 dB, 25–200 Hz	–
Recording medium and capacity[b]	Disk, semi-continuous	Disk, data download through pressure case	9 Gbyte disk, data download through pressure case	1 Gbyte DAT or 2 Gbyte semiconductor memory	–
Clock type and drift[c]	10 ms	Seascan Precision Timebase < 0.5 ms d^{-1}	Seascan Precision Timebase < 0.5 ms d^{-1}	Seascan Precision Timebase < 0.5 ms d^{-1}	–
Endurance	8 weeks–6 months	90 days, alkaline/ 1 year, lithium	80–180 days at 250/31.25 Hz sampling 5 days on NiCad rechargeable cells	300 days	–
Power consumption	550 mW	–	420 mW at 31.25 Hz, 1.6 W at 250 Hz for 2 channels	230–250 mW	–
Release type	Burnwire release, acoustically controlled, acoustic release, two backup timers	Burnwire, Edgetech acoustics	Double burnwire, Edgetech acoustics	Acoustic release with back-up timer	–
Mechanical configuration, launch/recovery weights	Single 43 cm diameter glass sphere, 85 kg/35 kg	63 kg/43 kg	110 kg/80 kg	Vertical cylinders, 175 kg/125 kg	Single 43 cm diameter glass sphere
Number available	37–39[d]	15 now, 40 more under construction	14 now, 74 more under construction	27 OBH + 11 OBS	100?
Total			51		

[a]Information on these instruments is incomplete.
[b]2 gigabytes is about 22 days of data sampling four channels at 128 Hz or 176 days sampling four channels at 16 Hz.
[c]1 ms d^{-1} is about 1 × 10 E-8.
[d]Includes instruments of the same design operated by IRD (formerly ORSTOM) and National Taiwan Ocean University.

Table 2 Characteristics of broadband ocean bottom seismometerrs

Parameter	SIO/ONR[a]	WHOI-BB	LDEO-BB[e]	SIO/IGPP-BB	ORI-BB[f]
Manager	Dorman/ Sauter http:// www-mpl. ucsd.edu/obs	Detrick/ Collins http:// www.obsip.org	Webb/ http:// www.obsip.org	Orcutt/Babcock http://www. obsip.org	–
Seismic sensor	1 Hz Mark Products L4C-3D or PMD 2123	Guralp CMG-3ESP	1 Hz Mark Products L4C-3D	Kinemetrics	PMD 2023
Frequency response	0.033–32 Hz[b]	0.033–50 Hz[b]	0.005–30 Hz[b]	0.02–50 Hz	0.033–50 Hz
Hydrophone	Cox-Webb DPG	HighTech	Cox-Webb DPG for low frequency hydrophone for high frequency	High Tech HYI-90-U or Cox-Webb DPG	–
Frequency response	0.001–5 Hz	0.033–X Hz	0.001–60 Hz	0.05–15 kHz or 0.01–32 Hz	–
Digitizer type, Dynamic range, sample rates	16-bit + gain-ranging, 126 dB, X-128 sps	Quanterra QA330 24-bit, 135 dB	nominal 24-bit ~ 135 dB, 1, 20, 40, 100, 200 sps	24-bit Cirrus/ Crystal CS 5321–CS-5322, 130 dB	20-bit
Recording medium and capacity[c]	Disk, 9–27 Gbyte	Disk, 2 Gbyte	Disk, 18 Gb, 72 Gb planned	Disk, 9 Gbyte	4 × 6.4 Gbyte disks
Clock drift[d]	< 1 ms d^{-1}	< 1 ms d^{-1}	0.5 ms d^{-1}	< 1 ms d^{-1}	–
Endurance	6–12 months	6–12 + months	Up to 15 months	6–12 months	9 months?
Power consumption	400 mW	1.5 W at 20 sps	–	–	~ 600 mW
Release type	Two EG&G 8242 acoustic releases	One EG&G 8242 acoustic relase with back-up acoustic burnwire release	Acoustically controlled burnwire?	EG&G acoustically controlled burnwire	Acoustically controlled burn-plate
Mechanical configuration, launch/ recovery weights	Fiberglass frame, aluminum pressure cases, glass flotation (ONR OBS-style)	ONR OBS-style, floating above anchor, 570 kg/ 472 kg	Aluminum pressure cases, plastic plate frame, 215/145 kg	178/138 kg	50 cm diameter, titanium sphere
Number available	14	25 under construction	64 under construction	15–20 (using recording package from SP instrument)	15
Total			~ 133		

[a]These instruments incorporate a fluid flowmeter/sampler in the instrument frame (see Tryon *et al.* 2001).
[b]Seismometers are free from spurious resonances below 20 Hz.
[c]2 gigabytes is about 22 days of data sampling four channels at 128 Hz or 176 days sampling four channels at 16 Hz.
[d]1 ms is about 1×10^{-8}.
[e]1 See Webb *et al.*, 2001.
[f]1 See Shiobara *et al.*, 2000.

seismology. One change over the past decade has been the disappearance of analog recording systems.

A side effect of this rapid change is that a review such as this provides a snapshot of the technology, rather than a long-lasting reference. The technical details reported below are for instruments at two stages of development: existing instruments (UTIG, SIO/ONR, SIO/IGPP-SP, GEOMAR, LDEO-BB) and instruments still in design and construction (WHOI-SP, WHOI-BB SIO/IGPP-BB) (see **Tables 1** and **2**; **Figures 1–6**. The latter construction project has the acronym 'OBSIP' for OBS instrument pool, and sports a polished, professionally designed web site at http://www.obsip.org.

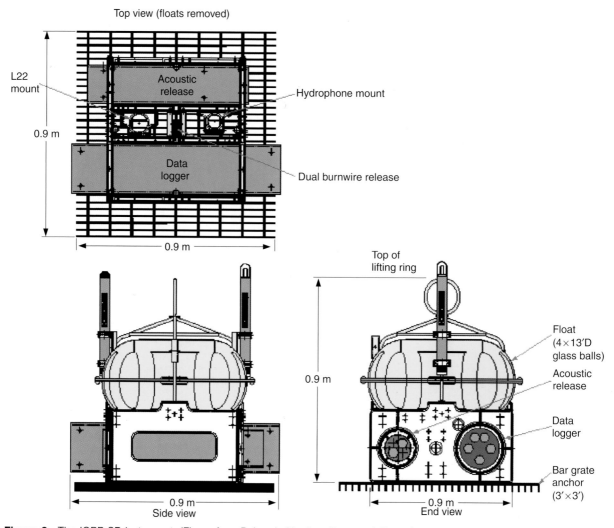

Top view (floats removed)

L22 mount

0.9 m

0.9 m

Acoustic release

Hydrophone mount

Data logger

Dual burnwire release

Top of lifting ring

0.9 m

0.9 m
Side view

Float (4×13′D glass balls)

Acoustic release

Data logger

Bar grate anchor (3′×3′)

0.9 m
End view

Figure 2 The IGPP-SP instrument. (Figure from Babcock, Harding, Kent, and Orcutt.)

D2 OBH/S

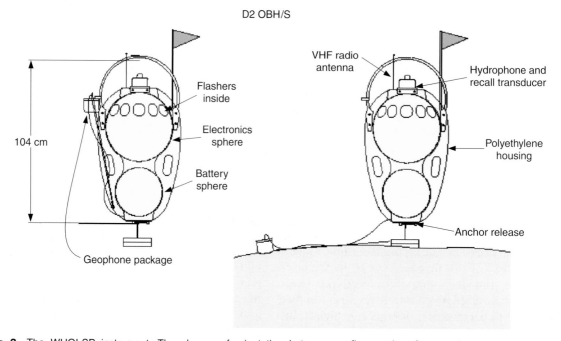

104 cm

Flashers inside

Electronics sphere

Battery sphere

Geophone package

VHF radio antenna

Hydrophone and recall transducer

Polyethylene housing

Anchor release

Figure 3 The WHOI-SP instrument. The change of orientation between seafloor and surface modes allows the acoustic transducer an unobstructed view of the surface while on the seafloor and permits acoustic ranging while the instrument is on the surface. (Figure by Beecher Wooding and John Collins.)

Figure 4 The GEOMAR OBH/S. The shipping/storage container is equipped with an overhead rail so that it serves as an instrument dispenser. The OBS version is shown here. (Photograph by Michael Tryon, UCSD.)

OBS designs are roughly divided into two categories which for brevity will be called 'short-period' (SP) and 'broadband' (BB). The distinction blurs at times because some instruments of both classes use a common recording system, a possibility which emerges when a high data-rate digitizer has the capability of operation in a low-power, high endurance mode.

Short Period (SP) Instruments

The SP instruments (**Table 1**) are light in weight and easy to deploy, typically use 4.5 Hz geophones, commonly only the vertical component, and/or hydrophones, may have somewhat limited recording capacity and endurance, and are typically used in active-source seismic experiments and for

Figure 5   The SIO-ONR OBSs (Jacobson *et al.* 1991, Sauter *et al*, 1990) being launched in Antarctica. The anchor serves as a collector for the CAT fluid fluxmeter (Tryon *et al.* 2001). The plumbing for the flowmeter is in the light-colored box at the right-hand end of the instrument. (Photograph by Michael Tryon, UCSD.)

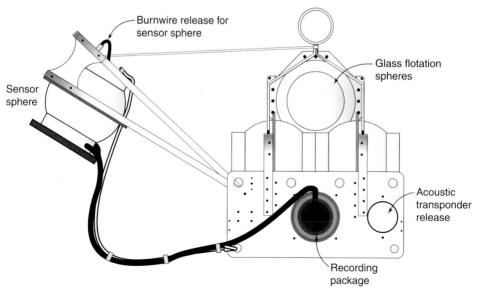

Figure 6   The LDGO-BB OBS. This is based on the Webb design in use during the past few years. The earlier version established a reputation for high reliability and was the lowest noise OBS and lightest of its time period. The main drawback of the earlier version was its limited (16-bit) dynamic range. (Figure from S. Webb, LDEO.)

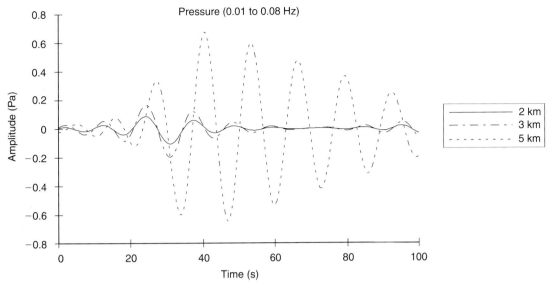

Figure 7 Synthetic seismograms of pressure at three ocean depths.

micro-earthquake studies with durations of a week to a few months. Two types (UTIG and JAMSTEC) are single-sphere instruments.

Broadband (BB) Instruments

The BB instruments (**Table 2**) provide many features of land seismic observatories, relatively high dynamic range, excellent clock stability – < 1 ms d^{-1} drift. This class of instruments can be equipped with hydrophones useful down to a millihertz. The BB instruments are designed in two parts, the main section contains the recording package, and release and recovery aids, while the sensor package is physically separated from the main section. This configuration allows isolation from mechanical noise and and permits tuning of the mechanical resonance of the sensor–seafloor system.

Sensor Considerations

Emplacement of a sensor on the seafloor is almost always suboptimal in comparison with land stations. Instruments dropped from the sea surface can land tens of meters from the desired location. The seafloor material is almost always softer than the surficial sediments (these materials have shear velocities as low as a few tens of meters per second.

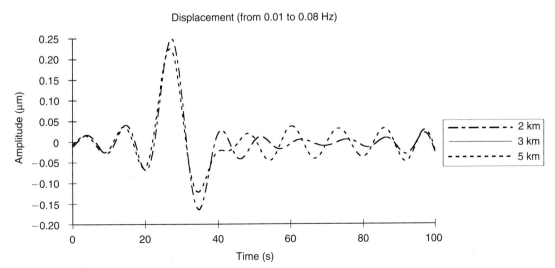

Figure 8 Synthetic seismograms of vertical motion at the same depths as **Figure 7**; note the reduction in the effects of the water column reverberations. (From Lewis and Dorman, 1998).

The sensor is thus almost always poorly coupled to the seafloor and sensor resonances can occur within the frequency range of interest for short period sensors. Fortunately, these resonances have little effect on lower frequencies. However, lower frequency sensors are affected, since a soft foundation permits tilt either in response to sediment deformation by the weight of the sensor or in response to water currents. The existing Webb instruments combat this problem by periodic releveling of the sensor gimbals. The PMD sensors have an advantage here in that the mass element is a fluid and the horizontal components self-level to within 5°.

Why not use hydrophones then? These make leveling unnecessary and are more robust mechanically. In terms of sensitivity, they are comparable to seismometers. The disadvantage of hydrophones lies in the physics of reverberation in the water layer. A pulse incident from below is reflected from the sea surface completely, and when it encounters the seafloor it is reflected to a significant degree. Since the seafloor has a higher acoustic impedance than water, the reflected pressure pulse has the same sign as the incident pulse and the signal is large. However, the seafloor motion associated with pressure pulses traveling in opposite directions is opposite in sign, so cancellation occurs. Unfortunately, the frequency range in which these reverberations are troublesome is in the low noise region. **Figures 7** and **8** show synthetic seismograms of pressure and vertical motion illustrating this effect.

See also

Mid-Ocean Ridge Seismic Structure. Seismic Structure.

Further Reading

Barash TW, Doll CG, Collins JA, Sutton GH and Solomon SC (1994) Quantitative evaluation of a passively leveled ocean bottom seismometer. *Marine Geophysical Researches* 16: 347–363.

Dorman LM (1997) Propagation in marine sediments. In: Crocker MJ ed. *Encyclopedia of Acoustics*, pp. 409–416. New York, John Wiley.

Dorman LM, Schreiner AE and Bibee LD (1991) The effects of sediment structure on sea floor noise. In: Hovem J *et al.* (eds) *Proceedings of Conference on Shear Waves in Marine Sediments*, pp. 239–245. Dordrecht: Kluwer Academic Publishers.

Duennebier FK and Sutton GH (1995) Fidelity of ocean bottom seismic observations. *Marine Geophysical Researches* 17: 535–555.

Jacobson RS, Dorman LM, Purdy GM, Schultz A and Solomon S (1991) *Ocean Bottom Seismometer Facilities Available*, EOS, Transactions AGU, 72, pp. 506, 515.

Sauter AW, Hallinan J, Currier R *et al.* (1990) *Proceedings of the MTS Conference on Marine Instrumentation*, pp. 99–104.

Tryon M, Brown K, Dorman L and Sauter A (2001) A new benthic aqueous flux meter for very low to moderate discharge rates. *Deep Sea Research* (in press).

Webb SC (1998) Broadband seismology and noise under the ocean. *Reviews in Geophysics* 36: 105–142.

SENSORS FOR MEAN METEOROLOGY

K. B. Katsaros, Atlantic Oceanographic and Meteorological Laboratory, NOAA, Miami, FL, USA

Copyright © 2001 Academic Press

doi:10.1006/rwos.2001.0329

Introduction

Basic mean meteorological variables include the following: pressure, wind speed and direction, temperature, and humidity. These are measured at all surface stations over land and from ships and buoys at sea. Radiation (broadband solar and infrared) is also often measured, and sea state, swell, wind sea, cloud cover and type, and precipitation and its intensity and type are evaluated by an observer over the ocean. Sea surface temperature and wave height (possibly also frequency and direction of wave trains) may be measured from a buoy at sea; they are part of the set of parameters required for evaluating net surface energy flux and momentum transfer. Instruments for measuring the quantities described here have been limited to the most common and basic. Precipitation is an important meteorological variable that is measured routinely over land with rain gauges, but its direct measurement at sea is difficult because of ship motion and wind deflection by ships' superstructure and consequently it has been measured routinely over the ocean only from ferry boats. However, it can be estimated at sea by satellite techniques, as can surface wind and sea surface temperature. Satellite methods are included in this article, since they are increasing in importance and provide the only means for obtaining complete global coverage.

Pressure

Several types of aneroid barometers are in use. They depend on the compression or expansion of an evacuated metal chamber for the relative change in atmospheric pressure. Such devices must be compensated for the change in expansion coefficient of the metal material of the chamber with temperature, and the device has to be calibrated for absolute values against a classical mercury in glass barometer, whose vertical mercury column balances the weight of the atmospheric column acting on a reservoir of mercury. The principle of the mercury barometer was developed by Evangilista Toricelli in the 17th century, and numerous sophisticated details were worked out over a period of two centuries. With modern manufacturing techniques, the aneroid has become standardized and is the commonly used device, calibrated with transfer standards back to the classical method. The fact that it takes a column of about 760 mm of the heavy liquid metal mercury (13.6 times as dense as water) illustrates the substantial weight of the atmosphere. Corrections for the thermal expansion or contraction of the mercury column must be made, so a thermometer is always attached to the device. Note that the word 'weight' is used, which implies that the value of the earth's gravitational force enters the formula for converting the mercury column's height to a pressure (force/unit area). Since gravity varies with latitude and altitude, mercury barometers must be corrected for the local value of the acceleration due to gravity.

Atmospheric pressure decreases with altitude. The balancing column of mercury decreases or the expansion of the aneroid chamber increases as the column of air above the barometer has less weight at higher elevations. Conversely, pressure sensors can therefore be used to measure or infer altitude, but must be corrected for the variation in the atmospheric surface pressure, which varies by as much as 10% of the mean (even more in case of the central pressure in a hurricane). An aneroid barometer is the transducer in aircraft altimeters.

Wind Speed and Direction

Wind speed is obtained by two basic means, both depending on the force of the wind to make an object rotate. This object comprises either a three- or four-cup anemometer, half-spheres mounted to horizontal axes attached to a vertical shaft (**Figure 1A**). The cups catch the wind and make the shaft rotate. In today's instruments rotations are counted by the frequency of the interception of a light source to produce a digital signal.

Propeller anemometers have three or four blades that are turned by horizontal wind (**Figure 1B**). The propeller anemometer must be mounted on a wind vane that keeps the propeller facing into the wind. For propeller anemometers, the rotating horizontal shaft is inserted into a coil. The motion of the shaft generates an electrical current or a voltage difference that can be measured directly. The signal is large enough that no amplifiers are needed.

Both cup and propeller anemometers, as well as vanes, have a threshold velocity below which they do not turn and measure the wind. For the propeller anemometer, the response of the vane is also crucial, for the propeller does not measure wind speed off-axis very well. These devices are calibrated in wind tunnels, where a standard sensor evaluates the speed in the tunnel. Calibration sensors can be fine cup anemometers or pitot tubes.

The wind direction is obtained from the position of a wind vane (a vertical square, triangle, or otherwise shaped wind-catcher attached to a horizontal shaft, **Figures 1A and B**). The position of a sliding contact along an electrical resistance coil moved by the motion of the shaft gives the wind direction relative to the zero position of the coil. The position is typically a fraction of the full circle (minus a small gap) and must be calibrated with a compass for absolute direction with respect to the Earth's north.

Other devices such as sonic anemometers can determine both speed and direction by measuring the modification of the travel time of short sound pulses between an emitter and a receiver caused by the three-dimensional wind. They often have three sound paths to allow evaluation of the three components of the wind (**Figure 1C**). These devices have recently become rugged enough to be used to measure mean winds routinely, and have a high enough frequency response to also determine the turbulent fluctuations. The obvious advantage is that the instrument has no moving parts. Water on the sound transmitter or receiver causes temporary difficulties, so a sonic anemometer is not an all-weather instrument. The sound paths can be at arbitrary angles to each other and to the natural vertical. Processing of the data transforms the measurements into an Earth-based coordinate system. The assumptions of zero mean vertical velocity and zero mean cross-wind velocity allow the relative orientation between the instrument axes and the Earth-based coordinate system to be found. Difficulties arise if the instrument is experiencing a steady vertical velocity at its location due to flow distortion around the measuring platform, for instance.

Cup and propeller anemometers are relatively insensitive to rain. However, snow and frost are problematic to all wind sensors, particularly the ones described above with moving parts. Salt

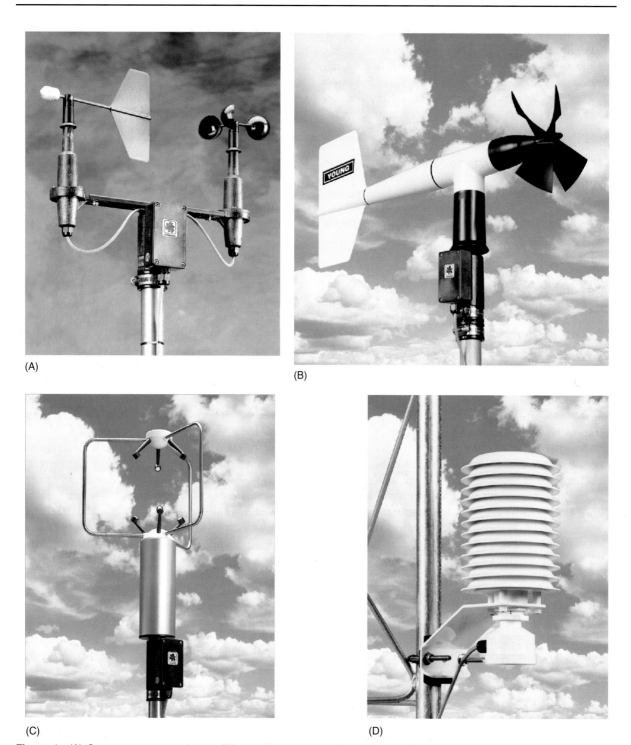

Figure 1 (A) Cup anemometer and vane; (B) propeller vane assembly; (C) three-dimensional sonic anemometer; (D) radiation shield for temperature and humidity sensors. (Photographs of these examples of common instruments were provided courtesy of R.M. Young Company.)

contamination over the ocean also causes deterioration of the bearings in cup and propeller anemometers. Proper exposure of wind sensors on ships is problematic because of severe flow distortion by increasingly large ships. One solution has been to have duplicate sensors on port and starboard sides of the ship and selecting the valid one on the basis of the recording of the ship's heading and the relative wind direction.

Temperature

The measurements of both air and water temperature will be considered here, since both are important in air–sea interaction. Two important considerations for measuring temperature are the exposure of the sensor and shielding from solar radiation. The axiom that a 'thermometer measures its own temperature' is a good reminder. For the thermometer to represent the temperature of the air, it must be well ventilated, which is sometimes assured by a protective housing and a fan pulling air past the sensor. Shielding from direct sunlight has been done traditionally over land and island stations by the use of a 'Stephenson screen,' a wooden-roofed box with slats used for the sides, providing ample room for air to enter. Modern devices have individual housings based on the same principles (**Figure 1D**).

The classic measurements of temperature were done with mercury in glass or alcohol in glass thermometers. For sea temperature, such a thermometer was placed in a canvas bucket of water hauled up on deck. Today, electronic systems have replaced most of the glass thermometers. **Table 1** lists some of these sensors (for details see the Further Reading section).

The sea surface temperature (SST) is an important aspect of air–sea interaction. It enters into bulk formulas for estimating sensible heat flux and evaporation. The temperature differences between the air at one height and the SST is also important for determining the atmospheric stratification, which can modify the turbulent fluxes substantially compared with neutral stratification.

The common measure of SST is the temperature within the top 1 or 2 m of the interface, obtained with any of the contact temperature sensors described in **Table 1**. On ships, the sensor is typically placed in the ship's water intake, and on buoys it may even be placed just inside the hull on the bottom, shaded side of the buoy. Because the heat losses to the air occur at the air–sea interface, while solar heating penetrates of the order of tens of meters (depth depending on sun angle), a cool skin, 1–2 mm in depth and 0.1–0.5°C cooler than the lower layers, is often present just below the interface. Radiation thermometers are sometimes used from ships or piers to measure the skin temperature directly (*see* **Radiative Transfer in the Ocean**).

Humidity

The Classical Sling Psychrometer

An ingenious method for evaluating the air's ability to take up water (its deficit in humidity with respect to the saturation value, *see* **Evaporation and Humidity**) is the psychrometric method. Two thermometers (of any kind) are mounted side by side, and one is provided with a cotton covering (a wick) that is wetted with distilled water. The sling psychrometer (**Figure 2**) is vigorously ventilated by swinging it in the air. The air passing over the sensors changes their temperatures to be in equilibrium with the air; the dry bulb measures the actual air temperature, the wet bulb adjusts to a temperature that is intermediate between the dew point and air temperature. As water from the wick is evaporated, it takes heat out of the air passing over the wick until an equilibrium is reached between the heat supplied to the wet bulb by the air and the heat lost due to evaporation

Table 1 Electronic devices for measuring temperature in air or water

Name	Principle	Typical use
Thermocouple	Thermoelectric junctions between two wires (e.g. Copper-Constantan) set up a voltage in the circuit, if the junctions are at different temperatures. The reference junction temperature must be measured as well	Good for measuring differences of temperature
Resistance thermometer	$R = R_{Ref}(1 + \alpha T)$ Where R is the electrical resistance, R_{Ref} is resistance at a reference temperature, and α is the temperature coefficient of resistance	Platinum resistance thermometers are used for calibration and as reference thermometers
Thermistor	$R = a \exp(b/T)$ Where R is resistance, T is absolute temperature, and a and b are constants	Commonly used in routine sensor systems
Radiation thermometer	Infrared radiance in the atmospheric window, 8–12 μm, is a measure of the equivalent black body temperature	Usually used for measuring water's skin temperature

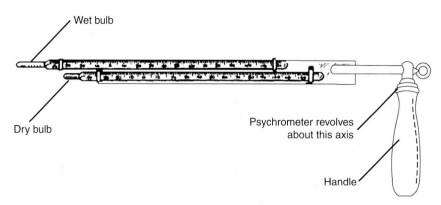

Figure 2 Sling pyschrometer. (Reproduced with permission from Parker, 1977.)

of water from the wick. This is the wet bulb temperature. The *Smithsonian Tables* provide the dew point temperature (and equivalent saturation humidity) corresponding to the measured 'wet bulb temperature depression,' i.e. the temperature difference between the dry bulb and the wet bulb thermometers at the existing air temperature.

Resistance Thermometer Psychrometer

A resistance thermometer psychrometer consists of stainless steel-encased platinum resistance thermometers housed in ventilated cylindrical shields. Ventilation can be simply due to the natural wind (in which case errors at low wind speeds may develop), or be provided by a motor and a fan (typically an air speed of $3\,\mathrm{m\,s^{-1}}$ is required). A water reservoir must be provided to ensure continuous wetting of the wet bulb. The reservoir should be mounted below the psychrometer so that water is drawn onto the wet bulb with a long wick. (This arrangement assures that the water has had time to equilibrate to the wet bulb temperature of the air.)

If these large wet bulbs collect salt on them over time, the relative humidity may be in error. This is not a concern for short-term measurements. A salt solution of 3.6% on the wet bulb would result in an overestimate of the relative humidity of approximately 2%.

Capacitance Sensors of Humidity

The synoptic weather stations often use hygrometers based on the principle of capacitance change as the small transducer absorbs and desorbs water vapor. To avoid contamination of the detector, special filters cover the sensor. Dirty filters (salt or other contaminants) may completely mask the atmospheric effects. Even the oil from the touch of a human hand is detrimental. Two well-known sensors go under the names of Rotronic and Humicap. Calibration with mercury in glass psychrometers is useful.

Exposure to Salt

As for wind and temperature devices, the humidity sensors are sensitive to flow distortion around ships and buoys. Humidity sensors have an additional problem in that salt crystals left behind by evaporating spray droplets, being hygroscopic, can modify the measurements by increasing the local humidity around them. One sophisticated, elegant, and expensive device that has been used at sea without success is the dew point hygrometer. It depends on the cyclical cooling and heating of a mirror. The cooling continues until dew forms, which is detected by changes in reflection of a light source off the mirror, and the temperature at that point is by definition the dew point temperature. The problem with this device is that during the heating cycle sea salt is baked onto the mirror and cannot be removed by cleaning.

Several attempts to build devices that remove the spray have been tried. Regular Stephenson screen-type shields provide protection for some time, the length of which depends both on the generation of spray in the area, the height of the measurement, and the size of the transducer (i.e. the fraction of the surface area that may be contaminated). One of the protective devices that was successfully used in the Humidity Exchange over the Sea (HEXOS) experiment is the so-called 'spray flinger.'

Spray-Removal Device

The University of Washington 'spray flinger' (**Figure 3**) was designed to minimize flow modification on scales important to the eddy correlation calculations of evaporation and sensible heat flux employing

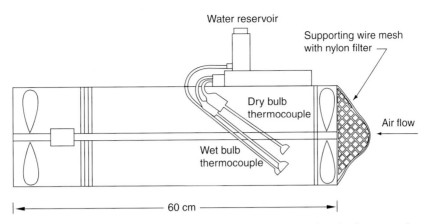

Figure 3 Sketch of aspirated protective housing, the 'spray flinger', used for the protection of a thermocouple psychrometer by the University of Washington group. The system is manually directed upwind. The spray flinger is a 60 cm long tube, 10 cm in diameter, with a rotating filter screen and fan on the upwind end, and an exit fan and the motor at the downwind end. The filter is a single layer of nylon stocking, which is highly nonabsorbent, supported by a wire mesh. Particles and droplets are intercepted by the rotating filter and flung aside, out of the airstream entering the tube. The rotation rate of the filter is about 625 rpm. Inspection of the filter revealed that this rate of rotation prevented build-up of water or salt. The nylon filter needs to be replaced at least at weekly intervals. (Reproduced with permission from Katsaros *et al.*, 1994.)

data from temperature and humidity sensors inside the housing. The design aims to ensure that the droplets removed from the airstream do not remain on the walls of the housing or filter where they could evaporate and affect the measurements. The device has been tested to ensure that there are no thermal effects due to heating of the enclosure, but this would be dependent on the meteorological conditions encountered, principally insolation. The housing should be directed upwind.

Although there is a slow draw of air through the unit by the upwind and exit fans ($1-2 \, \mathrm{m\,s}^{-1}$), it is mainly a passive device with respect to the airflow. Inside the tube, wet and dry thermocouples or other temperature and humidity sensors sample the air for mean and fluctuating temperature and humidity. Wind tunnel and field tests showed the airflow inside the unit to be steady and about one-half the ambient wind speed for wind directions $< 40°$ off the axis. Even in low wind speeds there is adequate ventilation for the wet bulb sensor. Comparison between data from shielded and unshielded thermocouples (respectively, inside and outside the spray flinger) show that the measurements inside are not noticeably affected by the housing.

A quantitative test of the effectiveness of the spray flinger in removing aerosols from the sample airstream was performed during HEXMAX (the HEXOS Main Experiment) using an optical particle counter to measure the aerosol content with diameters between 0.5 and 32 μm in the environmental air and at the rear of the spray flinger. Other devices have been constructed, but have had various difficulties, and the 'spray flinger' is not the final

answer. Intake tubes that protect sensors have also been designed for use on aircraft.

Satellite Measurements

With the global ocean covering 70% of the earth's surface, large oceanic areas cannot be sampled by *in situ* sensors. Most of the meteorological measurements are taken by Voluntary Observing Ships (VOS) of the merchant marine and are, therefore, confined to shipping lanes. Research vessels and military ships may be found in other areas and have contributed substantially to our knowledge of conditions in areas not visited by VOS. The VOS report their observations on a 3 hour or 6 hour schedule. Some mean meteorological quantities such as SST and wind speed and direction are observable by satellites directly, while others can be inferred from less directly related measurements. Surface insolation and precipitation depend on more complex algorithms for evaluation. Satellite-derived surface meteorological information over the ocean are mostly derived from polar-orbiting, sun-synchronous satellites. The famous TIROS and NOAA series of satellites carrying the Advanced Very High Resolution Radiometer (AVHRR) and its predecessors has provided sea surface temperature and cloud information for more than three decades (SST only in cloud-free conditions). This long-term record of consistent measurements by visible and infrared sensors has provided great detail with a resolution of a few kilometers of many phenomena such as oceanic eddy formation, equatorial Rossby and Kelvin waves, and the El Niño phenomenon.

Because of the wide swath of these short wavelength devices, of the order of 2000 km, the whole earth is viewed daily by either the ascending or descending pass of the satellite overhead, once in daytime and once at night.

Another mean meteorological variable observable from space is surface wind speed, with microwave radiometers and the wind vector from scatterometers, active microwave instruments. Both passive and active sensors depend on the changing roughness of the sea as a function of wind speed for their ability to 'sense' the wind. The first scatterometer was launched on the Seasat satellite in 1978, operating for 3 months only. The longest record is from the European Remote Sensing (ERS) satellites 1 and 2 beginning in 1991 and continuing to function well in 2000. Development of interpretation of the radar returns in terms of both speed and direction depends on the antennae viewing the same ocean area several times at different incidence angles relative to the wind direction. A recently launched satellite (QuikSCAT in 1999) carries a new design with a wider swath, the SeaWinds instrument. Scatterometers are providing surface wind measurements with accuracy of $\pm 1.6\,\mathrm{m\,s^{-1}}$ approximately in speed and $\pm 20°$ in direction at 50 km resolution for ERS and 25 km for SeaWinds. They view all of the global ocean once in 3 days for ERS and in approximately 2 days for QuikSCAT. Microwave radiometers such as the Special Sensor Microwave/Imager (SSM/I), operational since 1987 on satellites in the US Defense Meteorological Satellite Program, have wider swaths covering the globe daily, but they are not able to sense the ocean surface in heavy cloud or rainfall areas and do not give direction. They can be assimilated into numerical models where the models provide an initial guess of the wind fields, which are modified to be consistent with the details of the radiometer-derived wind speeds.

Surface pressure and atmospheric surface air temperature are not yet amenable to satellite observations, but surface humidity can be inferred from total column water content (*see* **Evaporation and Humidity**). From the satellite-observed cloudiness, solar radiation at the surface can be inferred by use of radiative transfer models. This is best done from geostationary satellites whose sensors sweep across the Earth's surface every 3 hours or more often, but only view a circle of useful data extending $\pm 50°$ in latitude, approximately.

Precipitation can also be inferred from satellites combining microwave data (from SSM/I) with visible and infrared signals. For tropical regions, the Tropical Rainfall Measuring Mission (TRMM) on a low-orbit satellite provides precipitation estimates on a monthly basis. This satellite carries a rain radar with 500 km swath in addition to a microwave radiometer.

Developments of multispectral sensors and continued work on algorithms promises to improve the accuracy of the satellite information on air–sea interaction variables. Most satellite programs depend on the simple *in situ* mean meteorological measurements described above for calibration and validation. A good example is the important SST record provided by the US National Weather Service and used by all weather services. The analysis procedure employs surface data on SST from buoys, particularly small, inexpensive, free-drifting buoys that are spread over the global oceans to 'tie-down' the correction for atmospheric interference for the satellite estimates of SST. The satellite-observed infrared radiances are modified by the transmission path from the sea to the satellite, where the unknown is the aerosol that can severely affect the interpretation. The aerosol signal is not directly observable yet by satellite, so the surface-measured SST data serve an important calibration function.

Future Developments

New measurement programs are being developed by international groups to support synoptic definition of the ocean's state similarly to meteorological measurements and to provide forecasts. The program goes under the name of the Global Ocean Observing System (GOOS). It includes new autonomous buoys cycling in the vertical to provide details below the interface, a large surface drifter component, and the VOS program, as well as certain satellite sensors. The GOOS is being developed to support a modeling effort, the Global Ocean Data Assimilation Experiment (GODAE), which is an experiment in forecasting the oceanic circulation using numerical models with assimilation of the GOOS data.

See also

Evaporation and Humidity. Heat and Momentum Fluxes at the Sea Surface. Sensors for Micrometeorological Flux Measurements. Wind Driven Circulation.

Further Reading

Atlas RS, Hoffman RN, Bloom SC, Jusem JC and Ardizzone J (1996) A multiyear global surface wind velocity data set using SSM/I wind observations. *Bulletin of the American Meteorological Society* 77: 869–882.

Bentamy A, Queffeulou P, Quilfen Y and Katsaros KB (1999) Ocean surface wind fields estimated from satellite active and passive microwave instruments. *IEEE Transactions Geosci Remote Sens.* 37: 2469–2486.

de Leeuw G (1990) Profiling of aerosol concentrations, particle size distributions, and relative humidity in the atmospheric surface layer over the North Sea. *Tellus* 42B: 342–354.

Dobson F, Hasse L and Davies R (eds) (1980) *Instruments and Methods in Air–Sea Interaction*, pp. 293–317. New York: Plenum.

Geernaert GL and Plant WJ (eds) (1990) *Surface Waves and Fluxes*, vol. 2, pp. 339–368. Dordrecht: Kluwer Academic Publishers.

Graf J, Sasaki C, *et al.* (1998) NASA Scatterometer Experiment. *Asta Astronautica* 43: 397–407.

Gruber A, Su X, Kanamitsu M and Schemm J (2000) The comparison of two merged rain gauge-satellite precipitation datasets. *Bulletin of the American Meteorological Society* 81: 2631–2644.

Katsaros KB (1980) Radiative sensing of sea surface temperatures. In: Dobson F, Hasse L and Davies R (eds) *Instruments and Methods in Air–sea* Interaction, pp. 293–317. New York: Plenum Publishing Corp.

Katsaros KB, DeCosmo J, Lind RJ *et al.* (1994) Measurements of humidity and temperature in the marine environment. *Journal of Atmospheric and Oceanic Technology* 11: 964–981.

Kummerow C, Barnes W, Kozu T, Shiue J and Simpson J (1998) The tropical rainfall measuring mission (TRMM) sensor package. *Journal of Atmospheric and Oceanic Technology* 15: 809–817.

Liu WT (1990) Remote sensing of surface turbulence flux. In: Geenaert GL and Plant WJ (eds) *Surface Waves and Fluxes*, vol. 2, pp. 293–309. Dordrecht: Kluwer Academic Publishers.

Parker SP (ed.) (1977) *Encyclopedia of Ocean and Atmospheric Science*. New York: McGraw-Hill.

Pinker RT and Laszlo I (1992) Modeling surface solar irradiance for satellite applications on a global scale. *Journal of Applied Meteorology* 31: 194–211.

Reynolds RR and Smith TM (1994) Improved global sea surface temperature analyses using optimum interpolation. *Journal of Climate* 7: 929–948.

List RJ (1958) *Smithsonian Meteorological Tables*, 6th edn. City of Washington: Smithsonian Institution Press.

van der Meulen JP (1988) On the need of appropriate filter techniques to be considered using electrical humidity sensors. In: *Proceedings of the WMO Technical Conference on Instruments and Methods of Observation (TECO-1988)*, pp. 55–60. Leipzig, Germany: WMO.

Wentz FJ and Smith DK (1999) A model function for the ocean-normalized radar cross-section at 14 GHz derived from NSCAT observations. *Journal of Geophysical Research* 104: 11 499–11 514.

SENSORS FOR MICROMETEOROLOGICAL FLUX MEASUREMENTS

J. B. Edson, Woods Hole Oceanographic Institution, Woods Hole, MA, USA

Copyright © 2001 Academic Press

doi:10.1006/rwos.2001.0330

Introduction

The exchange of momentum, heat, and mass between the atmosphere and ocean is the fundamental physical process that defines air–sea interactions. This exchange drives ocean and atmospheric circulations, and generates surface waves and currents. Marine micrometeorologists are primarily concerned with the vertical exchange of these quantities, particularly the vertical transfer of momentum, heat, moisture, and trace gases associated with the momentum, sensible heat, latent heat, and gas fluxes, respectively. The term flux is defined as the amount of heat (i.e., thermal energy) or momentum transferred per unit area per unit time.

Air–sea interaction studies often investigate the dependence of the interfacial fluxes on the mean meteorological (e.g., wind speed, degree of stratification or convection) and surface conditions (e.g., surface currents, wave roughness, wave breaking, and sea surface temperature). Therefore, one of the goals of these investigations is to parametrize the fluxes in terms of these variables so that they can be incorporated in numerical models. Additionally, these parametrizations allow the fluxes to be indirectly estimated from observations that are easier to collect and/or offer wider spatial coverage. Examples include the use of mean meteorological measurements from buoys or surface roughness measurements from satellite-based scatterometers to estimate the fluxes.

Direct measurements of the momentum, heat, and moisture fluxes across the air–sea interface are crucial to improving our understanding of the coupled atmosphere–ocean system. However, the operating requirements of the sensors, combined with the often harsh conditions experienced over the ocean, make this a challenging task. This article begins with a description of desired measurements and the operating requirements of the sensors. These requirements involve adequate response time, reliability, and survivability. This is followed by a description of the sensors used to meet these requirements, which includes examples of some of the obstacles that marine researchers have had to overcome. These obstacles include impediments caused by environmental conditions and engineering challenges that are unique to the marine environment. The discussion is limited to the measurement of velocity, temperature, and humidity. The article concludes with a description of the state-of-the-art sensors currently used to measure the desired fluxes.

Flux Measurements

The exchange of momentum and energy a few meters above the ocean surface is dominated by turbulent processes. The turbulence is caused by the drag (i.e., friction) of the ocean on the overlying air, which slows down the wind as it nears the surface and generates wind shear. Over time, this causes faster-moving air aloft to be mixed down and slower-moving air to be mixed up; the net result is a downward flux of momentum. This type of turbulence is felt as intermittent gusts of wind that buffet an observer looking out over the ocean surface on a windy day.

Micrometeorologists typically think of these gusts as turbulent eddies in the airstream that are being advected past the observer by the mean wind. Using this concept, the turbulent fluctuations associated with these eddies can be defined as any departure from the mean wind speed over some averaging period (eqn [1]).

$$u(t) = U(t) - \bar{U} \qquad [1]$$

In eqn [1], $u(t)$ is the fluctuating (turbulent) component, $U(t)$ is the observed wind, and the overbar denotes the mean value over some averaging period. The fact that an observer can be buffeted by the wind indicates that these eddies have some momentum. Since the eddies can be thought to have a finite size, it is convenient to consider their momentum per unit volume, given by $\rho_a U(t)$, where ρ_a is the density of air. In order for there to be an exchange of momentum between the atmosphere and ocean, this horizontal momentum must be transferred downward by some vertical velocity. The mean vertical velocity associated with the turbulent flux is normally assumed to be zero. Therefore, the turbulent transfer of this momentum is almost exclusively via the turbulent vertical velocity, $w(t)$, which we associate with overturning air.

The correlation or covariance between the fluctuating vertical and horizontal wind components is the most direct estimate of the momentum flux. This approach is known as the eddy correlation or direct covariance method. Computation of the covariance involves multiplying the instantaneous vertical velocity fluctuations with one of the horizontal components. The average of this product is then computed over the averaging period.

Because of its dependence on the wind shear, the flux of momentum at the surface is also known as the shear stress defined by eqn [2], where $\hat{\mathbf{i}}$ and $\hat{\mathbf{j}}$ are unit vectors, and v is the fluctuating horizontal component that is orthogonal to u.

$$\tau_0 = \hat{\mathbf{i}}\overline{uw} - \hat{\mathbf{j}}\overline{vw} \qquad [2]$$

Typically, the coordinate system is rotated into the mean wind such that u, v, and w denote the longitudinal, lateral, and vertical velocity fluctuations, respectively. Representative time series of longitudinal and vertical velocity measurements taken in the marine boundary layer are shown in **Figure 1**. The velocities in this figure exhibit the general trend that downward-moving air (i.e., $w < 0$) is transporting eddies with higher momentum per unit mass (i.e., $\rho_a u > 0$) and vice versa. The overall correlation is therefore negative, which is indicative of a downward flux of momentum.

Close to the surface, the wave-induced momentum flux also becomes important. At the interface, the turbulent flux actually becomes negligible and the momentum is transferred via wave drag and viscous shear stress caused by molecular viscosity.

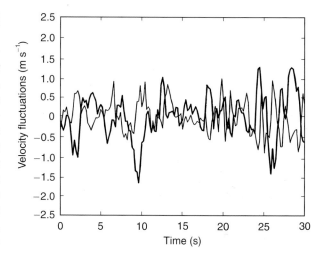

Figure 1 Time-series of the longitudinal (thick line) and vertical velocity (thin line) fluctuations measured from a stable platform. The mean wind speed during the sampling period was $10.8\,\mathrm{m\,s^{-1}}$.

The ocean is surprisingly smooth compared to most land surfaces. This is because the dominant roughness elements that cause the drag on the atmosphere are the wind waves shorter than 1 m in length. Although the longer waves and swell give the appearance of a very rough surface, the airflow tends to follow these waves and principally act to modulate the momentum flux supported by the small-scale roughness. Therefore, the wave drag is mainly a result of these small-scale roughness elements.

Turbulence can also be generated by heating and moistening the air in contact with the surface. This increases the buoyancy of the near-surface air, and causes it to rise, mix upward, and be replaced by less-buoyant air from above. The motion generated by this convective process is driven by the surface buoyancy flux (eqn [3]).

$$B_0 = \rho_a c_p \overline{w \theta_v} \qquad [3]$$

In eqn [3] c_p is the specific heat of air at constant pressure and θ_v is the fluctuating component of the virtual potential temperature defined in eqn [4], where $\overline{\Theta}$ and θ are the mean and fluctuating components of the potential temperature, respectively, and q is the specific humidity (i.e., the mass of water vapor per unit mass of moist air).

$$\theta_v = \theta + 0.61 \overline{\Theta} q \qquad [4]$$

These quantities also define the sensible heat (eqn [5]) and latent heat (eqn [6]) fluxes.

$$H_0 = \rho_a c_p \overline{w \theta} \qquad [5]$$

$$E_0 = L_e \overline{w \rho_a q} \qquad [6]$$

where L_e is the latent heat of evaporation. The parcels of air that are heated and moistened via the buoyancy flux can grow into eddies that span the entire atmospheric boundary layer. Therefore, even in light wind conditions with little mean wind shear, these turbulent eddies can effectively mix the marine boundary layer.

Conversely, when the air is warmer than the ocean, the flow of heat from the air to water (i.e., a downward buoyancy flux) results in a stably stratified boundary layer. The downward buoyancy flux is normally driven by a negative sensible heat flux. However, there have been observations of a downward latent heat (i.e., moisture) flux associated with the formation of fog and possibly condensation at the ocean surface. Vertical velocity fluctuations have to work to overcome the stratification since upward-moving eddies are trying to bring up denser air and vice versa. Therefore, stratified boundary layers tend to dampen the turbulent fluctuations and reduce the flux compared to their unstable counterpart under similar mean wind conditions. Over the ocean, the most highly stratified stable boundary layers are usually a result of warm air advection over cooler water. Slightly stable boundary conditions can also be driven by the diurnal cycle if there is sufficient radiative cooling of the sea surface at night.

Sensors

Measurement of the momentum, sensible heat, and latent heat fluxes requires a suite of sensors capable of measuring the velocity, temperature, and moisture fluctuations. Successful measurement of these fluxes requires instrumentation that is rugged enough to withstand the harsh marine environment and fast enough to measure the entire range of eddies that transport these quantities. Near the ocean surface, the size of the smallest eddies that can transport these quantities is roughly half the distance to the surface; i.e., the closer the sensors are deployed to the surface, the faster the required response. In addition, micrometeorologists generally rely on the wind to advect the eddies past their sensors. Therefore, the velocity of the wind relative to a fixed or moving sensor also determines the required response; i.e., the faster the relative wind, the faster the required response. For example, planes require faster response sensors than ships but require less averaging time to compute the fluxes because they sample the eddies more quickly.

The combination of these two requirements results in an upper bound for the required frequency response (eqn [7]).

$$\frac{fz}{U_r} \approx 2 \qquad [7]$$

Here f is the required frequency response, z is the height above the surface, and U_r is the relative velocity. As a result, sensors used on ships, buoys, and fixed platforms require a frequency response of approximately 10–20 Hz, otherwise some empirical correction must be applied. Sensors mounted on aircraft require roughly an order of magnitude faster response depending on the sampling speed of the aircraft.

The factors that degrade sensor performance in the marine atmosphere include contamination, corrosion, and destruction of sensors due to sea spray

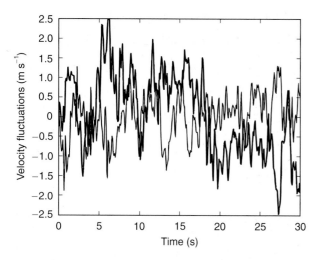

Figure 2 Time-series of the longitudinal (thick line) and vertical velocity (thin line) fluctuations measured from a 3 m discus buoy. The mean wind speed during the sampling period was 10.9 m s^{-1}. The measured fluctuations are a combination of turbulence and wave-induced motion of the buoy.

and salt water; and fatigue and failure caused by long-term operation that is accelerated on moving platforms. Additionally, if the platform is moving, the motion of the platform will be sensed by the instrument as an additional velocity and will contaminate the desired signal (**Figure 2**). Therefore, the platform motion must be removed to accurately measure the flux. This requires measurements of the linear and angular velocity of the platform. The alternative is to deploy the sensors on fixed platforms or to reduce the required motion correction by mounting the sensors on spar buoys, SWATH vessels, or other platforms that are engineered to reduced the wave-induced motion.

Shear Stress

The measurement of momentum flux or shear stress has a long history. The earliest efforts attempted to adapt many of the techniques commonly used in the laboratory to the marine boundary layer. A good example of this is the use of hot-wire anemometers that are well-suited to wind tunnel studies of turbulent flow. Hot-wire anemometry relies on very fine platinum wires that provide excellent frequency response and satisfy eqn [7] even close to the surface. The technique relies on the assumption that the cooling of heated wires is proportional to the flow past the wire. Hot-wire anemometers are most commonly used in constant-temperature mode. In this mode of operation, the current heating the wire is varied to maintain a constant temperature using a servo loop. The amount of current or power

required to maintain the temperature is a measure of the cooling of the wires by the wind.

Unfortunately, there are a number of problems associated with the use of these sensors in the marine environment. The delicate nature of the wires (they are typically 10 μm in diameter) makes them very susceptible to breakage. Hot-film anemometers provide a more rugged instrument with somewhat slower, but still excellent, frequency response. Rather than strands of wires, a hot-film anemometer uses a thin film of nickel or platinum spread over a small cylindrical quartz or glass core. Even when these sensors are closely monitored for breakage, aging and corrosion of the wires and films due to sea spray and other contaminants cause the calibration to change over time. Dynamic calibration in the field has been used but this requires additional sensors. Therefore, substantially more rugged anemometers with absolute or more stable calibrations have generally replaced these sensors in field studies.

Another laboratory instrument that meets these requirements is the pitot tube; this uses two concentric tubes to measure the difference between the static pressure of the inner tube, which acts as a stagnation point, and the static pressure of the air flowing past the sensor. The free stream air also has a dynamic pressure component. Therefore, the difference between the two pressure measurements can be used to compute the dynamic pressure of the air flow moving past the sensor using Bernoulli's equation (eqn [8]).

$$\Delta p = \frac{1}{2}\rho_a \alpha U^2 \qquad [8]$$

Here α is a calibration coefficient that corrects for departures from Bernoulli's equations due to sensor geometry. A calibrated pitot tube can then be used to measure the velocity.

The traditional design is most commonly used to measure the streamwise velocity. However, three-axis pressure sphere (or cone) anemometers have been used to measure fluxes in the field. These devices use a number of pressure ports that are referenced against the stagnation pressure to measure all three components of the velocity. This type of anemometer has to be roughly aligned with the relative wind and its ports must remain clear of debris (e.g., sea spray and other particulates) to operate properly. Consequently, it has been most commonly used on research aircraft where the relative wind is large and particulate concentrations are generally lower outside of clouds and fog.

The thrust anemometer has also been used to directly measure the momentum flux in the marine

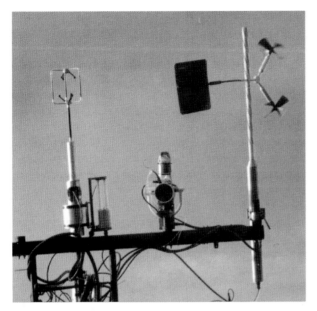

Figure 3 The instrument at the far right is a K-Gill anemometer shown during a deployment on a research vessel. The instrument on the left is a sonic anemometer that is shown in more detail in **Figure 4**. Photograph provided by Olc Persson (CIRES/NOAA/ETL).

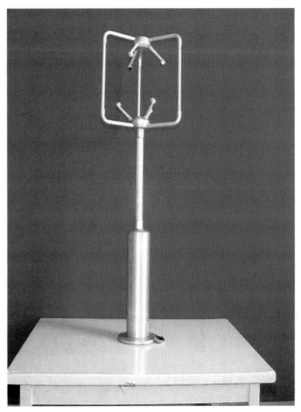

Figure 4 A commercially available pulse-type sonic anemometer. The three sets of paired transducers are cabable of measuring the three components of the velocity vector. This type of device produced the time-series in **Figure 1** and **Figure 2**. The sonic anemometer measures 0.75 m from top to bottom.

atmosphere. This device measures the frictional drag of the air on a sphere or other objects. The most successful design uses springs to attach a sphere and its supporting structure to a rigid mount. The springs allow the sphere to be deflected in both the horizontal and vertical directions. The deflection due to wind drag on the sphere is sensed by proximity sensors that measure the displacement relative to the rigid mount. Carefully calibrated thrust anemometers have been used to measure turbulence from fixed platforms for extended periods. They are fairly rugged and low-power, and have adequate response for estimation of the flux. The main disadvantages of these devices are the need to accurately calibrate the direction response of each sensor and sensor drift due to aging of the springs.

A very robust sensor for flux measurements, particularly for use on fixed platforms, relies on a modification of the standard propellor vane anemometer used to measure the mean wind. The modification involves the use of two propellers on supporting arms set at 90° to each other. The entire assembly is attached to a vane that keeps the propeller pointed into the wind. The device is known as a K-Gill anemometer from the appearance of the twin propeller-vane configuration (**Figure 3**). The twin propellers are capable of measuring the instantaneous vertical and streamwise velocity, and the vane reading allows the streamwise velocity to be broken down into its u and v components. This device is

also very robust and low power. However, it has a complicated inertial response on moving platforms and is therefore most appropriate for use on fixed platforms. Additionally, the separation between the propellers (typically 0.6 m) acts as a spatial filter (i.e., it cannot detect eddies smaller than the separation), so it cannot be used too close to the surface and still satisfy eqn [7]. This is generally not a problem at the measurement heights used in most field deployments.

Over the past decade, sonic anemometers have become the instrument of choice for most investigations of air–sea interaction. These anemometers use acoustic signals that are emitted in either a continuous or pulsed mode. At present, the pulse type sonic anemometers are most commonly used in marine research. Most commercially available devices use paired transducers that emit and detect acoustic pulses (**Figure 4**). One transducer emits the pulse and the other detects it to measure the time of flight between them. The functions are then reversed to measure the time of flight in the other direction. The basic concept is that in the absence of any wind the time of flight in either direction is the same. However, the times of flight differ if there is a component of the wind velocity along the path between the transducers. The velocity is directly computed from the two time of flight measurements, t_1 and t_2, using eqn [9], where L is the distance between the transducers.

$$U = \frac{L}{2}\left(\frac{1}{t_1} - \frac{1}{t_2}\right) \qquad [9]$$

Three pairs of transducers are typically used to measure all three components of the velocity vector. These devices have no moving parts and are therefore far less susceptible to mechanical failure. They can experience difficulties when rain or ice covers the transducer faces or when there is a sufficient volume of precipitation in the sampling volume. However, the current generation of sonic anemometers have proven themselves to be remarkably reliable in long-term deployments over the ocean; so much so that two-axis versions of sonic anemometers are also beginning to replace cup and propellor/vane anemometers for mean wind measurements over the ocean.

Motion Correction

The measurement of the fluctuating velocity components necessary to compute the fluxes is complicated by the platform motion on any aircraft, sea-going research vessel, or surface mooring. This motion contamination must be removed before the fluxes can be estimated. The contamination of the signal arises from three sources: instantaneous tilt of the anemometer due to the pitch, roll, and yaw (i.e., heading) variations; angular velocities at the anemometer due to rotation of the platform about its local coordinate system axes; and translational velocities of the platform with respect to a fixed frame of reference. Therefore, motion sensors capable of measuring these quantities are required to correct the measured velocities. Once measured, these variables are used to compute the true wind vector from eqn [10].

$$\mathbf{U} = T(\mathbf{U}_m + \boldsymbol{\Omega}_m \times \mathbf{R}) + \mathbf{U}_p \qquad [10]$$

Here $\mathbf{U}$ is the desired wind velocity vector in the desired reference coordinate system (e.g., relative to water or relative to earth); $\mathbf{U}_m$ and $\boldsymbol{\Omega}_m$ are the measured wind and platform angular velocity vectors respectively, in the platform frame of reference; T is the coordinate transformation matrix from the platform coordinate system to the reference coordi-

nates; $\mathbf{R}$ is the position vector of the wind sensor with respect to the motion sensors; and $\mathbf{U}_p$ is the translational velocity vector of the platform measured at the location of the motion sensors.

A variety of approaches have been used to correct wind sensors for platform motion. True inertial navigation systems are standard for research aircraft. These systems are expensive, so simpler techniques have been sought for ships and buoys, where the mean vertical velocity of the platform is unambiguously zero. These techniques generally use the motion measurements from either strapped-down or gyro-stabilized systems.

The strapped-down systems typically rely on a system of three orthogonal angular rate sensors and accelerometers, which are combined with a compass to get absolute direction. The high-frequency component of the pitch, roll, and yaw angles required for the transformation matrix are computed by integrating and highpass filtering the angular rates. The low-frequency component is obtained from the lowpass accelerometer signals or, more recently, the angles computed from differential GPS. The transformed accelerometers are integrated and highpass filtered before they are added to lowpass filtered GPS or current meter velocities for computation of $\mathbf{U}_p$ relative to earth or the sea surface, respectively. The gyro-stabilized system directly computes the orientation angles of the platform. The angular rates are then computed from the time-derivative of the orientation angles.

Heat Fluxes

The measurement of temperature fluctuations over the ocean surface has a similar history to that of the velocity measurements. Laboratory sensors such as thermocouples, thermistors, and resistance wires are used to measure temperature fluctuations in the marine environment.

Thermocouples rely on the Seebeck effect that arises when two dissimilar materials are joined to form two junctions: a measuring junction and a reference junction. If the temperature of the two junctions is different, then a voltage potential difference exists that is proportional to the temperature difference. Therefore, if the temperature of the reference junction is known, then the absolute temperature at the measuring junction can be determined. Certain combinations of materials exhibit a larger effect (e.g., copper and constantan) and are thus commonly used in thermocouple design. However, in all cases the voltage generated by the thermoelectric effect is small and amplifiers are often used along with the probes.

Thermistors and resistance wires are devices whose resistance changes with temperature. Thermistors are semiconductors that generally exhibit a large negative change of resistance with temperature (i.e., they have a large negative temperature coefficient of resistivity). They come in a variety of different forms including beads, rods, or disks. Microbead thermistors are most commonly used in turbulence studies; in these the semiconductor is situated in a very fine bead of glass. Resistance wires are typically made of platinum, which has a very stable and well-known temperature–resistance relationship. The trade-off is that they are less sensitive to temperature change than thermistors. The probe supports for these wires are often similar in design to hot-wire anemometers, and they are often referred to as cold-wires.

All of these sensors can be deployed on very fine mounts (**Figure 5**), which greatly reduces the adverse effects of solar heating but also exposes them to harsh environments and frequent breaking. Additionally, the exposure invariably causes them to become covered with salt from sea spray. The coating of salt causes spurious temperature fluctuations due to condensation and evaporation of water vapor on these hygroscopic particles. These considerations generally require more substantial mounts and some sort of shielding from the radiation and spray.

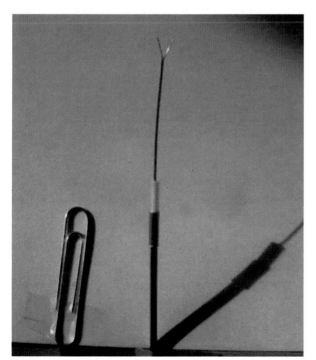

Figure 5 A thermocouple showing the very fine mounts used for turbulence applications. The actual thermocouple is situated on fine wires between the probe supports and is too small to be seen in this photograph.

While this is acceptable for mean temperature measurements, the reduction in frequency response caused by the shields often precludes their use for turbulence measurements.

To combat these problems, marine micrometeorologists have increasingly turned to sonic thermometry. The time of flight measurements from sonic anemometers can be used to measure the speed of sound c along the acoustic path (eqn [11]).

$$c = \frac{L}{2}\left(\frac{1}{t_1} + \frac{1}{t_2}\right) \qquad [11]$$

The speed of sound is a function of temperature and humidity and can be used to compute the sonic temperature T_s defined by eqn [12], where U_N is velocity component normal to the transducer path.

$$T_s = T(1 + 0.51q) = \frac{c^2 + U_N^2}{403} \qquad [12]$$

The normal wind term corrects for lengthening of the acoustic path by this component of the wind. This form of velocity crosstalk has a negligible effect on the actual velocity measurements, but has a measurable effect on the sonic temperature.

Sonic thermometers share many of the positive attributes of sonic anemometers. Additionally, they suffer the least from sea salt contamination compared to other fast-response temperature sensors. The disadvantage of these devices is that velocity crosstalk must be corrected for and they do not provide the true temperature signal as shown by eqn [12]. Fortunately, this can be advantageous in many investigations because the sonic temperature closely approximates the virtual temperature in moist air $T_v = T(1 + 0.61q)$. For example, in many investigations over the ocean an estimate of the buoyancy flux is sufficient to account for stability effects. In these investigations the difference between the sonic and virtual temperature is often neglected (or a small correction is applied), and the sonic anemometer/thermometer is all that is required. However, due to the importance of the latent heat flux in the total heat budget over the ocean, accurate measurement of the moisture flux is often a crucial component of air–sea interaction investigations.

The accurate measurement of moisture fluctuations required to compute the latent heat flux is arguably the main instrumental challenge facing marine micrometeorologists. Sensors with adequate frequency response generally rely on the ability of water vapor in air to strongly absorb certain wavelengths of radiation. Therefore, these devices require a narrowband source for the radiation and a detector to measure the reduced transmission of that radiation over a known distance.

Early hygrometers of this type generated and detected ultraviolet radiation. The Lyman α hygrometer uses a source tube that generates radiation at the Lyman α line of atomic hydrogen which is strongly absorbed by water vapor. This device has excellent response characteristics when operating properly. Unfortunately, it has proven to be difficult to operate in the field due to sensor drift and contamination of the special optical windows used with the source and detector tubes. A similar hygrometer that uses krypton as its source has also been used in the field. Although the light emitted by the krypton source is not as sensitive to water vapor, the device still has more than adequate response characteristics and generally requires less maintenance than the Lyman α. However, it still requires frequent calibration and cleaning of the optics. Therefore, neither device is particularly well suited for long-term operation without frequent attention.

Commercially available infrared hygrometers are being used more and more in marine micrometeorological investigations (**Figure 6**). Beer's

law (eqn [13]) provides the theoretical basis for the transmission of radiation over a known distance.

$$T = e^{-(\gamma + \delta)D} \qquad [13]$$

Here T is the transmittance of the medium, D is the fixed distance, and γ and δ are the extinction coefficients for scattering and absorption, respectively. This law applies to all of the radiation source described above; however, the use of filters with infrared devices allows eqn [13] to be used more directly. For example, the scattering coefficient has a weak wavelength dependence in the spectral region where infrared absorption is strongly wavelength dependent. Filters can be designed to separate the infrared radiation into wavelengths that exhibit strong and weak absorption. The ratio of transmittance of these two wavelengths is therefore a function of the absorption (eqn [14]), where the subscripts s and w identify the variables associated with the strongly and weakly absorbed wavelengths.

$$\frac{T_s}{T_w} \approx e^{-\delta_s D} \qquad [14]$$

Calibration of this signal then provides a reliable measure of water vapor due to the stability of current generation of infrared sources.

Infrared hygrometers are still optical devices and can become contaminated by sea spray and other airborne contaminants. To some extent the use of the transmission ratio negates this problem if the contamination affects the two wavelengths equally. Obviously, this is not the case when the optics become wet from rain, fog, or spray. Fortunately, the devices recover well once they have dried off and are easily cleaned by the rain itself or by manual flushing with water. Condensation on the optics can also be reduced by heating their surfaces. These devices require longer path lengths (0.2–0.6 m) than Lyman α or krypton hygrometers to obtain measurable absorption (**Figure 6**). This is not a problem as long as they are deployed at heights $\gg D$.

Conclusions

The state of the art in sensor technology for use in the marine surface layer includes the sonic anemometer/thermometer and the latest generation of infrared hygrometers (**Figure 7**). However, the frequency response of these devices, mainly due to spatial averaging, precludes their use from aircraft. Instead, aircraft typically rely on gust probes for

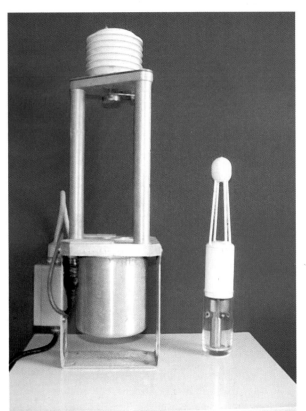

Figure 6 Two examples of commercially available infrared hygrometers. The larger hygrometer is roughly the height of the sonic anemometer shown in **Figure 4**.

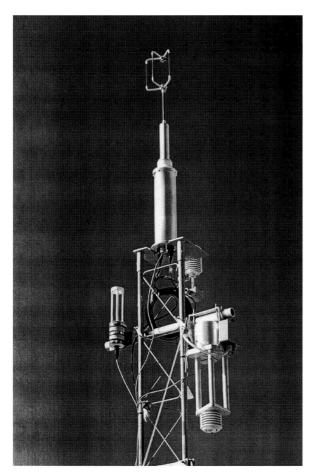

Figure 7 A sensor package used to measure the momentum, sensible heat, and latent heat fluxes from a moving platform. The cylinder beneath the sonic anemometer/thermometer holds 3-axis angular rate sensors and linear accelerometers, as well as a magnetic compass. Two infrared hygrometers are deployed beneath the sonic anemometer. The radiation shield protects sensors that measure the mean temperature and humidity. Photograph provided by Wade McGillis (WHOI).

deployments with careful power management. The use of these sensor packages is expected to continue owing to the desirability of these measurements and technological advances leading to improved power sources and reduced power consumption by the sensors.

See also

Air–Sea Gas Exchange. Breaking Waves and Near-surface Turbulence. Heat and Momentum Fluxes at the Sea Surface. Moorings. Rigs and Offshore Structures. Satellite Remote Sensing Microwave Scatterometers. Sensors for Mean Meteorology. Ships. Surface, Gravity and Capillary Waves. Turbulence Sensors. Upper Ocean Heat and Freshwater Budgets. Upper Ocean Mixing Processes. Wave Generation by Wind. Wind and Buoyancy-forced Upper Ocean. Wind Driven Circulation.

Further Reading

Ataktürk SS and Katsaros KB (1989) The K-Gill, a twin propeller-vane anemometer for measurements of atmospheric turbulence. *Journal of Atmospheric and Oceanic Technology* 6: 509–515.

Buck AL (1976) The variable path Lyman-alpha hygrometer and its operating characteristics. *Bulletin of the American Meteorological Society* 57: 1113–1118.

Crawford TL and Dobosy RJ (1992) A sensitive fast-response probe to measure turbulence and heat flux from any airplane. *Boundary-Layer Meteorology* 59: 257–278.

Dobson FW, Hasse L and Davis RE (1980) *Air–Sea Interaction; Instruments and Methods.* New York: Plenum Press.

Edson JB, Hinton AA, Prada KE, Hare JE and Fairall CW (1998) Direct covariance flux estimates from mobile platforms at sea. *Journal of Atmospheric and Oceanic Technology* 15: 547–562.

Fritschen LJ and Gay LW (1979) *Environmental Instrumentation.* New York: Springer-Verlag.

Kaimal JC and Gaynor JE (1991) Another look at sonic thermometry. *Boundary-Layer Meteorology* 56: 401–410.

Larsen SE, Højstrup J and Fairall CW (1986) Mixed and dynamic response of hot wires and measurements of turbulence statistics. *Journal of Atmospheric and Oceanic Technology* 3: 236–247.

Schmitt KF, Friehe CA and Gibson CH (1978) Humidity sensitivity of atmospheric temperature sensors by salt contamination. *Journal of Physical Oceanography* 8: 141–161.

Schotanus P, Nieuwstadt FTM, de Bruin HAR (1983) Temperature measurement with a sonic anemometer and its application to heat and moisture fluxes. *Boundary-Layer Meteorology* 26: 81–93.

measurement of the required velocity fluctuations, thermistors for temperature fluctuations, and Lyman α hygrometers for moisture fluctuations. Hot-wire and hot-film anemometers along with the finer temperature and humidity devices are also required to measure directly the viscous dissipation of the turbulent eddies that occurs at very small spatial scales.

Instruments for measuring turbulence are generally not considered low-power when compared to the mean sensors normally deployed on surface moorings, so past deployments of these sensors were mainly limited to fixed platforms or research vessels with ample power. Recently, however, sensor packages mounted on spar and discus buoys have successfully measured motion-corrected momentum and buoyancy fluxes on month- to year-long

SHELF-SEA AND SLOPE FRONTS

J. Sharples, Southampton University School of Ocean and Earth Science, Southampton, UK
J. H. Simpson, University of Wales, Bangor, UK

Copyright © 2001 Academic Press

doi:10.1006/rwos.2001.0152

Introduction

Horizontal gradients in temperature and salinity in the ocean are generally very weak. Regions of enhanced horizontal gradient are referred to as fronts. The scalar gradients across a front indicate concomitant changes in the physical processes that determine water column structure. Fronts are important oceanographic features because, corresponding to the physical gradients, they are also sites of rapid chemical and biological changes. In particular, fronts are often sites of enhanced standing stock of primary producers and primary production, with related increases in zooplankton biomass, fish, and foraging seabirds. These frontal aggregations of fish are also an important marine resource, targeted specifically by fishing vessels.

This article discusses three types of fronts. In shelf seas fronts can be generated by the influence of freshwater runoff, or by surface heat fluxes interacting with horizontal variations of tidal mixing. At the edge of the continental shelves, fronts are often seen separating the inherently contrasting temperature–salinity characteristics of shelf and open ocean water. These three types are illustrated schematically in **Figure 1**.

Fresh Water Fronts in Shelf Seas

The lateral input of fresh water from rivers results in coastal waters having a lower salinity than the ambient shelf water. In the absence of any vertical mixing, this low-salinity water would spread out above the saltier shelf water as a density-driven current, eventually turning anticyclonically (i.e., to the right in the Northern Hemisphere, to the left in the Southern Hemisphere) to form a buoyancy current. Parallel to the coastline there will be a thermohaline front separating the low-salinity water from the shelf water. If there is stronger vertical mixing, supplied by tidal currents or wind stress, or if the fresh water input is very strong, then the front can extend from the surface to the seabed.

Determining the Position of a Fresh Water Front

The distance offshore at which this front lies depends on whether or not the buoyancy current 'feels' the seabed. If the buoyancy flow is confined to the surface, then the coast-parallel region containing the low-salinity surface layer will be approximately one internal Rossby radius thick (i.e. the distance traveled seaward by the surface buoyancy current before the effect of the Earth's rotation drives it parallel to the coastline). Typically, in temperate regions, this distance will be a few tens of kilometers.

When the buoyancy current is in contact with the seabed, the situation is altered by the breakdown of geostrophy within the bottom Ekman layer of the flow. Within this Ekman layer there is a component of transport perpendicular to the front, pushing the bottom front offshore and driving low-density water beneath the higher-density shelf water. Thus the frontal region becomes convectively unstable, and overturns rapidly. The effect of this is to shift the position of the front further offshore. Numerical modeling studies have suggested that this continual offshore movement of the front is halted as a result of the vertical shear in the alongshore buoyancy current. As the water deepens, this vertical shear results in a reduction of the offshore bottom flow. Eventually the offshore flow is reversed, so that low-density water is no longer transported underneath the shelf water, and the front becomes fixed at that particular isobath.

Mixing and Frontogenesis in ROFIs

Fronts associated with the lateral buoyancy flux from rivers are affected by the amount of vertical mixing. In regions of fresh water influence (ROFIs) the modulation of tidal mixing over the spring–neap cycle has a dramatic effect on frontal dynamics. Strong vertical mixing at spring tides results in a vertically mixed water column, with salinity increasing offshore and often only a weak horizontal front. As the mixing then decreases toward neap tides, a point is reached when the vertical homogeneity of the water column cannot be maintained against the tendency for the low-density coastal water to flow offshore above the denser shelf water. This surface density-driven offshore current then rapidly establishes vertical stratification, with the offshore progression eventually being halted by the earth's rotation.

Fluid dynamics experiments have shown that such a relaxation of the initially vertically mixed density structure will produce a strong front within any nonlinear region of the initial horizontal density gradient. Furthermore, a periodic modulation of the

mixing about the level required to prevent this frontogenesis will result in a similar periodic variation in the density-driven mass flux. Spring–neap control of frontogenesis has been observed in a number of shelf seas; for instance, Liverpool Bay (eastern Irish Sea), the Rhine outflow (southern North Sea), and Spencer Gulf (South Australia). The physical switching between vertically mixed and stratified conditions is well established (**Figure 2**), though the biological responses within these dynamic environments have yet to be determined.

Tidal Mixing Fronts in Shelf Seas

In summer, away from sources of fresh water, temperate shelf seas are partitioned into thermally stratified and vertically well-mixed regions. Such partitioning is clearly visible in satellite remote sensing images of sea surface temperature (SST, **Figure 3**).

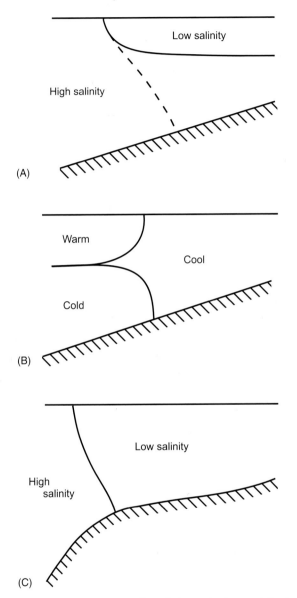

Figure 1 Schematic illustration of the main types of fronts found in shelf seas. (A) A fresh water front, caused by the input of fresher estuarine water into the coastal zone. The front can either be confined to the surface (weak vertical mixing) or extend from the surface to the seabed (strong vertical mixing). (B) A shelf sea tidal mixing front, caused by competition between surface heating and tidal mixing. The stratified water on the left occurs because of weak tidal mixing being unable to counter the stratification generated by surface heating. The mixed water on the right is the result of strong tidal mixing being able to prevent thermal stratification. (C) A shelf break front. The low-salinity water on the shelf results from the combination of all the estuarine inputs from the coast. There can also be a cross-shelf edge contrast in temperature.

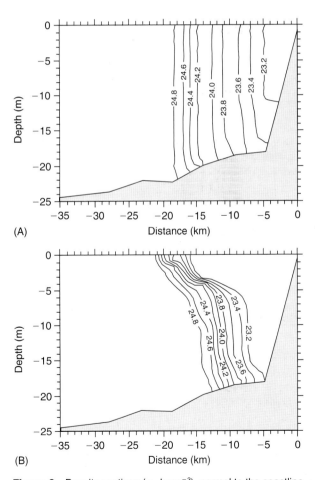

Figure 2 Density sections (σ_t, kg m^{-3}), normal to the coastline through the Rhine outflow. (A) Spring tide section, showing vertically mixed water with a fresh water-induced horizontal density gradient. (B) Neap tide section, showing the relaxation of the horizontal gradient and stratification caused by the reduction in mixing. (After Souza and Simpson (1997) *Journal of Marine Systems* 12: 311–323. Courtesy of Elsevier Science.)

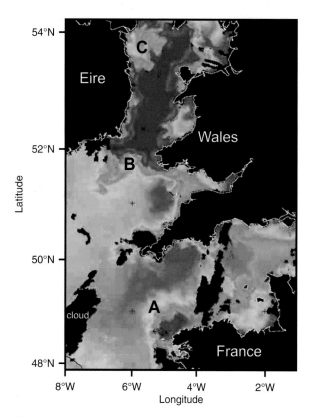

Figure 3 Sea surface temperature image from the Advanced Very High Resolution Radiometer (AVHRR). The image was taken at 0419 GMT on 12 July 1999. Violet/blue represents a temperature of 13–14°C, and shows regions of shelf sea that are vertically mixed. Red represents 18–19°C, indicating the surface temperature of strongly stratified water. Green/yellow represents 16–17°C, and shows the regions of weak stratification at the tidal mixing fronts. The regions of strong horizontal temperature gradient separating the mixed and stratified areas are the tidal mixing fronts (A, Ushant front; B, Celtic Sea front; C, Western Irish Sea front). (Image courtesy of the Dundee Satellite Receiving Station, and the Remote Sensing Group, Plymouth Marine Laboratory.)

Warm SST indicates the temperature of the surface mixed layer of a stratified water column, while cool SST shows the temperature of the entire, vertically homogeneous water column. The transition region between these stratified and well-mixed regions, with horizontal temperature gradients of typically $1°C \, km^{-1}$, are the shelf sea tidal mixing fronts.

Physical Control of Fronts and h/u^3

The suggestion that the intensity of tidal mixing was responsible for controlling the vertical structure of shelf seas was first made by Bigelow in the late 1920s, with reference to the variations in vertical temperature structure on and off Georges Bank. The first quantitative link between shelf sea fronts and tidal mixing was made by Simpson and Hunter in 1974. Surface heating, which is absorbed rapidly within the upper few meters of the ocean, acts to stabilize the water column by expanding the near-surface water and thus reducing its density. Friction between tidal currents and the seabed generates turbulence. Most of this turbulence is dissipated as heat (although the heating produced is insignificant) but a small fraction of it (typically 0.3%) is available for working against the thermal stratification near the sea surface. This seemingly low conversion rate of turbulence to mixing arises because turbulence is dissipated very close to where it is generated (the 'equilibrium hypothesis'). Most of the vertical current shear (and hence turbulence production and dissipation) in a tidal flow is close to the seabed. Current shear higher in the water column near the thermocline (where turbulence can work against stratification) is much weaker, leading to a low overall efficiency.

Thus, there is a competition between the rate at which the water column is being stratified by the surface heating and the ability of the tidal turbulence to erode and prevent stratification. If the magnitude of the heating component exceeds that of the tidal mixing term, then the water column will stratify. Alternatively, a stronger tidal current, and therefore more mixing, results in a situation where the heat input is continuously being well distributed through the entire depth and the water column is kept vertically mixed. A shelf sea front marks the narrow transition between these two conditions, with equal contributions from the heating and tidal mixing. This simple analysis led Simpson and Hunter to predict that tidal mixing fronts should follow lines of a critical value of h/u^3, with h (m) the total water depth and u ($m \, s^{-1}$) a measure of the amplitude of the tidal currents. Subsequent analysis of satellite SST images in comparison with maps of h/u^3 confirmed the remarkable power of this simple theory: shelf sea front positions are controlled by a local balance between the vertical physical processes of tidal mixing and sea surface heating (**Figure 4**).

Modifications to h/u^3

A prediction of the Simpson–Hunter h/u^3 hypothesis is that a shelf sea front should change position periodically with the spring–neap tidal cycle, owing to the fortnightly variation in tidal currents (**Figure 5**). In NW European shelf seas, spring tidal current amplitudes are typically twice those at neap tides. However, predicting the horizontal displacement of a shelf sea front using $h/u^3_{Springs}$ and h/u^3_{Neaps} leads to a substantial overestimate, typically suggesting a transition distance of 40–50 km compared to satellite-derived observations of only 2–4 km. Two

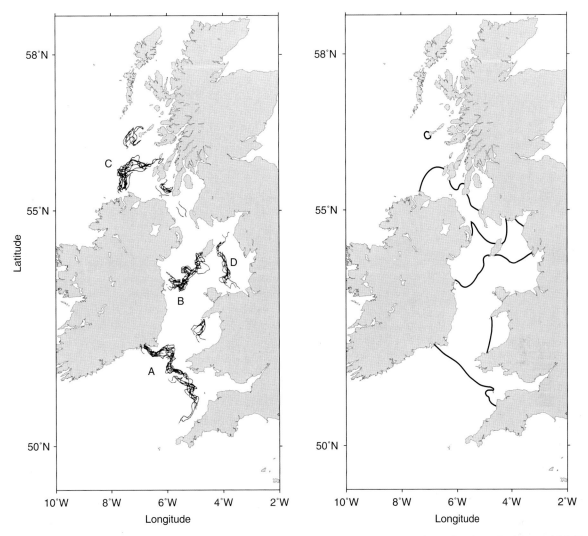

Figure 4 Left, mean positions of tidal mixing fronts observed in SST images, May 1978. A, Celtic Sea front; B, Western Irish Sea front; C, Islay front. Right, contours of $\log_{10} (h/u^3) = 2.7$. Note the correspondence between fronts A, B, and C, and the contours of constant h/u^3. Front D is caused by fresh water inputs from the estuaries of NW England, and so does not conform to the h/u^3 hypothesis. (After Simpson and James (1986) *Coastal and Estuarine Sciences*, 3: 63–93. Courtesy of the American Geophysical Union.)

modifications to the theory were subsequently made. First, as tidal turbulence increases from neap to spring tides, the mixing not only has to counteract the instantaneous heat supply but it must also break down the existing stratification that has developed as a result of the previous neap tide. Incorporating this behavior into the theory reduced the amplitude of the adjustment region to 10–20 km. Second, stratification inhibits vertical mixing, so the mixing efficiency would be expected to be lower as the existing stratification was being eroded. Simpson and Bowers used a simple parametrization linking mixing efficiency to the strength of the stratification, and showed that the predicted spring–neap adjustment was then similar to that observed. More recently the use of a turbulence closure model,

providing a less arbitrary link between stability and mixing, has provided further confirmation of the need to include variable mixing efficiency.

The only source of mixing accounted for in the h/u^3 theory is tidal friction with the seabed, and the success of the theory in NW European shelf seas is arguably a result of the dominance of the tides in these regions. A better prediction of frontal position could be made by incorporating wind-driven mixing, so that the competition becomes one of surface heating versus the sum of tidal + wind mixing. Again, only a small fraction of the wind-driven turbulence is available to work against the stratification, about 2–3%. This is significantly larger than the tidal mixing efficiency because the thermocline is generally nearer to the sea surface than the

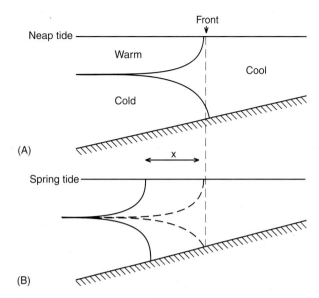

Figure 5 Schematic illustration of the adjustment of a tidal mixing front as a consequence of the spring–neap variation of tidal currents. At neap tides (A) the weaker tidal mixing allows stratification to develop in shallower water. At spring tides (B) the stronger tidal currents re-mix the shallow stratification. The adjustment distance, x, is typically 2–4 km.

seabed, and hence in a region of wind-driven current shear.

A debate arose in the late 1980s concerning the validity of the h/u^3 theory. Loder and Greenberg, and subsequently Stigebrandt, put forward an alternative hypothesis based on a more realistic description of tidal turbulence that includes the effect of the bottom rotational boundary layer as a control on the vertical extent of tidal turbulence away from the seabed. The position of the shelf sea front would, in this theory, simply reflect the position at which the tidal boundary layer was thick enough to reach over the entire depth, and fronts should follow a critical value of h/u. Moreover, in temperate latitudes the similar values of Coriolis and tidal frequencies suggests that there should be a very significant rotational constraint on frontal position as the tidal currents become cyclonically polarized.

Observations of frontal positions were not precise enough to determine which of the two theories was correct. However, use of a numerical model showed that both mechanisms contributed. For anticyclonically polarized tidal currents, boundary layer limitation is not a significant factor, and the frontal position is well described by the h/u^3 theory. As currents become more cyclonic, the vertical limitation of turbulence due to the reducing thickness of the boundary layer does alter the frontal position away from that predicted using h/u^3, but by less

than predicted using the boundary layer theory alone.

Circulation at Shelf Sea Fronts

The density gradients associated with shelf sea fronts drive a weak residual circulation, superimposed on the dominant tidal flows (**Figure 6**). A surface convergence of flow at the front often leads to an accumulation of buoyant debris. This can form a clear visual indicator of a front. The convergence is associated with a downwelling, predicted by models to be around $4\,\mathrm{cm\,s^{-1}}$. On the stratified side of the front a surface, geostrophic jet is predicted to flow parallel to the front. Models have predicted this flow to be of the order of $10\,\mathrm{cm\,s^{-1}}$. Direct observations of such flows against the background of strong tidal currents is difficult, but both drogued buoys and high frequency radar have been used successfully to observe along-front speeds of 10–$15\,\mathrm{cm\,s^{-1}}$. These frontal jets are prone to baroclinic instability, with meanders forming along the front, growing, and eventually producing baroclinic eddies that transfer water between the two sides of the front.

Biological Implications

The physical structure of a shelf sea front controls associated biochemical gradients. From the mid-1970s, alongside the physical oceanographic studies

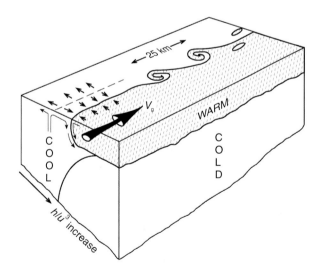

Figure 6 Pattern of circulation at a tidal mixing front. V_g is an along-front surface geostrophic jet, typically about $10\,\mathrm{cm\,s^{-1}}$. There is a surface current convergence at the front, followed by downwelling, with a compensatory upwelling and surface divergence in the mixed region. Baroclinic eddies develop along the front, with typical wavelengths of around 25 km. (After Simpson and James (1986) *Coastal and Estuarine Sciences*, 3: 63–93. Courtesy of the American Geophysical Union.)

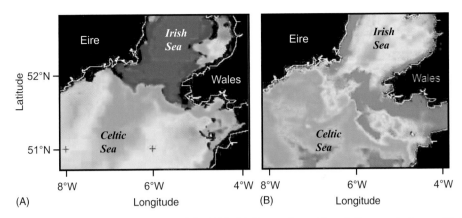

Figure 7 (A) SST image of the Celtic Sea front, 12 July 1999. (B) Sea surface chlorophyll concentration on 12 July in the same region, derived from the SeaWIFS sensor on NASA's SeaStar satellite. The mixed water of the Irish Sea is associated with chlorophyll concentrations of $1-2\,\text{mg}\,\text{m}^{-3}$. The strongly stratified water in the Celtic Sea has surface chlorophyll concentrations of less than $0.5\,\text{mg}\,\text{m}^{-3}$. At the front there is a clear signature of enhanced chlorophyll concentration, reaching about $5\,\text{mg}\,\text{m}^{-3}$. (Image courtesy of the Dundee Satellite Receiving Station, and the Remote Sensing Group, Plymouth Marine Laboratory.)

of fronts, it was recognized that enhanced levels of chlorophyll (phytoplankton biomass) were often seen in the frontal surface water. The availability of satellite remote sensing of surface chlorophyll (in particular the SeaWIFS sensor) now allows dramatic evidence of these frontal accumulations of phytoplankton (**Figure 7**). It is conceivable that the convergence of flow at a front could lead to enhancement of surface chlorophyll, by concentrating the biomass from the mixed and stratified water on either side. However, the spatial extent of the observed frontal chlorophyll ($\sim 1\text{--}10\,\text{km}$) is typically at least an order of magnitude greater than the horizontal extent of the convergence region (of order $100\,\text{m}$). More recently, at the Georges Bank frontal system, the enhanced frontal chlorophyll has been observed directly associated with an increase in rates of primary production compared to the waters on either side of the front. Thus, it appears that locally enhanced concentrations of frontal phytoplankton biomass are a result of locally increased production, and so require some source of nitrate to be mixed into the region.

For primary production the shelf sea front marks the transition between a nutrient-replete but light-limited, environment and a stratified water column with a well-lit but nutrient-deficient surface layer (**Figure 8**). Highest nutrient levels are usually found in the bottom mixed layer on the stratified side of the front, owing to negligible utilization and the contribution from detritus sinking down from the surface layer. Enhanced primary production in the frontal surface waters requires a mechanism to transport nutrients into the region, from the deep, high-nutrient water (vertical nutrient flux) and/or from the moderate nutrient-containing waters on the mixed side of the front (horizontal nutrient flux).

Four supply mechanisms have been suggested (**Figure 8**). First, the surface outcropping of the front is a region of gradually reducing vertical stratification, and thus a region where the inhibition of vertical mixing is reduced. The increased turbulent flux of nutrients will be available for surface primary production, as long as the residence time of the phytoplankton cells in the photic zone is still sufficient to allow net growth. Second, the spring–neap adjustment of a front's position results in the 2–4 km adjustment region undergoing a fortnightly mixing–stratification cycle. Thus, toward spring tides the region becomes vertically mixed and replenished with nutrients throughout the water column, and as the water restratifies toward neap tides the new surface nutrients become available for primary production. The predicted fortnightly pulses in surface frontal biomass have been reported at the Ushant shelf sea front, in the Western English Channel. A third nutrient supply mechanism is weak diapycnal flux, transferring water from the mixed side of the front into the surface frontal water. Finally, baroclinic eddies will transfer pools of water from the mixed side into the stratified side, though at the cost of a similar flux of water containing phytoplankton in the opposite direction.

Shelf Slope Fronts

Typical seabed slopes in shelf seas are about $0.5°\text{--}0.8°$. At the edge of the shelf seas this slope increases to $1.3°\text{--}3.2°$, a transition that occurs at a depth typically between 100 and 200 m. This region of steeper bathymetry, just seaward of the shelf edge, is the shelf slope, and is often associated with sharp horizontal gradients in temperature and/or salinity (for example, see **Figure 9**). The difference in the water characteristics of shelf seas and the open ocean arises as the result of several mechanisms. Coastal and shelf waters tend to have lower salinity than the open ocean, owing to the input of fresh water from land runoff. The fresh water input also alters the temperature of the shelf water, as does the seasonal heating/cooling cycle, which will generate more pronounced temperature fluctuations within the shallow water. Offshore, the open ocean is part of a larger, basin-scale circulation that, for instance, brings much warmer water from

equatorial regions past the shelf edge (e.g., consider the along-slope circulation of the world's western boundary currents). There is often a marked seasonality in the form of the shelf break front. Shelf waters can be more buoyant than oceanic waters during summer, but surface cooling in winter can reverse this to leave a denser water mass on the shelf that has the potential for cascading off the shelf and down the shelf slope.

The Position of a Shelf Slope Front

While the reasons for the contrast in water characteristics are straightforward, an explanation is required of why these differences between shelf and oceanic waters are maintained across such sharp fronts at the shelf slope. This is not as straightforward as for the case of the tidal mixing front. The limited number of processes governing the vertical structure of shelf seas resulted in a testable prediction for the position of a tidal mixing front in terms of water depth and tidal current amplitude. At the shelf break there are a number of potential controlling factors on a front's position, and a corresponding difficulty in producing an unambiguous, testable hypothesis. Numerical modeling provides the best technique for investigating frontal dynamics, allowing simultaneous consideration of several physical processes. However, a major problem with the assessment of any description of controls on shelf slope fronts is that there are considerable logistic difficulties in collecting current and scalar observations of sufficient quality and resolution to compare with the model outputs. The following arguments are based on both analytical and numerical models of shelf slope fronts.

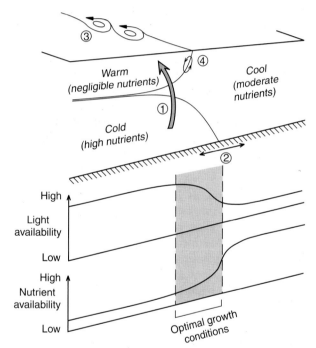

Figure 8 Nutrient supply mechanisms that are thought to fuel primary production at a tidal mixing front. (1) Vertical turbulent nutrient flux through the weaker stratification at the front. (2) The spring–neap adjustment of the front, causing a fortnightly replenishment of surface nutrients within the adjustment region. (3) Baroclinic eddy shedding along the front, transferring nutrient-rich water from the mixed side of the front into the stratified side. (4) A weak cross-frontal circulation caused by friction between the residual flows within the front (see **Figure 6**). In the surface layer on the stratified side of the front, the algae receive plenty of light but are prevented from growing because new nutrients cannot be supplied through the strong thermocline. On the mixed side of the front, nutrients are plentiful but growth is limited by a lack of light as the algae are mixed throughout the entire depth of the water column. The problem of lack of light on the mixed side is compounded by tidal re-suspension of bed sediments, acting to increase the opacity of the water. As the transition zone between these two extremes, the front provides optimal conditions for algal growth. Processes (1) and (2) are thought to be capable of supplying about 80% of the nutrient requirements at a typical front.

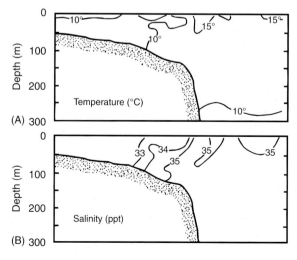

Figure 9 (A) Temperature and (B) salinity structure across the shelf break south of Cape Cod, eastern North America. (After Wright (1976) *J Marine Research* 34: 1–14. Courtesy of the Sears Foundation for Marine Research.)

A fundamental dynamic constraint on the exchange of water masses across the shelf slope lies with the geostrophic behavior of the oceanic flows. Geostrophic currents cannot cross steep bathymetry. Instead, they are forced to flow along isobaths, parallel to the topography of the shelf slope and shelf edge. Both the oceanic flow seaward of the shelf edge, and the buoyancy-driven flows of shelf water close to the shelf edge, behave geostrophically. This basic topographic constraint on these geostrophic flows often forms the basis of descriptions of shelf slope fronts (e.g., work by Csanady, Ou, and Hsueh). This topographic constraint on offshore movement of shelf water has been shown to be more important when the shelf water is denser than the oceanic water, with the shelf edge controlling the cascading of the denser water down the shelf slope. When shelf water is less dense, the internal Rossby deformation radius appears to have the dominant influence.

Such gravitational relaxation of the horizontal density structure across the shelf slope only explains the formation of the front. The resulting strong along-slope flows and current shear suggest that the frontal signature should be rapidly mixed and dissipated, and yet observations clearly show that the fronts exist for prolonged periods. This implies that the dynamics of the fronts must also act to maintain frontal structure, in addition to causing its initial formation. One suggestion by Ou is that the front can be maintained, paradoxically, by the action of wind stress at the sea surface. This wind mixing generates a surface mixed layer, which still contains a cross-shelf horizontal density gradient and so continues to relax under gravity and feed the along-slope current.

The above mechanisms for frontal formation and maintenance explicitly use a cross-shore density gradient as the pivotal dynamical process. However, it has been noted (e.g., by Chapman), that in the Middle Atlantic Bight in summer the combined frontal structures of temperature and salinity compensate to produce no horizontal density gradient. In other words, a shelf slope front can exist in the scalar fields without any apparent horizontal density structure to maintain them. Chapman, again utilizing a numerical model, showed that such a situation could be supported if there is a strong along-shore flow on the shelf and a distinct shelf break. Friction with the seabed in the shallower shelf water causes a cross-shelf component of the flow. Above the shelf slope, in deeper water, the effect of friction is reduced, and so there is a convergence of the cross-shelf flow close to the shelf break. The existence of the temperature and/or salinity front is then dependent on the relative contributions of advection and diffusion. Seaward of the shelf edge, diffusion is the dominant process, smoothing out any horizontal gradients. The convergence at the shelf edge concentrates the cross-shelf scalar gradients into a front, and the dominance of advection moves this structure along the shelf edge faster than diffusive processes can erode it. Thus, the front can be visible along several hundred kilometers of the Middle Atlantic Bight.

Implications of Shelf Slope Fronts

As with the tidal mixing fronts, shelf slope fronts in summer are often associated with concentrations of relatively high chlorophyll biomass, compared to the oceanic and shelf surface waters on either side. Fundamentally, this is again likely to be due to the diffusion of bottom water nutrients through the weaker stratification just at the surface front. Evidence from some shelf edge regions indicate the areas to be influenced by energetic internal waves on the thermocline, driven by the dissipation of the internal tidal wave (which is itself generated on the steep slope bathymetry seaward of the shelf edge). There has been some suggestion that secondary production can be more clearly linked to the primary production at shelf slope fronts than at shelf sea tidal mixing fronts, due to the temporal variability of the tidal fronts (e.g., spring–neap adjustment). Certainly many shelf slope regions are places of intense fishing activity.

The shelf slope is recognized as a region key to the global cycling of carbon. The shelf seas and slope areas are highly productive, and thus have a high capacity for uptake of atmospheric carbon. Atmospheric carbon is drawn into the ocean as the result of algal growth extracting carbon from the sea water. The fate of some of this carbon uptake is to sink when the algae die, and become buried in the shelf and slope sediments. This flux of carbon to the seabed is an important carbon removal process, with shelf sea and slope regions currently thought to be responsible for about 90% of the global oceanic removal of carbon. Thus, one of the important questions in oceanography concerns the transfer of water across the shelf edge, between the slope and shelf seas. This transfer controls both the rate at which carbon is transferred to the shelf slope from the shelf seas and the rate at which new nutrients from the slope waters are supplied to the shelf waters ready to fuel new carbon uptake.

Summary

Locally enhanced regions of horizontal salinity, temperature, and density gradient occur across the coastal and shelf seas, driven by a variety of mechanisms. The dynamics controlling the structure and position of fresh water fronts and shelf sea tidal mixing fronts are relatively well understood. Fresh water fronts result from the relaxation of a horizontal density gradient, arrested either by the diversion of flow caused by the Earth's rotation or by the interaction between nearbed cross-frontal flows and a sloping seabed. Tidal mixing fronts are controlled by the competition between the rate of supply of mixing energy (supplied either by tidal current stress against the seabed or by wind stress against the sea surface) and the rate of stratification (produced by surface heating). For fronts at the shelf edge/slope region, the change in the slope of the seabed must play a pivotal role, but the full dynamics controlling the fronts are less clear. Partially this is due to the difficulty in collecting observations of sufficient resolution, in both time and space, against which to test hypotheses. Also, in particular contrast with the tidal mixing fronts, there appears to be no dominant process controlling these fronts.

All fronts are observed to be regions of enhanced surface primary production. The common feature causing this production is likely to be the reduced stability close to surface fronts allowing increased vertical turbulent mixing of nutrients into the well-lit surface water. Fronts close to the shelf edge, or other regions of steep bathymetry, have the additional feature of locally generated internal waves providing enhanced mixing across the shallowing pycnocline. This increased primary production is often seen to be associated with increases in zooplankton and larger fish, ultimately supporting populations of sea birds and providing an important fisheries resource for people.

There are still important questions that remain to be answered concerning the physics of fronts. For instance, direct measurements of turbulent mixing have only recently become possible, so the potential for horizontal gradients in rates of vertical turbulent exchange still needs to be addressed. Shelf slope fronts are perhaps the most lacking in terms of a coherent theory of their dynamics (assuming such a general approach is possible), and have particular questions related to cross-frontal transfers that still require attention. The link between the physics of fronts and the closely coupled biology and chemistry is perhaps the area of greatest research potential. Oceanographic instrumentation is developing rapidly to allow the biological and chemical environment to be observed at the same spatial and temporal scales as the controlling physics.

See also

Carbon Cycle. Dispersion in Shallow Seas. Ekman Transport and Pumping. Primary Production Distribution. Primary Production Methods. Primary Production Processes. Tides. Turbulence in the Benthic Boundary Layer.

Further Reading

Chapman DC (1986) A simple model of the formation and maintenance of the shelf/slope front in the Middle Atlantic Bight. *Journal of Physical Oceanography* 16: 1273–1279.

Hill AE (1998) Buoyancy effects in coastal and shelf seas. In: Brink KH and Robinson R (eds) *The Sea*, vol. 10, pp. 21–62. New York: Wiley.

Holligan PM (1981) Biological implications of fronts on the northwest European continental shelf. *Philosophical Transactions of the Royal Society of London A* 302: 547–562.

Mann KH and Lazier JRN (1996) *Dynamics of Marine Ecosystems*, 2nd edn. Boston: Blackwell Science.

Simpson JH (1998) Tidal processes in shelf seas. In: Brink KH and Robinson R (eds) *The Sea*, vol. 10, pp. 113–150. New York: Wiley.

Simpson JH and Hunter JR (1974) Fronts in the Irish Sea. *Nature* 250: 404–406.

SHELF-SEA MODELS

See **REGIONAL AND SHELF SEA MODELS**

SHIPPING AND PORTS

H. L. Kite-Powell, Woods Hole Oceanographic Institution, Woods Hole, MA, USA

doi:10.1006/rwos.2001.0495

Introduction

Ships and ports have been an important medium for trade and commerce for thousands of years. Today's maritime shipping industry carries 90% of the world's 5.1 billion tons of international trade. Total seaborne cargo movements exceeded 21 trillion ton-miles in 1998. Shipping is an efficient means of transport, and becoming more so. World freight payments as a fraction of total import value were 5.24% in 1997, down from 6.64% in 1980. Modern container ships carry a 40-foot container for < 10 cents per mile – a fraction of the cost of surface transport.

Segments of the shipping industry exhibit varying degrees of technological sophistication and economic efficiency. Bulk shipping technology has changed little in recent decades, and the bulk industry is economically efficient and competitive. Container shipping has advanced technologically, but economic stability remains elusive. The shipping industry has a history of national protectionist measures and is governed by a patchwork of national and international regulations to ensure safety and guard against environmental damage.

Units

Two standard measures of ship size are deadweight tonnage (dwt) and gross tonnage (gt). Deadweight describes the vessel's cargo capacity; gross tonnage describes the vessel's enclosed volume, where 100 cubic feet equal one gross ton. In container shipping, the standard unit of capacity is the twenty-foot equivalent unit (TEU), which is the cargo volume available in a container 20 feet long, 8 feet wide, and 8 feet (or more) high. These containers are capable of carrying 18 metric tons but more typically are filled to 13 tons or less, depending on the density of the cargo.

World Seaborne Trade

Shipping distinguishes between two main types of cargos: bulk (usually shiploads of a single commodity) and general cargos (everything else). Major dry bulk trades include iron ore, coal, grain, bauxite, sand and gravel, and scrap metal. Liquid bulk or tanker cargos include crude oil and petroleum products, chemicals, liquefied natural gas (LNG), and vegetable oil. General cargo may be containerized, break-bulk (noncontainer packaging), or 'neo-bulk' (automobiles, paper, lumber, etc.). Table 1 shows that global cargo weight has grown by a factor of 10 during the second half of the twentieth century, and by about 4% per year since 1986. Tanker cargos (mostly oil and oil products) make up about 40% of all cargo movements by weight.

Both the dry bulk and tanker trades have grown at an average rate of about 2% per year since 1980.

Table 1 World seaborne dry cargo and tanker trade volume (million tons) 1950–1998

Year	Dry cargo	Tanker cargo	Total	% change
1950	330	130	460	—
1960	540	744	1284	—
1970	1165	1440	2605	—
1986	2122	1263	3385	3
1987	2178	1283	3461	2
1988	2308	1367	3675	6
1989	2400	1460	3860	5
1990	2451	1526	3977	3
1991	2537	1573	4110	3
1992	2573	1648	4221	3
1993	2625	1714	4339	3
1994	2735	1771	4506	4
1995	2891	1796	4687	4
1996	2989	1870	4859	4
1997	3163	1944	5107	5
1998	3125	1945	5070	− 1

Current bulk trades total some 9 trillion ton-miles for dry bulk and 10 trillion ton-miles for tanker cargos.

Before containerization, general cargo was transported on pallets or in cartons, crates, or sacks (in general cargo or break-bulk vessels). The modern container was first used in US domestic trades in the 1950s. By the 1970s, intermodal services were developed, integrating maritime and land-based modes (trains and trucks) of moving containers. Under the 'mini land-bridge' concept, for example, containers bound from Asia to a US east coast port might travel across the Pacific by ship and then across the USA by train. In the 1980s, intermodal services were streamlined further with the introduction of double-stack trains, container transportation became more standardized, and shippers began to treat container shipping services more like a commodity. Today, > 70% of general cargo moves in containers. Some 58.3 million TEU of export cargo moved worldwide in 1999; including empty and domestic movement, the total was over 170 million TEU. The current total container trade volume is some 300 billion TEU-miles. Container traffic has grown at about 7% per year since 1980.

Table 2 lists the major shipping trade routes for the most significant cargo types, and again illustrates the dominance of petroleum in total global cargo volume. The most important petroleum trade routes are from the Middle East to Asia and to Europe, and within the Americas. Iron ore and coal trades are dominated by Australian exports to Asia, followed by exports from the Americas to Europe and Asia. The grain trades are dominated by American exports to Asia and Europe, along with

Table 2 Cargo volume on major shipping routes (million tons) 1998 (for containers: millions of TEU, 1999)

Cargo type	Route	Cargo volume	% of total for cargo
Petroleum	Middle East–Asia	525.0	26
	Intra-Americas	326.5	16
	Middle East–Europe	229.5	12
	Intra-Europe	142.7	7
	Africa–Europe	141.4	7
	Middle East–Americas	137.7	7
	Africa–Americas	106.1	5
Dry bulk: iron ore	Australia–Asia	116.8	28
	Americas–Europe	87.1	21
	Americas–Asia	63.3	15
Dry bulk: coal	Australia–Asia	118.4	25
	Americas–Europe	61.3	13
	S. Africa–Europe	37.8	8
	Americas–Asia	35.2	7
	Intra-Asia	29.3	6
	Australia–Europe	27.3	6
Dry bulk: grain	Americas–Asia	60.2	31
	Americas–Europe	20.6	10
	Intra-Americas	33.1	17
	Americas–Africa	15.2	8
	Australia–Asia	15.1	8
Container	Trans-Pacific (Asia/N. America)	7.3	14
	Asia/Europe	4.5	8
	Trans-Atlantic (N. America/Europe)	4.0	8

intra-American routes. Container traffic is more fragmented; some of the largest routes connect Asia with North America to the east and with Europe to the west, and Europe and North America across the Atlantic.

The World Fleet

The world's seaborne trade is carried by an international fleet of about 25 000 ocean-going commercial cargo ships of more than 1000 gt displacement. As trade volumes grew, economies of scale (lower cost per ton-mile) led to the development of larger ships: ultra-large crude oil carriers (ULCCs) of > 550 000 dwt were launched in the 1970s. Structural considerations, and constraints on draft (notably in US ports) and beam (notably in the Panama Canal) have curbed the move toward larger bulk vessels. Container ships are still growing in size. The first true container ships carried around 200 TEU; in the late 1980s, they reached 4000 TEU; today they are close to 8000 TEU. The world container fleet capacity was 4.1 million TEU in 1999.

Table 3 shows the development of the world fleet in dwt terms since 1976. Total fleet capacity has grown by an average of 1.5% per year, although it contracted during the 1980s. The significant growth in the 'other vessel' category is due largely to container vessels; container vessel dwt increased faster

than any other category (nearly 9% annually) from 1985 to 1999. Specialized tankers and auto carriers have also increased rapidly, by around 5% annually. Combos (designed for a mix of bulk and general cargo) and general cargo ships have decreased by > 5% per year.

Four standard vessel classes dominate both the dry bulk and tanker fleets, as shown in **Table 4**. Panamax bulkers and Suezmax tankers are so named because they fall just within the maximum dimensions for the Panama and Suez Canals, respectively. Capesize bulkers and VLCCs (very large crude carriers) are constrained in size mostly by the limitations of port facilities (draft restrictions).

In addition to self-propelled vessels, ocean-going barges pulled by tugs carry both bulk and container cargos, primarily on coastal routes.

Dry Bulk Carriers

The dry bulk fleet is characterized by fragmented ownership; few operators own more than 100 vessels. Most owners charter their vessels to shippers through a network of brokers, under either long-term or single-voyage ('spot') charter arrangements (see **Table 5**). The top 20 dry bulk charterers account for about 30% of the market. The dry bulk charter market is, in general, highly competitive.

Table 3 World fleet development (million dwt) 1976–1998

Year	Oil tankers	Dry bulk carriers	Other vessels	Total world fleet	% change
1976	320.0	158.1	130.3	608.4	–
1977	335.3	174.4	139.1	648.8	7
1978	339.1	184.5	146.8	670.4	3
1979	338.3	188.5	154.7	681.5	2
1980	339.8	191.0	160.1	690.9	1
1981	335.5	199.5	162.2	697.2	1
1982	325.2	211.2	165.6	702.0	1
1983	306.1	220.6	167.8	694.5	− 1
1984	286.8	228.4	168.1	683.3	− 2
1985	268.4	237.3	168.0	673.7	− 1
1986	247.5	235.2	164.9	647.6	− 4
1987	245.5	231.8	163.5	640.8	− 1
1988	245.0	230.1	162.0	637.1	− 1
1989	248.4	231.4	166.9	646.7	2
1990	257.4	238.9	170.5	666.8	3
1991	264.2	244.0	176.1	684.3	3
1992	270.6	245.7	178.3	694.6	2
1993	270.2	251.3	194.4	715.9	3
1994	269.4	254.3	220.3	744.0	4
1995	265.8	259.8	241.5	767.1	3
1996	271.2	269.6	252.2	793.0	4
1997	272.5	277.8	263.0	813.3	3
1998	279.7	271.6	274.0	825.3	2

Capesize bulkers are used primarily in the Australian and South American bulk export trades. Grain shipments from the USA to Asia move mainly on Panamax vessels. Some bulk vessels, such as oil/bulk/ore (OBO) carriers, are designed for a variety of cargos. Others are 'neo-bulk' carriers, specifically designed to carry goods such as automobiles, lumber, or paper.

Tankers

Ownership of the tanker fleet is also fragmented; the average tanker owner controls fewer than three vessels; 70% of the fleet is owned by independent owners and 25% by oil companies. Like dry bulk ships, tankers are chartered to shippers on longer time charters or spot voyage charters (see **Table 5**). The top five liquid bulk charterers account for about 25% of the market. Like the dry bulk market, the tanker charter market is highly competitive.

VLCCs are used primarily for Middle East exports to Asia and the USA. In addition to liquid tankers, a fleet of gas carriers (world capacity: 17 million gt) moves liquefied natural gas (LNG) and petroleum gas (LPG).

General Cargo/Container Ships

Ownership is more concentrated in the container or 'liner' fleet. ('Liner' carriers operate regularly

Table 4 Major vessel categories in the world's ocean-going cargo ship fleet, 2000

Cargo type	Category	Typical size	Number in world fleet
Dry bulk	Handysize	27 000 dwt	2700
	Handymax	43 000 dwt	1000
	Panamax	69 000 dwt	950
	Capesize	150 000 dwt	550
Tanker	Product tanker	45 000 dwt	1400
	Aframax	90 000 dwt	700
	Suezmax	140 000 dwt	300
	VLCC	280 000 dwt	450
General cargo	Container	> 2000 TEU	800
	Container feeder	< 2000 TEU	1800
	Ro/Ro	< 2000 TEU	900
	Semi-container	< 1000 TEU	2800

VLCC, very large crude carrier; RoRo, roll-on/roll-off.

Table 5 Average daily time charter rates, 1980–2000

Cargo type	Vessel type	$/day
Dry bulk	Handysize	6000
	Handymax	8000
	Panamax	9500
	Capesize	14000
Tanker	Product tanker	12000
	Aframax	13000
	Suezmax	16500
	VLCC	22000
Container	400 TEU geared	5000
	1000 TEU geared	9000
	1500 TEU geared	13500
	2000 TEU gearless	18000

VLCC, very large crude carrier.

scheduled service between specific sets of ports.) Of some 600 liner carriers, the top 20 account for 55% of world TEU capacity. The industry is expected to consolidate further. Despite this concentration, liner shipping is competitive; barriers to entry are low, and niche players serving specialized cargos or routes have been among the most profitable. Outside ownership of container vessels and chartering to liner companies increased during the 1990s.

Large ships (> 4000 TEU) are used on long ocean transits between major ports, while smaller 'feeder' vessels typically carry coastwise cargo or serve niche markets between particular port pairs. In addition to container vessels, the general cargo trades are served by roll-on/roll-off (RoRo) ships (world fleet capacity: 23 million gt) that carry cargo in wheeled vehicles, by refrigerated ships for perishable goods, and by a diminishing number of multi-purpose general cargo ships.

Passenger Vessels

The world's fleet of ferries carries passengers and automobiles on routes ranging from urban harbor crossings to hundreds of miles of open ocean. Today's ferry fleet is increasingly composed of so-called fast ferries, > 1200 of which were active in 1997. Many of these are aluminum-hull catamarans, the largest capable of carrying 900 passengers and 240 vehicles at speeds between 40 and 50 knots. Fast ferries provide an inexpensive alternative to land and air travel along coastal routes in parts of Asia, Europe, and South America.

The cruise ship industry was born in the 1960s, when jet aircraft replaced ships as the primary means of crossing oceans, and underutilized passenger liners were converted for recreational travel. The most popular early cruises ran from North America to the Caribbean. Today, the Caribbean remains the main cruise destination, with more than half of all

passengers; others are Alaska and the Mediterranean. The US cruise market grew from 500 000 passengers per year in 1970 to 1.4 million in 1980 and 3.5 million in 1990. Today, more than 25 cruise lines serve the North American cruise market. The three largest lines control over 50% of the market, and further consolidation is expected; 149 ships carried 6 million passengers and generated $18 billion in turnover in 1999. The global market saw 9 million passengers. Cruise vessels continue to grow larger due to scale economies; the largest ships today carry more than 2000 passengers. Most cruise lines are European-owned and fly either European or open registry flags (see below).

Financial Performance

Shipping markets display cyclical behavior. Charter rates (see **Table 5**) are set by the market mechanism of supply and demand, and are most closely correlated with fleet utilization (rates rise sharply when utilization exceeds 95%). Shipping supply (capacity) is driven by ordering and scrapping decisions, and sometimes by regulatory events such as the double hull requirement for tankers (see below). Demand (trade volume) is determined by national and global business cycles, economic development, international trade policies, and trade disruptions such as wars/conflicts, canal closures, or oil price shocks. New vessel deliveries typically lag orders by 1–2 years. Excessive ordering of new vessels in times of high freight rates eventually leads to overcapacity and a downturn in rates; rates remain low until fleet contraction or trade growth improves utilization, and the cycle repeats.

Average annual returns on shipping investments ranged from 8% (dry bulk) to 11% (container feeder ships) and 13% (tankers) during 1980–2000, with considerable volatility. Second-hand ship asset prices fluctuate with charter rates, and many ship owners try to improve their returns through buy-low and sell-high strategies.

Law and Regulation

Legal Regimes

International Law of the Sea The United Nations Convention on the Law of the Sea (UNCLOS) sets out an international legal framework governing the oceans, including shipping. UNCLOS codifies the rules underlying the nationality (registry) of ships, the right of innocent passage for merchant vessels through other nations' territorial waters, etc. (see the entry in this volume on the Law of the Sea for details).

International Maritime Organization The International Maritime Organization (IMO) was established in 1948 and became active in 1959. Its 158 present member states have adopted some 40 IMO conventions and protocols governing international shipping. Major topics include maritime safety, marine pollution, and liability and compensation for third-party claims. Enforcement of these conventions is the responsibility of member governments and, in particular, of port states. The principle of port state control allows national authorities to inspect foreign ships for compliance and, if necessary, detain them until violations are addressed.

In 1960, IMO adopted a new version of the International Convention for the Safety of Life at Sea (SOLAS), the most important of all treaties dealing with maritime safety. IMO next addressed such matters as the facilitation of international maritime traffic, load lines, and the carriage of dangerous goods. It then turned to the prevention and mitigation of maritime accidents, and the reduction of environmental effects from cargo tank washing and the disposal of engine-room waste.

The most important of all these measures was the Marine Pollution (MARPOL) treaty, adopted in two stages in 1973 and 1978. It covers accidental and operational oil pollution as well as pollution by chemicals, goods in packaged form, sewage, and garbage. In the 1990s, IMO adopted a requirement for all new tankers and existing tankers over 25 years of age to be fitted with double hulls or a design that provides equivalent cargo protection in the event of a collision or grounding.

IMO has also dealt with liability and compensation for pollution damage. Two treaties adopted in 1969 and 1971 established a system to provide compensation to those who suffer financially as a result of pollution.

IMO introduced major improvements to the maritime distress communications system. A global search and rescue system using satellite communications has been in place since the 1970s. In 1992, the Global Maritime Distress and Safety System (GMDSS) became operative. Under GMDSS, distress messages are transmitted automatically in the event of an accident, without intervention by the crew.

An International Convention on Standards of Training, Certification and Watchkeeping (STCW), adopted in 1978 and amended in 1995, requires each participating nation to develop training and certification guidelines for mariners on vessels sailing under its flag.

National Control and Admiralty Law The body of private law governing navigation and shipping in each country is known as admiralty or maritime law. Under admiralty, a ship's flag (or registry) determines the source of law. For example, a ship flying the American flag in European waters is subject to American admiralty law. This also applies to criminal law governing the ship's crew.

By offering advantageous tax regimes and relatively lax vessel ownership, inspection, and crewing requirements, so-called 'flags of convenience' or 'open registries' have attracted about half of the world's tonnage (see **Table 6**). The open registries include Antigua and Barbuda, Bahamas, Bermuda, Cayman Islands, Cyprus, Gibraltar, Honduras, Lebanon, Liberia, Malta, Mauritius, Oman, Panama, Saint Vincent, and Vanuatu. Open registries are the flags of choice for low-cost vessel operation, but in some instances they have the disadvantage of poor reputation for safety.

Protectionism/Subsidies

Most maritime nations have long pursued policies that protect their domestic flag fleet from foreign

Table 6 Fleets of principal registries (in 1000s of gt) 1998 (includes ships of 100 gt)

Country of registry	Tankers	Dry bulk	General cargo	Other	Total	% of world
Panama	22 680	40 319	26 616	8608	98 223	18.5
Liberia	26 361	16 739	8803	8590	60 493	11.4
Bahamas	11 982	4990	7143	3601	27 716	5.2
Greece	12 587	8771	1943	1924	25 225	4.7
Malta	9848	8616	4658	952	24 074	4.5
Cyprus	3848	11 090	6981	1383	23 302	4.4
Norway	8994	4041	4153	5949	23 137	4.4
Singapore	8781	4585	5596	1408	20 370	3.8
Japan	5434	3869	3111	5366	17 780	3.3
China	2029	6833	6199	1443	16 504	3.1
USA	3436	1268	3861	3286	11 851	2.2
Russia	1608	1031	3665	4786	11 090	2.1
Others	33 448	46 414	57 350	34 917	172 129	32.4
World total	151 036	158 566	140 079	82 213	531 894	100.0

competition. The objective of this protectionism is usually to maintain a domestic flag fleet and cadre of seamen for national security, employment, and increased trade.

Laws that reserve domestic waterborne cargo (cabotage) for domestic flag ships are common in many countries, including most developed economies (although they are now being phased out within the European Union). The USA has a particularly restrictive cabotage system. Cargo and passengers moving by water between points within the USA, as well as certain US government cargos, must be carried in US-built, US-flag vessels owned by US citizens. Also, US-flag ships must be crewed by US citizens.

Vessels operating under restrictive crewing and safety standards may not be competitive in the international trades with open registry vessels due to high operating costs. In some cases, nations have provided operating cost subsidies to such vessels. The US operating subsidy program was phased out in the 1990s.

Additional subsidies have been available to the shipping industry through government support of shipbuilding (see below).

Liner Conferences

The general cargo trades have traditionally been served by liner companies that operate vessels between fixed sets of ports on a regular, published schedule. To regulate competition among liner companies and ensure fair and consistent service to shippers, a system of 'liner conferences' has regulated liner services and allowed liner companies, within limits, to coordinate their services.

Traditionally, so-called 'open' conferences have been open to all liner operators, published a single rate for port-to-port carriage of a specific commodity, and allowed operators within the conference to compete for cargo on service. The US foreign liner trade has operated largely under open conference rules. By contrast, many other liner trades have been governed by closed conferences, participation in which is restricted by governments. In 1974, the United Nations Conference on Trade and Development (UNCTAD) adopted its Code of Conduct for Liner Conferences, which went into effect in 1983. Under this code, up to 40% of a nation's trade can be reserved for its domestic fleet.

In the 1970s, price competition began within conferences as carriers quoted lower rates for intermodal mini land-bridge routes than the common published port-to-port tariffs. By the 1980s, intermodal rates were incorporated fully in the liner conference scheme. The 1990s saw growing deregu-lation of liner trades (for example, the US 1998 Ocean Shipping Reform Act), which allowed carriers to negotiate confidential rates separately with shippers. The effect has been, in part, increased competition and a drive for consolidation among liner companies that is expected to continue in the future.

Environmental Issues

With the increased carriage of large volumes of hazardous cargos (crude oil and petroleum products, other chemicals) by ships in the course of the twentieth century, and especially in the wake of several tanker accidents resulting in large oil spills, attention has focused increasingly on the environmental effects of shipping. The response so far has concentrated on the operational and accidental discharge of oil into the sea. Most notably, national and international regulations (the Oil Pollution Act (OPA) of 1990 in the US; IMO MARPOL 1992 Amendments, Reg. 13F and G) are now forcing conversion of the world's tanker fleet to double hulls. Earlier, MARPOL 1978 required segregated ballast tanks on tankers to avoid the discharge of oil residue following the use of cargo tanks to hold ballast water.

Recently, ballast and bilge water has also been identified as a medium for the transport of non-indigenous species to new host countries. A prominent example is the zebra mussel, which was brought to the USA from Europe in this way.

Liability and Insurance

In addition to design standards such as double hulls, national and international policies have addressed the compensation of victims of pollution from shipping accidents through rules governing liability and insurance. The liability of ship owners and operators for third-party claims arising from shipping accidents historically has been limited as a matter of policy to encourage shipping and trade. With the increased risk associated with large tankers, the international limits were raised gradually in the course of the twentieth century. Today, tanker operators in US waters face effectively unlimited liability under OPA 90 and a range of state laws.

Most liability insurance for the commercial shipping fleet is provided through a number of mutual self-insurance schemes known as P&I (protection and indemnity) clubs. The International Group of P&I Clubs includes 19 of the largest clubs and collectively provides insurance to 90% of the world's ocean-going fleet. Large tankers operating in US waters today routinely carry liability insurance coverage upward of $2 billion.

Shipbuilding

Major technical developments in twentieth century shipbuilding began with the introduction of welding in the mid-1930s, and with it, prefabrication of steel sections. Prefabrication of larger sections, improved welding techniques, and improved logistics were introduced in the 1950s. Mechanized steel prefabrication and numerically controlled machines began to appear in the 1970s. From 1900 to 2000, average labor hours per ton of steel decreased from 400–500 to fewer than 100; and assembly time in dock/berth decreased from 3–4 years to less than 6 months.

South Korea and Japan are the most important commercial shipbuilding nations today. In the late 1990s, each had about 30% of world gross tonnage on order; China, Germany, Italy, and Poland each accounted for between about 4 and 6%; and Taiwan and Romania around 2%. The US share of world shipbuilding output fell from 9.5% in 1979 to near zero in 1989.

Shipbuilding is extensively subsidized in many countries. Objectives of subsidies include national security goals (maintenance of a shipbuilding base), employment, and industrial policy goals (shipyards are a major user of steel industry output). Direct subsidies are estimated globally around $5–10 billion per year. Indirect subsidies include government loan guarantees and domestic-build requirements (for example, in the US cabotage trade).

An agreement on shipbuilding support was developed under the auspices of the Organization for Economic Cooperation and Development (OECD) and signed in December 1994 by Japan, South Korea, the USA, Norway, and the members of the European Union. The agreement would limit loan guarantees to 80% of vessel cost and 12 years, ban most direct subsidy practices, and limit government R&D support for shipyards. Its entry into force, planned for 1996, has been delayed by the United States' failure to ratify the agreement.

Ports

Commercial ships call on thousands of ports around the world, but global cargo movement is heavily concentrated in fewer than 100 major bulk cargo ports and about 24 major container ports. The top 20 general cargo ports handled 51% of all TEU movements in 1998 (see **Table 7**).

Most current port development activity is in container rather than bulk terminals. Port throughput efficiency varies greatly. In 1997, some Asian container ports handled 8800 TEU per acre per year, while European ports averaged 3000 and US ports only 2100 TEU per acre per year. The differences are due largely to the more effective use of automation and lesser influence of organized labor in Asian ports. Because container traffic is growing rapidly and container ships continue to increase in size, most port investment today is focused on the improvement and new development of container terminals.

In many countries, the public sector plays a more significant role in port planning and development than it does in shipping. Most general cargo or

Table 7 Top 20 container ports by volume, 1998

Port	Country	Throughput (1000s TEU)
Singapore	Singapore	15 100
Hong Kong	China	14 582
Kaohsiung	Taiwan	6271
Rotterdam	Netherlands	6011
Busan	South Korea	5946
Long Beach	USA	4098
Hamburg	Germany	3547
Los Angeles	USA	3378
Antwerp	Belgium	3266
Shanghai	China	3066
Dubai	UAE	2804
Manila	Philippines	2690
Felixstowe	UK	2524
New York/New Jersey	USA	2500
Tokyo	Japan	2169
Tanjung Priok	Indonesia	2131
Gioia Tauro	Italy	2126
Yokohama	Japan	2091
San Juan	Puerto Rico	2071
Kobe	Japan	1901

'commercial' ports are operated as publicly controlled or semi-public entities, which may lease space for terminal facilities to private terminal operators. Bulk cargo or 'industrial' ports for the export or import/processing of raw materials traditionally have been built by and for a specific industry, often with extensive public assistance. The degree of national coordination of port policy and planning varies considerably. The USA has one of the greatest commercial port densities in the world, particularly along its east coast. US commercial ports generally compete among each other as semi-private entities run in the interest of local economic development objectives. The US federal government's primary role in port development is in the improvement and maintenance of navigation channels, since most US ports (apart from some on the west coast) are shallow and subject to extensive siltation.

Future Developments

A gradual shift of the 'centroid' of Asian manufacturing centers westward from Japan and Korea may in the future shift more Asia–America container cargo flows from trans-Pacific to Suez/trans-Atlantic routes. Overall container cargo flows are expected to increase by 4–7% per year. Bulk cargo volumes are expected to increase as well, but at a slower pace.

Container ships will continue to increase in size, reaching perhaps 12 000 TEU capacity in the course of the next decade. Smaller, faster cargo ships may be introduced to compete for high-value cargo on certain routes. One concept calls for a 1400 TEU vessel capable of 36–40 knots (twice the speed of a conventional container ship), making scheduled Atlantic crossings in 4 days.

Port developments will center on improved container terminals to handle larger ships more efficiently, including berths that allow working the ship from both sides, and the further automation and streamlining of moving containers to/from ship and rail/truck terminus.

Conclusions

Today's maritime shipping industry is an essential transportation medium in a world where prosperity is often tied to international trade. Ships and ports handle 90% of the world's cargo and provide a highly efficient and flexible means of transport for a variety of goods. Despite a history of protectionist regulation, the industry as a whole is reasonably efficient and becoming more so. The safety of vessels and their crews, and protection of the marine environment from the results of maritime accidents, continue to receive increasing international attention. Although often conservative and slow to adopt new technologies, the shipping industry is poised to adapt and grow to support the world's transportation needs in the twenty-first century.

See also

International Organizations. Law of the Sea. Marine Policy Overview. Ships.

Further Reading

Containerization Yearbook 2000. London: National Magazine Co.

Shipping Statistics Yearbook. Bremen: Institute of Shipping Economics and Logistics.

Lloyd's Shipping Economist. London: Lloyd's of London Press.

United Nations Conference on Trade and Development (UNCTAD) (Annual) *Review of Maritime Transport*. New York: United Nations.

SHIPS

R. P. Dinsmore, Woods Hole Oceanographic Institution, Woods Hole, MA, USA

Copyright © 2001 Academic Press

doi:10.1006/rwos.2001.0299

Introduction

Oceanographic research vessels are shipboard platforms which support the conduct of scientific research at sea. Such research may include mapping and charting, marine biology, fisheries, geology and geophysics, physical processes, marine meteorology, chemical oceanography, marine acoustics, underwater archaeology, ocean engineering, and related fields.

Unlike other types of vessels (i.e., passenger, cargo, tankers, tugs, etc.) oceanographic research vessels (RVs) are a highly varied group owing to the diverse disciplines in which they engage. However, characteristics common to most RVs are relatively small size (usually 25–100 m length overall); heavy outfit of winches, cranes, and frames for overboard work; spacious working decks; multiple laboratories; and state-of-the-art instrumentation for navigation, data acquisition and processing.

Categories of RVs may vary according to the geographic areas of operations as well as the nature and sponsorship of work. Examples are:

1. By Region:

Coastal, usually smaller vessels of limited endurance and capability

Ocean going, larger vessels usually with multipurpose capabilities, ocean-wide or global range

Polar, ice reinforced with high endurance and multipurpose capability

2. By discipline:

Mapping and charting, emphasis on bathymetry

Fisheries, stock surveys, gear research, environmental studies

Geophysics, seismic and magnetic surveys

Support, submersibles, buoys, autonomous vehicles, diving

Multiple purpose, biological, chemical, and physical oceanography; marine geology; acoustics; ocean engineering; student training; may also include support services.

3. By sponsor:

Federal, usually mission oriented and applied research

Military, defense, mapping and charting, acoustics

Academic, basic research, student practicum

Commercial, exploration, petroleum, mining and fisheries

According to information available from the International Research Ship Operators Meeting (ISOM), in 2000 there were approximately 420 RVs (over 25 m length overall) operated by 49 coastal nations.

History

The history of RVs is linked closely to the cruises and expeditions of early record. Most of these were voyages seeking new lands or trade routes. Oceanic studies at best were limited to the extent and boundaries of the seas. Little is known of the ships themselves except it can be assumed that they were typical naval or trading vessels of the era. A known example is an expedition sent by Queen Hatshepsut of Egypt in 1500 BC. Sailing from Suez to the Land of Punt (Somaliland) to seek the source of myrrh, the fleet cruised the west coast of Africa. Pictures of the ships can be seen on reliefs in the temple of Deir-el-Bahri.

These ships represent the high-water mark of the Egyptian shipbuilder and although handsome vessels, they had serious weaknesses and did not play a role in the development of naval architecture.

The earliest known voyage of maritime exploration comes from Greek legend and sailors' tales. It is that of Jason dispatched from Iolcus on the northeast coast of Greece to the Black Sea c.1100 BC with the ship *Argo* and a celebrated crew of Argonauts in quest of the Golden Fleece. Although embellished by Greek mythology, this may have a factual basis; by sixth century BC there were Greek settlements along the shores explored by Jason, and the inhabitants traditionally collected gold dust from rams fleeces lain in river beds.

Better recorded is an expedition commissioned by Pharaoh Necho II of Egypt in 609 BC. A Phoenician fleet sailed south from Suez and circumnavigated Africa on a four-year voyage. Information from this and other Phoenician voyages over the next 300 years (including Hanno to the west coast of Africa in 500 BC, and Pytheas to the British Isles in 310 BC), although not specifically oceanographic research, contributed to a database from which philosophers and scholars would hypothesize the shape and extent of the world's oceans. In today's terminology these vessels would be characterized as ships of opportunity.

Under the Roman Empire, sea-trade routes were forged throughout the known world from Britain to the Orient. Data on ocean winds and currents were compiled in early sailing directions which added to the growing fund of ocean knowledge. Aristotle, although not a seafarer, directed much of his study to the sea both in physical processes and marine animals. From samples and reports taken from ships he named and described 180 species of fish and invertebrates Pliny the Elder cataloged marine animals into 176 species, and searched available ocean soundings proclaiming an average ocean depth of 15 stadia (2700 m). Ptolemy (AD 90–168), a Greek mathematician and astronomer at Alexandria employed ship reports in compiling world maps which formed the basis of mapping new discoveries through the sixteenth century.

With the fall of the Roman Empire came the dark ages of cartography and marine exploration slowed to a standstill except for Viking voyages to Greenland and Newfoundland, and Arab trade routes in the Indian Ocean and as far as China. Arabian sea tales such as 'Sinbad the Sailor' or the 'Wonders of India' by Ibn Shahryar (905) rival those of Homer and Herodotus. The art of nautical surveying began with the Arabs during medieval times. They had the compass and astrolabe and made charts of the coastlines which they visited.

In the West the city-state of Venice became in the ninth century the most important maritime power in the Mediterranean. It was from Venice that Marco

Polo began his famous journeys. In the fifteenth century, maritime exploration resumed, much of it under the inspiration and patronage of Prince Henry of Portugal (1394–1460), known as the Navigator, who established an academy at Sagres, Portugal, and attracted mathematicians, astronomers, and cartographers. This led to increasingly distant voyages leading up to those of Columbus, Vasco da Gama, and Magellan. With the results of these voyages, the shape of the world ocean was beginning to emerge, but little else was known. From the tracks of ships and their logs, information on prevailing winds and ocean currents were made into sailing directions. Magellan is reported to have made attempts at measuring depths, but with only 360 m of sounding line, he achieved little.

By the late seventeenth and early eighteenth centuries instruments were being devised for deep soundings, water samples, and even subsurface temperatures. However, any ship so engaged either a naval or trading vessel remained on an opportunity basis. In the mid to late eighteenth century the arts of navigation and cartography were amply demonstrated in the surveys and voyages of Captains James Cook and George Vancouver. The latter carried out surveys to especially high standards, and his vessel HMS *Discovery* might well be considered one of the first oceanographic research vessels.

Hydrographic departments were set up by seafaring nations: France in 1720, Britain in 1795, Spain in 1800, the US in 1807. Most were established as a navy activity, and most remain so to this day; exceptions were the French Corps of Hydrographic Engineers, and the US Coast Survey (now the National Ocean Survey). A date significant to this history is 1809 when the British Admiralty assigned a vessel permanently dedicated to survey service. Others followed suit; the first such US ship was the Coast Survey Schooner *Experiment* (1831). By now hydrographic survey vessels were a recognized type of vessel.

In the early to mid-nineteenth century expeditions were setting to sea with missions to include oceanographic investigations. Scientists (often termed 'naturalists') were senior members of the ships' complement. These include the vessels: *Astrolobe*, 1826–29 (Dumont D'Urville); *Beagle*, 1831–36 (Charles Darwin); US Exploring Expedition, 1838–42 (James Wilkes and James Dana); *Erebus & Terror*, 1839–43 (Sir James Ross); *Beacon*, 1841 (Edward Forbes); *Rattlesnake*, 1848–50 (Thomas Huxley). Most of the vessels participating in these expeditions were navy ships, but the nature of the work, their outfitting, and accomplishments mark them as oceanographic vessels of their time.

A new era in marine sciences commenced with the voyage of the HMS *Challenger*, 1872–76, a 69 m British Navy steam corvette. Equipped with a capable depth sounding machine and other instruments for water and bottom sampling, the *Challenger* under the scientific direction of Sir Charles Wyville Thomson obtained data from 362 stations worldwide. More than 4700 new species of marine life were discovered, and a sounding of 8180 m was made in the Marianas Trench. Modern oceanography is said to have begun with the Challenger Expedition. This spurred interest in oceanographic research, and many nations began to field worldwide voyages. These included the German *Gazelle* (1874–76); Russian *Vitiaz* (1886–89); Austrian *Pola*; USS *Blake* (1887–1880); and the Arctic cruise of the Norwegian *Fram* (1893–96).

The late 1800s and early 1900s saw a growing interest in marine sciences and the founding of both government- and university-sponsored oceanographic institutions. These included: Stazione Zoologia, Naples; Marine Biological Laboratory, Woods Hole, USA; Geophysical Institute, Bergen, Norway; Deutsche Seewarte, Hamburg; Scripps Institution of Oceanography, La Jolla, USA; Oceanographic Museum, Monaco; Plymouth Laboratory of Marine Biology, UK. These laboratories acquired and outfitted ships, which although mostly still conversions of naval or commercial vessels, became recognized oceanographic research vessels. The first vessel designed especially for marine research was the US Fish Commission Steamer *Albatross* built in 1882 for the new laboratory at Woods Hole. This was a 72 m iron-hulled, twin-screw steamer. It also was the first vessel equipped with electric generators (for lowering arc lamps to attract fish and organisms at night stations). Based at the Woods Hole Laboratory, the *Albatross* made notable deep sea voyages in both the Atlantic and Pacific Oceans. As with the *Challenger*, the *Albatross* became a legend in its own time and continued working until 1923.

In the early 1900s research voyages were aimed at strategic regions. Nautical charting and resources exploration were concentrated in areas of political significance: the Southern Ocean, South-east Asia, and the Caribbean. Many of the ships were designed and constructed as research vessels, notably the *Discovery* commanded by R. F. Scott on his first Antarctic voyage 1901–04. Another was the nonmagnetic brig *Carnegie* built in 1909 which carried out investigations worldwide. The *Carnegie* was the first research vessel to carry a salinometer, an electrical conductivity instrument for determining the chlorinity of sea water thus avoiding arduous chemical titrations. Research voyages by now were highly systematic on the pattern set by the *Challenger* expedition.

Another era of oceanic investigations began in 1925 with the German Atlantic Expedition voyage of the *Meteor*, a converted 66 m gunboat, that made transects of the South Atlantic Ocean using acoustic echo sounders, modern sampling bottles, bottom corers, current meters, deep-sea anchoring, and meteorological kites and balloons. Unlike the random cruise tracks of most earlier expeditions, the *Meteor* worked on precise grid tracklines. Ocean currents, temperatures, and salinities, and bathymetry of the Mid-Atlantic Ridge were mapped with great accuracy.

After World War II interest in oceanography and the marine sciences increased dramatically. As echo sounding after World War I was a milestone in oceanography, the advent of electronic navigation in the 1950s was another. Loran (LOng RAnge Navigation), a hyperbolic system using two radio stations transmitting simultaneously, provided ships with continuous position fixing. Universities and government agencies both embarked on new marine investigations, the former concentrating on basic science and the latter on applied science. In addition, commercial interests were becoming active in resource exploration. The ships employed were mostly, ex-wartime vessels: tugs, minesweepers, patrol boats, salvage vessels, etc. A 1950 survey of active research vessels showed 155 ships operated by 34 nations.

The first International Geophysical Year, 1957–58, brought together research ships of many nations working on cooperative projects. This decade also saw the formation of international bodies including: the International Oceanographic Congress; Intergovernmental Oceanographic Commission (IOC); the Special Committee on Oceanic Research (SCOR) of the International Council of Scientific Unions, all of which added to the growing pace of oceanic investigations.

The decade of the 1960s is often referred to as the 'golden years of oceanography.' Both public and scientific awareness of the oceans increased at an unprecedented rate. Funding for marine science both by government and private sources was generous, and the numbers of scientists followed suit. New shipboard instrumentation included the Chlorinity–Temperature–Depth sounder which replaced the old method of lowering water bottles and reversing thermometers. Computers were available for shipboard data processing. Advances in instrumentation and the new projects they generated were making the existing research ships obsolete; and the need for new and more capable ships became a pressing

issue. As a result, shipbuilding programs were started by most of the larger nations heavily involved in oceanographic research. One of the most ambitious was the construction of fourteen AGOR-3 Class ships by the US Navy. These ships, especially designed for research and surveys, were 70 m in length, 1370 tonnes, and incorporated quiet ship operation, centerwells, multiple echo sounders, and a full array of scientific instrumentation. Of the fourteen in the class, 11 were retained in the US and three were transferred to other nations. During this period, the Soviet Union also embarked on a major building program which resulted by the mid-1970s in probably the world's largest fleet – both by vessel size and numbers. In 1979 there was an estimated total of 720 research vessels being operated by 72 nations, the USSR (194 ships), USA (115), and Japan (94) being the leading three.

By the mid-1980s, the shipbuilding boom which started in the late 1950s had dwindled, but many of those vessels themselves were becoming obsolete. New ships were planned to meet the growing needs of shipboard investigators. This resulted in larger ships with improved maneuverability, seakeeping, and data-acquisition capabilities. The new ships built to meet these requirements plus improvements to selected older vessels constitute today's oceanographic research vessels. The worldwide fleet is now smaller in terms of numbers than 25 years ago, but the overall tonnage is greater and the capability vastly superior.

The Nature of Research Vessels

The term 'oceanographic research vessel' is relatively new; earlier ships with limited roles were 'hydrographic survey vessels' or 'fisheries vessels.' As marine science evolved to include biological, chemical, geological, and physical processes of the ocean and its floor and the air–sea interface above, the term 'oceanography' and 'oceanographic research vessel' has come to include all of these disciplines.

When research vessels became larger and more numerous it was inevitable that regulations governing their construction and operation would come into force. Traditionally, ships were either commercial, warships, or yachts. Research ships with scientific personnel fit into none of these, and it became necessary to recognize the uniqueness of such ships in order to preclude burdensome and inapplicable laws. Most nations have now established a definition of an oceanographic research vessel. The United Nations International Maritime Organization (IMO) has established a category of 'Special Purpose Ship' which includes 'ships engaged in research, expeditions, and survey'. Scientific personnel are defined as '... all persons who are not passengers or members of the crew ... who are carried on board in connection with the special purpose ...'

United States law is more specific; it states

> The term oceanographic research vessel means a vessel that the Secretary finds is being employed only in instruction in oceanography or limnology, or both, or only in oceanographic or limnological research, including those studies about the sea such as seismic, gravity meter, and magnetic exploration and other marine geophysical or geological surveys, atmospheric research, and biological research.

The same law defines scientific personnel as those persons who are aboard an oceanographic research vessel solely for the purpose of engaging in scientific research, or instructing or receiving instruction in oceanography, and shall not be considered seamen.

The specific purposes of research vessels include: hydrographic survey (mapping and charting); geophysical or seismic survey; fisheries; general purpose (multidiscipline); and support vessels. Despite their differences, there are commonalities that distinguish a research vessel from other ships. These are defined in the science mission requirements which set forth the operational capabilities, working environment, science accommodations and outfit to meet the science role for which the ship type is intended. The science mission requirements are the dominant factors governing the planning for a new vessel or the conversion of an existing one. The requirements can vary according to the size, area of operations, and type of service, but the composition of the requirements is a product of long usage.

A typical set of scientific mission requirements for a large high-endurance general purpose oceanographic research ship is as follows.

1. General: the ship is to serve as a large general purpose multidiscipline oceanographic research vessel. The primary requirement is for a high endurance ship capable of worldwide cruising (except in close pack ice) and able to provide both overside and laboratory work to proceed in high sea states. Other general requirements are flexibility, vibration and noise free, cleanliness, and economy of operation and construction.

2. Size: size is ultimately determined by requirements which probably will result in a vessel larger than existing ships. However, the length-over-all (LOA) should not exceed 100 m.

3. Endurance: sixty days; providing the ability to transit to remote areas and work 3–4 weeks on station; 15 000 mile range at cruising speed.

4. Accommodations: 30–35 scientific personnel in two-person staterooms; expandable to 40 through the use of vans. These should be a science-library lounge with conference capability and a science office.

5. Speed: 15 knots cruising; sustainable through sea state 4 (1.25–2.5 m); speed control ± 0.1 knot in the 0–6 knot range, and ± 0.2 knot in the range of 6–15 knots.[1]

6. Seakeeping: the ship should be able to maintain science operations in the following speeds and sea states:

15 knots cruising through sea state 4 (1.25–2.5 m);
13 knots cruising through sea state 5 (2.5–4 m);
8 knots cruising through sea state 6 (4–6 m);
6 knots cruising through sea state 7 (6–9 m).

7. Station keeping: the ship should be able to maintain science operations and work in sea states through 5, with limited work in sea state 7. There should be dynamic positioning, both relative and absolute, at best heading in 35-knot wind, sea state 5, and 3-knot current in depths to 6000 m, using satellite and/or bottom transponders; $\pm 5°$ heading and 50 m maximum excursion. It should be able to maintain a precision trackline (including towing) at speeds as low as 2 knots with a 45° maximum heading deviation from the trackline under controlled conditions (satellite or acoustic navigation) in depths to 6000 m, in 35 knot wind; and 3-knot current. Speed control along track should be within 0.1 knot with 50 m of maximum excursion from the trackline.

8. Ice strengthening: ice classification sufficient to transit loose pack ice. It is not intended for ice-breaking or close pack work.

9. Deck working area: spacious stern quarter area – 300 m² minimum with contiguous work area along one side 4×15 m minimum. There should be deck loading up to 7000 kg m^{-2} and there should be overside holddowns on 0.5 m centers. The area should be highly flexible to accommodate large, heavy, and portable equipment, with a dry working deck but not more than 2–3 m above the waterline. There should be a usable clear foredeck area to accommodate specialized towers and booms extending beyond the bow wave. All working decks should be accessible for power, water, air, and data and voice communication ports.

10. Cranes: a suite of modern cranes:
(a) to reach all working deck areas and offload vans and heavy equipment up to 9000 kg;

---

[1] 1 (Nautical) mile $= 1.853$ km, 1 knot $= 1.853 \text{ km h}^{-1} = 0.515 \text{ ms}^{-1}$.

(b) articulated to work close to deck and water surface;

(c) to handle overside loads up to 2500 kg, 10 m from side and up to 4500 kg closer to side;

(d) overside cranes to have servo controls and motion compensation;

(e) usable as overside cable fairleads at sea.

The ship should be capable of carrying portable cranes for specialized purposes such as deploying and towing scanning sonars, photo and video devices, remotely operated vehicles (ROVs), and paravaned seismic air gun arrays.

11. Winches: oceanographic winch systems with fine control (0.5 m min^{-1}; constant tensioning and constant parameter; wire monitoring systems with inputs to laboratory panels and shipboard data systems; local and remote controls including laboratory auto control.

Permanently installed general purpose winches should include:

- two winches capable of handling 10 000 m of wire rope or electromechanical conducting cables having diameters from 0.6 mm to 1.0 cm.

- a winch complex capable of handling 12 000 m of 1.5 cm trawling or coring wire, and 10 000 m of 1.75 cm electromechanical conducting cable (up to 10 kVA power transmission and fiberoptics); this can be two separate winches or one winch with two storage drums.

Additional special purpose winches may be installed temporarily at various locations along working decks. Winch sizes may range up to 40 mtons and have power demands up to 250 kW. (See also multichannel seismics.) Winch control station(s) should be located for optimum operator visibility with communications to laboratories and ship control stations.

12. Overside handling: various frames and other handling gear able to accommodate wire, cable, and free-launched arrays; matched to work with winch and crane locations but able to be relocated as necessary. The stern A-frame must have 6 m minimum horizontal and 10 m vertical clearances, 5 m inboard and outboard reaches, and safe static working load up to 60 mtons. It must be able to handle, deploy and retrieve very long, large-diameter piston cores up to 50 m length, 15 mtons weight and 60 mtons pullout tension. There should be provision to carry additional overside handling rigs along working decks from bow to stern. (See also multichannel seismics).

13. Towing: capable of towing large scientific packages at 4500 kg tension at 6 knots, and 10 000 kg at 2.5 knots in sea state 5; 35 knots of wind and 3 knot current.

14. Laboratories: approximately 400 m^2 of laboratory space including: main lab (200 m^2) flexible for subdivision providing smaller specialized labs; hydro lab (30 m^2) and wet lab (40 m^2) both located contiguous to sampling areas; bio-chem analytical lab (30 m^2); electronics/computer lab and associated users space (60 m^2); darkroom (10 m^2); climate-controlled chamber (15 m^2); and freezer(s) (15 m^2).

Labs should be arranged so that none serve as general passageways. Access between labs should be convenient. Labs, offices, and storage should be served by a man-rated lift having clear inside dimensions not less than 3×4 m.

Labs should be fabricated of uncontaminated and 'clean' materials. Furnishings, doors, hatches, ventilation, cable runs, and fittings should be planned for maximum lab cleanliness. Fume hoods should be installed permanently in wet and analytical labs. Main lab should have provision for temporary installation of fume hoods.

Cabinetry should be of high-grade laboratory quality with flexibility for arrangements through the use of bulkhead, deck and overhead holddown fittings.

Heating, ventilation, and air conditioning (HVAC) should be appropriate to labs, vans and other science spaces being served. Laboratories should be able to maintain a temperature of 20–23°C, 50% relative humidity, and 9–11 air changes per hour. Filtered air should be provided to the analytical lab. Each lab should have a separate electric circuit on a clean bus and continuous delivery capability of at least 250 VA m^{-2} of lab area. Total estimated laboratory power demand is 100 kVA. There should be an uncontaminated seawater supply to most laboratories, vans, and several key deck areas.

15. Vans: carry four standardized 2.5×6 m portable vans which may have laboratory, berthing, storage, or other specialized use. With hook-up provision for power, HVAC, fresh water, uncontaminated sea water, compressed air, drains, communications, data and shipboard monitoring systems. There should be direct van access to ship interior at key locations. There should be provision to carry up to four additional vans on working and upper decks with supporting connections at several locations. The ship should be capable of loading and off-loading vans with its own cranes.

16. Workboats: at least one and preferably two inflatable boats located for ease of launching and recovery. There should be a scientific workboat 8–10 m LOA fitted out for supplemental operation at sea including collecting, instrumentation and wide-angle signal measurement. It should have 12 h

endurance, with both manned and automated operation, be of 'clean' construction and carried in place of one of the four van options above.

17. Science storage: total of $600\,m^3$ of scientific storage accessible to labs by lift and weatherdeck hatch(es). Half should have suitable shelving, racks and tiedowns, and the remainder open hold. A significant portion of storage should be in close proximity to science spaces (preferably on the same deck).

18. Acoustical systems: ship to be as acoustically quiet as practicable in the choice of all shipboard systems and their installation. The design target is operationally quiet noise levels at 12 knots cruising in sea state 5 at the following frequency ranges:

- $4\,Hz$–$500\,Hz$ seismic bottom profiling
- $3\,kHz$–$500\,kHz$ echo sounding and acoustic navigation
- $75\,kHz$–$300\,kHz$ Doppler current profiling

The ship should have $12\,kHz$ precision and $3.5\,kHz$ sub-bottom echo sounding systems, and provision for additional systems. There should be a phased array, wide multibeam precision echo sounding system; transducer wells ($0.5\,m$ diam) one located forward and two midships; pressurized sea chest ($1.5 \times 3\,m$) located at the optimum acoustic location for afloat installation and servicing of transducers and transponders.

19. Multichannel seismics: all vessels shall have the capability to carry out multichannel seismic profiling (MCS) surveys using large source arrays and long streamers. Selected vessels should carry an MCS system equivalent to current exploration industry standards.

20. Navigation/positioning: There shall be a Global Positioning System (GPS) with appropriate interfaces to data systems and ship control processors; a short baseline acoustic navigation system; a dynamic positioning system with both absolute and relative positioning parameters.

21. Internal communications: system to provide high quality voice communications throughout all science spaces and working areas. Data transmission, monitoring and recording systems should be available throughout science spaces including vans and key working areas. There should be closed circuit television monitoring and recording of all working areas including subsurface performance of equipment and its handling. Monitors for all ship control, environmental parameters, science and overside equipment performance should be available in most science spaces.

22. Exterior communications: reliable voice channels for continuous communications to shore stations (including home laboratories), other ships, boats and aircraft. This includes satellite, VHF and UHF. There should be: facsimile communications to transmit high-speed graphics and hard copy on regular schedules; high-speed data communications links to shore labs and other ships on a continuous basis.

23. Satellite monitoring: transponding and receiving equipment including antennae to interrogate and receive satellite readouts of environmental remote sensing.

24. Ship control: the chief requirement is maximum visibility of deck work areas during science operations and especially during deployment and retrieval of equipment. This would envision a bridge-pilot house very nearly amidships with unobstructed stem visibility. The functions, communications, and layout of the ship control station should be designed to enhance the interaction of ship and science operations; ship course, speed, attitude, and positioning will often be integrated with science work requiring control to be exercised from a laboratory area.

These science mission requirements are typical of a large general-purpose research vessel. Requirements for smaller vessels and specialized research ships can be expected to differ according to the intended capability and service.

Design Characteristics of Oceanographic Vessels

In general the design of an oceanographic ship is driven by the science mission requirements described above. In any statement of requirements an ordering of priorities is important for the guidance of the design and construction of the ship. In the case of research vessels the following factors have been ranked by groups of practicing investigators from all disciplines.

1. Seakeeping: station keeping
2. Work environment: lab spaces and arrangements; deck working area; overside handling (winches and wire); flexibility
3. Endurance: range; days at sea
4. Science complement
5. Operating economy
6. Acoustical characteristics
7. Speed: ship control
8. Pay load: science storage; weight handling

These priorities are not necessarily rank ordered although there is general agreement among oceanographers that seakeeping, particularly on station, and work environment are the two top priorities. The remaining are ranked so closely together that they are of equal importance. The science mission requirements set for each of these areas become threshold levels, and any characteristic which falls below the threshold becomes a dominant priority.

General Purpose Vessels

Ships of this type (also termed multidiscipline) constitute the classic oceanographic research vessels and are the dominant class in terms of numbers today. They have outfitting and laboratories to support any of the physical, chemical, biological, and geological ocean science studies plus ocean engineering. The science mission requirements given above describe a large general purpose ship. Smaller vessels can be expected to have commensurately reduced requirements.

Current and future multidiscipline oceanographic ships are characterized as requiring significant open deck area and laboratory space. Accommodations for scientific personnel are greater than for single purpose vessels due to the larger science parties carried. Flexibility is an essential feature in a general purpose research vessel. A biological cruise may be followed by geology investigations which can require the reconfiguration of laboratory and deck equipment within a short space of time.

In addition to larger scientific complements, the complexity and size of instrumentation now being deployed at sea has increased dramatically over the past half century. As a result, the size of general purpose vessels has increased significantly. A research vessel of 60 m in length was considered to be large in 1950; the same consideration today has grown to 100 m and new vessels are being built to that standard. Even existing vessels have been lengthened to meet the growing needs.

The majority of oceanographic research vessels, however, are smaller vessels, 25–50 m in length, and limited to coastal service.

Mapping and Charting Vessels

Mapping and charting ships were probably the earliest oceanographic vessels, usually in conjunction with an exploration voyage. Incident to the establishment of marine trade routes, nautical charting of coastal regions became routine and the vessels so engaged were usually termed hydrographic survey ships. Surveys were (and still are) carried out using wire sounding, drags, and launches. Survey vessels are characterized by the number of boats and launches carried and less deck working space than general purpose vessels. Modern survey vessels, however, are often expected to carry out other scientific disciplines, and winches, cranes and frames can be observed on these ships.

Recent developments have affected the role (and therefore design) of this class of vessel. As a result of the International Law of the Sea Conferences, coastal states began to exercise control over their continental shelves and economic zones 200 miles (321.8 km) from shore. This brought about interest in the resources (fishery, bottom and sub-bottom) of these newly acquired areas, and research vessels were tasked to explore and map these resources. The usual nautical charting procedures are not applicable in the open ocean regions, and modern electronic echo-sounding instruments have supplanted the older wire measurements. This involves large hull-mounted arrays of acoustic projectors and hydrophones which can map a swath of ocean floor with great precision up to five miles in width, and at cruising speeds. The design of vessels to carry this equipment requires a hull form to optimize acoustic transmission and reception, and to minimize hull noise from propulsion and auxiliary machinery. Further, such new ships also may be outfitted to perform other oceanographic tasks incident to surveys. Their appearance, therefore, may come closer to a general purpose vessel.

Fisheries Research Vessels

Fisheries research generally includes three fields of study: (1) environmental investigations, (2) stock assessment, and (3) gear testing and development. The first of these are surveys and analyses of sea surface and water column parameters; both synoptic and serial. These are biological, physical, and chemical investigations (as well as geological if bottom fisheries are considered); and can be accomplished from a general purpose oceanographic research vessel.

Ships engaged in fish stock assessment and exploratory fishing, or development work in fishing methods and gear, fish handling, processing, and preservation of fish quality on board, are specialized types of vessels closely related to actual fishing vessels. Design characteristics include a stern ramp and long fish deck for bringing nets aboard, trawling winches, and wet labs for analyses of fish sampling. Newer designs also include instrumentation and laboratories for environmental investigations, and extensive electronic instrumentation for acoustic fish finding, biomass evaluation, and fish identification and population count.

As with mapping and charting ships, most fishery research vessels are operated by government agencies.

Geophysical Research Vessels

The purpose of marine geophysical research vessels is to investigate the sea floor and sub-bottom, oceanic crust, margins, and lithosphere. The demanding design aspect for these ships is the requirement for a MCS system used to profile the deep geologic structure beneath the seafloor. The missions range from basic research of the Earth's crust (plate tectonics) to resources exploration.

The primary components of an MCS system are the large air compressors needed to 'fire' multiple towed airgun sound source arrays and a long towed hydrophone streamer which may reach up to 10 km in length. The supporting outfit for the handling and deployment of the system includes large reels and winches for the streamer, and paravanes to spread the sound source arrays athwart the ship's track. This latter results in the need for a large stern working deck close to the water with tracked guide rails and swingout booms. Electronic and mechanical workshops are located close to the working deck. The design incorporates a large electronics room for processing the reflected signals from the hydrophones and integrating the imagery with magnetics, gravity and navigation data.

The highly specialized design requirements for a full-scale marine geophysics ship usually precludes work in other oceanographic disciplines. On the other hand, large general purpose research ships often carry compressors and portable streamer reels sufficient for limited seismic profiling.

Polar Research Vessels

Whereas most oceanographic research vessels are classed by the discipline in which they engage, polar research vessels are defined by their area of operations. Earlier terminology distinguished between polar research vessels and icebreakers with the former having limited icebreaking capability, and the latter with limited or no research capability. The more current trend is to combine full research capability into new icebreaker construction.

Arctic and Antarctic research ships in the nineteenth and early twentieth centuries were primarily ice reinforced sealing vessels with little or no icebreaking capability. World War II and subsequent Arctic logistics, and the International Antarctic Treaty (1959) brought about increased interest in polar regions which was furthered by petroleum exploration in the Arctic in the late twentieth century. Icebreakers with limited research capability early in this period became full-fledged research vessels by the end of the century.

The special requirements defining a polar research vessel include increased endurance, usually set at 90 days, helicopter support, special provisions for cold weather work, such as enclosed winch rooms and heated decks, and icebreaking capability. Other science mission requirements continue the same as for a large general purpose RVs. Of special concern is seakeeping in open seas. Past icebreaker hull shapes necessarily resulted in notoriously poor seakindliness. Newer designs employing ice reamers into the hull form also offer improved seakeeping.

Ice capability is usually defined as the ability to break a given thickness of level ice at 3 knots continuous speed, and transit ice ridges by ramming. Current requirements for polar research vessels have varied from 0.75 m to 1.25 m ice thickness in the continuous mode, and 2.0 to 3.0 m ridge heights in the ramming mode. These correspond to Polar Class 10 of Det Norske Veritas or Ice Class A3–A4 of the American Bureau of Shipping.

Support Vessels

Ships that carry, house, maintain, launch and retrieve other platforms and vehicles have evolved into a class worthy of note. These include vessels that support submersibles, ROVs, buoys, underwater habitats, and scientific diving. Earlier ships of this class were mostly converted merchant or fishing vessels whose only function was to launch and retrieve and supply hotel services. Recent vessels, especially those dedicated to major programs such as submersible support, are large ships and fully outfitted for general purpose work.

Other Classes of Oceanographic Research Vessels

In addition to the above types, there are research ships which serve other purposes. These include ocean drilling and geotechnical ships, weather ships, underwater archaeology, and training and education vessels. The total number of these ships is relatively small, and many of them merge in and out of the category and serve for a limited stretch of time.

Often ships will take on identification as a research vessel for commercial expediency or other fashion not truly related to oceanographic research. Such roles may include treasure hunting, salvage, whale watching, recreational diving, ecology tours, etc. These vessels may increase the popular awareness of oceanography but are not bona fide oceanographic research vessels.

Research Vessel Operations

Oceanographic cruises are usually the culmination of several years of scientific and logistics planning. Coastal vessels, typically 25–50 m length, will usually remain in a home, or adjacent regions on cruises of 1–3 weeks' duration. Ocean-going vessels, 50–100 m length, may undertake voyages of 1–2 years away from the home port, with cruise segments of 25–35 days working out of ports of opportunity. New scientific parties may join the ship at a port call and the nature of the following cruise leg can change from, for instance biological to physical oceanography. This involves complex logistics and careful planning and coordination, within a typical 4–5 day turnaround.

From the time of the *Meteor* expedition of 1925, cruise plans are usually highly systematic with work concentrating along preset track lines and grids, or confined to a small area of intense investigation. Work can take place continuously while underway using hull-mounted and/or towed instrumentation, or the ship will stop at a station and lower instruments for water column or bottom sampling. Typical stations are at 15–60-mile intervals and can last 1–4 h. Measurements or observations that are commonly made are shown in **Table 1**. In addition, work may include towed vehicles along a precise trackline at very slow controlled speeds making many of the observations, chiefly acoustic, photographic, and video.

Cooperative projects among research vessels including different nations have become common-place. These share a common scientific goal, and cruise tracks, times, methodology, data reduction, and archiving are assigned by joint planning groups.

A significant factor affecting oceanographic research cruises today is the permission required by a research vessel to operate in another nation's 200-mile economic zone. As a result of the United Nations Law of the Sea Conventions and the treaties resulting therefrom (1958–1982), coastal nations were given jurisdiction over the conduct of marine scientific research extending 200 nautical miles out from their coast (including island possessions). This area is termed exclusive economic zone (EEZ). As of 2000 there were approximately 151 coastal states (this number varies according to the world political makeup), and 36% of the world ocean falls within their economic zones. Most coastal nations have prescribed laws governing research in their zones. The rules include requests for permission, observers, port calls, sharing data, and penalties. This often poses a burden on the operator of a research vessel when permission requirements are arbitrary and untimely. It is not, however, an unworkable burden if done in an orderly manner, and does have desirable features for international cooperation. Problems arise when requests are not submitted within the time specified, or resulting data are not forthcoming. These are complicated by unrealistic requirements on the part of the coastal state, delay in acknowledging or acting on a request, or ignoring it totally. These can have a profound effect on scientific research and need to be addressed in future Law of the Sea Conventions.

World Oceanographic Fleet

The precise number of oceanographic research vessels worldwide is difficult to ascertain. Few nations maintain lists specific to research vessels, and numbers are available chiefly by declarations on the part of the operator. Some ships move in and out of a research status from another classification, e.g., fishing, passenger, yacht, etc. Also, some operators keep hydrographic survey, seismic exploration, and even fisheries research ships as categories separate from oceanographic research; here they are included within the general heading of oceanographic research vessels.

Based on the best available information, 48 nations or international agencies operated 420 oceanographic research vessels of size greater than 25 m LOA in 2000. Of these, 310 ships were from nine nations operating 10 or more ships each (**Table 2**).

A significant step in international cooperation affecting oceanographic research vessels and the

Table 1 Common measurements and observations made from RVs

Underway	On station
Single channel echo sounding	Echo sounding
Multichannel echo sounding	Sub-bottom profiling
Sub-bottom profiling (3.5 kHz)	Acoustic Doppler profiling
Acoustic Doppler profiling	Surface to any depth
Sea surface:	Temperature
Temperature	Salinity
Salinity	Sound velocity
Fluorometry	Dissolved oxygen
Dissolved oxygen	Water sample
Towed magnetometer	Bottom sampling
Gravimeter	Bottom coring
Meteorological	Bottom photography and video
Wind speed and direction	Bottom dredging
Barometric pressure	Geothermal bottom probe
Humidity	Biological net tows and trawls
Solar radiation	Biological net tows
Towed plankton recorder	

Table 2 Countries operating 10 or more oceanographic research ships

Russia	86
United States	84
Japan	66
China	17
Ukraine	14
Korea	11
Germany	11
United Kingdom	10
Canada	10
All others (39)	111
Total	420

marine sciences they support has been the International Ship Operators Meeting, an intergovernmental association, founded in 1986, comprising representatives from various ship operating agencies that meets periodically to exchange information on ship operations and schedules, and work on common problems affecting research vessels. In 1999, 21 ship-operating nations were represented and extended membership by other states is ongoing.

Future Oceanographic Ships

The interest in, and growth of, marine science over the past half century shows little or no indication of diminishing. The trend in oceanographic investigations has been to carry larger and more complex instrumentation to sea. The size and capability of research vessels in support of developing projects has also increased.

Future oceanographic research ships can be expected to become somewhat larger than their counterparts today. This will result from demands for more sizeable scientific complements and laboratory spaces. Workdeck and shops will be needed for larger equipment systems such as buoy arrays, bottom stations, towed and autonomous vehicles. Larger overside handling systems incorporating motion compensation will make demands for more deck space.

There will be fewer differences between basic science, fisheries, and hydrographic surveying vessels so that one vessel can serve several purposes. This may result in fewer vessels, but overall tonnage and capacity can be expected to increase.

New types of craft may take a place alongside conventional ships. These include submarines, 'flip'-type vessels which transit horizontally and flip vertically on station, and small waterplane-area twin hull ships (SWATH). SWATH, or semisubmerged ships, are a relatively recent development in ship design. Although patents employing this concept show up in 1905, 1932, and 1946, it was not until 1972 that the US Navy built a 28 m, 220 ton prototype model. The principle of a SWATH ship is that submerged hulls do not follow surface wave motion, and thin struts supporting an above water platform which have a small cross-section (waterplane) are nearly transparent to surface waves, and have longer natural periods and reduced buoyancy force changes than a conventional hull. The result is that SWATH ships, both in theory and performance, demonstrate a remarkably stable environment and platform configuration which is highly attractive for science and engineering operations at sea.

See also

Coastal Zone Management. Fishing Methods and Fishing Fleets. International Organizations. Law of the Sea. Maritime Archaeology. Shipping and Ports.

SILICA

See **MARINE SILICA CYCLE**

SINGLE COMPOUND RADIOCARBON MEASUREMENTS

T. I. Eglinton and A. Pearson, Woods Hole Oceanographic Institution, Woods Hole, MA, USA

doi:10.1006/rwos.2001.0171

Introduction

Many areas of scientific research use radiocarbon (carbon-14, ^{14}C) measurements to determine the age of carbon-containing materials. Radiocarbon's ~5700-year half-life means that this naturally occurring radioisotope can provide information over decadal to millennial timescales. Radiocarbon is uniquely suited to biogeochemical studies, where much research is focused on carbon cycling at various spatial and temporal scales. In oceanography, investigators use the ^{14}C concentration of dissolved inorganic carbon (DIC) to monitor the movement of water masses throughout the global ocean. In marine sediment geochemistry, a major application is the dating of total organic carbon (TOC) in order to calculate sediment accumulation rates. Such chronologies frequently rely on the premise that most of the TOC derives from marine biomass production in the overlying water column.

However, the ^{14}C content of TOC in sediments, as well as other organic pools in the ocean (dissolved and particulate organic matter in the water column) often does not reflect a single input source. Multiple components with different respective ages can contribute to these pools and can be deposited concurrently in marine sediments (**Figure 1**). This is particularly true on the continental margins, where fresh vascular plant debris, soil organic matter, and fossil carbon eroded from sedimentary rocks can contribute a significant or even the dominant frac-

tion of the TOC. This material dilutes the marine input and obscures the true age of the sediment. Although such contributions from multiple organic carbon sources can complicate the development of TOC-based sediment chronologies, these sediment records hold much important information concerning the cycling of organic carbon both within and between terrestrial and marine systems. The challenge, then, is to decipher these different inputs by resolving them into their individual parts.

Most of the allochthonous, or foreign, sources represent carbon with lower ^{14}C concentrations ('older' radiocarbon ages) than the fraction of TOC originating from phytoplanktonic production. The only exception is the rapid transport and sedimentation of recently synthesized terrestrial plant material, which is in equilibrium with the ^{14}C concentration of atmospheric CO_2. Other sources of nonmarine carbon typically are of intermediate (10^3–10^4 years) or 'infinite' (beyond the detection limit of 50–60 000 years) radiocarbon age, depending on the amount of time spent in other reservoirs such as soils, fluvial deposits, or carbon-rich rocks.

It is only at the molecular level that the full extent of this isotopic heterogeneity resulting from these

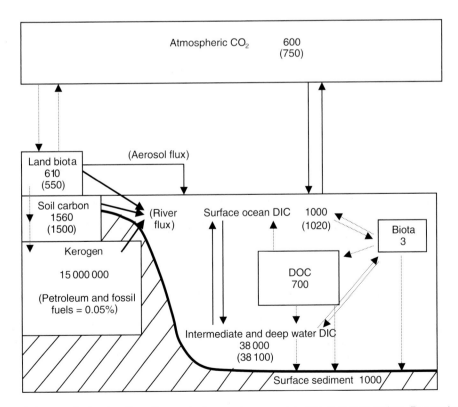

Figure 1 Major global reservoirs involved in active production, exchange and cycling of organic carbon. Reservoir sizes are shown in Gt carbon (1 GtC = 10^{15} g C). Numbers in parentheses are based on 1980s values; numbers without parentheses are estimates of the pre-anthropogenic values. Fluxes primarily mediated by biological reactions are shown with dashed arrows; physical transport processes are shown with solid arrows. (Modified after Siegenthaler and Sarmiento (1993) and Hedges and Oades (1997).)

diverse organic carbon inputs is expressed. Isotopic analysis of individual biomarker compounds was employed originally to study the stable carbon isotope (^{13}C) distribution in lipids of geological samples. It proved to be a useful tool to describe the diversity of carbon sources and metabolic pathways as well as to link specific compounds with their biological origins. Recently, this approach was expanded into a second isotopic dimension by the development of a practical method to achieve compound-specific ^{14}C analysis. Not only do these new ^{14}C analyses of individual biomarker molecules provide a tool for dating sediments, but they are another source of fundamental information about biogeochemical processes in the marine environment.

Carbon Isotopes

Carbon in the geosphere is composed of the stable isotopes ^{12}C (98.9%) and ^{13}C (1.1%), and the cosmogenic radionuclide, ^{14}C (radiocarbon). Upon production, ^{14}C is incorporated quickly into atmospheric CO_2, where it occurs as approximately 10^{-10}% of the total atmospheric abundance of CO_2. The distribution of the minor isotopes relative to ^{12}C is governed by thermodynamic and kinetic fractionation processes[1], in addition to the radioactive decay associated with ^{14}C.

^{14}C Systematics

Today, most radiocarbon data are obtained through the use of accelerator mass spectrometry (AMS) rather than by counting individual decay events. In particular, the advantage of AMS is its small carbon requirement (micrograms to milligrams); this ability to analyze small samples is critical to the compound-specific ^{14}C approach, where sample sizes typically range from tens to hundreds of micrograms. These sample sizes are dictated by natural concentrations of the analytes in geochemical samples (often $< 1\,\mu g\,g^{-1}$ dry sediment), and by the capacity of the techniques used to isolate the individual compounds in high purity.

Raw AMS data are reported initially as fraction modern (f_m) carbon (eqn [1]).

$$f_m = \frac{R_{sn}^{14/12}}{R_{std}^{14/12}} \qquad [1]$$

---

[1] This article assumes the reader is familiar with the conventions used for reporting stable carbon isotopic ratios, i.e., δ^{13}C (ppt) $= 1000[(R/R_{PDB}) - 1]$ where $R \equiv ^{13}C/^{12}C$. For further explanation, see the additional readings listed at the end of this article.

$R^{14/12} \equiv ^{14}C/^{12}C$ (some laboratories use $R = ^{14}C/^{13}C$), sn indicates the sample has been normalized to a constant ^{13}C fractionation equivalent to $\delta^{13}C = -25$ppt, and std is the oxalic acid I (HOxI) or II (HOxII) modern-age standard, again normalized with respect to ^{13}C.

For geochemical applications, data often are reported as Δ^{14}C values (eqn [2]).

$$\Delta^{14}C = \left[f_m \left(\frac{e^{\lambda(y-x)}}{e^{\lambda(y-1950)}} \right) - 1 \right] \times 1000 \qquad [2]$$

Here, $\lambda = 1/8267(y^{-1})(= t_{1/2}/\ln 2)$, y equals the year of measurement, and x equals the year of sample formation or deposition (applied only when known by independent dating methods, for example, by the use of ^{210}Pb). This equation standardizes all Δ^{14}C values relative to the year AD 1950. In oceanography, Δ^{14}C is a convenient parameter because it is linear and can be used in isotopic mass balance calculations of the type shown in eqn [3].

$$\Delta^{14}C_{total} = \sum_i (\chi_i \Delta^{14}C_i) \qquad \sum_i \chi_i = 1 \qquad [3]$$

The 'radiocarbon age' of a sample is defined strictly as the age calculated using the Libby half-life of 5568 years (eqn [4]).

$$\text{Age} = -8033\,ln(f_m) \qquad [4]$$

For applications in which a calendar date is required, the calculated ages subsequently are converted using calibration curves that account for past natural variations in the rate of formation of ^{14}C. However, the true half-life of ^{14}C is 5730 years, and this true value should be used when making decay-related corrections in geochemical systems.

^{14}C Distribution in the Geosphere

Natural Processes

Atmospheric $^{14}CO_2$ is distributed rapidly throughout the terrestrial biosphere, and living plants and their heterotrophic consumers (animals) are in equilibrium with the Δ^{14}C value of the atmosphere. Thus, in radiocarbon dating, the ^{14}C concentration of a sample is strictly an indicator of the amount of time that has passed since the death of the terrestrial primary producer. When an organism assimilates a fraction of pre-aged carbon, an appropriate 'reservoir age' must be subtracted to correct for the deviation of this material from the age of the atmosphere. Therefore, reservoir time must be considered

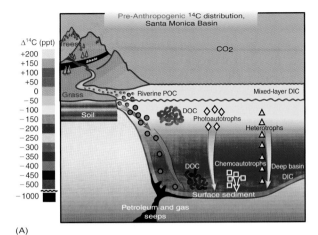

(A)

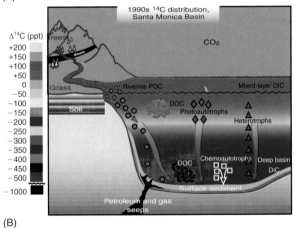

(B)

Figure 2 $\Delta^{14}C$ values for bulk carbon reservoirs in the region of Santa Monica Basin, California, USA: (A) prior to human influence ('pre-bomb'), and (B) contemporary values ('post-bomb'). Modified after Pearson (2000).)

when interpreting the ^{14}C 'ages' of all of the global organic carbon pools other than the land biota.

For example, continuous vertical mixing of the ocean provides the surface waters with some abyssal DIC that has been removed from contact with atmospheric CO_2 for up to 1500 years. This process gives the ocean an average surface water reservoir age of about 400 years ($\Delta^{14}C = -50\,$ppt). A constant correction factor of 400 years often is subtracted from the radiocarbon dates of marine materials (both organic and inorganic). There are regional differences, however, and in upwelling areas the true deviation can approach 1300 years.

An example of the actual range of $\Delta^{14}C$ values found in the natural environment is shown in **Figure 2A**. This figure shows the distribution of ^{14}C in and around Santa Monica Basin, California, USA, prior to significant human influence. The basin sediments are the final burial location for organic matter derived from many of these sources, and

the TOC $\Delta^{14}C$ value of $-160\,$ppt represents a weighted average of the total organic carbon flux to the sediment surface.

Anthropogenic Perturbation

In addition to natural variations in atmospheric levels of ^{14}C, anthropogenic activity has resulted in significant fluctuations in ^{14}C content. The utilization of fossil fuels since the late nineteenth century has introduced ^{14}C- (and ^{13}C-) depleted CO_2 into the atmosphere (the 'Suess effect'). In sharp contrast to this gradual change, nuclear weapons testing in the 1950s and 1960s resulted in the rapid injection of an additional source of ^{14}C into the environment. The amount of ^{14}C in the atmosphere nearly doubled, and the $\Delta^{14}C$ of tropospheric CO_2 increased to greater than $+900\,$ppt in the early 1960s. Following the above-ground weapons test ban treaty of 1962, this value has been decreasing as the excess ^{14}C is taken up by oceanic and terrestrial sinks for CO_2. This anthropogenically derived $^{14}CO_2$ 'spike' serves as a useful tracer for the rate at which carbon moves through its global cycle. Any carbon reservoir currently having a $\Delta^{14}C$ value $>0\,$ppt has taken up some of this 'bomb-^{14}C'. Carbon pools that exhibit no increase in $\Delta^{14}C$ over their 'pre-bomb' values have therefore been isolated from exchange with atmospheric CO_2 during the last 50 years. This contrast between 'pre-bomb' and 'post-bomb' $\Delta^{14}C$ values can serve as an excellent tracer of biogeochemical processes over short timescales. These changes in ^{14}C concentrations can be seen in the updated picture of the Santa Monica Basin regional environment shown in **Figure 2B**, where bomb-^{14}C has invaded everywhere except for the deep basin waters and older sedimentary deposits.

In general, the global distribution of organic ^{14}C is complicated by these interreservoir mixing and exchange processes. The more end-member sources contributing organic carbon to a sample, the more complicated it is to interpret a measured $\Delta^{14}C$ value, especially when trying to translate that value to chronological time. Source-specific ^{14}C dating is needed, and this requires isotopic measurements at the molecular level.

Compound-specific ^{14}C Analysis: Methods

The ability to perform natural-abundance ^{14}C measurements on individual compounds has only recently been achieved. This capability arose from refinements in the measurement of ^{14}C by AMS that allow increasingly small samples to be measured, and from methods that resolve the complex mixtures encountered in geochemical samples into their individual components. Here we describe the methods that are currently used for this purpose.

Selection of Compounds for ^{14}C Analysis

The organic matter in marine sediments consists of recognizable biochemical constituents of organisms (carbohydrates, proteins, lipids, and nucleic acids) as well as of more complex polymeric materials and nonextractable components (humic substances, kerogen). Among the recognizable biochemicals, the lipids have a diversity of structures, are comparatively easy to analyze by gas chromatographic and mass spectrometric techniques, and are resistant to degradation over time. These characteristics have resulted in a long history of organic geochemical studies aimed at identifying and understanding the origins of 'source-specific' lipid 'biomarker' compounds. Frequently, lipids from several organic compound classes are studied within the same sample (**Figures 3** and **4**).

Although many of the most diagnostic compounds are polar lipids that are susceptible to modification during sediment diagenesis (e.g., removal of functional groups, saturation of double bonds), several retain their marker properties through the preservation of the carbon skeleton (**Figure 5**). Thus sterols (e.g., cholesterol) are transformed to sterenes and ultimately steranes. The isotopic integrity of the compound is also preserved in this way.

It is sometimes the case that families of compounds can also be characteristic of a particular source. For example, plant waxes comprise homologous series of *n*-alkanes, *n*-alkanols, and *n*-alkanoic acids (**Figure 3**). As a result, ^{14}C measurements of a compound class can yield information with similar specificity to single compound ^{14}C analysis, with the benefits of greater total

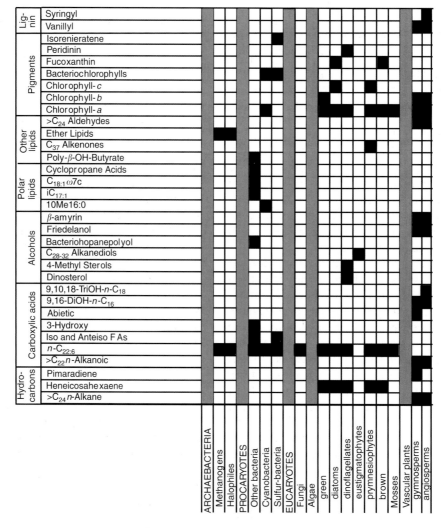

Figure 3 Common source assignments of lipid biomarkers (Modified from Hedges and Oades (1997).)

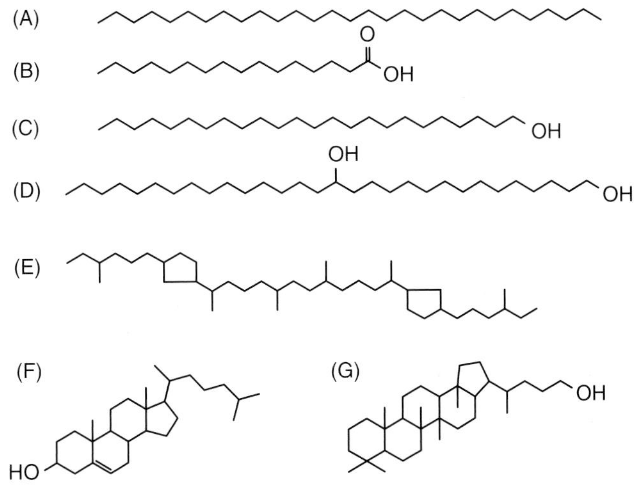

Figure 4 Selected example structures (carbon skeletons and functional groups) for biomarkers shown in Table 1: (A) n-C_{29} alkane; (B) n-$C_{16:0}$ alkanoic acid; (C) n-C_{24} alkanol; (D) C_{30} alkanediol; (E) $C_{40:2cy}$ isoprenoid; (F) C_{27} Δ^5-sterol (cholesterol); (G) C_{32} hopanol.

analyte abundance and, potentially, simpler isolation schemes.

Compound Separation and Isolation

Procedures for single-compound ^{14}C analysis are quite involved, requiring extraction, purification, modification and isolation of the target analytes (**Figure 5**). For lipid analyses, the samples are processed by extracting whole sediment with solvents such as methylene chloride, chloroform, or methanol to obtain a total lipid extract (TLE). The TLE is then separated into compound classes using solid–liquid chromatography. The compound classes elute on the basis of polarity differences, from least polar (hydrocarbons) to most polar (free fatty acids) under normal-phase chromatographic conditions. Individual compounds for ^{14}C analysis are then isolated from these polarity fractions. Additional chromatographic steps or chemical manipulations may be included to reduce the number of components in each fraction prior to single compound isolation, or to render the compounds amenable to isolation by the method chosen. These steps may include silver nitrate-impregnated silica gel chromatography (separation of saturated from unsaturated compounds), 'molecular sieving' (e.g., urea adduction, for separation of branched/cyclic compounds from straight-chain compounds), and derivatization (for protection of functional groups, such as carboxyl or hydroxyl groups, prior to gas chromatographic separation).

For ^{14}C analysis by AMS, tens to hundreds of micrograms of each individual compound must be isolated from the sample of interest. Isolation of individual biomarkers from geochemical samples

such as marine sediments and water column particulate matter requires separation techniques with high resolving power. To date, this has been most effectively achieved through the use of automated preparative capillary gas chromatography (PCGC; **Figure 6**).

A PCGC system consists of a commercial capillary gas chromatograph that is modified for work on a semipreparative, rather than analytical, scale. Modifications include a large-volume injection system; high-capacity, low-bleed 'megabore' (e.g., 60 m length × 0.53 mm inner diameter × 0.5 μm stationary phase film thickness) capillary columns; an effluent splitter; and a preparative trapping device in which isolated compounds are collected in a series of cooled U-tube traps. Approximately 1% of the effluent passes to a flame ionization detector (FID) and the remaining 99% is diverted to the collection system. The traps are programmed to receive compounds of interest on the basis of chromatographic retention time windows determined from the FID trace. Computerized synchronization of the trapping times permits collection of multiple identical runs (often > 100 consecutive injections). Using PCGC, baseline resolution of peaks can be achieved at concentrations > 100-fold higher than typical analytical GC conditions, allowing up to 5 μg of carbon per chromatographic peak, per injection, to be separated (to achieve greater resolution, typical loadings are usually about 1 μg of carbon per peak). An example of a typical PCGC separation is shown in **Figure 7**, where ∼ 40–130 μg of individual sterols (as their acetate derivatives) were resolved and isolated from a total sterol fraction obtained from Santa Monica Basin surface sediment.

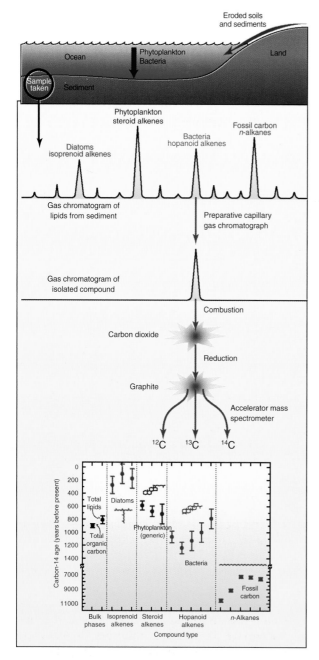

Figure 5 Schematic diagram showing steps for the isolation and ^{14}C analysis of individual sedimentary lipids.

Another practical means of isolating individual components from compound mixtures is high-performance liquid chromatography (HPLC). While the resolving power of HPLC is lower, this technique is particularly suited to polar, nonvolatile, or thermally unstable analytes that are difficult to separate by GC. It also offers higher loading capacity than capillary GC.

In addition to chromatographic resolution and capacity, two additional aspects that require consideration are the potential for contamination of the analytes during the isolation procedure, and corrections for carbon associated with any derivative groups that have been appended to the molecule of interest. Regarding the former, entrainment 'bleed' of chromatographic stationary phase can result in significant carbon contamination of the isolated compound, unless steps are taken to avoid this problem (e.g., use of ultra-low bleed GC columns, removal of contaminants after the chromatographic isolation). This problem is likely to be most acute in HPLC when reversed-phase chromatographic phases are used. Comparison of yields and the $\delta^{13}C$ compositions of the isolated compound and the CO_2 resulting from its combustion are effective means of assessing potential contamination problems.

AMS Measurement of ^{14}C

The purified compounds are sealed in evacuated quartz tubes with CuO as an oxidant. The material is combusted to CO_2, purified, and then reduced to graphite over cobalt or iron catalyst. The mixture of graphite and catalyst is loaded into a cesium sputter ion source. ^{14}C-AMS analysis is performed using special methods necessary for the accurate determination of $\Delta^{14}C$ in samples containing only micrograms, rather than milligrams, of carbon. AMS targets containing $< 150\,\mu g$ of carbon are prone to machine-induced isotopic fractionation, which appears to be directly related to the lower levels of carbon ion current generated by these samples. Therefore, small samples are analyzed with identically prepared, size-matched small standards to compensate for these effects. The f_m values that are calculated relative to these standards no longer show a size-dependent fractionation.

Examples of Applications

Lipid Biomarkers in Santa Monica Basin Sediments

As one example of the application of single-compound radiocarbon analysis, we show a detailed data set for a range of lipid biomarkers extracted from marine sediments. This work focused on the upper few centimeters of a core from Santa Monica Basin. The basin has a high sedimentation rate, and its suboxic bottom waters inhibit bioturbation. As a result, laminated cores recovered from the basin depocenter allow decadal resolution of recent changes in the ^{14}C record. On the timescale of radiocarbon decay, these samples are contemporary and have no in situ ^{14}C decay. However, the $\Delta^{14}C$ values of the end-member carbon sources have

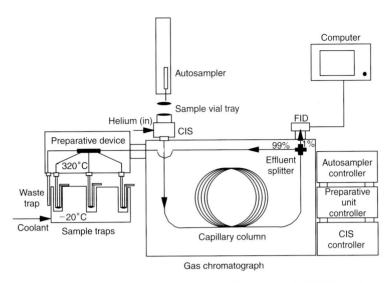

Figure 6 Diagrammatic representation of a preparative capillary gas chromatograph (PCGC) system.

changed (**Figure 2A, B**). 'Bomb-^{14}C' has invaded the modern surface ocean phytoplankton and the terrestrial biota, and through subsequent sedimentation of their organic detritus, this bomb-^{14}C is carried to the underlying sediments. The contrast between 'pre-bomb' and 'post-bomb' Δ^{14}C values, or the relative rate of bomb-^{14}C uptake, therefore is a useful tracer property. It can help distinguish biogeochemical

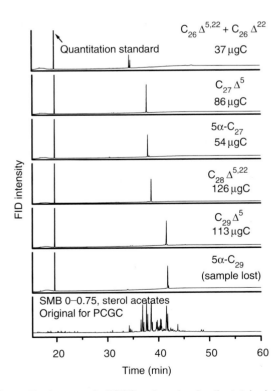

Figure 7 An example PCGC series, showing the total original mixture and the six individual, trapped compounds. In this case the analytes are sterols (as their acetate derivatives). (From Pearson (2000).)

processes that transfer carbon within years or decades (source-specific lipids that now contain bomb-^{14}C) from biogeochemical processes that do not exchange with atmospheric CO_2 on a short timescale (lipids that remain free from bomb-^{14}C).

Compound-specific Δ^{14}C values for 31 different lipid biomarker molecules are shown in **Figure 8** for sedimentary horizons corresponding to pre-bomb (before AD 1950) and post-bomb (1950–1996) eras. These organic compounds represent phytoplanktonic, zooplanktonic, bacterial, archaeal, terrestrial higher plant, and fossil carbon sources. The lipid classes include long-chain n-alkanes, alkanoic (fatty) acids, n-alcohols, C_{30} mid-chain ketols and diols, sterols, hopanols, and C_{40} isoprenoid side chains of the ether-linked glycerols of the *Archaea*.

The data show that the carbon source for the majority of the analyzed biomarkers is marine euphotic zone production. Most of the lipids from 'pre-bomb' sediments have Δ^{14}C values equal to the Δ^{14}C of surface water DIC at this time (dotted line), while most of the lipids from 'post-bomb' sediments have Δ^{14}C values equal to the Δ^{14}C of present-day surface water DIC (solid line).

However, it is clear that two of the lipid classes do not reflect carbon originally fixed by marine photoautotrophs. These are the n-alkanes, for which the Δ^{14}C data are consistent with mixed fossil and contemporary terrestrial higher plant sources, and the archaeal isoprenoids, for which the Δ^{14}C data are consistent with chemoautotrophic growth below the euphotic zone. This is just one example of the way in which compound-specific ^{14}C analysis can distinguish carbon sources and biogeochemical processes simultaneously. The large number of compounds that appear to record the Δ^{14}C of surface

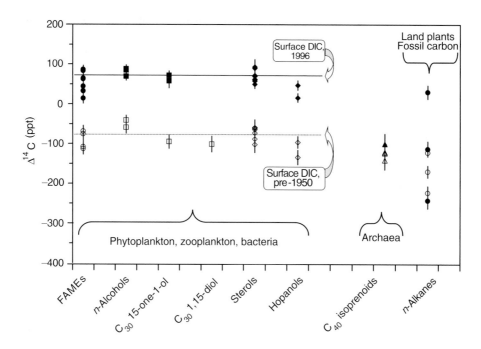

Figure 8 $\Delta^{14}C$ data for individual lipids extracted from Santa Monica Basin sediments. The solid symbols represent compounds extracted from the post-bomb sedimentary horizon (AD 1950–1996). The hollow symbols represent compounds extracted from the pre-bomb sedimentary horizon (deposited prior to AD 1950). (Modified after Pearson (2000).)

water DIC, and therefore marine primary production, points to the potential for numerous tracers of marine biomass; these are the target compounds of interest when developing refined sediment chronologies. In particular, the sterols appear to be particularly effective tracers of surface ocean DIC, and hence suitable for this purpose.

Monosaccharides in Oceanic High-Molecular-Weight Dissolved Organic Matter

The second example illustrates the utility of single compound, as well as compound class, ^{14}C measurements as ocean process tracers. In this case, the process of interest is the cycling of dissolved organic matter (DOM) in the ocean. Much progress has been made in characterizing this large carbon pool. A significant fraction of the DOM pool is composed of high-molecular-weight (HMW) compounds (> 1 kDa), and a substantial fraction of this HMW DOM is known to be comprised of complex polysaccharides. Evidence suggests that these polysaccharides are produced in the surface ocean as a result of primary productivity, and/or attendant heterotrophic activity, and should therefore carry a bomb-influenced ^{14}C signature. Similar polysaccharides have been detected in HMW DOM well below the surface mixed layer, implying that these compounds are transported to the deep ocean. Two possible mechanisms can explain these observations:

(1) advection of DOM associated with ocean circulation, and/or (2) aggregation and vertical transport followed by disaggregation/dissolution at depth. Because the timescales of aggregation and sinking processes are short relative to deep water formation and advective transport, ^{14}C measurements on polysaccharides in HMW DOM provide means of determining which mechanism is dominant.

Figure 9 shows vertical ^{14}C profiles for DIC and DOM as well as ^{14}C results for selected samples of sinking and suspended particulate organic matter (POM), HMW DOM, and monosaccharides isolated from selected depths at a station in the North-east Pacific Ocean. Individual monosaccharides were obtained by hydrolysis of HMW DOM, and purified and isolated by HPLC. The similarity of $\Delta^{14}C$ values of individual monosaccharides implies that they derive from a common polysaccharide source. Such similarities in ^{14}C lend support to the utility of ^{14}C measurements at the compound class level. Furthermore, the similarity between $\Delta^{14}C$ values of these compounds and surface ocean DIC indicates that they are either directly or indirectly the products of marine photoautotrophy. The deep ocean (1600 m) data shows the presence of bomb-radiocarbon in the monosaccharides. Their enrichment in ^{14}C relative to DIC, and similarity to suspended POC at the same depth, suggests that this component of HMW DOM is

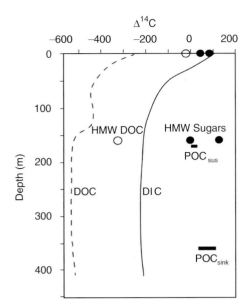

Figure 9 $\Delta^{14}C$ values (ppt) for different fractions of carbon in the North-East Pacific: solid bars labeled POC_{sus} and POC_{sink} correspond to suspended and sinking POC, respectively; solid and dashed lines show depth profiles for DIC and DOC, respectively; open circles are total HMW DOC and closed circles are individual sugars (monosaccharides) isolated from the HMW DOC fractions. (Modified from Aluwihare (1999).)

injected into the deep ocean by vertical transport as particles.

Summary

The ability to perform single-compound ^{14}C measurements has only recently been realized, and as a consequence its application as a tracer in ocean sciences remains in its infancy. The above examples highlight potential applications of single-compound ^{14}C measurements as tools for understanding the biogeochemical cycling of organic matter in the ocean. There are several other areas of study where this approach holds great promise. For example, ^{14}C measurements of vascular plant biomarkers (e.g., plant waxes, lignin-derived phenols) in continental shelf sediments provide constraints on the timescales over which terrestrial organic matter is delivered to the ocean. The 'infinite ^{14}C age' signature that polycyclic aromatic hydrocarbons and other fossil fuel-derived contaminants carry provides an effective means of tracing their inputs to the

coastal ocean relative to contributions from natural processes (e.g., biomass burning). As methods are streamlined, it is anticipated that single compound ^{14}C measurements will find increasing application in marine biogeochemistry.

See also

Ocean Carbon System, Modelling of. Radiocarbon. Stable Carbon Isotope Variations in the Ocean.

Further Reading

Aluwihare LI (1999) *High Molecular Weight (HMW) Dissolved Organic Matter (DOM) in Seawater: Chemical Structure, Sources and Cycling.* PhD thesis, Massachusetts Institute of Technology/Woods Hole Oceanographic Institution.

Eglinton TI, Aluwihare LI, Bauer JE, Druffel ERM and McNichol AP (1996) Gas chromatographic isolation of individual compounds from complex matrices for radiocarbon dating. *Analytical Chemistry* 68: 904–912.

Eglinton TI, Benitez-Nelson BC, Pearson A *et al.* (1997) Variability in radiocarbon ages of individual organic compounds from marine sediments. *Science* 277: 796–799.

Faure G (1986) *Principles of Isotope Geology.* New York: Wiley.

Hayes JM (1993) Factors controlling the ^{13}C content of sedimentary organic compounds: principles and evidence. *Marine Geology* 113: 111–125.

Hedges JI (1992) Global biogeochemical cycles: progress and problems. *Marine Chemistry* 39: 67–93.

Hedges JI and Oades JM (1997) Comparative organic geochemistries of soils and marine sediments, *Organic Geochemistry* 27: 319–361.

Hoefs J (1980) *Stable Isotope Geochemistry.* New York: Springer-Verlag.

Pearson A (2000) *Biogeochemical Applications of Compound-Specific Radiocarbon Analysis.* PhD thesis, Massachusetts Institute of Technology/Woods Hole Oceanographic Institution.

Siegenthaler U and Sarmiento JL (1993) Atmospheric carbon dioxide and the ocean. *Nature* 365: 119–125.

Tuniz C, Bird JR, Fink D and Herzog GF (1998) *Accelerator Mass Spectrometry: Ultrasensitive Analysis for Global Science.* Boca Raton, FL: CRC Press.

Volkman JK, Barrett SM, Blackburn SI *et al.* Microalgal biomarkers: a review of recent research developments. *Organic Geochemistry* 29: 1163–1179.

SINGLE POINT CURRENT METERS

P. Collar and G. Griffiths, Southampton Oceanography Centre, Southampton, UK

doi:10.1006/rwos.2001.0326

Introduction

A current meter estimates the speed and direction of water moving relative to the instrument. The single point current meter is, therefore, only part of a measurement system that includes the mooring or mounting hardware or technique. This article begins with a discussion of the interaction between the current meter, the method of mounting and the characteristics of the currents within the environment being studied. This is followed by an introduction to the principles of current meter design, which are largely independent of the chosen implementation technology. Some examples of commonly used instruments follow, with an assessment of the strengths and weaknesses of the different types of sensor. The importance of direction measurement and calibration are discussed as prerequisites to making accurate observations. (For typical current meter moorings *see* **Moorings**). This article concludes with a note on the future for current measurement systems.

The first self-recording current meters were ingenious mechanical devices such as the Pillsbury instrument (first used in 1884) and the Ekman current meter, available in 1904. However, the slowness of progress during the first half of the twentieth century is reflected in the view of the German hydrographer Bohnecke in 1954 that 'The subject of current measurements has kept the oceanographers busy for more than a hundred years without having found – this must honestly be admitted – an entirely satisfactory solution.' In the 1960s and 1970s the growing need for current measurements in the deep ocean provided a stimulus for the development of robust, self-contained recording instruments capable of deployment over periods of months.

Measurement of current in the open sea is usually achieved by mounting the instrument on a mooring. Movement of the mooring makes true fixed-point, or Eulerian, measurement impossible, although careful attention to mooring design can generally provide an acceptably good approximation to a fixed point. In some circumstances a fixed measurement platform can be used, for example in shallow seas, or at the deep ocean floor. Care then needs to be taken to avoid, as far as possible, disturbance to the flow by the sensor itself, and by any supporting structure. In the case of moored current meters the design of the mooring must minimize vibration, which can lead to the sensor sampling in its own turbulent wake, thereby generating significant errors. With proper attention to design of the mooring or platform, and selection of an appropriate current meter, it should be possible to make most deep-sea measurements to within about $1\,\mathrm{cm\,s^{-1}}$ in speed and 2–$5°$ in direction, and with rather better precision in the case of bottom-mounted instruments. Many of the issues of current meter data quality have been reviewed in the literature.

Particular problems arise in the case of near-surface measurements. Wave orbital motion decays exponentially with depth but may be considered significant, if somewhat arbitrarily, to a depth equivalent to half the wavelength of the dominant surface waves. In the open ocean the influence of surface waves can thus easily extend to a depth of several tens of meters. Within this region the difficulty presented by the lack of a fixed Eulerian frame of reference for current measurement is compounded by the presence of three-dimensional wave orbital velocities. These can be large compared with the horizontal mean flow, making it difficult to avoid flow obstruction by the sensor itself and necessitating a linear response over a large dynamic range. If instruments are suspended some way beneath a surface buoy large errors can result from vertical motion induced by the surface buoy relative to the local water mass.

However, a more fundamental problem arises in the surface wave zone. This may be illustrated by reference to a water particle undergoing progressive wave motion in a simple small amplitude wave. Neglecting any underlying current, a particle at depth z experiences a net Lagrangian displacement, or Stokes drift, in the direction of wave travel of $O[a^2\sigma k\exp(-2kz)]$, where a is the wave amplitude, k is the wavenumber and σ is its angular frequency. In $10\,\mathrm{s}$ waves of amplitude $2\,\mathrm{m}$ this amounts to about $4.5\,\mathrm{cm\,s^{-1}}$ at a depth of $10\,\mathrm{m}$. However, a fixed instrument, even if perfect in all respects will not be able to detect the Stokes drift. Nevertheless an instrument that is moving in a closed path in response to wave action, but unable to follow drifting particles, will record some value of current speed related in a complex but generally unknown fashion to the drift. Close to the surface, where the path of a current sensor over a wave cycle can be more easily arranged to approximate the path of a water particle, the value recorded by the instrument should more closely resemble the surface value of the local Stokes drift. This has been verified in laboratory measurements involving simple waves, but is not easily tested in the open sea. The reader is referred to the bibliography for discussion of these points.

Current Meter Design

Fluid motion can be sensed in a number of ways: techniques most frequently employed nowadays include the rotation of a mechanical rotor, electromagnetic sensing, acoustic travel time measurement, and measurement of the Doppler frequency of back-scattered acoustic energy. **Figure 1** shows examples of practical current meters based on these techniques, and **Table 1** shows the main characteristics of some commonly used instruments. The evolution in design of experimental and commercial instruments from 1970 to 2000 can be traced by comparing the descriptions of the current sensors and *in situ* processing to the 3-D current mapping discussed in contemporary literature. Acoustic Doppler and correlation back-scatter techniques can also measure current profiles, as discussed elsewhere in this volume (*see* **Profiling Current Meters**).

In quasi-steady flow, relatively unsophisticated instruments often produce acceptable results. However, in circumstances in which an instrument may need to cope with a broad frequency band of fluid motions, as in the wave zone or when subject to appreciable mooring motion, there are implications for the design of the sampling system.

If the sensor is to determine horizontal current it should be completely insensitive to any vertical component, while responding linearly to horizontal components across a frequency band which includes the wave spectrum.

Then if, as is usual, the sensor output is sampled in a discrete manner, the provisions of the Nyquist sampling theorem must be observed, i.e., the sampling rate must be at least twice the highest frequency component of interest, whereas negligible spectral content should exist at frequencies above the highest frequency of interest. The highest frequencies that need to be measured are encountered in velocity fluctuations in small-scale turbulence, for example in measurements of Reynolds stress from the time-averaged product of a horizontal velocity component with the vertical velocity. A frequency response to at least 50 Hz is generally required, perhaps even

higher frequencies if the measurements are being made from a moving platform. Satisfying spatial sampling criteria is as important as satisfying temporal sampling requirements. Hence, the sampling path length, or sampling volume of the sensor must be less than the spatial scale corresponding to the highest frequencies of interest. In this case, specialist turbulence dissipation probes that employ miniature sensors measuring velocity shear are used (*see* **Turbulence Sensors**).

Experiments involving the use of laser back-scatter instruments have been carried out at sea, for example to measure fine-scale turbulence near the ocean floor, but their characteristics are generally better matched to high resolution studies in fluid dynamics in the laboratory.

Vector Averaging

Apart from the study of turbulence, the existence of significant wave energy and instrument motion down to periods of 1 s means that a sampling rate (f_s) of ≥ 2 Hz is often used. At this frequency substantial amounts of data are generated and, unless the high frequency content is specifically of interest, it is usual to average before storing data. If done correctly this involves the summation of orthogonal Cartesian components individually prior to computation of the magnitude. Any other form of averaging can produce erroneous results.

If the instrument makes a polar measurement, for example if it measures flow by determining instantaneous rotor speed V_i and the instrument is aligned with the current using a vane whose measured angle relative to north is θ_i the averages are formed:

$$\bar{E} = \frac{1}{n}\sum_{i=1}^{n} V_i \sin \theta_i$$

$$\bar{N} = \frac{1}{n}\sum_{i=1}^{n} V_i \cos \theta_i$$

If on the other hand the instrument measures orthogonal velocity components X_i, Y_i directly, as for example in electromagnetic or acoustic sensors, it forms:

$$\bar{E} = \frac{1}{n}\sum_{i=1}^{n}(X_i \cos \theta_i + Y_i \sin \theta_i)$$

$$\bar{N} = \frac{1}{n}\sum_{i=1}^{n}(-X_i \sin \theta_i + Y_i \cos \theta_i)$$

where θ_i is the instantaneous angle between the Y axis and north; n is chosen so as to reduce noisy

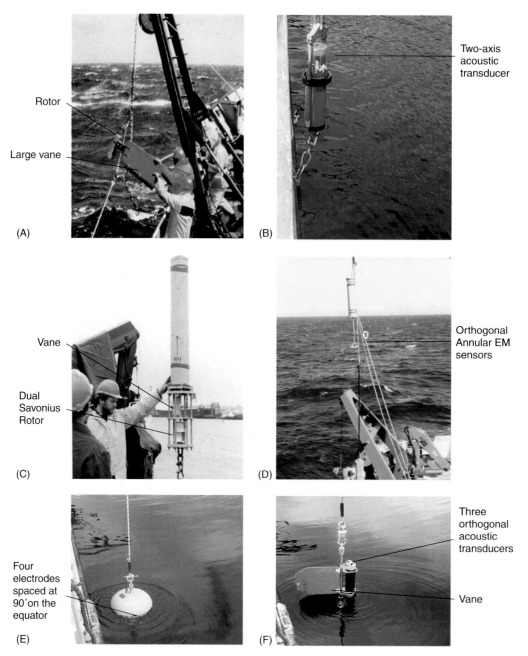

Figure 1 Current meters based on different sensors. (A) Aanderaa RCM4 deep ocean rotor-vane instrument; (B) Aanderaa RCM9 single cell Doppler current meter; (C) EG&G Vector averaging current meter with dual Savonius rotor (at the base) and small vane (immediately above); (D) Vector averaging electromagnetic current meter based on an annular sensor; (E) Interocean S4 electromagnetic current meter; (F) Nortek Aquadopp high precision single cell Doppler instrument, capable of measuring horizontal and vertical currents.

contributions from, for example, the wave spectrum; a value of $nf_s^{-1} > 50\,s$ is usual.

The averaged magnitude and direction are then given by:

$$\bar{U} = ((\bar{E})^2 + (\bar{N})^2)^{0.5}$$

$$\bar{\theta} = \tan^{-1}\frac{\bar{E}}{\bar{N}}$$

Mechanical Current Meters

At first, mechanical current meters were relatively simple. For example, the early, mechanically encoded Aanderaa current meter of the 1960s combined a scalar average of speed with a spot measurement of the direction (**Figure 1A**). Speed was measured by a rotor consisting of six impellers

Table 1 Main characteristics of some contemporary current meters and a vector averaging current meter (VACM) from the 1970s

Type	Speed accuracy	Resolution (cm s^{-1})	Range (cm s^{-1})	Direction accuracy	Depth rating (m)	Weight (kg)	Data capacity (records)
Aanderaa RCM9 MkII single-point Doppler	±0.5 cm s^{-1}	0.3	0–300	±5° for 0–15° tilt	2000	Air: 17 Water: 12	up to 36 100
Aanderaa RCM8 vector averaging rotor-vane	±1.0 cm s^{-1} or ±2% of speed, whichever greater	Not specified	2–295	±5° for speeds 5–100 cm s^{-1} ±7.5° for 2.5–5 cm s^{-1} and 100–200 cm s^{-1}	6000	Air: 29.3 Water: 22.7	up to 43 600
InterOcean S4 electromagnetic sensor	±1.0 cm s^{-1} or ±2% of speed, whichever greater	0.03–0.43 depending on range	0–350	±2° for 0–5° tilt ±4° for 15–25° tilt	S4 1000 m S4Deep 6000 m	S4 Air: 11 S4 Water: 1.5 S4Deep Air: 34.5 S4Deep Water: 10.5	S4 348 000 S4A 7 000 000
Sontek Argonaut-ADV acoustic travel time	±0.5 cm s^{-1}	0.01	0–600	optional extra, at ±2°	60 m	Air: 3.2 Water: 0.45.	> 100 000
FSI 3D ACM acoustic travel time	±1.0 cm s^{-1} or ±2% of speed, whichever greater	0.01	0–300	±2.5° at unspecified tilt	1000 m	Not specified	200 000
EG & G VACM (1970s design)	2.6 cm s^{-1} threshold accuracy not stated	Not stated	2.6–309 cm s^{-1}	± 2.8°	6096 m	Air: 72.5 kg Water: 34.9 kg	50 925–76 388

Source: Company specification sheets.

of cylindrical shape mounted between circular end plates. The rotor shaft ran in ball-race bearings at each end, and at the lower end two magnets communicated the rotation to an internal recording device. The large plastic vane, with a counterweight at the rear end, aligned the instrument with the current.

As experience in a range of deployment conditions and types of mooring widened, such sampling schemes were found to be unsuitable when the sensor experienced accelerating flow as a result of wave motion or mooring movement. The introduction of vector averaging schemes followed, initially in the vector averaging current meter (VACM) (**Figure 1C**), and provided a substantial improvement in accuracy in such conditions. Improved sampling regimes were facilitated in later instruments by low power microprocessor technology. It was also realized that it is necessary to understand fully the behavior of speed/velocity and direction sensors in unsteady flow conditions.

By the time the dual orthogonal propeller vector measuring current meter (VMCM) was developed in the late 1970s sufficient was understood about the pitfalls of near-surface current measurement to realize that rotor design required a combination of modeling and experimental testing in order to ensure a linear response. For example, the propellor in the VMCM was designed to avoid nonlinearity due to the different response times to accelerating and decelerating flows that had been found in the 'S'-shaped Savonius rotor of the VACM. Today, mechanical current meter development might be regarded as mature.

Electromagnetic Current Meters

In electromagnetic current meters an alternating current (a.c.) or switched direct current (d.c.) magnetic field is imposed on the surrounding sea water using a coil buried in the sensing head, and measurements of the potential gradients arising from the Faraday effect are made using orthogonally mounted pairs of electrodes, as illustrated in **Figure 2**. Some electromagnetic techniques make use of the Earth's field, but in self-contained instruments simple d.c. excitation is avoided. This is because unwanted potential differences arising, for example, from electrochemical effects can exceed flow-induced potential differences, which are typically between 20 and $100\,\mu V\,m^{-1}\,s^{-1}$, by two orders of magnitude. Flow-field characteristics around the sensor head, including hydrodynamic boundary layer thickness and flow separation, are of critical importance in determining the degree of sensor

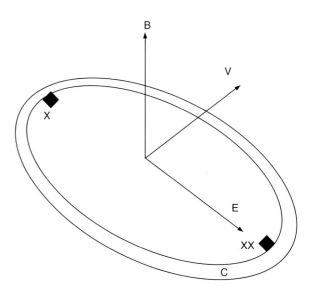

Figure 2 Sketch showing the Faraday effect, which forms the basis of the electomagnetic current meter. The effect results in a potential difference $E = BVL$ induced between two electrodes (X and XX) with a separation L when a conductor (sea water) moves at a resolved velocity V perpendicular to the line X–XX and perpendicular to a magnetic field with a flux density of B induced by coil C.

linearity as well as the directional response. Modeling techniques can help to evaluate specific cases.

Forms of sensor head that have been considered or used include various solids of revolution, such as spheres, cylinders, and ellipsoids. Although hydrodynamic performance weighs heavily in choice of shape, this may be balanced by consideration of ease of fabrication and robustness. One neat solution incorporates the entire instrument within a spherical housing that can be inserted directly into a mooring line (**Figure 1E**). For a smooth sphere, the resulting instrument dimensions would normally give rise to a transition from a laminar to a turbulent boundary layer over the instrument at some point within its working velocity range, at a Reynolds number of $\sim 10^5$, but this is forestalled by use of a ribbed surface so as to introduce a fully turbulent boundary layer at all measurable current speeds. Good linearity is thereby achieved.

Lower flow disturbance can be achieved using an open form of head construction (**Figure 1D**) which has been shown to provide excellent linearity and off-axis response, the only disadvantage relative to solid heads being greater complexity in construction and perhaps some reduction in robustness.

Unlike mechanical current meters electromagnetic instruments have no zero velocity threshold. In the past zero stability has presented a problem, but with modern electronics, and care in head design and fabrication, stability to within a few $mm\,s^{-1}$ over many months of immersion should be achieved.

Acoustic Travel Time (ATT) Current Meters

ATT systems are based on the valid assumption that the resultant velocity of an acoustic pressure wave propagating at any point in a moving fluid is the vector sum of the fluid velocity at that point and the sound velocity in the fluid at rest. The method involves the measurement of the difference in propagation time of an acoustic pulse along reciprocal paths of known length in the moving fluid, although the principle can be realized equally in terms of measurement of phase or of frequency difference. Using reciprocal paths removes the need to know the precise speed of sound. The three techniques present differing design constraints. Typically an acoustic path length l may be of order 10 cm. For resolution of currents Δv to $1\,cm\,s^{-1}$, the required time discrimination of acoustic pulse arrivals can be calculated from:

$$\Delta t = \Delta v.l/c^2$$

or about $4 \times 10^{-10}\,s$ (since the sound speed, c, is about $1500\,m\,s^{-1}$) requiring stable, wide-band detection in the electronic circuitry. In contrast, phase measurement, made on continuous wave signals, is effected within a narrow bandwidth, thereby relaxing the front-end design in the receiver. Phase measurement provides good zero stability and low power consumption, but the pathlength may be constrained by the need to avoid phase ambiguity.

Whichever method is chosen, hydrodynamic considerations are important in achieving accuracy: rigid mounting arrangements which do not disturb the flow significantly are required for the transducers at each end of the acoustic path. Techniques for minimizing flow obstruction have included the use of mirrors to route sound paths away from wakes and, with the development of substantial *in situ* processing, the use of redundant acoustic paths. For a given instrument orientation, the least-disturbed paths can be selected for processing.

ATT techniques have been implemented in various forms for a range of applications, including miniature probes for laboratory tanks, profiling instruments and self-recording current meters. Of the three basic methods, the measurement of frequency difference seems to have been the least exploited, although it has been successfully used in such diverse applications as a miniature profiling sensor

for turbulence measurement, and a buoy-mounted instrument with 3 m path length providing surface current measurements.

ATT current meters offer well-defined spatial averaging, high resolution of currents (better than $1\,mm\,s^{-1}$), potentially good linearity and high frequency response. The main disadvantage, tackled with varying degrees of success in individual types of instrument, is associated with disturbance of flow in the acoustic path by transducers, support struts, and the instrument housing.

Remote Sensing Single-Point Current Meters

One current measurement technique that avoids flow obstruction altogether is that of acoustic back-scatter, using either Doppler shift or spatial or temporal cross-correlation. In the past, these computationally intensive techniques were restricted to use in current profilers, where the relatively expensive instrument could nevertheless substitute for an array of less-expensive single point current meters. Nowadays, the availability of low cost, low-power yet high-performance digital signal processing circuits has made it possible and economic to produce single-point acoustic back-scatter current meters (**Figure 1B & F**). Such instruments provide a combination of several desirable specifications, including: rapid data output rate, with 25 Hz being common; a dynamic range extending from $1\,mm\,s^{-1}$ to several $m\,s^{-1}$; an accuracy of $\pm 1\%$ or $\pm < 5\,mm\,s^{-1}$; a typical sampling volume of a few cubic centimeters and the capability of operating within a few millimeters of a boundary. These characteristics make this class of instrument almost ideal for current measurement within boundary layers, e.g., in the surf zone, while also enabling the collection of concurrent velocity and directional wave spectrum information through sensing the wave orbital velocity components.

Directional Measurement

The directional reference for measurement of current is invariably supplied by a magnetic compass, two main types of which are in common use. The first type is the traditional bar magnet, often mounted on an optically read encoded disk. The entire assembly is mounted on jeweled bearings, with arrangements for damping and gimballing. In the fluxgate compass, the second type of sensor, a soft magnetic core is driven into saturation by an a.c. signal. Orthogonal secondary windings detect the out-of-balance harmonic signals caused by the polarizing effect of the Earth's field and, from an appropriately summed output, the orientation of the sensor relative to the Earth's field can be determined. In current meters a gimballed two-component system may be used, but as in the case of the magnet compass, this does require that the system will respond correctly to any rotational and translational motions arising from mooring or platform motion.

Calibration, Evaluation and Intercomparison

The calibration, evaluation, and intercomparison of current measuring instruments are closely related and are central to the issue of data quality assurance. Basic velocity calibration can be carried out in a tank of nominally still water by moving the instrument, usually suspended from a moving carriage, at a constant, independently measured velocity. Compass calibration is done, typically to a precision of ~ 1°, in an area free from stray magnetic fields either using a precisely orientated compass table equipped with a vernier scale or by invoking a self-calibration program built into the instrument that obviates the need for an accurate heading reference.

Modern instruments can correct for heading-dependent errors in real time as well as correcting for a user-supplied magnetic variation. However, older instruments usually require the corrections to be applied at the post-processing stage.

There is a variety of practices relating to routine calibration, ranging from checks before and after every deployment to almost complete lack of checks. It has been argued that sensitivities of acoustic and electromagnetic sensors are determined by invariant physical dimensions and stable electronic gains, whereas mechanical instruments require only a simple in-air test to ensure free revolution of the rotor. However, good practice is represented by regular calibration checks in water.

Current meters generally behave well in steady flows but, as remarked above, in the near surface zone, or in the presence of appreciable mooring or platform motion, substantial differences can occur in data recorded by different instruments at the same nominal place and time. The fact is that no amount of simple rectilinear calibration in steady flow conditions can reveal the instrument response to the complex broadband fluid motions experienced in the sea and as yet there are no standard instruments or procedures for more comprehensive

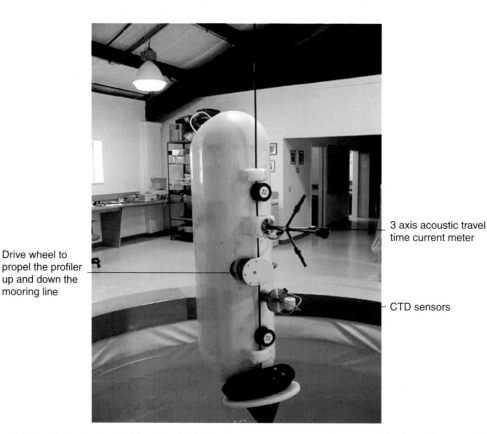

Drive wheel to propel the profiler up and down the mooring line

3 axis acoustic travel time current meter

CTD sensors

Figure 3 Acoustic travel time current meter as one instrument among many on a package capable of crawling up and down a wire mooring to obtain profiles of properties in water depths of up to 5000 m. (Illustration courtesy of McLane Research Inc.)

calibration. Some efforts have, however, been made to model the errors incurred in some specific instruments, with a view to the prediction of performance at sea from dynamic simulation data acquired in the laboratory test tank.

Laboratory tests in controlled conditions thus provide a necessary, though insufficient basis for judging performance, and when a new instrument, or technique, is first used at sea considerable effort is put into intercomparisons with other, longer established instruments or techniques. Not surprisingly, most of the impetus for testing and intercomparison has come from the scientific community; the costs of providing anything other than basic performance data in controlled flow conditions is, with some justification, considered prohibitive by manufacturers. Extensive information on the performance at sea of instruments of many types is, therefore, to be found in the scientific literature, although cheaper instruments are generally less well represented.

Evolutionary Trends

As a result of the advances in electronics and battery technology in recent years, and the painstaking evaluation work accompanying the introduction of new instrument types, sufficient is now known about current measurement that it can in this sense at least be regarded as a relatively mature technology. Yet clear evolutionary trends are in evidence, driven by an increasing operational need for data in support of large-scale monitoring programmes. A further factor is the growing commercial involvement in data gathering. The tendency is towards cheaper, lighter instruments which are more easily handled at sea, and which can be deployed in larger numbers. An example of changes in size, recording capacity and weight that have taken place over the past 25 years is shown by comparing the Vector Averaging Current Meter from the 1970s with a modern acoustic or electromagnetic current meter of similar performance (**Figure 1** and **Table 1**).

Another trend brought about by the growth of processing capability *in situ* is towards the incorporation of current measurement within a complete measurement system embracing a range of physical, chemical, and biological parameters (**Figure 3**). Operational requirements for current data may also in time result in the routine deployment of telemetering systems. At present, satellite telemetry of surface and near-surface measurements is well established, but telemetry of midwater measurements is not yet common practice.

See also

Ocean Circulation. Moorings. Profiling Current Meters. Sonar Systems. Three-dimensional (3D) Turbulence. Turbulence Sensors.

Further Reading

Appell GF and Curtin TB (eds) (1991) Special issue on current measurement. *IEEE Journal of Oceanic Engineering* 16(4): 305–414.

Collar PG, Carson RM and Griffiths G (1983) Measurement of near-surface current from a moored wave-slope follower. *Deep-Sea Research* 30A(1): 63–75.

Dobson F, Hasse L and Davis R (eds) (1980) *Air–Sea Interaction: Instruments and Methods.* New York: Plenum Press.

Hine A (1968) *Magnetic Compasses and Magnetometers.* London: Adam Hilger.

Howarth MJ (1989) *Current Meter Data Quality.* Cooperative Research Report No. 165, Copenhagen: International Council for the Exploration of the Sea.

Myers JJ, Holm CH and McAllister RF (1969) *Handbook of Ocean and Underwater Engineering.* New York: McGraw-Hill.

Pinkel R and Smith JA (1999) *Into the Third Dimension: Development of Phased Array Doppler Sonar.* Proceedings of the IEEE Sixth Working Conference on Current Measurement. San Diego, March 1999.

Shercliff JA (1962) *The Theory of Electromagnetic Flow Measurement.* Cambridge: Cambridge University Press.

SIRENIANS

T. J. O'Shea, Midcontinent Ecological Science Center, Fort Collins, Colorado, USA
J. A. Powell, Florida Marine Research Institute, St Petersburg, FL, USA

Copyright © 2001 Academic Press

doi:10.1006/rwos.2001.0433

Introduction

The Sirenia are a small and distinctive Order of mammals. They evolved from ancient terrestrial plant feeders to become the only fully aquatic, large mammalian herbivores. This distinctive mode of life is accompanied by a suite of adaptations that make the Sirenia unique among marine mammals in anatomical and physiological features, distribution, ecology, and behavior. Although the Sirenia have a long history of interaction with humans, some of their biological attributes now render them vulnerable to extinction in the face of growing human populations throughout their coastal and riverine habitats.

Evolution and Classification

Fossil History

The order Sirenia arose in the Paleocene from the Tethytheria, a group of hoofed mammals that also gave rise to modern elephants (Order Proboscidea). The beginnings of the Sirenia were probably in the ancient Tethys Sea, near what is now the area joining Africa and Asia. By the early Eocene, sirenians had reached the New World, as evidenced by fossils from the primitive family Prorastomidae from Jamaica. The prorastomids and the sirenian family Protosirenidae were restricted to the Eocene. Protosirenids and prorastomids were amphibious and could walk on land, but probably spent most of the time in water. The fully aquatic Dugongidae also arose around the Eocene, and persisted to the Recent, as represented by the modern dugong (subfamily Dugonginae) and Steller's sea cow. The dugongines were the most diverse lineage, particularly in the Miocene, with greatest radiation in the western Atlantic and Caribbean but also spreading back into the Old World. Two other subfamilies of dugongids also differentiated but became extinct: the Halitherinae, which disappeared around the late Pliocene, and the Hydrodamalinae, which was lost with the extinction of Steller's sea cow in 1768. The hydrodamalines are noteworthy for escaping the typically tropical habitats of sirenians and occupying colder climates of the North Pacific. The family Trichechidae (manatees) arose from dugongids around the Eocene–Oligocene boundary.

Two subfamilies of trichechids have been delineated: the Miosireninae, which became extinct in the Miocene, and the Trichechinae, which persists in the three living species of manatees. Early trichechids arose in estuaries and rivers of an isolated South America in the Miocene, where building of the Andes Mountains provided conditions favorable to the flourishing of aquatic vegetation, particularly the true grasses. The abrasiveness of these plants resulted in natural selection for the indeterminate tooth replacement pattern unique to trichechids. When trichechids returned to the sea about a million years ago, this persistent dentition may have allowed them to be more efficient at feeding on seagrasses and outcompete the dugongids that had remained in the Atlantic. Thus dugongids disappeared from the Atlantic, while forms of manatees probably very similar to the modern West Indian manatee spread throughout the tropical western Atlantic and Caribbean, and since the late Pliocene had also dispersed by transoceanic currents to reach West Africa. Beginning in the late Miocene, Amazonian manatees evolved in isolation in what was then a closed interior basin of South America.

Classification, Distribution, and Status of Modern Sirenians

There are four living and one recently extinct species of modern Sirenia. They are classified in two families, the Dugongidae (Steller's sea cow and the dugong) and the Trichechidae (three species of manatees). All four extant species are designated as vulnerable (facing a high risk of extinction in the wild in the medium-term future) by the International Union for the Conservation of Nature and Natural Resources–World Conservation Union.

Steller's sea cow Steller's sea cow, *Hydrodamalis gigas* (**Figure 1**), is placed in the subfamily Hydrodamalinae of the family Dugongidae. These largest of sirenians were found in shallow waters of the Bering Sea around Bering and Copper Islands in the

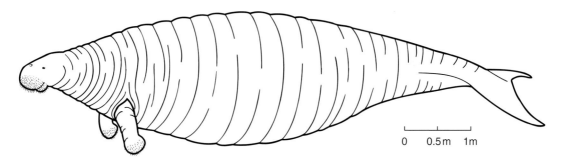

Figure 1 Steller's sea cow (*Hydrodamalis gigas*).

Commander Islands. The first scientist to discover these sea cows was Georg Steller, who first observed them in 1741 during expeditions into the North Pacific led by Vitus Bering. They were hunted to extinction for meat to provision fur hunters about 27 years after their discovery.

Dugongs The dugong (*Dugong dugon*) (**Figure 2**) is the only living member of the family Dugongidae, and is placed in the subfamily Dugonginae. There are no recognized subspecies. Dugongs occur in coastal waters in limited areas of the Indian and western Pacific Oceans. They are currently known from the island of Malagasy and off the east coast of Africa from Mozambique northward to the Red Sea and Persian Gulf; along the Indian subcontinent; off south-east Asia through southern China north to the island of Okinawa in Japan; and through the Phillipines, Malaysia, Indonesia, New Guinea, and most of northern Australia from Shark Bay in Western Australia to Moreton Bay in southern Queensland. Dugongs also occur in very low numbers around the Micronesian islands of Palau. Dugong populations are disjunct and in most areas depleted. Australia provides the major exception, and harbors most of the world's remaining dugongs. One conservative estimate suggests that 85 000 dugongs occur in Australian waters. Dugongs are classified as endangered under the US Endangered Species Act.

West Indian manatees There are two subspecies of West Indian manatees (**Figure 3**): the Florida manatee (*T. manatus latirostris*) (**Figure 4**) of the southeastern USA, and the Antillean manatee (*T. manatus manatus*) of the Caribbean, Central and South America. West Indian manatees are classified as endangered under the US Endangered Species Act. The Florida subspecies is found year-round in nearshore waters, bays, estuaries, and large rivers of Florida, with summer movements into other states bordering the Gulf of Mexico and Atlantic Ocean. Excursions as far north as Rhode Island have been documented. Winter stragglers occur outside of Florida, sometimes dying from cold exposure. It has not been possible to obtain rigorous population estimates, but it has been estimated that there are between 2500 and 3000 manatees in Florida.

Antillean manatee populations are found in coastal areas and large rivers around the Greater Antilles (Hispaniola, Cuba, Jamaica, and Puerto Rico), the east coast of Mexico, and coastal central America through the Lake Maracaibo region in western Venezuela. There do not appear to be resident manatees along the steep, high-energy

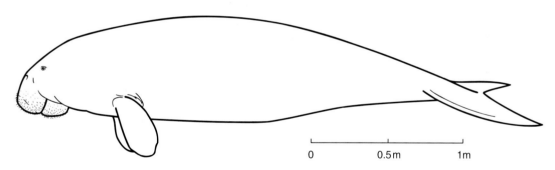

Figure 2 Dugong (*Dugong dugon*).

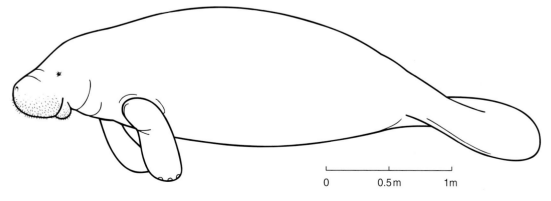

Figure 3 West Indian manatee (*Trichechus manatus*).

Figure 4 Florida manatee (*Trichechus manatus latirostris*).

Caribbean coastline of Venezuela, but populations are found along the Atlantic coast and far inland in large rivers from eastern Venezuela southward to below the mouth of the Amazon near Recife in Brazil. They have not been documented in the lower Amazon. Estimates of Antillean manatee population sizes are unavailable, but they are thought to have declined throughout their range.

West African manatees West African manatees (*Trichechus senegalensis*) (**Figure 5**) are found in Atlantic coastal waters of Africa from Angola in the south to Senegal and Mauritania in the north. They extend far inland in large rivers such as the Senegal, Gambia and Niger, into landlocked and desert countries such as Mali and Chad. There are no recognized subspecies. There are no rigorous data on populations, but they have probably suffered widespread declines due to hunting. West African manatees are classified as threatened under the US Endangered Species Act.

Amazonian manatees There are no recognized subspecies of Amazonian manatees (*Trichechus inunguis*) (**Figure 6**). This species is found in the Amazon River system of Brazil, as far inland as upper tributaries in Peru, Ecuador, and Colombia. Areas occupied include seasonally inundated forests. They apparently do not overlap with West Indian manatees near the mouth of the Amazon, and do not occur outside of fresh water. Amazonian manatee populations have declined this century, but there are

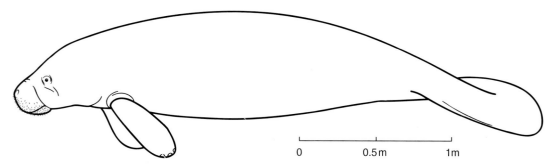

Figure 5 West African manatee (*Trichechus senegalensis*).

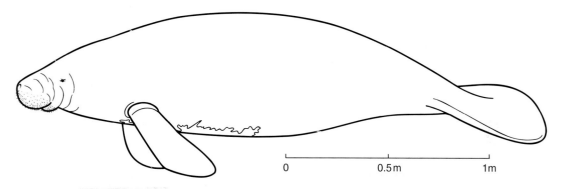

Figure 6 Amazonian manatee (*Trichechus inunguis*).

no firm estimates of numbers. They are classified as endangered under the US Endangered Species Act.

Morphology and Physiology

Adult dugongs range to 3.4 m in length and 420 kg in mass. Florida manatees are larger, ranging to 3.9 m and up to 1655 kg. Morphometric data on West African manatees and Antillean manatees are limited, but they are similar or slightly smaller in size than Florida manatees. Amazonian manatees are the smallest trichechids, ranging up to 2.8 m in length and 480 kg in mass. Steller's sea cow was the largest of all known sirenians, reaching a length of about 9–10 m, and a body mass estimated as high as 10 metric tonnes. The earliest sirenians were about the size of pigs, and had legs, narrow snouts, and the bulky (pachyostotic), dense (osteosclerotic) bones lacking marrow cavities also typical of dugongs and manatees. The heavy skeleton serves as ballast to keep the animals submerged in shallow water. Adaptations for an existence as fully aquatic herbivores also led to rapid loss of hind limbs and development of a broadened down-turned snout. The forelimbs became paddle-like flippers that were positioned relatively close to the head through shortening of the neck, allowing great leverage for

steering. Forelimbs are also used to grasp and manipulate objects, including food plants, and to 'walk' along the bottom. The flippers of West African and West Indian manatees have nails at their distal ends, but nails are not present in Amazonian manatees or dugongs. Forward propulsion is attained by vertical strokes of the horizontal tail, which is spatulate in manatees, but deeply notched like the flukes of whales in dugongs and Steller's sea cow. Horizontal orientation is important for feeding on plants, and is facilitated by the long, horizontally oriented, and unilobular lungs. The lungs are separated from the other internal organs by a horizontal diaphragm.

Manatees lack notable sexual dimorphism, although female Florida manatees attain larger sizes than males and the vestigial pelvic bones differ in size and shape. Mammary glands of manatees and dugongs consist of a single teat located in each axilla. Incisors erupt in male dugongs (and occasionally in old females) and take the form of small tusks used in mating-related behavior. The dorsal surfaces of dugongs often bear parallel scar marks left from wounds inflicted by tusks of males. The Amazonian manatee has a very dark gray-black appearance and many have white ventral patches. West Indian and West African manatees and

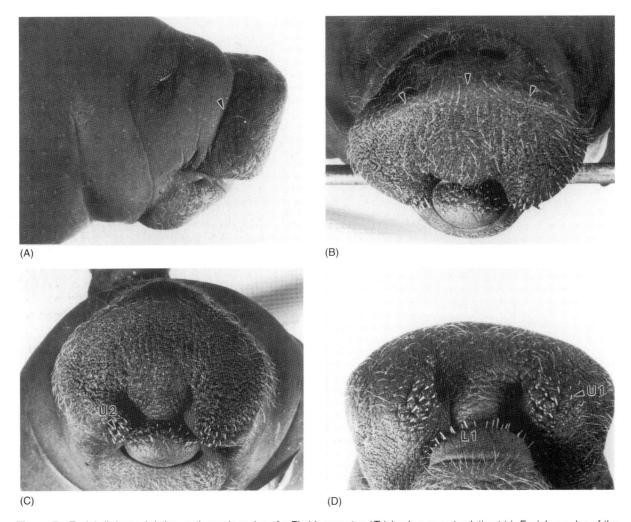

Figure 7   Facial vibrissae, bristles, and muzzle region of a Florida manatee (*Trichechus manatus latirostris*). Facial muscles of the oral disk and bristles combine to form the only prehensile bristles in mammals. (A) Arrow shows postnasal crease which bounds the supradisk of the prehensile muzzle-bristle apparatus. (B) Arrows show the orofacial ridge of the oral disk. (C) The oral disk is seen face forward, with U2 denoting one of the grasping bristle groups. (D) Bristles are everted at U1 and U2 and on the lower lip pad (L1). (Photographs reproduced with permission of *Marine Mammal Science* and Dr Roger Reep.)

dugongs are gray in color, but may also appear brown. Ventral patches are rare and there is little counter-shading in these three species. Hairs are sparsely distributed along the sides and dorsum, and are especially dense and sensitive on the muzzle. These serve a tactile function, which may be particularly important in turbid water and at night. Hairs inside the upper muzzle and lower lip pad are modified into stiff bristles that together with unique facial muscle arrangements form the only prehensile vibrissae known in mammals (**Figure 7**). Steller's sea cow had a small head in relation to the body, and the skin was described as bark-like, dark brown, and in some instances flecked or streaked with white. The forelimbs were reduced to stumps for locomotion along rocky shorelines. They had no teeth and instead relied on horny plates in the mouth to crush kelp. Dugongs have a strongly downwardly deflected snout and are obligate bottom feeders. Dugongs produce a total of six cheek (molariform) teeth in each quadrant of the jaw but also have tough, horny plates in the roof of the mouth. These plates play a significant role in mastication. Dugong teeth are resorbed anteriorly and fall out as they wear, except that the last two molars in each quadrant have persistent pulps and continue to grow throughout life. Manatee cheek teeth, in contrast, continually erupt at the rear of the jaw and move forward as they wear through resorption and reworking of bone. There are no fixed numbers of teeth in manatees, and they erupt throughout life.

Compared with most terrestrial and marine mammals of the same body size, sirenians have unusually low metabolic rates. Metabolic rates of Amazonian manatees are approximately one-third those expected based on measurements of other mammals of similar size. Florida manatee metabolic rates are 15–22% of rates predicted for mammals of their size. They are also poorly insulated, and cannot maintain positive energy balance in cool water, restricting distribution to tropical and subtropical areas. The large size of the Steller's sea cow was in part an adaptation to maintain thermal inertia in the cool North Pacific. Low metabolic rates in the Sirenia can be viewed as adaptations to a diet of aquatic vegetation, which is low in nutritional quality compared with the foods of other marine mammals (such as fish, krill, or squid). Sirenians are hind-gut digesters that may consume up to 10% of body weight per day. The stomach includes a specialized 'cardiac' gland in which special secretory cells are concentrated to avoid abrasion from coarse vegetation. The large intestine is up to 25 m long in dugongs, and passage time of ingesta may take 5–7 days.

Amazonian manatees have an unusual capacity to persist for long periods without food. When food is readily available during the flood season large quantities of fat are deposited, and when individuals are left in isolated pools with receding waters of the dry season they subsist by drawing on this stored energy. Fat deposits coupled with a low metabolic rate may allow Amazonian manatees to go without feeding for up to 7 months. Brains of adult sirenians are very small for mammals of their body size. This may be due to a combination of factors, including low metabolic rates and natural selection for lengthy postnatal growth in body size. Brains also lack marked cerebral convolutions. However, other aspects of neuroanatomy (elaboration of cortical centers and cerebral cytoarchitecture) suggest that sirenians are fully suited for complex behavior. Manatees in captivity can learn a variety of conditioned tasks.

Behavior and Social Organization

Manatees lack strong circadian (24 h) rhythms in activity, although in the more temperate climate of Florida, winter activity can depend on diel changes in water temperature. This lack of circadian activity correlates with the absence of a pineal gland. Seasonal migrations occur within Florida to avoid cool water in winter. Some of these are local movements to constant temperature artesian springs, whereas others span one-way distances of 500 km or more along a north–south gradient. Florida manatees can travel at rates of 50 km per day. Their seasonal

destinations can consist of core areas used annually. The methods by which sirenians orient and navigate accurately and directly between destinations through murky waters is unknown. They have small eyes, no external ear, and minute ear openings.

Mating behavior in Florida manatees consists of groups of up to about 20 males actively pursuing females in estrus. The pursuits can last up to 2 weeks, cover distances up to 160 km, and involve vigorous jostling and chasing. Copulations with more than one male have been observed. Dugongs show greater variability in mating systems, and during violent encounters males will inflict cuts with their tusks. In eastern Australia dugongs may exhibit mating behavior similar to that of Florida manatees. In western Australia, however, dugongs also form leks in which single males set up small territories that are visited by individual females. Males advertise their presence on these leks with complex audible underwater vocalizations. Florida manatees also produce audible underwater sounds, but these vocalizations function more as contact calls, may signal simple motivational states, and can serve in individual recognition. Vocal communication is most pronounced between females and calves. West African and Amazonian manatees also produce underwater communication sounds, but these have been less well studied. No stable social organization has been observed in sirenians, other than the long bond between females and calves during lactation. Manatees appear to be primarily solitary but form small transient groups of about 1–20. They may also aggregate in larger numbers around concentrated resources such as freshwater seeps and winter thermal refugia in Florida. Dugongs can aggregate in herds of a few hundred. Herding behavior may have advantages in cultivation grazing, in which feeding activities keep seagrass beds in stages of succession that favor certain food plants.

Ecology and Population Biology

Sirenians are usually found in shallow waters, as aquatic plants do not grow at significant depth where light is restricted. Dugongs feed exclusively on marine angiosperms, the seagrasses of the families Potamogetonaceae and Hydrocharitaceae. The historic distribution of dugongs coincides with the distribution of these food plants. Around tropical Australia, dugongs can be found wherever adequate seagrass beds occur, including distances as far as 60 km offshore and waters to 37 m deep. More delicate, sparsely dispersed deep-water species of seagrasses are fed upon under the latter conditions.

Manatees feed on a much wider variety of plants than dugongs, including seagrasses, overhanging mangrove leaves, true grasses along banks and in floating mats, and various rooted, submerged, and floating plants. Predation on sirenians has only rarely been observed. There are uncommon reports of sharks preying on manatees and of the presence of shark bite marks on surviving manatees, but manatees do not typically occur in regions occupied by large sharks. Uncommon reports also exist of predation or possible predation by jaguars on Amazonian manatees, and by sharks, killer whales, and crocodiles on dugongs.

Florida manatees and dugongs have life history characteristics that allow only modest population growth rates, and these characteristics are probably similar in Amazonian and West African manatees. Florida manatees can live 60 years. Litter size is one (with twin births occurring in < 2% of pregnancies) after an imprecisely known gestation estimated at about 1 year or slightly longer. Age at first reproduction for females is 3–5 years. Calves suckle for variable periods of 1–2 years, and adult females give birth at about 2.5 year intervals. Survival rates for Florida manatee calves and subadults are not well known. Adult survival, however, has the greatest impact on manatee population growth rates. If adult annual survival is 96%, manatee populations with the healthiest observed reproductive rates can increase at about 7% per year. Growth rates decline by about 1% for every 1% decrease in adult survival. Even when reproductive output is at its maximum, populations cannot remain stable with less than about 90 % adult survival. Dugongs have even slower growth rates than manatees. The oldest age attained by adult female dugongs from Australia was 73 years, litter size is one (twins have not been reliably documented) after a gestation period that is probably similar to that of the Florida manatee, and age at first reproduction for females may be 9–10 years or older. The period of lactation is at least 1.5 years, and adult females give birth on a schedule of 3–7 years. Data to support calculations of dugong survival rates are not available, but modeling of life history traits suggest that under best observed reproduction, populations cannot grow by more than about 5–6% annually.

Conservation and Interactions with Humans

Low population growth rates make sirenians vulnerable to modern agents of mortality. Intensifying human activities in coastal areas produce additional sources of death, injury, and habitat change (Figure 8). Throughout their recent history, humans have hunted sirenians for meat and fat. People in many indigenous tropical cultures use oil and powders from bones as folk remedies for numerous ailments. Hides have been used for leather, whips, shields for warfare, and even as machine belting during shortages of other materials in World War II. Numerous cultures ascribe magical powers to manatee and dugong bones and body parts. Jewelry and intricate carvings are also made from the dense bones. Ingenious means were employed to hunt sirenians, including the use of box traps in parts of West Africa (Figure 9). In most indigenous tropical cultures through the mid-1900s, however, sirenians were hunted principally by hand with harpoons and spear points, and were recognized to be elusive and difficult quarry. Human populations were more sparse, and typically only a few people in any region acquired skills needed to hunt sirenians. Mortality under such conditions may have been sustainable. With the advent of firearms, motors for boats and canoes, and burgeoning numbers of people, overexploitation has occurred. Sirenians have been hunted as a source of bush meat at frontier markets as well as commercially. In Brazil manatee meat was legally sold in processed form, and during peak exploitation in the 1950s as many as 7000 per year were killed for market. In addition, growth in artesanal fisheries has introduced many inexpensive synthetic gill-nets throughout areas used by sirenians. Death due to incidental entanglement has become an additional and significant mortality factor globally. Gill-net commercial fisheries have been excluded from several areas in Australia that are important for dugongs, but few efforts to manage gill-netting for sirenian protection have been instituted elsewhere. Nets deployed to protect bathers from sharks on the coast of Queensland resulted in the deaths of hundreds of dugongs from the 1960s to 1996. Many of these nets have since been replaced with drum lines and deaths have dropped. The vulnerability of sirenians to overexploitation was most markedly illustrated in the case of the Steller's sea cow. Unable to submerge, they were easy quarry as provisions of meat for fur traders on their long voyages into the North Pacific. This caused extirpation of this sea cow within 27 years of discovery by western science. Today sirenians are legally protected in nearly every country in which they occur, as well as by international treaties and agreements. However, few nations provide active law enforcement or enduring conservation programs for manatees.

Overexploitation of manatees by hunting is no longer an issue in Florida. Instead, extensive coastal development accompanied by technological

Figure 8 A Florida manatee showing massive wounds inflicted by a boat propeller. Wounds occur on the dorsum. The nostrils are in the lower center of the photograph, breaking the surface slightly as the animal rises to breathe. (Photograph courtesy of Sara Shapiro, Staff Biologist, Florida Fish and Wildlife Conservation Commission.)

advances in boating and water diversions have created major new sources of mortality. Overall human-related mortality accounts for about 30% of manatee deaths in Florida each year. Collisions with watercraft account for about 24% of all manatee mortality and have increased more rapidly than overall mortality in recent years (**Figure 10**). About 4% of known manatee deaths are caused by crushing or drowning in flood control structures and canal locks. Other human-related causes of manatee mortality such as entanglement in fishing gear represent approximately 3% of the total.

The east and south-west coasts of Florida have lost between 30 and 60% of former seagrass habitat due to development, dredging, filling, and scarring of seagrass beds caused by motor boats. Some human activities, however, such as dredging of canals and the construction of power-generating plants (that provide warm water for refuge at northern limits of the manatee's winter range), may have opened up previously unavailable habitat. Destruction of quality seagrass habitat through activities that disturb bottoms (*e.g.* dredging and mining) or increase sedimentation and turbidity can also impact

Figure 9 A box trap used to capture manatees in West Africa. Walls are made of poles bound by fibrous leaves, with a door at one end that is propped open. Pieces of cassava tubers are left around and in the box as bait. After a manatee swims inside to feed on the starchy cassava, it will jar loose the stick propping the door, which descends and traps the manatee alive.

dugongs. Loss of seagrasses due to a cyclone and flooding in Hervey Bay, Queensland, was accompanied by numerous dugong deaths, and a reduction in the estimated population in the bay from about 1500–2000 to < 100 dugongs. Direct mortality, reduced immune function, or impaired reproduction of sirenians in relation to environmental contaminants have not been observed. Low positions in marine food webs reduce exposure to many of the persistent organic contaminants that build up in tissues of other marine mammals, but sirenians may be more likely to be exposed to toxic elements that accumulate in sediments and are taken up by plants.

Outlook for the Future

In 1999, at least 268 manatees died in Florida and at least 82 (30%) of these were killed by watercraft. The total number of deaths in 1999 was the highest recorded, except for 1996 when a large red tide event increased manatee deaths in south-west Florida. Watercraft-related mortality is increasing more rapidly than overall mortality. However,

aerial survey counts and monitoring of identifiable individual manatees indicate that numbers in particular regions of Florida have been slowly increasing. Some argue that there are more manatees now than ever before, so we should expect a higher proportion to die each year as the population grows. Alternatively, population models suggest that the current level of manatee mortality and the low survival of adults, will result or has already resulted in a manatee population decline. Because of inherent variation in counts due to survey conditions, several years of survey data are required to determine conclusively whether a change in population trend has actually occurred. However, the current human population in Florida is about 15 million people, a doubling in the past 25 years. By 2025 the population is expected to reach 20 million. Florida's waterways are a major source of recreation and revenue. Increasing boat traffic and coastal development resulting from Florida's growth will probably accelerate human-related manatee mortality. Other factors that complicate the future of manatees are the loss of warm-water refugia as spring flows decrease from increased ground-water extraction and drought

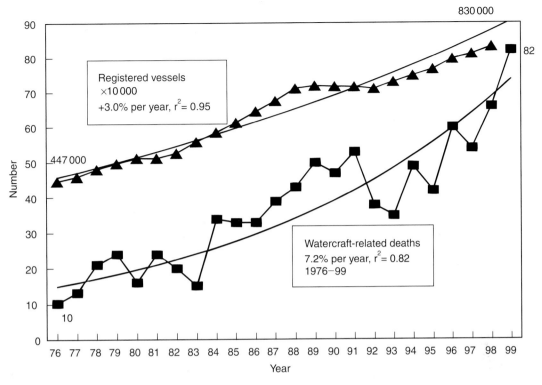

Figure 10 Patterns of increase in the annual numbers of boats registered in Florida and annual totals of boat-killed Florida manatees, 1976–99. Numbers of registered vessels have increased from 447 000 to 830 000. Numbers of boat-killed manatee carcasses recovered have increased from 10 annually to over 80.

conditions, and the phasing out of coastal power-generating plants because of industry deregulation and competition as more efficient inland plants come on-line. Florida manatees are thus faced with a very uncertain future. Proper planning, good information, and effective management are critical to the long-term survival of this species in Florida, where societal concern, dedicated financial resources, and formal protection for manatees are among the strongest in the world. West Indian manatees elsewhere in the Caribbean and South America probably have an even less optimistic future. This is because of past suppression or local elimination of populations from hunting and net entanglement, as well as habitat loss and ever-increasing pressure on resources from human populations. This is particularly true for manatees around island countries. On the mainland the outlook is more mixed. In some areas hunting pressure has declined, and conservation efforts have been enhanced. West Indian manatees are likely to persist in very remote and undeveloped areas of South America as long as such conditions prevail.

The outlook for dugongs in Australia is more guardedly optimistic, but recovery of populations and indeed the continued existence of dugongs in most of the rest of their range are much less likely. This is due to severe reductions from hunting and fishing activities in the past, and continued degradation of coastal habitat as human populations burgeon in these areas. Although the distribution of Amazonian manatees has remained similar to historical records, populations have been reduced in many areas of former abundance. Illegal capture for commercial sale of meat and incidental take in fishing nets continue, habitat degradation is increasing, and there is concern about heavy metal contamination of aquatic food plants from mining.

In West Africa the manatee's range has not changed appreciably from historical accounts. However, it is believed that numbers have been reduced due to illegal hunting. Several countries are particularly important for manatees, including Senegal, The Gambia, Guinea-Bissau, Sierra Leone, Ivory Coast, Nigeria, Cameroon, Gabon, and Angola. In these countries manatees are not uncommon, probably because there are extensive areas of optimal habitat located in relatively remote areas where hunting pressure is reduced. Hunting is the primary threat to manatees in Africa. For example, in Guinea-Bissau a single fisherman had sternums from over 40 manatees in his hut. Although manatees are protected throughout their range, the lack of enforcement of hunting laws and the need for supplemental protein in many areas contributes to hunting pressure. Damming of rivers poses an emerging threat, both directly and indirectly. Crushing in dam structures has been reported from Senegal, Ghana and Nigeria (and also in Florida). Dams on many rivers in the manatee's range may cut off needed seasonal migrations. Killing of manatees as they aggregate at freshwater overflows of dams has also been reported. In recent years, there has been increasing trade in West African manatees for commercial display facilities. Accelerating habitat destruction and cutting of mangroves for construction and firewood will have negative effects. There is considerable cause for concern for the manatee's future in several regions of Africa unless hunting is reduced. However, there has been increasing interest in West African manatee conservation, and several countries are moving towards increased protection involving coastal sanctuaries and law enforcement. To prevent future loss of manatees from all but a few remote and protected areas of West Africa, law enforcement and protection must become regional in scale, improvements must be made in economic conditions that contribute to hunting, and modifications will be needed on dams and other structures that kill manatees.

See also

Marine Mammal Evolution and Taxonomy. Marine Mammal Overview. Marine Mammals, History of Exploitation.

Further Reading

Bryden M, Marsh H and Shaughnessy P (1998) *Dugongs, Whales, Dolphins and Seals*. St. Leonards, NSW, Australia: Allen & Unwin.

Domning DP (1996) Bibliography and index of the Sirenia and Desmostylia. *Smithsonian Contributions to Paleobiology* 80: 1–611.

Domning DP (1999) Fossils explained 24: sirenians (seacows). *Geology Today* (March–April): 75–79.

Hartman DS (1979) *Ecology and Behavior of the Manatee (Trichechus manatus) in Florida*. American Society of Mammalogists Special Publication 5: 1–153.

O'Shea TJ (1994) Manatees. *Scientific American* 271 (1): 66–72.

O'Shea TJ, Ackerman BB and Percival HF (eds) (1995) *Population Biology of the Florida Manatee*. US Department of Interior, National Biological Service Information and Technology Report 1.

Reynolds JE III and Odell DK (1991) *Manatees and Dugongs*. New York: Facts on File, Inc.

Rosas FCW (1994) Biology, conservation and status of the Amazonian manatee, *Trichechus inunguis*. *Mammal Review* 24: 49–59.

SLOPE FRONTS

See **SHELF-SEA AND SLOPE FRONTS**

SMALL PELAGIC SPECIES FISHERIES

R. L. Stephenson, St. Andrews Biological Station, St. Andrews, New Brunswick, Canada
R. K. Smedbol, Dalhousie University, Halifax, Nova Scotia, Canada

doi:10.1006/rwos.2001.0447

Introduction

The so-called 'pelagic' fish are those that typically occupy the midwater and upper layers of the oceans, relatively independent of the seabed (*see* **Pelagic Fishes**). Small pelagic species include herrings and sprats, pilchards and anchovies, sardines, capelin, sauries, horse mackerel, mackerels, and whiting. Most of these fish have a high oil content and are characterized by a strong tendency to school and to form large shoals. These features have contributed to the development of large fisheries, using specialized techniques, and to a variety of markets for small pelagic species.

Two of the major pelagic species (herring and capelin) are found in polar or boreal waters, but most are found in temperate or subtropical waters. The sardine, anchovy, mackerel, and horse mackerel are each represented by several species around the world, with the largest concentrations being found in the highly productive coastal upwelling areas typically along western continental coasts. Intense fisheries have developed on a few small pelagic species, which occur in very dense aggregations in easily accessible areas near the coast. In recent years the largest fisheries have been on Atlantic herring, Japanese anchovy and Chilean Jack mackerel, although the dominance of individual fisheries has changed as the abundance of these species has fluctuated due to both natural factors and intense fishing. Small pelagic fish species with recent annual landings exceeding 100 000 t are listed in **Table 1**. Over the past three decades, fisheries for small pelagic species have made up almost half the total annual landings of all marine fin-fish species (**Figure 1**).

Catch Techniques in Pelagic Fisheries

Pelagic fish, by definition, are found near the ocean surface or in middle depths. As a result, pelagic fisheries must search larger volumes than demersal fisheries. However, most pelagic fish species exhibit behaviors that increase their catchability. The most import characteristic is shoaling, in which individuals of the same species form and travel in aggregations. Several pelagic species also exhibit clear patterns of vertical migration, often staying deep in the water column, or near bottom, by day but migrating to surface waters at dusk. In some cases fishers have used techniques such as artificial light sources to enhance shoaling behavior and improve fishing.

Pelagic fish shoals can be very large, which greatly increases their detectability. Surface shoals can be located visually, using spotters from shore or aboard vessels, and aerial searches are used in the fishery for some species. The location and depth of shoals can be determined using hydroacoustics, which has developed greatly in association with pelagic fisheries in recent decades. Echo sounders and sonars transmit an acoustic signal from a transducer associated with the vessel and receive echoes from objects within the path of the beam such as fish and

Table 1 Main species of pelagic fish with world catches greater than 100 000 tonnes in 1998. Catch information from FAO 1999

Order	Family	Species		Catch (10³ tonnes)
Beloniformes	Scomberesocidae	Pacific saury	Coloabis saira	181
Clupeoidei	Clupeidae	Gulf menhaden	Brevoortia patronus	497
		Atlantic menhaden	Brevoortia tyrannus	276
		Atlantic herring	Clupea harengus	2419
		Pacific herring	Clupea pallasi	498
		Bonga shad	Ethmalosa fimbriata	157
		Round sardinella	Sardinella aurita	664
		Goldstripe sardinella	Sardinella gibbosa	161
		Indian oil sardine	Sardinella longiceps	282
		California pilchard	Sardinops caeruleus	366
		Japanese pilchard	Sardinops melanostictus	296
		Southern African pilchard	Sardinops ocellatus	197
		South American pilchard	Sardinops sagax	937
		Araucanian herring	Strangomera bentincki	318
	Engraulidae	Pacific anchoveta	Cetengraulis mysticetus	181
		European anchovy	Engraulis encrasicolus	492
		Japanese anchovy	Engraulis japonicus	2094
		Peruvian anchoveta	Engraulis rigens	1792
		European pilchard	Sardina pilchardus	941
		European sprat	Sprattus sprattus	696
Gadiformes	Gadidae	Southern blue whiting	Micromesistius australis	184
		Blue whiting	Micromesistius poutassou	1191
Percoidei	Carangidae	Cape horse mackerel	Trachurus capensis	184
		Japanese jack mackerel	Trachurus japonicus	341
		Chilean jack mackerel	Trachurus murphyi	2056
		Atlantic horse mackerel	Trachurus trachurus	388
Salmonoidei	Osmeridae	Capelin	Mallotus villosus	988
Scombroidei	Scombridae	Indian mackerel	Rastrelliger kanagurta	284
		Chub mackerel	Scomber japonicus	1910
		Atlantic mackerel	Scomber scombrus	657
		Japanese Spanish mackerel	Scomberomorus niphonius	552
Scorpaeniformes	Hexagrammidae	Atka mackerel	Pleurogrammus azonus	344

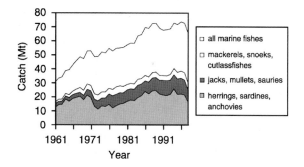

Figure 1 World catch (million tonnes, Mt) of groups of small pelagics in relation to the total world catch of all marine fish species. Catch information from FAO (1999).

the seafloor. When sector scanning and multibeam sonars are mounted on vessels within a moveable housing, they can be used to search the water column ahead and to the sides of the vessel. This permits improved detection of schools and targeting of shoals throughout the water column.

Methods for catching small pelagic fishes range from very simple to highly sophisticated. This section deals only with those catch methods and specialized gear that have been developed specifically for pelagic fisheries, and are widely used around the globe. In general, these methods can be placed in several broad categories:

1. static gear (trap, lift net and gill net)
2. towed gear (trawl net)
3. surrounding gear (purse seine, ring net and lampara net).

Static Gear (Trap, Lift Net and Gill Net)

The earliest fisheries for small pelagic species relied on shoals of fish encountering and becoming entrapped in nets that were fixed to the bottom or floating passively. A variety of forms of trap nets have been used historically where small pelagic species such as herring and mackerel occur near shore. Typically these nets have a lead and wings which intercept and direct a school of fish into a pot or pound where they remain alive until dipped or seined from the net. In the Bay of Fundy and Gulf of Maine, for example, juvenile herring have been caught in heart-shaped traps known locally as weirs (made of fine mesh covering poles driven into the seabed), or even trapped in coves by closing off the entrance with a seine. These catches have formed the basis of the canned sardine industry for over a century.

Unlike trap nets, a lift net functions as a moveable container that traps the target fish. The shoal swims over the net, and the fish are caught by lifting the net and thereby trapping the shoal within. Fish are

then usually removed by dipnet. This type of gear is used commonly in South-east Asia. The nets may be operated from shore or from specialized vessels. These vessels often use bright lights to lure the target shoals over their lift nets.

The gill net catches fish by serving as a barrier or screen through which fish larger than the mesh size cannot pass, but become entangled, usually by their gills, and are removed when the net is retrieved. The gill net is constructed of a section of fine netting with a border of stronger netting. The bottom of the net is weighted with a leadline, and the top or headline has floats so that the net hangs vertically in the water. Fishing typically involves a string of nets that can be set to a desired depth. Gillnets may be moored in a specific location (set nets) or may be allowed to float with the prevailing winds and currents (drift nets).

Towed Gear (Trawl Net)

The midwater trawl was developed specifically to capture pelagic species. Like the bottom trawl commonly used in demersal fisheries, it is towed behind the vessel. Trawling may be undertaken from either single or paired vessels, although trawling by one vessel is much more common. The net is set to the depth for a particular, targeted school of fish, and is towed only long enough to pass through (or to encompass) the school.

The trawl net is usually conical in form and may be very large. Typically, the net mouth, which may be circular or square, is held open during a tow by doors which spread the wings of the net, together with the buoyancy of floats placed on the headline, and a weighted footrope attached to the bottom of the net mouth. A transducer is placed on the headline so that the depth of the net can be monitored during a tow, and the school of fish can be seen passing the mouth into the net. Additional sensors on the trawl provide information about the state of the gear and potential catch.

Surrounding Gears (Purse Seine, Ring Net and Lampara Net)

The characteristic of pelagic fishes to occur in large shoals, right at the surface, makes them vulnerable to capture with surrounding gears. The purse seine is used worldwide and accounts for an appreciable fraction of the total annual catch by global fisheries. In general, this fishing method involves setting out a long net hanging vertically in the water to surround a school of fish. The top of the net usually floats at the surface. Following envelopment of the school, the bottom of the net is pulled together to trap the fish in a cup or 'purse' of netting. The net is hauled gradually, such that the purse shrinks in size until the school of fish is alongside the vessel. The fish are then brought aboard using pumps or a lifting net. The seine may be very large, up to 730 m long and 180 m deep. The main area of the net is usually of constant mesh size and material. Only the section of the gear wherein the catch will be concentrated during retrieval is strengthened. Purse seining usually targets fish found at the surface to a depth of about 130 m.

Purse seining may use one or two vessels. The gear is set beginning with one end and the vessel deploys the gear as it slowly steams around the fish school. When two vessels are used, the tasks of setting and retrieving the seine are shared between the vessels, and each vessel carries part of the gear. During deployment, each vessel sets the gear, beginning with the middle of the seine net, and then they proceed to envelope the school. Most often retrieval is with the aid of a power block.

A variation of the purse seine is the ring net. This encircling net is generally smaller than a purse seine and is used in relatively shallow waters. Ring net gear is usually used by pairs of vessels in the 12–19 m range.

A third type of surrounding gear is the lampara net. Although similar in shape and use to a purse seine, the bottom of this net is not pursed during retrieval. The float line on a lampara net is much longer than the weighted bottom line, which causes the middle portion of the net to form a bag. When the net is hauled, both sides (wings) of the gear are retrieved evenly. The leadlines meet, effectively closing off the bag of the net. The bag is brought alongside the vessel and the catch transferred aboard.

Products Derived from Pelagic Fisheries

Some small pelagic species (notably Atlantic herring) have been important as food, and have sustained coastal communities for centuries. Principal products (those used for human consumption) are listed in **Table 2**. The main reason for the underutilization of small pelagics as food is a lack of consumer demand. The development of new products and uses may increase this demand. One such innovation is the use of small pelagics as raw material in processed seafoods such as surimi.

More than half (approximately 65% during the period 1994–1998) of the harvest of small pelagic species in recent years has not been used directly for food, but rather has been processed into a variety of by-products. Some of these by-products, such as fish protein concentrate, may then be used in the production of foodstuffs for human consumption. The main by-products of the pelagic fisheries are:

1. fish meal and oil
2. fertilizer
3. silage and hydrolysate
4. compounds for industrial, chemical and pharmaceutical products.

Fish Meal and Fish Oil

Fish meal and fish oil are the most important by-products produced from pelagic fish. Meal is used mainly as a protein additive in animal feeds in both agriculture and aquaculture. During the 1980s the annual world production of fish meal was in the order of 6–7 million tonnes (Mt), requiring landings of 35–40 Mt of whole fish. Production has declined somewhat during the mid 1990s due to fluctuations in abundance of key target species.

Fish meal and oil are usually produced in tandem. The fish is cooked and then pressed to separate the solids (presscake) and liquids (pressliquor). After separation of the oil from the pressliquor, the rest of the soluble components are remixed with the presscake and this mixture is dried to form the finished solid product.

Table 2 World production of the main primary products derived from small pelagics

Product	Live-weight equivalent of product (10^3 tonnes)			
	1995	1996	1997	1998
Fresh or chilled	836	948	950	1084
Frozen	6138	6935	7481	6911
Prepared or preserved	1759	1743	2072	2013
Smoked	61	66	65	63
Dried or salted	464	433	391	360
Fats and oils of fish	3399	3090	2732	1709
Fish meal fit for human consumption	162	118	182	140

Source: FAO

Fish oil is an important by-product of the production process for fish meal. Annual production of fish oil is approximately 1.5 Mt, making up 1–2% of the total production of fats and oils worldwide. The oil is used mainly in margarines and shortenings, but there are also technical and industrial uses, such as in detergents, lubricants, water repellents, and as fuel. Fish oil consists mainly of triglycerides. The oils derived from pelagics comprise two main groups. The first group of oils contains a relatively large amount of monoenic fatty acids and a moderate proportion of polyunsaturated fatty acids. This group of oils is derived from fish caught in the North Atlantic. The second group has a high iodine number and exhibits a relatively high content of polyunsaturated fatty acids. These oils are from fish originating in the Pacific Ocean, and tropical and southern region of the Atlantic Ocean.

Fertilizer

Fish offal and soluble compounds are used as fertilizer on farms in coastal areas. Fertilizers are also produced through industrial processes, but the industry is relatively small compared to the economic value of other fish by-products.

Fish Silage and Hydrolysates

Fish silage is used as an ingredient in animal feed, and is treated separately from fish meal due mainly to differences in the production process. Fish silage is used mainly in fish feeds and moist feed pellets. During production, the fish is usually minced and mixed with materials that inhibit bacterial growth. The product is stored for a period during which the fish is liquefied by digestive enzymes. Some of the water may be removed via evaporation.

The production process for hydrolysates is similar to that of fish silage. Water is added to the minced fish, and suspended solids are liquefied through the action of proteolytic enzymes. Oils and the remaining solids are removed, leaving a hydrolysate. The product may be treated further, depending on how the product is to be used. Fish hydrolysate may be used as an ingredient in compound feeds much like fish silage. It is also used as a flavor additive. A third use for fish hydrolysate is as a component of growth media for industrial fermentation.

Pelagic Fish as a Resource for Biotechnological Products

A number of organic compounds found within small pelagic fish are of interest to the chemical and pharmaceutical industries. These compounds include, among others, fatty acids, lipids, hormones, nucleic acids, and organometallics. Scales have been used as a source of pearl essence.

Management Issues for Small Pelagic Fisheries

The exploitation of many stocks of pelagic fishes has exhibited a pattern of sharply increasing catches followed by an even more rapid decline, leading in several cases to closure of the fishery (**Figure 2**). The rapid increases in catches have been largely due to increasing technical developments, increasing markets and movement into new fishing areas of high abundance. The rapid decline has been considered, in some cases, to have a link with environmental change, but has been attributed generally to heavy fishing pressure, recruitment failure, and ineffective management. Most stocks have recovered after reduction/termination of fishing, and some fisheries have resumed following recovery.

Assessment and management of pelagic fisheries has been complicated by a number of issues. The long history of information of some pelagic stocks prior to mechanized fisheries demonstrates that these stocks are prone to pronounced fluctuations in abundance in the absence of fishing (**Figure 3**). Although such natural fluctuations are undoubtedly linked to the environment, there are likely several mechanisms and these are not well understood. In some cases (e.g., sardines and anchovies, herring, and pilchard) there have been apparent shifts in dominance over time, where periods of high abundance have alternated for the different species groups (**Figure 4**). It has been hypothesized that these switches in dominance relate to shifts in environmental conditions favoring one species over another, possible interactions with other species, and the effects of harvesting. Some species or species groups

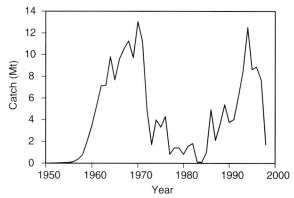

Figure 2 Catch (million tonnes; Mt) of Peruvian anchoveta (*Engraulis ringens*) showing fluctuations in abundance attributed largely to fishing. (Data from FAO, 1999.)

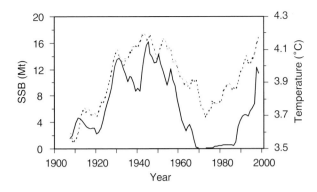

Figure 3 An example of possible environmentally induced biomass changes. Estimated spawning stock biomass (million tonnes; Mt) of Norwegian spring spawning herring (*Clupea harengus*), 1907–1998 and associated long-term temperature fluctuations. —, spawning stock, - - -, temperature. Adapted from Toresen and Østvedt (2000).

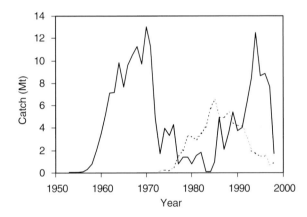

Figure 4 Catch (million tonnes; Mt) of Peruvian anchoveta (*Engraulis ringens*; —), and South American sardine (or pilchard) (*Sardinops sagax*; - - -), showing a shift in species dominance. Data from FAO (1999).

have exhibited coherence in high abundance periods over broad geographical scales (e.g., high abundance of sardines throughout the North Pacific basin during the same time periods).

The shoaling behavior of pelagic species can result in high fishery catch rates when there are relatively few fish remaining in the stock. As a result catch per unit effort, which is commonly used as an indicator of stock decline in demersal fisheries, is less useful in evaluating the state of pelagic fisheries. High catch rates can be maintained, in spite of declining stock size, and this has contributed to sudden stock collapse.

The biological basis for assessment and estimation of stock size for many small pelagic species is complicated by migration, mixing of stocks, lack of fishery-independent abundance indices, and poor analytical performance of common assessment methods. Management attempts have been further complicated by the rapid rate of decline of some small pelagic stocks.

The economic importance of herring to coastal European nations contributed to the prominent consideration of small pelagic fisheries in the evolution of fisheries science and management. Several developments, including the stock concept, recognition of year-to-year differences in year-class strength, development of hydroacoustic survey methods, and early fishery management systems have been based on fisheries for small pelagic species. Most fisheries for small pelagic species are regulated by some form of quota, although in some cases management is complicated by the occurrence of these fisheries outside or across areas of national jurisdiction.

Continued increase in mechanization of fisheries and expansion of markets increases the fishing pressure on several pelagic stocks. Technical developments such as improved hydroacoustic (sonars and sounders) detection of fish, global positioning systems, improved net design and materials, and improved vessel characteristics (e.g., larger size and refrigeration) have increased the efficiency of the fishery. These developments have contributed to increasing fishing effort, in spite of management attempts in some cases to regulate fishing capacity. The historical exploitation of several pelagic fish stocks can be shown to be directly linked to changes in markets. The increasing demand for fishmeal, for example, resulted in large increases in the landings from some pelagic fish stocks in the 1970s and 1980s.

Current research areas of particular interest for small pelagic species include the link between stock abundance, environmental fluctuation (particularly in areas of fronts and upwelling) and climate change. Most small pelagic species feed on zooplankton, and are in turn important food items for other fish, marine mammals, and seabirds. Assessment and management of fisheries for small pelagic fisheries are increasingly becoming concerned with multispecies interactions, including both the estimation of mortality of the pelagic species caused by predation and maintaining sufficient stock size of small pelagic fish as forage for other species.

See also

Acoustics, Arctic. Acoustics, Deep Ocean. Acoustics, Shallow Water. Demersal Fishes. Demersal Species Fisheries. Dynamics of Exploited Marine Fish Populations. Fish Migration, Vertical. Fisheries and Climate. Fishery Management. Fishing Methods and Fishing Fleets. Pelagic Fishes. Plankton.

Further Reading

Burt JR, Hardy R and Whittle KJ (eds) (1992) *Pelagic Fish: The Resource and its Exploitation*. Oxford: Fishing News Books.

FAO (1999) *The State of World Fisheries and Aquaculture 1998*. Rome: Food and Agriculture Organization of the United Nations.

FAOSTAT Website. http://apps.fao.org/default.htm The FAO website contains information from the FAO Yearbook of Fishery Statistics.

Patterson K (1992) Fisheries for small pelagic species: an empirical approach to management targets. *Review of Fish Biology and Fisheries* 2: 321–338.

Saville A (ed.) (1980) The assessment and management of pelagic fish stocks. *Rapp. P-v Reun., Cons. Int. Explor. Mer* 177: 517.

Toresen R and Østvedt OJ (2000) Variation in abundance of Norwegian spring-spawning herring (*Clupea harengus*, Clupeidae) throughout the 20th century and the influence of climatic fluctuations. *Fish and Fisheries* 1: 231–256.

SMALL-SCALE PATCHINESS, MODELS OF

D. J. McGillicuddy Jr, Woods Hole Oceanographic Institution, Woods Hole, MA, USA

doi:10.1006/rwos.2001.0405

Introduction

Patchiness is perhaps the most salient characteristic of plankton populations in the ocean. The scale of this heterogeneity spans many orders of magnitude in its spatial extent, ranging from planetary down to microscale (**Figure 1**). It has been argued that patchiness plays a fundamental role in the functioning of marine ecosystems, insofar as the mean conditions may not reflect the environment to which organisms are adapted. For example, the fact that some abundant predators cannot thrive on the mean concentration of their prey in the ocean implies that they are somehow capable of exploiting small-scale patches of prey whose concentrations are much larger than the mean. Understanding the nature of this patchiness is thus one of the major challenges of oceanographic ecology.

The patchiness problem is fundamentally one of physical–biological–chemical interactions. This interconnection arises from three basic sources: (1) ocean currents continually redistribute dissolved and suspended constituents by advection; (2) space–time fluctuations in the flows themselves impact biological and chemical processes; and (3) organisms are capable of directed motion through the water. This tripartite linkage poses a difficult challenge to understanding oceanic ecosystems: differentiation between the three sources of variability requires accurate assessment of property distributions in space and time, in addition to detailed knowledge of organismal repertoires and the processes by which ambient conditions control the rates of biological and chemical reactions.

Various methods of observing the ocean tend to lie parallel to the axes of the space/time domain in which these physical–biological–chemical interactions take place (**Figure 2**). Given that a purely observational approach to the patchiness problem is not tractable with finite resources, the coupling of models with observations offers an alternative which provides a context for synthesis of sparse data with articulations of fundamental principles assumed to govern functionality of the system. In a sense, models can be used to fill the gaps in the space/time domain shown in **Figure 2**, yielding a framework for exploring the controls on spatially and temporally intermittent processes.

The following discussion highlights only a few of the multitude of models which have yielded insight into the dynamics of plankton patchiness. Examples have been chosen to provide a sampling of scales which can be referred to as 'small' – that is, smaller than the planetary scale shown in **Figure 1A**. In addition, this particular collection of examples is intended to furnish some exposure to the diversity of modeling approaches which can be brought to bear on the problem. These approaches range from abstract theoretical models intended to elucidate specific processes, to complex numerical formulations which can be used to actually simulate observed distributions in detail.

Formulation of the Coupled Problem

A general form of the coupled problem can be written as a three-dimensional advection-diffusion-reaction equation for the concentration C_i of any particular organism of interest:

$$\underbrace{\frac{\partial C_i}{\partial t}}_{\text{local rate of change}} + \underbrace{\nabla \cdot (\mathbf{v}C_i)}_{\text{advection}} - \underbrace{\nabla \cdot (K \nabla C_i)}_{\text{diffusion}}$$

$$= \underbrace{R_i}_{\text{biological sources/sinks}} \qquad [1]$$

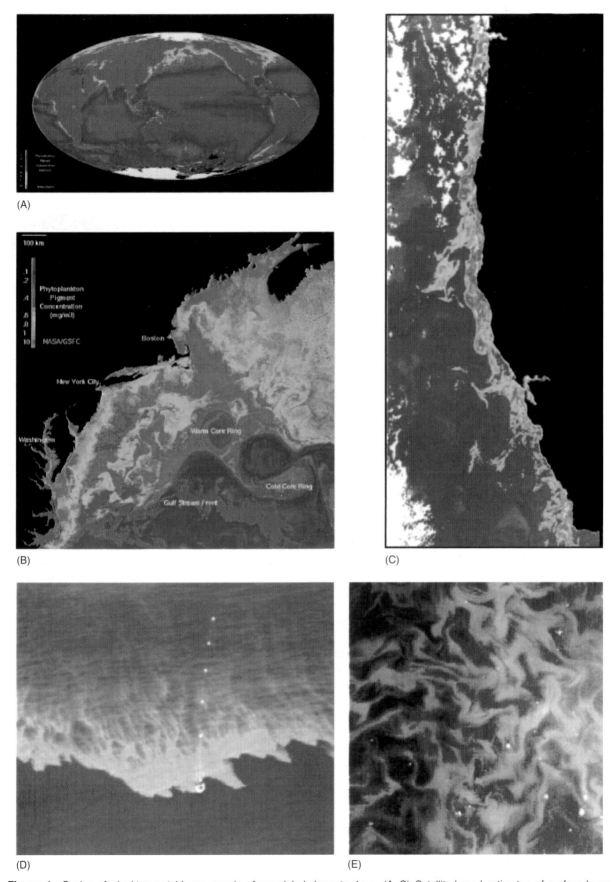

Figure 1 Scales of plankton patchiness, ranging from global down to 1 cm. (A–C) Satellite-based estimates of surface-layer chlorophyll computed from ocean color measurements. Images courtesy of the Seawifs Project and Distributed Active Archive Center at the Goddard Space Flight Center, Sponsored by NASA. (D) A dense stripe of *Noctiluca scintillans*, 3 km off the coast of La Jolla. The boat in the photograph is trailing a line with floats spaced every 20 m. The stripe stretched for at least 20 km parallel to the shore (photograph courtesy of P.J.S. Franks). (E) Surface view of a bloom of *Anabaena flos-aquae* in Malham Tarn, England. The area shown is approximately 1 m² (photograph courtesy of G.E. Fogg).

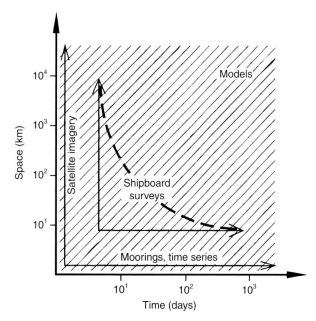

Figure 2 Space–time diagram of the scales resolvable with current observational capabilities. Measurements tend to fall along the axes; the dashed line running between the 'shipboard survey' axes reflects the trade-off between spatial coverage and temporal resolution inherent in seagoing operations of that type. Models can be used to examine portions of the space–time continuum (shaded area).

where the vector v represents the fluid velocity plus any biologically induced transport through the water (e.g., sinking, swimming), and K the turbulent diffusivity. The advection term is often written simply as $v \cdot \nabla C_i$ because the ocean is an essentially incompressible fluid (i.e., $\nabla \cdot v = 0$). The 'reaction term' R_i on the right-hand side represents the sources and sinks due to biological activity.

In essence, this model is a quantitative statement of the conservation of mass for a scalar variable in a fluid medium. The advective and diffusive terms simply represent the redistribution of material caused by motion. In the absence of any motion, eqn [1] reduces to an ordinary differential equation describing the biological and/or chemical dynamics. The reader is referred to the review by Donaghay and Osborn for a detailed derivation of the advection-diffusion-reaction equation, including explicit treatment of the Reynolds decomposition for biological and chemical scalars (see Further Reading).

Any number of advection–diffusion–reaction equations can be posed simultaneously to represent a set of interacting state variables C_i in a coupled model. For example, an ecosystem model including nutrients, phytoplankton, and zooplankton (an 'NPZ' model) could be formulated with $C_1 = dN$, $C_2 = P$ and $C_3 = Z$. The biological dynamics link-

ing these three together could include nutrient uptake, primary production, grazing, and remineralization. R_i would then represent not only growth and mortality, but also terms which depend on interactions between the several model components.

Growth and Diffusion – the 'KISS' Model

Some of the earliest models used to investigate plankton patchiness dealt with the competing effects of growth and diffusion. In the early 1950s, models developed independently by Kierstead (KI) and Slobodkin (S) and Skellam (S) – the so-called 'KISS' model – were formulated as a one-dimensional diffusion equation with exponential population growth and constant diffusivity:

$$\frac{\partial C}{\partial t} - K \frac{\partial^2 C}{\partial x^2} = \alpha C \qquad [2]$$

Note that this model is a reduced form of eqn [1]. It is a mathematical statement that the tendency for organisms to accumulate through reproduction is counterbalanced by the tendency of the environment to disperse them through turbulent diffusion. Seeking solutions which vanish at $x = 0$ and $x = L$ (thereby defining a characteristic patch size of dimension L), with initial concentration $C(x, 0) = f(x)$, one can solve for a critical patch size $L = \pi(K/\alpha)^{\frac{1}{2}}$ in which growth and dispersal are in perfect balance. For a specified growth rate α and diffusivity K, patches smaller than L will be eliminated by diffusion, while those that are larger will result in blooms. Although highly idealized in its treatment of both physical transport and biological dynamics, this model illuminates a very important aspect of the role of diffusion in plankton patchiness. In addition, it led to a very specific theoretical prediction of the initial conditions required to start a plankton bloom, which Slobodkin subsequently applied to the problem of harmful algal blooms on the west Florida shelf.

Homogeneous Isotropic Turbulence

The physical regime to which the preceding model best applies is one in which the statistics of the turbulence responsible for diffusive transport is spatially uniform (homogeneous) and has no preferred direction (isotropic). Turbulence of this type may occur locally in parts of the ocean in circumstances where active mixing is taking place, such as in a wind-driven surface mixing layer. Such motions might produce plankton distributions such as those shown in **Figure 1E**.

The nature of homogeneous isotropic turbulence was characterized by Kolmogoroff in the early 1940s. He suggested that the scale of the largest eddies in the flow was set by the nature of the external forcing. These large eddies transfer energy to smaller eddies down through the inertial subrange in what is known as the turbulent cascade. This cascade continues to the Kolmogoroff microscale, at which viscous forces dissipate the energy into heat. This elegant physical model inspired the following poem attributed to L. F. Richardson:

Big whorls make little whorls
which feed on their velocity;
little whorls make smaller whorls,
and so on to viscosity...

Based on dimensional considerations, Kolmogoroff proposed an energy spectrum E of the form

$$E(k) = A\varepsilon^{\frac{2}{3}}k^{-\frac{5}{3}}$$

where k is the wavenumber, ε is the dissipation rate of turbulent kinetic energy, and A is a dimensionless constant. This theoretical prediction was later borne out by measurements, which confirmed the 'minus five-thirds' dependence of energy content on wavenumber.

In the early 1970s, Platt published a startling set of measurements which suggested that for scales between 10 and 10^3 m the variance spectrum of chlorophyll in the Gulf of St Lawrence showed the same $-5/3$ slope. On the basis of this similarity to the Kolmogoroff spectrum, he argued that on these scales, phytoplankton were simply passive tracers of the turbulent motions. These findings led to a burgeoning field of spectral modeling and analysis of plankton patchiness. Studies by Denman, Powell, Fasham, and others sought to formulate more unified theories of physical–biological interactions using this general approach. For example, Denman and Platt extended a model for the scalar variance spectrum to include a uniform growth rate. Their theoretical analysis suggested a breakpoint in the spectrum at a critical wavenumber k_c (**Figure 3**), which they estimated to be in the order of $1\,\mathrm{km}^{-1}$ in the upper ocean. For wavenumbers lower than k_c, phytoplankton growth tends to dominate the effects of turbulent diffusion, resulting in a k^{-1} dependence. In the higher wavenumber region, turbulent motions overcome biological effects, leading to spectral slopes of -2 to -3. Efforts to include more biological realism in theories of this type have continued to produce interesting results, although Powell and others have cautioned that spectral characteristics may not be sufficient in and of themselves

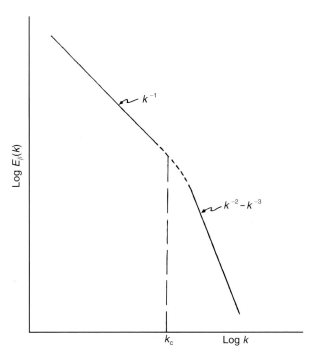

Figure 3 A theoretical spectrum for the spatial variability of phytoplankton, $E_\beta(k)$, as a function of wavenumber, k, displayed on a log-log plot. To the left of the critical wavenumber k_c, biological processes dominate, resulting in a k^{-1} dependence. The high wavenumber region to the right of k_c where turbulent motions dominate, has a dependence between k^{-2} and k^{-3}. (Reproduced with permission from Denman KL and Platt T (1976). The variance spectrum of phytoplankton in a turbulent ocean. *Journal of Marine Research* 34: 593–601.)

to resolve the underlying physical–biological interactions controlling plankton patchiness in the ocean.

Vertical Structure

Perhaps the most ubiquitous aspect of plankton distributions which makes them *anisotropic* is their vertical structure. Organisms stratify themselves in a multitude of ways, for any number of different purposes (e.g., to exploit a limiting resource, to avoid predation, to facilitate reproduction). For example, consider the subsurface maximum which is characteristic of the chlorophyll distribution in many parts of the world ocean (**Figure 4**). The deep chlorophyll maximum (DCM) is typically situated below the nutrient-depleted surface layer, where nutrient concentrations begin to increase with depth. Generally this is interpreted to be the result of joint resource limitation: the DCM resides where nutrients are abundant and there is sufficient light for photosynthesis. However, this maximum in chlorophyll does not necessarily imply a maximum in phytoplankton biomass. For example, in the

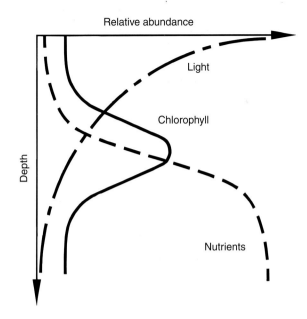

Figure 4 Schematic representation of the deep chlorophyll maximum in relation to ambient light and nutrient profiles in the euphotic zone (typically 10s to 100s of meters in vertical extent).

nutrient-impoverished surface waters of the open ocean, much of the phytoplankton standing stock is sustained by nutrients which are rapidly recycled; thus relatively high biomass is maintained by low ambient nutrient concentrations. In such situations, the DCM often turns out to be a pigment maximum, but not a biomass maximum. The mechanism responsible for the DCM in this case is photoadaptation, the process by which phytoplankton alter their pigment content according to the ambient light environment. By manufacturing more chlorophyll per cell, phytoplankton populations in this type of DCM are able to capture photons more effectively in a low-light environment.

Models have been developed which can produce both aspects of the DCM. For example, consider the nutrient, phytoplankton, zooplankton, detritus (NPZD) type of model (**Figure 5**) which simulates the flows of nitrogen in a planktonic ecosystem. The various biological transformations (such as nutrient uptake, primary production, grazing, excretion, etc.) are represented mathematically by functional relationships which depend on the model state variables and parameters which must be determined empirically. Doney *et al.* coupled such a system to a one-dimensional physical model of the upper ocean (**Figure 6**). Essentially, the vertical velocity (w) and diffusivity fields from the physical model are used to drive a set of four coupled advection-diffusion-reaction equations (one for each ecosystem state variable) which represent a subset of the full three-dimensional eqn [1]:

$$\frac{\partial C_i}{\partial t} + w\frac{\partial C_i}{\partial z} - \frac{\partial}{\partial z}\left(K\frac{\partial C_i}{\partial z}\right) = R_i \qquad [3]$$

The R_i terms represent the ecosystem interaction terms schematized in **Figure 5**. Using a diagnostic photoadaptive relationship to predict chlorophyll from phytoplankton nitrogen and the ambient light and nutrient fields, such a model captures the overall character of the DCM observed at the Bermuda Atlantic Time-series Study (BATS) site (**Figure 6**).

Broad-scale vertical patchiness (on the scale of the seasonal thermocline) such as the DCM is accompanied by much finer structure. The special volume of *Oceanography* on 'Thin layers' provides an excellent overview of this subject, documenting small-scale vertical structure in planktonic populations of many different types. One particularly striking example comes from high-resolution fluorescence measurements (**Figure 7A**). Such profiles often show strong peaks in very narrow depth intervals, which presumably result from thin layers of phytoplankton. A mechanism for the production of this layering was identified in a modeling study by P.J.S. Franks, in which he investigated the impact of near-inertial wave motion on the ambient horizontal and vertical patchiness which exists at scales much larger than the thin layers of interest. Near-inertial waves are a particularly energetic component in the internal wave spectrum of the ocean. Their horizontal velocities can be described by:

$$u = U_0\cos(mz - \omega t) \quad v = U_0\sin(mz - \omega t) \qquad [4]$$

where U_0 is a characteristic velocity scale, m is the vertical wavenumber, and ω the frequency of the wave. This kinematic model prescribes that the velocity vector rotates clockwise in time and counterclockwise with depth; its phase velocity is downward, and group velocity upward. In his words, 'the motion is similar to a stack of pancakes, each rotating in its own plane, and each slightly out of phase with the one below'. Franks used this velocity field to perturb an initial distribution of phytoplankton in which a Gaussian vertical distribution (of scale σ) varied sinusoidally in both x and y directions with wavenumber k_P. Neglecting the effects of growth and mixing, and assuming that phytoplankton are advected passively with the flow, eqn [1] reduces to:

$$\frac{\partial C}{\partial t} + u\frac{\partial C}{\partial x} + v\frac{\partial C}{\partial y} = 0 \qquad [5]$$

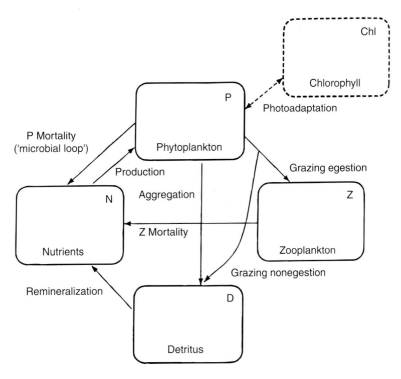

Figure 5 A four-compartment planktonic ecosystem model showing the pathways for nitrogen flow. (Reproduced with permission from Doney SC, Glover DM and Najjar RG (1996) A new coupled, one-dimensional biological–physical model for the upper ocean: applications to the JGOFS Bermuda Atlantic Time-series Study (BATS) site. *Deep-Sea Research II* 43: 591–624.)

Plugging the velocity fields [4] into this equation, the initial phytoplankton distribution can be integrated forward in time. This model demonstrates the striking result that such motions can generate vertical structure which is much finer than that present in the initial condition (**Figure 7B**). Analysis of the simulations revealed that the mechanism at work here is simple and elegant: vertical shear can translate horizontal patchiness into thin layers by stretching and tilting the initial patch onto its side (**Figure 7C**).

Mesoscale Processes: The Internal Weather of the Sea

Just as the atmosphere has weather patterns that profoundly affect the plants and animals that live on the surface of the earth, the ocean also has its own set of environmental fluctuations which exert fundamental control over the organisms living within it. The currents, fronts, and eddies that comprise the oceanic mesoscale, sometimes referred to as the 'internal weather of the sea', are highly energetic features of ocean circulation. Driven both directly and indirectly by wind and buoyancy forcing, their characteristic scales range from tens to hundreds of kilometers with durations of weeks to months. Their

space scales are thus smaller and timescales longer than their counterparts in atmospheric weather, but the dynamics of the two systems are in many ways analogous. Impacts of these motions on surface ocean chlorophyll distributions are clearly visible in satellite imagery (**Figure 1B**).

Mesoscale phenomenologies accommodate a diverse set of physical–biological interactions which influence the distribution and variability of plankton populations in the sea. These complex yet highly organized flows continually deform and rearrange the hydrographic structure of the near-surface region in which plankton reside. In the most general terms, the impact of these motions on the biota is twofold: not only do they stir organism distributions, they can also modulate the rates of biological processes. Common manifestations of the latter are associated with vertical transports which can affect the availability of both nutrients and light to phytoplankton, and thereby the rate of primary production. The dynamics of mesoscale and submesoscale flows are replete with mechanisms that can produce vertical motions.

Some of the first investigations of these effects focused on mesoscale jets. Their internal mechanics are such that changes in curvature give rise to horizontal divergences which lead to very intense

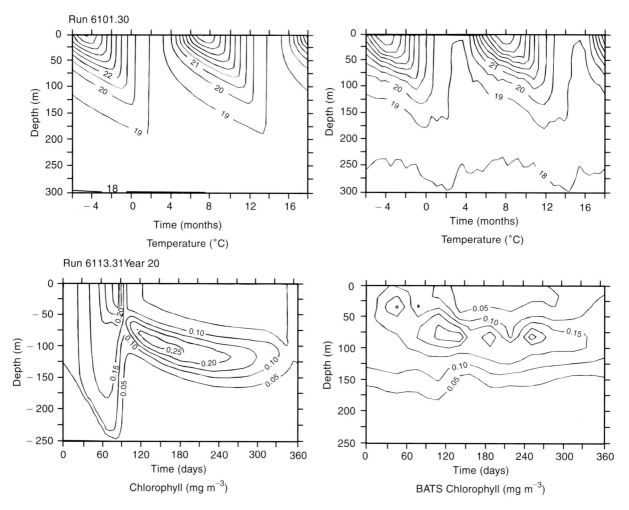

Figure 6 Simulated (left) and observed (right) seasonal cycles of temperature and chlorophyll at the BATS site. (Reproduced with permission from Doney SC, Glover DM and Najjar RG (1996) A new coupled, one-dimensional biological–physical model for the upper ocean: applications to the JGOFS Bermuda Atlantic Time-series Study (BATS) site. *Deep-Sea Research II* 43: 591–624.)

vertical velocities along the flanks of the meander systems (**Figure 8**). J.D. Woods was one of the first to suggest that these submesoscale upwellings and downwellings would have a strong impact on upper ocean plankton distributions (see his article contained in the volume edited by Rothschild; see Further Reading). Subsequent modeling studies have investigated these effects by incorporating planktonic ecosystems of the type shown in **Figure 5** into three-dimensional dynamical models of meandering jets. Results suggest that upwelling in the flank of a meander can stimulate the growth of phytoplankton (**Figure 9**). Simulated plankton fields are quite complex owing to the fact that fluid parcels are rapidly advected in between regions of upwelling and downwelling. Clearly, this complicated convolution of physical transport and biological response can generate strong heterogeneity in plankton distributions.

What are the implications of mesoscale patchiness? Do these fluctuations average out to zero, or are they important in determining the mean characteristics of the system? In the Sargasso Sea, it appears that mesoscale eddies are a primary mechanism by which nutrients are transported to the upper ocean. Numerical simulations were used to suggest that upwelling due to eddy formation and intensification causes intermittent fluxes of nitrate into the euphotic zone (**Figure 10A**). The mechanism can be conceptualized by considering a density surface with mean depth coincident with the base of the euphotic zone (**Figure 10B**). This surface is perturbed vertically by the formation, evolution, and destruction of mesoscale features. Shoaling density surfaces lift nutrients into the euphotic zone which are rapidly utilized by the biota. Deepening density surfaces serve to push nutrient-depleted water out of the well-illuminated surface layers. The asymmetric

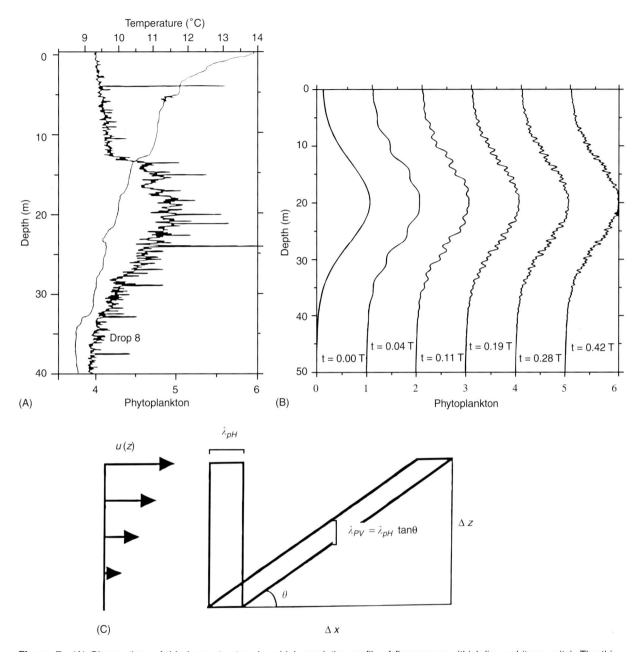

Figure 7 (A) Observations of thin layer structure in a high-resolution profile of fluorescence (thick line, arbitrary units). The thin line shows the corresponding temperature structure. (Data courtesy of Dr T. Cowles.) (B) Simulated vertical profiles of phytoplankton concentration (arbitrary units) at six sequential times. Each profile is offset from the previous by 1 phytoplankton unit. The times are given as fractions of the period of the near-inertial wave used to drive the model. (C) A schematic diagram of the layering process. Vertical shear stretches a vertical column of a property horizontally through an angle θ, creating a layer in the vertical profile. (Reproduced with permission from Franks PJS (1995) Thin layers of phytoplankton: a model of formation by near-inertial wave shear. *Deep-Sea Research I* 42: 75–91.)

light field thus rectifies vertical displacements of both directions into a net upward transport of nutrients, which is presumably balanced by a commensurate flux of sinking particulate material. Several different lines of evidence suggest that eddy-driven nutrient flux represents a large portion of the annual nitrogen budget in the Sargasso Sea. Thus, in this instance, plankton patchiness appears to be an essential characteristic that drives the mean properties of the system.

Coastal Processes

Of course, the internal weather of the sea is not limited to the eddies and jets of the open ocean. Coastal regions contain a similar set of phenomena,

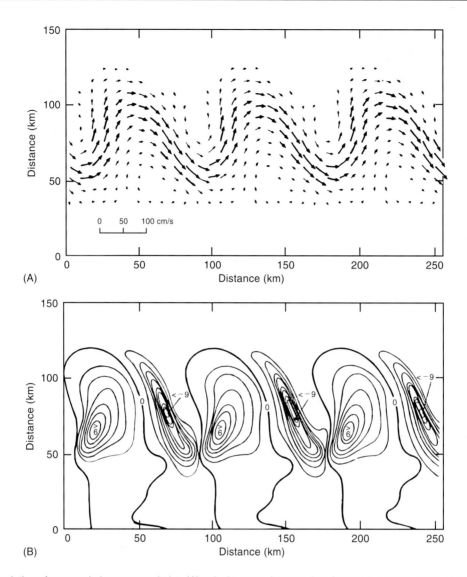

Figure 8 Simulation of a meandering mesoscale jet: (A) velocity on an isopycnal surface with a mean depth of 20 m; (B) vertical velocity (m d⁻¹) on the same isopycnal surface as in (A). Note the consistent pattern of the vertical motion with respect to the structure of the jet. (Reproduced with permission from Woods JD (1988) Mesoscale upwelling and primary production. In: Rothschild BJ (ed) *Toward a Theory on Physical–Biological Interactions in the World Ocean.* London: Kluwer Academic.)

in addition to a suite of processes in which the presence of a land boundary plays a key role. A canonical example of such a process is coastal upwelling, in which the surface layer is forced offshore when the wind blows in the alongshore direction with the coast to the left (right) in the northern (southern) hemisphere. This event triggers upwelling of deep water to replace the displaced surface water. The biological ramifications of this were explored in the mid-1970s by Wroblewski with one of the first coupled physical–biological models to include spatial variability explicitly. Configuring a two-dimensional advection-diffusion-reaction model in vertical plane cutting across the Oregon shelf, he studied the response of an NPZD-type ecosystem model to tran-

sient wind forcing. His 'strong upwelling' case provided a dramatic demonstration of mesoscale patch formation (**Figure 11**). Deep, nutrient-rich waters from the bottom boundary layer drawn up toward the surface stimulate a large increase in primary production which is restricted to within 10 km of the coast. The phytoplankton distribution reflects the localized enhancement of production, in addition to advective transport of the resultant biogenic material. Note that the highest concentrations of phytoplankton are displaced from the peak in primary production, owing to the offshore transport in the near-surface layers.

Although Wroblewski's model was able to capture some of the most basic elements of the

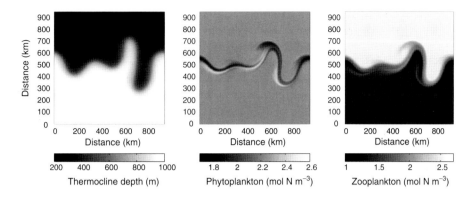

Figure 9   Results from a coupled model of the Gulf Stream: thermocline depth (left), phytoplankton concentration (middle), and zooplankton concentration (right). (Courtesy of GR Flierl, Massachusetts Institute of Technology.)

biological response to coastal upwelling, its two-dimensional formulation precluded representation of alongshore variations which can sometimes be as dramatic as those in the cross-shore direction. The complex set of interacting jets, eddies, and filaments characteristic of such environments (as in **Figure 1C**) have been the subject of a number of three-dimensional modeling investigations. For example, Moisan *et al.* incorporated a food web and bio-optical model into simulations of the Coastal Transition Zone off California. This model showed how coastal filaments can produce a complex biological response through modulation of the ambient light and nutrient fields (**Figure 12**). The simulations suggested that significant cross-shelf transport of carbon can occur in episodic pulses when filaments meander offshore. These dynamics illustrate the tremendous complexity of the processes which link the coastal ocean with the deep sea.

Behavior

The mechanisms for generating plankton patchiness described thus far consist of some combination of fluid transport and physiological response to the physical, biological, and chemical environment. The fact that many planktonic organisms have behavior (interpreted narrowly here as the capability for directed motion through the water) facilitates a diverse array of processes for creating heterogeneity in their distributions. Such processes pose particularly difficult challenges for modeling, in that their effects are most observable at the level of the population, whereas their dynamics are governed by interactions which occur amongst individuals. The latter aspect makes modeling patchiness of this type particularly amenable to individual-based models, in contrast to the concentration-based model described by eqn [1]. For example, many species of marine plankton are

known to form dense aggregations, sometimes referred to as swarms. Okubo suggested an individual-based model for the maintenance of a swarm of the form:

$$\frac{d^2x}{dt^2} = -k\frac{dx}{dt} - \omega^2 x - \phi(x) + A(t) \qquad [6]$$

where x represents the position of an individual. This model assumes a frictional force on the organism which is proportional to its velocity (with frictional coefficient k), a random force $A(t)$ which is white noise of zero mean and variance B, and attractive forces. Acceleration resulting from the attractive forces is split between periodic (frequency ω) and static ($\phi(x)$) components. The key aspect of the attractive forces is that they depend on the distance from the center of the patch. A Fokker-Planck equation can be used to derive a probability density function:

$$p(x) = p_0 \exp\left(-\frac{\omega^2}{2B}x^2 - \int \frac{\phi(x)}{b}\, dx \right) \qquad [7]$$

where p_0 is the density at the center of the swarm. Thus, the macroscopic properties of the system can be related to the specific set of rules governing individual behavior. Okubo has shown that observed characteristics of insect swarms compare well with theoretical predictions from this model, both in terms of the organism velocity autocorrelation and the frequency distribution of their speeds. Analogous comparisons with plankton have proven elusive owing to the extreme difficulty in making such measurements in marine systems.

The foregoing example illustrates how swarms can arise out of purely behavioral motion. Yet another class of patchiness stems from the joint effects of behavior and fluid transport. The paper by Flierl *et al.* is an excellent reference on this general topic

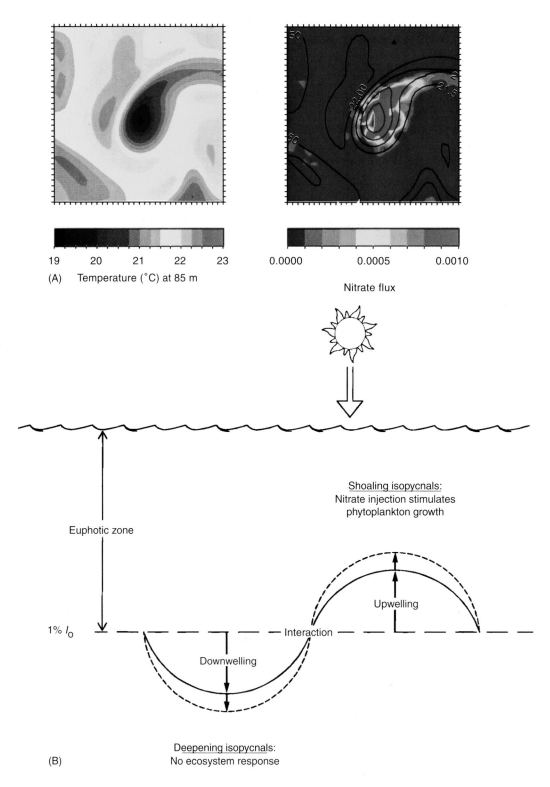

Figure 10 (A) A simulated eddy-driven nutrient injection event: snapshots of temperature at 85 m (left column, °C) and nitrate flux across the base of the euphotic zone (right column, moles of nitrogen $m^{-2}d^{-1}$). For convenience, temperature contours from the left-hand panels are overlayed on the nutrient flux distributions. The area shown here is a 500 km on a side domain. (The simulation is described in McGillicuddy DJ and Robinson AR (1997) Eddy induced nutrient supply and new production in the Sargasso Sea. *Deep-Sea Research I* 44(8): 1427–1450.) (B) A schematic representation of the eddy upwelling mechanism. The solid line depicts the vertical deflection of an individual isopycnal caused by the presence of two adjacent eddies of opposite sign. The dashed line indicates how the isopycnal might be subsequently perturbed by interaction of the two eddies. (Reproduced with permission from McGillicuddy DJ *et al.* (1998) Influence of mesoscale eddies on new production in the Sargasso Sea. *Nature* 394: 263–265).

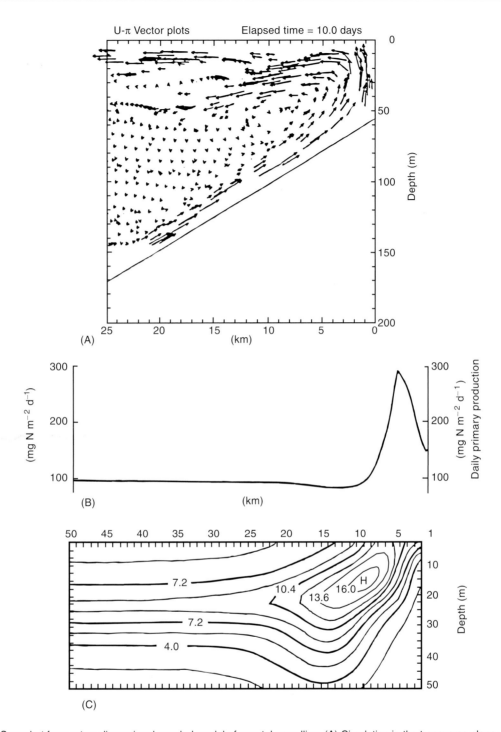

Figure 11 Snapshot from a two-dimensional coupled model of coastal upwelling. (A) Circulation in the transverse plane normal to the coast (maximum horizontal and vertical velocities are −6.1 and 0.05 cm s^{-1}, respectively); (B) daily gross primary production; (C) phytoplankton distribution (contour interval is 1.6 μg) at N l^{-1}. (Adapted with permission from Wroblewski JS (1977) A model of plume formation during variable Oregon upwelling. *Journal of Marine Research* 35(2): 357–394.)

(see Further Reading). One of the simplest examples of this kind of process arises in a population which is capable of maintaining its depth (either through swimming or buoyancy effects) in the presence of convergent flow. With no biological sources or sinks, eqn [1] becomes:

$$\frac{\partial C}{\partial t} + \mathbf{v} \cdot \nabla_H C + C \nabla_H \cdot \mathbf{v} - \nabla \cdot (K \nabla C) = 0 \quad [8]$$

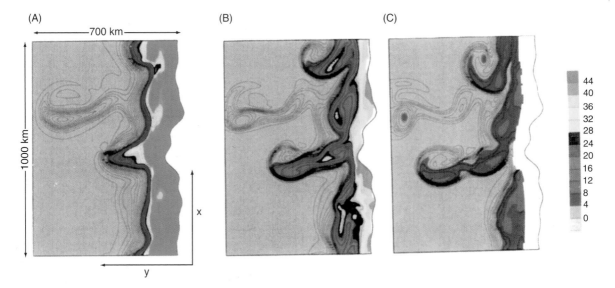

Figure 12 Modeled distributions of phytoplankton (color shading, mg nitrogen m^{-3}) in the Coastal Transition Zone off California. Instantaneous snapshots in panels (A–C) are separated by time intervals of 10 days. Contour lines indicate the depth of the euphotic zone, defined as the depth at which photosynthetically available radiation is 1% of its value at the surface. Contours range from 30 to 180 m, 40 to 180 m, and 60 to 180 m in panels (A), (B), and (C) respectively. (Reproduced with permission from Moisan *et al.* (1996) Modeling nutrient and plankton processes in the California coastal transition zone 2. A three-dimensional physical-bio-optical model. *Journal of Geophysical Research* 101(C10): 22 677–22 691.

where ∇_H is the vector derivative in the horizontal direction only. Because vertical fluid motion is exactly compensated by organism behavior (recall that the vector **v** represents the sum of physical and biological velocities), two advective contributions arise from the term $\nabla \cdot (\mathbf{v}C)$ in eqn [1]: the common form with the horizontal velocity operating on spatial gradients in concentration, *plus* a source/sink term created by the divergence in total velocity (fluid + organism). The latter term provides a mechanism for accumulation of depth-keeping organisms in areas of fluid convergence. It has been suggested that this process is important in a variety of different oceanic contexts. In the mid-1980s, Olson and Backus argued it could result in a 100-fold increase in the local abundance of a mesopelagic fish *Benthosema glaciale* in a warm core ring. Franks modeled a conceptually similar process with a surface-seeking organism in the vicinity of a propagating front (**Figure 13**). Simply stated, upward swimming organisms tend to accumulate in areas of downwelling. This mechanism has been suggested to explain spectacular accumulations of motile dinoflagellates at fronts (**Figure 1D**).

Conclusions

The interaction of planktonic population dynamics with oceanic circulation can create tremendously complex patterns in the distribution of organisms. Even an ocean at rest could accommodate significant inhomogeneity through geographic variations in environmental variables, time-dependent forcing, and organism behavior. Fluid motions tend to amalgamate all of these effects in addition to introducing yet another source of variability: space–time fluctuations in the flows themselves which impact biological processes. Understanding the mechanisms responsible for observed variations in plankton distributions is thus an extremely difficult task.

Coupled physical–biological models offer a framework for dissection of these manifold contributions to structure in planktonic populations. Such models take many forms in the variety of approaches which have been used to study plankton patchiness. In theoretical investigations, the basic dynamics of idealized systems are worked out using techniques from applied mathematics and mathematical physics. Process-oriented numerical models offer a conceptually similar way to study systems that are too complex to be solved analytically. Simulation-oriented models are aimed at reconstructing particular data sets using realistic hydrodynamic forcing pertaining to the space/time domain of interest. Generally speaking, such models tend to be quite complex because of the multitude of processes which must be included to simulate observations made in the natural environment. Of course, this complexity makes diagnosis of the coupled system more challenging. Nevertheless, the combination of models and observations provides a unique context for the synthesis of necessarily sparse data: space–time continuous representations of the real

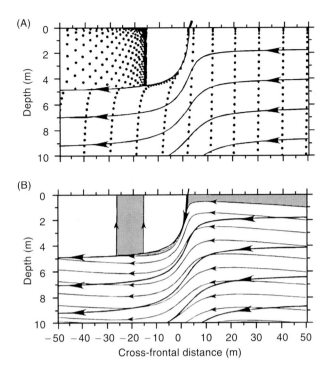

Figure 13 Surface-seeking organisms aggregating at a propagating front. Modeled particle locations (dots, panel (A)) and particle streamlines (thin lines, panel (B)) in the cross-frontal flow. The front is centered at $x = 0$, and the coordinate system translates to the right with the motion of the front. Flow streamlines are represented in both panels as bold lines; they differ from particle streamlines due to propagation of the front. The shaded area in (B) indicates the region in which cells are focused into the frontal zone, forming a dense band at $x = -20$ m. (Reproduced with permission from Franks, 1997.)

ocean which can be diagnosed term-by-term to reveal the underlying processes. Formal union between models and observations is beginning to occur through the emergence of inverse methods and data assimilation in the field of biological oceanography. **Biogeochemical Data Assimilation** provides an up-to-date review of this very exciting and rapidly evolving aspect of coupled physical–biological modeling.

Although the field is more than a half-century old, modeling of plankton patchiness is still in its infancy. The oceanic environment is replete with phenomena of this type which are not yet understood. Fortunately, the field is perhaps better poised than ever to address such problems. Recent advances in measurement technologies (e.g., high-resolution acoustical and optical methods, miniaturized biological and chemical sensors) are beginning to provide direct observations of plankton on the scales at which the coupled processes operate. Linkage of such measurements with models is likely to yield important new insights into the mechanisms controlling plankton patchiness in the ocean.

See also

Biogeochemical Data Assimilation. Coastal Circulation Models. Continuous Plankton Recorders. Fishery Management. Fish Migration, Vertical. Fluorometry for Biological Sensing. Fossil Turbulence. General Circulation Models. Krill. Mesoscale Eddies. Ocean Circulation. Ocean Color from Satellites. Patch Dynamics. Pelagic Biogeography. Phytoplankton Blooms. Small-scale Physical Processes and Plankton Biology. Three-dimensional (3D) Turbulence. Ships. Upper Ocean Mixing Processes.

Further Reading

Denman KL and Gargett AE (1995) Biological–physical interactions in the upper ocean: the role of vertical and small scale transport processes. *Annual Reviews of Fluid Mechanics* 27: 225–255.

Donaghay PL and Osborn TR (1997) Toward a theory of biological–physical control on harmful algal bloom dynamics and impacts. *Limnology and Oceanography* 42: 1283–1296.

Flierl GR, Grunbaum D, Levin S and Olson DB (1999) From individuals to aggregations: the interplay between behavior and physics. *Journal of Theoretical Biology* 196: 397–454.

Franks PJS (1995) Coupled physical–biological models in oceanography. *Reviews of Geophysics* (supplement): 1177–1187.

Franks PJS (1997) Spatial patterns in dense algal blooms. *Limnology and Oceanography* 42: 1297–1305.

Levin S, Powell TM and Steele JH (1993) *Patch Dynamics*. Berlin: Springer-Verlag.

Mackas DL, Denman KL and Abbott MR (1985) Plankton patchiness: biology in the physical vernacular. *Bulletin of Marine Science* 37: 652–674.

Mann KH and Lazier JRN (1996) *Dynamics of Marine Ecosystems: Biological–Physical Interactions in the Oceans*. Oxford: Blackwell Scientific Publications.

Okubo A (1980) *Diffusion and Ecological Problems: Mathematical Models*. Berlin: Springer-Verlag.

Okubo A (1986) Dynamical aspects of animal grouping: swarms, schools, flocks and herds. *Advances in Biophysics* 22: 1–94.

Robinson AR, McCarthy JJ and Rothschild BJ (2001) *The Sea: Biological–Physical Interactions in the Ocean*. New York: John Wiley and Sons.

Rothschild BJ (1988) *Toward a Theory on Biological–Physical Interactions in the World Ocean*. Dordrecht: D. Reidel.

Oceanography Society (1998) *Oceanography* 11(1): Special Issue on Thin Layers. Virginia Beach, VA: Oceanography Society.

Steele JH (1978) *Spatial Pattern in Plankton Communities*. New York: Plenum Press.

Wroblewski JS and Hofmann EE (1989) U.S. interdisciplinary modeling studies of coastal–offshore exchange processes: past and future. *Progress in Oceanography* 23: 65–99.

SMALL-SCALE PHYSICAL PROCESSES AND PLANKTON BIOLOGY

J. F. Dower, University of British Columbia, Vancouver, BC, Canada

K. L. Denman, University of Victoria, Victoria BC, Canada

Copyright © 2001 Academic Press

doi:10.1006/rwos.2001.0209

Introduction

By definition, plankton are aquatic organisms (including plants, animals, and microbes) that drift in the water and which cannot swim against any appreciable current. Most plankton are also very small, usually much less than 1 cm in size. Thus, it should come as no surprise that the behavior of, and the interactions between, individual plankton are strongly influenced by small-scale physical processes. Although this may seem intuitive, the fact is that oceanographers have only been aware of the importance of small-scale physical processes to plankton ecology for about the past 20 years, and have only been able to directly study plankton biology at these scales for the past 10 years or so. The primary reason for this is that the space and timescales relevant to individual plankton (millimeters → meters, and seconds → hours) are quite small, making it extremely difficult to sample properly in the oceans. Thus, much of what we know about the effect of small-scale physical processes on plankton biology is necessarily based on empirical and theoretical studies.

One might well ask why we need to understand the behavior of individual plankton at all, especially given that traditional plankton ecology (at least as conducted in field studies) generally involves comparisons between averages. For instance, we usually compare differences in the average zooplankton density or the average rate of primary productivity at several sites. However, in comparing averages, researchers make many implicit (though often unstated) assumptions about the behavior of the individual plankton that comprise these populations. Specifically, researchers assume that differences in average population-level responses are merely the sum of many individual responses. But is this a realistic assumption? To explore this idea, let us consider what individual planktonic organisms actually 'do' and examine how they are affected by small-scale physical processes.

Life in the Plankton

First, let us consider a typical phytoplankton cell. We will assume it is a large (e.g., $100 \, \mu m$) diatom. In the simplest sense, the biological processes of prime importance to this organism are that of finding sufficient light and nutrients to photosynthesize, grow, and reproduce, while at the same time trying to avoid sinking and predation. Next, let us consider a typical zooplanktonic organism. We will assume it is a copepod. What does it actually 'do'? On a daily basis it spends much of its time searching for food. Once food is encountered it must then be captured and ingested. In the meantime, this copepod must also try to avoid its own predators. Toward these goals of predator avoidance and feeding, certain zooplankton (including many copepods) also undertake diel vertical migrations of hundreds of meters (for more details on diel vertical migrations). Assuming our copepod reaches maturity (having successfully avoided being eaten) it must then find a mate in order to reproduce. Finally, like the diatom cell, our copepod must also try to avoid sinking. How might these various biological processes be affected by small-scale physics? There are two main factors that must be considered: viscosity and turbulence. Interestingly, both are related to the interaction between the very small size of most plankton, and the nature of the fluid environment that they inhabit.

Effects of Viscosity

The first thing to consider is that, although they live in open water and are transported by the background flow, the world probably feels quite 'sticky' to most plankton. This is due to the fact that at very small spatial scales and at the relatively slow swimming speeds or sinking rates of most plankton (i.e., of order $mm \, s^{-1}$), viscous forces dominate over inertial forces. The relative importance of viscous to inertial forces can be determined by calculating a dimensionless quantity known as the Reynolds number:

$$Re = \ell U / v \qquad [1]$$

where ℓ is the characteristic length (m) of the object in question, U is the speed ($m \, s^{-1}$) at which the object is moving, and v is the kinematic viscosity ($m^2 \, s^{-1}$) of the medium in which the object is

Table 1 Approximate value of Reynolds numbers for the swimming speeds of various animals. Note that the kinematic viscosity was taken as $1.05 \times 10^{-6}\,m^2\,s^{-1}$, which approximates that of sea water at 20°C

Animal	Length (m)	Speed (m s^{-1})	Re
Whale	20	10	200 000 000
Tuna	2	10	20 000 000
Human	2	2	4 000 000
Small adult fish	0.5	0.5	240 000
Post-metamorphosis fish	0.05	0.05	2400
Larval fish	0.005	0.005	24
Adult copepod	0.002	0.02	4

moving. For our purposes we will take $v = 1.05 \times 10^{-6}$, a value typical of sea water at about 20°C.

When $Re > 2000$, inertial forces dominate and the flow is turbulent. For values of $Re < 2000$ viscous forces become progressively more important. Note that Re is not meant to be a precise measure, serving merely as a 'ballpark measure' for comparing the flow regimes that apply to different objects. **Table 1** lists some estimates of Re that apply to the swimming motions of various animals. Note that for most plankton, Re is generally less than 100, meaning that flow conditions are strongly viscous. Thus, once active swimming ceases, instead of continuing to glide (as does an adult fish once it ceases swimming), the low Reynolds numbers that apply to most plankton ensure that they come to an almost immediate stop.

There is at least one notable exception to this, however; the escape responses initiated by certain copepods. Copepods are equipped with antennae that serve as highly sensitive mechanoreceptors. In addition to detecting food and potential mates, these mechanoreceptors can also warn of approaching predators. Recent work has shown that some copepods can initiate escape responses (a series of rapid hops in a direction away from the perceived threat) in $< 10\,ms$. Moreover, they can achieve burst speeds of several hundred body lengths per second. In some cases, the acceleration achieved is sufficient to enable the copepod to break through the 'viscous barrier' and enter the inertial world, albeit temporarily. Some copepod species have even evolved myelinated axons in their antennae to boost signal conduction along the nerves responsible for initiating the escape response.

Effects of Turbulence

Although occasionally occurring in very dense patches most zooplankton are actually rather dilute. Concentrations of $10–100\,l^{-1}$ are typical for many neritic copepods, and concentrations of planktonic organisms such as jellyfish and larval fish are usually orders of magnitude lower still (e.g. 1 per $10–100\,m^3$). Thus, given these low concentrations, it has long been (and continues to be) widely held that food concentrations in the ocean are limiting to growth and biological production. Throughout the 1960s and 1970s this belief was strengthened by the observation that successfully rearing copepods and larval fish in the lab often required food concentrations several times higher than those encountered in the oceans.

The point to consider, however, is that at the sub-meter scales relevant to most plankton, water motions tend to be dominated by random turbulence rather than directional flow. Turbulence is a ubiquitous feature of the ocean, and exists at all scales. In the surface layer of the ocean turbulent energy usually comes from the mixing effects of winds, or from the current shear between layers of water moving in different directions. In shallow coastal waters a second source of turbulence is tidal friction with the seafloor, which stirs the water column from the bottom up. From the perspective of our individual plankton, the key point is that in the surface layer of the ocean these small-scale turbulent velocities tend to be of the same order of magnitude as (or larger than) the typical swimming and/or sinking velocities of most plankton. How will this affect interactions between individual plankton?

Turbulence and Predator–Prey Interactions

Until the late 1980s, the contact rate between planktonic predators and prey was usually expressed as a simple function of (1) the relative swimming velocities of the predator and its prey, and (2) the prey concentration. Put simply, the faster predator and prey swim, and the higher the prey concentration, the more often should the predator and prey randomly encounter each other. Numerically, this can be represented as:

$$Z = D \times A \qquad [2]$$

where Z is the encounter rate, D is the prey density, and A is the relative velocity term. However, in 1988 researchers first theorized that small-scale turbulence might play an important role in predator–prey interactions in the plankton. Specifically, it was hypothesized that small-scale turbulent motions can randomly bring predator and prey together in the water column and, thus, that D and A from eqn. 2 should be rewritten as follows:

$$D = \pi R^2 N \qquad [3]$$

and

$$A = (u^2 + 3v^2 + 4w^4)/\sqrt{3(v^2 + w^2)} \qquad [4]$$

where R is the distance at which the predator can detect the prey (m), N is the prey concentration (m^{-3}), u and v are the prey and predator swimming speeds $(m\,s^{-1})$, and w is the turbulent velocity $(m\,s^{-1})$. Eqn. [3] assumes that the predator searches a circular area in front of itself, the result being that as it swims forward its 'search volume' assumes the shape of a cylindrical tube of radius R. In fact, this is only one of many search geometries displayed by visual planktonic predators. For instance, although some larval fish (e.g., herring larvae) are 'cruise predators' and scan a cylinder of water as described above, laboratory experiments have also shown that other species (e.g., cod larvae) are better described as 'pause–travel' predators. In this case, the predator searches for prey only during short pauses. These are followed by short bursts of swimming, during which there is no searching. Furthermore, the geometry of the volume searched also appears to be species specific, although in general it assumes the shape of a 'pie-shaped wedge' centered along the predator's line of vision.

There are also many predatory zooplankton that do not rely on vision at all. These include mechanoreceptor predators (e.g., raptorial copepods, chaetognaths) which detect the vibrations and hydrodynamic disturbances created by approaching prey, and contact predators (e.g., jellyfish, ctenophores) armed with stinging or entangling tentacles and which essentially rely on prey bumping into them. There are also a wide variety of filter-feeding (e.g., copepods, larvaceans) and suspension feeding zooplankton (e.g., heteropods) which generally feed on phytoplankton and protozoans that are either strained out of a feeding current or which are ingested after becoming trapped on mucus coated surfaces.

Regardless of the search geometry and mode of feeding, however, the general hypothesis is that as turbulence increases so, too, should encounter rates between predator and prey. Among the chief reasons plankton ecologists are interested in this idea is the longstanding belief that food availability is an important regulator of the growth and survival of zooplankton and larval fish. Throughout the 1990s, researchers sought empirical evidence of this phenomenon. A number of laboratory studies did show that copepods encounter more prey under increased turbulence. These experiments also showed that many copepods initially respond to increased turbulence by initiating escape responses. The explanation of this result was that, being mechanoreceptor (rather than visual) predators, copepods initially interpret an abrupt increase in turbulence as signaling the approach of a potential predator. Other laboratory studies have since demonstrated that larval fish also initiate more attacks and have higher levels of gut fullness under increased turbulence.

Of course, there are limitations to what can be modeled realistically in any laboratory study, especially those studying small-scale physical processes. For instance, the experiments described above typically involved videotaping the behavior of individual copepods that had been tethered in a flow field in which the turbulence could be varied. It remains to be seen whether the behavior of such tethered animals is the same as that of free-swimming copepods. Likewise, experiments on the effect of turbulence on larval fish feeding ecology usually involve offering only a single type and size of prey at a time, and usually at unrealistically high prey concentrations (e.g. 1000s of prey per liter). In the ocean, of course, larval fish encounter a wide range of prey types and prey sizes, many of which typically occur at relatively low concentrations.

Logistically, the biggest challenge is that of trying to create a realistic turbulent field under laboratory conditions. The small size of most plankton necessitates the use of rather small volume aquaria for experimentation, especially if (as is often the case) the goal is to use videographic techniques to follow individual plankton. Such experiments generally rely on variable speed oscillating grids or paddles to generate different levels of turbulence. Thus, although empirical studies have taught us a lot about the potential importance of turbulence in plankton ecology, the question remains as to whether this process is actually important in the ocean.

Further refinements of the theory have since suggested that the relationship between turbulence and feeding success should be dome-shaped (rather than linear) since, at some point, the predator will be unable to react to prey before they are carried away by high levels of turbulence (**Figure 1**). However,

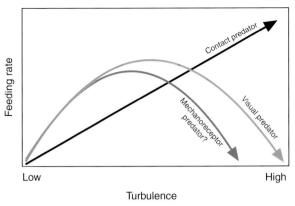

Figure 1 Possible responses of a visual predator, a mechanoreceptor predator, and a contact predator to turbulence. The domed response of visual predators such as larval fish has been supported by laboratory studies. The other two lines are more speculative. The linear increase proposed for contact predators (e.g. jellyfish and ctenophores) is based on the premise that these predators rely on prey bumping into them. However, given that mechanoreceptor predators such as raptorial copepods actually 'hear' the turbulence, it may well be that their ability to separate the signal of an approaching prey item from that of the background turbulent noise declines at higher turbulence levels, leading to a lower optimum level of turbulence than for the other types of predators.

this theory is based largely on the assumption of a visual predator. It remains to be seen how other types of planktonic predators should respond to turbulence. For instance, will the functional response of a mechanoreceptor predator (e.g., a raptorial copepod) be dome-shaped, too? Perhaps such a predator should have a lower 'optimum' level of turbulence than a visual predator since, at high turbulence levels, the background turbulence might make it more difficult to sense prey? Similarly, perhaps a linear response should be expected for 'contact predators' such as jellyfish or ctenophores. Further laboratory work will be needed to clarify this matter.

There have been some attempts to quantify the effect of turbulence on zooplankton and larval fish in the field. To date, although the evidence broadly suggests that turbulence increases encounter rates and even gut fullness (**Figure 2**), its effect on the growth and survival rates of larval fish and zooplankton remains uncertain. What is known is that not all species respond to turbulence in the same way. For instance, although gut fullness in larval Atlantic cod (*Gadus morhua*) and radiated shanny (*Ulvaria subbifurcata*) increases in response to increased turbulence, other species such as herring (*Clupea harengus*) and walleye pollock (*Theragra chalcogramma*) actively move deeper in the water column to avoid turbulence during windy condi-

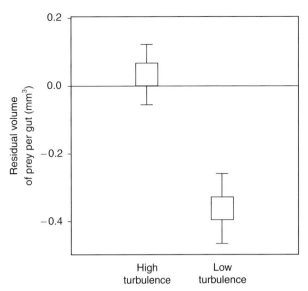

Figure 2 The effect of turbulence on gut fullness in larval radiated shanny *(Ulvaria subbifurcata)* during a 23-day time-series in July/August 1995 in coastal Newfoundland. Despite the fact that prey concentrations remained relatively constant throughout, the figure shows that the average volume of food in the guts of larval fish was significantly higher on high turbulence days than on low turbulence days. Note that the data have been detrended, since larger larvae will generally always have more food in their guts than smaller larvae. The figure thus shows the residual volume of food per gut (i.e., after the size effect has been removed). (Redrawn from Dower *et al.*, 1998.).

tions. Similarly, field observations demonstrate that the vertical segregation of at least two congeneric copepod species (*Neocalanus cristatus* and *Neocalanus plumchrus*) is determined by species-specific preferences for different turbulent regimes. Such responses make it extremely difficult to generalize whether turbulence is of net positive benefit to zooplanktonic predators.

Turbulence and Reproductive Ecology

In addition to its effect on feeding ecology, turbulence also appears to play a role in zooplankton reproduction. Recent work has shown that the males of some copepod species find mates by following chemical 'odor trails' left by the females. High-resolution videographic observations show the male swimming back and forth until he crosses the odor trail, at which point he immediately reverses direction to pick up the trail again. Having locked onto the trail the male then follows it back to the female (**Figure 3**). Should the male initially head the wrong way down the trail, he quickly reverses direction and goes the right way until reaching the female.

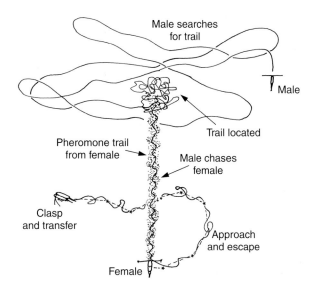

Figure 3 Conceptual interpretation of mate-attraction and mate-searching behavior in the copepod *Calanus marshallae*. The sequence of events first involves the female producing a pheromone trail, which alerts the male that a female is in the area. The male initially swims in smooth horizontal loops until he crosses the trail. At this point the male follows the trail back to the female. (Redrawn from Tsuda and Miller (1998) and used with the kind permission of the authors.).

Turbulence comes into play because the odor trails persist for only a few seconds before being dissipated by small-scale turbulent motions, making it harder for the males to find females. Although it has yet to be confirmed in the lab, this may partly explain why many zooplankton form dense swarms during mating. If the males are to successfully find females via odor trails it might be predicted that mating swarms should be dense enough that the time taken to randomly encounter an odor trail is shorter than the time taken for the trails to dissipate. It is also known that copepod species that usually inhabit the near-surface layers of the ocean often descend to depths of hundreds of meters to reproduce. Is it possible that this behavior is partly a response to the need for relatively quiescent waters in order to allow males to track females before odor trails are dissipated by small-scale turbulent motions?

Turbulence and Phytoplankton Ecology

Recall our typical diatom. Like other phytoplankton, it needs to stay relatively close to the surface in order to capture sufficient light for photosynthesis. Phytoplankton have evolved a variety of adaptations to reduce sinking, including the production of spines and other external ornamentations (to increase drag), the inclusion of oil droplets (to increase buoyancy), and the evolution of hydrodynamic geometries that induce 'side-slipping' (as opposed to vertical sinking). Despite all these adaptations, however, many phytoplankton, particularly the larger diatoms with their relatively heavy siliceous frustules, still rely on turbulent mixing to remain suspended in the water column. It has also been shown that turbulence plays an important role in determining the taxonomic composition of phytoplankton communities. For instance, the phytoplankton communities of upwelling zones are usually dominated by large diatoms, whereas less turbulent regions (e.g., open ocean) are more often dominated by much smaller phytoplankton taxa. Similarly, and particularly at temperate latitudes where there is a strong seasonality in wind-mixing, there is often a seasonal succession of phytoplankton species in coastal waters whereby large diatoms dominate during the spring bloom when winds, and thus near-surface turbulent mixing, are strongest. As the summer progresses, and stratification of the upper water column proceeds, large diatoms are often succeeded by more motile forms (e.g., dinoflagellates and coccolithophores) and smaller phytoplankton (e.g., cyanobacteria) that rely less on turbulence to remain near the surface.

A growing body of literature suggests that turbulence can also have novel effects on dinoflagellates. Many dinoflagellate species are bioluminescent, producing a flash of light when disturbed. Recent work has revealed that such bioluminescence may provide an indirect means of reducing predation. Experiments show that some dinoflagellates only bioluminesce in response to strong current shear, such as that induced by breaking surface waves or by predators attempting to capture them. The so-called 'burglar alarm' theory suggests that, by flashing in response to strong shear and lighting up the water around themselves, the dinoflagellates make their potential predators more visible, and thereby more prone to predation from *their* predators. This effect has been demonstrated in the lab, where species of squid and fish (feeding on mysids, shrimp, and other fish species) have been observed to increase their rates of attack and capture success when the water in which they are foraging contains bioluminescent dinoflagellates. Apparently, prey movements induce the dinoflagellates to flash, thereby illuminating the prey and making it easier for the predator to capture them.

Turbulence has also been shown to have negative effects on dinoflagellates. It has long been known that dinoflagellate blooms are most common when conditions are calm. However, recent work has shown that even moderate amounts of turbulent mixing can actually inhibit the growth of certain dinoflagellates. The effect seems to be that of preventing cell division, since other cellular processes (e.g., photosynthesis, pigment synthesis, nucleic acid synthesis) appear to continue as usual. If the period of turbulent mixing is short-lived, the cells recover and quickly begin dividing, possibly going on to form a bloom. If, however, the turbulence continues for longer periods, the cells will be prevented from dividing, or they may even die. Of particular interest is that the growth of at least two dinoflagellate species that form harmful algal blooms *(Lingulodinium polyedrum* and *Heterosigma carterae)* seems to be inhibited by turbulence.

Conclusions

Although our basic view of plankton as being organisms that 'go with the flow' still holds true, over the past 20 years we have also learned that interactions between plankton and their physical environment can be quite complex. Perhaps the most important result of this research has been the realization that small-scale physical processes affect so many different aspects of plankton ecology including feeding, predator–prey interactions, swimming and buoyancy, nutrient diffusion, mate selection, and even patterns of community composition. That many of these discoveries are rather new is largely due to the fact that, until relatively recently, oceanographers were simply unable to conduct experiments and observe plankton at appropriately small scales. Now that we have that capability, however, the challenge in the coming years will be to find ways to integrate what has been learned about the behavior of individual plankton to further develop our understanding of population-level processes. Are population-level processes merely the sum of innumerable individual interactions, or are

there other physical processes that affect populations at larger space and timescales? Alas, we do not yet know the answer to this question. However, finding new ways to extend what has been learned in the laboratory into more realistic field settings may prove one step in the right direction.

See also

Fish Larvae. Plankton.

Further Reading

Berdalet E (1992) Effects of turbulence on the marine dinoflagellate *Gymnodinium nelsonii. Journal of Phycology* 28: 267–272.

Dower JF, Miller TJ and Leggett WC (1997) The role of microscale turbulence in the feeding ecology of larval fish. *Advances in Marine Biology* 31: 170–220.

Dower JF, Pepin P and Leggett WC (1998) Enhanced gut fullness and an apparent shift in size selectivity by radiated shanny *(Ulvaria subbifurcata)* larvae in response to increased turbulence. *Canadian Journal of Fisheries and Aquatic Sciences* 55: 128–142.

Fuiman LA and Batty RS (1997) What a drag it is getting cold: partitioning the physical and physiological effects of temperature on fish swimming. *Journal of Experimental Biology* 200: 1745–1755.

Kiorboe T (1993) Turbulence, phytoplankton cell size, and the structure of pelagic food webs. *Advances in Marine Biology* 29: 1–72.

Mensinger AF and Case JF (1992) Dinoflagellate luminescence increases susceptibility of zooplankton to teleost predation. *Marine Biology* 112: 207–210.

Rothschild BJ and Osborn TR (1988) Small-scale turbulence and plankton contact rates. *Journal of Plankton Research* 10: 465–474.

Tsuda A and Miller CB (1998) Mate-finding behaviour in *Calanus marshallae* Frost. *Philosophical Transactions of the Royal Society of London,* B 353: 713–720.

Vogel S (1989) *Life in Moving Fluids: The Physical Biology of Flow*, 3rd edn. Princeton, NJ: Princeton University Press.

SOMALI CURRENT

M. Fieux, Université-Pierre et Marie Curie, Paris, France

Copyright © 2001 Academic Press

doi:10.1006/rwos.2001.0364

Introduction

The western Indian Ocean is the only region of the world where a large boundary current, as strong as the Gulf Stream, reverses twice a year in response to the wind reversals during the north-east winter monsoon and the south-west summer monsoon. This region of the Somali current is known to undergo the highest variability of the world ocean circulation.

Along the Somali coast, the reversals of winds and currents, known for many centuries, have been used by the Arabic traders for their navigation along the African coast and towards India. The term 'monsoon' comes from the Arabic word 'mawsin', which means seasonal.

Far from the large oceanographic research centers, the Indian Ocean used to be relatively poorly observed. The first large-scale international experiment was set up during the 1960s: the International Indian Ocean Experiment (IIOE), whose results were gathered into the Oceanographic IIOE Atlas. Until recently, the Somali current was known to reverse just twice a year with the direction of the monsoon, flowing south-westward during the NE monsoon and north-eastward during the SW monsoon. It is only with the international 'INDEX' experiment in 1979, under the initiative of Henry Stommel, that a careful survey of the response of the Somali current to the onset of the SW monsoon has been carried out.

Winds

The winds are the main driver of currents, in particular near the surface; therefore, we first recall their main characteristics. Their seasonal variability over the western boundary of the Indian Ocean can be described in four periods: the winter monsoon period, the summer monsoon period, and the two transition periods between the two monsoons (*see* Figure 2 of **Current Systems in the Indian Ocean**).

North of the Equator, the winter (NE) monsoon blows from the north east, with moderate strength, between December to March–April. At the Equator the winds are weak and usually from the north.

During the transition period between the end of the NE monsoon and the beginning of the SW monsoon, in April–May, the winds north of the Equator calm down. At the Equator moderate eastward winds blow, which contrast with the westward winds over the equatorial Pacific and Atlantic Oceans. As early as the end of March or early April, the SE monsoon starts south of the Equator, between the ITCZ (Inter-Tropical Convergence Zone) about 10°S and the Equator, with southerly winds along the East African Coast.

In most years, north of the Equator, the onset of the SW monsoon develops in two phases. The onset of the SW monsoon involves a reversal of the winds, which reach the Equator in early May as weak winds and progress northward along the Somali coast. Then a strong increase in wind occurs typically in late May to mid-June. In some years the onset can be gradual with no intermediate decrease between the first phase and the full-strength winds. They reach their full strength over the Arabian Sea usually in June–July. The summer (SW) monsoon blows steadily from the south west from June to September–October and is much stronger than the winter monsoon. Along the high orography of the east African coast, a low-level wind jet, called the Findlater jet (a kind of atmospheric western boundary flow) develops, bringing the strongest winds along the Somali coast toward the Arabian Sea, particularly north-east of Cape Guardafui (the horn of Africa). These are the strongest steady surface wind flows in the world, with mean July wind speed of $12 \, \mathrm{m \, s^{-1}}$ and peaks exceeding $20 \, \mathrm{m \, s^{-1}}$. At the Equator, the winds are moderate from the south and decrease eastward. In the southern Indian Ocean, during the southern winter (July), the SE Trades intensify and penetrate farther north than during the southern summer (January); they reach the Equator in the western part of the ocean and are the strongest in the three oceans. During that season the air masses transported by the SE Trade Winds cross the Equator in the west and flow, loaded with moisture, toward the Asian continent where they bring the awaited monsoon rainfall (for the Indian subcontinent, 'monsoon' means the wet monsoon, i.e., the SW monsoon).

October–November corresponds to the second transition period between the end of the SW monsoon and the beginning of the NE monsoon. North of the Equator, the winds vanish and the sea surface temperature can exceed 30°C. At the Equator moderate eastward winds blow again as during the first transition period.

This particular wind regime is dominated off the Equator by a strong annual period. At the Equator, the zonal wind component is dominated by a semiannual period associated with the transition westerly winds, while the meridional wind component has a strongly annual periodicity associated with the monsoon reversals (*see* Figure 1 of **Indian Ocean Equatorial Currents**).

The Western Boundary Current System

As in the other oceans, the strongest currents are found close to the western shores of the ocean and are called the western boundary current system. In the western boundary region influenced by the monsoonal winds, i.e., north of 10°S, there are two western boundary currents: the East African Coastal Current (EACC), also called the Zanzibar current, which always flows north-eastward; and the Somali Current (SC), which is highly variable in contrast to other western boundary currents (*see* Figure 3 of **Current Systems in the Indian Ocean**). The Somali Current is the more intense, and reverses twice a year owing to the complete seasonal reversal of

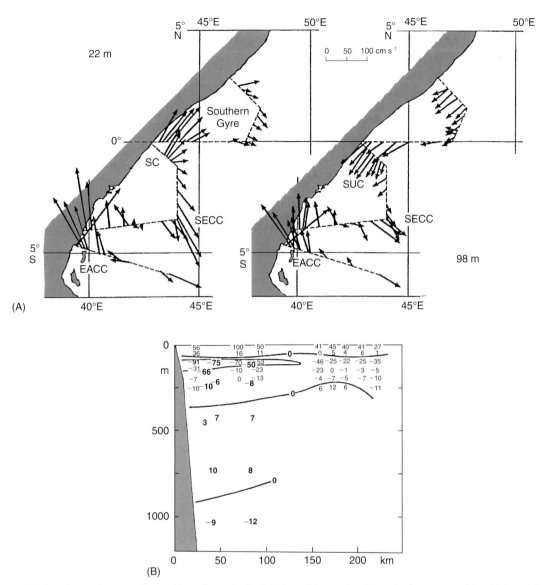

Figure 1 (A) Circulation along the East African Coast in April 1985 at 22 m and at 98 m depth, measured by shipboard Acoustic Doppler Current Profiler (ADCP). showing the EACC, the SECC, the reversal to the north at the equator of the Somali Current (SC), the Southern Gyre (SG) and the southward undercurrent (SUC) at the Equator. (B) Northward component of current along the equator in April 1985, from shipboard ADCP, in cm s⁻¹. Bold figures are mean northward components from moored currentmeters for the same period. (From Swallow et al., 1991, Structure and transport of the East African Coastal Current, *Journal of Geophysical Research* 96, C12, 22 245-22 257, 1991, copyright by the American Geophysical Union.)

the winds. It has been particularly studied during the Indian Ocean Experiment (INDEX) that started in the 1970s and the SINODE (Surface Indian Ocean Dynamic Experiment) in the 1980s. In 1990–1996, during the WOCE (World Ocean Circulation Experiment), a large number of new data were collected over the whole Indian Ocean and particularly in the Somali Basin.

The East African Coastal Current (Also Called the Zanzibar Current)

The EACC is fed by the branch of the South Equatorial Current (SEC) that passes north of Madagascar and splits northward around 11°S. It runs northward throughout the year between latitudes 11°S and 4°S. The location of its northern end depends on the season. In the northern winter, during the NE monsoon, the EACC converges around 3°–4°S with the southgoing Somali Current to form the eastward South Equatorial Counter Current (SECC). It flows against light winds and is then the weakest (**Figure 1A**).

Direct current measurements at 2°S during March–April 1970 and 1971 indicated that the current reversed to the north at least one month before the onset of the SW monsoon over the interior of the north Indian Ocean, immediately after the southerly winds began along the East African coast at the beginning of April (15–20 knots) and the onset of the SE monsoon to the south. At that time, the current is very sensitive to small variations in the local wind direction. The boundary between the northward (EACC) and the southward (SC) flow was distinctly marked by changes in fauna and water properties with lower salinities in the EACC and higher salinities in the south-westward Somali Current. At that time the EACC is strengthened by the winds. By the end of April, at 2°S, the current was about 100 nautical miles wide with peak speeds of 2 m s^{-1} within few miles offshore.

In April 1985, of the 10 Sv passing northward at 5°S between 0 and 100 m, 4.5 Sv continued across the Equator and 5.5 Sv join the SECC from the south (**Figure 1A**). The total transport down to 300 m of the SECC was 23 Sv of which 17 Sv came from the EACC. Most of the subsurface transport of the SECC moved eastward between 2.5°S and 6°S across 45°E and most of the northward near surface boundary current crossing the Equator was turning eastward south of 1.5°N. At the Equator, within the layer 0–100 m, below the surface northward transport, 2.5 Sv still flowed south across the Equator (**Figure 1B**).

Under the onset of the SW monsoon, the EACC merges into the reversing Somali Current, which progresses northward. Its surface speed can then exceed 2 m s^{-1} and its transport amounts to 20 Sv in the upper 500 m with 14 Sv in the upper 100 m at 1°S. Below the surface, the deeper EACC current flows northward across the Equator at all seasons. During the NE monsoon, it becomes an undercurrent under the southward-flowing Somali current.

The Somali Current

North of the Equator, the Somali current develops in different phases in response to the onset of the monsoon winds. **Figure 2** shows the evolution of the circulation from the 1979 SW monsoon observations. **Figure 3** gives a schematic representation of the western boundary current system for the different seasons for the surface layer.

During the transition period at the end of the NE monsoon, in April–early May, the Somali current flows south-westward along the coast from 4°–5°N to the Equator, whereas south of the Equator the EACC flows north-eastward (see above). At that time, the two currents converge and turn offshore to the south east to form the SECC. North of 4°–5°N, the current is already north-eastward, fed by the NE monsoon current, which brings waters from the interior Arabian Sea driven by the wind stress curl,

splitting into a northward boundary surface current between 5°N and 10°N associated with a southward subsurface current underneath, and a southward surface current towards the Equator.

During the early phase of the SW monsoon, in early May, the Somali current responds rapidly to the onset of the southerly winds at the Equator and reverses northward in continuity with the northward EACC, which crosses the Equator. It develops as a shallow cross-equatorial inertial current, turning offshore at about 3°N, where a cold upwelling wedge develops north of the turnoff latitude near the coast. As part of it recirculates southward across the Equator, it forms the anticyclonic Southern Gyre (**Figure 2**).

By mid-May, the SW wind onset propagates northward along the coast and the southern offshore-flowing branch, at 1°N to 3°N is strongly developed, with westward equatorial flow across 50°E indicating the recirculation of the Southern Gyre. North of that branch along the coast, the upwelling wedge spreads out bringing cold and enriched waters at the surface about 4°–5°N. Further to the north, the current is already northward from March. With southerly winds blowing parallel to the coast, a typical upwelling regime develops with northward surface flow, an undercurrent below and cold water along the coast.

When the onset of the strong summer monsoon winds occurs at these latitudes in June, the southern branch increases in strength and extends farther north (5°N) and a strong anticyclonic gyre, called the 'Great Whirl' develops between 5°N and 9°N with velocities at the surface higher than 2.5 m s^{-1} and transports around 90 Sv between the surface and 1000 m, where currents exceeding 0.1 m s^{-1} have been observed (**Figures 2 and 3**). Between the Somali coast and the northern branch of the Great Whirl a second strong upwelling wedge forms at its north-western flank where the flow turns offshore (**Figure 3**).

In August–September–October, depending on the year, when the winds decrease, it has been observed that the southern cold wedge propagates northward along the coast and coalesces with the northern one, which moves slightly northward (**Figure 4**). It is only at that time that the Somali current is continuous from the Equator up to the horn of Africa and brings fresher waters from the Southern Hemisphere into the Arabian Sea (**Figures 2, 5A–C, 6A,B**). The breakdown of the two-gyre system can occur at speeds of up to 1 m s^{-1}, replacing a 100 m thick and 100 km wide band of high-salinity water with lower-salinity water from south of the Equator, which represents a transport of 10 Sv.

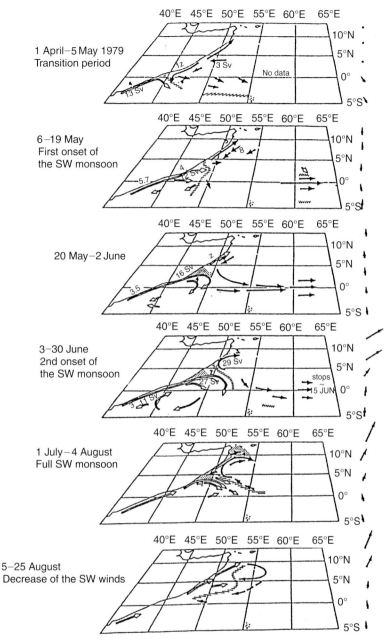

Figure 2 Evolution of the circulation during the onset of the SW monsoon from the 1979 observations during INDEX (speed in knots; transport in Sv = $10^6 \, m^3 \, s^{-1}$; open arrow = low salinity, solid arrow = high salinity, front, upwelling). The arrows on the right of the figure represent the wind stress observed at different latitudes along the coast; the full strength of the SW monsoon is reached in June (From M. Fieux, 1987, Circulation dans l'océan Indien occidental, Actes Colloque sur la Recherche Française dans les Terres Australes, Strasbourg, unpublished manuscript).

At the end of the summer monsoon and during the transition to the winter monsoon, in October, the continuous Somali Current no longer exists; instead the cross-equatorial flow, characterized by low surface salinities (35–35.2) with a transport of 12 Sv in the upper 100 m, turns offshore south of 2.5°N. The northward current component through the equatorial section has a subsurface maximum of more than $1.50 \, m \, s^{-1}$ near the coast at 40 m depth and velocities of more than $0.5 \, m \, s^{-1}$ at 200 m depth. The cross-equatorial transport in the upper 100 m was comparable to the 14 Sv transport at 1°S in the period May–June 1979. This means that the cross-equatorial transport of the Somali current in late autumn is very similar to that during the onset of the SW monsoon. The local winds are quite

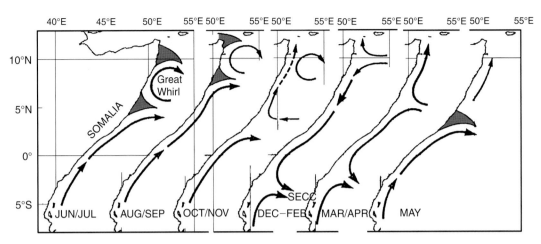

Figure 3 Somali Current flow patterns for the layer 0–100 m for different seasons with upwelling in grey (Reprinted from *Deep Sea Research*, 37(12), F. Schott et al., 1990, The Somali Current at the Equator: annual cycle of currents and transports in the upper 1000 m and connection to neighbouring latitudes, 1825–1848, copyright 1990, with permission from Elsevier Science).

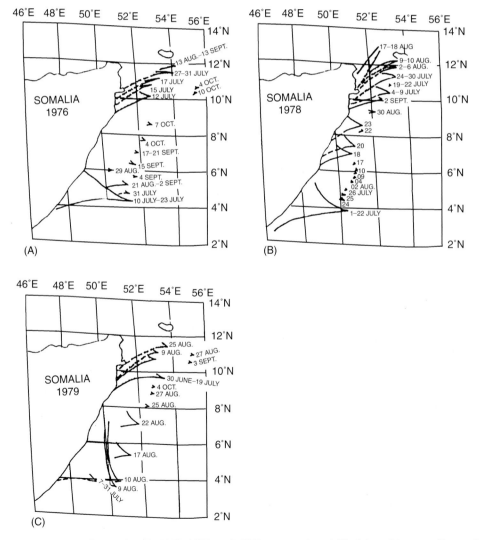

Figure 4 Propagation of upwelling wedges in 1976, 1978, and 1979, as seen in satellite infrared imagery (the northern one is in gray). (Reprinted from *Progress in Oceanography* 12, F. Schott, Monsoon response of the Somali Current and associated upwelling, 357–381, copyright 1983, with permission from Elsevier Science).

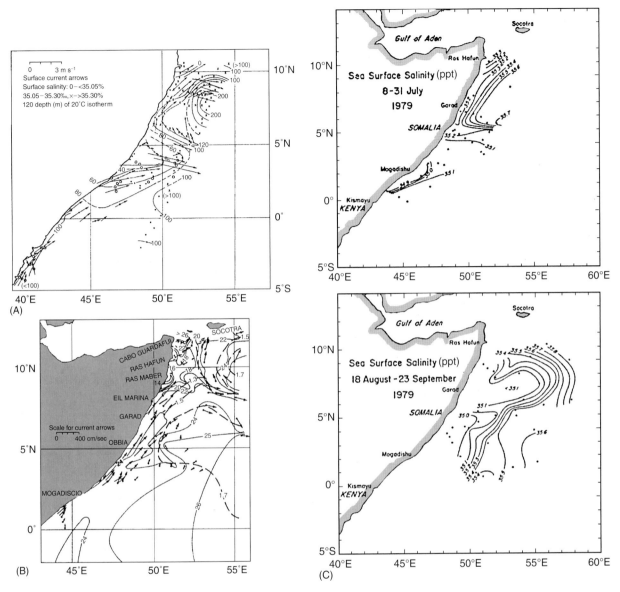

Figure 5 Variability of the circulation and salinity along the Somali coast. (A) 1 July–4 August 1979: surface currents (arrows), depth of the 20°C isotherm and surface salinity range (Reprinted from Swallow et al., 1983, Development of near-surface flow pattern and water mass distribution in the Somali Basin in response to the southwest monsoon of 1979, *Journal of Physical Oceanography*, 13, 1398–1415, with permission from American Meteorological Society). (B) 4 August–4 September 1964: currents at 10 m depth (arrows), dynamic heights of the sea surface relative to 1000 dbars in dyn. meters (heavy dashed) and sea surface temperatures in °C (solid) (Reprinted from *Progress in Oceanography*, 12, F. Schott, Monsoon response of the Somali Current and associated upwelling, 12, 357–381, copyright 1983, with permission from Elsevier Science (redrawn from J. C. Swallow and J. G. Bruce, 1966 and Warren et al., 1966)). (C) Surface salinities during the existence of the two gyre system, 8–31 July 1979, and during and after the northward propagation of the southern cold wedge, 18 August–23 September 1979 (Reprinted from Progress in Oceanography, 12, F. Schott, Monsoon response of the Somali Current and associated upwelling, 357–381, copyright 1983, with permission from Elsevier Science). At that time the Somali Current is continuous along the coast and brings fresher water from the south at the end of the SW monsoon season.

different during the two periods, whereas the Trade Winds over the subtropical south Indian ocean are similar, which suggests that, in October, the cross-equatorial flow is driven primarily by remote forcing through the inflow from the South Equatorial Current. Between 6°N and 11°N, the anticyclonic Great Whirl, marked by relatively high surface salinities (35.6–35.8), persists with transport of 33 Sv westward in the upper 250 m between 6°N and 8.5°N, and 32 Sv eastward between 8.5°N

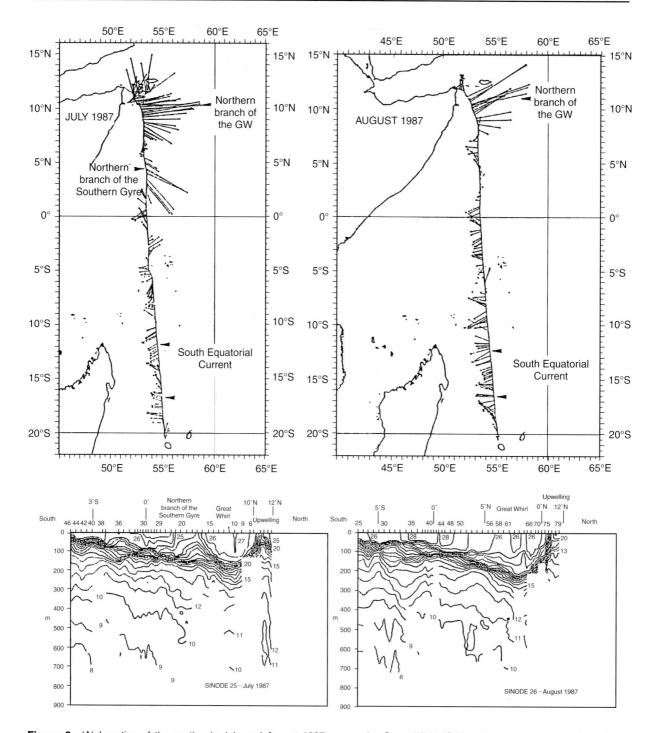

Figure 6 (A) Location of the section in July and August 1987 across the Great Whirl (GW) with corresponding surface drifts. (B) Corresponding temperature sections (°C) showing the strengthening of the northern front of the Great Whirl in August and the disappearance of the southern front in August compared to July at that longitude. (From M. Fieux, 1987, Circulation dans l'océan Indien occidental, Actes Colloque sur la Recherche Française dans les Terres Australes, Strasbourg, unpublished manuscript.)

and 11.5°N. Then a southward undercurrent is established during the transition period while the Great Whirl weakens.

The northern gyre is not symmetrical, as can be seen from the depth and slope of the 20°C isotherm, which is much steeper on the northern flank of the

gyre (**Figure 6B**). This can be explained by the strong nonlinearity of that system, causing a shift of high currents into the northern corner.

The northern Somali Current was found to be disconnected from the interior of the Arabian Sea in the latitude range 4°–12°N in terms of both water mass properties and current fields. Communication predominantly occurs through the passages between Socotra and the horn of Africa.

During the late phase of the SW monsoon, a third anticyclonic gyre appears north-east of the island of Socotra that is called the Socotra Gyre (**Figure 7**).

In summer 1993, a significant northward flow of 13 Sv was observed through the passage between the island of Abd al Kuri (west of Socotra) and Cape Guardafui (the horn of Africa). East of the Great Whirl, a band of northward warm water flow that provided low-latitude waters to the Socotra Gyre and the Socotra Passage separated the Great Whirl and the interior of the Arabian Sea. The net transport through the Socotra Passage is northward throughout most of the year.

During the transition period in October–November, the northward Somali circulation decreases, with a branch turning offshore south of 2.5°N.

In December–February, during the NE winter monsoon, the Somali current reverses southward from 10°N to 4°–5°S, where it converges with the northward EACC to form the SECC flowing eastward (see above).

Nonlinear reduced gravity numerical models driven by observed monthly mean winds are very successful in simulating the observed features of the circulation in this region, such as the formation and decay of the two-gyre system of the Somali Current during the SW monsoon. With interannually varying winds they also simulate a large interannual variability in the circulation.

Currents at Depth

Direct measurements made in 1984–1986 at the Equator, near Africa, show that the southward reversal implies only a thin surface layer below which, between 120 m and 400 m, there is a northward undercurrent, remnant of the SW monsoon season circulation (**Figures 8** and **9**), followed again by a southward current below 400 m. It results in a large variability of the cross-equatorial transport, which amounts to 21 Sv for the upper 500 m during the summer monsoon season and is close to zero for the winter monsoon mean transport. The annual mean cross-equatorial transport in the upper 500 m is 10 Sv northward, with very little transport in either season in the depth range 500–1000 m (**Figure 9**). Comparison of current profiles at 4°S in the EACC, at the Equator, and at 5°N in the Somali Current in both seasons shows that at 4°S the subsurface profile stays fairly constant while

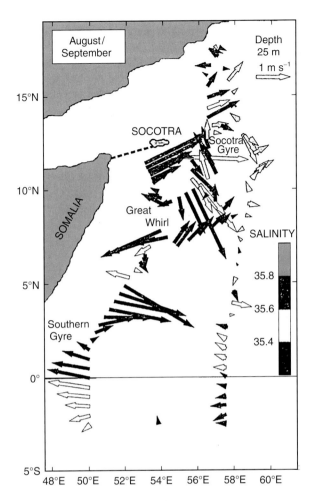

Figure 7 Near surface circulation at 25 m in August–September 1995, showing the Southern Gyre (SG), the Great Whirl (GW) and the Socotra Gyre (SG) together with surface salinities. (Reprinted from F. Schott et al., Summer monsoon response of the northern Somali Current, 1995, *Geophysical Research Letters*, 24(21), 2565–2568, 1997, copyright by the American Geophysical Union).

at 5°N drastic changes occur between the seasons as well as at the Equator (**Figure 10**). North of 5°N, there is less variability at subsurface, with the presence of a southward coastal undercurrent during most of the year except during the full strength of the deep-reaching Great Whirl in July–August to more than 1000 m, involving large deep transports. The Somali Current flow patterns for the different seasons in the 100–400 m layers are shown in **Figure 8**.

Deeper, below the Somali current, at the Equator during October 1984 to October 1986 at 1000 m, 1500 m, 2000 m, and 3000 m depth, the measured mean currents were very small. However, the only clear seasonal signal was observed at the 3000 m level, with a seasonal current parallel to the coast

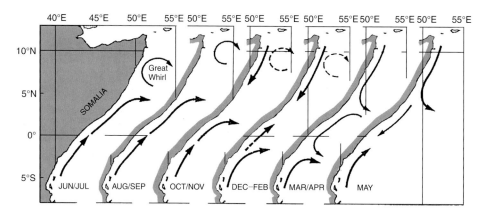

Figure 8 Schematic representation of Somali Current circulation patterns in the layer 100–400 m for different seasons. (Reprinted from *Deep Sea Research*, 37(12), F. Schott *et al.*, 1990, The Somali Current at the equator: annual cycle of currents and transports in the upper 1000 m and connection to neighbouring latitudes, 1825–1848, copyright 1990, with permission from Elsevier Science).

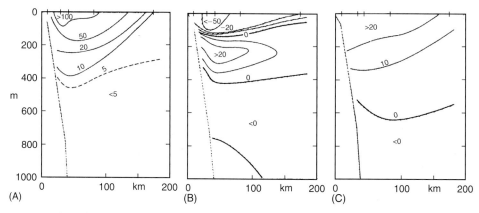

Figure 9 Mean sections of northward current component at the Equator (positive northward): (A) for winter monsoon (1 June–13 September); (B) for winter monsoon (15 December–15 February; (C) for overall mean, in cm s^{-1}. (Reprinted from *Deep Sea Research*, 37, (12), F. Schott *et al.*, 1990, The Somali Current at the equator: annual cycle of currents and transports in the upper 1000 m and connection to neighbouring latitudes, 1825–1848, copyright 1990, with permission from Elsevier Science).

approximately in phase with the local surface winds. This reached a north-eastward mean of 0.10 m s^{-1} between June and September, and a south-westward mean of 0.06 m s^{-1} between November and February. This variability seems in agreement with salinity distribution near that level along the coastal boundary, with slightly higher salinity at the end of the NE monsoon season and slightly lower salinity during the SW monsoon. Higher up, at 2000 m, 1500 m, and 1000 m, the currents are dominated by events of 1–2 months duration.

From the long-term current measurements, it seems that at the Equator the semiannual variability is stronger than the annual variability even near the coast. Off the equator, the annual component dominates.

Numerical models have shown that the location and motion of eddies are influenced by the distribu-

tion and strength of the wind forcing; an increase in the winds leads to a southward displacement of the offshore turning of the southern branch; the northward motion of eddies is very dependent upon the coastal geometry; and the onset of the northward Somali current depends on local winds forcing and on the wind forcing far out at sea. Baroclinic Rossby waves generated by the strong offshore anticyclonic windstress curl have been found to be the generation mechanism of the Great Whirl.

Conclusion

The western boundary of the northern Indian Ocean is a remarkable natural laboratory for studying the effect of the wind on the oceanic circulation, as regularly twice a year the winds change direction rapidly and are particularly strong. It is in the

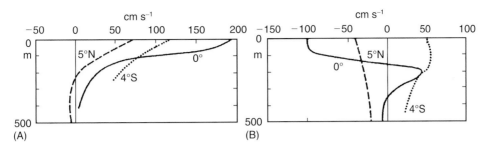

Figure 10 Current profiles for different seasons at 4°S, on the Equator, and at 5°N: (A) for summer monsoon; (B) for winter monsoon. (Reprinted from *Deep Sea Research*, 37(12), F. Schott *et al.*, 1990, The Somali Current at the equator: annual cycle of currents and transports in the upper 1000 m and connection to neighbouring latitudes, 1825–11848, copyright 1990, with permission from Elsevier Science).

Somali current that the highest variability as well as the highest current speeds in the world ocean are found.

See also

Current Systems in the Indian Ocean. Elemental Distribution: Overview. Indian Ocean Equatorial Currents. Thermohaline Circulation. Water Types and Water Masses. Wind Driven Circulation.

Further Reading

Fein JS and Stephens PL (eds) (1987) *Monsoons*. Washington, DC: Wiley Interscience.

Monsoon (1987) Fein JS and Stephens PL (ed.) NSF. A Wiley-Interscience Publication. Washington, USA: John Wiley & Sons.

Open University Oceanography Course Team (1993) *Ocean Circulation*. Oxford: Pergamon Press.

Schott F (1983) Monsoon response of the Somali Current and associated upwelling. *Progress in Oceanography* 12: 357–381

Schott F, Swallow JC and Fieux M (1990) The Somali Current at the equator: annual cycle of currents and transports in the upper 1000 m and connection to neighbouring latitudes. *Deep Sea Research* 37(12): 1825–1848.

Schott F, Fischer J, Garternicht U and Quadfasel D (1997) Summer monsoon response of the northern Somali Current, 1995. *Geophysical Research Letters* 24(21): 2565–2568.

Swallow JC, Molinari RL, Bruce JG, Brown OB and Evans RH (1983) Development of near-surface flow pattern and water mass distribution in the Somali Basin in response to the southwest monsoon of 1979. *Journal of Physical Oceanography* 13: 1398–1415.

Tomczak M and Godfrey S (1994) *Regional Oceanography: An Introduction*. Oxford: Pergamon Press.

SONAR SYSTEMS

A. B. Baggeroer, Massachusetts Institute of Technology, Cambridge, MA, USA

doi:10.1006/rwos.2001.0317

Introduction and Short History

Sonar (Sound Navigation and Ranging) systems are the primary method of imaging and communicating within the ocean. Electromagnetic energy does not propagate very far since it is attenuated by either absorption or scattering – visibility beyond 100 m is exceptional. Conversely, sound propagates very well in the ocean especially at low frequencies; consequently, sonars are by far the most important systems used by both man and marine life within the ocean for imaging and communication.

Sonars are classified as being either active or passive. In active systems an acoustic pulse, or more typically a sequence of pulses, is transmitted and a receiver processes them to form an 'image' or to decode a data message if operating as a communication system. The image can be as simple as the presence of a discrete echo or as complex as a visual picture. The receiver may be coincident with the transmitter – a monostatic system, or separate – a bistatic system. Both the waveform of the acoustic pulse and the beamwidths of both the transmitter and receiver are important and determine the performance of an active system. One typically associates an active sonar with the popular perception of sonar systems. Many marine mammals use active sonar for navigation and prey localization, as well as communication in ways which we are still attempting to understand. Many of the signals used by modern sonars have some of the same features as those of marine mammals.

Passive systems only receive. They sense ambient sound made by a myriad of sources in the ocean such as ships, submarines, marine mammals, volcanoes. These systems have been, and still are, especially important in anti-submarine warfare (ASW) where stealth is an important issue, and an active ping would reveal the location of the source.

The use of sound for detecting underwater objects was first introduced in a patent by Richardson in June 1912 for the 'sonic detection of icebergs,' 2 months after the sinking of the *Titanic*.[1] This was soon followed by the development of the Fessenden oscillator in 1914 which eventually led to the development of fathometers, an acoustic system for measuring the depth to the seabed. The French physicist/chemist Paul Langevin was the first to detect a submarine using sonar in 1918, motivated by the extensive damage of German U-boats. Between World Wars I and II both Britain and the US sponsored sonar research, especially on transducers. The former was conducted under the Antisubmarine Detection Investigation Committee, or ASDIC as sonar is still often referred to within the British military, and the latter was performed at the Naval Research Laboratory.

The re-emergence of the German U-boat stimulated the modern era of sonars and the physics of sound propagation in the ocean where major research programs were chartered in the USA (Columbia, Harvard, Scripps Institution of Oceanography, Woods Hole Oceanographic Institution), UK, and Russia. A very comprehensive summary was compiled by the US National Defense Research Council after World War II, which still remains a valuable reference (*see* Further Reading Section).

The development of the nuclear submarine, both as an attack boat (SSN) or as a missile carrier (SSBN) provided a major emphasis for sonar throughout the cold war. The USA, UK, Russia, and France all had substantial research programs on sonar for many applications, but ASW certainly had a major priority. The nuclear submarine could deny use of the oceans but could also unleash massive

destruction with nuclear missiles. With the end of the cold war, ASW now has a lower priority; however, the submarine still remains the platform of choice for many countries since modern diesel/electric submarines operating on batteries are extremely hard to detect and localize. Undoubtedly, the most extensively used reference was compiled by Urick (1975), which is frequently referenced as a handbook for sonar engineers.

While military operations have dominated the development of sonars, they are now used extensively for both scientific and commercial applications. The use of fathometers and closely related seismic methods provided much of the important data validating plate tectonics. There is also a lot of overlap between geophysical exploration for hydrocarbons and modern sonars. High resolution and multibeam systems are extensively used for charting the seabed and its sub-bottom characteristics, fish finding, current measurements exploiting Doppler, as well as archaeological investigations.

Active Sonar systems

The major components of an active system are indicated in **Figure 1**. A waveform generator forms a pulse or 'ping', which is then modulated, or frequency shifted, to an operating frequency, f_o which may be as low as tens of Hertz for very long-range systems, or as high as 1 MHz, for high resolution short-range imaging sonars. Next, the signal is often 'beamformed' by an array of transducers, that focuses the signal in specific directions either by mechanically rotating the array or by introducing appropriate time delays or phase shifts. The signals are amplified and then converted from an electrical signal to a sound wave by the transmit transducers. Efficient transduction, the conversion of electric power to sound power, and even a modest amount of directivity of the transmitter requires that the transducer have dimensions on the scale of the wavelength of the operating frequency; hence, low frequency transmitters are typically large and not very efficient, whereas high frequency transmitters are smaller and very efficient.

The pinging rate, usually termed the pulse repetition frequency (PRF) is determined by the duration over which strong echos (called 'returns') from the previously transmitted pulse can be expected, so that one return does not overlap and become confused with another. With some systems with well confined response durations, several pulses may be in transit at the same time.

The ocean introduces three important components before it is detected by a receiver.

---

[1] Much of this material in the history has been extracted from Beyer, 1999.

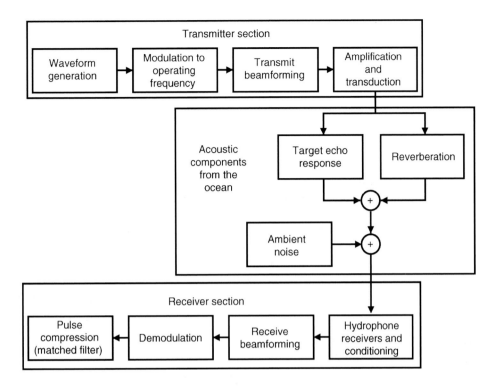

Figure 1 Active sonar system components.

- There is the desired echo from the target itself. This may be a simple echo, especially if the target is close, but it may also include many multipaths and/or modes as a result of reflections of the ocean surface and bottom as well as paths refracted completely within the ocean itself (*see* **Acoustics, Deep Ocean**).
- The ocean is filled with spurious, or unwanted reflectors which produce reverberation. The dominant source of this is the sea bottom, but the sea surface and objects (e.g. fish) can be important as well. Typically, the bottom is characterized in terms of a scattering strength per unit area insonified.
- Finally, the ocean is filled with ambient noise which is created by both natural and man-made sources. At low frequencies, 50–500 Hz, shipping tends to dominate the noise in the Northern Hemisphere, especially near shipping lanes. Wind and wave processes as well as rain can also be important. In specific areas, marine life may be a very important component.

The sonar receiver implements operations similar to the transmitter. Hydrophones convert the acoustic signal to an electric one whereupon it usually undergoes some 'signal conditioning' to amplify it to an appropriate level. In modern sonars the signal is digitized, since most of the subsequent operations are more easily implemented by digital signal processors. Next, a receiver beamformer, which may be quite different from the transmitter, focuses energy arriving from specific directions for spatial imaging. This is also done either by mechanically steering the array or by introducing time delays or phase shifts (if the processing is done in the frequency domain). The signal is then usually demodulated to a low frequency band which simplifies further electronics and signal processing steps. Finally it is 'pulse compressed,' or 'matched filtered,' which is a process that maximizes energy arriving at the travel time that corresponds to the range to a target. The matched filter is simply a correlation operation which seeks the best replica of the transmitted signal among all the signal components introduced by the ocean. In the simplest form of processing a sequence of 'pings' is rastered, i.e. the echo time-series are displayed one after the other, to construct an image. This is typical of a sidescan sonar system. In more sophisticated systems, especially those operating at low frequencies where phase coherence can be preserved or extracted, a sequence of outputs from the pulse compression filter is processed to form images. This is typical of synthetic aperture sonars. In both types of systems display algorithms, sometimes termed 'normalizers,' are important for emphasizing certain features by improving contrast and controlling the dynamic range of the output.

The performance of an active sonar system is captured in the active sonar equation; while imperfect in its details, it is very useful in assessing gross performance. It is expressed in logarithmic units, or decibels which are referenced to a standard level. For a monostatic system it is:

$$SE = SL - 2*TL + TS - \max(NL, RL) + DI_t$$
$$+ AG_r - DT$$

where[2,3] SE, is the signal excess at the receiver output; SL, is the source level referenced to a pressure level of $1\,\mu Pa$; TS, is the target strength, which is a function of aspect angle; NL, is the level of the ambient noise at a single hydrophone; RL, is the reverberation which is determined by the area insonified, the scattering strength, and the signal level; DI_t, is the directivity index of the transmitter, which is a measure of the gain compared omnidirectional radiation; AG_r, is the array gain of the receiver in the direction of the target (often this is described as a receiver); directivity index, DI_r; DT, is the direction threshold for a target to be seen on the output display (this can be a complicated function of the complexity of the environment and the sophistication of an operator); TL, is the transmission loss, i.e. the loss in signal energy as it transmits to the target and returns. If the $SE > 0$, then a target is discernible on a display.

The notation $\max(NL, RL)$ distinguishes the two important regimes for an active sonar. When $NL > RL$, the sonar is operating in a noise-limited environment; conversely, when $RL > NL$ the environment is reverberation-limited which is the case in virtually all applications.

Reverberation is the result of unwanted echoes from the sea surface, seafloor, and volume. In the simplest formulation its level for surfaces both bottom and top is usually characterized by a scattering strength per unit area, so the level is given by the product of the resolved area multiplied by the scattering strength (or sum if using a decibel formulation). Often, the Rayleigh parameter $\frac{2\pi\sigma_s}{\lambda}\sin(\phi)$ is used, where σ_s is the rms surface roughness and ϕ is the incident angle as a measure of when surface roughness becomes important, i.e. when the Rayleigh parameter is greater than 1. Similarly with volume scattering, a scattering strength per unit volume is used.

The operating frequency of a sonar is an important parameter in its design and performance. The important design issues are:

- Resolution. There are two aspects of resolution – 'cross-range' resolution and 'in range' resolution. 'Cross-range' resolution is determined by the dimensions of the transmit and receive apertures relative to the wavelength, λ, of the acoustic signal.[4] It is given by $R\lambda/L$ where R is the range and L is the transmitter and/or receiving aperture. Higher resolution requires higher operating frequencies since these result in smaller $R\lambda/L$ ratios.

  'In range' resolution is determined by the bandwidth of the signal and is given approximately by $c/2W$, where W is the available bandwidth and c is the sonar speed. Since W is usually proportional to the operating frequency, f_o, one tends to try to use higher frequencies, which limits practical ranges. Also the sonar channel is often very band-limited. As a result in most sonar systems the 'in range' resolution is typically significantly smaller than the 'cross-range' resolution, so care needs to be taken in interpreting images.

- Maximum operating range. Acoustical signals can propagate over very long ranges, but there are a number of phenomena which can both enhance and attenuate the signal power. These include the geometrical spreading, the stratification of the sound speed versus depth, absorption, and scattering processes. The first two are essentially independent of frequency while the latter two have strong dependencies. **Figure 2** indicates the absorption loss in $dB\,km^{-1}$ of sound at $20°$ and 35 apt salinity.[5] Essentially, the absorption loss factor increases quadratically with frequency. The two 'knees' in the figure relate to the onset of losses introduced by ionic relaxation phenomena. The net implication is that efforts are made to minimize the operating frequency so that it contributes $< 10\,dB$ of loss for the desired range, i.e $\alpha(f_o)R < 10$ where $\alpha(f_o)$ is attenuation per unit distance in dB.

The penalty of this, however, is a less directive signal, so it is more difficult to avoid contact with the ocean boundaries and consequent scattering

---

[2] There are many versions of the sonar equation and the nomenclature differs among them (see Urick, 1975).

[3] If the system is bistatic wherein the transmitter and receiver are not colocated, then the sonar equation is significantly more complicated (Cox, 1989).

[4] The wavelength in uncomplicated media is given by $\lambda = c/f_o$, where c is the sound speed, nominally $1500\,m\,s^{-1}$ and f_o is the operating frequency.

[5] $db\,km^{-1}$ represents 10 log of the fractional loss in power per kilometer representing an exponential decay versus range.

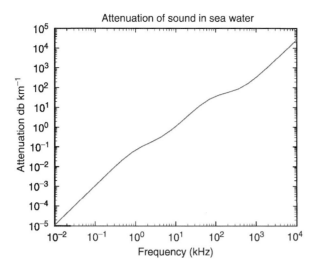

Figure 2 Attenuation of sound vs frequency.

losses, especially the bottom is shallow-water environments. As a result, the actual operating frequency of a sonar is a compromise based on the sound speed profile, the directivity of transmitter, receiver beamformers, and desired operating range and isolation.

- Target cross-section. The physics of sound reflecting from a target can be very complicated and there are few exact solutions, most of which involve long, complicated mathematical functions. In addition, the geometry of the target can introduce a significant aspect dependence. The important scaling number is $2\pi a/\lambda$ where a is a characteristic length scale presented to the incoming sound wave. Typically, if the number is less than unity, the reflected target strength depends upon the fourth power of frequency, so the target strength is quite small; this is the so-called Rayleigh region of scattering. Conversely, if this number is larger than unity, the target strength normalized by the presented area is typically be-

tween 1 and 10. There are two additional features to consider: (i) large, flat surfaces lead to large returns, often termed specular, and (ii) in a bistatic sonar, the forward scattering, essentially the shadow, is determined almost solely by the intercepted target shape, often called Babinet's principle.

The net effect is a desire for higher operating frequencies in order to stay in the Rayleigh region where there is significant target strength.

Overall high frequencies produce better resolution and higher target strengths. However, at high frequencies acoustic propagation is more complicated and absorption limits the range.

Table 1 indicates the operating frequencies of some typical active sonars.

Sonar System Components

The components in **Figure 1** all have significant impact on the performance of an active sonar. The signal processing issues are complex and there is a large sonar-related literature as well as radar where the issues are similar. The essential problem is to separate a target amidst the reverberation and ambient noise in either of two operating realms – 'noise-limited' and 'reverberation'-limited environments. In a 'noise-limited' environment the ambient background noise limits the performance of a system, so increasing the transmitter output power improves performance. A 'reverberation-limited' environment is one where the noise is composed of mostly unwanted reflections from objects other than the target, so increasing the transmitted power simply increases both the target and reverberation returns simultaneously with no net gain in signal to noise ping. Most active sonars operate in a reverberation limited environment. Effective design of an active sonar depends upon controlling reverberation through a combination of waveform design and beamforming.

Table 1 Features of some typical active sonars

Sonar system	Operating frequency range	Wavelength	Nominal range
Long-range, low frequency	50–500 Hz	30–3 m	1000 km
Military ASW sonars	3–4 kHz	0.5–0.75 m	100 km
Bottom-mapping echosounders	3–4 kHz	0.5–0.75 m	vertical
High-resolution fathometers	10–15 kHz	15–10 cm	vertical
Acoustic communications	10–30 kHz	15–5 cm	10 km
Sidescan sonar (long-range)	50–100 kHz	6–1.5 cm	5 km
Sidescan sonar (short range)	500–1000 kHz	3–1.5 mm	100 m
Acoustic localization nets	10–20 kHz	15–5 cm	vertical
Fish-finding sonars	25–200 kHz	6–1 cm	1–5 km
Recreational	100–250 km	15–6 mm	vertical

Waveform design There are two basic approaches to waveform design for resolving targets – 'range gating' and 'Doppler gating.' The simplest approach to 'range gating' is a short, high powered pulse. This essentially resolves every reflector and an image is constructed by successive pulses and then the returns are rastered. While this is the simplest waveform it has limitations when operating in environments with high noise and reverberation levels, since the peak power of most sonars, both man-made and marine mammal, is limited. This shortcoming can often be mitigated by exploiting bandwidth (resolution α'/β). This has led to a large literature on waveform design with the most popular being frequency modulated (FM) and coded (PRN) signals. With these signals[6] the center frequency is swept, or 'chriped' across a frequency band at the transmitter and correlated, or 'compressed' at the receiver. This class of signals is commonly used by marine mammals including whales and dolphins for target localization.

'Doppler gating' is based on differences in target motion. A moving target imparts a Doppler shift to the reflected signal which is proportional to operating frequency and the ratio v/c, where v is the target speed and c is the sound speed. 'Doppler gating' is particularly useful in some ASW contexts since it is difficult to keep a submarine stationary, thus a properly designed signal, one that resolves Doppler, can distinguish it against a fixed reverberant background. The ability to resolve Doppler frequency depends upon the duration of a signal, with a dependence of 1/duration, so good 'Doppler gating' waveforms are long.

There has been a lot of research on the topic of optimal waveform design. Ideally, one wants a waveform which can resolve range and Doppler simultaneously which implies long duration, and wide bandwidth. These requirements are difficult to satisfy simultaneously.

Beamforming Both the transmitter and the receiver beamformers provide spatial resolution for the sonar system. The angular resolution in degrees is approximately $\Delta\theta \approx 60\frac{\lambda}{L}$, where λ is the wavelength and L is the aperture length. Since acoustic wavelengths are large when compared with optical wavelengths, the angular resolutions tend to be large especially at low frequencies.[7] Beamformers, often termed array processors in receivers, have been an important research topic for several decades with the advent of digital signal processing which permitted increasingly more sophistication, especially in the realm of adaptive methods. One of the simplest transmit beamformers consists of a line array of transducers each radiating the same signal. The simple receiver and also a line of transducers adds all the signals together. This resolves the paths perpendicular or broadside to the array. If one wants to 'steer' the array, or resolve another direction, the array must be mechanically rotated. This method of beamforming is still used by many systems since it is quite robust. Another simple beamformer is a planar array of transducers.

Digital signal processing has led to more sophisticated array processing, especially for receivers. Beamformers which steer beams electronically by introducing delays, or phase shifts, shape beams to control sidelobes, place nulls to control strong reflectors, and reduce jamming are now practical because these features can be practical electronically rather than mechanically.

Examples of Active Sonar Images

This section describes two examples of sonars used for mapping seafloor bathymetry. In the first the sonar is carried on an unmanned underwater vehicle (UUV) close to the seafloor. The operating frequency is 675 kHz and the beam is mechanically steered from port to starboard as well as fore and aft as the UUV proceeds along its track, so the beams are steered forward and directly below the UUV (**Figure 3**). The onset time of the first echo return is the parameter of interest. It is converted to the depth of the seafloor after including the vehicle position, the direction of the beam and possibly refraction effects in the water itself. Usually straight-line acoustic propagation is assumed.

The signals are combined to generate a high resolution map of the seafloor. The processing to achieve this includes editing for spurious responses, registration of the rasters or images from successive transmission using the navigation sensors on the UUV (or more generally any vehicle) and normalization to improve the contrast so that weak features can be detected amidst strong ones.

---

[6] Pseudo Random Noise (PRN) are coded signals which appear to be random noise. Well designed signals have useful mathematical constructs which led to good outputs at the output of the pulse compression, or matched filter, processor.

[7] Sonars with angular resolutions of 1° are generally considered to have high resolution. Compare this with that of the human eye with a nominal diameter of 4 mm and the wavelength of light in the visible region is 0.4 μm leading to a resolution scale of 0.1 ms.

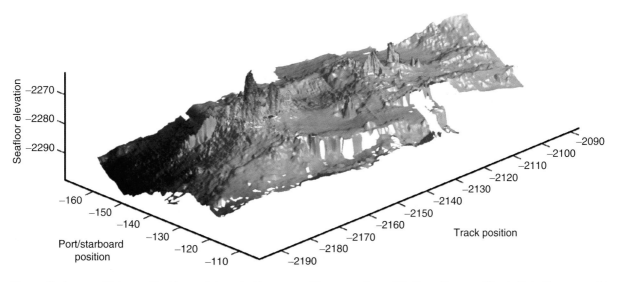

Figure 3 Image with forward-looking and down-looking sonar. (Figure courtesy of Dr Dana Yoerger, Woods Hole Oceanographic Institution.)

The second example of an active sonar is a multibeam bathymetric mapper. Most of these systems for deep water operate at a 12 kHz center frequency. The transmit beam is produced by a linear array running fore to aft along the bottom of the ship, thwartships beam, which produces a swath which resolves the seafloor along track (**Figure 4**). The receiver array is oriented port to starboard. The signals from this array are beamformed electronically, so the seafloor is resolved port to starboard within the transmitted swath, since the patch is the product of the transmit and receive beamwidth. This configuration allows two-dimensional resolution with two linear arrays instead of a full planar array. The depths from each of the multibeams are measured by combining the travel time and the ray refraction from the sound speed profile to obtain a depth. Subsequent processing edits anomalous returns and interpolates all the data to generate the contour map. Active sonar systems with additional features have been developed for special applications, but they all use the basic principles described above.

Passive Sonars

Passive sonars that only listen and do not transmit are used in a variety of applications including the military for antisubmarine warfare (ASW), tracking and classification of marine mammals, earthquake detection, and nuclear test ban monitoring.

Since the signals are passive there is no pulse compression, or matched filtering, so a passive sonar design primarily focuses upon the 'short-term' frequency wavenumber spectrum, or the directional spectrum and the power density spectrum and how it evolves in time. The data are nonstationary and inhomogeneous, but many of the processing algorithms are based upon stationary and homogeneous assumptions; hence the term 'short-term.' The performance of a passive system is characterized by the passive sonar equation:

$$SE = SL - TL - NL + AG_r - DT$$

where the terms are essentially the same as for an active sonar. In some applications the arrays are so large that the coherence of the received signal is important, and it is necessary to separate the array gain, AG_r into two terms, or:

$$AG_r = AG_{r,n} - SGD_s$$

where $AG_{r,n}$ is the array gain against the ambient noise and SGD_s is the signal gain degradation due to lack of coherence. $SGD_s = 0$ for a signal that is coherent across the entire array.

Passive Sonar Beamforming

The signals received by the sonar's hydrophone are preconditioned, which might include editing bad data channels, calibration, and filtering. They are then beamformed, either in the time domain by introducing delays to compensate for the travel time across the array, or in the frequency domain. With digital signals the former usually requires upsampling or interpolation of the data to avoid distortion.

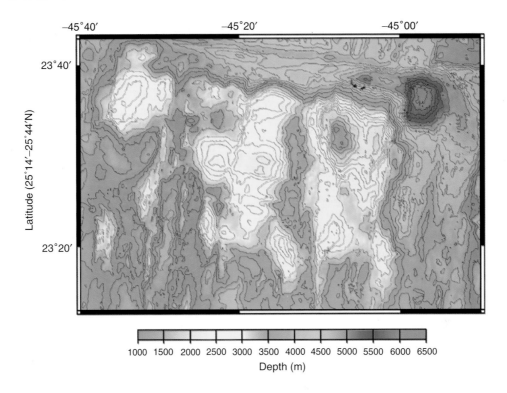

Figure 4 High resolution bathymetric map of the seafloor near the Mid-Atlantic Ridge: (A) contour map; (B) isometric projection (from top-right). (Figures courtesy of Dr Brian Tucholke, Woods Hole Oceanographic Institution.)

The latter is accomplished by FFT (fast Fourier transforms), phase shifting to compensate for the delays, and then IFFT (inverse fast Fourier transforming). Frequency domain beamforming allows simpler implementation of adaptive techniques that are useful in cases where the ambient field has many discrete components. Adaptive algorithms form beams with notches, i.e. poor response in the direction of interferers, thereby suppressing them. Many

algorithms have been designed to accomplish this, but the MVDR (minimum variance distortion filter – first introduced by Capon) and related algorithms have been used most extensively in practice.

Passive Sonar Display Formats

The output of the beamformer is a time-series for each beam. In certain applications the time-series itself may be of interest, however, in most cases the

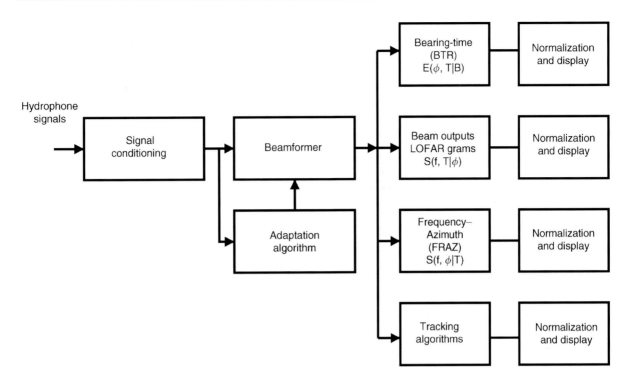

Figure 5 Passive sonar with modes of displaying output.

time-series is further processed to assist in extracting weak signals from the background noise. The parameter for signal processing schemes and display formats includes time (the epoch for data processing, T), angle (azimuth and elevation), and frequency (the spectral content of the data) (**Figure 5**).

Bearing-time Recording

Bearing time processing takes the beam outputs over a specified frequency band and plots the output versus time. Two modes of processing are often used: (i) energy detection, which forms an average of the beam outputs, or (ii) cross-correlation detection, where the array is split at the beamformer and then the two outputs are cross-correlated versus the direction. The processing is often classified according to the width of the band used, passive broadband (PBB) or passive narrowband (PNB).

The data for each time epoch are normalized to improve the contrast for signals of interest and each raster is plotted. Over a sequence of epochs, the directional components in the ambient field which are associated with shipping are observed. By maneuvering the array one can triangulate to obtain a range to each source as well.

Low Frequency Acoustic Recording and Analysis (LOFAR) grams Once a bearing or direction of interest has been determined, spectral analysis of the selected beam is used to produce a LOFAR gram, which is a plot of the signal spectrum for each analysis epoch, T, versus time. By examining the features of the LOFAR gram, such as frequency, peaks, harmonic signals, and their changes as a function of time, the source of the signal and some of its characteristics, such as speed, can be deduced. As in the case of the bearing-time display, the LOFAR gram is normalized to enhance features of interest.

FRAZ displays Frequency-azimuth, or FRAZ displays plot the spectral content as a function of frequency and azimuth for each epoch, T. Often a number of FRAZ outputs are averaged to improve signal to noise. FRAZ displays allow connection of the spectral content of a single source along a given bearing since the display contains a number of lines for each source at a given azimuth. As with the previous display, normalization algorithms to improve contrast are usually employed.

Trackers The objective of an ASW passive sonar is to detect, classify and track sources of radiating sound. Trackers are used to follow sources through direction and frequency space. There are a number of tracking algorithms build upon various signal models. Some separate the direction and frequency dimensions while others are coupled models. Since the ambient field can often have a number of sources, some targets and some interferers and since there are complicated propagation effects, the design of trackers is difficult. Most involve some form of Kalman filtering and some have integral propagation models.

Advanced beamforming

Most passive sonars are based upon a plane wave model for the ambient field components. Plane wave beamforming is robust, has several computational advantages, and is well understood. However, in many sonars this model is not adequate because the arrays are so long that wavefront curvature becomes an issue or the arrays are used vertical where acoustic propagation introduces multipaths. Advanced beamforming concepts can address these issues.

Wavefront curative becomes important when the target is in the near field of the array (usually given by the Fresnel number $2L^2/\lambda$) which is a consequence of either very long arrays or high frequencies. A quadratic approximation to the curvature is often used and the array is actually designed to focus it at specific range (rather than at infinite range which is the case for plane waves). When focused at short ranges, long-range targets are attenuated. This introduces the focal range as another parameter for displaying the sonar output.

At long ranges and low frequencies or in shallow water acoustic signals have complex multipath or multimode propagation which leads to coherent interference along a vertical array or a very long horizontal array. The appropriate array processing is to determine the full field Green's function for the signal and match the beamforming to it, a technique known as matched field processing (MFP). MFP requires knowledge of the sound speed profile along the propagation path, so its performance depends upon the accuracy of environmental data. It is a computationally intensive process, but has the powerful advantage of being able to resolve both target depth and range, as well as azimuth. MFP is an active subject of passive sonar research.

See also

Acoustic Scattering by Marine Organisms. Acoustics, Deep Ocean. Acoustics, Shallow Water.

Further Reading

Baggeroer AB (1978) Sonar signal processing. In: Oppenheim AV (ed.) *Applications of Digital Signal Processing*. Englewood Cliffs, NJ: Prentice-Hall.

Baggeroer A, Kuperman WA and Mikhalevsky PN (1993) An overview of Matched Field Processing. *IEEE Journal of Oceanic Engineering* 18: 401–424.

Beyer RT (1999) *Sounds of Our Times: 200 Years of Acoustics*. New York: Springer-Verlag.

Capon J (1969) High resolution frequency-wavenumber spectrum analysis. *Proceedings of the IEEE* 57: 1408–1418.

Clay C and Medwin H (1998) *Fundamentals of Acoustical Oceanography*. Chestnut Hill, MA: Academic Press.

Cox H (1989) Fundamentals of bistatic active sonar. In: Chan YT (ed.) *Underwater Acoustic Data Processing*. Kluwer Academic Publishers.

National Defense Research Council (1947) *Physics of Sound on the Sea: Parts 1–6*. National Defense Research Council, Division 6, Summary Technical Report.

Urick RJ (1975) *Principles of Underwater Sound for Engineers*, 2nd edn. McGraw-Hill Book Co.

SOUTHERN OCEAN FISHERIES

I. Everson, British Antarctic Survey, Cambridge, UK

doi:10.1006/rwos.2001.0451

History

The history of harvesting living resources in the Southern Ocean goes back two centuries. Soon after South Georgia was discovered by Captain Cook, hunters arrived in search of fur seals (*Arctocephalus gazella*). Between 1801 and 1822, over 1 200 000 skins were taken there before the sealers moved on to the South Shetlands. The extent of the carnage was such as to bring the species close to extinction by the 1870s. In the early twentieth century elephant seals (*Mirounga leonina*) were harvested at South Georgia because their blubber provided higher-grade oil than that from whales. A strong management regime was instituted in 1952, which allowed the population to recover from earlier overexploitation. Elephant seals were also harvested at Macquarie Island, although there was less of a link to the whaling industry.

Whaling has been the largest fishing operation in the Southern Ocean. Initially this was shore-based but, with the introduction of floating factories in 1925, catching vessels could search much of the

Southern Ocean. Overcapitalization of the industry and, in the early years, a lack of robust population models to provide management advice, meant that the industry declined. Research at that time, particularly by Discovery Investigations, provided much valuable information not only on whales but also on their food, krill.

With the decline in shore-based whaling, fishing fleets from Japan and the former Soviet Union turned their attention to harvesting fish and krill. Scientists expressed several concerns over this development: First that there was no international fishery regime in place for the region; second that an unregulated fishery on krill could lead to overexploitation as had happened with the whales; and third that significant fishing of krill might adversely affect the recovery of baleen whales.

Member states of the Antarctic Treaty, an agreement that arose directly from scientific collaboration during the International Geophysical Year, agreed that a management regime needed to be established for Southern Ocean resources. The first step was the Convention for the Conservation of Antarctic Seals, which came into force in 1972 even though there was at that time no harvesting of seals or any intention voiced to that end. Scientific advice underpinning this agreement come from the Scientific Committee for Antarctic Research (SCAR). The establishment of the BIOMASS (Biological Investigations of Marine Antarctic Systems and Stocks) program in 1976, again through support from SCAR, in collaboration with the Scientific Committee for Oceanic Research (SCOR), International Association of Biological Oceanography (IABO) and Advisory Committee on Marine Resources Research (ACMRR), led naturally to the Convention for the Conservation of Antarctic Marine Living Resources, negotiated in 1980 coming into force in 1982.

Commission for the Conservation of Antarctic Marine Living Resources (CCAMLR)

The overriding concern leading to the establishment of CCAMLR was that overexploitation of krill might adversely affect dependent species. This is encapsulated in Article II as a series of general principles of conservation that in summary form set out in paragraph 3 state that any harvesting shall be conducted in such a way as to

(a)   prevent the decrease in the size of the harvested population to levels below those which ensure its stable recruitment;

(b)   maintain the ecological relationships between harvested, dependent, and related populations; and

(c)   prevent changes or minimize the risk of changes in the marine ecosystem that are not reversible over two or three decades.

Subparagraph (a) is essentially a statement of the traditional single-species approach to fisheries management and setting an upper limit to the total catch that can be taken for any individual species. This maximal level – which, it should be noted, is a limit and not a target – must then be reduced to take account of the dependent species as required by subparagraph (b). Finally, there is the requirement in subparagraph (c) that any ecological changes should be reversible within a finite and limited period.

These three components of Article II encapsulate what has come to be known as the ecosystem approach to management of marine living resources. CCAMLR was the first international fisheries commission to make this a central plank of its management regime, and even though the approach has received wide approval it is only now, twenty years later, that it is beginning to be considered for implementation in other forums.

The Antarctic Treaty covers the land and ice-shelves south of 60°S with an agreement that all territorial claims within that region are 'frozen.' Essentially this means that anything other than sovereignty can be discussed. At the time of the negotiation of CCAMLR it was known that the distribution of Antarctic krill and some finfish stocks extended some way north of the Treaty zone. Accordingly, the CCAMLR region was designated to cover the area south of the Antarctic Convergence (now called the Antarctic Polar Frontal Zone, APFZ). This area is designated by lines joining parallels of latitude and meridians of longitude, that are set out in Article I paragraph 5, which approximate the Antarctic Convergence. For the most part the fit is good, but for political reasons some small parts were excluded. Within the region between the Convergence and latitude 60°S lie several Sub-Antarctic islands for which sovereignty is claimed by member states of CCAMLR. Prior to the conclusion of negotiation of the CCAMLR Convention, France had declared a 200 mile exclusive economic zone (EEZ) around Kerguelen and Crozet and obtained agreement that, through what has become to be known as 'The Chairman's Statement' of 19 May 1980, within the zone France would take note of CCAMLR advice but would be free to implement its own policy. Subsequently, other claimant states of Sub-Antarctic islands have implemented a similar policy to France.

Krill

Although originally applied to 'fish fry,' the term krill is now taken to refer to euphausiids, of which six species occur in the Southern Ocean. Of these the only one that is targeted commercially is the Antarctic krill, *Euphausia superba*. Antarctic krill are small shrimplike crustacea that grow to a maximum length of 65 mm. They are widely distributed in the Southern Ocean south of the APFZ. On the continental shelf they are replaced by their smaller congener *E. crystallorophias*. Concentrations are normally found close to islands and the continental shelf break and have in the past been considered as constituting separate populations or management units. Although they are carried around the continent on the Antarctic Circumpolar Current (ACC), the extent of mixing between groups of krill from different regions is not clear.

The sheer size of the Southern Ocean, allied to the fact that much of it is covered in sea ice, means that it has never been possible to estimate the standing stock over the whole region. Synoptic surveys have been undertaken over part of the region in the Atlantic and Indian Ocean sectors. The most recent in the Atlantic sector, CCAMLR Area 48 (see **Figure 1**), gave an estimated standing stock of 44 Mt, while a survey in the Indian Ocean sector in 1996 gave estimated values of 3.04 Mt between 80° and 115°E and 1.79 Mt between 115° and 150°E.

In common with many zooplankters, krill undergo diurnal vertical migration within the top 200 m, although the pattern is not consistent with time or locality. They are often found at the surface at night, although daytime surface concentrations are not infrequent. They frequently occur in swarms with a density of several thousand individuals per cubic meter and it is these swarms that are targeted by commercial fishers.

Antarctic krill are filter feeders dependent primarily on phytoplankton. Arising from this, in the early season 'green' krill, with large amounts of chlorophyll in the hepatopancreas, predominate. These green krill are associated with a 'grassy' flavour to products manufactured for human consumption. Later in the season, colorless 'white' krill predominate.

Estimates of production have been based on consumption of krill by dependent species and, even allowing for considerable uncertainty over the conversion factors, indicate that production is over 100 Mt a year.

Commercial fishing began in the 1960s and had reached 1000 t a year by 1970. Initially these catches were reported to FAO in the category 'Unspecified Marine Crustacea' and, although they are

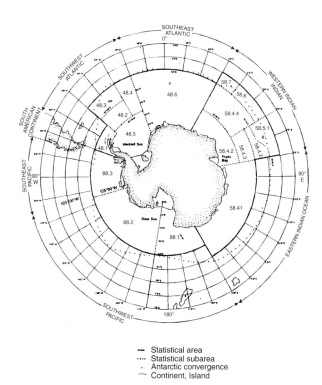

Figure 1 CCAMLR Statistical Areas, subareas, and divisions.

— Statistical area
···· Statistical subarea
· Antarctic convergence
⌒ Continent, Island

generally thought to refer to krill, the figures do not match those subsequently reported to CCAMLR as krill. By the late 1970s the total reported catch had risen to over 300 000 t, giving cause for concern that a very rapid and large-scale expansion of the fishery might be imminent. The main fishing nations at that time were the former Soviet Union, Japan, and Poland. With the collapse of the former Soviet Union, the total reported catch is now dominated by Japan, which has continued to take around 60 000–70 000 t per year. Apart from Japan and Poland, small catches have been reported in recent years by Bulgaria, Chile, Korea, Russia, Ukraine, and the United Kingdom (see **Table 1**).

In the early years of the fishery, catches were reported from the Indian and Atlantic Ocean sectors with relatively small amounts coming from the Pacific sector. During the 1990s the pattern has changed such that virtually all catches are taken from the Atlantic sector. In the summer months this is predominantly the Antarctic Peninsula (Subarea 48.1) and South Orkneys (Subarea 48.2); and in the winter, because the more southerly grounds are closed owing to sea ice, around South Georgia (Subarea 48.3).

In the early years a variety of fishing methods were attempted. Early attempts at developing surface trawls, with a codend pump, towed alongside the fishing vessel failed because of the cumbersome

Table 1 Annual reported catches (tonnes) of Antarctic krill by nation

Year	Japan	Poland	USSR	Russia	Ukraine	Argentina	Bulgaria[a]	Chile	E. Germany	Korea	Spain	Latvia[a]	Panama[a]	UK
1970														
1971														
1972														
1973	59													
1974	646		19 139											
1975	2 677		41 352											
1976	4 750		609					276						
1977	12 802		71 656					92						
1978	25 219	6 966	106 991				94		8	511				
1979	36 961		295 508				46		102	0				
1980	36 275	226	440 516					0	0	0				
1981	27 698	0	420 434					0	0	1 429				
1982	35 116	360	491 656					0	0	1 959				
1983	42 282	0	180 290					3 752	0	5 314				
1984	49 531	0	74 381					1 649	0	0				
1985	38 274	2 065	150 538					2 598	0	0				
1986	61 074	1 726	379 270					3 264	0	0				
1987	78 360	5 215	290 401					4 063	0	1 527	379			
1988	73 112	6 997	284 873					5 938	0	1 525				
1989	78 928	1 275	301 498					5 329	396	1 779				
1990	62 187	9 571	302 376					4 501		4 040				
1991	67 582	8 607	275 495					3 679		1 211				
1992	74 325	15 911		151 725	61 719			6 066		519				
1993	59 272	7 915		4 249	6 083			3 261		0				
1994	62 322	9 384		965	8 852			3 834		0		71		
1995	60 303	20 610			48 886			0		0			141	
1996	60 546	19 156			20 056			0		0			495	
1997	58 798	15 312			4 246			0		0			0	308
1998	63 233	18 554			0			0		1 618			0	634
1999	71 318				5 694	6 524		0		1 228			0	

[a]Not a member state of CCAMLR when the catches were made.

nature of the gear and the relative infrequency of daytime surface swarms. Likewise, purse seines have proved impractical. Currently, midwater trawls with a mouth area of 500–700 m^2 and a fine mesh throughout are used. Krill swarms are detected using directional sonar and the fishing depth of the net is adjusted by reference to a netsounder. When regions of high concentration have been found, the vessels continue to fish on them without spending a great deal of time in searching farther afield. Haul duration is generally controlled to ensure a maximum catch of 7–10 t. This is because with large catches the krill tend to became crushed in the cod end and product quality suffers.

Over several seasons Japanese fleets have changed their fishing patterns, delaying the commencement of fishing until late in the austral summer and continuing through to the winter so as to avoid catching 'green' krill; products made from green krill have a lower market value than white krill. There is also a preference for larger krill, which, owing to diurnal vertical migration, tend to be deeper by day than small krill, a factor reflected in the fishing depth.

Much effort has been expended in developing krill products for direct human consumption, but much of the current catch has been used for domestic animal feed and, particularly in recent years, for aquaculture feed. The Japanese fishery produces four types of product: fresh frozen (46% of catch), boiled frozen (10% of catch), peeled krill meat (10% of catch), and meal (34% of catch). These are used for aquaculture and aquarium feed (43% of catch), for sport fishing bait (45% of catch), and for human consumption (12% of catch). There are plans for further products such as freeze-dried krill and krill hydrolysates, although currently these are only at an early developmental stage.

In the 1970s high contents of fluoride were reported from krill. This was found to be localized to the exoskeleton, where it can reach 3500 μg F per g dry mass, although concentrations in the muscle appear to be less than 100 μg F per g dry mass. Providing the krill are peeled soon after capture or the whole krill are frozen at a temperature lower than $-30°C$, the fluoride concentration in the muscle remains low.

Finfish

Fisheries for finfish developed very quickly. In 1969 and 1970 around 500 000 t of fish were taken from the South Georgia, and in 1971 and 1972 a further 300 000 t were taken from Kerguelen. Although the catches were initially reported to FAO as 'Unspecified Demersal Percomorphs' it is widely assumed that the dominant species in the catches was *Notothenia rossii*. Smaller catches in following years were reported from South Shetland and South Orkney Islands. These catch rates were unsustainable and the species remains at a low stock level. Other species that were taken at this time included grey rockcod, *Lepidonotothen (Notothenia) squamifrons*, and mackerel icefish, *Champsocephalus gunnari*. During the 1980s and coincident with the start of CCAMLR, catches of other species were reported including *Chaenocephalus aceratus*, *Chaenodraco wilsoni*, *Channichthys rhinoceratus*, *Chionodraco rastrospinosus Pseudochaenichthys georgianus*, *Dissostichus eleginoides*, *Gobionotothen (Notothenia) gibberifrons* and *Patagonotothen guntheri*. There was also a fishery for Myctophidae, principally *Electrona carlsbergi*, which reached a peak of 78 000 t in 1991 but ceased two years later coincident with the collapse of the former Soviet Union. In recent years the only target species have been mackerel icefish and Patagonian toothfish.

Mackerel Icefish

The mackerel icefish is a Channichthyid; these 'white blooded fish' are the only group of vertebrates that do not have erythrocytes or any effective blood pigment. They occur on the shelf in waters rarely more than 500 m deep. Spawning is thought to occur in fiords and bays during April and May and the eggs, > 3 mm diameter, hatch in the spring coincident with the spring primary production bloom and new generations of copepods in the plankton. The fish become sexually mature at around 3 years of age when they are 25 cm in total length. Their diet is predominantly krill, although they will take a variety of other species, particularly when krill are scarce.

At Kerguelen it has been possible to follow growth through several years by analysis of length distributions. Of particular interest is a three year cycle in the population. It is not clear whether this is a natural phenomenon or one induced by fishing pressure. At South Georgia there have been similar fluctuations in stock size, although in that case they have been attributed to enhanced predation pressure by fur seals at times of krill scarcity.

Initially bottom trawls were used in the fishery but these were banned by CCAMLR in the early 1990s to reduce damage to the benthic environment. The use of midwater trawls, as well as having a much reduced impact on benthic biota, is also accompanied by reduced by-catches. Currently there are limited trawl fisheries at Kerguelen and South Georgia. The fish are generally headed and gutted and then frozen; they fetch a price similar to that of

Table 2 Toothfish catches (tonnes) by area[a]

Subarea/Division	Estimated total catch	Reported catch 1996/97	Estimated unreported catch from catch/ effort data	Unreported catch as % of the estimated total catch
South Georgia (48.3)	2 389	2 389	Probably low	Probably low
Prince Edward Is. (58.7)	14 286	2 386	11 900	83.3
Crozet Is. (58.6)	19 233	333	18 900	98.2
Kerguelen (58.5.1)	6 681	4 681	2 000	29.9
Heard Is. (58.5.2)	8 037–12 837	837	7 200–12 000	89.6–93.4
All subareas	48 856–53 656	10 856	38 000–42 800	77.8–79.8

[a]Reported in relation to unreported catch of toothfish during the 1996/97 season. The unreported catch is estimated from market landings, numbers of vessels seen, and observed catch rates. (Anon, 1997.)

medium-small to medium sized cod. The fillets contain slightly more fat than average whitefish.

At Kerguelen the management regime has been operated exclusively through the French authorities. Around South Georgia a Total Allowable Catch (TAC) has been set based on production estimates using trawl survey results. The most recent TACs have been based on the lower 95% confidence limit from the surveys.

Toothfish

There are two species of the genus *Dissostichus*, which are broadly separated geographically. *D. mawsoni* is present in the high Antarctic close to the continent, whereas *D. eleginoides* is present around sub-Antarctic islands and extends outside of the Antarctic zone to Patagonia and the west coast of South America. Toothfish are the largest Nototheniid fish in the Southern Ocean, growing to a maximum length of over 2 m and mass of around 100 kg. They are relatively slow-growing reaching sexual maturity at around 90–100 cm. Spawning is thought to occur in the latter part of the winter.

At Kerguelen a trawl fishery has become established in various parts of the shelf, with trawlers mainly from France and Russia. The management regime is implemented by French authorities.

In the early years at South Georgia, toothfish appeared as a by-catch in the trawl fishery, sometimes to the extent of a tonne of fish in a haul. Toward the end of the 1980s, long-liners from the former Soviet Union began fishing at the shelf slope around South Georgia and obtained good catch rates. With the increasing fishing effort being applied, concern was expressed at the sustainability of the resource. Owing to the high value of the catch, other nations have joined the fishery; these include Argentina, Bulgaria, Chile, Korea, Ukraine, and the United Kingdom. Two types of long-line system are used: Mustad Autoliner and the Spanish system with

a ground line and parallel fishing line. The lines are baited with squid, horse mackerel or sprats and fished on the bottom for around 12 h in water around 1000 m deep.

It is possible that the by-catch of rays and grenadiers may be significant, although at present data are scarce. Of much greater concern is the incidental mortality caused to seabirds that attack the baits as the lines are shot. Very low recruitment of albatrosses to the breeding populations at South Georgia has been directly attributed to mortality caused by long-line fishing, not only for toothfish within the Southern Ocean but also due to tuna long-lining over much of the Southern Hemisphere. It has been demonstrated that restricting the setting of lines to the hours of darkness together with weighting the lines is adequate to eliminate this mortality, if applied diligently.

Toothfish fetch a high price on international markets and this has been a major cause of illegal and unregulated fishing within the Southern Ocean. This reached a peak in 1997 when over 90% of catches in some areas were thought to have been taken outside of CCAMLR regulations. To combat this activity, CCAMLR has introduced a catch documentation scheme whereby only fish that have been taken in compliance with CCAMLR Conservation Measures can be issued with a valid certificate. This measure is having a positive effect as currently certificated catches attract a price of around $10 000 per tonne whereas uncertificated catches fetch only around $3000 per tonne (see **Table 2**).

Crabs

There are two species of crab in the waters around South Georgia, *Paralomis spinosissima* and *P. formosa*. In the early 1990s there was an exploratory study to fish them. In view of the lack of knowledge of either species, CCAMLR imposed a rigorous fishing plan for the study. Following fishing in the area during 1993, when 299 t were caught, and 1996, when 497 t were caught, the fishing company decided that these catches were uneconomic and did not pursue the matter. Subsequently there have been expressions of interest by other companies, but no further fishing activity.

Squid

For many years it has been known that squid figure in the diets of many predators in the Southern Ocean. Regurgitated stomach content samples from albatrosses and petrels indicated that in many cases these squid were quite fresh, leading to the conclusion that they had been taken from adjacent waters. Exploratory ventures in 1989 and 1996 by Korean squid jiggers found small concentrations of *Martialia hyadesi* close to the APFZ in summer and on the South Georgia shelf in winter. World market prices of squid have meant that further ventures are likely to be only marginally profitable. However, advances in processing whereby the second tunic can be easily removed may change this situation in the near future.

Southern Ocean Management Regime

The sections of Article II of the CCAMLR Convention provide a clear framework for the management regime. In the first instance, harvested species need to be managed such that their long-term sustainability is not threatened by commercial activities. This requirement can be met through the application of traditional fisheries models and others developed within CCAMLR such as the Krill Yield Model (KYM) and Generalized Yield Model, the underlying principles of which are set out below.

1. To aim to keep the krill biomass at a level higher than would be the case for single-species harvesting considerations.
2. Given that krill dynamics have a stochastic component, to focus on the lowest biomass that might occur over a future period, rather than on the average biomass at the end of that period, as might be the case in a single species context.
3. To ensure that any reduction of food to predators that might arise out of krill harvesting is not such that land-breeding predators with restricted foraging ranges are disproportionately affected compared to predators in pelagic habitats.
4. To examine what levels of krill escapement are sufficient to meet the reasonable requirements of predators.

The KYM has the general formula [1], where Y = yield, B_0 = median unexploited biomass, and γ = proportionality coefficient.

$$Y = \gamma B_0 \qquad [1]$$

Currently two values of γ are calculated. The first, γ_1, is chosen such that the probability of the spawning biomass dropping below 20% of its pre-exploitation median level over a 20-year harvesting period is 10%, and the second γ_2 is chosen so that the median krill escapement over a 20-year period is 75%. The lower of γ_1 and γ_2 is selected for the calculation of krill yield. These principles are applied so as to account for sustained consistency in catch with time while at the same time taking account of uncertainties in the estimators. This implies a conservative approach to the application of fishery models in pursuit of the precautionary principle; this approach is followed in the advice that comes forward from the Scientific Committee and is implemented by the Commission. Within this overall framework, due regard has to be taken of the requirements of dependent species – the ecosystem approach.

Early in the history of CCAMLR it became apparent that a wide range of interpretations can be put on the definition of an 'ecosystem approach.' At one extreme is an endeavor to understand all interactions in the food web in order to formulate management advice. Such an approach is favored by idealists on the one hand and those who did not wish to see any form of control on the other. It is impracticable because no advice would emerge in a timely manner. Recognizing this, CCAMLR has set up an Ecosystem Monitoring Program (CEMP) whereby certain features of a small suite of dependent species are monitored. Currently the CEMP species are Adelie, chinstrap, gentoo, and macaroni penguins; black-browed albatross; fur seal and crabeater seal. Key parameters associated with the ecology of each of these species are monitored according to a series of agreed CEMP protocols. The aim of the program is to determine how the dependent species perform in response to krill availability and how this is affected by fishing activities. CEMP parameters integrate krill availability over different time and space scales, varying from months and hundreds of kilometers in the case of the total mass of individual penguins on arrival at a colony at the start of breeding, to days and tens of kilometers in the case of foraging trip duration while feeding chicks. These provide indicators of overlap with commercial fishing. The other components monitored relate to the vital rates of dependent species – changes in population size, mortality, and recruitment – and progress is currently in hand to integrate these into the overall management process.

Recent papers have linked variations in krill distribution and standing stock to climate change and El Niño–Southern Oscillations through their effects on the ACC and sea ice regime. The CCAMLR management scheme implicitly takes account of such long-term environmental change because its regime can be adjusted to compensate as the monitored species and variables change with time.

See also

Crustacean Fisheries. Current Systems in the Southern Ocean. International Organizations. Krill. Marine Mammals, History of Exploitation.

Further Reading

Information on the status of Southern Ocean resources and management decisions is provided in the

Reports and Statistical Bulletins of the Commission and Scientific Committee published by CCAMLR, Hobart, Tasmania 7000, Australia.

Anon (1997) Estimates of catches of *Dissostichus eleginoides* inside and outside the CCAMLR Area. SC-CAMLR-XVI, Annex 5, Appendix D.

Constable AJ, de la Mare WK, Agnew DJ, Everson I and Miller D (2000) Managing fisheries to conserve the Antarctic marine ecosystem: practical implementation of the Convention on the Conservation of Antarctic Marine Living Resources (CCAMLR). *ICES Journal of Marine Science* 57: 778–791.

El-Sayed SZ (ed.) (1994) *Southern Ocean Ecology: The BIOMASS Perspective*. Cambridge: Cambridge University Press.

Everson I (1978) Antarctic fisheries. *Polar Record* 19(120): 233–251.

Everson I (ed.) (2000) *Krill: Biology, Ecology and Fisheries*. Oxford: Blackwell Science.

Kock K-H (1992) *Antarctic Fish and Fisheries*. Cambridge: Cambridge University Press.

SPERM WHALES AND BEAKED WHALES

S. K. Hooker, Natural Environment Research Council, Cambridge, UK

doi:10.1006/rwos.2001.0430

Introduction

Sperm whales and beaked whales are among the largest and most enigmatic of the odontocetes (toothed whales). These species tend to live far offshore in regions of deep water, and perform long, deep dives in search of their squid prey. This has generally made the study of these animals much more difficult than that of more accessible, nearshore cetacean species. In addition, the pygmy and dwarf sperm whales, and many species of beaked whale, have superficially similar external morphology, and so are often difficult to identify to species level in the wild. The study of many of these species has therefore been based primarily on examination of stranded and beachcast animals. As a result, we currently know little about many of these relatively large mammals. For example, one species of beaked whale, Longman's beaked whale, has been identified only from two skulls in Australia and Somalia. Another putative species, *Mesoplodon* species 'A' has only ever been observed at sea, and knowledge of its morphological characteristics remains far from complete. New species of beaked whales are still being discovered. For example, the pygmy beaked whale

and Bahamonde's beaked whale were only identified in the last decade from specimens collected in Peru and Chile, respectively. Likewise, the dwarf and pygmy sperm whale were only recognized as separate species in the 1960s.

The sperm and beaked whale species about which we know most are the sperm whale, the northern bottlenose whale and Baird's beaked whale. Much of the information about these species has come from scientific research programs conducted in conjunction with historic whaling operations. Longer-term, nonlethal studies of wild populations only began in the early 1980s. These focused initially on sperm whales, and today include research on populations of northern bottlenose whales and dense-beaked whales. Such studies help provide important behavioral information about these species which was previously not available from studies of dead animals.

Taxonomy and Phylogeny

There are three superfamilies within the odontocetes: the Physeteroidea (sperm whales), the Ziphioidea (beaked whales), and the Delphinoidea (river dolphins, oceanic dolphins, porpoises, and monodontids). The superfamily Physeteroidea encompasses two families: the Physeteridae which contains the sperm whale, and the Kogiidae which contains the pygmy sperm whale and the dwarf

Table 1 Sperm and beaked whale species, approximate demographic distribution and size

Species		General location	Adult size (m)	Notes
Family Physeteridae				
Sperm whale	*Physeter macrocephalus*	Global	12–18	
Family Kogiidae				
Dwarf sperm whale	*Kogia simus*	Tropical and temperate oceanic	2.7–3.4	
Pygmy sperm whale	*Kogia breviceps*	Tropical and temperate, continental shelf and slope	Up to 2.7	
Family Ziphiidae				
Baird's beaked whale	*Berardius bairdii*	North Pacific	11.9–12.8	
Arnoux's beaked whale	*Berardius arnuxii*	Southern Ocean	Up to 9.7	
Cuvier's beaked whale	*Ziphius cavirostris*	Global, common in eastern tropical Pacific	7–7.5	
Northern bottlenose whale	*Hyperoodon ampullatus*	North Atlantic	8.7–9.8	
Southern bottlenose whale	*Hyperoodon planifrons*	Southern Ocean	7.2–7.8	
Tropical bottlenose whale	*Hyperoodon sp.*	Tropical Indian and Pacific Oceans	4–9	Sightings only
Shepherd's beaked whale	*Tasmacetus shepherdi*	Southern temperate	6.6–7	Few stranded specimens
Longman's beaked whale	*Indopacetus pacificus*	Australia, Somalia	Over 6	Two skulls
Blainville's beaked whale	*Mesoplodon densirostris*	Temperate global	Up to 4.7	
Gray's beaked whale	*Mesoplodon grayi*	Southern temperate circumglobal	Up to 5.6	
Ginkgo-toothed beaked whale	*Mesoplodon ginkgodens*	Temperate/tropical Indian and Pacific Oceans	Up to 4.9	
Hector's beaked whale	*Mesoplodon hectori*	Southern temperate, extralimital in S. California	4.3–4.4	
Hubbs' beaked whale	*Mesoplodon carlhubbsi*	North Pacific	Up to 5.3	
Sowerby's beaked whale	*Mesoplodon bidens*	Northern North Atlantic	5.1–5.5	
Gervais' beaked whale	*Mesoplodon europaeus*	Temperate/tropical Atlantic	4.5–5.2	
True's beaked whale	*Mesoplodon mirus*	Temperate N. Atlantic and temperate Southern Ocean	Up to 5	
Strap-toothed beaked whale	*Mesoplodon layardii*	Southern temperate	5.9–6.2	
Andrews' beaked whale	*Mesoplodon bowdoini*	South Indian and Pacific Oceans	4.6–4.7	
Stejneger's beaked whale	*Mesoplodon stejnegeri*	North Pacific	Up to 5.3	
Pygmy beaked whale	*Mesoplodon peruvianus*	Eastern tropical Pacific	Up to 3.7	Few strandings; tentative sightings
Bahamonde's beaked whale	*Mesoplodon bahamondii*	Peru	Estimated 5–5.5	Partial skull
Mesoplodon species 'A'	*Mesoplodon sp.*	Eastern tropical Pacific	5.5	Sightings only

Compiled from Jefferson *et al.*, 1993.

sperm whale. The Ziphioidea encompasses only the family Ziphiidae, which includes at least 20 species of beaked whales (**Table 1**).

Although some genetic studies have challenged the relationship of the sperm whales to other toothed whales, the analytical methods used to determine this have been questioned, and there is general agreement between morphological and other molecular data that the sperm whales and beaked whales are basal odontocetes (**Figure 1**). Physeterids appeared in the fossil record in the early Miocene deposits of Argentina (approximately 25–30 Ma). In the past this family included a diverse array of

genera, but today it is represented only by the sperm whale. The kogiids are thought to have diverged from the physeterids in late Miocene and early Pleiocene (approximately 5–10 Ma). The earliest ziphiids have been found in deposits from the middle Miocene (10–15 Ma). Relationships among the beaked whales are not clear. The six genera in this family have previously been separated into two tribes grouping *Berardius* and *Ziphius*, and grouping *Tasmacetus*, *Indopacetus*, *Hyperoodon* and *Mesoplodon*. However, it has also been suggested that *Tasmacetus*, with a full set of teeth in upper and lower jaws, may be the sister group to all other

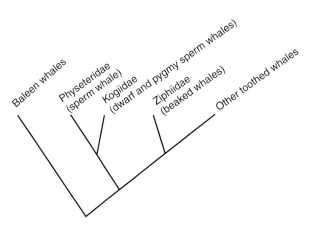

Figure 1 Phylogenetic diagram showing relationship of sperm whales and beaked whales to other cetaceans. (Based on Heyning, 1997.)

living species. Current work investigating the systematics of this group by DNA sequence data should shed further light on their phylogenetic relationships.

Anatomy and Morphology

Beaked whales are characterized by the possession of a long and slender rostrum resulting in a prominent beak in most species. An evolutionary trend in ziphiids has led to the reduction in number of teeth in all genera except *Tasmacetus*. Most species have retained only one or two pairs of teeth, set in varying positions in the lower jaw (**Figure 2**). In most beaked whale species these teeth only erupt in adult males. From observations of scarring patterns on the animals, these teeth appear to function as weapons in intra-specific combat, and have become much enlarged in some species. Other features which distinguish beaked whales from other groups include the possession of two conspicuous throat grooves or creases which form a forward-pointing V-shape, and the lack of a notch in the flukes. The skull morphology of beaked whales is also unique, exhibiting elevated maxillary ridges behind the nasals.

The Physeteroidea are characterized by several features of the skull, including a large supracranial basin. This basin holds the 'spermaceti organ', a fat-filled structure, which lies behind the melon in the forehead, and is unique to these species. This structure was named for the presence of spermaceti, an oily substance thought to resemble semen (after which it was named). It is generally thought that this organ functions in sound transmission.

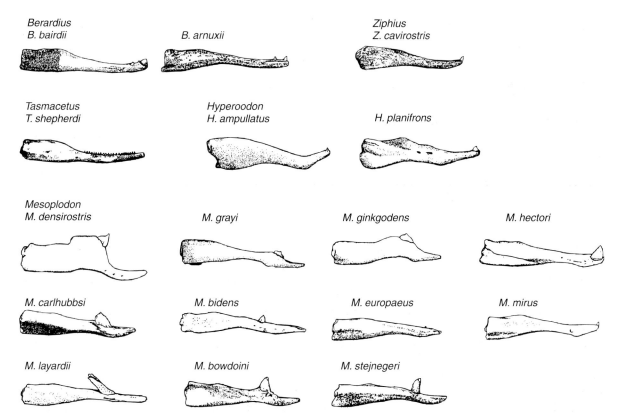

Figure 2 Variation in position, size, and morphology of the lower jaw teeth of adult male beaked whales shown for the majority of recorded beaked whale species. (Reproduced with permission from Jefferson *et al.*, 1993.)

Externally, sperm whales can be differentiated from other species by their narrow lower jaw, and an upper jaw which extends well past the lower. This group also has reduced dentition compared with many other odontocetes. The sperm whale has teeth (18–25 pairs) only in the lower jaw. The dwarf and pygmy sperm whales have reduced numbers of teeth in the lower jaw (generally 12–16 pairs in the pygmy sperm whale, and 8–11 pairs in the dwarf sperm whale). Some teeth may be present in the upper jaw of both dwarf and pygmy sperm whales, although this is less common in the pygmy sperm whale.

The sperm whales show highly pronounced asymmetry of the skull. Most beaked whale species also have asymmetrical skulls, although *Berardius* and *Tasmacetus* have nearly symmetrical cranial characteristics. This suggests that such symmetry is the ancestral characteristic and that cranial asymmetry in beaked whales has evolved independently from that in sperm whales.

Distribution and Abundance

All sperm whales and beaked whales are deep-water oceanic species. However, there is wide variation in species coverage. The sperm whale is found throughout the world's oceans, from the tropics to the poles. The pygmy and dwarf sperm whales also have fairly cosmopolitan distributions, and are found in temperate and tropical waters worldwide. In contrast, many beaked whale species have quite limited distributions and only two species, the dense-beaked whale and Cuvier's beaked whale, show similar ranges to the sperm whales (**Table 1**). Many other beaked whale species are limited to a single ocean basin, and several species pairs show an antitropical distribution, e.g. *Hyperoodon* species (*H. ampullatus* in the North Atlantic and *H. planifrons* in the Southern Ocean) and *Berardius* species (*B. bairdii* in the North Pacific and *B. arnuxii* in the Southern Ocean).

There are very few estimates of abundance for sperm and beaked whales. Many of these species are difficult to detect and identify at sea, and so are probably more common than sighting records would suggest. The status of all sperm and beaked whales as currently listed by the IUCN (International Union for the Conservation of Nature) Red Book is 'insufficiently known'.

Foraging Ecology

The majority of sperm and beaked whales are thought to feed primarily on squid. The reduced dentition of these species is thought to be due to this dietary specialization. One exception to this is Shepherd's beaked whale, in which both sexes possess a full set of functional teeth, and the diet appears to consist primarily of fish.

The reduced dentition of the beaked whales, together with their narrow jaws and throat grooves, have been suggested to function in suction feeding. Among the males of species such as the strap-toothed whale, the elaborate growth of the strap-like teeth may limit the aperture of the gape to a few centimeters, and it is difficult to see how prey capture techniques other than suction feeding could be successful. The same mechanism is thought to be used by sperm whales, which also have a comparatively small mouth area. Anecdotal evidence suggests that the lower jaw teeth of the sperm whale are not required for feeding, as apparently healthy animals have been seen with broken and badly set lower jaws resulting from past injuries.

Sperm whales and beaked whale species are known to be excellent divers. Dives of up to an hour have been recorded from several beaked whale species, although many of these records have been based on surface observations of diving whales. Similarly dives of up to 25 minutes have been recorded from *Kogia* species. Two studies have used time-depth recorders to monitor dives in more detail from these species. Acoustic transponder tags on two sperm whales recorded repeated dives to depths of 400–600 m (including a dive to 1185 m), for durations of 35–45 minutes. Similarly deployments of time-depth data loggers on two northern bottlenose whales showed regular dives to depths of over 800 m (maximum 1452 m) for durations of 30–40 minutes (maximum 70 minutes), with dives approximately every 80 minutes. Both these species are thought to forage at these depths in search of deep-water squid.

The similarity of ecological niches among beaked and sperm whales might be expected to lead to competition between these species. The relatively discrete distributions of many beaked whale species may have resulted from this. For example, several *Mesoplodon* species coexist in the North Atlantic, but have separate centers of distribution, with little overlap in range: Sowerby's beaked whale has a more northerly distribution than True's beaked whale, which in turn is found to the north of Gervais' beaked whale. On a much smaller spatial scale, there is some suggestion of competitive exclusion between sperm whales and northern bottlenose whales in habitat use of a submarine canyon area off eastern Canada. Prey species (mainly squid) identified from the stomachs of stranded *Kogia* specimens suggest that these species occur primarily along the continental shelf and slope in the epi- and meso-pelagic zones. Although the diets of both species overlap, the relative contribution of prey types suggests that the dwarf sperm whale feeds on smaller squid in shallower waters and thus occurs further inshore than the pygmy sperm whale.

Social Organization

The social organization of the majority of sperm and beaked whale species is poorly known, with the exception of the sperm whale. The social system of the sperm whale appears quite unlike that known for other cetaceans. Groups of females and juveniles are found in temperate and tropical latitudes (**Figure 3**). Males become segregated from these female groups at or before puberty, and migrate to higher latitudes. Younger males are found in 'bachelor schools', which consist of animals of approximately the same age. These schools decrease in size with increasing age of the members, to the point at which large mature animals are typically solitary. Sexually mature males return to the tropical waters inhabited by females in order to breed. There, these males were traditionally viewed as 'harem masters', each remaining with a single group of females throughout the breeding season. However, recent mark-recapture data, relative parasite loads, and indications of synchronous estrus suggest that instead, males rove between groups of females, and remain with any one group for only a few hours, although they may revisit groups on consecutive days.

Female sperm whales are found in groups of 20 or so individuals. These groups appear to consist of two or more stable units that associate for periods of approximately 10 days. Genetic evidence has suggested that these groups are composed of one or more matrilines. However, there are also suggestions of paternal relatedness between grouped matrilines, and recent photo-identification studies suggest that some animals occasionally switch groups, and thus may not be of the same maternal lineages as other group members.

Whalers observed that sperm whale groups often exhibited epimeletic behavior, with individuals supporting and staying with harpooned, injured, and even dead group members. It is thought that this may be a result of the close genetic ties between the individuals in a group. Sperm whales have also been observed to exhibit allomaternal care (babysitting behavior). Calves remain on the surface when a group is feeding, presumably since they are unable to dive to the depths at which adults forage, and

Figure 3 Group of female and juvenile sperm whales off the Galapagos Islands. (Photograph by Sascha K. Hooker.)

adults have been observed to stagger their dives such that it is more likely for there to be an adult at the surface with the calf, and the proportion of time that the calf is alone is reduced.

Observations of wild *Kogia* suggest that they typically form small groups of one to four animals, with occasional groups of up to 10 reported. However, almost nothing is known of the composition of these groups or of the behavior of these species at sea.

Many beaked whale species appear to show intra-specific aggression between adult males, presumably for access to females. The prominent and elaborate teeth of many beaked whale species are thought to be used in this male–male conflict (**Figure 2**), resulting in the extensive scarring seen on adult males. However, in other beaked whale species, males possess only comparatively small lower jaw teeth, and these do not appear to be used for fighting. The northern bottlenose whale is an example of this. Instead this species shows marked sexual dimorphism in skull structure and associated forehead or melon shape, which is relatively small in females, but is enlarged and flattened in adult males. Recent observations have suggested that this melon morphology is also associated with male–male competition, as adult males have been observed to head-butt each other.

Among beaked whales, the composition of social groups is not well known. The two beaked whales for which most data have been collected are the northern bottlenose whale and Baird's beaked whale. Long-term photo-identification studies of individual bottlenose whales have in fact suggested stronger associations between males than between females. However, the aggression observed between some associated males makes further interpretation difficult. Anatomical studies of groups of Baird's beaked whales taken in the continuing fishery off Japan are suggestive of a different type of social structure for this species. Among this species, both males and females possess erupted teeth, and females are slightly larger than males. Males appear to reach sexual maturity at an average of 4 years earlier than females and may live for up to 30 years longer. This has led to speculation that males may be providing parental care in this species, although further work is needed to confirm this.

Acoustics, Sound Production, and Sound Reception

The acoustic behavior of sperm whales is relatively well documented. These whales produce broad-band clicks in the frequency range of 1–12 kHz. These clicks are thought to function primarily in echolocation, although some repetitive patterned clicks (termed codas) also appear to be used in a social context. Adult male sperm whales also produce especially loud resonant clicks (termed 'clangs') that may function in female choice or as a threat to other males. Neither *Kogia* species appears to be highly vocal, although high-frequency clicks have been recorded from the pygmy sperm whale. The social whistles characteristic of other odontocete species are absent from the physeterids, and may be absent from some beaked whale species, although they have been recorded in others. No whistles were documented in several hours of recordings from northern bottlenose whales, which appear to produce primarily echolocation-type clicks instead. These were superficially similar to sperm whale clicks, although often at ultrasonic frequencies (∼ 20–30 kHz). Recordings made from Baird's beaked whale and Arnoux's beaked whale included frequency-modulated whistles, burst-pulse clicks, and discrete clicks in rapid series. Only a few other records of beaked whale acoustic behavior exist and the majority of these were obtained from stranded animals.

The sound production mechanism used by both sperm and beaked whales for echolocation is homologous with that of other odontocetes, consisting of a sound-producing complex (the 'monkey lips'/dorsal bursae) in the upper nasal passages. Sound is thought to propagate into the water through the melon, a low-density lipid-filled structure which has been hypothesized to act as an acoustic lens to focus high-frequency sound ahead of the animal. The echoes of this sound are then thought to be received via the fat body in the lower jaw which connects with the bulla of the middle ear.

The sperm whale head is unique in comparison with other odontocetes in that the blowhole and sound production mechanism is situated at the front of the head rather than above the eyes. The signal generated is reverberated within the front of the head, generating a decaying series of pulses, with the time interval between these pulses related to the size of the head. It is thought that female sperm whales may select males based on this inter-pulse interval (indicating the size of the male). The extreme sexual dimorphism seen in male sperm whales in both the increased ratio of head to body size in addition to the overall much larger size of adult males (up to 18 m length compared with up to 12 m length of adult females) may therefore be based on this selection pressure associated with the sound production mechanism.

Predation

It was previously thought that large size was adequate defense against predators, but female sperm whales (of 10–12 m) have been observed under lethal attack by *transient* (mammal-eating) killer whales. Additionally large sharks are thought to be a threat to these species, and particularly to juvenile animals. Various methods of defense may be employed. For such a deep-diving species, it is surprising that deep dives are not used as a method of escape from predators. This may be because young calves are unable to dive to the depths or for the same duration as adults. Instead sperm whales appear to show a behavioral response to the threat of predation and form a 'marguerite', with the adults forming a circle (heads innermost) around the calves.

Pygmy and dwarf sperm whales evacuate reddish-brown intestinal fluid when startled, in a similar manner to squids. The lower intestine is expanded in both species, forming a balloon-like structure filled with up to 12 liters (in large specimens) of this liquid. Additionally, these species possess a crescent-shaped light-colored mark, often called a 'false gill', on the side of the head behind the eye and before the flippers. Along with the underslung mouth, this can lead to the mistaken identification of these animals as sharks. However, whether this patterning functions as camouflage against predation is unknown.

Conservation

The larger species of sperm and beaked whales were all targeted by whaling operations in the past. The sperm whale was the most heavily hunted, primarily due to the prized spermaceti oil that it contained. This fishery spanned the seventeenth to twentieth centuries and at its peak (in the 1960s) average annual catches reached 25 000 animals. Northern bottlenose whales were also quite severely depleted by whaling. The northern bottlenose whale fishery began in the late nineteenth century and between 1880 and 1920 approximately 60 000 bottlenose whales were caught. The other species in this group which has been taken in relatively large numbers is Baird's beaked whale. This species has been hunted in Japan since at least the seventeenth century, but has generally been taken in relatively low numbers (a maximum annual catch of 322 in 1952, but recently averaging 40 whales per year). This fishery still continues today.

Current threats faced by these species range from other factors potentially causing immediate death, such as ship-strikes, to the more insidious threats of ocean plastic, chemical, and acoustic pollution. Since many sperm and beaked whales feed primarily on squid, they are very susceptible to the ingestion of plastics, apparently mistaking it for prey. Stranded animals from several sperm and beaked whale species have been found with plastic in their stomachs and in some cases, this appears to have blocked the normal function of the stomach, causing severe emaciation and probably contributing to their death. The ecological role of odontocetes as long-lived top predators also exposes these animals to increased levels of chemical pollutants. Cetaceans store energy (and pollutants) in their blubber, and have a lower capacity to metabolize some PCB isomers than many other mammals. Foreign and toxic substances are therefore often biomagnified in odontocete species, and even species living offshore in relatively pristine environments have been found to contain high levels of pollutants. These high pollutant levels can have two major effects: (1) inhibition of immune system capacity to respond to naturally occurring diseases, and (2) potentially causing reproductive failure.

There is also increasing concern about the effect of anthropogenic noise in the marine environment. Sperm whales and beaked whales appear to be particularly susceptible to the effects of such noise. Sperm whales have been observed to react to several types of underwater noise including sonar, seismic activity, and low-frequency sound. Beaked whales also appear to be susceptible to high-intensity underwater sound. Several beaked whale stranding events appear to have coincided with military naval exercises, and associated increased noise levels. As sound seems essential to their foraging and social behavior, increasing levels of underwater noise are therefore of particular concern for these species.

See also

Bioacoustics. Marine Mammal Diving Physiology. Marine Mammal Evolution and Taxonomy. Marine Mammal Migrations and Movement Patterns. Marine Mammal Overview. Marine Mammal Social Organization and Communication. Marine Mammal Trophic Levels and Interactions. Marine Mammals, History of Exploitation.

Glossary

echolocation the production of high-frequency sound and reception of its echoes, used to navigate and locate prey
epimeletic care-giving behavior
mandible lower jaw
matriline descendants of a single female

rostrum anterior portion or beak region of the skull that is elongated in most cetaceans

sexual dimorphism morphological differences between males and females of a species

ultrasonic high-frequency sounds, beyond the upper range of human hearing

Further Reading

Berta A and Sumich JL (1999) *Marine Mammals: Evolutionary Biology*. London: Academic Press.

Best PB (1979) Social organisation in sperm whales. In: Winn HE and Olla BL (eds) *Behaviour of Marine Mammals*. vol 3: *Cetaceans*, pp. 227–289. New York: Plenum Press.

Heyning JE (1997) Sperm whale phylogeny revisited: analysis of the morphological evidence. *Marine Mammal Science* 13: 596–613.

Jefferson TA, Leatherwood S and Webber MA (1993) *Marine Mammals of the World*. Rome: United Nations Environment Program, FAO.

Moore JC (1968) Relationships among the living genera of beaked whales with classification, diagnoses and keys. *Fieldiana: zoology* 53: 209–298.

Rice DW (1998) *Marine Mammals of the World: Systematics and Distribution*. Society for Marine Mammalogy, Special Publication Number 4. Lawrence, Kansas: Allen Press.

Ridgway SH and Harrison R (1989) *Handbook of Marine Mammals*, vol. 4: *River Dolphins and the Larger Toothed Whales*. London: Academic Press.

Whitehead H and Weilgart L (2000) The sperm whale: social females and roving males. In: Mann J, Connor RC, Tyack P and Whitehead H (eds) *Cetacean Societies: Field Studies of Dolphins and Whales*. Chicago: Chicago University Press.

SPHENISCIFORMES

L. S. Davis, University of Otago, Dunedin, New Zealand

doi:10.1006/rwos.2001.0227

What is a Penguin?

Penguins, with their upright stance and dinner-jacket plumage, constitute a distinct and unmistakable order of birds (Sphenisciformes). Granted there are a few embellishments here and there – the odd crest, a black line or two on the chest – but otherwise, penguins conform to a very conservative body plan. The design of penguins is largely constrained by their commitment to an aquatic lifestyle. Penguins have essentially returned to the sea from which their ancestors, and those of all tetrapods, came. In that sense, they share more in common with seals and sea turtles than they do with other birds. Their spindle-shaped bodies and virtually everything about them have evolved in response to the demands of living in water (**Table 1**).

The loss of flight associated with their aquatic makeover is the penguins' most telling modification. While isolated examples of flightlessness can be found in virtually all other groups of waterbirds, penguins are the only group in which all members cannot fly. Among birds generally, they share that distinction only with the ratites (the kiwis, ostriches, emus, and their ilk), where flight has been sacrificed for large size and running speed.

Despite earlier claims to the contrary, it is clear that penguins have evolved from flying birds. The evidence from morphological and molecular studies suggests that penguins are closely related to loons (Gaviiformes), petrels, and albatrosses (Procellariiformes), and at least some families of the Pelicaniformes, most notably frigate-birds. Despite this, the exact nature of the relationship between penguins and these groups remains unresolved: at the moment it would seem to be a dead heat between loons and petrels as to which group is the sister taxon of penguins (**Figure 1**). (On the surface, loons may seem strange candidates to be so closely allied to penguins – penguins are found in the Southern Hemisphere, loons in the Northern Hemisphere; penguins are wing-propelled divers, loons are foot-propelled divers. However, it seems that loons, or their ancestors, were wing-propelled divers in their past.)

If the relationship of penguins to other birds seems confusing and controversial, the relationships among penguins themselves are no less so. Penguins are confined to the Southern Hemisphere and the distribution of fossilized penguin bones discovered to date mirrors their present-day distribution. Fossils have been found in New Zealand, Australia, South America, South Africa, and islands off the Antarctic Peninsula. The oldest confirmed fossil penguins have been described from late Eocene deposits in New Zealand and Australia, dating back some 40 million years. However, fossils from Waipara, New Zealand, unearthed from late Paleocene/early Eocene deposits that are about 50–60 million years

Table 1 Some adaptations of penguins for an aquatic lifestyle

Attribute	How modified	Purpose
Wings	Shorter, more rigid. Flipper acts as paddle/propeller. Feathers much reduced to decrease resistance	Increase diving capacity. By eliminating flight, the birds no longer need to keep body light.
Body shape	Spindle shape. Very low coefficient of drag	Reduce drag and increase efficiency of swimming. Water is more dense and offers more resistance than does air.
Bones	Nonpneumatic	Unlike flying birds, which have spaces in their bones to make them light (pneumatic), penguins have solid bones to increase strength and density.
Feathers	Short, rigid and interlocking	Feathers trap air beneath them, creating a feather survival suit that provides most of the insulation necessary for a warm-blooded animal to exist in water.
Fat	Subdermal layer of fat	Whereas flying birds cannot afford to carry too much fat, the subdermal layer of fat in penguins contributes a little to their insulation and also enables them to endure long periods of fasting on land necessary for incubation and molting.
Coloration	Dark back, light belly	To aid concealment in the open ocean, like many pelagic predators, penguins have a dark back, so that they merge with the bottom when viewed from above, and a light belly, so that they merge with the surface when viewed from below.
Legs	Placed farther back. Very short tarsometatarsus. Upright stance on land	To reduce drag in water. Feet act as rudder.
Eyesight	Variable	The eye is able to be altered to accommodate refractive differences when moving between water and land.
Circulation	Countercurrent blood system	Allows penguins to reduce heat loss in the water or very cold environments, while aiding heat dissipation when hot.

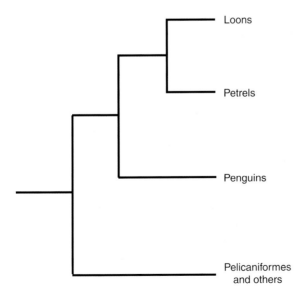

Figure 1 Nearest living relatives of penguins (adapted from Davis and Renner (in prep.)).

old, represent possibly the earliest penguin remains. These have still to be described fully, but they show a mixture of attributes from flying birds and those of penguins (the bones are heavy and nonpneumatic; the wing bones are flattened in a way that is consistent with being a wing-propelled diver). The Waipara fossils may very well be near the base of the penguin radiation, when the transition was being made from flyer to swimmer. In any case, by 40 million years ago, penguins were already very specialized, in much the same way as modern penguins, for underwater swimming.

Diving versus Flying

There is a trade-off between diving performance and flying. Even so, it would be wrong to conclude that flying birds, as a matter of course, cannot dive well; that somehow the riches that the sea has to offer are denied to them much beyond the surface. The truth is that some flying birds – for example, the diving-petrels and, especially, the alcids (auks, auklets, and murres) – can literally fly underwater as well as above. Although they are not closely related to penguins, auks are often considered to be the Northern Hemisphere's ecological equivalent of penguins, and in proportion to their size, auks can dive as deeply as many penguins. Nevertheless, the requirements for efficient flight – light bodies and flexible wings with a large surface area/small wing loadings – are not the same as those needed for diving efficiency – large, heavy bodies and stiff, powerful wings. Wingloadings are a measure of the body mass of the animal relative to the surface area of the wing. They give an indication of the lift that can be provided by the wing. If the linear measurements of a bird were simply scaled up, because body mass is a function of volume, the loading on the wing would become greater. This means that, to generate the same lift, larger birds must have wings with a disproportionately larger surface area.

Birds with relatively light bodies expend energy simply counteracting the buoyancy of their bodies in water. The time a bird can stay underwater and the depth to which it can travel are physiologically related to the size of the bird, providing another advantage to large birds. But as body mass increases, the surface area of wings needs to increase to a much greater extent just to maintain the same lift. There appears to be a cut-off point at about 1 kg beyond which it is not possible to both dive and fly efficiently. Indeed, auks that exceeded this size threshold, such as the extinct great auk, were flightless like penguins.

Logically, then, it seems that the transition from flighted to flightless would have taken place in birds around the 1 kg threshold, which is roughly the size of today's little penguins (*Eudyptula minor*) and consistent with the size of the smallest fossil penguins. However, once penguins were freed of the need to balance the requirements of flight with the requirements for underwater diving, they radiated rapidly and were able to attain much greater sizes. Many fossil penguins were larger than living penguins, with the tallest being up to 1.7 m and well over 100 kg.

Living Penguins

The species living today represent but a small remnant of the penguin diversity from times past. Exactly how modern penguins relate to the fossil penguins is not at all clear, as most of the extant (living) genera show up in the fossil record only relatively recently (within the last 3 million years or so). While there is argument about the precise number of species living today (most authorities list between 16 and 18 species), there is agreement that the extant penguins fall into six distinct genera (**Table 2**).

Sphensicus

In contrast to the popular misconception that penguins are creatures of the snow and ice, representatives of the *Spheniscus* penguins breed in tropical to temperate waters, with one species, the Galapagos penguin (*S. mendiculus*), breeding right on the equator.

There are four species: the African (*S. demersus*), Humboldt (*S. humboldti*) and Magellanic penguins (*S. magellanicus*), in addition to the Galapagos penguin. They were the first penguins to be discovered by Europeans. After Vasco de Gama's ships sailed around the Cape of Good Hope in 1497, they encountered African penguins in Mossel Bay, South Africa. However, the account of this discovery was not published until 1838 and the first announcement to the world at large concerning penguins was to come from another famous voyage, that of Magellan's circumnavigation of the globe (1519–1522). A passenger, Pigafetta, described in his diary a great number of flightless 'geese' seen on two islands near Punta Tombo, Argentina, home to a large concentration of Magellanic penguins.

The *Spheniscus* penguins are characterized by black lines on their chests and distinct black and white bands on their faces (as a consequence, they are sometimes referred to as the 'ringed' or 'banded' penguins) (**Figure 2**). Their faces are also distinguished by having patches of bare pink skin, which help to radiate heat. Warm-blooded animals like penguins and seals must be well insulated if they are to maintain a constant body temperature when in the water, because the sea acts as a huge heat sink; but this creates problems of overheating for penguins when ashore, especially in the hot climates that the *Spheniscus* penguins inhabit. For this reason, all these species nest in burrows or, where available, in caves, in clefts in rocks, or under vegetation.

Penguins, with their limited foraging range due to their being flightless, require a consistent and good food supply near to their breeding areas. Tropical waters are typically not very productive and are unable to support the dense swarms of fish or krill on which penguins depend. This has probably acted as a barrier, restricting penguins to the Southern Hemisphere. The *Spheniscus* penguins are able to breed as far north as they do because of wind-driven upwelling of nutrient-rich water and because the cold-water Benguela and Humboldt Currents, which run up the sides of southern Africa and South America, bring nutrient-rich waters from farther south. *Spheniscus* penguins feed mainly on small pelagic schooling fish such as sprats and anchovies. The top mandible is hooked at its far end, which helps the penguin to catch and hold fish.

With the exception of the Magellanic penguin, which is at the southern extreme of their range, the *Spheniscus* penguins are inshore foragers making

Table 2 Species of living penguins[a]

Species	Scientific name	Principal location	Latitude (°S)	Body mass (kg)	Foraging type	Migratory	Nest type	Diet
Galapagos	*Spheniscus mendiculus*	Galapagos	0	2.1 (M), 1.7 (F)	▣	▣	▣	▣
Humboldt	*Spheniscus humboldti*	Peru, Chile	5–42	4.9 (M), 4.5 (F)	▣	▣	▣	▣
African	*Spheniscus demersus*	South Africa	24–35	3.3 (M), 3.0 (F)	▣	▣	▣	▣
Magellanic	*Spheniscus magellanicus*	Argentina, Chile, Falklands Is.	29–54	4.9 (M), 4.6 (F)	■	■	▣	▣
Little	*Eudyptula minor*	Australia, New Zealand	32–47	1.2 (M), 1.0 (F)	□	□	▣	□
Yellow-eyed	*Megadyptes antipodes*	New Zealand, Auckland Is.	46–53	5.5 (M), 5.1 (F)	▣	▣	□	□
Fiordland	*Eudyptes pachyrhynchus*	New Zealand	44–47	4.1 (M), 3.7 (F)	■	■	□	□
Snares	*Eudyptes robustus*	Snares Islands	48	3.3 (M), 2.8 (F)	■	■	■	□
Erect-crested	*Eudyptes sclateri*	Antipodes Is., Bounty Is.	47–49	6.4 (M), 5.4 (F)	■	■	■	□
Rockhopper	*Eudyptes chrysocome*	Subantarctic	37–53	2.5 (M), 2.3 (F)	■	■	■	□
Macaroni/ Royal	*Eudyptes chrysolophus*	Subantarctic	46–65	5.2 (M), 5.3 (F)	■	■	■	□
Gentoo	*Pygoscelis papua*	Subantarctic, Antarctic	46–65	5.6 (M), 5.1 (F)	▣	□	■	■
Chinstrap	*Pygoscelis antarctica*	Antarctic	54–69	5.0 (M), 4.8 (F)	■	■	■	■
Adelie	*Pygoscelis adeliae*	Antarctic	54–77	5.4 (M), 4.8 (F)	■	■	■	■
King	*Aptenodytes patagonicus*	Subantarctic	45–55	16.0 (M), 14.3 (F)	■	■	■	▣
Emperor	*Aptenodytes forsteri*	Antarctic	66–78	36.7 (M), 28.4 (F)	■	■	■	□

Key

	Foraging type	Migratory	Nest type	Diet
▣	inshore	resident	burrow	fish
□	either	either	forest	fish, cephalopods, crustaceans
■	offshore	migratory	open	crustaceans

[a]M, Male; F, Female.

foraging trips of relatively short duration (1–2 days) throughout the breeding period. They lay two similar-sized eggs. Breeding can occur in all months of the year (African and Galapagos), at two peak times (Humboldt), or just once (Magellanic): a pattern that roughly corresponds with the more pronounced seasonality the farther south they breed.

Eudyptula

Little penguins (*Eudyptula minor*) are the smallest living penguins and are found only in Australia (often called fairy penguins) and New Zealand (often called little blue or blue penguins). While most authorities recognize only the single species, whether this really is a monotypic genus is a matter of debate. Six subspecies have been described and it

has been suggested in some quarters that at least one of these, the white-flippered penguin (*E. m. albosignata*), is deserving of separate species status in its own right. However, the latter freely hybridize with other subspecies and analyses of isozymes and DNA do not support such a conclusion at this stage. More extensive genetic studies of this genus are warranted.

Apart from their small size, little penguins are characterized by being nocturnal (coming ashore after dark) and nesting in burrows, although they will breed in caves or under suitable vegetation in some locations. Morphologically they are quite similar to *Spheniscus* penguins and also feed on small schooling fish. The plumage of all penguins conforms to the dark top and white undersides common to pelagic marine predators, but in little

Figure 2 Magellanic penguins. (Photograph: L.S. Davis).

Figure 4 Yellow-eyed penguin on nest. (Photograph: J. Darby.)

Figure 3 Little penguin chick, almost ready to fledge, begging for a meal from its parent. (Photograph: M. Renner.)

penguins the plumage on the backs has more of a blue hue compared to the blackish coloration of other penguins (**Figure 3**).

Patterns of breeding are very variable throughout their range in both the duration of the breeding period and the number of clutches per year (either one or two). Little penguins lay clutches of two similarly-sized eggs.

Typically they are described as inshore foragers (with foraging trips lasting 1–2 days), but there is evidence of plasticity in their feeding strategies. In some locations or in years of poor food supply, they may feed considerably farther offshore, with a concomitant increase in the duration of their foraging trips (up to 7 days or more).

Megadyptes

The representative of another monotypic genus, the yellow-eyed penguin (*Megadyptes antipodes*), is often touted as the world's rarest penguin. (The Galapagos penguin may well be rarer, and the Fiordland penguin is not much better off.) It breeds on the south-east coast of New Zealand and the sub-Antarctic Auckland Islands. The yellow-eyed is a medium-sized penguin, distinguished by having pale yellow eyes and a yellow band of plumage that runs around the back of its head from its eyes (**Figure 4**). But perhaps its most unique characteristic is a behavioral one: it nests under dense vegetation and typically nests are visually isolated from each other. Whether this visual isolation is an absolute requirement or simply a consequence of their need for dense cover to escape the sun's heat, there is little doubt that yellow-eyed penguins are among the least overtly social of the penguins. Although they nest in loose colonies, individual nests can be from a few meters to several hundred meters apart. Perhaps partly as a consequence of this, mate fidelity is higher than in the cheek-by-bill colonies of other species where opportunities for mate switching and extra-pair copulations are rife.

Another factor contributing to the apparent faithfulness of yellow-eyed penguins is that adults remain more-or-less resident at the breeding site throughout the year, even though breeding occurs only between September to February. They are inshore foragers, feeding mainly on fish and cephalopods, and typically foraging trips are from 1 to 2 days during both incubation and chick rearing. The clutch consists of two equal-sized eggs and there appears to be little in the way of evolved mechanisms for brood reduction: if both chicks hatch, they have an equally good chance of fledging.

Eudyptes

In contrast, the members of the genus *Eudytptes*, which is closely related to *Megadyptes* and may have been derived from ancestors that were very like the yellow-eyed penguin, are famous for their obligate brood reduction (i.e., they lay two eggs but only ever fledge one chick). Collectively, the six Eudyptid species are known as the crested penguins. They are distinguished by plumes of yellow or orange feathers, of varying length, that arise above their eyes like out-of-control eyebrows (**Figure 5**).

For the most part, they are penguins of the sub-Antarctic, although Fiordland penguins (*Eudyptes pachyrhynchus*) breed on the south-west corner of New Zealand (though some would argue that conditions there are not too dissimilar from the sub-Antarctic, especially as the Fiordland penguin breeds during winter). Other crested penguins also have quite restricted distributions: the Snares penguin (*E. robustus*) breeds only on the Snares Islands south of New Zealand; the erect-crested penguin (*E. sclateri*) breeds on the Bounty and Antipodes Islands, also near New Zealand; and the royal penguin (*E. schlegeli*) breeds only on Macquarie Island. However, there is a considerable body of evidence and opinion that argues that the royal penguin is really only a pale-faced subspecies of the more ubiquitously distributed macaroni penguin (*E. chrysolophus*). The rockhopper penguin (*E. chrysocome*), the smallest of the crested species, has a circumpolar distribution, but there are three recognized subspecies and preliminary analysis of DNA would suggest that they are at least as distinct as royal penguins are from macaroni penguins.

All the crested penguins are migratory, spending two-thirds of the year at sea (their whereabouts during this period are largely unknown) and, with the exception of the Fiordland penguin, returning to their breeding grounds in spring. Despite laying

Figure 5   Erect-crested penguins. (Photograph: L.S. Davis.)

Figure 6   Extreme egg-size dimorphism in erect-crested penguin eggs. (Photoraph: L.S. Davis.)

a single clutch of two eggs, they only ever fledge one chick. Moreover, the eggs are of dramatically different sizes, with the egg laid second being substantially bigger than that laid first (this form of egg-size dimorphism is unique among birds). In those species with the most extreme egg-size dimorphism (erect-crested, royal/macaroni), the second egg can be double the size of the first egg (**Figure 6**). Furthermore, although the laying interval between the first and second eggs is the longest recorded for penguins (4–7 days), the chick from the large second-laid egg usually hatches first.

The evolutionary reasons for this bizarre breeding pattern have long been matters of debate and, it is fair to say, have yet to be determined satisfactorily. Crested penguins are probably unable adequately to feed two chicks. They are all offshore foragers (feeding on a mixture of fish, cephalopods, and swarming crustaceans). The logistics of provisioning chicks from a distant food source probably make it too difficult to bring back enough food to fledge two chicks; but this idea has not been tested properly yet. While in some species it can be argued that the small first egg provides a measure of insurance for the parents in case the second egg is lost (or the chick from the second-laid egg dies), this cannot be a universal explanation. Studies of erect-crested and royal penguins have found that first-laid eggs are usually lost before or on the day the second egg is laid and it has even been suggested that the parents may actively eject the first eggs. While the evidence for egg ejection must be viewed as equivocal, first eggs cannot provide much of an insurance policy in these species. In any case, none of this explains why it is the first egg that is the smallest and that usually does not produce a chick.

For penguins breeding in higher latitudes, over-heating when on land is less of a problem, permitting them to breed in colonies in the open. Some, like macaroni penguins, form vast colonies, while Fiordland penguins nest in small colonies in forest or caves.

Pygoscelis

The three members of this genus are the classic dinner-jacketed penguins of cartoons. They all breed in the Antarctic to varying degrees. The gentoo penguin (*P. papua*) consists of two subspecies, one of which breeds on the Antarctic Peninsula and associated islands (*P. p. ellsworthii*) and the other (*P. p. papua*) on several sub-Antarctic islands. Chinstrap penguins (*P. antarctica*) breed below the Antarctic Convergence, mainly on the Antarctic Peninsula, but also on a few sub-Antarctic islands. The Adélie penguin (*P. adeliae*), with very few exceptions (Bouvetoya Island), breeds only on the Antarctic continent and its offshore islands.

The three species are medium-sized penguins with white undersides and black backs. They differ most obviously in the markings on their faces. Adélie penguins have black faces and throats, and prominent white eye-rings. The proximal ends of the mandibles are covered by feathers, giving the impression that the bill is short (**Figure 7**). Gentoo penguins also have black faces and white eye-rings, but have white patches above the eyes and a bright orange-red bill. Chinstrap penguins have white faces with a black crown, and a black line running under the chin, from ear-to-ear, producing the effect from which they derive their name.

All three species breed in colonies in the open and nest just once per year during the austral summer. A clutch of two eggs is laid, with the first slightly larger than the second. The northern subspecies of gentoo penguins (*P. p. papua*) lay their eggs in nests made from vegetation. They are inshore foragers and remain resident at the breeding site throughout most of the year. They nest in colonies, but the location of these colonies can shift from season to season as the area becomes soiled. Despite this, mate fidelity is high. In contrast, the other Pygoscelid penguins are migratory, make their nests out of stones, and the nest sites are more permanent. Chinstrap and Adélies are offshore foragers, with foraging trips commonly lasting two or more weeks during incubation. All the Pygoscelid penguins feed on a type of swarming crustacean known as krill (*Euphausia* spp.).

Aptenodytes

This genus is comprised of the two largest of the living penguins: emperor (*A. forsteri*) and king (*A. patagonicus*). Apart from their large size, these species are distinguished by their bright yellow (emperor) or orange (king) auricular patches and the fact that they lay a single large egg. Both are offshore foragers, and it seems unlikely that they could possibly bring back enough food to successfully rear two of their very large chicks (**Figure 8**). Each of them has a truly remarkable pattern of breeding.

Emperor penguins are notable for breeding during the heart of the Antarctic winter. They breed on ice in dense colonies: there are no nests, the single egg (or small chick) is carried upon the parent's feet. The male incubates the egg entirely by himself, while the female goes off to feed for about 2 months. Together with the period they are at the colony during courtship, that means the males go without food for up to 3.5 months. Emperor penguins feed on fish, cephalopods, or crustaceans, with the relative importance of each varying according to the location or time. To help reduce heat loss during the Antarctic winter, the penguins stand in a tight huddle.

In some ways the breeding of king penguins is even more remarkable. It takes them over a year

Figure 7 Adélie penguin with chick. (Photograph: L.S. Davis.)

Figure 8 King penguin chicks on the Falkland Islands. (Photograph: L.S. Davis.)

(14–16 months) to provide enough food for their chicks to reach a sufficient size to fledge. As a consequence, king penguins can breed only twice every three years. During the winter months the parents must forage at great distances and the chicks are left to fast, sometimes for as long as 5 months! In anyone's book, that is a long time between dinners. The king penguin breeds on sub-Antarctic islands. It does not build a nest, carrying the large single egg on its feet like an emperor penguin, but it does defend a territory within the colony. Some colonies can be enormous, with 100 000 or more birds. Mate fidelity is low in both king and emperor species. During incubation, foraging trips can last 2 weeks or more. They feed largely on Myctophid fish.

Life in the Sea

Tests using life-size plastic models of penguin bodies have shown that they have lower coefficients of drag than those for any cars or planes that scientists and engineers have been able to manufacture. When swimming, penguins move their flippers to generate vortex-based lift forces, in essence creating a form of underwater jet propulsion. The usual underwater traveling speed of most penguins is around $2 \, \mathrm{m \, s^{-1}}$. However, penguins are capable of high-speed movements known as 'porpoising', whereby they clear the water surface to breathe. Porpoising speeds can be $3 \, \mathrm{m \, s^{-1}}$ or higher. The speed at which a penguin must be traveling to porpoise is related to its body size, and it is likely that for large penguins, like emperor and king penguins, porpoising requires just too much effort. All smaller penguins use porpoising to escape from predators. Porpoising is energetically expensive compared to underwater swimming and, consequently, inshore foraging penguins tend not to use porpoising simply for traveling to and from their foraging areas. For offshore foragers, however, the time savings gained from porpoising may outweigh the energetic costs, and it seems likely that porpoising may sometimes, at least, be used in these species as an expedient means of travel and not just an escape reaction.

Maximum dive depth and duration are correlated with body mass. Emperor penguins have been recorded diving to over 500 m and all but little penguins are readily capable of diving to depths of 100 m or more. Even so, most foraging takes place at much shallower depths: penguins are visual predators and so light levels in the water column will limit their effectiveness as hunters (albeit bioluminescent prey, such as squid, enable some penguins to hunt at very low light levels, either at night or at great depth). The prey of penguins tends to vary with latitude and phylogenetic grouping (**Table 2**).

A Life in Two Worlds

An important determinant of the breeding success of penguins is how they manage to balance the time at sea needed for foraging against the time required ashore for breeding and molting. Nesting duties are shared, and the male and female of a pair take alternating turns being in attendance at the nest or at sea foraging. The foraging penguin needs to find enough food to sustain itself, to cover the costs of getting the food and, once chicks hatch, to ensure the growth of its chicks. However, if it spends too long at sea either the partner on the nest might desert or, if the chicks have hatched, the chicks might starve to death.

For inshore foragers, which make short and frequent trips to sea, the risks of desertion and starvation are not as acute as for those that need to forage farther afield. All adult penguins are capable of enduring relatively long periods of fasting, yet there are finite limits to how long an incubating bird can live off its fat reserves (e.g., male Adelie penguins arriving at the colony at the start of the breeding season will endure just over a month without food). For many offshore foragers, foraging trips during the incubation period tend to be from 10 to 20 days (albeit female emperor penguins are away for about 2 months). The situation changes dramatically once the chicks hatch, as chicks of all species need to be fed frequently, requiring that foraging trips be relatively short during at least the early stages of chick rearing. For offshore foragers that take long foraging trips during incubation, this creates a dilemma: they must switch over to short foraging trips by the time the chicks hatch or risk losing them to starvation. There is a growing body of evidence that indicates that penguins have an internal mechanism that measures the duration of the incubation period, precipitating the early return of a foraging bird if its chicks are about to hatch. The remnants of the yolk sac provide a little insurance against a tardy parent by permitting penguin chicks to survive about 3–6 days without being fed from the time they first hatch. In the case of the emperor penguin – in which the male undertakes all the incubation – if the female has not returned by the time the chick hatches, the male is able to feed it an oesophageal 'milk' made from breaking down his own tissues even though he has not eaten himself for over 3 months.

In contrast to flying birds, which cannot afford to store much fat because of the weight, the ability of penguins to store enough fat to withstand long fasts is important not only when breeding but also when molting. Although a small amount of their insulation is provided by the subdermal layer of fat, most insulation comes from the feathers. But feathers wear and, to provide an effective barrier against heat-stealing waters, they must be renewed to maintain their integrity. This is an energetically expensive process, whereby a new suit of feathers is grown beneath the old one. During this process, which can take 2 or 3 weeks, the birds typically remain on land and are unable to eat.

Threats and Conservation

Desertions and starvations can be major causes of breeding failure for some species of penguins and the likelihood of these occurring will be exacerbated by anything that increases the time penguins must spend foraging. This makes penguins especially vulnerable to perturbations in the marine environment. Threats to penguins come from anything that reduces their food supply (causing a concomitant increase foraging times), such as commercial fisheries or pollution.

The major threat facing penguins may be from global warming. Warmer water associated with ENSO (El Niño Southern Oscillation) events can reduce the amount of upwelling of nutrients and seriously affect the availability of prey in various areas. ENSO events can be especially critical for nonmigratory species like Galapagos and yellow-eyed penguins, which rely on a persistent and steady food supply in a localized area. Breeding dates of little penguins throughout Australia and New Zealand are correlated with sea surface temperatures, with laying commencing later and fewer birds attempting to breed or being successful when water temperatures are warmer.

The commitment penguins made to the sea by becoming flightless has made them doubly vulnerable. Not only are they at risk from factors affecting their food supply, but their lack of agility on land makes them potentially easy targets for predators. It is for this reason that flightlessness in water birds has usually evolved on offshore islands or isolated places relatively free of predators. However, humans have managed to undo much of that by introducing exotic predators to many areas where penguins breed. Mustelids, rats, and feral cats have had serious impacts upon penguins when they have been introduced to places like New Zealand, South Africa, and the sub-Antarctic islands. During the nineteenth and early twentieth centuries, humans were often significant predators themselves, killing adult penguins for their oil and their skins and taking eggs for food. It has been estimated that in one area alone, over 13 million eggs were harvested from African penguins during a 30-year period. Humans also have an impact on penguins by reducing the availability of suitable habitat through harvesting guano (this is especially so of African and Humboldt Penguins in South Africa and South America, respectively) or deforestation. The latter dramatically reduced the numbers of yellow-eyed penguins breeding on mainland New Zealand: this trend has been reversed in recent years by extensive replantings.

Until fairly recently, the species inhabiting the Subantarctic and Antarctic have been able to rely on their isolation for protection. However, as humans reach ever more into these nether regions of the planet, they are no longer immune to the threats from overfishing, pollution, disturbance from tourists, introduced predators, and introduced diseases. On a positive note, however, the design of penguins has enabled them to survive, if not flourish, for over 40 million years. They have been through periods of vast climatic change in the earth's history. While they have had to conform to a design shaped by the requirements of living in water, paradoxically, this has given them a great deal of versatility with respect to the environments they can exploit on land. They are the only 100-degree birds: the only birds capable of breeding at temperatures from $-60°C$ (midwinter in the Antarctic) to $+40°C$ (midsummer in Peru). It is to be hoped that these qualities will serve them well for the next 40 million years.

See also

El Niño Southern Oscillation (ENSO). El Niño Southern Oscillation (ENSO) Models. Seabird Conservation. Seabird Foraging Ecology. Seabird Population Dynamics.

Further Reading

Ainley DG, LeResche RE and Sladen WJL (1983) *Breeding Biology of the Adelie Penguin.* Berkeley: University of California Press.

Dann P, Norman I and Reilly P (eds) (1995) *The Penguins: Ecology and Management.* Chipping Norton, NSW, Australia: Surrey Beatty & Sons.

Davis LS (1993) *Penguin: A Season in the Life of the Adélie Penguin.* London: Pavilion.

Davis LS and Darby JT (eds) (1990) *Penguin Biology.* San Diego: Academic Press.

Stonehouse B (ed.) (1975) *The Biology of Penguins.* London: Macmillan.

Williams TD (1995) *The Penguins.* Oxford: Oxford University Press.

STABLE CARBON ISOTOPE VARIATIONS IN THE OCEAN

K. K. Turekian, Yale University, New Haven, CT, USA

Copyright © 2001 Academic Press

doi:10.1006/rwos.2001.0167

The two stable isotopes of carbon, ^{12}C and ^{13}C, vary in proportions in different reservoirs on earth. The ratio of ^{13}C to ^{12}C is commonly given relative to a standard (a belemnite from the Peedee formation in South Carolina and therefore called PDB). On the basis of this standard δ^{13}C is defined as:

$$\left[\frac{^{13}C/^{12}C_{sample}}{^{13}C/^{12}C_{standard}} - 1 \right] \times 1000$$

The values for some major carbon reservoirs are: marine limestones, δ^{13}C = 0; C-3 plants, δ^{13}C = − 25; air CO_2, δ^{13}C = − 7. The inorganic carbon in the surface ocean is in isotopic equilibrium with atmospheric CO_2 and has a value of about 2. Organic matter in the shallow ocean ranges from − 19 at high latitudes to − 28 at low latitudes. The midlatitude value is around − 21. The transport of organic matter to depth and subsequent metabolism adds inorganic carbon to the water. The isotopic composition of dissolved inorganic carbon then reflects the amount of addition of this metabolic carbon. **Figure 1** is a profile of δ^{13}C for the North Pacific. It is typical of other profiles in the oceans.

Carbon isotope measurements in all the oceans were made on the GEOSECS expedition. These values are given in the *GEOSECS Atlas* (1987).

See also

Carbon Cycle. Carbon Dioxide (CO₂) Cycle.

Further Reading

Chesselet R, Fontagne M, Buat-Menard P, Ezat U and Lambert CE (1981) The origin of particulte organic matter in the marine atmosphere as indicated by its stable carbon isotopic composition. *Geophysical Research Letters* 8: 345–348.

GEOSECS Atlantic, Pacific, and Indian Ocean Expeditions, vol 7: *Shorebased Data and Graphics*. National Science Foundation. 200 pp. (1987).

Kroopnick P, Deuser WG and Craig H (1970) Carbon-13 measurements on dissolved inorganic carbon at the North Pacific (1969) GEOSECS station. *Journal of Geophysical Research* 75: 7668–7671.

Sackett WM (1964) The depositional history and isotopic organic composition of marine sediments. *Marine Geology* 2: 173–185.

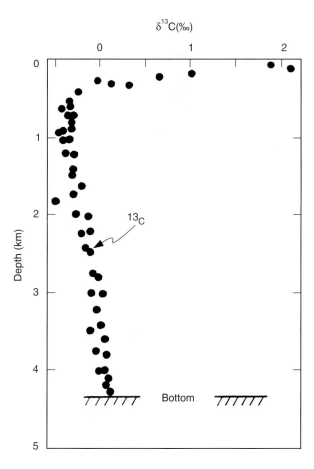

Figure 1 Variation of δ^{13}C in dissolved inorganic carbon with depth in the Pacific Ocean at GEOSECS Station 346 (28°N, 121°W) (Kroopnick, Deuser and Craig, 1970).

STERNIDAE

See **LARIDAE, STERNIDAE AND RYNCHOPIDAE**

STORM SURGES

R. A. Flather, Proudman Oceanographic Laboratory, Bidston Hill, Prenton, UK

doi:10.1006/rwos.2001.0124

Introduction and Definitions

Storm surges are changes in water level generated by atmospheric forcing; specifically by the drag of the wind on the sea surface and by variations in the surface atmospheric pressure associated with storms. They last for periods ranging from a few hours to 2 or 3 days and have large spatial scales compared with the water depth. They can raise or lower the water level in extreme cases by several meters; a raising of level being referred to as a 'positive' surge, and a lowering as a 'negative' surge. Storm surges are superimposed on the normal astronomical tides generated by variations in the gravitational attraction of the moon and sun. The storm surge component can be derived from a time-series of sea levels recorded by a tide gauge using:

$$\text{surge residual} = (\text{observed sea level})$$
$$- (\text{predicted tide level}) \qquad [1]$$

producing a time-series of surge elevations. **Figure 1** shows an example.

Sometimes, the term 'storm surge' is used for the sea level (including the tidal component) during a storm event. It is important to be clear about the usage of the term and its significance to avoid confusion. Storms also generate surface wind waves that have periods of order seconds and wavelengths, away from the coast, comparable with or less than the water depth.

Positive storm surges combined with high tides and wind waves can cause coastal floods, which, in terms of the loss of life and damage, are probably the most destructive natural hazards of geophysical origin. Where the tidal range is large, the timing of the surge relative to high water is critical and a large surge at low tide may go unnoticed. Negative surges reduce water depth and can be a threat to navigation. Associated storm surge currents, superimposed on tidal and wave-generated flows, can also contribute to extremes of current and bed stress responsible for coastal erosion. A proper understanding of storm surges, the ability to predict them and measures to mitigate their destructive effects are therefore of vital concern.

Storm Surge Equations

Most storm surge theory and modeling is based on depth-averaged hydrodynamic equations applicable to both tides and storm surges and including nonlinear terms responsible for their interaction. In vector form, these can be written:

$$\frac{\partial \zeta}{\partial t} + \nabla.(D\mathbf{q}) = 0 \qquad [2]$$

$$\frac{\partial \mathbf{q}}{\partial t} + \mathbf{q}.\nabla \mathbf{q} - f\mathbf{k} \times \mathbf{q} = -g\nabla(\zeta - \bar{\zeta}) - \frac{1}{\rho}\nabla p_a$$
$$+ \frac{1}{\rho D}(\tau_s - \tau_b) + A\nabla^2 \mathbf{q} \qquad [3]$$

where t is time; ζ the sea surface elevation; $\bar{\zeta}$ the equilibrium tide; $\mathbf{q}$ the depth-mean current; τ_s the wind stress on the sea surface; τ_b the bottom stress; p_a atmospheric pressure on the sea surface; D the total water depth ($D = h + \zeta$, where h is the undisturbed depth); ρ the density of sea water, assumed to be uniform; g the acceleration due to gravity; f the Coriolis parameter ($= 2\omega \sin\varphi$, where ω is the angular speed of rotation of the Earth and φ is the latitude); $\mathbf{k}$ a unit vector in the vertical; and A the coefficient of horizontal viscosity. Eqn [2] is the continuity equation expressing conservation of volume. Eqn [3] equates the accelerations (left-hand side) to the force per unit mass (right-hand side).

In this formulation, bottom stress, τ_b is related to the current, $\mathbf{q}$, using a quadratic law:

$$\tau_b = k\rho \mathbf{q}|\mathbf{q}| \qquad [4]$$

where k is a friction parameter (~ 0.002). Similarly, the wind stress, τ_s, is related to $\mathbf{W}$, the wind velocity at a height of 10 m above the surface, also using a quadratic law:

$$\tau_s = c_D \rho_a \mathbf{W}|\mathbf{W}| \qquad [5]$$

where ρ_a is the density of air and c_D a drag coefficient. Measurements in the atmospheric boundary layer suggest that c_D increases with wind speed, W, accounting for changes in surface roughness associated with wind waves. A typical form due to J. Wu is:

$$10^3 c_D = 0.8 + 0.065W \qquad [6]$$

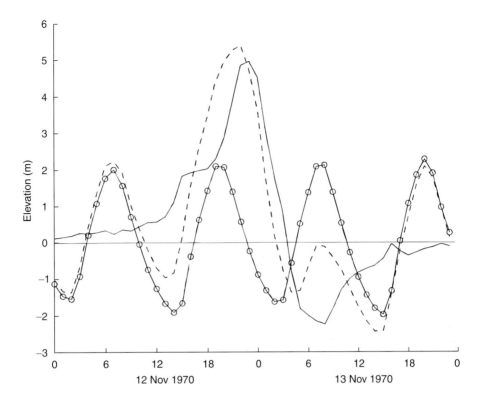

Figure 1 Water level (dashed line), predicted tide (line with ○), and the surge residual (continuous line) at Sandwip Island, Bangladesh, during the catastrophic storm surge of 12–13 November 1970 (times are GMT).

Alternatively, from dimensional analysis, H. Charnock obtained $gz_0/u_*^2 = \alpha$, where z_0 is the aerodynamic roughness length associated with the surface wavefield, u_* is the friction velocity ($u_*^2 = \tau_s/\rho_a$), and α is the Charnock constant. So, the roughness varies linearly with surface wind stress. Assuming a logarithmic variation of wind speed with height z above the surface, $W(z) = (u_*/\kappa) \ln(z/z_0)$, where κ is von Kármán's constant. It follows that for $z = 10$ m:

$$c_D = [(1/\kappa)\ln\{gz/(\alpha c_D W^2)\}]^{-2} \qquad [7]$$

Estimates of α range from 0.012 to 0.035.

Generation and Dynamics of Storm Surges

The forcing terms in eqn [3] which give rise to storm surges are those representing wind stress and the horizontal gradient of surface atmospheric pressure. Very simple solutions describe the basic mechanisms. The sea responds to atmospheric pressure variations by adjusting sea level such that, at depth, pressure in the water is uniform, the hydrostatic approximation. Assuming in eqn [3] that $\mathbf{q} = 0$ and

$\tau_s = 0$, then $g\nabla\zeta + (1/\rho)\nabla p_a = 0$, so $\rho g\zeta + p_a =$ constant. This gives the 'inverse barometer effect' whereby a decrease in atmospheric pressure of 1 hPa produces an increase in sea level of approximately 1 cm. Wind stress produces water level variations on the scale of the storm. Eqns [5] and [6] imply that the strongest winds are most important since effectively $\tau_s \propto W^3$. Both pressure and wind effects are present in all storm surges, but their relative importance varies with location. Since wind stress is divided by D whereas ∇p_a is not, it follows that wind forcing increases in importance in shallower water. Consequently, pressure forcing dominates in the deep ocean whereas wind forcing dominates in shallow coastal seas. Major destructive storm surges occur when extreme storm winds act over extensive areas of shallow water.

As well as the obvious wind set-up, with the component of wind stress directed towards the coast balanced by a surface elevation gradient, winds parallel to shore can also generate surges at higher latitudes. Wind stress parallel to the coast with the coast on its right will drive a longshore current, limited by bottom friction. Geostrophic balance gives (in the Northern Hemisphere) a surface gradient raising levels at the coast (*see* **Tides**).

Amplification of surges may be caused by the funneling effect of a converging coastline or estuary and by a resonant response; e.g. if the wind forcing travels at the same velocity as the storm wave, or matches the natural period of oscillation of a gulf, producing a seiche.

As a storm moves away, surges generated in one area may propagate as free waves, contributing as externally generated components to surges in another area. Generally, away from the forcing center, the response of the ocean consists of longshore propagating coastally trapped waves. Examples include external surges in the North Sea (see **Figure 2**), which are generated west and north of Scotland and propagate anticlockwise round the basin like the diurnal tide; approximately as a Kelvin wave. Low mode continental shelf waves have been identified in surges on the west coast of Norway, in the Middle Atlantic Bight of the US, in the East China Sea, and on the north-west shelf of Australia. Currents associated with low mode continental shelf waves generated by a tropical cyclone crossing the north-west shelf of Australia have been observed and explained by numerical modeling. Edge waves can also be generated by cyclones travelling parallel to the coast in the opposite direction to shelf waves.

From eqns [3] and [4], bottom stress (which dissipates surges) also depends on water depth and is non-linear; the current including contributions from tide and surge, $q = q_T + q_s$. Consequently, dissipation of surges is stronger in shallow water and where tidal currents are also strong. In deeper water and where tides are weak, free motions can persist for long times or propagate long distances. For example, the Adriatic has relatively small tides and seiches excited by storms can persist for many days.

In areas with substantial tides and shallow water, non-linear dynamical processes are important, resulting in interactions between the tide and storm surge such that both components are modified. The main contribution arises from bottom stress, but time-dependent water depth, $D = h + \zeta(t)$, can also be significant (e.g. $\tau_s/(\rho D)$ will be smaller at high tide than at low tide). An important consequence is that the linear superposition of surge and tide without accounting for their interaction gives substantial errors in estimating water level. For example, for surges propagating southwards in the North Sea into the Thames Estuary, surge maxima tend to occur on the rising tide rather than at high water (**Figure 3**).

Interactions also occur between the tide–surge motion and surface wind waves (see below).

Areas Affected by Storm Surges

Major storm surges are created by mid-latitude storms and by tropical cyclones (also called hurricanes and typhoons) which generally occur in geographically separated areas and differ in their scale. Mid-latitude storms are relatively large and evolve slowly enough to allow accurate predictions of their wind and pressure fields from atmospheric forecast models. In tropical cyclones, the strongest winds occur within a few tens of kilometers of the storm center and so are poorly resolved by routine weather prediction models. Their evolution is also rapid and much more difficult to predict. Consequently prediction and mitigation of the effects of storm surges is further advanced for mid-latitude storms than for tropical cyclones.

Tropical cyclones derive energy from the warm surface waters of the ocean and develop only where the sea surface temperature (SST) exceeds 26.5°C. Since their generation is dependent on the effect of the local vertical component of the Earth's rotation, they do not develop within 5° of the equator. **Figure 4** shows the main cyclone tracks. Areas affected include: the continental shelf surrounding the Gulf of Mexico and on the east coast of the US (by hurricanes); much of east Asia including Vietnam, China, the Philippines and Japan (by typhoons); the Bay of Bengal, in particular its shallow north-east corner, and northern coasts of Australia (by tropical cyclones). Areas affected by mid-latitude storms include the North Sea, the Adriatic, and the Patagonian Shelf. Inland seas and large lakes, including the Great Lakes, Lake Okeechobee (Florida), and Lake Biwa (Japan) also experience surges.

The greatest loss of life due to storm surges has occurred in the northern Bay of Bengal and Meghna Estuary of Bangladesh. A wide and shallow continental shelf bounded by extensive areas of low-lying poorly protected land is impacted by tropical cyclones. Cyclone-generated storm surges on 12–13 November 1970 and 29–30 April 1991 (**Figure 5**) killed approximately 250 000 and 140 000 people, respectively, in Bangladesh.

A severe storm in the North Sea on 31 January–1 February 1953 generated a large storm surge, which coincided with a spring tide to cause catastrophic floods in the Netherlands (**Figure 6**) and south-east England, killing approximately 2000 people. Subsequent government enquiries resulted in the 'Delta Plan' to improve coastal defences in Holland, led to the setting up of coastal flood warning authorities, and accelerated research into storm surge dynamics.

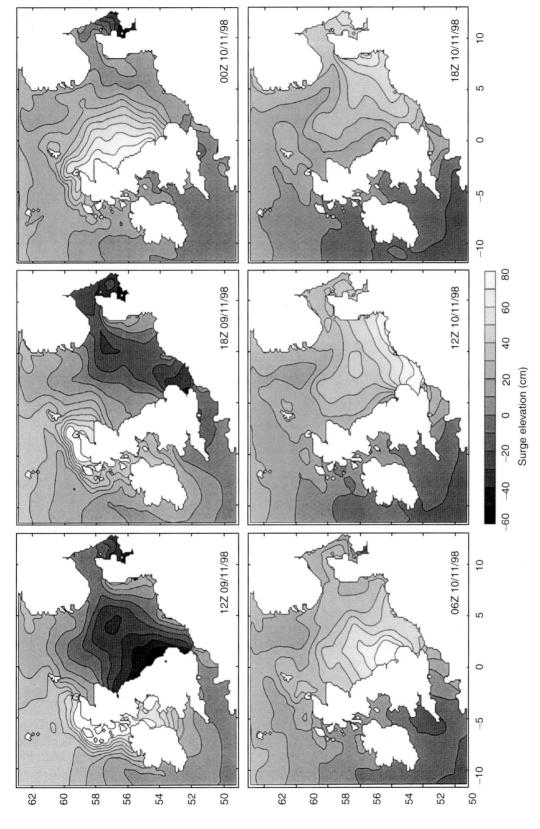

Figure 2 Propagation of an external surge in the North Sea from a numerical model simulation.

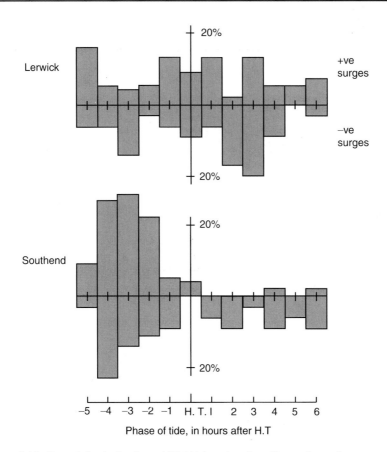

Lerwick
+ve surges
−ve surges

Southend

Phase of tide, in hours after H.T

Figure 3 Frequency distribution relative to the time of tidal high water of positive and negative surges at Lerwick (northern North Sea) and Southend (Thames Estuary). The phase distribution at Lerwick is random, whereas due to tide–surge interaction most surge peaks at Southend occur on the rising tide (re-plotted from Prandle D and Wolf J, 1978, The interaction of surge and tide in the North Sea and River Thames. *Geophys. J. R. Astr. Soc.*, 55: 203–216, by permission of the Royal Astronomical Society).

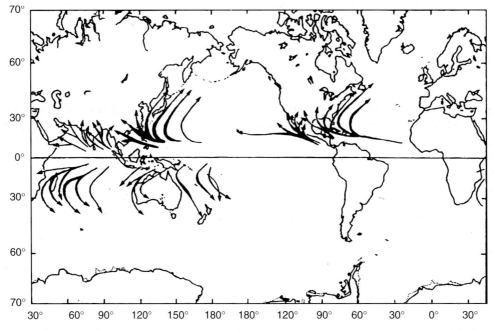

Figure 4 Tropical cyclone tracks (from Murty, 1984, reproduced by permission of the Department of Fisheries and Oceans, Canada).

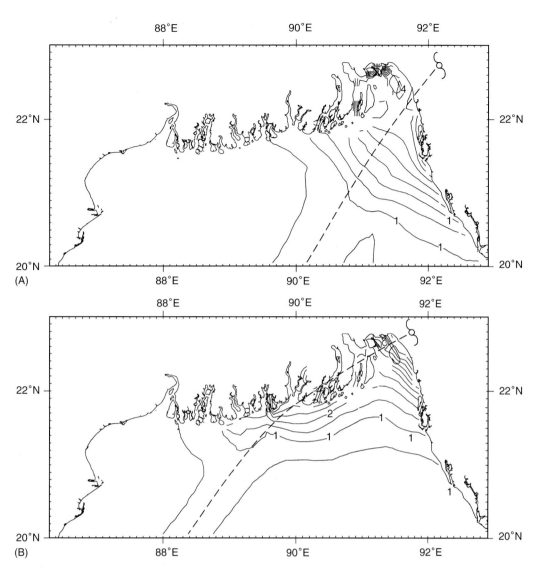

Figure 5 Cyclone tracks (dashed lines) and maximum computed surge elevation measured in meters in the northern Bay of Bengal during the cyclones of (A) 1970 and (B) 1991 from numerical model simulations (contour interval 0.5 m). (Re-plotted from Flather, 1994, by permission of the American Meteorological Society.)

The city of Venice, in Italy, suffers frequent 'acqua alta' which flood the city, disrupting its life and accelerating the disintegration of the unique historic buildings.

In 1969 Hurricane Camille created a surge in the Gulf of Mexico which rose to 7 m above mean sea level, causing more than 100 deaths and about 1 billion dollars worth of damage. An earlier cyclone in the region, in 1900, flooded the island of Galveston, Texas, with the loss of 6000 lives.

Storm Surge Prediction

Early research on storm surges was based on analysis of observations and solution of simplified – usually linearized – equations for surges in idealized channels, rectangular gulfs, and basins with uniform depth. The first self-recording tide gauge was installed in 1832 at Sheerness in the Thames Estuary, England, so datasets for analysis were available from an early stage. Interest was stimulated by events such as the 1953 storm surge in the North Sea, which highlighted the need for forecasts.

First prediction methods were based on empirical formulae derived by correlating storm surge elevation with atmospheric pressure, wind speed and direction and, where appropriate, observed storm surges from a location 'upstream'. Long time-series of observations are required to establish reliable correlations. Where such observations existed, e.g. in the North Sea, the methods were quite successful.

Figure 6 Breached dyke in the Netherlands after the 1953 North Sea storm surge. (Reproduced by permission of RIKZ, Ministry of Public Works, The Netherlands.)

From the 1960s, developments in computing and numerical techniques made it possible to simulate and predict storm surges by solving discrete approximations to the governing equations (eqns [1] and [2]). The earliest and simplest methods, pioneered in Europe by W. Hansen and N.S. Heaps and in the USA by R.O. Reid and C.P. Jelesnianski, used a time-stepping approach based on finite difference approximations on a regular grid. Surge–tide interaction could be accounted for by solving the non-linear equations and including tide. Effects of inundation could also be included by allowing for moving boundaries; water levels computed with a fixed coast can be O(10%) higher than those with flooding of the land allowed.

Recent developments have revolutionized surge modeling and prediction. Among these, coordinate transformations, curvilinear coordinates and grid nesting allow better fitting of coastal boundaries and enhanced resolution in critical areas. A simple example is the use of polar coordinates in the SLOSH (Sea, Lake and Overland Surges from Hurricanes) model focusing on vulnerable sections of the US east coast. Finite element methods with even greater flexibility in resolution (e.g. **Figure 7**) have also been used in surge computations in recent years.

There has also been increasing use of three-dimensional (3-D) models in storm surge studies. Their main advantage is that they provide information on the vertical structure of currents and, in particular, allow the bottom stress to be related to flow near the seabed. This means that in a 3-D formulation the bottom stress need not oppose the direction of the depth mean flow and hence of the

water transport. Higher surge estimates result in some cases.

In the last decades many countries have established and now operate model-based flood warning systems. Although finite element methods and 3-D models have been developed and are used extensively for research, most operational models are still based on depth-averaged finite-difference formulations.

A key requirement for accurate surge forecasts is accurate specification of the surface wind stress. Surface wind and pressure fields from numerical weather prediction (NWP) models are generally used for mid-latitude storms. Even here, resolution of small atmospheric features can be important, so preferably NWP data at a resolution comparable with that of the surge model should be used. For tropical cyclones, the position of maximum winds at landfall is critical, but prediction of track and evolution (change in intensity, etc.) is problematic. Presently, simple models are often used based on basic parameters: p_c, the central pressure; W_m, the maximum sustained 10 m wind speed; R, the radius to maximum winds; and the velocity, $\mathbf{V}$, of movement of the cyclone's eye. Assuming a pressure profile, e.g. that due to G.J. Holland:

$$p_a(r) = p_c + \Delta p \exp[-(R/r)^B] \qquad [8]$$

where r is the radial distance from the cyclone center, Δp the pressure deficit (difference between the ambient and central pressures), and B is a 'peakedness' factor typically $1.0 < B < 2.5$. Wind fields can then be estimated using further assumptions and approximations. First, the gradient or

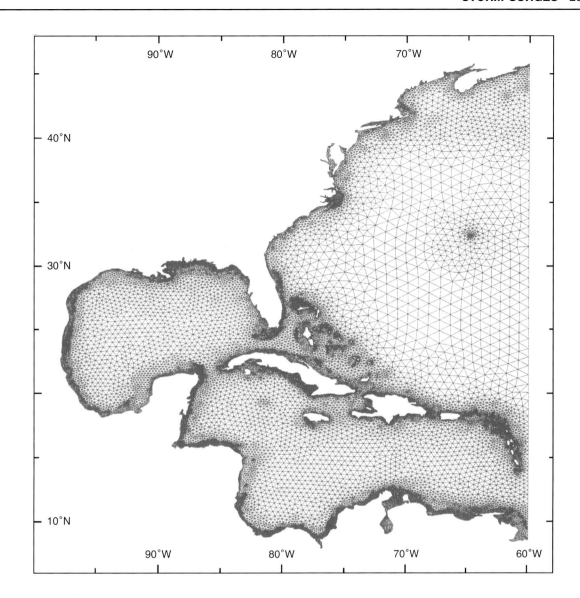

Figure 7 A finite element grid for storm surge calculations on the east coast of the USA, the Gulf of Mexico, and Caribbean (from Blain CA, Westerink JJ and Leuttich RL (1994) The influence of domain size on the response characteristics of a hurricane storm surge model. *J. Geophys. Res.*, 99(C9) 18467–18479. Reproduced by permission of the American Geophysical Union.)

cyclostrophic wind can be calculated as a function of r. An empirical factor (~ 0.8) reduces this to W, the 10 m wind. A contribution from the motion of the storm (maybe 50% of V) can be added, introducing asymmetry to the wind field, and finally to account for frictional effects in the atmospheric boundary layer, wind vectors may be turned inwards by a cross-isobar angle of $10°$–$25°$. Such procedures are rather crude, so that cyclone surges computed using the resulting winds are unlikely to be very accurate. Simple vertically integrated models of the atmospheric boundary layer have been used to compute winds from a pressure distribution such as eqn [8], providing a more consistent approach. In reality, cyclones interact with the ocean. They generate wind waves, which modify the sea surface roughness, z_0, and hence the wind stress generating the surge. Wind- and wave-generated turbulence mixes the surface water changing its temperature and so modifies the flux of heat from which the cyclone derives its energy. Progress requires improved understanding of air–sea exchanges at extreme wind speeds and high resolution coupled atmosphere–ocean models.

Interactions with Wind Waves

As mentioned above, observations suggested that Charnock's α was not constant but depended on water depth and 'wave age', a measure of the state

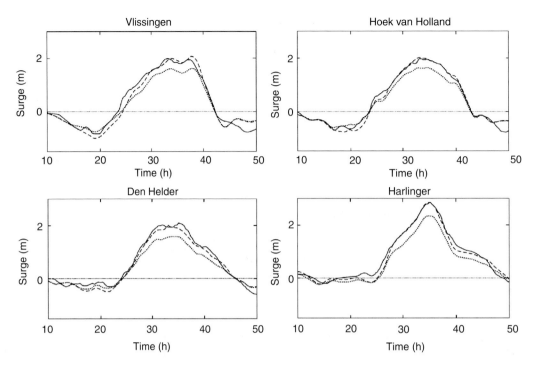

Figure 8 Computed surge elevations during 13–16 February 1989 with (dashed line) and without (dotted) wave stress compared with observations (continuous line). (Re-plotted from Mastenbroek *et al.*, 1994 by permission of the American Geophysical Union.)

of development of waves. Young waves are steeper and propagate more slowly, relative to the wind speed, than fully developed waves and so are aerodynamically rougher enhancing the surface stress. These effects can be incorporated in a drag coefficient which is a function of wave age, wave height, and water depth and agrees well with published datasets over the whole range of wave ages. Further research, considering the effects of waves on airflow in the atmospheric boundary layer, led P.A.E.M. Janssen to propose a wave-induced stress enhancing the effective roughness. Application of this theory requires the dynamical coupling of surge and wave models such that friction velocity and roughness determine and are determined by the waves. Mastenbroek *et al.* obtained improved agreement with observed surges on the Dutch coast by including the wave-induced stress in a model experiment (**Figure 8**). However, they also found that the same improvement could be obtained by a small increase in the standard drag coefficient.

In shallow water, wave orbital velocities also reach the seabed. The bottom stress acting on surge and tide is therefore affected by turbulence introduced at the seabed in the wave boundary layer. With simplifying assumptions, models describing these effects have been developed and can be used in storm surge modeling. Experiments using both 2-D and 3-D surge models have been carried out. 3-D

modeling of surges in the Irish Sea using representative waves shows significant effects on surge peaks and improved agreement with observations. Bed stresses are much enhanced in shallow water. Because the processes depend on the nature of the bed, a more complete treatment should take account of details of bed types.

Non-linear interactions give rise to a wave-induced mean flow and a change in mean water depth (wave set-up and set-down). The former has contributions from a mean momentum density produced by a non-zero mean flow in the surface layer (above the trough level of the waves), and from wave breaking. Set-up and set-down arise from the 'radiation stress', which is defined as the excess momentum flux due to the waves (*see* **Waves on Beaches**). Mastenbroek *et al.* showed that the radiation stress has a relatively small influence on the calculated water levels in the North Sea but cannot be neglected in all cases. It is important where depth-induced changes in the waves, as shoaling or breaking, dominate over propagation and generation, i.e. in coastal areas. The effects should be included in the momentum equations of the surge model.

Although ultimately coupled models with a consistent treatment of exchanges between atmosphere and ocean and at the seabed remain a goal, it appears that with the present state of understanding the benefits may be small compared with other

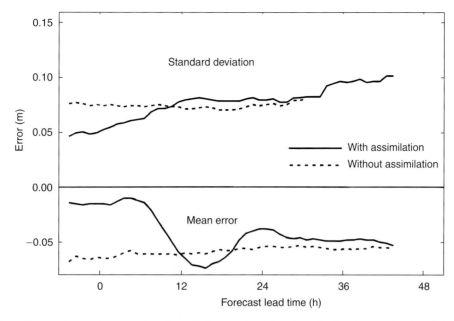

Figure 9 Variation of forecast errors in surge elevation with forecast lead-time from the Dutch operational model with (continuous line) and without (dashed line) assimilation of tide gauge data. (Replotted from Flather, 2000, by permission of Elsevier Science BV.)

inherent uncertainties. In particular, accurate definition of the wind field itself and details of bed types (rippled or smooth, etc.) are not readily available.

Data Assimilation

Data assimilation plays an increasing role, making optimum use of real-time observations to improve the accuracy of initial data in forecast models. Bode and Hardy (see Further Reading section) reviewed two approaches, involving solution of adjoint equations and Kalman filtering. The Dutch operational system has used Kalman filtering since 1992, incorporating real-time tide gauge data from the east coast of Britain. Accuracy of predictions (**Figure 9**) is improved for the first 10–12 h of the forecast.

Related Issues

Extremes

Statistical analysis of storm surges to derive estimates of extremes is important for the design of coastal defenses and safety of offshore structures. This requires long time-series of surge elevation derived from observations of sea level where available or, increasingly, from model hindcasts covering O(50 years) forced by meteorological analyses.

Climate Change Effects

Climate change will result in a rise in sea level and possible changes in storm tracks, storm intensity and frequency, collectively referred to as 'storminess'. Changes in water depth with rising mean sea level (MSL) will modify the dynamics of tides and surges, increasing wavelengths and modifying the generation, propagation, and dissipation of storm surges. Increased water depth implies a small reduction in the effective wind stress forcing, suggesting smaller surges. However, effects of increased storminess may offset this. It has been suggested, for example, that increased temperatures in some regions could raise sea surface temperatures resulting in more intense and more frequent tropical cyclones. Regions susceptible to tropical cyclones could also be extended. Research is in progress to assess and quantify some of these effects, e.g. with tide–surge models forced by outputs from climate GCMs. An important issue is that of distinguishing climate-induced change from the natural inter-annual and decadal variability in storminess and hence surge extremes.

See also

Fish Feeding and Foraging. Tides. Waves on Beaches. Wind Driven Circulation.

Further Reading

Bode L and Hardy TA (1997) Progress and recent developments in storm surge modelling. *Journal of Hydraulic Engineering* 123(4): 315–331.

Flather RA (1994) A storm surge prediction model for the northern Bay of Bengal with application to the cyclone disaster in April 1991. *Journal of Physical Oceanography* 24: 172–190.

Flather RA (2000) Existing operational oceanography. *Coastal Engineering* 41: 13–40.

Heaps NS (1967) Storm surges. In: Bavnes H (ed.) *Oceanography and Marine Biology Annual Review 5*: 11–47. London, Allen and Unwin.

Mastenbroek C, Burgers G and Janssen PAEM (1993) The dynamical coupling of a wave model and a storm surge model through the atmospheric boundary layer. *Journal of Physical Oceanography* 23: 1856–1866.

Murty TS (1984) Storm surges – meteorological ocean tides. *Canadian Bulletin of Fisheries and Aquatic Sciences* 212: 1–897.

Murty TS, Flather RA and Henry RF (1986) The storm surge problem in the Bay of Bengal. *Progress in Oceanography* 16: 195–233.

Pugh DT (1987) *Tides, Surges, and Mean Sea Level.* Chichester: John Wiley & Sons.

World Meteorological Organisation (1978) *Present Techniques of Tropical Storm Surge Prediction.* Report 13. Marine science affairs, WMO No. 500. Geneva, Switzerland.

SUB ICE-SHELF CIRCULATION AND PROCESSES

K. W. Nicholls, British Antarctic Survey, Cambridge, UK

doi:10.1006/rwos.2001.0010

Introduction

Ice shelves are the floating extension of ice sheets (*see* **Ice-shelf Stability**). They extend from the grounding line, where the ice sheet first goes afloat, to the ice front, which usually takes the form of an ice cliff dropping down to the sea. Although there are several examples on the north coast of Greenland, the largest ice shelves are found in the Antarctic where they cover 40% of the continental shelf. Ice shelves can be up to 2 km thick and have horizontal extents of several hundreds of kilometers. The base of an ice shelf provides an intimate link between ocean and cryosphere. Three factors control the oceanographic regime beneath ice shelves: the geometry of the sub-ice shelf cavity, the oceanographic conditions beyond the ice front, and tidal activity. These factors combine with the thermodynamics of the interaction between sea water and the ice shelf base to yield various glaciological and oceanographic phenomena: intense basal melting near deep grounding lines and near ice fronts; deposition of ice crystals at the base of some ice shelves, resulting in the accretion of hundreds of meters of marine ice; production of sea water at temperatures below the surface freezing point, which may then contribute to the formation of Antarctic Bottom Water (*see* **Bottom Water Formation**); and the upwelling of relatively warm Circumpolar Deep Water. Although the presence of the ice shelf itself makes measurement of the sub-ice shelf environment difficult, various field techniques have been used to study the processes and circulation within sub-ice shelf cavities. Rates of basal melting and freezing affect the flow of the ice and the nature of the ice–ocean interface, and so glaciological measurements can be used to infer the ice shelf's basal mass balance. Another indirect approach is to make ship-based oceanographic measurements along ice fronts. The properties of in-flowing and out-flowing water masses give clues as to the processes needed to transform the water masses. Direct measurements of oceanographic conditions beneath ice shelves have been made through natural access holes such as rifts, and via access holes created using thermal (mainly hot-water) drills. Numerical models of the sub-ice shelf regime have been developed to complement the field measurements. These range from simple one-dimensional models following a plume of water from the grounding line along the ice shelf base, to full three-dimensional models coupled with sea ice models, extending out to the continental shelf-break and beyond.

The close relationship between the geometry of the sub-ice shelf cavity and the interaction between the ice shelf and the ocean implies a strong dependence of the ice shelf/ocean system on the state of the ice sheet. During glacial climatic periods the geometry of ice shelves would have been radically different to their geometry today, and ice shelves probably played a different role in the climate system.

Geographical Setting

By far the majority of the world's ice shelves are found fringing the Antarctic coastline (**Figure 1**). Horizontal extents vary from a few tens to several hundreds of kilometers, and maximum thickness at the grounding line varies from a few tens of meters to 2 km. By area, the Ross Ice Shelf is the largest at around 500 000 km². The most massive, however, is the very much thicker Filchner-Ronne Ice Shelf in the southern Weddell Sea. Ice from the Antarctic Ice Sheet flows into ice shelves via fast-moving ice streams (**Figure 2**). As the ice moves seaward, further nourishment comes from snowfall, and, in some cases, from accretion of ice crystals at the ice shelf base. Ice is lost by melting at the ice shelf base and by calving of icebergs at the ice front. Current estimates suggest that basal melting is responsible for around 25% of the ice loss from Antarctic ice shelves; most of the remainder calves from the ice fronts as icebergs.

Over central Antarctica the weight of the ice sheet depresses the lithosphere such that the seafloor beneath many ice shelves deepens towards the ground-ing line. The effect of the lithospheric depression has probably been augmented during glacial periods by the scouring action of ice on the seafloor: at the glacial maxima the grounding line would have been much closer to the continental shelf-break. Since ice shelves become thinner towards the ice front and float freely in the ocean, a typical sub-ice shelf cavity has the shape of a cavern that dips down-wards towards the grounding line (**Figure 2**). This geometry has important consequences for the ocean circulation within the cavity.

Oceanographic Setting

The oceanographic conditions over the Antarctic continental shelf depend on whether relatively warm, off-shelf water masses are able to cross the continental shelf-break.

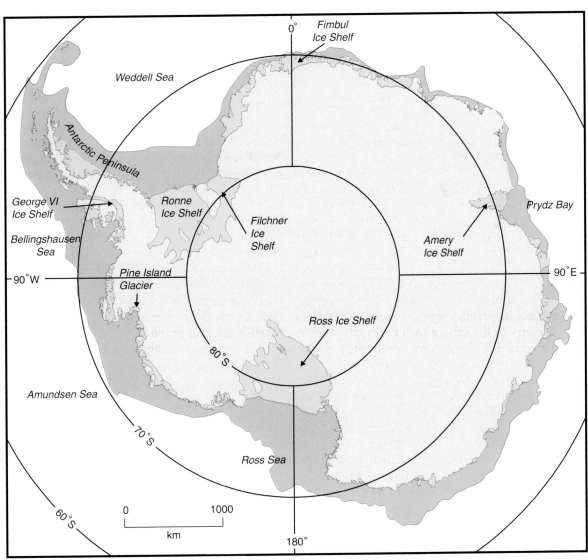

Figure 1 Map showing ice shelves (blue) covering about 40% of the continental shelf (dark gray) of Antarctica.

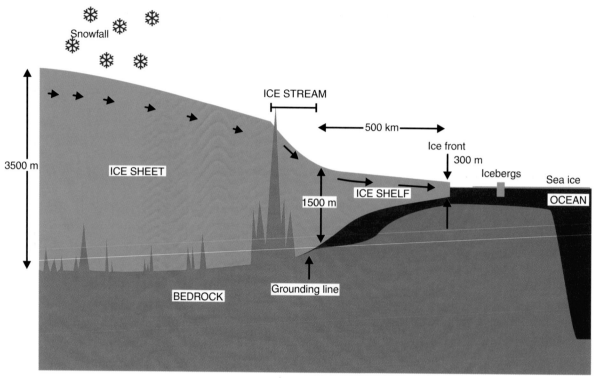

Figure 2 Schematic cross-section of the Antarctic ice sheet showing the transition from ice sheet to ice stream to ice shelf. Also shown is the depression of the lithosphere that results in the deepening of the seabed towards the continental interior.

For much of Antarctica a dynamic barrier at the shelf-break prevents advection of circumpolar deep water (CDW) onto the continental shelf itself. In these regions the principal process determining the oceanographic conditions is production of sea ice (*see* **Sea Ice**) in coastal polynyas (*see* **Polynyas**) and leads, and the water column is largely dominated by high salinity shelf water (HSSW). Long residence times over some of the broader continental shelves, for example in the Ross and southern Weddell seas, enable HSSW to attain salinities of over 34.8 PSU. HSSW has a temperature at or near the surface freezing point (about − 1.9°C), and is the densest water mass in Antarctic waters. Conditions over the continental shelves of the Bellingshausen and Amundsen seas (**Figure 1**) represent the other extreme. There, the barrier at the shelf-break appears to be either weak or absent. At a temperature of about 1°C, CDW floods the continental shelf.

Between these two extremes there are regions of continental shelf where tongues of modified warm deep water (MWDW) are able to penetrate the shelf-break barrier (**Figure 3**), in some cases reaching as far as ice fronts. MWDW comes from above the warm core of CDW: the continental shelf effectively skims off the shallower and cooler part of the water column.

What Happens When Ice Shelves Melt into Sea Water?

The freezing point of fresh water is 0°C at atmospheric pressure. When the water contains dissolved salts, the freezing point is depressed: at a salinity of around 34.7 PSU the freezing point is − 1.9°C. Sea water at a temperature above − 1.9°C is therefore capable of melting ice. The freezing point of water is also pressure dependent. Unlike most materials, the pressure dependence for water is negative: increasing the pressure decreases the freezing point. The freezing point T_f of sea water is approximated by:

$$T_f = aS + bS^{3/2} - cS^2 - dp$$

where $a = - 5.75 \times 10^{-2} °C\,PSU^{-1}$, $b = 1.710523 \times 10^{-3} °C\,PSU^{-3/2}$, $c = - 2.154996 \times 10^{-4} °C\,PSU^{-2}$ and $d = - 7.53 \times 10^{-4} °C\,dbar^{-1}$. S is the salinity in PSU, and p is the pressure in dbar. Every decibar increase in pressure therefore depresses the freezing point by 0.75 m°C. The depression of the freezing point with pressure has important consequences for the interaction between ice shelves and the ocean. Even though HSSW is already at the surface freezing point, if it can be brought into contact with an ice

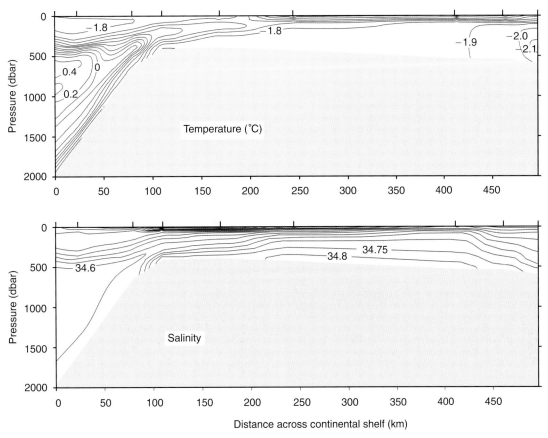

Figure 3 Hydrographic section over the continental slope and across the open continental shelf in the southern Weddell Sea, as far as the Ronne Ice Front. Water below the surface freezing point ($-1.9°C$) can be seen emerging from beneath the ice shelf. The majority of the continental shelf is dominated by HSSW, although in this location a tongue of warmer MWDW penetrates across the shelf-break. The station locations are shown by the heavy tick marks along the upper axes.

shelf base, melting will take place. As the freezing point at the base of deep ice shelves can be as much as 1.5°C lower than the surface freezing point, the melt rates can be high.

When ice melts into sea water the effect is to cool and freshen. Consider unit mass of water at temperature T_0, and salinity S_0 coming into contact with the base of an ice shelf where the *in situ* freezing point is T_f. The water first warms m kg of ice to the freezing point, and then supplies the latent heat necessary for melting. The resulting mixture of melt and sea water has temperature T and salinity S. If the initial temperature of the ice is T_i, the latent heat of melting is L, the specific heat capacity of sea water and ice, c_w and c_i, then heat and salt conservation requires that:

$$(T - T_f)(1 + m)c_w + m(c_i(T_f - T_i) + L)$$
$$= (T_o - T_f)c_w$$
$$S(1 + m) = S_o$$

Eliminating m, and then expressing T as a function of S reveals the trajectory of the mixture in T–S space as a straight line passing through (S_0, T_0), with a gradient given by:

$$\frac{dT}{dS} = \frac{L}{S_o c_w} + \frac{(T_f - T_i)c_i}{S_o c_w} + \frac{(T_o - T_f)}{S_o}$$

The gradient is dominated by the first term, which evaluates to about $2.4°C\,PSU^{-1}$. In polar waters the third term is two orders of magnitude lower than the first; the second term results from the heat needed to warm the ice, and, at about a tenth the size of the first term, makes a measurable contribution to the gradient. This relationship allows the source water for sub-ice shelf processes to be found by inspection of the T–S properties of the resultant water masses. Examples of T–S plots from beneath ice shelves in warm and cold regimes are shown in **Figure 4**.

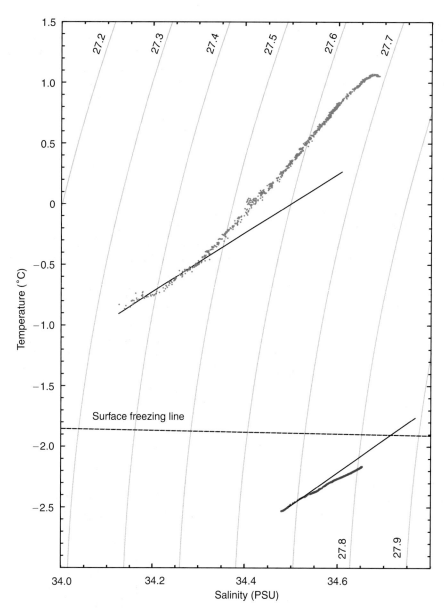

Figure 4 Temperature and salinity trajectories from CTD stations through the George VI Ice Shelf (red) and Ronne Ice Shelf (blue). The cold end of each trajectory corresponds to the base of the ice shelf. The straight lines are at the characteristic gradient for ice melting into sea water. For the Ronne data, as the source water will be HSSW at the surface freezing point, the intersection of the characteristic with the broken line gives the temperature and salinity of the source water. The isopycnals (gray lines) are referenced to sea level.

Two important passive tracers are introduced into sea water when glacial ice melts. When water evaporates from the ocean, molecules containing the lighter isotope of oxygen, ^{16}O, evaporate preferentially. Compared with sea water the snow that makes up the ice shelves is therefore low in ^{18}O. By comparing the $^{18}O/^{16}O$ ratios of the outflowing and inflowing water it is possible to calculate the concentration of melt water, provided the ratio is known for the glacial ice. Helium contained in the air bubbles in the ice is also introduced into the sea

water when the ice melts. As helium's solubility in water increases with increasing water pressure, the concentration of dissolved helium in the melt water can be an order of magnitude greater than in ambient sea water, which has equilibrated with the atmosphere at surface pressure.

Modes of Sub-ice Shelf Circulation

Various distinguishable modes of circulation appear to be possible within a sub-ice shelf cavity. Which mode is active depends primarily on the oceanographic forcing from seaward of the ice front, but also on the geometry of the sub-ice shelf cavity. Thermohaline forcing drives three modes of circulation, although the tidal activity is thought to play an important role by supplying energy for vertical mixing. Another mode results from tidal residual currents.

Thermohaline Modes

Cold regime external ventilation Over the parts of the Antarctic continental shelf dominated by the production of HSSW, such as in the southern Weddell Sea, the Ross Sea, and Prydz Bay, the circulation beneath large ice shelves is driven by the drainage of HSSW into the sub-ice shelf cavities. The schematic in **Figure 5** illustrates the circulation mode. HSSW drains down to the grounding line where tidal mixing brings it into contact with ice at depths of up to 2000 m. At such depths HSSW is up to 1.5°C warmer than the freezing point, and relatively rapid melting ensues (up to several meters of ice per year). The HSSW is cooled and diluted, converting it into ice shelf water (ISW), which is defined as water with a temperature below the surface freezing point.

ISW is relatively buoyant and rises up the inclined base of the ice shelf. As it loses depth the *in situ* freezing point rises also. If the ISW is not entraining sufficient HSSW, which is comparatively warm, the reduction in pressure will result in the water becoming *in situ* supercooled. Ice crystals are then able to form in the water column and possibly rise up and

accrete at the base of the ice shelf. This 'snowfall' at the ice shelf base can build up hundreds of meters of what is termed 'marine ice'. Entrainment of HSSW, and the possible production of ice crystals, often result in the density of the ISW finally matching the ambient water density before the plume has reached the ice front. The plume then detaches from the ice shelf base, finally emerging at the ice front at midwater depths.

The internal Rossby radius beneath ice shelves is typically only a few kilometers, and so rotational effects must be taken into account when considering the flow in three dimensions. HSSW flows beneath the ice shelf as a gravity current and is therefore gathered to the left (in the Southern Hemisphere) by the Coriolis force. As an organized flow, it then follows bathymetric contours. Once converted into ISW, the flow is again gathered to the left, following either the coast, or topography in the ice base. If the ISW plume fills the cavity, conservation of potential vorticity would demand that it follow contours of constant water column thickness. The step in water column thickness caused by the ice front then presents a topographic obstacle for the outflow of the ISW. However, the discontinuity can be reduced by the presence of trenches in the seafloor running across the ice front. This has been proposed as the mechanism that allows ISW to flow out from beneath the Filchner Ice Shelf, in the southern Weddell Sea (**Figure 1**).

Initial evidence for this mode of circulation came from ship-based oceanographic observations along

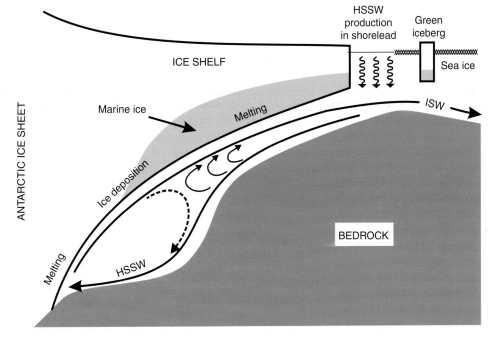

Figure 5 Schematic of the two thermohaline modes of sub-ice shelf circulation for a cold regime ice shelf.

the ice front of several of the larger ice shelves. Water with temperatures up to 0.3°C below the surface freezing point indicated interaction with ice at a depth of at least 400 m, and the $^{18}O/^{16}O$ ratio confirmed the presence of glacial melt water at a concentration of several parts per thousand. Nets cast near ice fronts for biological specimens occasionally recovered masses of ice platelets, again from depths of several hundred meters. The ISW flowing from beneath the Filchner Ice Shelf has been traced overflowing the shelf-break and descending the continental slope, ultimately to mix with deep waters and form bottom water.

Evidence from the ice shelf itself comes in the form of glaciological measurements. By assuming a steady state (the ice shelf neither thickening nor thinning with time at any given point) conservation arguments can be used to derive the basal mass balance at individual locations. The calculation needs measurements of the local ice thickness variation, the horizontal spreading rate of the ice as it flows under its own weight, the horizontal speed of the ice, and the surface accumulation rate. This technique has been applied to several ice shelves, but is time-consuming, and has rarely been used to provide a good areal coverage of basal mass balance. However, it has demonstrated that high basal melt rates do indeed exist near deep grounding lines; that the melt rates reduce away from the grounding line; that further still from the grounding line, melting frequently switches to freezing; and that the balance usually returns to melting as the ice front is approached.

One-dimensional models have been to study the development of ISW plumes from the grounding line to where they detach from the ice shelf base. The most sophisticated includes frazil ice dynamics, and suggests that the deposition of ice at the base depends not only on its formation in the water column, but also on the flow regime being quiet enough to allow the ice to settle at the ice base. As the flow regime usually depends on the basal topography, the deposition is often highly localized. For example, a reduction in basal slope reduces the forcing on the buoyant plume, thereby slowing it down and possibly allowing any ice platelets to be deposited.

Deposits of marine ice become part of the ice shelf itself, flowing with the overlying meteoric ice. This means that, although the marine ice is deposited in well-defined locations, it moves towards the ice front with the flow of the ice and may or may not all be melted off by the time it reaches the ice front. Icebergs that have calved from Amery Ice Front frequently roll over and reveal a thick layer of marine ice. Impurities in marine ice result in different optical properties, and these bergs are often termed 'green icebergs'.

Ice cores obtained from the central parts of the Amery and Ronne ice shelves have provided other direct evidence of the production of marine ice. The interface between the meteoric and marine ice is clearly visible – the ice changes from being white and bubbly, to clear and bubble-free. Unlike normal sea ice, which typically has a salinity of a few PSU, the salinity of marine ice was found to be below 0.1 PSU. The salinity in the cores is highest at the interface itself, decreasing with increasing depth. A different type of marine ice was found at the base of the Ross Ice Shelf. There, a core from near the base showed 6 m of congelation ice with a salinity of between 2 and 4 PSU. Congelation ice differs from marine ice in its formation mechanism, growing at the interface directly rather than being created as an accumulation of ice crystals that were originally formed in the water column.

Airborne downward-looking radar campaigns have mapped regions of ice shelf that are underlain by marine ice. The meteoric (freshwater) ice/marine ice interface returns a characteristically weak echo, but the return from marine ice/ocean boundary is generally not visible. By comparing the thickness of meteoric ice found using the radar with the surface elevation of the freely floating ice shelf, it is possible to calculate the thickness of marine ice accreted at the base. In some parts of the Ronne Ice Shelf basal accumulation rates of around $1\,m\,a^{-1}$ result in a marine ice layer over 300 m thick, out of a total ice column depth of 500 m. Accumulation rates of that magnitude would be expected to be associated with high ISW fluxes. However, cruises along the Ronne Ice Front have been unsuccessful in finding commensurate ISW outflows.

Internal recirculation Three-dimensional models of the circulation beneath the Ronne Ice Shelf have revealed the possibility of an internal recirculation of ISW. This mode of circulation is driven by the difference in melting point between the deep ice at the grounding line, and the shallower ice in the central region of the ice shelf. The possibility of such a recirculation is indicated in **Figure 5** by the broken line. Intense deposition of ice in the freezing region salinifies the water column sufficiently to allow it to drain back towards the grounding line. In three dimensions, the recirculation consists of a gyre occupying a basin in the topography of water column thickness. The model predicts a gyre strength of around one Sverdrup ($10^6\,m^3\,s^{-1}$).

This mode of circulation is effectively an 'ice pump' transporting ice from the deep grounding line regions to the central Ronne Ice Shelf. The mechanism does not result in a loss or gain of ice overall. The heat used to melt the ice at the grounding line is later recovered in the freezing region. The external heat needed to maintain the recirculation is therefore only the heat to warm the ice to the freezing point before it is melted. Ice leaves the continent at a temperature of around $-30°C$, and has a specific heat capacity of around $2010\,J\,kg^{-1}°C^{-1}$. As the latent heat of ice is $335\,kJ\,kg^{-1}$, the heat required for warming is less than 20% of that required for melting. To support an internal redistribution of ice therefore requires a small fraction of the external heat that would be needed to melt and remove the ice from the system entirely. A corollary is that a recirculation of ISW effectively decouples much of the ice shelf base from external forcings that might be imposed, for example, by climate change.

Apart from the lack of a sizable ISW outflow from beneath the Ronne Ice Front, evidence in support of an ISW recirculation deep beneath the ice shelf is scarce, as it would require observations beneath the ice. Direct measurements of conditions beneath ice shelves are limited to a small number of sites. Fissures through George VI and Fimbul ice shelves (**Figure 1**) have allowed instruments to be deployed with varying degrees of success. The more important ice shelves, such as the Ross, Amery and Filchner-Ronne system have no naturally occurring access points. Instead, access holes have to be created using hot water, or other thermal-type drills. In the late 1970s researchers used various drilling techniques to gain access to the cavity at one location beneath the Ross Ice Shelf before deploying various instruments. During the 1990s several access holes were made through the Ronne Ice Shelf, and data from these have lent support both to the external mode of circulation, and most recently, to the internal recirculation mode first predicted by numerical models.

Warm regime external ventilation The flooding of the Bellingshausen and Amundsen seas' continental shelf by barely modified CDW results in very high basal melt rates for the ice shelves in that sector. The floating portion of Pine Island Glacier (**Figure 1**) has a mean basal melt rate estimated to be around $12\,m\,a^{-1}$, compared with estimates of a few tens of centimeters per year for the Ross and Filchner-Ronne ice shelves. Basal melt rates for Pine Island Glacier are high even compared with other ice shelves in the region. George VI Ice Shelf on the west coast of the Antarctic Peninsula, for example, has an estimated mean basal melt rate of $2\,m\,a^{-1}$. The explanation for the intense melting beneath Pine Island Glacier can be found in the great depth at the grounding line. At over $1100\,m$, the ice shelf is $700\,m$ thicker than George VI Ice Shelf, and this results in not only a lower freezing point, but also steeper basal slopes. The steep slope provides a stronger buoyancy forcing, and therefore greater turbulent heat transfer between the water and the ice.

The pattern of circulation in the cavities beneath warm regime ice shelves is significantly different to its cold regime counterpart. Measurements from ice front cruises show an inflow of warm CDW $(+1.0°C)$, and an outflow of CDW mixed with glacial melt water. **Figure 6** shows a two-dimensional schematic of this mode of circulation. Over the open

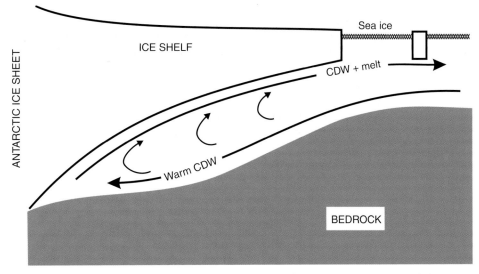

Figure 6 Schematic of the thermohaline mode of sub-ice shelf circulation for a warm regime ice shelf.

continental shelf the ambient water column consists of CDW overlain by colder, fresher water left over from sea ice production during the previous winter. Although the melt water-laden outflow is colder, fresher, and of lower density than the inflow, it is typically warmer and saltier than the overlying water, but of similar density. Somewhat counter-intuitively, therefore, the products of sub-glacial melt are often detected over the open continental shelf as relatively warm and salty intrusions in the upper layers. Again, measurements of oxygen isotope ratio, and also helium, provide the necessary confirmation that the upwelled CDW contains melt water from the base of ice shelves. In the case of warm regime ice shelves, melt water concentrations can be as high as a few percent.

Tidal Forcing

Except for within a few ice thicknesses of grounding lines, ice shelves float freely in the ocean, rising and falling with the tides. Tidal waves therefore propagate through the ice shelf-covered region, but are modified by three effects of the ice cover: the ice shelf base provides a second frictional surface, the draft of the ice shelf effectively reduces the water column thickness, and the step change in water column thickness at the ice front presents a topographic feature that has significant consequences for the generation of residual tidal currents and the propagation of topographic waves along the ice front.

Conversely, tides modify the oceanographic regime of sub-ice shelf cavities. Tidal motion helps transfer heat and salt beneath the ice front. This is a result both of the regular tidal excursions, which take water a few kilometers into the cavity, and of residual tidal currents which, in the case of the Filchner-Ronne Ice Shelf, help ventilate the cavity far from the ice front. The effect of the regular advection of potentially seasonally warmed water from seaward of the ice shelf is to cause a dramatic increase in basal melt rates in the vicinity of the ice front. Deep beneath the ice shelf, tides and buoyancy provide the only forcing on the regime. Tidal activity contributes energy for vertical mixing, which brings the warmer, deeper waters into contact with the base of the ice shelf. **Figure 7A** shows modeled tidal ellipses for the M_2 semidiurnal tidal constituent for the southern Weddell Sea, including the sub-ice shelf domain. A map of the modeled residual currents for the area of the ice shelf is shown in **Figure 7B**. Apart from the activity near the ice front itself, a residual flow runs along the west coast of Berkner Island, deep under the ice shelf. However, this flow probably makes only a minor contribution to the ventilation of the cavity.

How Does the Interaction between Ice Shelves and the Ocean Depend on Climate?

The response to climatic changes of sub-ice shelf circulation depends on the response of the oceanographic conditions over the open continental shelf. In the case of cold regime continental shelves, a reduction in sea ice would lead to a reduction in HSSW production. Model results, together with the implications of seasonality observed in the circulation beneath the Ronne Ice Shelf, suggest that drainage of HSSW beneath local ice shelves would then reduce, and that the net melting beneath those ice shelves would decrease as a consequence. Some general circulation models predict that global climatic warming would lead to a reduction in sea ice production in the southern Weddell Sea. Reduced melting beneath the Filchner-Ronne Ice Shelf would then lead to a thickening of the ice shelf. Recirculation beneath ice shelves is highly insensitive to climatic change. The thermohaline driving is dependent only on the difference in depths between the grounding lines and the freezing areas. A relatively small flux of HSSW is required to warm the ice in order to allow this mode to operate.

The largest ice shelves are in a cold continental shelf regime. If intrusions of warmer off-shelf water were to become more dominant in these areas, or if the shelf-break barrier were to collapse entirely and the regime switch from cold to warm, then the response of the ice shelves would be a dramatic increase in their basal melt rates. There is some evidence from sediment cores that such a change might have occurred at some point in the last few thousand years in what is now the warm regime Bellingshausen Sea. Evidence also points to the possibility that one ice shelf in that sector, the floating extension of Pine Island Glacier (**Figure 1**), might be a remnant of a much larger ice shelf (*see* **Ice-shelf Stability**).

During glacial maxima the Antarctic ice sheet thickens and the ice shelves become grounded. In many cases they ground as far as the shelf-break. There are two effects. The continental shelf becomes very limited in extent, and so there is little possibility for the production of HSSW; and where the ice shelves overhang the continental shelf-break, the only possible mode of circulation will be the warm regime mode. Substantial production of ISW during glacial conditions is therefore unlikely.

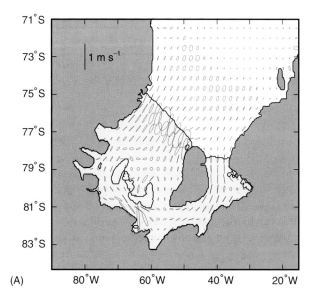

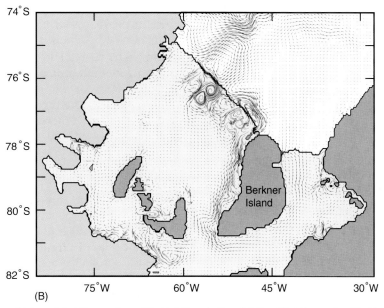

Figure 7 Results from a tidal model of the southern Weddell Sea, in the vicinity of the Ronne Ice Shelf. (A) The tidal ellipses for the dominant M_2 species. (B) Tidally induced residual currents.

See also

Bottom Water Formation. Current Systems in the Southern Ocean. Holocene Climate Variability. Ice–Ocean Interaction. Ice-shelf Stability. Internal Tidal Mixing. Polynyas. Sea Ice: Overview. Shelf-sea and Slope Fronts. Thermohaline Circulation. Tides. Under-ice Boundary Layer. Water Types and Water Masses. Weddell Sea Circulation.

Further Reading

Jenkins A and Doake CSM (1991) Ice–ocean interactions on Ronne Ice Shelf, Antarctica. *Journal of Geophysical Research* 96: 791–813.

Jacobs SS, Hellmer HH, Doake CSM, Jenkins A and Frolich RM (1992) Melting of ice shelves and the mass balance of Antarctica. *Journal of Glaciology* 38: 375–387.

Nicholls KW (1997) Predicted reduction in basal melt rates of an Antarctic ice shelf in a warmer climate. *Nature* 388: 460–462.

Oerter H, Kipfstuhl J, Determann J *et al.* (1992) Ice-core evidence for basal marine shelf ice in the Filchner-Ronne Ice Shelf. *Nature* 358: 399–401.

Williams MJM, Jenkins A and Determann J (1998) Physical controls on ocean circulation beneath ice shelves revealed by numerical models. In: Jacobs SS and Weiss RF (eds) *Ocean, Ice, and Atmosphere: Interactions at the Antarctic Continental Margin, Antarctic Research Series* 75: pp. 285–299. Washington DC: American Geophysical Union.

SUBMERSIBLES

See **MANNED SUBMERSIBLES, DEEP WATER; MANNED SUBMERSIBLES, SHALLOW WATER**

SUB-SEA PERMAFROST

T. E. Osterkamp, University of Alaska,
Alaska, AK, USA

doi:10.1006/rwos.2001.0008

Introduction

Sub-sea permafrost, alternatively known as submarine permafrost and offshore permafrost, is defined as permafrost occurring beneath the seabed. It exists in continental shelves in the polar regions (**Figure 1**). When sea levels are low, permafrost aggrades in the exposed shelves under cold subaerial conditions. When sea levels are high, permafrost degrades in the submerged shelves under relatively warm and salty boundary conditions. Sub-sea permafrost differs from other permafrost in that it is relic, warm, and generally degrading. Methods used to investigate it include probing, drilling, sampling, drill hole log analyses, temperature and salt measurements, geological and geophysical methods (primarily seismic and electrical), and geological and geophysical models. Field studies are conducted from boats or, when the ocean surface is frozen, from the ice cover. The focus of this article is to review our understanding of sub-sea permafrost, of processes ocurring within it, and of its occurrence, distribution, and characteristics.

Sub-sea permafrost derives its economic importance from current interests in the development of offshore petroleum and other natural resources in the continental shelves of polar regions. The presence and characteristics of sub-sea permafrost must be considered in the design, construction, and operation of coastal facilities, structures founded on the seabed, artificial islands, sub-sea pipelines, and wells drilled for exploration and production.

Scientific problems related to sub-sea permafrost include the need to understand the factors that control its occurrence and distribution, properties of warm permafrost containing salt, and movement of heat and salt in degrading permafrost. Gas hydrates that can occur within and under the permafrost are a potential abundant source of energy. As the sub-sea permafrost warms and thaws, the hydrates destabilize, producing gases that may be a significant source of global carbon.

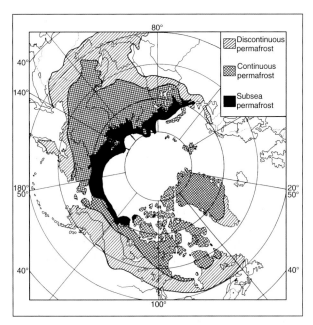

Figure 1 Map showing the approximate distribution of sub-sea permafrost in the continental shelves of the Arctic Ocean. The scarcity of direct data (probing, drilling, sampling, temperature measurements) makes the map highly speculative, with most of the distribution inferred from indirect measurements, primarily water temperature, salinity, and depth (100 m depth contour). Sub-sea permafrost also exists near the eroding coasts of arctic islands, mainlands, and where seabed temperatures remain negative. (Adapted from Pewe TL (1983). *Arctic and Alpine Research* 15(2):145–156 with the permission of the Regents of the University of Coloroado.)

Nomenclature

'Permafrost' is ground that remains below 0°C for at least two years. It may or may not contain ice. 'Ice-bearing' describes permafrost or seasonally frozen soil that contains ice. 'Ice-bonded' describes ice-bearing material in which the soil particles are mechanically cemented by ice. Ice-bearing and ice-bonded material may contain unfrozen pore fluid in addition to the ice. 'Frozen' implies ice-bearing or ice-bonded or both, and 'thawed' implies non-ice-bearing. The 'active layer' is the surface layer of sediments subject to annual freezing and thawing in areas underlain by permafrost. Where seabed temperatures are negative, a thawed layer ('talik') exists near the seabed. This talik is permafrost but does not contain ice because soil particle effects, pressure, and the presence of salts in the pore fluid can depress the freezing point 2°C or more. The boundary between a thawed region and ice-bearing permafrost is a phase boundary. 'Ice-rich' permafrost contains ice in excess of the soil pore spaces and is subject to settling on thawing.

Formation and Thawing

Repeated glaciations over the last million years or so have caused sea level changes of 100 m or more (**Figure 2**). When sea levels were low, the shallow continental shelves in polar regions that were not covered by ice sheets were exposed to low mean annual air temperatures (typically − 10 to − 25°C). Permafrost aggraded in these shelves from the exposed ground surface downwards. A simple conduction model yields the approximate depth (X) to the bottom of ice-bonded permafrost at time t, (eqn [1]).

$$X(t) = \sqrt{\frac{2K(T_e - T_g)t}{h}} \qquad [1]$$

K is the thermal conductivity of the ice-bonded permafrost, T_e is the phase boundary temperature at the bottom of the ice-bonded permafrost, T_g is the long-term mean ground surface temperature during emergence, and h is the volumetric latent heat of the sediments, which depends on the ice content. In eqn [1], K, h, and T_e depend on sediment properties. A rough estimate of T_g can be obtained from information on paleoclimate and an approximate value for t can be obtained from the sea level history (**Figure 2**). Eqn [1] overestimates X because it neglects geothermal heat flow except when a layer of ice-bearing permafrost from the previous transgression remains at depth. Timescales for permafrost growth are such that hundreds of meters of permafrost could have aggraded in the shelves while they were emergent (**Figure 3**).

Cold onshore permafrost, upon submergence during a transgression, absorbs heat from the seabed above and from the geothermal heat flux rising from below. It gradually warms (**Figures 4 and 5**), becoming nearly isothermal over timescales up to a few millennia (**Figure 4**, time t_3). Substantially longer times are required when unfrozen pore fluids are present in equilibrium with ice because some ice must thaw throughout the permafrost thickness for it to warm.

A thawed layer develops below the seabed and thawing can proceed from the seabed downward, even in the presence of negative mean seabed temperatures, by the influx of salt and heat associated with the new boundary conditions. Ignoring seabed erosion and sedimentation processes, the thawing rate at the top of ice-bonded permafrost during submergence is given by eqn [2].

$$\dot{X}_{top} = \frac{J_t}{h} - \frac{J_f}{h} \qquad [2]$$

J_t is the heat flux into the phase boundary from above and J_f is the heat flux from the phase boundary into the ice-bonded permafrost below. J_t depends on the difference between the long-term mean

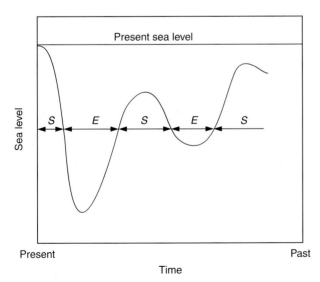

Figure 2 Schematic sea level curve during the last glaciation. The history of emergence (E) and submergence (S) can be combined with paleotemperature data on sub-aerial and sub-sea conditions to construct an approximate thermal boundary condition for sub-sea permafrost at any water depth.

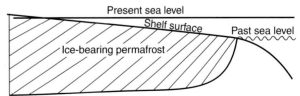

Figure 3 A schematic illustration of ice-bearing sub-sea permafrost in a continental shelf near the time of minimum sea level. Typical thicknesses at the position of the present shoreline would have been about 400–1000 m with shelf widths that are now typically 100–600 km.

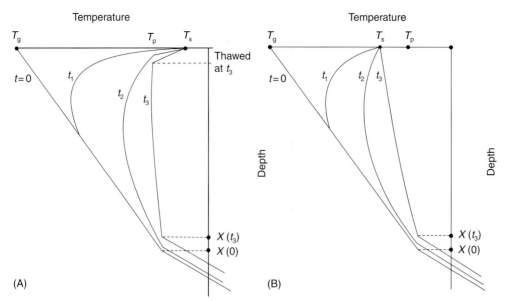

Figure 4 Schematic sub-sea permafrost temperature profiles showing the thermal evolution at successive times (t_1, t_2, t_3) after submergence when thawing occurs at the seabed (A) and when it does not (B). T_g and T_s are the long-term mean surface temperatures of the ground during emergence and of the seabed after submergence. T_p is the phase boundary temperature at the top of the ice-bonded permafrost. $X(0)$ and $X(t_3)$ are the depths to the bottom of ice-bonded permafrost at times $t = 0$ and t_3. (Adapted with permission from Lachenbruch AH and Marshall BV (1977) Open File Report 77-395. US Geological Survey, Menlo Park, CA.)

temperature at the seabed, T_s, and phase boundary temperature, T_p, at the top of the ice-bonded permafrost. For $J_t = J_f$, the phase boundary is stable. For $J_t < J_f$, refreezing of the thawed layer can occur from the phase boundary upward. For thawing to occur, T_s must be sufficiently warmer than T_p to make $J_t > J_f$. T_s is determined by oceanographic conditions (currents, ice cover, water salinity, bathymetry, and presence of nearby rivers). T_p is determined by hydrostatic pressure, soil particle effects, and salt concentration at the phase boundary (the combined effect of *in situ* pore fluid salinity, salt transport from the seabed through the thawed layer, and changes in concentration as a result of freezing or thawing).

Nearshore at Prudhoe Bay, $\dot{X}_{top}$ varies typically from centimeters to tens of centimeters per year while farther offshore it appears to be on the order of millimeters per year. The thickness of the thawed layer at the seabed is typically 10 m to 100 m, although values of less than a meter have been observed. At some sites, the thawed layer is thicker in shallow water and thinner in deeper water.

Sub-sea permafrost also thaws from the bottom by geothermal heat flow once the thermal disturbance of the transgression penetrates there. The approximate thawing rate at the bottom of the ice-bonded permafrost is given by eqn [3].

$$\dot{X}_{bot} \cong \frac{J_g}{h} - \frac{J_f'}{h} \qquad [3]$$

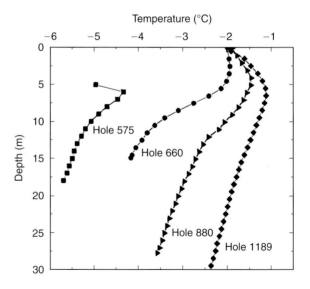

Figure 5 Temperature profiles obtained during the month of May in sub-sea permafrost near Barrow, Alaska, showing the thermal evolution with distance (equivalently time) offshore. Hole designation is the distance (m) offshore and the shoreline erosion rate is about 2.4 m y^{-1}. Sea ice freezes to the seabed within 600 m of shore. (Adapted from Osterkamp TE and Harrison WD (1985) Report UAGR-301. Fairbanks, AK: Geophysical Institute, University of Alaska.)

J_g is the geothermal heat flow entering the phase boundary from below and J_f' is the heat flow from the phase boundary into the ice-bonded permafrost above. J_f' becomes small within a few millennia except when the permafrost contains unfrozen pore fluids. $\dot{X}_{bot}$ is typically on the order of centimeters per year. Timescales for thawing at the permafrost table and base are such that several tens of thousands of years may be required to completely thaw a few hundred meters of sub-sea permafrost.

Modeling results and field data indicate that impermeable sediments near the seabed, low T_s, high ice contents, and low J_g favor the survival of ice-bearing sub-sea permafrost during a transgression. Where conditions are favorable, substantial thicknesses of ice-bearing sub-sea permafrost may have survived previous transgressions.

Characteristics

The chemical composition of sediment pore fluids is similar to that of sea water, although there are detectable differences. Salt concentration profiles in thawed coarse-grained sediments at Prudhoe Bay (**Figure 6**) appear to be controlled by processes occurring during the initial phases of submergence. There is evidence for highly saline layers within ice-bonded permafrost near the base of gravels overlying a fine-grained sequence both onshore and offshore. In the Mackenzie Delta region, fluvial sand units deposited during regressions have low salt concentrations (**Figure 7**) except when thawed or when lying under saline sub-sea mud. Fine-grained mud sequences from transgressions have higher salt concentrations. Salts increase the amount of unfrozen pore fluids and decrease the phase equilibrium temperature, ice content, and ice bonding. Thus, the sediment layering observed in the Mackenzie Delta region can lead to unbonded material (clay) between layers of bonded material (fluvial sand).

Thawed sub-sea permafrost is often separated from ice-bonded permafrost by a transition layer of ice-bearing permafrost. The thickness of the ice-bearing layer can be small, leading to a relatively sharp (centimeters scale) phase boundary, or large, leading to a diffuse boundary (meters scale). In general, it appears that coarse-grained soils and low salinities produce a sharper phase boundary and fine-grained soils and higher salinities produce a more diffuse phase boundary.

Sub-sea permafrost consists of a mixture of sediments, ice, and unfrozen pore fluids. Its physical and mechanical properties are determined by the individual properties and relative proportions. Since ice and unfrozen pore fluid are strongly temperature dependent, so also are most of the physical and mechanical properties.

Ice-rich sub-sea permafrost has been found in the Alaskan and Canadian portions of the Beaufort Sea and in the Russian shelf. Thawing of this permafrost can result in differential settlement of the seafloor that poses serious problems for development.

Processes

Submergence

Onshore permafrost becomes sub-sea permafrost upon submergence, and details of this process play a major role in determining its future evolution. The rate at which the sea transgresses over land is determined by rising sea levels, shelf topography, tectonic setting, and the processes of shoreline erosion, thaw settlement, thaw strain of the permafrost, seabed erosion, and sedimentation. Sea levels on the polar continental shelves have increased more than 100 m in the last 20 000 years or so. With shelf widths of 100–600 km, the average shoreline retreat rates would have been about 5 to 30 m y^{-1}, although maximum rates could have been much larger. These average rates are comparable to areas with very rapid shoreline retreat rates observed today on the Siberian and North American shelves. Typical values are 1–6 m y^{-1}.

It is convenient to think of the transition from sub-aerial to sub-sea conditions as occurring in five regions (**Figure 8**) with each region representing different thermal and chemical surface boundary conditions. These boundary conditions are successively applied to the underlying sub-sea permafrost during a transgression or regression. Region 1 is the onshore permafrost that forms the initial condition for sub-sea permafrost. Permafrost surface temperatures range down to about $-15°C$ under current sub-aerial conditions and may have been 8–10°C colder during glacial times. Ground water is generally fresh, although salty lithological units may exist within the permafrost as noted above.

Region 2 is the beach, where waves, high tides, and resulting vertical and lateral infiltration of sea water produce significant salt concentrations in the active layer and near-surface permafrost. The active layer and temperature regime on the beach differ from those on land. Coastal banks and bluffs are a trap for wind-blown snow that often accumulates in insulating drifts over the beach and adjacent ice cover.

Region 3 is the area where ice freezes to the seabed seasonally, generally where the water depth is less than about 1.5–2 m. This setting creates unique thermal boundary conditions because, when the ice freezes into the seabed, the seabed becomes conductively coupled to the atmosphere and thus very cold. During summer, the seabed is covered with shallow, relatively warm sea water. Salt concentrations at the seabed are high during winter because of salt rejection from the growing sea ice

and restricted circulation under the ice, which eventually freezes into the seabed. These conditions create highly saline brines that infiltrate the sediments at the seabed.

Region 4 includes the areas where restricted under-ice circulation causes higher-than-normal sea water salinities and lower temperatures over the

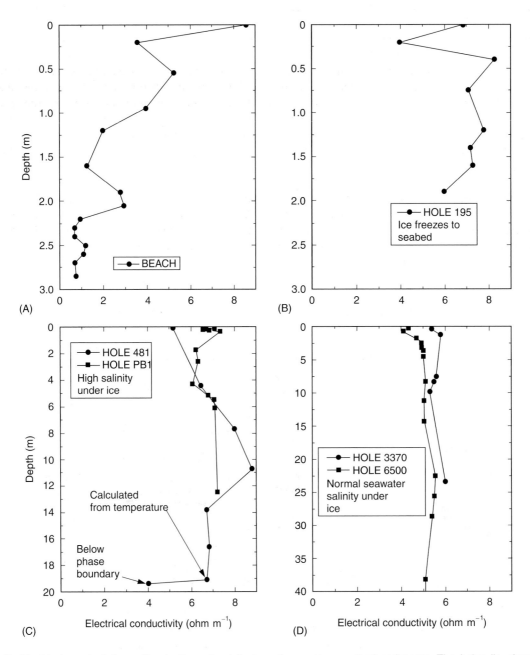

Figure 6 Electrical conductivity profiles in thawed, relatively uniform coarse-grained sediments. The holes lie along a line extending offshore near the West Dock at Prudhoe Bay, Alaska except for PB1, which is in the central portion of Prudhoe Bay. Hole designation is the distance (m) offshore. On the beach (A), concentrations decrease by a factor of 5 at a few meters depth. There are large variations with depth and concentrations may be double that of normal seawater where ice freezes to the seabed (B) and where there is restricted circulation under the ice (C). Farther offshore (D), the profiles tend to be relatively constant with depth with concentrations about the same or slightly greater than the overlying seawater. (Adapted from Iskandar IK *et al.* (1978) *Proceedings, Third International Conference on Permafrost*, Edmonton, Alberta, Canada, pp. 92–98. Ottawa, Ontario: National Research Council.)

sediments. The ice does not freeze to the seabed or only freezes to it sporadically. The existence of this region depends on the ice thickness, on water depth, and on flushing processes under the ice. Strong currents or steep bottom slopes may reduce its extent or eliminate it.

The setting for region 5 consists of normal sea water over the seabed throughout the year. This results in relatively constant chemical and thermal boundary conditions.

There is a seasonal active layer at the seabed that freezes and thaws annually in both regions 3 and 4. The active layer begins to freeze simultaneously with the formation of sea ice in shallow water. Brine drainage from the growing sea ice increases the water salinity and decreases the temperature of the water at the seabed because of the requirement for phase equilibrium. This causes partial freezing of the less saline pore fluids in the sediments. Thus, it is not necessary for the ice to contact the seabed for the seabed to freeze. Seasonal changes in the pore fluid salinity show that the partially frozen active layer redistributes salts during freezing and thawing, is infiltrated by the concentrated brines, and influences the timing of brine drainage to lower depths in

the sediments. These brines, derived from the growth of sea ice, provide a portion or all of the salts required for thawing the underlying sub-sea permafrost in the presence of negative sediment temperatures.

Depth to the ice-bonded permafrost increases slowly with distance offshore in region 3 to a few meters where the active layer no longer freezes to it (**Figure 8**) and the ice-bonded permafrost no longer couples conductively to the atmosphere. This allows the permafrost to thaw continuously throughout the year and depth to the ice-bonded permafrost increases rapidly with distance offshore (**Figure 8**).

The time an offshore site remains in regions 3 and 4 determines the number of years the seabed is subjected to freezing and thawing events. It is also the time required to make the transition from sub-aerial to relatively constant sub-sea boundary conditions. This time appears to be about 30 years near Lonely, Alaska (about 135 km southeast of Barrow) and about 500–1000 years near Prudhoe Bay, Alaska. **Figure 9** shows variations in the mean seabed temperatures with distance offshore near Prudhoe Bay where the shoreline retreat rate is about $1\,\mathrm{m\,y^{-1}}$.

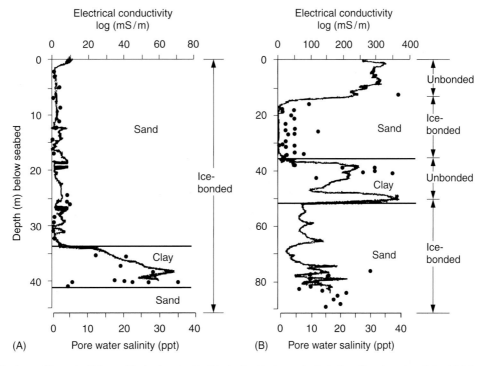

Figure 7 Onshore (A) and offshore (B) (water depth 10 m) electrical conductivity log (lines) and salinity (dots) profiles in the Mackenzie Delta region. The onshore sand–clay–sand sequence can be traced to the offshore site. Sand units appear to have been deposited under sub-aerial conditions during regressions and the clay unit under marine conditions during a transgression. At the offshore site, the upper sand unit is thawed to the 11 m depth. (Adapted with permission from Dallimore SR and Taylor AE (1994). *Proceedings, Sixth International Conference on Permafrost*, Beijing, vol. 1, pp.125–130. Wushan, Guangzhou, China: South China University Press.)

The above discussion of the physical setting does not incorporate the effects of geology, hydrology, tidal range, erosion and sedimentation processes, thaw settlement, and thaw strain. Regions 3 and 4 are extremely important in the evolution of sub-sea permafrost because the major portion of salt infiltration into the sediments occurs in these regions. The salt plays a strong role in determining T_p and, thus, whether or not thawing will occur.

Heat and Salt Transport

The transport of heat in sub-sea permafrost is thought to be primarily conductive because the observed temperature profiles are nearly linear below the depth of seasonal variations. However, even when heat transport is conductive, there is a coupling with salt transport processes because salt concentration controls T_p. Our lack of understanding of salt transport processes hampers the application of thermal models.

Thawing in the presence of negative seabed temperatures requires that T_p be significantly lower than T_s, so that generally salt must be present for thawing to occur. This salt must exist in the permafrost on submergence and/or be transported from the seabed to the phase boundary. The efficiency of salt transport through the thawed layer at the seabed appears to be sensitive to soil type. In clays, the salt transport process is thought to be diffusion, a slow process; and in coarse-grained sands and gravels, pore fluid convection, which (involving motion of fluid) can be rapid.

Diffusive transport of salt has been reported in dense overconsolidated clays north of Reindeer Island offshore from Prudhoe Bay. Evidence for convective transport of salt exists in the thawed coarse-grained sediments near Prudhoe Bay and in the layered sands in the Mackenzie Delta region. This includes rapid vertical mixing as indicated by large seasonal variations in salinity in the upper 2 m of sediments in regions 3 and 4 and by salt concentration profiles that are nearly constant with depth and decrease in value with distance offshore in region 5 (**Figures 5–7**). Measured pore fluid pressure profiles (**Figure 10**) indicate downward fluid motion. Laboratory measurements of downward brine drainage velocities in coarse-grained sediments indicate that these velocities may be on the order of 100 m y^{-1}.

The most likely salt transport mechanism in coarse-grained sediments appears to be gravity-

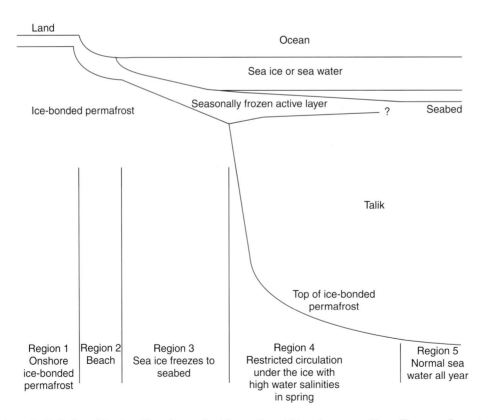

Figure 8 Schematic illustration of the transition of permafrost from sub-aerial to sub-sea conditions. There are five potential regions with differing thermal and chemical seabed boundary conditions. Hole 575 in **Figure 5** is in region 3 and the rest of the holes are in region 4. (Adapted from Osterkamp TE (1975) Report UAGR-234. Fairbanks, AK: Geophysical Institute, University of Alaska.)

driven convection as a result of highly saline and dense brines at the seabed in regions 3 and 4. These brines infiltrate the seabed, even when it is partially frozen, and move rapidly downward. The release of relatively fresh and buoyant water by thawing ice at the phase boundary may also contribute to pore fluid motion.

Occurrence and Distribution

The occurrence, characteristics, and distribution of sub-sea permafrost are strongly influenced by regional and local conditions and processes including the following.

1. Geological (heat flow, shelf topography, sediment or rock types, tectonic setting)
2. Meteorological (sub-aerial ground surface temperatures as determined by air temperatures, snow cover and vegetation)
3. Oceanographic (seabed temperatures and salinities as influenced by currents, ice conditions, water depths, rivers and polynyas; coastal erosion and sedimentation; tidal range)
4. Hydrological (presence of lakes, rivers and salinity of the ground water)
5. Cryological (thickness, temperature, ice content, physical and mechanical properties of the onshore permafrost; presence of sub-sea permafrost that has survived previous transgressions; presence of ice sheets on the shelves)

Lack of information on these conditions and processes over the long timescales required for permafrost to aggrade and degrade, and inadequacies in the theoretical models, make it difficult to formulate reliable predictions regarding sub-sea permafrost. Field studies are required, but field data are sparse and investigations are still producing surprising results indicating that our understanding of sub-sea permafrost is incomplete.

Pechora and Kara Seas

Ice-bonded sub-sea permafrost has been found in boreholes with the top typically up to tens of meters below the seabed. In one case, pure freshwater ice was found 0.3 m below the seabed, extending to at least 25 m. These discoveries have led to difficult design conditions for an undersea pipeline that will cross Baydaratskaya Bay transporting gas from the Yamal Peninsula fields to European markets.

Laptev Sea

Sea water bottom temperatures typically range from −0.5°C to −1.8°C, with some values colder than

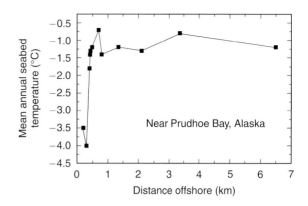

Figure 9 Variation of mean annual seabed temperatures with distance offshore. Along this same transect, about 6 km offshore from Reindeer Island in 17 m of water, the mean seabed temperature was near −1.7°C. Data on mean annual seabed salinities in regions 3 and 4 do not appear to exist. (Adapted from Osterkamp TE and Harrison WD (1985) Report UAGR-301. Fairbanks, AK: Geophysical Institute, University of Alaska.)

−2°C. A 300–850 m thick seismic sequence has been found that does not correlate well with regional tectonic structure and is interpreted to be ice-bonded permafrost. The extent of ice-bonded permafrost appears to be continuous to the 70 m isobath and widespread discontinuous to the 100 m isobath. Depth to the ice-bonded permafrost ranges from 2 to 10 m in water depths from 45 m to the shelf edge. Deep taliks may exist inshore of the 20 m isobath. A shallow sediment core with ice-bonded material at its base was recovered from a water depth of 120 m. Bodies of ice-rich permafrost occur under shallow water at the locations of recently eroded islands and along retreating coastlines.

Bering Sea

Sub-sea permafrost is not present in the northern portion except possibly in near-shore areas or where shoreline retreat is rapid.

Chukchi Sea

Seabed temperatures are generally slightly negative and thermal gradients are negative, indicating ice-bearing permafrost at depth within 1 km of shore near Barrow, Rabbit Creek and Kotzebue.

Alaskan Beaufort Sea

To the east of Point Barrow, bottom waters are typically −0.5°C to −1.7°C away from shore, shoreline erosion rates are rapid (1–10 m y^{-1}) and sediments are thick. Sub-sea permafrost appears to be thicker in the Prudhoe Bay region and thinner west of Harrison Bay to Point Barrow. Ice layers up

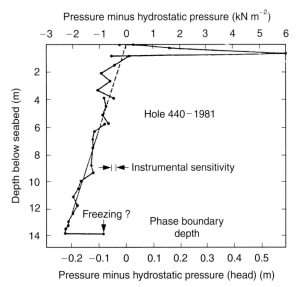

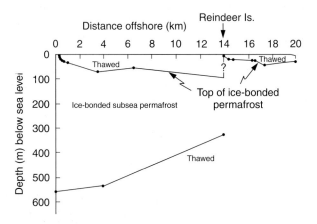

Figure 10 Measured *in situ* pore fluid pressure minus calculated hydrostatic pressure through the thawed layer in coarse-grained sediments at a hole 440 m offshore in May 1981 near Prudhoe Bay, Alaska. High pressure at the 0.54 m depth is probably related to seasonal freezing and the solid line is a least-squares fit to the data below 5.72 m. The negative pressure head gradient (−0.016) indicates a downward component of pore fluid velocity. (Adapted from Swift DW *et al.* (1983) *Proceedings, Fourth International Conference on Permafrost*, Fairbanks, Alaska, vol. 1, pp. 1221-1226. Washington DC: National Academy Press.)

Figure 11 Sub-sea permafrost profile near Prudhoe Bay, Alaska, determined from drilling and well log data. The sediments are coarse-grained with deep thawing inshore of Reindeer Island and fine-grained with shallow thawing farther offshore. Maximum water depths are about 8 m inshore of Reindeer Island and 17 m about 6 km north. (Adapted from Osterkamp TE *et al.* (1985) *Cold Regions Science and Technology* 11: 99-105, 1985, with permission from Elsevier Science.)

to 0.6 m in thickness have been found off the Sagavanirktok River Delta.

Surface geophysical studies (seismic and electrical) have indicated the presence of layered ice-bonded sub-sea permafrost. A profile of sub-sea permafrost near Prudhoe Bay (**Figure 11**) shows substantial differences in depth to ice-bonded permafrost between coarse-grained sediments inshore of Reindeer Island and fine-grained sediments farther offshore. Offshore from Lonely, where surface sediments are fine-grained, ice-bearing permafrost exists within 6–8 m of the seabed out to 8 km offshore (water depth 8 m). Ice-bonded material is deeper (∼15 m). In Elson Lagoon (near Barrow) where the sediments are fine-grained, a thawed layer at the seabed of generally increasing thickness can be traced offshore.

Mackenzie River Delta Region

The layered sediments found in this region are typically fluvial sand and sub-sea mud corresponding to regressive/transgressive cycles (see **Figure 12**). Mean seabed temperatures in the shallow coastal areas are generally positive as a result of warm river water, and negative farther offshore. The thickness of ice-bearing permafrost varies substantially as a result of a complex history of transgressions and regressions, a complex history of transgressions and regressions,

discharge from the Mackenzie River, and possible effects of a late glacial ice cover. Ice-bearing permafrost in the eastern and central Beaufort Shelf exceeds 600 m. It is thin or absent beneath Mackenzie Bay and may be only a few hundred meters thick toward the Alaskan coast. The upper surface of ice-bearing permafrost is typically 5–100 m below the seabed and appears to be under control of seabed temperatures and stratigraphy.

The eastern Arctic, Arctic Archipelago and Hudson Bay regions were largely covered during the last glaciation by ice that would have inhibited permafrost growth. These regions are experiencing isostatic uplift with permafrost aggrading in emerging shorelines.

Antarctica

Negative sediment temperatures and positive temperature gradients to a depth of 56 m below the seabed exist in McMurdo Sound where water depth is 122 m and the mean seabed temperature is −1.9°C. This sub-sea permafrost did not appear to contain any ice.

Models

Modeling the occurrence, distribution and characteristics of sub-sea permafrost is a difficult task. Statistical, geological, analytical, and numerical models are available. Statistical models attempt to combine geological, oceanographic, and other information into algorithms that make predictions about sub-sea permafrost. These statistical models have not been very successful, although new GIS methods could potentially improve them.

Geological models consider how geological processes influence the formation and development of sub-sea permafrost. It is useful to consider these models since some sub-sea permafrost could potentially be a million years old. A geological model for the Mackenzie Delta region (**Figure 12**) has been

developed that provides insight into the nature and complex layering of the sediments that comprise the sub-sea permafrost there.

Analytical models for investigating the thermal regime of sub-sea permafrost include both one- and two-dimensional models. All of the available analytical models have simplifying assumptions that limit their usefulness. These include assumptions of one-dimensional heat flow, stable shorelines or

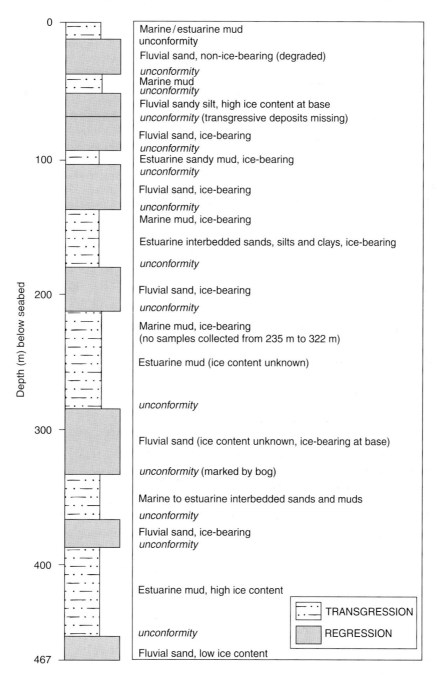

Figure 12 Canadian Beaufort Shelf stratigraphy in 32 m of water near the Mackenzie River Delta. Eight regressive/transgressive fluvial sand/marine mud cycles are shown. It is thought that, except for thawing near the seabed, the ice-bearing sequence has been preserved through time to the present. (Adapted with permission from Blasco S (1995) GSC Open File Report 3058. Ottawa, Canada: Geological Survey of Canada.)

shorelines that undergo sudden and permanent shifts in position, constant air and seabed temperatures that neglect spatial and temporal variations over geological timescales, and constant thermal properties in layered sub-sea permafrost that is likely to contain unfrozen pore fluids. Neglect of topographical differences between the land and seabed, geothermal heat flow, phase change at the top and bottom of the sub-sea permafrost, and salt effects also limits their application. An analytical model exists that addresses the coupling between heat and salt transport but only for the case of diffusive transport with simplifying assumptions. Nevertheless, these analytical models appear to be applicable in certain special situations and have shaped much of the current thinking about sub-sea permafrost.

Two-dimensional numerical thermal models have addressed most of the concerns related to the assumptions in analytical models except for salt transport. Models have been developed that address salt transport via the buoyancy of fresh water generated by thawing ice at the phase boundary. Models for the infiltration of dense sea water brines derived from the growth of sea ice into the sediments do not appear to exist.

Successful application of all models is limited because of the lack of information over geological timescales on initial conditions, boundary conditions, material properties, salt transport and the coupling of heat and salt transport processes. There is also a lack of areas with sufficient information and measurements to fully test model predictions.

Symbols used

h	Volumetric latent heat of the sediments (1 to $2 \times 10^8 \, J \, m^{-3}$)
J_g	Geothermal heat flow entering the bottom phase boundary from below
J_f	Heat flow from the bottom phase boundary into the ice-bonded permafrost above
J_t	Heat flux into the top phase boundary from above
J_f	Heat flux from the top phase boundary into the ice-bonded permafrost below
K	Thermal conductivity of the ice-bonded permafrost (1 to $5 \, W \, m^{-1} \, K^{-1}$)
t	Time
T_s	Long-term mean temperature at the seabed
T_p	Phase boundary temperature at the top of the ice-bonded permafrost (0 to $-2°C$)
T_e	Phase boundary temperature at the bottom of the ice-bonded permafrost (0 to $-2°C$)
T_g	Long-term mean ground surface temperature during emergence

$X(t)$	Depth to the bottom of ice-bonded permafrost at time, t
$\dot{X}_{top}$	Thawing rate at top of ice-bonded permafrost during submergence
$\dot{X}_{bot}$	Thawing rate at the bottom of ice-bonded permafrost during submergence

See also

Arctic Basin Circulation. Coastal Circulation Models. Glacial Crustal Rebound, Sea Levels and Shorelines. Heat Transport and Climate. Holocene Climate Variability. Methane Hydrates and Climatic Effects. Mid-ocean Ridge Tectonics, Volcanism and Geomorphology. Millenial Scale Climate Variability. Penetrating Shortwave Radiation. Polynyas. River Inputs. Sea Level Change. Sea Level Variations Over Geologic Time. Sea Ice: Overview; Variations in Extent and Thickness. Sub Ice-shelf Circulation and Processes. Under-ice Boundary Layer. Upper Ocean Heat and Freshwater Budgets.

Further Reading

Dallimore SR and Taylor AE (1994) Permafrost conditions along an onshore–offshore transect of the Canadian Beaufort Shelf. *Proceedings, Sixth International Conference on Permafrost*, Beijing, vol. 1, pp. 125–130. Wushan, Guangzhou, China: South China University Press.

Hunter JA, Judge AS, MacAuley HA *et al.* (1976) *The Occurrence of Permafrost and Frozen Sub-sea Bottom Materials in the Southern* Beaufort Sea. Beaufort Sea Project, Technical Report 22. Ottawa: Geological Survey Canada.

Lachenbruch AH, Sass JH, Marshall BV and Moses TH Jr (1982) Permafrost, heat flow, and the geothermal regime at Prudhoe Bay, Alaska. *Journal of Geophysical Research* 87(B11): 9301–9316.

Mackay JR (1972) Offshore permafrost and ground ice, Southern Beaufort Sea, Canada. *Canadian Journal of Earth Science* 9(11): 1550–1561.

Osterkamp TE, Baker GC, Harrison WD and Matava T (1989) Characteristics of the active layer and shallow sub-sea permafrost. *Journal of Geophysical Research* 94(C11): 16227–16236.

Romanovsky NN, Gavrilov AV, Kholodov AL *et al.* (1998). Map of predicted offshore permafrost distribution on the Laptev Sea Shelf. *Proceedings, Seventh International Conference on Permafrost*, Yellowknife, Canada, pp. 967–972. University of Laval, Quebec: Center for Northern Studies.

Sellmann PV and Hopkins DM (1983) Sub-sea permafrost distribution on the Alaskan Shelf. *Final Proceedings, Fourth International Conference on Permafrost*, Fairbanks, Alaska, pp. 75–82. Washington, DC: National Academy Press.

SURFACE FILMS

W. Alpers, University of Hamburg, Hamburg, Germany

doi:10.1006/rwos.2001.0065

Introduction

Surface films floating on the sea surface are usually attributed to anthropogenic sources. Such films consist, for example, of crude oil discharged from tankers during cleaning operations or accidents. However, much more frequently surface films that are of natural origin are encountered at the sea surface. These natural surface films consist of surface-active compounds that are secreted by marine plants or animals. According to their physicochemical characteristics the film-forming substances tend to be either enriched at the sea surface (more hydrophobic character, sometimes referred to as 'dry surfactants') or they prevail within the upper water layer (more hydrophobic character 'wet surfactants'). The first type of surface-active substances ('dry surfactants') are able to form monomolecular slicks at the air–water interface and damp the short-scale surface waves (short-gravity capillary waves) much more strongly than the second type. This implies that they have a strong effect on the mass, energy, and momentum transfer processes at the air–water interface. They also affect these transfer processes by reducing the turbulence in the subsurface layer which is instrumental in transporting water from below to the surface.

Both types of surface films are easily detectable by radars because radars are roughness sensors and surface films strongly reduce the short-scale sea surface roughness.

Orgin of Surface Films

Surface films at the sea surface can be either of anthropogenic or natural origin. Anthropogenic surface films consist, e.g., of crude or petroleum oil spilled from ships or oil platforms ('spills'), or of surface-active substances discharged from municipal or industrial plants ('slicks'). Natural surface films may also consist of crude oil which is leaking from oil seeps on the seafloor, but usually they consist of surface-active substances, which are produced by biogenic processes in the sea ('biogenic slicks'). In general, the biogenic surface slicks consisting of suf-

ficiently hydrophobic substances ('dry surfactants') are only one molecular layer thick (approximately 3 nm). This implies that it needs only few liters of surface-active material to cover an area of $1 \, km^2$. The prime biological producers of natural surface films in the sea are algae and some bacteria. Also zooplankton and fish produce surface-active materials, but the amount is usually small in comparison with the primary biological production. Primary production depends on the quantity of light energy available to the organism and the availability of inorganic nutrients. In the higher latitudes light energy depends strongly on the season of the year which results in a seasonal variation of the primary biological production in the ocean and thus of the slick coverage by natural surface films. At times when the biological productivity is high, i.e., during plankton blooms, the probability of encountering natural biogenic surface films is strongly enhanced.

In other regions of the world's ocean where the primary production is not mainly determined by the quantity of light energy available to the organisms, but by the nutrient levels, the seasonal variation of the slick coverage is smaller, however, still observable. Surface slicks of biogenic origin are mainly encountered in sea regions where the nutrition factor favors productivity. This is the case on continental shelves, slopes, and in upwelling regions where nutrition-rich cold water is transported to the sea surface.

The areal extent, the concentration and the composition of the surface films vary strongly with time. At high wind speeds (typically above $8-10 \, m \, s^{-1}$) breaking waves disperse the films by entrainment into the underlying water such that they disappear from the sea surface. In general, the probability of encountering surface films of biogenic origin decreases with wind speed. Furthermore, after storms, enhanced coverage of the sea surface with biogenic slicks consisting of 'dry surfactants' is often observed which is due to the fact that, firstly, the secretion of surface-active material by plankton is being increased during higher wind speed periods, and secondly, the surface-active substances are being transported to the sea surface from below by turbulence and rising air bubbles generated by breaking waves. The composition of the sea slicks varies also with time because constituents of the surface films are selectively removed by dissolution, evaporation, enzymatic degradation and photocatalytic oxidation.

Modifications of Air–Sea Interaction by Surface Films

Numerous processes that take place at the air–sea interface are affected by surface films. Among other things, surface films: (1) attenuate the surface waves; (2) reduce wave breaking; (3) reduce gas transfer; (4) increase the sea-surface temperature; (5) change the reflection of sunlight; and (6) reduce the intensity of the radar backscatter.

Attenuation of Surface Waves

Two main factors contribute to the damping of short-scale surface waves by surface films:

1. the enhanced viscous dissipation in a thin water layer below the water surface caused by strong velocity gradients induced by the presence of viscoelastic films at the water surface; and
2. the decrease in energy transfer from the wind to the waves due to the reduction of the aerodynamic roughness of the sea surface.

The enhanced viscous dissipation caused by the surface films results from the fact that, due to the different boundary condition imposed by the film at the sea surface, strong vertical velocity gradients are encountered in a thin layer below the water surface. This layer, also called shear layer, has a thickness of the order of 10^{-4} m. Here strong viscous dissipation takes place. In the case of mineral oil films floating on the sea surface, the shear layer may lie completely within the oil layer, but more often it extends also into the upper water layer since the thickness of mineral oil films is typically in the range of 10^{-3}–10^{-6} m.

In the case of biogenic monomolecular surface films accumulating at a rough sea surface, the surface films are compressed and dilated periodically, causing variations of the concentration of the molecules and thus of the surface tension. In this case not only the well-known gravity-capillary waves (surface waves) are excited, but also the so-called Marangoni waves. The Marangoni waves are predominantly longitudinal waves in the shear layer. They are heavily damped by viscous dissipitation; at a distance of only one wavelength from their source their amplitude has already decreased to less than one-tenth of the original value. This is the reason why Marangoni waves escaped detection until 1968. When these gravity-capillary waves and Marangoni waves are in resonance as given by linear wave theory, the surface waves experience maximum damping. Depending on the viscoelastic properties of the surface film, maximum damping of the surface waves usually occurs in the centimeter to decimeter wavelength region.

Reduction of Wave Breaking

Since surface films reduce roughness of the sea surface, the stress exerted by the wind on the sea surface is reduced. Furthermore, the steepness of the short-scale waves is decreased which leads to less wave breaking.

Reduction of Gas Transfer

Biogenic surface films do not constitute a direct resistance for gas transfer. However, they do have a major effect on the structure of subsurface turbulence and thus on the rate at which surface water is renewed by water from below. Furthermore, surface films reduce the air turbulence above the ocean surface and thus also the surface renewal. As a consequence, the gas transfer rate across the air–sea interface is reduced in the presence of surface films.

Change of Sea Surface Temperature

In infrared images the sea surface areas covered with biogenic surface films usually appear slightly warmer than the adjacent slick-free sea areas (typical temperature increase 0.2–0.5 K). This is due to the fact that surface films reduce the mobility of the near-surface water molecules and thus slow down the conventional overturn of the surface layer by evaporation.

Change of Reflection of Sunlight

Slick-covered areas of the sea surface are easily visible by eye when they lie in the sun-glitter area. This is an area where facets on the rough sea surface are encountered that have orientations that reflect the sunlight to the observer. When the surface is covered with a surface film, the sea surface becomes smoother and thus the orientation of the facets is changed such that the amount of light reflected to the observer is increased. Thus surface slicks become detectable in sun-glitter areas as areas of increased brightness. Outside the sun-glitter area they are sometimes also visible, but with a much fainter contrast. In this case they appear as areas of reduced brightness relative to the surrounding.

Radar Backscattering

Surface slicks floating on the sea surface also become visible on radar images because they reduce the short-scale sea surface roughness. Since the intensity of the radar backscatter is determined by the amplitude of short-scale surface waves, slick-covered sea surfaces appear on radar images as areas of reduced radar backscattering. Since radars have their own illumination source and transmit electromagnetic waves with wavelengths in the centimeter to decimeter range, radar images of the sea surface

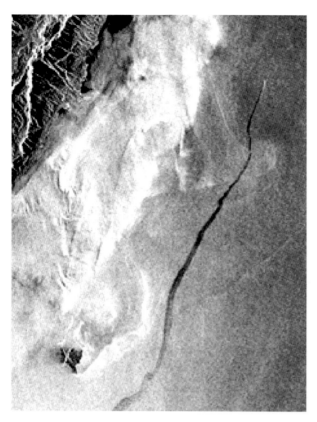

Figure 1 Radar image acquired by the synthetic aperture radar (SAR) aboard the First European Remote Sensing satellite (ERS-1) on 20 May 1994 over the coastal waters east of Taiwan. The imaged area is 70 km × 90 km. It shows a ship (the bright spot at the front of the black trail) discharging oil. The oil trail, which is approximately 80 km long, widens towards the rear because the oil disperses with time. Copyright © 2000, European Space Agency.

Figure 2 Radar image acquired by the SAR aboard the Second European Remote Sensing satellite (ERS-2) on 10 May 1998 over the Western Baltic Sea which includes the Bight of Lübeck (Germany). The imaged area is 90 km × 100 km. Visible are the lower left coastal areas of Schleswig-Holstein with the island of Fehmarn (Germany) and in the upper right part of the Danish island of Lolland. The black areas are sea areas covered with biogenic slicks which are particularly abundant in this region during the time of the spring plankton bloom. The slicks follow the motions of the sea surface and thus render oceanic eddies visible on the radar image. Copyright © 2000, European Space Agency.

can be obtained independent of the time of the day and independent of cloud conditions. This makes radar an ideal instrument for detecting oil pollution and natural surface films at the sea surface. Consequently, most oil pollution surveillance aircraft which patrol coastal waters for locating illegal discharges of oil from ships are equipped with imaging radars.

Unfortunately the reduction in backscattered radar intensity caused by mineral oil films is often of the same order (typically 5–10 decibels) as that of natural surface films. This makes it difficult by using the information contained in the reduction of the backscattered radar intensity to differentiate whether the black patches visible on radar images of the sea surface originate from one or the other type of film. However, in many cases the shape of the black patches on the radar images can be used for discrimination. A long elongated dark patch is indicative of an oil spill originating from a travelling ship. Examples of radar images on which both types of surface films are visible are shown in **Figures 1** and **2**.

See also

Air–Sea Gas Exchange. Oil Pollution. Phytoplankton Blooms. Satellite Remote Sensing of Sea Surface Temperatures. Surface, Gravity and Capillary Waves.

Further Reading

Alpers W and Hühnerfuss H (1989) The damping of ocean waves by surface films: a new look at an old problem. *Journal of Geophysical Research* 94: 6251–6265.

Levich VG (1962) *Physico-Chemical Hydrodynamics.* Englewood Cliffs, NJ: Prentice-Hall.

Lucassen J (1982) Effect of surface-active material on the damping of gravity waves: a reappraisal. *Journal of Colloid Interface Science* 85: 52–58.

Tsai WT (1996) Impact of surfactant on turbulent shear layer under the air–sea interface. *Journal of Geophysical Research* 101: 28557–28568.

SURFACE, GRAVITY AND CAPILLARY WAVES

W. K. Melville, Scripps Institution of Oceanography,
La Jolla, CA, USA

doi:10.1006/rwos.2001.0129

Introduction

Ocean surface waves are the most common oceanographic phenomena that are known to the casual
observer. They can at once be the source of inspiration and primal fear. It is remarkable that the complex, random wave field of a storm-lashed sea can
be studied and modeled using well-developed theoretical concepts. Many of these concepts are based
on linear or weakly nonlinear approximations to the
full nonlinear dynamics of ocean waves. Early contributors to these theories included such luminaries
as Cauchy, Poisson, Stokes, Lagrange, Airy, Kelvin
and Rayleigh. Many of the current challenges in the
study of ocean surface waves are related to nonlinear processes which are not yet well understood.
These include dynamical coupling between the
atmosphere and the ocean, wave–wave interactions,
and wave breaking.

For the purposes of this article, surface waves are
considered to extend from low frequency swell from
distant storms at periods of 10 s or more and
wavelengths of hundreds of meters, to capillary
waves with wavelengths of millimeters and frequencies of O(10) Hz. In between are wind waves with
lengths of O(1–100) m and periods of O(1–10) s.
Figure 1 shows a spectrum of surface waves measured from the Research Platform FLIP off the coast
of Oregon. The spectrum, Φ, shows the distribution
of energy in the wave field as a function of frequency. The wind wave peak at approximately 0.13
Hz is well separated from the swell peak at approximately 0.06 Hz.

Ocean surface waves play an important role in
air–sea interaction. Momentum from the wind goes
into both surface waves and currents. Ultimately the
waves are dissipated either by viscosity or breaking,
giving up their momentum to currents. Surface
waves affect upper-ocean mixing through both wave
breaking and their role in the generation of Langmuir circulations. This breaking and mixing influences the temperature of the ocean surface and thus
the thermodynamics of air–sea interaction. Surface
waves impose significant structural loads on ships
and other structures. Remote sensing of the ocean
surface, from local to global scales, depends on the
surface wave field.

Basic Formulations

The dynamics and kinematics of surface waves are
described by solutions of the Navier–Stokes equations for an incompressible viscous fluid, with appropriate boundary and initial conditions. Surface
waves of the scale described here are usually generated by the wind, so the complete problem would
include the dynamics of both the water and the air
above. However, the density of the air is approximately 800 times smaller than that of the water,
so many aspects of surface wave kinematics and
dynamics may be considered without invoking
dynamical coupling with the air above.

The influence of viscosity is represented by the
Reynolds number of the flow, $R_e = UL/v$, where
U is a characteristic velocity, L a characteristic
length scale, and $v = \mu/\rho$ is the kinematic viscosity,
where μ is the viscosity and ρ the density of the
fluid. The Reynolds number is the ratio of inertial
forces to viscous forces in the fluid and if $R_e \gg 1$, the
effects of viscosity are often confined to thin boundary layers, with the interior of the fluid remaining
essentially inviscid ($v = 0$). (This assumes a homogeneous fluid. In contrast, internal waves in a continuously stratified fluid are rotational since they
introduce baroclinic generation of vorticity in the
interior of the fluid). Denoting the fluid velocity by
$\mathbf{u} = (u, v, w)$, the vorticity of the flow is given by
$\zeta = \nabla \times \mathbf{u}$. If $\zeta = 0$, the flow is said to be irrotational. From Kelvin's circulation theorem, the
irrotational flow of an incompressible ($\nabla . \mathbf{u} = 0$)
inviscid fluid will remain irrotational as the flow
evolves. The essential features of surface waves may
be considered in the context of incompressible irrotational flows.

For an irrotational flow, $\mathbf{u} = \nabla \phi$ where the scalar
ϕ is a velocity potential. Then, by virtue of incom-
pressibility, ϕ satisfies Laplace's equation

$$\nabla^2 \phi = 0 \qquad [1]$$

We denote the surface by $z = \eta(x, y, t)$, where (x, y)
are the horizontal coordinates and t is time. The
kinematic condition at the impermeable bottom at
$z = -h$, is one of no flow through the boundary:

$$\frac{\partial \phi}{\partial z} = 0 \text{ at } z = -h \qquad [2]$$

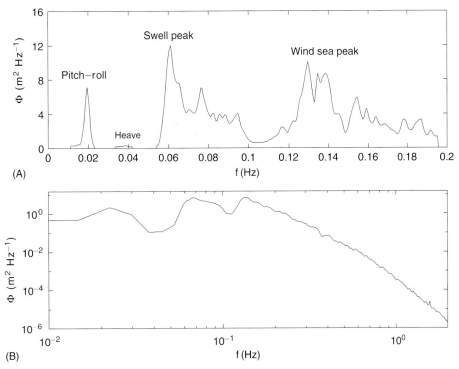

Figure 1 (A) Surface displacement spectrum measured with an electromechanical wave gauge from the Research Platform FLIP in 8 m s^{-1} winds off the coast of Oregon. Note the wind-wave peak at 0.13 Hz, the swell at 0.06 Hz and the heave and pitch and roll of FLIP at 0.04 and 0.02 Hz respectively. (B) An extension of (A) with logarithmic spectral scale, note that from the wind sea peak to approximately 1 Hz the spectrum has a slope like f^{-4}, common in wind-wave spectra. (Reproduced with permission from Felizardo FC and Melville WK (1995). Correlations between ambient noise and the ocean surface wave field. *Journal of Physical Oceanography* 25: 513–532.)

There are two boundary conditions at $z = \eta$:

$$\frac{\partial \eta}{\partial t} + u \frac{\partial \eta}{\partial x} + v \frac{\partial \eta}{\partial y} = w \qquad [3]$$

$$\frac{\partial \phi}{\partial t} + \frac{1}{2} \mathbf{u}^2 + g\eta = (p_a - p_-)/\rho. \qquad [4]$$

The first is a kinematic condition which is equivalent to imposing the condition that elements of fluid at the surface remain at the surface. The second is a dynamical condition, a Bernoulli equation, which is equivalent to stating that the pressure p_- at $z = \eta_-$, an infinitesimal distance beneath the surface, is just a constant atmospheric pressure, p_a, plus a contribution from surface tension. The effect of gravity is to impose a restoring force tending to bring the surface back to $z = 0$. The effect of surface tension is to reduce the curvature of the surface.

Although this formulation of surface waves is considerably simplified already, there are profound difficulties in predicting the evolution of surface waves based on these equations. Although Laplace's equation is linear, the surface boundary conditions are nonlinear and apply on a surface whose specification is a part of the solution. Our ability to accurately predict the evolution of nonlinear waves is limited and largely dependent on numerical techniques. The usual approach is to linearize the boundary conditions about $z = 0$.

Linear Waves

Simple harmonic surface waves are characterized by an amplitude a, half the distance between the crests and the troughs, and a wavenumber vector $\mathbf{k}$ with $|\mathbf{k}| = k = 2\pi/\lambda$, where λ is the wavelength. The surface displacement, (unless otherwise stated, the real part of complex expressions is taken)

$$\eta = a e^{i(\mathbf{k} \cdot \mathbf{x} - \sigma t)} \qquad [5]$$

where $\sigma = 2\pi/T$ is the radian frequency and T is the wave period. Then ak is a measure of the slope of the waves, and if $ak \ll 1$, the surface boundary conditions can be linearized about $z = 0$.

Following linearization, the boundary conditions become

$$\frac{\partial \eta}{\partial t} = w \qquad [6]$$

$$\frac{\partial \phi}{\partial t} + g\eta = \frac{\Gamma}{\rho}\left(\frac{\partial^2 \eta}{\partial x^2} + \frac{\partial^2 \eta}{\partial y^2}\right) \text{ at } z = 0 \qquad [7]$$

where the linearized Laplace pressure is

$$p_a - p_- = \Gamma\left(\frac{\partial^2 \eta}{\partial x^2} + \frac{\partial^2 \eta}{\partial y^2}\right) \qquad [8]$$

where Γ is the surface tension coefficient.

Substituting for η and satisfying Laplace's equation and the boundary conditions at $z = 0$ and $-h$ gives

$$\phi = \frac{ig'a}{\sigma} \frac{\cosh k(z+h)}{\cosh kh} \qquad [9]$$

where

$$\sigma^2 = g'k \tanh kh \qquad [10]$$

and

$$g' = g(1 + \Gamma k^2/\rho) \qquad [11]$$

Equations relating the frequency and wavenumber, $\sigma = \sigma(k)$, are known as dispersion relations, and for linear waves provide a fundamental description of the wave kinematics. The phase speed,

$$c = \sigma/k = \left(\frac{g'}{k} \tanh kh\right)^{1/2} \qquad [12]$$

is the speed at which lines of constant phase (e.g., wave crests) move.

For waves propagatinge in the x-direction, the velocity field is

$$u = \frac{g'ak}{\sigma} \frac{\cosh(z+h)}{\cosh kh} e^{i(kx - \sigma t)} \qquad [13]$$

$$v = 0 \qquad [14]$$

$$w = -\frac{ig'ak}{\sigma} \frac{\sinh(z+h)}{\cosh kh} e^{i(kx - \sigma t)} \qquad [15]$$

and the pressure

$$p = \rho g'\eta \frac{\cosh(z+h)}{\cosh kh} \qquad [16]$$

The velocity decays with depth away from the surface, and, to leading order, elements of fluid execute elliptical orbits as the waves propagate.

For shallow water, $kh \ll 1$,

$$(u, v, w, p) = \left(\frac{g'k}{\sigma}, 0, 0, \rho g'\right)\eta \qquad [17]$$

so that there is no vertical motion, just a uniform sloshing backwards and forwards in the horizontal plane in phase with the surface displacement η. The phase speed $c = (g'h)^{1/2}$, is independent of the wavenumber. Such waves are said to be nondispersive. Waves propagating towards shore eventually attain this condition, and, as the depth tends to zero, nonlinear effects become important as ak increases.

For very deep water, $kh \gg 1$,

$$(u, v, w, p) = \left(\frac{g'k}{\sigma}, 0, -\frac{ig'k}{\sigma}, \rho g'\right)e^{kz}\eta \qquad [18]$$

so that the water particles execute circular motions that decay exponentially with depth. The horizontal motion is in phase with the surface displacement, and the phase speed of the waves

$$c = (g'/k)^{1/2} = \left[\frac{g}{k}(1 + \Gamma k^2/\rho g)\right]^{1/2} \qquad [19]$$

These deep-water waves are dispersive; that is, the phase speed is a function of the wavenumber as shown in **Figure 2**. The influence of surface tension relative to gravity is determined by the value of the dimensionless parameter $\Sigma = \Gamma k^2/\rho g$. When $\Sigma = 1$, the wavelength $\lambda = 1.7\,\text{cm}$ and the phase speed is a minimum at $c = 23\,\text{cm s}^{-1}$. When $\Sigma \gg 1$, surface tension is the dominant restoring force, the wavelength is less than 1.7 cm, and the phase speed

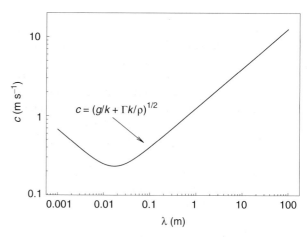

Figure 2 The phase speed of surface gravity-capillary waves as a function of wavelength λ. A minimum phase speed of 23 cm s^{-1} occurs for $\lambda = 0.017$ m. Shorter waves approach pure capillary waves, whereas longer waves become pure gravity waves. Note that there are both capillary and gravity waves for a given phase speed. This is the basis of the generation of parasitic capillary waves on the forward face of steep gravity waves.

increases as the wavelength decreases. When $\Sigma \ll 1$, gravity is the dominant restoring force, the wavelength is greater than 1.7 cm, and the phase speed increases as the wavelength increases.

The Group Velocity

Using the superposition principle over a continuum of wavenumbers a general disturbance (in two spatial dimensions) can be represented by

$$\eta(x, t) = \int_{-\infty}^{\infty} a(k) e^{i(kx - \sigma t)} dk \qquad [20]$$

where, as above, only the real part of the integral is taken. Assuming the disturbance is confined to wavenumbers in the neighbourhood of k_o, and expanding $\sigma(k)$ about k_o gives

$$\sigma(k) = \sigma(k_o) + (k - k_o) \frac{d\sigma}{dk}|_{k=k_0} + \ldots \qquad [21]$$

whence

$$\eta(x, t) \doteq e^{i(k_o x - \sigma(k_o) t)} \int_{-\infty}^{\infty} a(k) e^{i(k - k_o)(x - c_g t)} dk + \ldots$$

$$[22]$$

where

$$c_g = \left| \frac{d\sigma}{dk} \right|_{k=k_o} \qquad [23]$$

is the group velocity. Eqn [22] demonstrates that the modulation of the pure harmonic wave propagates at the group velocity. This implies that an isolated packet of waves centered around the wavenumber k_o will propagate at the speed c_g, so that an observer wishing to follow waves of the same length must travel at the group velocity. Since the energy density is proportional to a^2 (see below), it is also the speed at which the energy propagates. These properties of the group velocity apply to linear waves, and more subtle effects may become important at large slopes. In general, $c_g \neq c$.

For deep-water gravity waves,

$$c_g = \frac{1}{2} c = \frac{1}{2} \left(\frac{g}{k} \right)^{1/2} \qquad [24]$$

so the wave group travels at half the phase speed, with waves appearing at the rear of a group propagating forward and disappearing at the front of the group.

For deep-water capillary waves,

$$\sigma^2 = \Gamma k^3 / \rho, \quad c = (\Gamma k / \rho)^{1/2}, \quad c_g = \frac{3}{2} c \qquad [25]$$

so waves appear at the front of the group and disappear at the rear of the group as it propagates.

For shallow water gravity waves, $kh \ll 1$, $c_g = c$.

Second Order Quantities

The energy density (per horizontal surface area) of surface waves is

$$E = \frac{1}{2} \rho g' a^2 \qquad [26]$$

being the sum of the kinetic and potential energies. In the case of gravity waves, the potential energy results from the displacement of the surface about its equilibrium horizontal position. For capillary waves, the potential energy arises from the stretching of the surface against the restoring force of surface tension.

The mean momentum density **M** is given by

$$\mathbf{M} = \frac{1}{2} \rho \sigma a^2 \coth kh \mathbf{e} = \frac{E}{c} \mathbf{e} \qquad [27]$$

where the unit vector $\mathbf{e} = \mathbf{k}/k$.

To leading order, linear gravity waves transfer energy without transporting mass; however, there is a second order mass transport associated with surface waves. In a Lagrangian description of the flow it can be shown that for irrotational inviscid wave motion the mean horizontal Lagrangian velocity (Stokes drift) of a particle of fluid originally at $z = z_o$ is

$$\mathbf{u}_l = \sigma k a^2 \frac{\cosh 2k(z_o + h)}{2 \sinh^2 kh} \mathbf{e} \qquad [28]$$

which reduces to $(ak)^2 c e^{2kz_o} \mathbf{e}$ when $kh \gg 1$. This second order velocity arises from the fact that the orbits of the particles of fluid are not closed. Integrating eqn [28] over the depth it can be shown that this mean Lagrangian velocity accounts for the wave momentum **M** in the Eulerian description. The Stokes drift is important for representing scalar transport near the ocean surface, but this transport is likely to be significantly enhanced by the intermittent larger velocities associated with wave breaking.

Longer waves, or swell, from distant storms can travel great distances. An extreme example is the propagation of swell along great circle routes from storms in the Southern Ocean to the coast of California. For waves to travel so far, the effects of dissipation must be small. In deep water, where the wave motions have decayed away to negligible levels at depth, the contributions to the dissipation come from the thin surface boundary layer and the rate of strain of the irrotational motions in the bulk of the fluid. It can be shown that the integral is dominated by the latter contributions, and the time-scale for the decay of the wave energy is just

$$\tau_e = -\left(\frac{1}{E}\frac{dE}{dt}\right)^{-1} = (4vk^2)^{-1} \quad [29]$$

or $\sigma/8\pi v k^2$ wave periods. This gives negligible dissipation for long-period swell in deep water over scales of the ocean basins. More realistic models of wave dissipation must take into account breaking and near surface turbulence which is sometimes parameterized as a 'super viscosity' or eddy viscosity, several orders of magnitude greater than the molecular value. When waves propagate into shallow water, the dominant dissipation may occur in the bottom boundary layer.

Eqn [27] shows that dissipation of wave energy is concomitant with a reduction in wave momentum, but since momentum is conserved, the reduction of wave momentum is accompanied by a transfer of momentum from waves to currents. That is, net dissipative processes in the wave field lead to the generation of currents.

Waves on Currents: Action Conservation

Waves propagating in varying currents may exchange energy with the current, thus modifying the waves. Perhaps the most dramatic examples of this effect come when waves propagating against a current become larger and steeper. Examples occur off the east coast of South Africa as waves from the Southern Ocean meet the Aghulas Current; as North Atlantic storms meet the northward flowing Gulf Stream, or at the mouths of estuaries as shoreward propagating waves meet the ebb tide.

For currents $U = (U, V)$ that only change slowly on the scale of the wavelength, and a surface displacement of the form

$$\eta = a(x, y, t)e^{i\theta(x, y, t)} \quad [30]$$

where a is the slowly varying amplitude and θ is the phase. The absolute local frequency $\omega = -\partial\theta/\partial t$,

and the x- and y-components of the local wave number are given by $k = \partial\theta/\partial x$, $l = \partial\theta/\partial y$. The frequency seen by an observer moving with the current U is

$$-\left(\frac{\partial\theta}{\partial t} + U.\nabla\theta\right) \quad [31]$$

which is equal to the intrinsic frequency σ. Thus

$$\sigma = \omega - U.k \quad [32]$$

which is just the Doppler relationship.

We also have,

$$\frac{\partial k}{\partial t} + \nabla\omega = 0, \quad [33]$$

which can be interpreted as the conservation of wave crests, where k is the spatial density of crests and ω the wave flux.

The velocity of a wave packet along rays is

$$\frac{dx_i}{dt} = U_i + \frac{\partial\sigma}{\partial x_i} = U_i + c_{gi} \quad [34]$$

which is simply the vector sum of the local current and the group velocity in a fluid at rest. Furthermore, refraction is governed by

$$\frac{dk_i}{dt} = -k_j\frac{\partial U_j}{\partial x_i} - \frac{\partial\sigma}{\partial x_i} \quad [35]$$

where the first term on the right represents refraction due to the current and the second is due to gradients in the waveguide, such as changes in the depth. It is this latter term which results in waves, propagating from deep water towards a beach, refracting so that they propagate normal to shore.

For steady currents, the absolute frequency is constant along rays but the intrinsic frequency may vary, and the dynamics lead to a remarkable and quite general result for linear waves. If E is the energy density then the quantity $\mathscr{A} = E/\sigma$, the wave action, is conserved:

$$\frac{\partial\mathscr{A}}{\partial t} + \frac{\partial}{\partial x_i}[(U_i + c_{gi}).\mathscr{A}] = 0 \quad [36]$$

In other words, the variations in the intrinsic frequency σ and the energy density E, are such as to conserve the quotient.

This theory permits the prediction of the change of wave properties as they propagate into varying currents and water depths. For example, in the case of waves approaching an increasing counter current, the waves will move to shorter wavelengths (higher k), larger amplitudes, and hence greater slopes, ak. As the speed of the adverse current approaches the group velocity, the waves will be 'blocked' and be unable to propagate further. In this simplest theory, a singularity occurs with the wave slope becoming infinite, but higher order effects lead to reflection of the waves and the same blocking effect. This theory also forms the basis of models of long-wave–short-wave interaction that are important for wind-wave generation and the interpretation of remote sensing measurements of the ocean surface, including the remote sensing of long nonlinear internal waves.

Nonlinear Effects

The nonlinearity of surface waves is represented by the wave slope, ak. For typical gravity waves at the ocean surface the average slope may be $O(10^{-2}–10^{-1})$; small, but not negligibly so. Nonlinear effects may be weak and can be described as a perturbation to the linear wave theory, using the slope as an expansion parameter. This approach, pioneered by Stokes in the mid-nineteenth century, showed that for uniform approach deep-water gravity waves,

$$\sigma^2 = gk(1 + a^2k^2 + \cdots), \qquad [37]$$

and

$$\eta = a\cos\theta + \frac{1}{2}a^2k\cos 2\theta + \cdots. \qquad [38]$$

Weakly nonlinear gravity waves have a phase speed greater than linear waves of the same wavelength. The effect of the higher harmonics on the shape of the waves leads to a vertical asymmetry with sharper crests and flatter troughs.

The largest such uniform wave train has a slope of $ak = 0.446$ a phase speed of $1.11c$, and a discontinuity in slope at the crest containing an included angle of $120°$. This limiting form has sometimes been used as the basis for the models of wave breaking; however, uniform wave trains are unstable to side-band instabilities at significantly lower slopes, and it is unlikely that this limiting form is ever achieved in the ocean.

With the assumption of both weak nonlinearity and weak dispersion (or small bandwidth, $\delta k/k_o \ll 1$), it may be shown that if

$$\eta(x, y, t) = \mathscr{R}_e[A(x, y, t)e^{i(k_o x - \sigma_o t)}] \qquad [39]$$

where $\sigma_o = \sigma(k_o)$ and $\mathscr{R}_e$ means that the real part is taken, then the complex wave envelope $A(x, y, t)$ satisfies a nonlinear Schrödinger equation or one of its variants. Solutions of the nonlinear Schrödinger equation for initial conditions that decay sufficiently rapidly in space evolve into a series of envelope solitons and a dispersive tail. Solitons propagate as waves of permanent form and survive interactions with other solitons with just a change of phase. Attempts have been made to describe ocean surface waves as fields of interacting envelope solitons; however, instabilities of the two-dimensional soliton solutions, and the effects of higher-order nonlinearities, random phase and amplitude fluctuations in real wave fields give pause to the applicability of these idealized theoretical results.

Resonant Interactions

Modeling the generation, propagation, interaction, and dissipation of wind-generated surface waves is of great importance for a variety of scientific, commercial and social reasons. A rigorous theoretical foundation for all components of this problem does not yet exist, but there is a rational theory for weakly nonlinear wave–wave interactions.

For linear waves freely propagating away from a storm, the spectral content at any later time is explicitly defined by the initial storm conditions. For a nonlinear wave field, wave–wave interactions can lead to the generation of wavenumbers different from those comprising the initial disturbance. For surface gravity waves, these nonlinear effects lead to the generation of waves of lower and higher wavenumber with time. The timescale for this evolution in a random homogeneous wave field is of the order of $(ak)^4$ times a characteristic wave period; slow, but significant over the life of a storm.

The foundation of weakly nonlinear interactions between surface waves is the resonant interaction between waves satisfying the linear dispersion relationship. It is a simple consequence of quadratic nonlinearity that pairs of interacting waves lead to the generation of waves having sum and difference frequencies relative to the original waves. Thus

$$\mathbf{k}_3 = \pm\mathbf{k}_1 \pm \mathbf{k}_2, \quad \sigma_3 = \pm\sigma_1 \pm \sigma_2 \qquad [40]$$

If in addition, $\sigma_i(i = 1, 2, 3)$ satisfies the dispersion relationship, then the interaction is resonant. In the case of surface waves, the nonlinearities arise from the surface boundary conditions, and resonant triads are possible for gravity capillary waves, and gravity waves in water of intermediate depth.

For deep-water gravity waves, cubic nonlinearity is required before resonance occurs between a quartet of wave components:

$$\mathbf{k}_1 \pm \mathbf{k}_2 \pm \mathbf{k}_3 \pm \mathbf{k}_4 = 0,$$

$$\sigma_1 \pm \sigma_2 \pm \sigma_3 \pm \sigma_4 \pm \cdots = 0, \quad \sigma_i = (gk_i)^{1/2} \qquad [41]$$

These quartet interactions comprise the basis of nonlinear wave–wave interactions in operational models of surface gravity waves. Exact resonance is not required, since even with detuning significant energy transfer can occur across the spectrum. The formal basis of these theories may be cast as problems of multiple spatial and temporal scales, and higher-order interactions should be considered as these scales increase, and the wave slope increases.

Parasitic Capillary Waves

The longer gravity waves are the dominant waves at the ocean surface, but recent developments in air–sea interaction and remote sensing, have placed increasing importance on the shorter gravity-capillary waves. Measurements of gravity-capillary waves at sea are very difficult to make and much of the detailed knowledge is based on laboratory experiments and theoretical models.

Laboratory measurements suggest that the initial generation of waves at the sea surface occurs in the gravity-capillary wave range, initially at wavelengths of $O(1)$ cm. As the waves grow and the fetch increases, the dominant waves, those at the peak of the spectrum, move into the gravity-wave range. A simple estimate of the effects of surface tension based on the surface tension parameter Σ using the gravity wavenumber k would suggest that they are unimportant, but as the wave slope increases and the curvature at the crest increases, the contribution of the Laplace pressure near the crest increases. A consequence is that so-called parasitic capillary waves may be generated on the forward face of the gravity wave (**Figure 3**).

The source of these parasitic waves can be represented as a perturbation to the underlying gravity wave caused by the localized Laplace pressure component at the crest. This is analogous to the 'fish-line' problem of Rayleigh, who showed that due to the differences in the group velocities, capillary waves are found ahead of, and gravity waves

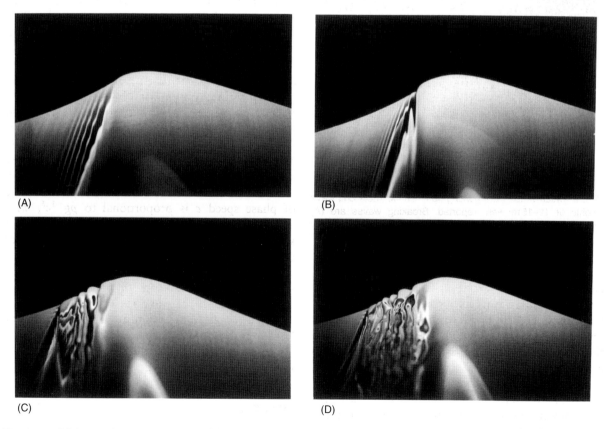

(A)

(B)

(C)

(D)

Figure 3 (A)–(D) Evolution of a gravity wave towards breaking in the laboratory. Note the generation of parasitic capillary waves on the forward face of the crest. (Reproduced with permission from Duncan JH *et al.* (1994) The formation of a spilling breaker. *Physics of Fluids* **6**: S2.)

(A)

(B)

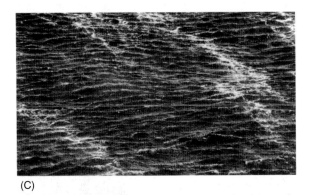

(C)

Figure 4 Waves in a storm in the North Atlantic in December 1993 in which winds were gusting up to 50–60 knots and wave heights of 12–15 m were reported. Breaking waves are (A) large, (B) intermediate and (C) small scale. (Photographs by E. Terrill and W.K. Melville; reproduced with permission from Melville, (1996).)

behind, a localized source in a stream. In this context the capillary waves are considered to be steady relative to the crest. The possibility of the direct resonant generation of capillary waves by perturbations moving at or near the phase speed of longer gravity waves is implied by the form of the dispersion curve in **Figure 2**. Free surfaces of large curvature, as in parasitic capillary waves, are not irrotational and so the effects of viscosity in transporting vorticity and dissipating energy must be accounted for. Theoretical and numerical studies show that the viscous dissipation of the longer gravity

waves is enhanced by one to two orders of magnitude by the presence of parasitic capillary waves. These studies also show that the observed high wavenumber cut-off in the surface wave spectrum that has been observed at wavelengths of approximately $O(10^{-3}-10^{-2})$ m can be explained by the properties of the spectrum of parasitic capillary waves bound to short steep gravity waves.

Wave Breaking

Although weak resonant and near-resonant interactions of weakly nonlinear waves occur over slow timescales, breaking is a fast process, lasting for times comparable to the wave period. However, the turbulence and mixing due to breaking may last for a considerable time after the event. Breaking, which is a transient, two-phase, turbulent, free-surface flow, is the least understood of the surface wave processes. The energy and momentum lost from the wave field in breaking are available to generate turbulence and surface currents, respectively. The air entrained by breaking may, through the associated buoyancy force on the bubbles, be dynamically significant over times comparable to the wave period as the larger bubbles rise and escape through the surface. The sound generated with the breakup of the air into bubbles is perhaps the dominant source of high frequency sound in the ocean, and may be used diagnostically to characterize certain aspects of air–sea interaction. **Figure 4** shows examples of breaking waves in a North Atlantic storm.

Since direct measurements of breaking in the field are so difficult, much of our understanding of breaking comes from laboratory experiments and simple modeling. For example, laboratory experiments and similarity arguments suggest that the rate of energy loss per unit length of the breaking crest of a wave of phase speed c is proportional to $\rho g^{-1} c^5$, with a proportionality factor that depends on the wave slope, and perhaps other parameters. Attempts are underway to combine such simple modeling along with field measurements of the statistics of breaking fronts to give an estimate of the distribution of dissipation across the wave spectrum. Recent developments in the measurement and modeling of breaking using optical, acoustical microwave and numerical techniques hold the promise of significant progress in the next decade.

See also

Acoustic Noise. Breaking Waves and Near-surface Turbulence. Bubbles. Heat and Momentum Fluxes at the Sea Surface. Internal Waves. Langmuir

Circulation and Instability. Surface Films. Wave Energy. Wave Generation by Wind. Whitecaps and Foam. Wind Driven Circulation.

Further Reading

Komen GJ, Cavaleri L, Donelan M *et al.* (1994) *Dynamics and Modelling of Ocean Waves*. Cambridge: Cambridge University Press.

Lamb H (1945) *Hydrodynamics*. New York: Dover Publications.

LeBlond PH and Mysak LA (1978) *Waves in the Ocean*. Amsterdam: Elsevier.

Lighthill J (1978) *Waves in Fluids*. Cambridge: Cambridge University Press.

Mei CC (1983) *The Applied Dynamics of Ocean Surface Waves*. New York: John Wiley.

Melville WK (1996) The role of wave breaking in air–sea interaction. *Annual Review of Fluid Mechanics* 28: 279–321.

Phillips OM (1977) *The Dynamics of the Upper Ocean*. Cambridge: Cambridge University Press.

Whitham GB (1974) *Linear and Nonlinear Waves*. New York: John Wiley.

Yuen HC and Lake BM (1980) Instability of waves on deep water. *Annual Review of Fluid Mechanics* 12: 303–334.

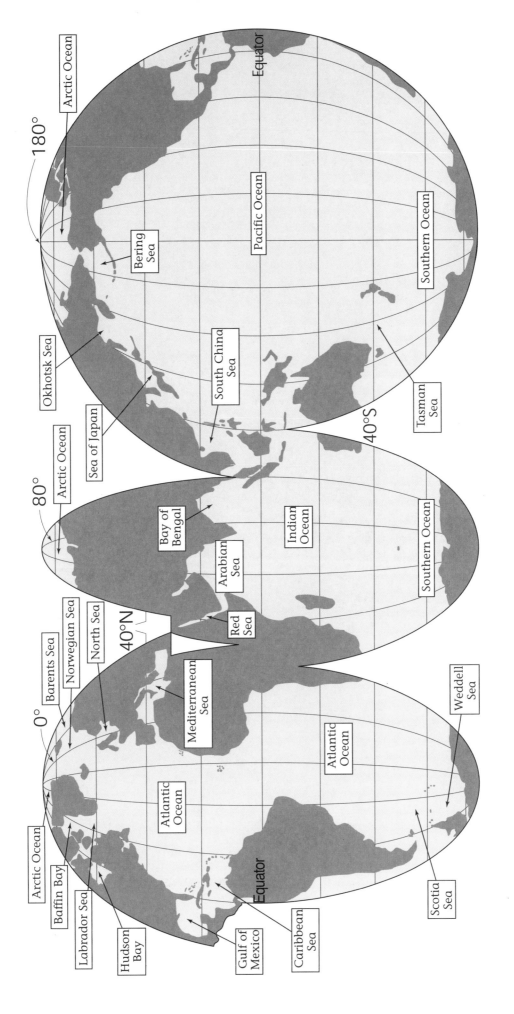